# Potential Microorganisms for Sustainable Agriculture

## A Techno-commercial Perspective

# Potential Microorganisms for Sustainable Agriculture
## A Techno-commercial Perspective

*Editors*
**D.K. Maheshwari**
**R. C. Dubey**

**I.K. International Publishing House Pvt. Ltd.**
NEW DELHI • BANGALORE

*Published by*
I.K. International Publishing House Pvt. Ltd.
S-25, Green Park Extension
Uphaar Cinema Market
New Delhi 110 016 (India)
E-mail: ik_in@vsnl.net

ISBN 978-81-907462-0-5



Published by Krishan Makhijani for I.K. International Publishing House Pvt. Ltd., S-25, Green Park Extension, Uphaar Cinema Market, New Delhi 110 016. Printed by Rekha Printers Pvt. Ltd., Okhla Industrial Area, Phase II, New Delhi 110 020.

# Preface

Modern system of agriculture is based on a substantial use of microorganisms. It aims at mutually reinforcing relationship between agricultural productivity and conservation of nature. Productivity advance in perpetuity without accompanying ecological harm necessarily involves pathways which do not adversely affect soil health, water quality, biodiversity, atmosphere and remarkable energy sources. Similar to other industries, microorganisms such as plant growth promoting rhizobacteria (PGPR) are as effective as pure chemicals on plant growth enhancement and disease control besides managing abiotic and other stresses in crop plants. *Bacillus*-based bioinoculant formulations have achieved success for improving the plant growth. The resistant life stage of the genera has paved the way for commercialization of PGPR-based products in various countries including India. The integrated use of organic and inorganic fertilizers has been found to be more effective on crop yield and product quality. A good number of endophytic diazotrophs viz. *Acetobacter spp. Herbaspirillum spp.*, *Azotobacter, Azospirillum* etc. and endophytic fungi may benefit crops like sugarcane, cereals, forage grass, oil palm trees and coffee plants. Genetic improvement of PGPR strains may contribute in understanding their beneficial traits. On the other hand, lowering of crop yield and quality cannot be ruled out due to lack of fungal and bacterial pathogens. It becomes easy to control their disease producing ability. The understanding of pathogen diversity can effectively contribute in plant growth promotion abatement of pathogens and plant parasitic nematodes but their prevention is desired to evolve such microorganisms, which effectively and efficiently colonize the root and rhizosphere. Such microorganisms are now alternative paradigms for commercialization. Seeing the importance of such microorganisms in the protection of plant health new biotechnological approaches are employed regularly to develop newer and much better microbial agents for management of the pathogen. Even viruses attack plants and the biotechnological approaches of their control needs to be evolved. These techniques have allowed genetic material to move between organisms and expressed in them.

Most of the modern agrochemicals are selective/specific in their action but microorganisms, such as *Trichoderma* spp. have multifarious agricultural industries. A number of successful products based on different species of *Trichoderma* have been commercialized. In addition, basic concepts on plant-microbe interaction mineral-nutrient exchange, and soil acidity-alkalinity requirements have been customized to fit into the track of emerging rhizomediation.

Microorganisms also act as food. Conception of direct fed microbial helps as additives in ruminant nutrition and production. Even costal sand dunes contain microbial resources, which are also being exploited in the benefits to agricultural based industries. All the articles of this volume depict the role of microorganisms to agricultural industries. The exploitation of such beneficial microorganism may improve agricultural systems with economically sound production of human food and animal feed.

**D.K. Maheshwari**
**R.C. Dubey**

# Contents

*Preface* *v*

*Contributors* *ix*

*The Editors* *xv*

1. **Coinoculation of PGPR for Plant Growth Promotion and Disease Control** **1**
   *Varun Kumar, R.C. Dubey and D.K. Maheshwari*

2. **PGPR- Biotechnology for Management of Abiotic and Biotic Stresses in Crop Plants** **26**
   *V. Govindasamy, M. Senthilkumar, Upendra Kumar and K. Annapurna*

3. **Nature and Applications of *Bacillus* Species for Improving Plant Growth** **49**
   *Pankaj Trivedi, Bhavesh Kumar and Anita Pandey*

4. **Biofertilizers for Sustainability, Agroresources and Technology** **67**
   *Sunita Gaind and B.D. Kaushik*

5. **Biodiversity of Plant Pathogenic *Fusarium* species** **88**
   *Vishwa Dhar, Subhojit Datta and Rita Rai*

6. **Biocontrol of Soil-Borne Phytopathogens by Fluorescent Pseudomonads as a Sustainable Alternative to Agrochemicals** **99**
   *N.K. Arora and Ekta Khare*

7. **Biological Control : Alternative Paradigms for Commercialization** **115**
   *Swati and Alok Adholeya*

8. **Screening, Evaluation and Formulation of Delivery Systems for Biocontrol Agents** **142**
   *R. Saravanamuthu*

9. **Molecular Basis of Plant Growth Promotion and Biocontrol Activities of Rhizobacteria** **169**
   *Yogesh Kumar Negi, S.K. Garg and J. Kumar*

10. **Rhizospheric Competent Microorganisms in Biological Control of Plant-Parasitic Nematodes** 195
*T. Kumar, S.C. Kang and D.K. Maheshwari*

11. **Recent Advances in Biotechnology of *Beauveria Bassiana* : A Mycopesticide Against Insect Pests** 218
*S.R. Niranjana, N. Haraprasad and H.S. Shetty*

12. **Endophytic Fungi and Their Role in Biological Control and Plant Growth Promotion** 242
*Jyoti Saxena, Anamika and Vinay Sharma*

13. **Microorganism-Mediated Management of Plant Pathogens** 267
*Vishal Prasad, B.K. Singh, K.S. Kumar, M. Ahmed and R.S. Upadhyay*

14. **Retrospects and Prospects of Tomato Leaf Curl Virus Research in India** 286
*J.P. Tewari and N.A. Ansari*

15. **Biotechnological Application of Microorganisms in the Management of Viral Diseases of Crops** 305
*L.P. Awasthi, P. Kumar and S. Singh*

16. ***Trichoderma* : Biology and Biotechnology** 319
*V.K. Gupta, A. Gupta, N. Modi and D.R. Modi*

17. ***Trichoderma*: A Potential Biocontrol Agent for Biopesticide Industries** 336
*H.B. Singh and Shishir Srivastava*

18. **Potential of *Trichoderma* for Biocontrol and its Mass Production** 355
*D.P. Singh*

19. **Rhizomediation in Soil for Sustainable Agriculture** 364
*Meenu Saraf, Neelam Tank and Rajendran Narayanan*

20. **Strategy to Improve Ruminant Productivity Through Direct-Fed Microbials** 394
*Ravinder Nagpal, A.K. Puniya, Monica Puniya, Tejpal Dhewa, J.P. Sehgal and Kishan Singh*

21. **Tripartite Symbiotic Association: Legume-Rhizobia-Mycorrhiza** 406
*D. Harwani, Preeti Choudhary, Sangeeta Dhaker, Kamal Prasad and S.K. Mahna*

22. ***Agrobacterium* as Natural Tool for Plant Genetic Engineering** 436
*R. Pati, V.K. Gupta, L.P. Yadava, N. Srivastava, A. Gupta, N. Modi and D.R. Modi*

23. **Coastal Sand Dune Vegetation and Microbial Resources : Benefits, Threats and Safeguards** 460
*K.R. Sridhar and A.B. Arun*

**Index** 477

# Contributors

**A. Gupta**, *Central Drug Research Institute, Lucknow (U.P.)*

**A.K. Puniya**, *Dairy Microbiology Division, Dairy Cattle Nutrition Division, National Dairy Research Institute, Karnal-132 001, Haryana (H.R.)*

**A.B. Arun**, *College of Agriculture and Natural Resources, Department of Soil Environmental Science, National Chung Hsing University, 250 Kuo-Khang Road Taichung, Taiwan-40227, Republic of China*

**Alok Adholeya**, *Centre for Mycorrhizal Research, Bioresources and Bioresources Management, The Energy and Resources Institute, Darbari Seth Block, Habitat Place, Lodhi Road, New Delhi-110 003*

**Anamika**, *Department of Bioscience and Biotechnology, Banasthali Vidyapith-304 022, Rajasthan, India.*

**Anita Pandey**, *Environmental Physiology and Biotechnology, G B Pant Institute of Himalayan Environment and Development, Kosi-Katarmal, Almora-263 643(Uttarakhand)*

**B.K. Singh**, *Laboratory of Plant pathology and Microbial Technology, Centre of Advanced Study in Botany, Banaras Hindu University, Varanasi-221 005, (U.P.)*

**B.D. Kaushik**, *Microbiology Division, Indian Agricultural Research Institute New Delhi-110 012*

**Bhavesh Kumar**, *Environmental Physiology and Biotechnology, G B Pant Institute of Himalayan Environment and Development, Kosi-Katarmal, Almora-263 643(Uttarakhand)*

**D.P. Singh**, *Crop Protection Programme, Directorate of Wheat Research, Karnal-132 001 (Haryana)*

**D.K. Maheshwari**, *Department of Botany and Microbiology, Gurukul Kangri University, Haridwar-249 404, (Uttarakhand)*

**D.R. Modi**, *Dept. of Biotechnology, B.B.A. University, Lucknow (U.P.)*

**Dharmesh Harwani**, *INCO- DEV Research Laboratory, Department of Botany, Maharshi Dayanand Saraswati University, Ajmer-305 009 (Rajasthan)*

**Ekta Khare**, *Department of Microbiology, Institute of Biosciences and Biotechnology, CSJM University, Kanpur-208 024 (U.P.)*

**H.B. Singh**, *Department of Plant Pathology, National Botanical Research Institute, Lucknow-226 001(U.P.)*

**H.S. Shetty**, *Department of Applied Botany & Biotechnology, Biocontrol Research Laboratory, Manasagangotri, University of Mysore, Mysore-570 006 (Karnatka)*

**J. Kumar**, *University of Horticulture and Forestry, Ranichauri, Tehri Garhwal, (Uttarakhand)*

**J.P. Sehgal**, *Dairy Microbiology Division, Dairy Cattle Nutrition Division, National Dairy Research Institute, Karnal-132 001 (Haryana)*

**J.P. Tewari**, *Plant Pathology Research Lab, Department of Botany, M.L.K. P.G. College, Balrampur-271 201 (U.P.)*

**Jyoti Saxena**, *Department of Bioscience and Biotechnology, Banasthali Vidyapith-304 022 (Rajasthan)*

**K. Annapurna**, *Division of Microbiology, Indian Agricultural Research Institute, PUSA, New Delhi-110 012*

**K.S. Kumar**, *Laboratory of Plant Pathology and Microbial Technology, Centre of Advanced Study in Botany, Banaras Hindu University, Varanasi-221 005 (U.P.)*

**K.R. Sridhar**, *Microbiology and Biotechnology, Department of Biosciences, Mangalore University, Mangalagangotri-574 199, Mangalore, (Karnataka)*

**Kamal Prasad**, *INCO- DEV Research Laboratory, Department of Botany, Maharshi Dayanand Saraswati University, Ajmer-305 009 Rajasthan (India)*

**Kishan Singh**, *Dairy Microbiology Division, Dairy Cattle Nutrition Division, National Dairy Research Institute, Karnal-132 001 (Haryana)*

**L.P. Awasthi**, *Department of Plant Pathology, N.D.University of Agriculture & Technology, Kumarganj, Faizabad-224 229 (U.P.)*

**L.P. Yadava**, *Central Institute for Subtropical Horticulture, Lucknow-227 107 (U.P.)*

**M. Ahmed**, *Laboratory of Plant Pathology and Microbial Technology, Centre of Advanced Study in Botany, Banaras Hindu University, Varanasi-221 005 (U.P.)*

**M. Senthilkumar**, *Division of Microbiology, Indian Agricultural Research Institute, New Delhi-110 012, India.*

**Meenu Saraf**, *Department of Microbiology, Gujarat University, Ahmedabad, (Gujarat)*

**Monica Puniya**, *Dairy Microbiology Division, Dairy Cattle Nutrition Division, National Dairy Research Institute, Karnal-132 001 (Haryana)*

**N.A. Ansari**, *Plant Pathology Research Lab, Department of Botany, M.L.K. P.G. College, Balrampur-271 201 (U.P.) India*

**N. Haraprasad**, *Department of Biotechnology, Sri Jayachamarajendra College of Engineering, Manasagangotri, Mysore-570 006 (Karnatka)*

**N. Modi**, *Dept. of Biotechnology, B.B.A. University, Lucknow (U.P.)*

**N. Srivastava**, *Central Institute for Subtropical Horticulture, Lucknow-227 107 (U.P.)*

**N.K. Arora**, *Department of Microbiology, Institute of Biosciences and Biotechnology, CSJM University, Kanpur-208 024 (U.P.)*

**Neelam Tank**, *Department of Microbiology, Gujarat University, Ahmedabad, Gujarat)*

**P. Kumar**, *Department of Plant Pathology, N.D.University of Agriculture & Technology, Kumarganj, Faizabad-224 229 (U.P.)*

**Pankaj Trivedi**, *Environmental Physiology and Biotechnology, G B Pant Institute of Himalayan Environment and Development, Kosi-Katarmal, Almora-263 643 (Uttarakhand)*

**Preeti Choudhary**, *INCO-DEV Research Laboratory, Department of Botany, Maharshi Dayanand Saraswati University, Ajmer-305 009 Rajasthan (India)*

**R. Pati**, *Central Institute for Subtropical Horticulture, Lucknow-227 107 (U.P.)*

**R.S. Upadhyay**, *Laboratory of Plant Pathology and Microbial Technology, Centre of Advanced Study in Botany, Banaras Hindu University, Varanasi-221 005 (U.P.)*

**R. Saravanamuthu**, *Department of Botany, A.V.C. College (Autonomous), Mannampandal, Mayiladuthurai-609 305 (Tamil Nadu)*

**R.C. Dubey**, *Department of Botany and Microbiology, Gurukul Kangri University, Haridwar-249 404 (Uttarakhand)*

**Rajendran Narayanan**, *Molecular Biology, Department of Biology, Kentucky State University, Frankfurt, Kentucky, USA.*

**Ravinder Nagpal**, *Dairy Microbiology Division, Dairy Cattle Nutrition Division, National Dairy Research Institute, Karnal-132 001 (Haryana)*

**Rita Rai**, *Indian Institute of Pulses research, Kanpur-208 024 (UP)*

**S.K. Garg**, *R.M.L. Avadh University, Faizabad, (U.P.)*

**S.R. Niranjana**, *Department of Applied Botany & Biotechnology, Biocontrol Research Laboratory, Manasagangotri, University of Mysore, Mysore-570 006 (Karnatka)*

**S. Singh**, *Department of Plant Pathology, N.D.University of Agriculture & Technology, Kumarganj, Faizabad-224 229 (U.P.)*

**S.C. Kang**, *Department of Food, Biological and Chemical Engineering, Daegu University, Gyungsan-712 714, Korea*

**Sangeeta Dhaker**, *INCO- DEV Research Laboratory, Department of Botany, Maharshi Dayanand Saraswati University, Ajmer-305 009 (Rajasthan)*

**Shishir Srivastava**, *Department of Plant Pathology, National Botanical Research Institute, Lucknow-226 001 (U.P.)*

**Subhojit Datta**, *Indian Institute of Pulses Research, Kanpur-208 024 (U.P.)*

**Sunita Gaind**, *Microbiology Division, Indian Agricultural Research Institute New Delhi-110 012*

**Suresh Kumar Mahna**, *INCO-DEV Research Laboratory, Department of Botany, Maharshi Dayanand Saraswati University, Ajmer-305 009 (Rajasthan)*

**Swati**, *TERI School of Advanced Studies, Bioresources and Biotechnology, the Energy and Resources Institute, Darbari Seth Block, Habitat Place, Lodhi Road, New Delhi-110 003*

**T. Kumar**, *Department of Botany and Microbiology, Gurukul Kangri University, Hardwar-249 404,(Uttarakhand)*

**Tejpal Dhewa**, *Dolphin (PG) Institute of Bio-Medical and Natural Sciences, Dehradun-248 007, (Uttarakhand)*

**Upendra Kumar**, *Division of Microbiology, Indian Agricultural Research Institute, New Delhi-110 012*

**Varun Kumar**, *Department of Botany and Microbiology, Gurukul Kangri University, Haridwar-*

*249 404, (Uttarakhand)*

**V. Govindasamy**, *Division of Microbiology, Indian Agricultural Research Institute, New Delhi-110 012.*

**V.K. Gupta**, *Central Institute for Subtropical Horticulture, Lucknow-227 107 (U.P.)*

**Vinay Sharma**, *Department of Bioscience and Biotechnology, Banasthali Vidyapith-304 022 (Rajasthan)*

**Vishal Prasad**, *Laboratory of Plant Pathology and Microbial Technology, Centre of Advanced Study in Botany, Banaras Hindu University, Varanasi-221 005 (U.P.)*

**Vishwa Dhar**, *Indian Institute of Pulses Research, Kanpur-208 024 (U.P.)*

**Yogesh Kumar Negi**, *Sardar Bhagwan Singh (PG) Institute of Biomedical Sciences and Research, Balawala, Dehradun (Uttarakhand)*

# The Editors

**D.K. Maheshwari** started teaching career as lecturer in the Department of Botany, D.A.V. (PG) College, Muzaffarnagar and thereafter served as Reader in the Department of Microbiology, Barkatullah University, Bhopal. He joined Gurukul Kangri University, Haridwar as Professor in the year 1990 and was Dean, Faculty of Life-Sciences (2004-2007). Prof. Maheshwari is an active member of several scientific bodies of international repute and in the board of panels of various academic and administrative bodies including UPSC and NAAC of Govt. of India. He was Editor of the *Journal of Indian Botanical Society* (1999-2002). He was awarded "Prof. Y.S. Murthy Medal" for his outstanding contribution.

As a young scientist, he was selected under UNESCO programmes and worked in Biological Research Centre, Szeged (Hungary) in the year 1983-84 and also visited Czechoslovakia, Belgium, Holland, Spain, Germany, Japan and S. Korea He was visiting Professor, University of Tokyo, in 1993 and 1998, and Guest Professor, in the University of Ulm (Germany). Prof. Maheshwari visited S. Korea in the year 2000, 2003 and 2006 under bilateral scientific exchange programmes sponsored by Indian National Science Academy (INSA), New Delhi.

Prof. Maheshwari has about 95 research papers to his credit in leading peer-reviewed journals both Indian as well as foreign. More than 28 candidates have been awarded Doctorate (Ph.D.) degree under him. He has completed 12 major research projects sponsored by various funding agencies. Prof. Maheshwari is the editor of *Microbes: Agriculture, Industry and Environment* (2000), *Innovative Approaches in Microbiology* (2002) and *Biotechnological Applications of Microorganisms: A Techno-Commercial Approach* (2006). He is co-author of *Practical Microbiology* (2nd ed., 2006), and *A Text Book of Microbiology* (2nd ed, 2005) and member Editorial board, *Korean J. Agric. Chem. and Biotechnology* and *Indian Phytopathology*.

He is a Fellow of the Indian Botanical Society and Indian Phytopathological Society and life-member of Indian Science Congress, Phytopathological Society of India, Indian Botanical Society and Association of Microbiologists in India.

**R.C. Dubey**, obtained his M.Sc. and Ph.D. degrees from the Banaras Hindu University in 1981 and 1986, respectively. He started his teaching career as Lecturer in Kumaun University (Nainital) (1987 to 1996) thereafter, he joined Gurukul Kangri University (Haridwar) in 1996. He became Professor in 2004.

His fields of interest are *soil microbiology* with *biological control* of soil-borne plant pathogens, *mycorrhizae*, *rhizosphere microbiology* and *botanical pesticides*, besides enzymology and heavy metal toxicity. He has published more than 75 research and review papers in the national and international journals of repute. He has produced 12 Ph.Ds and supervising seven research students on varied aspects of microbiology.

He has authored three books viz., *Practical Microbiology* (2nd ed., 2006), *A Text Book of Biotechnology* (4th ed, 2006) and *A Text Book of Microbiology* (2nd ed, 2005), and is Co-editor of four books viz., *Himalayan Microbial Diversity* (1997), *Microbes: Agriculture, Industry and Environment* (2000), *Innovative Approaches in Microbiology* (2002), and *Biotechnological Applications of Microorganisms : A Techno-Commercial Approach* (2006).

He has been the Organising Secretary of the National Seminar on *Bioinoculants for Holistic Sustainable Rural Development* jointly organized by the Department of Botany & Microbiology and DDU State Institute for Rural Development (U.P. Govt.) (1998).

He is the Life Member and Fellow of the Indian Botanical Society, Indian Phytopathological Society and the International Society for Conservation of Natural Resources. He was elected Associate Editor of the Journal of the Indian Botanical Society (2000-2004), and at present he is a member in the Editorial Board of the Journal of Environmental Biology and Conservation (2005-2009).

1

# Coinoculation of PGPR for Plant Growth Promotion and Disease Control

**Varun Kumar, R.C. Dubey and D.K. Maheshwari**

*Department of Botany and Microbiology, Gurukul Kangri University, Haridwar-249404, Uttarakhand*

## ABSTRACT

The increasing use of chemical fertilizers, pesticides and weed control agents has damaged soil structure, ecology, environment, and most importantly human health. By adopting a strategy using a combination of microorganisms, the soil can be protected from further impoverishment. Particular bacterial strains in certain natural environment prevent infectious diseases of plant roots. This article reviews the researches that could be explored in the future to strengthen the importance of the bioinoculants. The review includes the role of coinoculation of plant growth promotory rhizobacteria (PGPR) with special reference to rhizobia and pseudomonads in plant growth and disease suppression. Combinations of PGPR strains are expected to result in a higher level of potential to promote plant growth and suppression of multiple plant diseases. The complexity of the soil ecosystem is constraint that makes biological control of these root pathogens by introducing antagonists as a challenge. By combining several microorganisms, multiple beneficial activities are also combined and one may assume that at least one biocontrol mechanism will be functional under the conditions faced by the released biocontrol agents.

## INTRODUCTION

In recent years tonnes of chemicals in the form of inorganic fertilizers and synthetic pesticides with various adjuvants, solvents, carriers and diluents are applied in the soil or sprayed on to plants. These highly potent chemicals exert deleterious effects on beneficial soil microflora. Nitrogen (N), phosphorus (P), potassium (K) and sulphur (S) are the basic chemical nutrients required by the plant, (Ayala and Rao, 2002) but higher doses of these chemical fertilizers disturb microbial ecology and decrease soil fertility. Therefore, a shift towards non-chemical strategies has to be adopted which may likely to correct the imbalance with the approaches known as integrated nutrient management (INM), integrated pest management (IPM) or integrated disease management (IDM). Unfortunately, microorganisms as bioinoculants typically have a relatively narrow spectrum of activity as compared with synthetic pesticides and fertilizers (Janisiewicz, 1996) and often exhibit inconsistent performance in practical agriculture, resulting in limited commercial use of bio-inoculant approaches. Plant growth promoting rhizobacteria (PGPR) have been proved more effective in combinations for plant growth promotion as compared to single strain (Park *et al.*, 1988). Bacteria with additional traits of producing antibiotics, siderophores, HCN, phytohormones and phosphate solubilizing activity are known to serve as substitute for chemical fertilizers and pesticides (Pandey *et al.*, 2005). Naturally occurring rhizosphere microbial community plays an important role in improvement of root health and biological control of plant pathogens. Particularly, fluorescent pseudomonads and rhizobia have revolutionized the field of biocontrol of soil-borne plant pathogenic fungi (Arora *et al.*, 2001; Gupta *et al.*, 2002). Combinations of biocontrol strains are expected to result in a higher level of potential to suppress multiple plant diseases. Sheneviratne (2003) has advocated the use of microbial consortium for growth enhancement and elucidated that Coinoculation and co-culture of microorganisms yield better task than their individual cultures.

## PLANT GROWTH PROMOTING RHIZOBACTERIA (PGPR)

Bacteria are abundantly present in rhizosphere, most often organized in the form of micro-colonies. Some of these rhizobacteria not only derive benefits from the nutrients secreted by the plant root, but also influence the plants directly or indirectly, resulting in stimulation of their growth. The term 'rhizobacteria' is used to describe a subset of rhizosphere bacteria able to colonize the root environment. Beneficial root colonizing, rhizosphere bacteria i.e. the PGPR, are defined by three intrinsic characteristics:

(i) They must be able to colonize the root.

(ii) They must survive and multiply in microhabitats associated with the root surface in competition with other microbiota, at least for the time needed to express their plant promotion/protection activities.

(iii) They must promote plant growth.

These plant growth promoting rhizobacteria (PGPR) can be classified according to their beneficial effects. For instance, biofertilizers can fix nitrogen, which can subsequently be used by the plant, thereby improving plant growth when the amount of nitrogen in the soil is limiting. Phytostimulators can directly promote the growth of plants, usually through the production of hormones.

The bacteria useful to plants are characterized in two general types: bacteria forming a symbiotic relationship with the plant, and the free-living ones found in the soil, which are often found near, on or even within the plant tissues (Kloepper *et al.,* 1988; Frommel *et al.,* 1991). Beneficial free-living soil bacteria that enhance the plant growth are usually referred to as plant growth promoting rhizobacteria (PGPR) (Kloepper *et al.,* 1989) or yield increasing bacteria (YIB). PGPR, originally defined (Kloepper and Schrath, 1978) as root colonizing bacteria (rhizobacteria), cause either growth promotion or biological control of plant diseases. Bashan and Holguin (1997) proposed the division of plant growth promoting rhizobacteria into two categories: biocontrol-PGPB (plant growth promoting bacteria) and PGPB. Biocontrol agents are able to protect the plants from infection by phytopathogenic organisms. The large-scale application of PGPRs to crops as inoculants would be attractive as it would substantially reduce the use of chemical fertilizers and pesticides, which pollute the environment. In addition, the application of PGPRs would increase crop yield, thereby helping to feed the growing world population. A growing number of PGPRs is being marketed. These strains can only be used optimally, however, if the molecular basis of their beneficial effects, and the way these traits are influenced by biotic and abiotic factors, are understood.

PGPR's are usually in contact with the root surface or rhizoplane, and increase plant yield by one or more mechanisms such as improved mineral nutrition, disease suppression, or phytohormone production (Défago and Keel, 1995). A large number of plant growth promotory rhizobacteria are inhabitants of rhizosphere. In this article, mainly rhizobia and pseudomonads have been discussed .

## RHIZOBIA

Rhizobia are classically defined as symbiotic bacteria capable of eliciting and invading root and stem tissue, forming nodules on leguminous plants where they carry out symbiotic nitrogen fixation. Beijerinck (1888) had, for the first time, isolated a bacterium from the root nodules of legumes and named it *Bacillus radicicola.* This was subsequently renamed *Rhizobium* (Frank, 1889). The earliest classification of rhizobia was based on specificity of symbiotic plant range of bacterial species, recognized six species in genus *Rhizobium*, viz., *R. leguminosarum* (*Lathyrus*, *Lens*, *Pisum* and *Vicia*), *R. lupine* (*Lupinus*), *R. meliloti* (*Melilotus*, *Medicago, Trigonella*), *R. phaseoli* (*Phaseolus*) and *R. trifolii* (*Trifolium*) based on their host range, though also describe certain morphological and physiological properties of the identified species. Based on the growth rate, bacteria were grouped as fast growers and slow growers, but were still placed in the genus *Rhizobium* till Jordan (1982) coined the new genus, *Bradyrhizobium japonicum* described for isolate of *Glycine max.* In the first edition of Bergey's Manual of Systematic Bacteriology (Kreigand and Holt, 1984), two rhizobial genera (*Bradyrhizobium* and *Rhizobium*) and four species were described solely on the basis of growth rate and symbiotic host ranges. These include *Bradyrhizobium japonicum* (slow growing isolates) nodulating *Glycine max, Rhizobium leguminosarum* isolates nodulating *Lathyrus, Lens, Phaseolus vulgaris, P. augustifolius, P. multiflorus, Pisum* and *Trifolium* (Frank, 1979), *R. meliloti* nodulating *Medicago, Melilotus* and *Trigeonella* and *R. loti* isolates nodulating *Anthyllis, Caragana arborescens, Cicer arietinum, Luceana lucocephala, Lotus* spp., *Lupinus* and *Mimosa.*

The genera *Azorhizobium, Bradyrhizobium* and *Mesorhizobium*, which have been included in the family rhizobiaceae (van Berkum and Earstly, 1998), are allocated in the forthcoming second edition of the Bergey's Manual of Systematic Bacteriology in the new families Azorhizobiaceae, Bradyrhizobiaceae and Phyllobacteriaceae, respectively. The revised family Rhizobiaceae, having two genera *Rhizobium* and *Sinorhizobium*, is genotypically related to Bartonellaceae, Brucillaceae and Phyllobactericeae (Young *et al.*, 2001). Families Azorhizobiaceae and Bradyrhizobiaceae are closely related to each other, but are distant from other families in the group based on 16S rDNA sequence alignment.

Recently, proteobacteria from β-subclass have also been identified in the nodules of legumes. Bacterial isolates from nodules of *Aspalathus carnosa* and *Machaerium lunatum* have been identified as *Burkholderia* spp. (Barkum and Eardly, 1998; Moulin *et al.*, 2001). Subsequently, *Ralstonia taiwanensis* has been identified in the nodules of *Mimosa* nodules from China (Chen *et al.*, 2001) and *R. eutropha* in *Mimosa pudica* from India (Verma *et al.*, 2004). Recently, Pandey *et al.* (2005) have reported *Burkholderia* sp. from Indian *Mimosa* on the basis of polyphasic taxonomic approach.

Since early 21st century, sequence comparisons of 16S rRNA genes and genetic fingerprinting methods using PCR have been used extensively for characterizing rhizobia. The topology of the phylogenetic trees obtained from aligned sequences and those obtained from ARDRA patterns have been found to be well-correlated (Laguerre *et al.*, 1996). Phylogenies resulting by complete sequence of certain rhizobial species are congruent (Young *et al.*, 1991). Oligonucleotide probes are not reliable, as until now no single probe is specific for rhizobia (Ludwig *et al.*, 1998). The above arguments support the use of sequencing of full 16S rRNA gene and different PCR tools as most reliable to ascertain the phylogeny of rhizobial isolates. On the basis of current taxonomy of root nodule bacteria, greater than 70% DNA-DNA homology coupled to distinctive phenotypic characters is one of the main criteria for defining species. However, in a few cases, strains sharing lower homology (40-60%) have been found within a single species, viz. *R. tropici* type A and B (Martinez-Romero, 1991) and *Mesorhizobium plurifarium*. At present, 50 rhizobial species distributed among five genera of rhizobia are recognized based on the polyphasic approach.

The ways by which PGPR can influence plant growth directly may differ from species to species as well as strain to strains. Symbiotic plant colonizers and few free-living bacteria such as *Azotobacter, Azospirillum, Guconacetobacter, Herbaspirillum, Burkholderia* mostly contribute to plant growth by nitrogen fixation symbiotically and asymbiotically. All these bacteria form a host specific symbiosis with legumes and non-legumes. The symbiosis is initiated by the formation of root or stem nodules in response to the presence of the bacterium. Molecular signal molecules (lipo-oligosaccharide) secreted by these bacteria play a critical role in this process (Lang, 2000; Spaink, 2000; Perrot *et al.,* 2000). Association of rhizobia with legumes is well known, but it was Postgate (1987) and Trinik and Habdobas (1995) who highlighted the existence of rhizobia symbiosis beyond leguminaceae. Rhizobia isolated from legumes can be used as potential PGPR with non-legumes (Antaun *et al.*, 1998). Several rhizobia have also been proven by nodulating the non-legumes (Trinik and Hadobas, 1995).

## PSEUDOMONADS

Members of genus *Pseudomonas* are rod-shaped Gram-negative bacteria characterized by metabolic versatility, aerobic respiration (some strains also have anaerobic respiration with nitrate as a terminal electron acceptor and for arginine fermentation), motility owing to several polar flagella, and high genomic G+C content (59.68%) (Holt *et al.*, 1994). The term "pseudomonads"

(*Pseudomonas* like bacteria) is often used to describe strains for which the taxonomic affiliation has not been established in detail. Fluorescent pseudomonads produce the fluorescent pigment pyoverdin (Pvd) (also known as pseudobactin). This large and heterogenous group comprises most of them notably, *P. aeruginosa*, *P. putida*, *P. fluorescens* and *P. syringae* (recently reviewed by Kumar *et al.*, 2005).

*Pseudomonas* spp. are particularly suitable for application as agricultural biocontrol agents since they can use many exudate as a nutrient source, which is indicative of their adaptive potential (Lugtenberg *et al.*, 2004). *Pseudomonas* possess diverse mechanisms by which they can exert inhibitory activity towards phytopathogens and thereby mediate crop protection. They have a high growth rate relative to many other rhizosphere bacteria, are easy to grow *in vitro,* and can subsequently be reintroduced into the rhizosphere by seed bacterization. In most of the investigations, the efficacy of individual biocontrol strain or combination of different strains is assessed by screening these agents against a single isolate of the target pathogen. This could be one of the reasons of failure of commercial bioinoculant, as for evaluation no consideration of the genetic diversity of the target pathogen is considered (Mazzola *et al.*, 1995). Significant genotypic variation in a fungal pathogen population may exist among geographic region, within a given field or even among lesions of the same plant (Budge *et al.*, 1995; Mc Donald *et al.*, 1994). Enhancing biological control by using mixtures of antagonists may have advantages, as it may broaden the spectrum of activity, enhance the efficacy and reliability of the biological control, and allow the combination of various traits without employment of genetic engineering (Mc Donald *et al.*, 1994).

## PLANT-MICROBE INTERACTIONS IN RHIZOSPHERE

Naturally occurring rhizosphere microbial community plays an important role in improvement of root health and biological control of plant pathogens. Particularly, fluorescent pseudomonads and rhizobia have revolutionized the field of biocontrol of soil-borne plant pathogenic fungi (Arora *et al.*, 2001; Gupta *et al.*, 2002). Combinations of biocontrol strains are expected to result in a higher level of potential to suppress multiple plant diseases. Sheneviratne (2003) advocated the use of microbial consortium for growth enhancement and elucidated that coinoculation and co-culture of microorganisms yield better task than their individual microbial cultures. The rhizosphere supports a large and metabolically active microbial population that can be several orders of magnitude higher than in rhizosphere soil (Schloter *et al.*, 1997). Interactions between plant and microorganisms and between rhizosphere microorganisms are largely unknown, and recent works show that these interactions are complex and are dependent on multiple traits. The potential rhizosphere colonization for biotechnology is being investigated and applications include soil biofertilizer (Hoflich *et al.*, 1995), biocontrol (Dowling and O'Gara, 1994) and more recently rhizoremediation, i.e., the use of integrated plant/microorganism system for *in situ* soil rhizoremediation. In rhizoremediation systems, the increased population in the rhizosphere metabolizes or co-metabolizes organic pollutants, while the plant root provides an ecological niche to support these populations.

On the basis of current knowledge, indirect interactions between plants and microbes occur in the rhizosphere due to the loss of carbon by exudation, secretion or autolysis of older root cells of the plants. Root exudates have been shown to increase the biomass and activity of soil microbes and fauna found in the rhizosphere (Butler 2003). They have a profound qualitative and quantitative effect on the rhizosphere microflora. The chemical components released by roots are classified into six groups, depending on their mode of excretion: *(i)* water-soluble exudates, such as sugars, amino acids, organic acids, hormones and vitamins, which leak from the root without the involvement of metabolic energy (Lynch and Whipps, 1990), *(ii)* secreted substances, such as polymeric carbohydrates and enzymes, which depend upon metabolic processes for their release (Gramss and Rudeschko, 1998), *(iii)* lysates released when cells autolyse, including cell walls, and with time, whole roots (Curl and Truelove, 1986), *(iv)* gases such as ethylene, $CO_2$ and $CH_4$ (Mitra *et al*., 2005), *(v)* plant mucilage (Nagahashi and Douds, 2004), and *(vi)* mucigel (Curl and Truelove, 1986). Roots also may liberate antimicrobial agents. In some instances, these are antifungal substances (Ye *et al*., 2004), and a single root system may give rise to a number of such toxicants (Turrini *et al*., 2004).

Microbial populations can produce biologically active substances which affect plant growth. This may be positive when these substances promote growth, as seems to be the case when biotin and pantothenic acid are produced or when iron uptake by the root is enchanced by agrobactin production by *Agrobacterium tumefaciens* (Lynch 1990). The presence of homoserine lactone (HSL, a degradation product of N-acyl-homoserine lactone, a regulatory signal molecule that is commonly used by plant-associated bacteria) increased stomatal conductance and transpiration by 20-30%, and transpiration helps plants and rhizospheric bacteria to obtain diffusion-limited mineral nutrients such as P. It is known that *Bacillus* species are prevalent in the rhizosphere, where they can produce plant growth factors and synthesize antifungal agents (Wipat and Harwood 1999). Plant roots support the growth and activities of a wide variety of microorganisms that may have a profound effect on the growth and/or health of plants. Among these microorganisms, a high diversity of bacteria have been identified and categorized as deleterious, beneficial or neutral with respect to the plant. The beneficial bacteria, i.e., plant growth promoting rhizobacteria (PGPR), are widely studied by microbiologists and agronomists because of their potential in plant production (Barea *et al*., 2004).

PGPR may increase plant growth regulating ethylene production in roots, releasing phytohormones (Lynch 1990) and decreasing heavy metal toxicity. For example, *Azospirillum*, a genus of versatile PGPR, is able to enhance the growth and yield of a wide range of economically important crops in different soils. Beneficial effects of *Azospirillum* have been attributed mainly to the production of phytohormones, nitrate reduction, and nitrogen fixation (Somers 2004).

Most of the studies in which plants were grown in pots as a monoculture revealed that specific groups of microorganisms were associated with the rhizosphere. However, contradictory results were found from field studies with a mixed plant community (White *et al.* 2005). Some studies did not find such selection in the rhizosphere but other studies reported a strong correlation between plant and soil microbial community. At the functional level, microorganisms that colonize the rhizosphere help plants to aquire P and K, and some enhance N uptake from the soil by their effect on root morphology and physiology. Recent observations on a higher diversity of ammonia-oxidizing bacteria and nitrogen-fixing (*nifH*) genes in the rhizosphere compared with bulk suggest that plants might promote functional groups rather than taxonomic groups of microorganisms. There is accumulating evidence that biotic interactions, occurring in the rhizosphere, play an important role in determining plant diversity above ground by indirect effects on competing plants.

Soil microbial populations are multispecific communities, which fluctuate according to the environmental changes. In nature, every microbial cell is likely to interact with other cells, and the dynamics of every microbial population is partly a result of its interactions with other populations. Microbial life and microbial interactions are, therefore, inseparable and the latter are as ubiquitous as the former.

The interactions can be divided into direct interaction (the most important type of direct microbe-microbe interactions in the rhizosphere is antagonism, although this may simply reflect the fact that, in general, such activity is easier to detect and investigate than the other major types of interactions, such as mutualism and competition) (Curl and Truelove, 1986), and/or interactions mediated by a resource (*e.g.*, competition for a nutrient), and/or indirect interactions (*e.g.*, a strong competitor negatively affecting the predator of a weaker competitor by decreasing the food base of this predator).

## COINOCULATION OF PGPR IN PLANT GROWTH PROMOTION

### Biofertilization

At present, biofertilization accounts for approximately 65% of the nitrogen supply to crops worldwide. Legumes are often used as green fertilizers. The most efficient nitrogen fixers are bacterial strains belonging to the genera *Rhizobium*, *Sinorhizobium*, *Mesorhizobium*, *Bradyrhizobium*, *Azorhizobium* and *Allorhizobium*, and these strains have been studied in detail. These bacteria form a host-specific symbiosis with leguminous plants. The symbiosis is initiated by the formation of root or stem nodules in response to the presence of the bacterium. Lipooligosacharide signal molecules that are secreted by the bacterium play a crucial role in this process. The bacteria penetrate the cortex, induce root nodules, multiply and subsequently differentiate into bacteroids which produce the nitrogenase complex. Within the root nodules, the plant creates a low oxygen concentration, which allows bacterial nitrogenase to convert atmospheric nitrogen into ammonia. In return, the plant supplies the bacteria with a carbon source.

The gene products of both the *nif* and the *fix* genes, which are involved in nitrogen fixation, have been characterized and described in detail. A major challenge in rhizobial research is to understand how the bacterial signals, which trigger the formation of the nodule, are perceived by the plant. Genome analyses of model legumes such as *Lotus* and *Medicago* are in progress, and comparative genomics with other non-legumes may indicate which genes make leguminous plants susceptible to the establishment of the symbiosis. An exciting new finding is that rhizobial bacteria contain genes on the Sym (Symbiosis) plasmid or chromosomally in a symbiotic island that are homologous to those encoding type-III secretion systems, which are used by pathogenic bacteria to deliver virulence factors into host cells. Free-living nitrogen-fixing rhizobacteria such as *Azospirillum*, *Herbaspirillum*, *Acetobacter*, *Azotobacter* and *Azoarcus* are also able to fix atmospheric nitrogen. They use a nitrogenase complete that functions under low oxygen conditions and is not as specific in its interaction with the plant as are rhizobia. *Azospirillum* predominantly colonizes the rhizosphere, whereas the other bacteria are predominantly found as endophytes inside roots, stems and leaves. The genes involved in nitrogen fixation, nitrogen assimilation and nitrogen regulation have been described for *Azospirillum*. Several of the *nif* genes have also been described for the other free-living nitrogen fixers, which have similar nitrogenase complexes, except for *Azoarcus* that possesses three differently encoded nitrogenase complexes.

## Phytostimulation

Phytostimulators enhance plant growth in a direct way. Organic substances capable of regulating plant growth produced either endogenously or applied exogenously are called plant growth regulators. They regulate growth by affecting physiological and morphological processes at very low concentrations. A large number of studies done so far have indicated the involvement of plant hormones produced by bacteria. There are different types of plant hormones secreted by microorganisms such as indole acetic acid (Mordukhova *et al.,* 1991; Gupta *et al.*, 1999), gibberrelic acid (Mahmoud *et al.,* 1984), cytokinins and ethylene (Glick *et al.,* 1995). Phytohormone like IAA and cytokinin producing PGPR play a significant role in growth of non-leguminous plants (Noel *et al.*, 1996; Patten and Glick, 2002).

Phosphorus is one of the key plant nutrients next to nitrogen. The phosphorus content in average soil is about 0.05% (w/w); however, only 0.1% of this is available to plants (Illmer and Schinner, 1995a). Large amounts of phosphatic fertilizers added to soil enhance agro-productivity on a regular basis, but most of this remains immobilized and hence is not available to plants. Maintenance of available phosphorus remains a major challenge. The biogeochemical cycle of phosphorus is influenced by various microorganisms, which affect phosphorus metabolism in soil. The precipitated inorganic phosphate is solubilized by the action of minerals and organic acids produced by soil bacteria (Deshwal *et al.*, 2003) and other organisms.

Organic compounds containing phosphorus are decomposed and mineralized by bacteria (Aaronson and Patni, 1992) and other group of microorganisms. Several bacteria such as *Azotobacter, Bacillus, Clostridium*, fluorescent pseudomonads and rhizobia are reported to solubilize inorganic phosphorus (Gupta *et al.*, 1999; 2001; 2002; Arora *et al.*, 2001; Deshwal *et al.*, 2003).

## COINOCULATION OF PGPR FOR CONTROL OF PHYTOPATHOGENS

Interest in biological control arose during 1920s and 1930s, when some plant pathogens were suppressed by introducing some antibiotic producing microbes to the natural habitats. Microorganisms that can grow in the rhizosphere are ideal for use as biocontrol agents, since rhizosphere is the front-line defence for roots against attack by pathogenic fungi (Weller, 1988). In rhizosphere region, organic exudates are released from the roots and root metabolic activities such as respiration. Kloepper and Schorth (1978) and Suslow (1982) coined the term "rhizobacteria" for those bacteria in the rhizosphere that aggressively colonized the root system. The premier examples of bacterial biocontrol agents are *Actinoplanes, Agrobacterium, Alcaligens, Amorphosporangium, Arthrobacter, Azotobacter, Bacillus, Burkholderia, Cellulomonas, Enterobacter, Erwinia, Flavobacterium, Gluconacetobacter, Micromonospora, Pseudomonas,* Rhizobia, *Serratia, Streptomyces, Xanthomonas,* etc. as stated by large number of microbiologists (Bashan and Levanony, 1990; Tang, 1994; Genzalez, 1994; Weller and Thomashow, 1994; Glick *et al.*, 1999; M. Lucy *et al.*, 2004).

A number of other PGPR are being discovered day-by-day. Recently, a novel PGPR, *Delftia tsuruhatensis* HR4 has been reported both as diazotroph as well as potential biocontrol against various plant pathogens. Biological control agents colonize the rhizosphere, the site requiring protection, and leave no toxic residues, as opposed to chemicals. The antagonistic bacteria and fungi have been used extensively for biological control of *M. phaseolina*. Fluorescent pseudomonads, rhizobia and *Trichoderma* spp. are the most frequently used microorganisms for biological control and of fungal pathogens which indirectly promotes plant as summarized in a recent review by Kumar *et al.* (2005).

## ANTIBIOSIS

One of the most effective mechanisms that antagonists employ to prevent proliferation of phytopathogens is the synthesis of antibiotics. A large number of antibiotics have been reported from different fluorescent pseudomonads including agrocin-84, agrocin-434, 2,4-diacetyl phloroglucinol, herbicolin, oomycin, phenazine, pyoluteorin, pyrrolnitrin, pyrroles, etc. They have a role to play in inhibition of pathogens (Gutterson *et al.,* 1986; James and Gutterson 1986; Kerr, 1980). Fluorescent *Pseudomonas* spp. have been reported to produce 2,4-diacetylfluoroglucinol (2,4-DAPG) by which they provide biological control of soil-borne pathogens on a wide range of crops and have a key role in the suppressiveness of some plant pathogens (Raaijmakers and Weller, 2001; Weller *et al.*, 2002). Recently, isolated 2-Hydroxy-methyl-chromo-4-one produced from *Burkholderia* sp. MSSP, also showed antifungal activities. Antibiotics from rhizobia are not new, as several antibiotics have been reported such as anthranilic acid from *R. leguminosarum* (Rioux *et al.*, 1986), 4,3-Hydroxy-2,3,9-trimethoxy pterocarpan phytoalexin from *R. leguminosarum* bv *viciae,* trifolitoxin (TFX) from *R. leguminosarum* bv *trifoli* T 24 (Chakraborty and Chakraborty 1988) and rhizobitoxin from *Bradyrhizobium* spp (Chakraborty and Purkayastha, 1983; Deshwal *et al.*, 2003).

### Siderophore

Iron is present in the Earth's crust in a highly insoluble form of ferric hydroxide ($Fe^{3+}$), and thus unavailable to microorganisms and plants. Some bacteria have developed iron uptake systems (Neilands and Nakamura, 1997). These systems involve a siderophore (an iron binding legand) and an uptake protein needed to transport iron into the cell. Siderophores are low molecular weight (~ 400–1000 Da) iron-chelating compounds that bind $Fe^{3+}$, transport it back to the cell and make it available for the microbial cells (Neilands and Leong, 1986; Briat, 1992). The secreted siderophore molecules have a very high affinity (kd = $10^{-20}$ to $10^{-50}$) for iron and bind most of the $Fe^{3+}$ available in the rhizosphere and prevent the pathogens present in immediate vicinity from proliferation because of lack of iron (O'Sullivan and O' Gara 1992). Antagonists can prevent the proliferation of fungal phytopathogens by producing siderophores that bind most of the $Fe^{3+}$ in the rhizosphere.

Thus, paucity of iron caused by such microorganisms prevents any fungal pathogen from proliferating in their immediate vicinity. Kloepper *et al.*, (1988) have supported this mechanism and stated that the production of siderophores that chelate and, thereby, scavenge the ferric iron in the rhizosphere may result in growth inhibition of other microorganisms whose affinity for iron is lower. It has been suggested that the ability to produce specific siderophores, and/or to utilize a broad spectrum of siderophores may contribute to the root colonization ability of biocontrol strains.

In addition, siderophores also mediate the iron uptake by plant roots in iron limiting conditions. Siderophore production has been reported by a number of workers in different groups of plant growth promoting bacteria such as rhizobia (Arora *et al.*, 2001; Deshwal *et al.,* 2003), pseudomonads (Gupta *et al.,* 2002; Bhatia *et al.*, 2003), *Azotobacter* (Budzikiewicz *et al.*, 1992), *Azospirillum*, *Burkholderia* sp. (Pandey *et al.*, 2005). Liles *et al.*, (2000) discovered that *Legionella pneumophila* (an aquatic bacteria) produced a non-classical siderophore, legiobactin. Loss of structural integrity in *Fusarium udum* was observed by electron microscopy (Fig. 1a) due to *Pseudomonas aeruginosa* LPK2 (unpublished result). *P. aeruginosa* LPK2 produced siderophore on Chrome-Azurol S (CAS) medium (Fig. 1b). Recently, it was suggested that siderophore production was conserved among the members of legionellae family (Starkenburg *et al.*, 2004). Various authors have described the role of siderophores in the inhibition of several fungal pathogens such as *M. phaseolina* (Arora *et al.*, 2001; Deshwal *et al.*, 2003; Bhatia *et al.*, 2003; Pandey *et al.*, 2005).

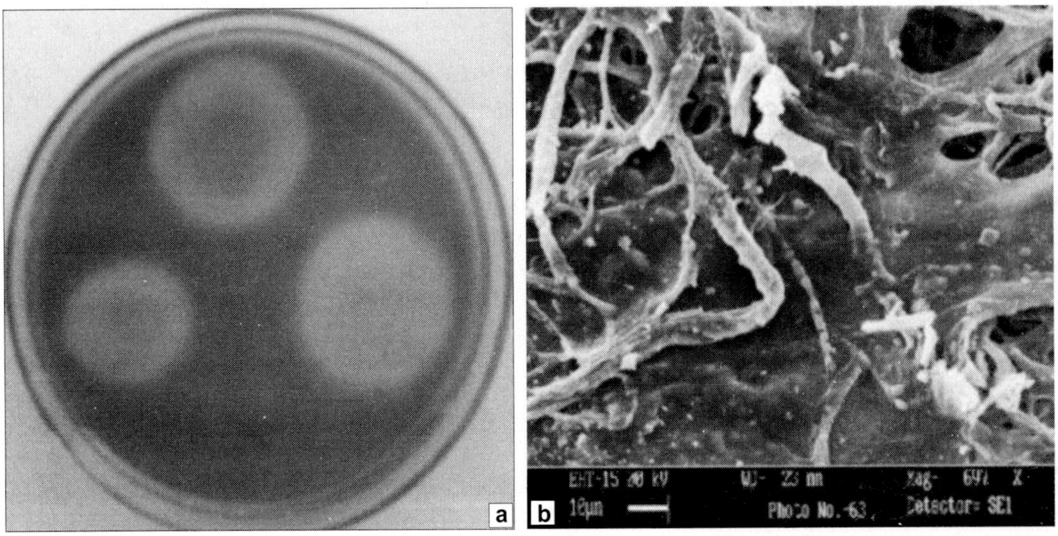


**Fig. 1** (a) Siderophore produced by *P. aeruginosa* LPK2, (b) Scanning electron photo micrograph of hyphal perforation and hyphal lysis during early stage of growth in *Fusarium udum* due to presence of *P. aeruginosa* LPK2 (unpublished results)

## HCN PRODUCTION

The ability of some antagonists to synthesize hydrogen cyanide (HCN) to which they are resistant may be linked to the ability of those strains to inhibit some pathogenic fungi. Cyanides play a role in suppression of root rot in tobacco caused by *Thilaviopsis basicola* (Ahl *et al.,* 1986). Cyanide produced by rhizosphere pseudomonads has detrimental effect on plant establishment in several crops, but it also suppresses root pathogens (Defago *et al.,* 1990). The production of volatiles in liquid cultures proved inhibitory to spore germination and mycelial growth as reported by numerous workers (Arora and Upadhyay, 1978; Schippers *et al.*, 1987; Defago and Hass, 1990; O'Sulliwan and O'Gara, 1992; Dowling and O'Gara, 1994). Cyanide also seems to play a role in the suppression of take all disease in wheat and *Rhizoctonia solani* in cotton. *Pseudomonas fluorescent* strain CHA0 resulted in increased weight of wheat and plants infected with *Rhizoctonia solani. P. fluorescens* and *P. putida* produced HCN that could inhibit the wilt of sunflower caused by *Sclerotinia* spp. (Epert and Digat, 1995). *Septoria tritici* blotch and leaf rust of wheat were suppressed by cyanide producing strains of *P. putida* (Flaishman *et al.*, 1996). HCN producing fluorescent strains of *P. aeruginosa* GRC1 and *P. aeruginosa* GRC2 (Gupta *et al.* 1999; 2001; Bhatia *et al.* 2003) inhibited charcoal rot of peanut and sclerotial rot of *Brassica* caused by *M. phaseolina* and *Sclerotiorum sclerotonium*, respectively. However, rhizobia do contain low amount of HCN in comparison to that of fluorescent pseudomonads, and only 12.5% (Beauchamp *et al.,*1991) and 3% (Antoun *et al.,*1998) strains were found to be HCN producers.

## CELL WALL DEGRADING ENZYMES

Recently, the hydrolytic enzymes have received considerable attention because they play a role in controlling diseases. The enzymatic digestion or deformation of cell wall components of phytopathogenic fungi by chitinase and β-1, 3-glucanase of several plant growth promoting bacteria has been discovered in recent years. Hyperparasitism occurs when a fungus exists in intimate association with another fungus, which derives some nutrients while conferring no benefit in return. Hyperparasitism and lysis of propagules in soil is a logically satisfying method of biological control of soil-borne plant pathogens by microbial antagonists. The cell wall degrading enzymes such as chitinase (EC 3.2.1.14), β-1, 3-glucanase (EC 3.2.1.39), proteases (EC 3.4.21.14) and lipase (EC 3.1.1.3) (Lim and Kim, 1995; Dunne *et al.,* 1997; Vivekanathan *et al.*, 2004) degrade fungal cell walls, resulting in lysis of wall material.

## EFFECT OF PGPR ON SEED GERMINATION

Charcoal rot due to *M. phaseolina* in *Arachis hypogea* could be checked when two plant growth promoting fluorescent pseudomonads PS1 and PS2 were applied through seed bacterization (Bhatia *et al.*, 2003). About 45% to 68% reduction in disease symptoms was recorded in comparison to *M. phaseolina* in peanut crop grown in infected soil. About 20% increase in crop yield was recorded. This clearly indicated the role of fluorescent pseudomonads in the disease suppression (Bhatia *et al.*, 2003). Earlier, Arora *et al.*, (2001) observed enhanced seed germination, seedling biomass and nodule weight with reduced disease incidence in groundnut crop. Increased seed germination by 47% and 53%, biomass by 16.5% and 22%, and fresh nodule weight by 36% and 56.6% were obtained with respective inoculation of *Sinorhizobium meliloti* RMP3 and *S. meliloti* RMP5. Gupta *et al.*, (2002) found reduced disease incident, better vegetative growth parameters and ultimately enhanced grain yield in peanut by the addition of *Pseudomonas aeruginosa* $GRC_2$ in *M. phaseolina*-infested field soil.

Seed germination of groundnut was enhanced by 83% in comparison to *M. phaseolina*-infested soil, while 11% in comparison to non-infested soil. Other plant parameters such as fresh nodule weight, shoot length, root length, and dry weight were enhanced, while charcoal rot disease was drastically reduced by 99% in *M. phaseolina*-infested field. The final yield of groundnut per plant got enhanced by 38% in case of infested soil, and 50% without infestation of *M. phaseolina* (Gupta *et al.*, 2002).

Fluorescent *Pseudomonas* spp. rapidly and aggressively colonizes the root system and suppresses pathogenic microorganisms, thus improving plant growth and grain yield (Schippers *et al.*, 1987; Weller, 1988; Gupta *et al.*, 2002). Fluorescent pseudomonads are reported for the biological control of different fungal species of various genera such as *Rhizoctonia*, *Fusarium*, *Sclerotinia*, *Pythium*, *Erwinia*, *Macrophomina phaseolina,* etc. (Defago *et al.*, 1990; Gupta *et al.*, 2001; Garbeva *et al.*, 2004).

Few rhizobia are also reported for biological control of deleterious phytopathogens such as *M. phaseolina* (Arora *et al.*, 2001; Deshwal *et al.*, 2003), *F. oxysporum*, *F. solani*, *Rhizoctonia solani, Pythium* spp., etc. (Chao, 1990), producing several secondary metabolites which were similar to fluorescent pseudomonads in their mode of action. Pseudomonads and rhizobia are also reported to direct plant growth promoting activities leading to yield promotion (Arora *et al.*, 2001; Gupta *et al.*, 2002; Deshwal *et al.*, 2003). Little work has been done to date concerning the beneficial relationship of *Rhizobium* and non-leguminous plants. Although, attachment of bacteria with maize, wheat, rice, oat, sunflower, mustard and asparagus has been reported (Triouchi and Syono, 1990) and improved growth of certain non-legumes inoculated with rhizobia or increased growth in legume soils has been reported. Recently, rhizospheric strains have been reported as potential promoters of plant growth for increased yield in landraces of African cereal crops. Rhizobia are able to produce plant growth regulators or stimulate production of these compounds by plant roots.

Various species of *Azotobacter* are reported for biological control of different phytopathogens such as *Alternaria, Helminthosporium, Fusarium,* etc. under *in vitro* conditions (Singh, 1999). *Azotobacter choococcum* AZO2 inhibited the growth of *Macrophomina phaseolina* in fungal pathogen causing charcoal rot in *Sesamum indicum*. Morphological observations of the fungi, obtained from the zone of interaction of *A. chroococcum* AZO2 and *M. phaseolina* during dual culture revealed by light microscopy and electron microscopy the loss of structural integrity of the mycelial in *M. phaseolina* in abundance (Fig. 2). *Azotobacter* spp. are also being applied as bioinoculant due to their several direct PGP activities including asymbiotic nitrogen fixation, phosphate solubilization, growth hormones production and vitamins production (Shende *et al.*, 1977).

## INCONSISTENCE PERFORMANCE

Although numerous microorganisms capable of suppressing plant diseases have been identified, but few have been commercialized as biopesticides (Table 1), primarily due to their inconsistent performance (Weller, 1988). The complexity of the soil ecosystem makes biological control of root pathogens by introduced bacteria particularly challenging. Many factors contribute to the inconsistent behaviour of seed or seed pieces-applied bacteria given the complex interaction among host, pathogen, bacteria, and soil environment. One major factor is variable root colonization by introduced bacteria (Weller, 1988). Multiple steps are involved in colonization. A

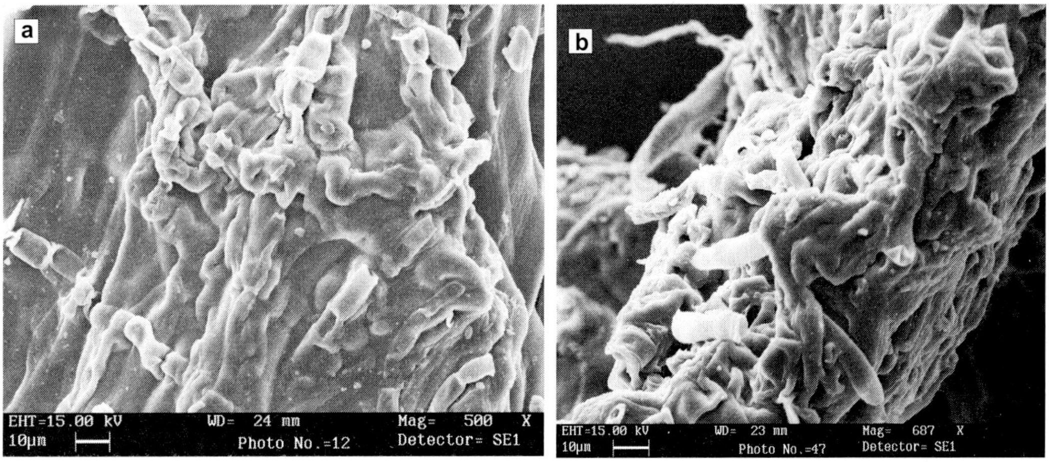


**Fig. 2** Scanning electron photomicrograph of post interaction events in *M. phaseolina* mycelial growth in the presence of *A. chroococcum* AZ02 (a, b) Degenerated, leakage of cytoplasm and lysed hyphal biomass of *M. phaseolina* (unpublished results)

variety of biotic and abiotic factors can affect the process of colonization (Howie *et al.*, 1987) and many bacterial traits may contribute to rhizosphere competence (Buell and Anderson, 1992; Mazzola *et al.*, 1992).

The demonstration by Bull *et al.*, (1991) of an inverse relationship between the population size of *P. fluorescens* 2-79 on wheat roots and the number of take all lesions demonstrated incomplete colonization (Bull *et al.*, 1991). Another factor that can contribute to inconsistent performance is variable production *in situ* of bacterial metabolites responsible for disease control. Kanner *et al.*, (1978) and Shanahan *et al.*, (1992) observed *in situ* production of PGP compounds sensitive to the physical and chemical environment in the rhizosphere. Inconsistent performance may also result from occurrence of diseases caused by non-target pathogens (Weller, 1988). If a biocontrol agent successfully suppresses a target pathogen but has no effect on second non-target pathogen that subsequently limits yield, then the treatment will appear ineffective (Weller, 1988).

An accurate evaluation of biocontrol activity against a target pathogen is nearly impossible when there is both a large amount of disease caused by a non-target pathogen(s) and yield is used as the measure of effectiveness. Finally, spatial and temporal variability in disease incidence or severity is another factor that can interfere with the ability to evaluate disease suppression effectively. Under environmental conditions that are unsuitable for disease development, the introduction of a biocontrol agent may 'fail' to provide an increase in crop yield simply because the disease was not severe enough to limit yield.

The complexity of the soil ecosystem is a constraint that makes biological control of these root pathogens by introducing antagonists as a challenge. Pierson and Weller (1994) have advanced that the establishement of threshold densities of an antagonist as a key factor in biological control, the rational of using a mixture of antagonists isolates is a logical approach, because mixtures more closely mimic microbial communities and have multiple mechanism of disease suppression

**Table 1** Biocontrol products available in the market

| *Agents/Product* | *Target Pathogen* | *Plant* | *Reference* |
|---|---|---|---|
| Intercept (*Burkholderia cepacia*) | *Fusarium spp., Rhizoctonia spp., Aspergillus* | Maize, vegetables, cotton | Soil Technologies corp. USA |
| Kodiak (*Bacillus subtilis* GB03) | *Rhizoctonia solani, Fusarium spp., Pythium spp.* | Cotton, legumes | Gustatson Inc. USA |
| *Candida oleophila* I-182 | *Botrytis* spp. | Citrus, pome fruits | Ecogen, Inc., (2005) Cabot Blvd. West, Langhome, PA, 19074 |
| *Trichoderma* spp. | Pathogenic fungi that cause wilt, take-all, root rot, and internal decay of wood products and decay in tree wounds | Flowers, fruit, ornamentals, turf and vegetables | Binab, Box 1s61, 546 22, Karlsbor, Sweden |
| *Fusarium oxysporum* | *Fusarium oxysporum, Fusarium molniforme* | Basil, carnation, cyclamen, tomato | www.biofox.com |
| Koni (*Coniothyrium minitans*) | *Sclerotinia sclerotiorum* and *S. minor* | Cucumber, lettuce, and ornamental flowers in greenhouse production | Bioved Ltd., Hungary |
| Messenger (*Erwinia amylovora* HrpN harpin) | *Rhizoctonia solani, Fusarium* spp. | Field, ornamental and vegetable crops | EDEN Biosicence Corp., USA |
| Mycostop (*Steptomyces griseoviridis* strain K61) | *Fusarium* spp., *Alternaria brassicola, Pythium* spp. | Field, ornamental and vegetable crops | Kemira Agro Oy, Finland |
| Nogall (*Agrobacterium radiobacter* K1026) | *Agrobacterium tumefaciens* | Fruit and nut trees, and other ornamental nursery stock | Bio care Technology, Australia |
| Companion (*Bacillus subtilis* GB03, other *Bacillus subtilis, B. licheniformis, B. megaterium*) | *Rhizoctonia, Phyophthora* | Greenhouse and nursery | www.growthproducts.com |
| *Burkholderia cepacia* type Wisconsin | *Rhizoctonia, Pythium,* and *Fusarium* | Alfalfa, barley, beans, clover, cotton, peas, grain shorgum, vegetable crops, and wheat | www.helenachemical.com |
| Polyversam (*Pythium oligandrum*) | *Pythium* spp., *Fusarium* spp., *Botrytis* spp., *Phytophthora* spp., *Gaeumanomyces graminis, Rhizoctonia solani* | Vegetables, fruits (grapes, strawberries, citrus), legumes, cereals, canola, and ornamental plants | Biopreparaty Ltd., Czech Republic |
| Actinovate (*Sreptomyces lydicus*) | *Rhizoctonia* spp. | Greenhouse and nursery crops, turf | Natural Industries Inc., USA |
| AQ10 Biofungicide (*Ampelomyces quisqualis* isolate M-10) | Powdery mildew | Apples, Cucurbits, grapes, ornamentals, strawberries, and tomatoes | Ecogen Inc., USA |
| Aspire (*Candida oleophila* I-182) | *Botrytis* spp., *Penicillum* spp. | Citrus, pome fruits | Ecogen Inc., USA |
| Binab (Various *Trichoderma* products) | *Rhizoctonia* spp., *Pythium* spp., *Fusarium* spp. | Flowers, fruits, ornamentals, turf, and vegetables | Ecogen Inc., USA |

*Table 1 Contd.*

*Table 1 Contd.*

| | | | |
|---|---|---|---|
| Biofox C (*Fusarium oxysporum*) | *Fusarium oxysporum, Fusarium moniliforme* | Basil, carnation, tomato | S.I.A.P.A., Italy |
| Bio-Fungus Supervisit (*Trichoderma* spp.) | *Sclerotinia, Phytopthora, Rhizoctonia solani, Pythium spp., Fusarium, Verticillum* | Flowers, trees, and vegetables | De Ceuster, Belgium |
| BioJect Spotless (*Pseudomonas aureofaciens*) | Dollar spot, Anthracenose, *Pythium aphanidermatum*, Michrochium patch (Pink snow mold) | Turf and others | Eco Soil Systems Inc., USA |
| Bio-save 10LP, 110 (*Pseudomonas syrangiae*) | *Botrytis cinerea, Penicillium* spp., *Mucor pyroformis, Geotrichum candidum* | Pome fruit, citrus, cherries, and potatoes | Village Farms, LLC, USA |
| Contans WG, Intercept WG (*Coniotyrum minitans*) | *Sclerotinia sclerotiorum* and *S. minor* | Field crops/vegetables | Prophyta Biologischer, Germany |
| Deny (*Barkholderia cepacia* type Wisconsin) | *Rhizoctonia, Pythium, Fusarium* and *Phytopthora* spp. | Legumes | www.helenachemical.com |
| Fusaclean (*Fusarium oxysporum* nonpathogenic) | *Fusarium oxysporum* | Asparagus, basil, carnation, cyclamen, gerbera, and tomato | Natural Plant Protection, France |
| *Bacillus licheniformis* | Dollar spot, moderate disease pressure | Turf | www.novozyme.com/ecoguard |
| *Fusarium oxysporum* | *Fusarium oxysporum* | Asparagus, basil, carnation, cyclamen, gerbera, and tomato | Natural plant Protection, Route d'Artix B.P. 80, 64150, Nogueres, France |
| RootShield, PlantSheild, T-22 Planter box *Trichoderma harzianum* Rifai strain KRL-AG2 (T-22) | *Pythium spp., Rhizoctonia solani, Fusarium* spp. | Trees, shrubs, transplants, cabbage, tomato, and cucumber | Bioworks Inc., USA |
| Rotstop (*Phelbia gigantea*) | *Heterobesidium annosum* | Trees | |
| SoilGard (*Gliocladium virens* GL-21) | *Rhizoctonia solani,* and *Pythium spp.* | Ornamentals and food crop plants grown in greenhouses | Certis Thermo Trilogy, USA |
| Trichodex (*Trichoderma harzianum*) | Seed and root rot, *Botrytis cinerea, Collectotrichum* spp., *Fulvia fulva, Monilia laxa, plasmopara viticola, Pseudoperonospora cubensis* | Cucumber, grape, soybean, strawberry, sunflower, tomato | Mycontrol Ltd., Israel |
| *Agrobacterium radiobacter* strain 84 | Crown gal disease caused by *Agrobactium tumifaciens* | Fruit, nut and ornamental nursery stock | www.crowngall.com |
| *Trichoderma* spp. | *Sclerotinia, Phytopthora, Rhizctonia solani, Pythium* spp. *Fusarium verticellum* | Flowers, strawberries, trees, vegetables | www.bioplant.dk |
| Galltrol (*Agrobacterium radiobactor strain*) | *Agrobacterium tumefaciens* | Fruit, nut and ornamental nursery stock | AgBioChem Inc., USA |
| Root Pro (*Trichoderma Harzianum*) | *Rhizoctonia solani, Pythium* spp., *Fusarium* spp., and *Sclerotium rolfsii* | All susceptible flowers and vegetable plants | Mycontrol Ltd., Israel |
| *Pseudomonas syringae* | *Botrytis cinerea, Penicillum* spp., *Mucor pyroformis Geotrichum candidum* | Fruit, citrus, cherries, and potatoes | www.villagefarms.com/biosave/index.html |
| *Pseudomonas fluorescens* A506 | Frost damage, *Erwinia amylovora* | Almond, apple, apricot, blue berry, cherry, pench, pear, potato, strawberry, and tomato | www.neufarm.com |

(Schisler *et al.*, 1997). Microorganisms as biocontrol agents typically have a relatively narrow spectrum of activity compared with synthetic pesticides as described by Baker (1991) and Janisiewicz (1996) who often exhibited inconsistent performance in practical agriculture, resulting in limited commercial use of biocontrol approaches for suppression of plant pathogens (Backman *et al.*, 1997). Most approaches for biological control of plant diseases have used single biocontrol agents as antagonists against a single pathogen. This has partially been achieved by: (i) modifying genetics of the biocontrol agent to add mechanisms of disease suppression that are operable against more than one pathogen, (ii) alteration of environment to favour the biological control agent and to disfavour competitive microflora, and (iii) to develop strain mixtures or consortia with superior biocontrol activity (Janisiewicz, 1988).

## WHY COINOCULATION?

A microbial consortium is a group of different species of microorganisms that act together as a community. For developing a consortium one can choose microorganisms that are resistant to environmental shock, fast acting, synergistically active, producing natural enzymatic activity, easy to handle, having long shelf life, good sustainability, non-pathogenic, non-corrosive of consistent quality and economical.

The combinations of biocontrol agents have been advocated by Duffy and Weller (1995). By combining several microorganisms, multiple antifungal activities are also got to be combined and one may assume that at least one biocontrol mechanism will be functional under the conditions faced by the released biocontrol agents. Moreover, combinations of biocontrol agents are expected to result in a higher level of protection (Dunne *et al.*, 1998) which has reduced variability of biological control. In fact, several authors have demonstrated that combining microorganisms can result in improved plant stand (Höflich *et al.*, 1994). Mixtures of biocontrol agents also resulted in enhanced control of several fungal plant diseases (Leeman *et al.*, 1996); Raupach and Kloepper, 1998; Singh *et al.*, 1999). Dunne *et al.* (1998) demonstrated that combining phloroglucinol-producing and proteolytic bacteria improve biocontrol *Pythium*-mediated damping off of sugar beat. On the other hand, Sung and Chung (1997) showed that a mixture of bacteria that produce chitinase and antibiotics could efficiently suppress rice sheath blight caused by *Rhizoctonia solani*.

Such enhanced disease suppression may involve not only different disease suppressive mechanisms but can also result from interaction between the introduced strains that positively influence growth, root colonization and/or activity of the strains. On the other hand, interactions between two or more introduced biocontrol agents can also negatively influence disease control. de Boer *et al.* (1999) demonstrated that *Pseudomonas putida* strain RE8 inhibits the growth of *P. fluorescens* strain RSIII *in vitro* and both the strains RE8 and RSIII did not suppress *Fusarium* wilt of radish, whereas the combination of RE8 with RSIII-9, i.e. a spontaneous mutant of RSIII, was no longer inhibited by RE8, enhanced disease suppression. Such results may be due to genetic modification of the strains RSIII that could yield the desirable results.

It is likely that in most cases of naturally occurring biological control result from mixtures of antagonists, rather than from high population of a single antagonist. For example, mixtures of antagonists are considered due to their action for protection in disease-suppressive soils (Lemanceau and Alabouvette, 1991; Schippers, 1992). Consequently, application of mixture of introduced bicontrol agents would more closely mimic to the natural situation and might broaden the spectrum of biocontrol activity, enhance the efficacy and reliability of various mechanisms without the need for genetic engineering (Hubbard *et al.*, 1983). Previous studies of combinations of biological agents for plant diseases have included mixtures of fungi (Budge *et al.*, 1995; Datnoff *et al.*, 1995; De Boer *et al.*, 1997), mixtures of fungi and bacteria (Duffy *et al.*, 1996; Hassan *et al.*, 1997; Janisiewicz, 1996; Leeman *et al.*, 1996; Leibinger *et al.*, 1997) and mixtures of bacteria (Pierson and weller, 1994; Wiehe *et al.*, 1996; De Boer *et al.*, 1997; Sung and Chung, 1997; Shanmugam *et al.*, 2002; de Boer *et al.*, 2003).

Most of these reported biological control agents do not result in improved suppression of disease compared with the individual antagonists (Dandurand and Knudsan, 1993; Hubbard *et al.*, 1983; Miller and May, 1991; Sneh *et al.*, 1984). This incompatibility of the co-inoculants is due to the action of biocontrol agents on each other as well as on the target pathogen(s) (Leeman *et al.*, 1996). Thus, it is an important prerequisite attribute of the co-inoculated microorganisms (Baker, 1990; de Boer *et al.*, 1997)

## MICROBIAL INTERACTIONS IN RHIZOSPHERE

For any disease suppressive mechanism to be effective it is important that the antagonist first be able to efficiently establish itself in the rhizosphere of that particular plant (Kloepper *et al.*, 1980). Many workers documented that inadequate colonization leads to decreased PGP activities (Schippers *et al.*, 1997; Antoun *et al.*, 1998). Root colonization is an important step in the interaction of antagonist group of bacteria with host plant reported that antagonistic microorganisms are ideal biocontrol agents because front line defence is provided by the rhizosphere against the pathogenic infection. *Pseudomonas* sp. was prominent in the rhizosphere, rhizoplane and ectorhizosphere (Curl and Truelove, 1986) because of their siderophore complexes (Baker *et al.*, 1986) and production of antibiotic compound (Howell and Stipanovic, 1980). *Rhizobium leguminosarum* bv. *trifolii* can multiply and survive under field conditions as well as in rhizosphere of non-legumes host (Wiehe and Hoflich, 1995). Rhizobia grew readily in the presence of germinating seeds and developing root systems in a similar manner with legumes and non-legumes (Pena-Cabriales and Alexander, 1983). Role of rhizobia in growth promotion of non-legumes has been well documented (Chabot *et al.*, 1996).

Root colonization has been the subject of considerable interest. The relative rates of utilization of root exudates may determine the degree of competence and the differences in the extent of bacterial attachment and the colonization when bacterial are closer to the source of exudates (Dazzo, 1980). This is because of the reason that root exudates offer a carbon rich diet to the rhizosphere microorganisms in the form of organic acids (citrate, malate, sucinate, pyruvate, fumarate, oxylate and acetone) and sugars (glucose, xylose, fructose, maltose, sucrose, galactose and ribose) constitute the 'main course', where variable amounts of $\alpha$-amino acids, nucleobases and vitamins (thiamine and biotin) proved the 'entrée' or 'dessert' (Lugtenberg and Bloemberg, 2004). In addition to root exudates, other materials such as root secretions, mucilages, mucigel and lysates from different portions of the plant contain different types of nutritional compounds available to rhizospheric microflora (Somers and Vanderleyden, 2004).

*Rhizobium* strain R39 colonized pea, maize and sugar beet rhizosphere better than *Pseudomonas fluorescens* (Hoflich *et al.*, 1994). Recently, several strains of *Rhizobium leguminosarum* bv *trifolii* were isolated from the roots of wheat cultivated in rotation with clover established that rhizobia colonized well with non-legumes. *Rhizobium leguminosarum* bv *trifolii* colonized the corn exclusively in the rhizosphere soil and on rhizoplane and only occasionally in the inner root tissues (Scholter *et al.*, 1997). *B. japonicum* was also found as an endophyte in the roots of sweet corn and cotton indicating that this bacterium is capable of colonizing internal plant niches (McInroy and Kloepper, 1995). Two major breakthroughs in plant inoculation technology occurred in the late 1970s by involvement of the use of biocontrol agents mainly *Pseudomonas fluorescens* and *P. putida* groups (Defago *et al.*, 1990; Glick *et al.*, 1995). A broad range of approaches has been used to investigate soil food webs. Recent developments in different methods are shedding some light on rhizosphere interactions. The increasing application of molecular techniques to study rhizosphere interactions will provide the basis for sensitive broad-scale (community level) to fine-scale (species level) investigations. Such studies coupled with use of well characterized experimental systems should provide important complimentary information using more traditional techniques.

## MONITORING OF FIELD PERFORMANCE THROUGH MOLECULAR MARKERS

Inoculants applied once in the soil irrespective of any means should be effective for further few years. Introduced strains are re-isolated from the applied field by using suitable markers viz., intrinsic resistance to high level of antibiotic, *Tn*5, *gfp* (green fluorescent proteins), *gus* marker, etc. (Sessitich *et al.*, 1998; Tomboline *et al.*, 1999; Hallmann *et al.*, 2001). Enumeration of applied bacterial strains tagged with any one of markers mentioned can be done above from the field soil by using serial dilution technique or standard plate count (Dubey and Maheshwari, 2002) is generally used. Other markers used in strain detection are antigenic molecules that locate the cell surface by reacting with specific antibodies. This immunological response can be detected by ELISA, fluorescent-labelled antibodies or immunodiffusion technique (Sessitch *et al.*, 1998). In recent years, molecular tools have been developed to identify and quantify specific microbial activity in mixed population. These tools are especially valuable for analysis of specific rhizobia in complex environmental samples (Topp, 2003). Research has concentrated on the development of marker genes for tapping a particular bacterial species of interest, so that the cells can be specially identified and monitored.

The reported *gfp* gene can be used as marker for monitoring the bioinoculants. The *gfp* plasmid is not suitable for fluorescence *Pseudomonas* spp. unlike other markers because of difficulty in discrimination of green fluorescent pigments produced by the bacterium and *gfp* gene. Moreover, being a non-conjugal plasmid, spread of the antibiotic marker enclosed on it is highly restricted. Therefore, *gfp* can be used as marker for monitoring the survival of *Rhizobium* sp. in bio-inoculant. (Saha *et al.*, 2001). Since no cofactor is required for the folded *gfp* protein to develop fluorescence, it has been possible to use the gene encoding *gfp* as a marker in many different organisms (Saha *et al.*, 2001). For instance, *gfp* has successfully been used to follow various aspects of the nodulation processes after expression in rhizobia. By screening and rational design, either together or separately, a number of *gfp* mutants have been derived that display altered excitation and emission properties. These *gfp* colour variants (collective called auto-fluorescent proteins; AFPs) allow the simultaneous expression of multiple fluorescent markers and using proper ways to excite the samples and detect the emitted light—they can easily be distinguished from each other (Saha *et al.*, 2001)

Use of marker genes has several advantages in studying rhizobial competition as compared to traditional approaches. Reporter genes such as the β-glucosonidase genes (*gus* A) or a thermo-stable β-glucosidase gene (*cel* B) allow detection of *Rhizobium* strains in non-nodulating plants (Sessitsch *et al.*, 1998; Wilson *et al.*, 1999). The genes encoding bacterial luciferase (Lux A, B) or firefly luciferase (Luc) have been found to be very useful markers, since tagged cells can be detected on the basis of their bioluminescent phenotype. Transposons such as *Tn*5 carrying drug resistance ($Kao^R$ and $Neo^R$) traits can be used to tag on rhizobacteria for their monitoring. The cells having *Tn*5 inserted into their genomes can be identified on specific media containing kanamycin and neomycin. An advantage of *Tn*5 insertion is that it gives the permanent kanamycin and neomycin resistance to the cells (Selvaraj and Iyer, 1984). Antibiotic resistance markers have been frequently used because of simplicity of the identification technique and cost-effective (Obaton, 1973). The intrinsic resistance of rhizobacteria towards antibiotic has been exploited to observe survival of introduced rhizobacteria in rhizosphere. Arora *et al.* (2001) and Gupta *et al.* (2002) used 100 μg $ml^{-1}$ of ampicillin and streptomycin to re-isolate the *Rhizobium* $RMP1^{amp+}$ and *Pseudomonas* $GRC1^{str+}$ and $GRC2^{neo+}$ from the un-inoculated rhizosphere. Though, the intrinsic resistance markers may show high background activity but still, it is the first choice of workers due to simplicity, non-expensive and time saving measures and a number of workers have successfully used such markers (Arora *et al.*, 2001; Gupta *et al.*, 2002; Deshwal *et al.*, 2003). Obaton *et al.* (2002) documented successful recovery of rhizobia antibiotic marker even after its twenty years of field inoculation.

## References

Aaronson, S. & Patni, N.J. (1992) *Limnol. Oceanogr. Limnol. Oceanogr.* **21**: 838-845.

Ahl, P., Volsard, C. & Defago, G. (1986) *J. Phytopathol.* **116**: 121-134.

Antoun, H., Beauchamp, C.J., Goussard, N., Chabot, R. & Lalinde, R. (1998) *Plant Soil.* **204**: 57-67.

Arora, D.K. & Upadhyaya, R.K. (1978) *Plant Soil.* **49**: 685-690.

Arora, N.K., Kang, S.C. & Maheshwari. D.K. (2001) *Curr. Sci.* **81**: 673-677.

Ayala, S. & Rao, E.V.S.P. (2002) *Curr. Sci.* **82**: 797-807.

Babich, H. & Stotzky, G. (1983) *Regul. Toxicol. Pharmacol.* **3** : 82-86.

Backman, P.A., Wilson, M., & Murphy, J.F. (1997) In: *Environmentally Safe Approaches to Crop Diseases Control.* (Eds. Rechigl, N. A. & Rechigl, J.E.), CRC Lewis Publishers, Boca Raton, FL., pp 95-109.

Baker, K. F. & Cook, R. J. (1974) In: *Biological Control of Plant Pathogens*, W. H. Freeman and Co., San Francisco, California. Amer. Phytopathol. Soc., St. Paul, Minnesota, pp 433.

Baker, R. (1990) In: *Biological control of soil-borne plant pathogens.* (Ed. Hormby, D.), CAB Inte. Wellinford, UK., pp 375-388.

Baker, R. (1991) *Crop Prot.* **10**: 85-94.

Balesdent J. & Balabane, M. (1992) *Soil Biol. Biochem.* **24**: 97-101.

Barea J.M., Azco'n, R., Azco'n-Aguilar, C. (2004) In: *Plant surface microbiology* (Eds. Varma, A., Abbott, L., Werner, D. & Hampp, R.) Heidelberg, Germany: Springer-Verlag, pp. 351-371.

Bashan, Y. & Holguin, G. (1997) *Can. J. Microbiol.* **43**: 102-121.

Beauchamp, C. J., Dion, P., Kloepper, J.W. D. & Antoun, H. (1991) *Plant Soil.* **132**: 273-279.

Beijerinck, M.W. (1888) *Botaniscle Zeitung.* **46**: 797-804.

Bhatia, S., Bhaita, S., Dubey, R.C. & Maheshwari D.K. (2003) *Ind. J. Expt. Biol.* **41**: 1442-1446.

Briat, J.F. (1992) *J. Gen. Microbiol.* **138**: 2475-2483.

Budge, S.P., McQuilken, M.P., Fenlon, P. & Whipps, J.M. (1995) *Biol. Control* **5**: 513-522.

Budzikiewicz, H., Schaffner, E. & Taraz, K. (1992) *Nat. Prod. Lett.* **1**: 9-14.

Buell, C.R. & Anderson, A.J. (1992) *Mol. Plant-Microbe Interact.* **5**: 154-162.

Bull, C.T., Weller, D. M. & Thomashow, L.S. (1991) *Phytopathol.* **81**: 954-959.

Butler JL *et al.* (2003) *Appl. Environ. Microbial.* **69**: 6793-6800

Chabot, R., Antoun, H. & Cescas, M.P. (1996) *Plant Soil.* **184**: 311-321

Chakraborty, U. & Chakraborty, B.N. (1988) *Can J. Bot.* **67**: 1698-1722.

Chakraborty, U. & Purkayastha, R.P. (1983) *Can. J. Microbiol.* **30**:285-289.

Chao, W.L. (1990) *Lett. Appl. Microbiol.* **10**: 213-215.

Chen, W.M., Laevens, S. Lee, T.M. Coeze, T., Vos P., Mergeay, M. & Vandamme, P. (2001) *Int. J. Syst. Evol. Microbial.* **51**: 1729-1735.

Curl, E.A. & Truelove, B. (1986) In: *The rhizosphere*, Springer-Verlag Berlin Heidelberg, Germany, pp. 52-54.

Dandurand, L.M. & Kundsen, G.R. (1993) *Phytopathol.* **83**: 265-270.

Datnoff, L.E., Nemec, S. & Pernezny, K. (1995) *Biol. Control.* **5**: 427-431.

Dazzo, F.B., Truchet, G.L., Sherwood, J.E., Hrabak, E.M., Abe, M. & Pankratz, S.H. (1984) *Appl. Environ. Microbiol.* **48**: 1140-1150.

de Boer, M., Sluis Van der, I., Loon Van, L. C. & Baker, P.A.H.M. (1999) *Eur. J. of Plant Pathol.***105**: 201-210.

de Boer, M., Bom, P., Kindt, F., Keurentjes, J.B., van Der Sluis, I., van Lun, L.C., Bakker, P.A.H.M. (2003) *Phytopathol.* **93**: 626-632.

de Boer, M., van der Siluis ,I., van Loon , L.C., Baker, P.AH.M. (1997) In: *Plant growth promoting rhizobacteria-Present status and future prospects.* (Eds. Ogoshi, A., Kobayashi, K., Homma, Y., Komadma, F., Kondo, N. & Akino, S.), Prock 4th Int. Workshop on Plant growth promoting rhizobacteria, Nakanishi printing Sapporo. Japan, pp. 380-382.

De Mot, R., Prowwst, P., Van Dame, J. & Vandarleyden, J. (1992) *Mol. Gen. Genet.* **231**: 489-493.

Defago, G. and Haas, D. (1990) In: *Soil Biochemistry* (Eds. Bollage, J.M. & Stotzky, G.), Vol 6, New York, Basel, pp 249-291.

Defago, G., Berling, C.H., Burger, U., Hass, D., Hahr, G., Keel, C.Voisard, C., Wirthner, P.H. & Wutrich, B. (1990) In: *Biological control of soil-borne plant pathogens.* (Ed. Hornby, D.), CAB international, pp. 93-108.

Deshmukh, M.R., Jain, H.C., Duhwn, S.S. & Goswami, U. (2002) *J. oilseed Res.* **19**: 73-75.

Deshwal, V.K., Dubey, R.C. & Maheshwari, D.K. (2003) *Curr. Sci.* **84**: 443-448.

Dowling, D.N. & O'Gara, F. (1994) *TIBTECH.* **12**: 133-141.

Dubey, R.C. & Maheshwari, D.K. (2001) In: *Practical Microbiology*. S Chand & Co. New Delhi.

Duffy, B.K. & Weller, D.M. (1995) *Plant Dis.* **79**: 907-911.

Duffy, B.K., Simon, A. & Weller, D.M. (1996) *Phytopathol.* **86**: 188-194.

Duhoon, S.S., Jain, H.C. Deshmukh, M.R. &

Goswami, U. (2001) *J. Oilseed Res.* **18**: 81-84.

Dunne, C., Moenne-Loccoz, Y, Mccarthy, J., Higgins, P., Powell, J., Dowling, D.N., & O'Gara , F. (1998) *Plant Pathol.* **47**: 299-307.

Epert, J.M. and Digat, B. (1995) *Can. J. Microbiol.* 41:685-691.

Flaishman, M.A., Eyal, Z., Zilberstein, A., Voisarsd, C. & Haas, D. (1996) *Mol. Plant-Microbe Interact.* **9**: 642-645.

Frank, B. (1879) *Bot. Ztg.* **37**: 376-387.

Frank, B. (1889) *Bot. Ges.* **7**: 332-346.

Frommel, M.I., Nowak, J., & Lazarovits,G. (1991) *Plant Physiol.* **96**: 928-936.

Galimov, E.M. (1985) In: *The biological fractionation of isotopes*. Academic Press, Orlando, FL.

Garbeva, P., van Veen, J.A. & van Elas, J.D. (2004) *FEMS Microbiol. Ecol.* **47**: 51-64.

Garcia de Salamone, I.E., Hynes, R.K. & Nelson, L.M. (2001) *Can. J. Microbiol.* **47**: 404-411.

Garret, S.D. (1965) In: *Prelude to Biological Control.* (Eds. Baker, K.F. & Synder, W.C.) California Univ. Press, Berkeley, pp 4-17.

Ghosh, D.C. & Das, A.K. (1998) *Ind. Agric.* **42**: 109-113.

Glick, B. R. (1995) *Can. J Microbiol.* **47**: 109-117.

Glick, B.R., Patten, C.L., Holguin, G. & Penrose, D.M. (1999) Biochemical and genetic mechanism used by plant growth-promoting bacteria. Imperial College Press, London, UK.

Glick, B.R., Penrose, D. M. & Li, J. (1998) *J. Theor. Bio.* **190**: 63-68.

Gramss, G. & Rudeschko, O. (1998) *New Phytologist.* **138 (3)**: 401-409.

Grayston, S.J. *et al.* (2001) *Soil Biol. Biochem.* **33**: 533-551.

Guetsky, R., Shtienberg, D., Elad, Y. & Dinoor, A. (2001) *Phytopathol.* **91**: 621-627.

Gupta, C.P., Dubey, R.C. & Maheshwari D. K. (2002) *Biol. Fert. Soils.* **35**: 399-405.

Gupta, C.P., Sharma, A., Dubey, R.C. & Maheshwari, D.K. (1999) *Cytobios.* **99**: 183-189.

Gutterson, N.I., Layton, T.J., Ziegle, J.S. & Warren, G.J. (1986) *J. Bacteriol.* **165**: 696-703.

Halder, S.A.K., Mishra, A.K., Bhhatacharya, P. & Chakrabarty, P.K. (1990) *J. Gen. Appl. Microbiol.* **36**: 81-92.

Hallmann, J., Quadt-Hallmann, A., Miller, W.G., Sikira, R.A. & Lindow, S.E. (2001) *Amer. Phytopathol. Soc.* **91**: 415-422.

Hassan, D.G., Zarjar, M. & Beigh, G.M. (1997) *Mol. Ecol.* **34**: 74-80.

Hegde, D.M. (1998) *J. Oilseed Res.* **15**: 1-17.

Hera, C. (1996) *Fertil. Environmen.* 131-149.

Hirsch, A.M., Fang, Y., Asad, S. & Kapulnik, Y. (1997) *Plant Soil.* **194**: 171-184.

Hoflich, G., Wiehe, W. & Khun, G. (1994) *Experientia* **50**:897-905.

Holt, J.G., Krieg, N.R., Sneath, P.H.A., Staley, J.T., & Williams, S.T. (1994) In: *Bergey's manual of Determinative Bacteriology.* Williams and Wilkins Press, Baltimore, U.S.A.

Howell, C.R. & Stipanovic, R.D. (1980) *Phytopathol.* **70**: 712-715.

Howie, W.J., Cook, R. J. & Weller, D.M. (1987) *Phytopathol.* **77**: 286-292.

Hubbard, J.P., Harman, G.E. & Hadar, Y. (1983) *Phytopathol.* **73**: 655-659.

Hwang, S.F. & Chatravarty, P. (1993) *Z. Pflanzenkr. Pflanzenschuz.* **100**: 308-318.

Illmer, P. & Sphinner, F. (1995a) *Soil Biol. Biochem.* **27**: 257-263.

James, D.W.Jr. & Guttorson, N. (1986) *Appl. Environ. Microbiol.* **52**: 1183-1189.

Janisiewicz, W.J. (1988) *Phytopathol.* **78**:194-198.

Janisiewicz, W.J. (1996) *Phytopathol.* **86**: 473-479.

Janisiewicz, W.J., & Bors, B. (1995) *Appl. Environ. Microbial.* **61**:3261-3267.

Jordan, D.C. (1982) *Int. J. Sys.t Bacteriol.* **32**: 136-139.

Kanner, D., Gerber, N. N. & Bartha, R. (1978) *J. Bacteriol.***134**:690-692.

Kerr, A. (1980) *Plant Dis.* **64**: 25-30.

Kloepper, J.W., Leong, J, Teintze, M. & Schroth, M.N. (1980) *Nature* **286**: 885-886.

Kloepper, J.W. and Schorth, M.N. (1978) In: *Proc. 4th Int. Conf. Plant Pathology Bacteria,* (Ed. Angers, I.N.R.A.), Vol. 2, staion de pathologie, Gibert-Clary, Tours, France, pp 879-882.

Kloepper, J.W., Hume, D.J, Scher, F.M., Singeleton, C., Tipping, B., Laliberte, M., Frauley, K., Kutchaw, T., Simonson, C., Lifshitzr, R., Zeleska, I. & Lee, L. (1988) *Plant Dis.* **72**: 42-46.

Kloepper, J.W., Leong, J., Teintze, M. & Schroth, M.N. (1980) *Curr. Microbiol.* **4**: 317-320.

Kloepper, J.W., Lifshitz, R. & Zablotowich, R.K. (1989) *Trends. Biotechnol.* **7**: 39-43.

Kreig, N.R. & Holt, J.G. (1984) In: *Bergie's manual of Systematic Bacteriology*, Williums and Wilkins Co., Baltimore, Md.

Kumar, B., Dubey, R.C. & Maheshwari, D.K. (2005) In: *Microbial diversity: Current perspectives and potential applications* (Eds. Satyanarayana, T. & Johri, B.N.), I.K. International Pvt. Ltd., New Delhi, pp 471-492.

Laguerre, G. *et al.* (1996) *Appl. Environ. Microbiol.* **62**: 2029-2036.

Lang, S.R. (2000) *Plant Physiol.* **125**: 69-72.

Leeman, M., Den Ouden , F.M. van Pelt , J.A., Cornelissen, C., Matamala -Garros, A., Bakker, P.A.H.M. & Schippers , B. (1996) *Eur. J. Plant Pathol.* **102**: 21-31.

Leibinger, W., Beuker,B., Hahn, M. & Mendgen, K. (1997) *Phytopathol.* **87**: 1103-1110

Lemanceau, P. & Alabouvette, C. (1991) *Crop. Prot.* **10**: 279-286.

Liles, M.R., Averscheel, T., C. & Cianciotto, N.P. (2000) *J. Bacteriol.* **182**: 749-757.

Lim, H.S. & Kim, S.D. (1995) *Curr. Microbiol.* **33**: 295-301

Loper, J.E. Suslow, T.V. & Schroth, M.N. (1984) *Phytopathol.* **74**: 1454-1460.

Ludwig, W., Amann, R., Martinez-Romero, E., Schonhubar, W., Bauer, S., Neef, A. & Schleifer, K.H. (1998) *Plant Soil.* **204**: 1-19.

Lugtenberg, B.J.J. and Bloemberg, G.B. (2004) In: *Pseudomonas Vol I*, (Ed. Ramos, J.L.) Kluwer, Academic Publishers, New York, pp. 403-430.

Lynch, J.M. (1990) In: *The rhizosphere* (Ed. Lynch, J.M.) A Wiley-Interscience publication pp. 177-207.

Lynch, J.M. & Whipps, J.M. (1990) *Plant and Soil* **129**: 1-10.

M. Lucy., Reed & Glick, R. (2004) *Antonie van Leeuwenhek* **86**: 1-25.

Mahmoud, S.A.Z., Ramadan, E.M., Thabet, F.M. & Khater, T. (1984) *Zbl. Mikrobiol.* **139**: 227-232.

Mandal, B.K., Mandal, B.B., Murmu, N.N., Tudu, B.C. & Khanda, C.M. (2003) *Ind. J. Plant Physiol.* **8**: 103-105.

Manefield, M. *et al.*, (2002) *Appl. Environ. Microbiol.* **68**: 5367-5373.

Martinez-Romero, E., Segovia, L., Mercante, F.M., Franco, A.A., Graham, P. & Pardo, M.A. (1991) *Ibid* **1**: 417-426.

Mazzola, M. Cook, R.J. Thomashaow, L.S. Weller, D.M. & Pierson, L.S. (1992) *Appl. Environ. Microbiol.* **58**: 2616-2624.

Mazzola, M. Fugimato, D.K., Thomshow, L.S. & Cook, R.J. (1995) *Appl. Environ, Microbiol.* **61**: 2554-2559.

Mc Donald, B.A., Miles, A.J., Nelson, L.R. & Pettway, R.E. (1994) *Phytopathol.* **84**: 250-255.

Mc Inroy, J.A. & Kloepper, J.W. (1995). *Plant Soil* **73**: 337-342.

Meikle, A., Killham, K., Prosser, J.I., & Glover, L.A. (1992) *FEMS Microbiol. Lett.* **99**: 217-220.

Miller, R.H. & May, S. (1991) In: *The rhizosphere and plant growth* (Eds. Keister, D.L. & Cregan, P.B.), Kluwer Academic Publishers, Dordrecht, The Netherlands, pp. 123-134.

Mitra, S., Aulakh, M.S., Wassmann, R., & Olk, D.C.(2005) *Soil Science Society of America Journal.* **69 (2)**: 563-570.

Mohuddin, M. D., Das, A. K. & Ghosh, D. C. (2000) *Ind. J. Plant Physiol.* **5**: 334-338.

Mordukhova, E.A., Skvortsova, N.P., Kochaetko, V.V., Dubeikovskii, A.N. & Boronin., A.M. (1991) *Mikrobiologiya* **60**: 494-500.

Moulin, L., Munive, A., Preybis, B. & Boivin-Mosson, C. (2001) *Nature* **411**: 948-950.

Mukhopadhyay, A.N., Shrestha, S.M. & Mukherjee, P. K. (1992) *FAQ Plant Protect. Bull.* **40**: 110.

Muntasir, I.M., Tauhiduzzaman, S. & Ishaque, M. (2001) *Indian J. Agric. Chem.* **XXXIV (1, 2)**: 15-21.

Nagahashi, G., & Douds, D.D. (2004) *Mycological Res.***108 (9):** 1079-1088.

Neilands, J.B. & Leong, S.A. (1986) *Ann. Rev. Plant. Physiol.* **37**: 187-208.

Neilands, J.B., Nakamura, K. (1997) In: CRC Handbook of Microbial Iron Chelates (Ed. Winkelmann, G.). CRC Press, Boca Raton pp1-14.

Noel, T.C., Sheng, C., York, C. K., Pharis, R. P. & Hynes, M. F. (1996) *Can J Microbiol.* **42**: 279-283.

O'Sullivan, D.J. & O'Gara, F. (1992) *Microbial. Rev.* **56**: 662-676.

Obaton, M. (1973) *Bull. Eco Comm. NFR.* **17**: 170-171.

Obaton, M., Bouniols, A., Piva, G. & Vadez, V. (2002) *Plant and Soil.* **245**: 315-326.

Oyaizu, H., Naruhashi, N. & Gamou, T. (1992) *Biodivers Conserv.* **1**: 237-249.

Pandey, P., Kang, S.C. & Maheshwari, D.K. (2005) *Curr Sci.* **89**: 177-180.

Pannase, S.K., Thakur, N.S. & Sawarkar, S.D. (1995) *Indian J. Agron.* **40**: 323-324.

Park, C., Paulitz, T.C. & Bakker, R. (1988) *Phytopathol.* **78**: 190-194.

Paterson, E. *et al.* (1997) *Global Change Biology* **3**: 363-377.

Patten, C.L. & Glick, B.R. (2002) *Appl. Environ. Microbiol.* **68**: 3795-3801.

Pena-Cabriales, J.J. & Alexander, M. (1983) *Soil. Sci. Soc. Ame. J.* **47**: 81-84.

Perrot, X., Stachelin, C. & Broughton, W.J. (2000) *Microbiol. Mol. Biol. Rev.* **64**: 180-201.

Pierson, E.A. & Weller, D.M. (1993) *Phytopathol.* **83**: 1365.

Pierson, E.A. & Weller, D.M. (1994) *Phytopathol.* **84**: 940-947.

Postgate, J.R. (1987) *J. App. Bacteriol.* 855-916.

Prosser, J.I., Killham, K., Glover, L.A., & Rattray, E.A.S. (1996) *Critical Rev. Biotech.***16**: 157-183.

Raaijmakers, J.M. & Weller, D.M. (2001) *Appl Environ. Microbil.* **67**: 2545-2554.

Ramirez, M.E., Israel, D.W. & Wollum, A.G. (1998) *Can. J. Microbiol.* **44**: 753-758.

Raupach, G.S. & Kloepper, J. W. (1998) *Phytopathol.* **8**: 1158-1164.

Rovira, A.D. (1969) *Bot. Rev.* **35**: 35-57.

Saha, A.K., Deshpande, M.V. & Kapadnis, B.P. (2001) *Curr. Sci.* **80**: 669-671.

Sahgal, M. & Johri, B.N. (2003) *Curr. Sci.* **84**: 43-48.

Saraf, M. & Sood, N. (2002) *J. Ind. Bot. Soc.* **81**: 154-157.

Scheffer, F. & Schachtschabel, P. (1992) In: *Lehrbuch der Bondenkunde*. Ferdinand Enke Verlag, Stuttgard, Germany.

Schippers, B. (1992) In: *Biological control of plant diseases, progress and challengers for the future* (Eds. Tjamos, EC., . Papavizas, G.C. & Cook, R.J.), NATO ASI Series A-Life Sciences, Vol 230. Plenum Press New York, pp. 21-34.

Schippers, B., Bakker, A.W. & Bakker, P.A.H.M. (1987) *Ann. Rev. Phytopathol.* **25**: 339-358.

Schisler, D.A., Slininger, P.J. & Botahst, R.J. (1997) *Phytopathol.* **87**: 177-183.

Schloter, M., Wiehe, W., Assmus, B., Steindl, H., Beckeh, H., Hoflich, G. & Hartmann, A. (1997) *Appl.Environ. Microbiol.* **63**: 2038-2046.

Selvaraj, G. & Iyer, N.N. (1984). *Int. J. Bacteriol.* **158**: 580-589.

Sessitsch, A., Haedarson, G., de Vos, W.M. & Wilson, K.J. (1998) *Plant Soil.* **204**: 35-45.

Shanahan, P., O'Sullivan, D.J., Simpson, P., Glennon, J.D. & O'Gara, G. (1992) *Appl. Environ. Microbiol.***58**: 353-358.

Shanmungam V, Senthil N, Raguchander T, Ramanathan A & Samiyappan R (2002) *Phytoparasitica* **30**: 169-176.

Shende, S.T., Apte, R.G., & Singh,T. (1977) *Curr. Sci.* **46**: 675.

Sheneviratne, G. (2003) *Curr. Sci.* **85**: 1395-96.

Singh, B.K., Millard, P., Whiteley, A.S. & Murrell, J.C. (2004) *Trends Microbiol.* **12 (8)**: 386-393.

Sneh, B., Dupler, M., Elad, Y. & Baker, R. (1984) *Phytopathol.* **74**: 1115-1124.

Somers, E., Vanderleyden, J. & Srinivasan, M. (2004) *Crit. Rev. Microbiol.* **30(4)**: 205-240

Somers, E. & Vanderleyden, J. (2004) *Ann Rev. Microbiol.* **30**: 205-240.

Spaink, H.P. (2000) *Ann. Rev. Microbiol.* **54**: 277-288.

Stanier, R. Y., Adelberg, E. A. & Ingraham, J. L. (1985) In: *General Microbiology*. McMilan Publication, London. pp. 273.

Starkenburg, S.R., Casey, J.N. & Cianciotto, N.P. (2004) *Curr. Microbiol.* **49**: 203-207.

Stephens, J.H.G & Rask, H.M. (2000) *Field Crop Res.* **65**: 249-258.

Sung, K.C. & Chung, Y.R. (1997) In: *Plant growth-promoting rhizobacteria-Present status and future prospects.* Proc Int. Workshop on Plant growth-promoting rhizobacteria, $4^{th}$. A. Ogosi, K.Obiayashi Printing, Sapporo, Japan, pp. 370-372.

Suslow, T.V. (1982) In: *Phytopathogenic Prokaryots*, (Eds. Mount, M.S. & Lacy, G.H.), London: Acedamic Press.

Tang, W.H. (1994) In: *improving plant productivity with rhizosphere bacteria* (Eds. Ryder, M.H., Stephens, P.M. & Bowen, G.D.), Commonwealth scientific and industrial research organization, Adelaide, Australia. pp. 267-278.

Tiwari, K.P., Namdev, K.V., Tomar, R.K.S. & Ragu J.S. (1995) *Ind. J. Agron.* **40**: 134-136.

Tomar, R.K.S., Namdeo, K.N., Raghu, J.S. & Tiwari, K.P. (1995) *Ind. J. Agric. Sci.* **65**: 256-259.

Tomboline, R., Vandergaag, D.J., Gerhardson, B. & Jansson, J.K. (1999) *Appl. Environ. Microbiol.* **65**: 3674-3680

Topp, E. (2003) *Cem. J. Soil* **83**: 803-809.

Trinick, M.J. & Hadobas, P.A. (1995) *Plant Soil.* **172**: 207-219.

Triouchi, N., & Syono, K. (1990) *Plant Cell Physiol.* **31**: 119-127.

Tu, J. C. (1989) *Plant Dis.* **73**: 281-285.

Turrini, A., Sbrana, C., Pitto, L., Castiglione, M.R., Giorgetti, L., Briganti, R., Bracci, T., Evangelista, M., Nuti, M.P. & Giovannetti, M. (2004) *New Phytologist* **163 (2):** 393-403.

Upadhyay, R. S. & Jayaswal, R. K. (1992) *Curr. Microbiol.* **24**: 181 - 187.

Valladares, F. Salvador, P. V., Dominguez, S., Fernandez, M., Penuelas, J. L. & Pugnaire, F. I. (2002) *Plant Soil.* **240**: 253-262.

Van Berkum, P. & Earldy, B. D. (1998) In: *The Rhizobiaceae-molecular biology of plant associated bacteria* (Eds. Spaionk, H.P., Kondorosi, A. & Hooykaas, P.J.J.) Kluwer, Dordrecht, pp. 1-24.

Vargas, M.A.T., Mandes, I.C. & Hungaria, M. (2000) *Biol. Fert. Soil.* **33**: 228-233.

Verma, S.C., Chowdhury, S.A. & Tripathi, A.K. (2004) *J. Microbiol.* **50**: 313-322.

Vivekanathan, R., Ravi, M., Ramanathan, A. & Samiyappan, R. (2004) *W. J. Microbiol. Biotechnol.* **20**: 235-244.

Vyas, S.C. (1994) *Ind. J. Mycol. Plant Pathol.* **24**: 132-134.

Weller, D.M. (1988) *Annu. Rev. Phytopathol* **26**: 279-407.

Weller, D.M. & Thomashao, L.S. (1994) In: *Molecular ecology of rhizosphere microorganisms: Biotechnology and the release of GMO's.* (Eds O'Gara,. F., Dowling, D.N. & Boesten, B.) Vch, Weinheim, Germany, pp. 1-18.

Weller, D.M., Raaijmakers, J.M., Gardner, B.B.M. & Thomashow, L.S. (2002) *Ann Rev. Phytopathol.* **40**: 309-348.

White, C., Tardif, J.C., Adkins, A. & Staniforth, R. (2005) *Soil Biol. Biochem.* 37 (7): 1359-1372.

Wiehe, W. & Hoflich, G. (1995) *Microbiol. Res.* **150**: 201-206.

Wilson, K.J., Parra, A. & Botera, L. (1999) *Can J. Microbil.* **45**: 678-685.

Wipat, A. & Harwood, C.R. (1999) *FEMS Microbiol Ecol.* **28**: 1-9.

Ye, S.F., Yu, J.Q., Peng, Y.H., Zheng, J.H. & Zou, L.Y. (2004) *Plant and Soil.* **263 (1-2)**: 143-150.

Young, J.M., Kuykendall. L.D., Martinez-Romero, E., Kerr, A. and Sawada, H. (2001) *Int. J. Syst Evol. Microbiol.* **51**: 89-103.

Young, J.P.W., Downer, H.L. & Eardly, B.D. (1991) *J. Bacteriol.* **173**: 2271-2277.

Zahir, Z.A., Arshad, M. & Frankenberger, W.T. (2004) *Advances in Agronomy* **81**: 97-168.

# 2

# PGPR-Biotechnology for Management of Abiotic and Biotic Stresses in Crop Plants

**V. Govindasamy, M. Senthilkumar, Upendra Kumar and K. Annapurna**
*Division of Microbiology, Indian Agricultural Research Institute, New Delhi-110 012, India.*

## ABSTRACT

Soil microbial populations are immersed in a framework of interactions, which are known to affect plant fitness and soil quality. The bacteria that provide some benefit to plants are of either symbiotic types or free-living in the soil. Beneficial free-living bacteria in the plant rhizosphere are usually referred as plant growth promoting rhizobacteria (PGPR) which promote the plant growth by means of direct and indirect mechanisms. Many plant growth promoting rhizobacteria contain the enzyme 1-aminocyclopropane -1- carboxylate (ACC) deaminase and promote plant growth by sequestering and cleaving plant produced ACC, the immediate precursor of the plant hormone ethylene and thereby lowering the level of ethylene in the plant. A wide range of abiotic stresses such as flooding, drought, salt, heavy metals and organic contaminants; high and low temperature can induce synthesis of stress ethylene which causes severe damage to the plants and affecting the crop production. ACC deaminase containing PGPR can be exploited as successful strategy for protecting the plants against the deleterious effects caused by these stresses. Amongst the biotic stresses, phytopathogens can reduce crop yield which is an enormous potential loss to crop productivity. PGPR protect plants from the damages caused by phytopathogens in a number of different mechanisms including; synthesis of antibiotics, antifungal metabolites and defence enzymes, exhibiting competition with phytopathogens, secretion of iron chelating siderophores and induction of systemic resistance in plants. Development of superior or novel PGPR strains by improving above traits can be possible using genetic manipulations. These PGPR-biotechnologies can be exploited as a low-input, sustainable and environment-friendly technology for the management of plant stresses.

## INTRODUCTION

The complexity of the soil system is determined by the numerous and diverse interactions among its physical, chemical, and biological components, as modulated by the prevalent environmental conditions (Buscot, 2005). In particular, the varied genetic and functional activities of the extensive microbial populations have a critical impact on soil functions, based on the fact that microorganisms are driving forces for fundamental metabolic processes involving specific enzyme activities (Nannipieri *et al.*, 2003). Many microbial interactions, which are regulated by specific molecules/signals (Pace, 1997), are responsible for key environmental processes, such as the biogeochemical cycling of nutrients and matter and the maintenance of plant health and soil quality (Barea *et al.*, 2004). Many studies have demonstrated that soil-borne microbes interact with plant roots and soil constituents at the root-soil interface (Lynch, 1990; Linderman, 1992; Glick, 1995a; Kennedy, 1998; Bowen and Rovira, 1999; Barea *et al.*, 2002b). The great array of root-microbe interactions results in the development of a dynamic environment known as the rhizosphere where microbial communities also interact. The differing physical, chemical, and biological properties of the root associated soil, compared with those of the root-free bulk soil, are responsible for changes in microbial diversity and for increased numbers and activity of micro-organisms in the rhizosphere microenvironment (Kennedy, 1998).

A large number of different microorganisms are commonly found in the soil including bacteria, fungi, actinomycetes, protozoa, and algae (Paul and Clark, 1989). Of these, bacteria are by far the most common type of soil microorganisms, possibly because they can grow rapidly and have the ability to utilize a wide range of substances as either carbon or nitrogen sources. While many of the bacteria found in soil are bound to the surface of soil particles and are found in soil aggregates, a number of soil bacteria interact specifically with the roots of plants. The interaction between bacteria and the roots of plants may be beneficial, harmful, or neutral for the plant, and sometimes the effect of a particular bacterium may vary as a consequence of soil conditions (Lynch, 1990). The bacteria that provide some benefit to plants are of two general types—those that form a symbiotic relationship with the plant, which has been studied extensively, and those that are free-living in the soil, but are often found near, on, or even within the roots of plants (Kloepper *et al.*, 1988; Van Peer and Schippers, 1989; Frommel *et al.*, 1991). The beneficial free-living soil bacteria are usually referred to as plant growth promoting rhizobacteria or PGPR (Kloepper *et al.*, 1989), or, by one group of workers in China, as yield increasing bacteria or YIB (Piao *et al.*, 1992; Tang, 1994). Beneficial, root colonizing, rhizosphere bacteria, the PGPR, are defined by three intrinsic characteristics: (i) they must be able to colonize the root, (ii) they must survive and multiply in microhabitats associated with the root surface, in competition with other microbiota, at least for the time needed to express their plant promotion/protection activities, and (iii) they must promote plant growth.

Plant growth promoting rhizobacteria can affect plant growth in a variety of ways (Glick, 1995a; Glick *et al.*, 1999). They increase the seed germination and viability, root respiration and formation; stimulate root growth and whole plant growth by accelerating cell division; and increase plant membrane permeability for most efficient nutrient uptake. Some of the mechanisms of PGPR can effectively reduce the damages caused by abiotic and biotic stresses on crop plants and thereby management of these stresses in crop production.

## PLANT ETHYLENE HORMONE AND ABIOTIC STRESSES

Ethylene is a gaseous plant hormone which is produced by almost all plants. It mediates a range of different plant responses and developmental steps (Abeles *et al.*, 1991; Mattoo and Suttle, 1991). Ethylene plays an active role in seed germination, tissue differentiation, the formation of root and shoots primordia, root elongation, lateral bud development, flowering initiation, anthocyanin synthesis, flower opening and senescence, pollination, fruit ripening and degreening, the production of volatile organic compounds responsible for aroma formation in fruits, storage product hydrolysis and leaf and fruit abscission. Ethylene is also involved in plant-microbial symbiotic interactions that are important for establishment of the legume-*Rhizobium* association (Lynch and Brown, 1997). Ethylene has the ability to trigger exaggerated disease symptoms and exacerbate an environmental pressure. Except for fruit ripening and lateral root initiation, high levels of ethylene are usually deleterious to plant growth and health.

Abiotic stresses are caused in living organisms by non-living environmental factors, such as drought, extreme temperatures, edaphic conditions and high winds. Plants are especially dependent on environmental factors, and continued abiotic stress can have harmful effects on them or to force them for natural selection. Abiotic stress is a major constraint in crop and food production. A wide range of abiotic stresses including flooding, drought, high and low temperature, heavy metals and salt can induce the synthesis of stress ethylene in plants. With various stresses in plants, the increase in ethylene synthesis serves as a common step in the chain of events leading to a variety of responses that are exhibited by plants.

One model that explains the paradoxical effects of stress ethylene on plants growing under abiotic stress conditions emphasizes the fact that in stressed plant tissues there is an initial small peak of ethylene close in time to the onset of stress, and then a second much larger peak some time later (Stearns and Glick, 2003). The first peak is only a small fraction of the magnitude of the second peak and is thought to initiate a protective response by the plant, such as transcription of stress and pathogenesis related genes (Ciardi *et al.*, 2000). The second peak is so large, however, that processes such as senescence, chlorosis and abscission are initiated, the overall effect of which is detrimental to plant survival. The first peak of ethylene production is thought to consume the existing pool of 1-aminocyclopropane-1-carboxylic acid (ACC), the immediate precursor to ethylene, after which ACC synthase genes are transcribed and more ACC begins to accumulate to fuel the second wave of ethylene production (Robison *et al.*, 2001). This occurs due to the regulation of ACC synthase genes by environmental and developmental cues and the enhancement of its enzymatic action during stress conditions (Yang and Hoffman, 1984).

## PGPR AND ACC DEAMINASE ENZYME

Plant growth promoting rhizobacteria include a diverse group of free-living soil bacteria that can stimulate the growth of plants by one or more direct or indirect mechanisms (Glick, 1995a). Indirect stimulation of plant proliferation includes preventing phytopathogens from inhibiting plant growth and development, while direct stimulation includes providing plants with compounds such as fixed nitrogen, phytohormones, or solubilized iron from the soil (Glick, 1995a; Glick *et al.*, 1999). A number of different bacteria may be considered to be PGPR, including *Azotobacter* species, *Azospirillum* species, Pseudomonads, *Acetobacter* species, *Burkholderia* species, and Bacilli (Klopper *et al.*, 1988; Bashan and Levanony, 1990; Tang, 1994).

Many plant growth promoting rhizobacteria contain the enzyme 1-aminocyclopropane-1-carboxylate (ACC) deaminase and this enzyme can cleave the ethylene precursor ACC to α-ketobutyrate and ammonia and thereby lower the level of ethylene in developing or stressed plants (Glick, 1995a; Glick *et al.*, 1998; Jacobson *et al.*, 1994). When ACC deaminase-containing plant growth-promoting bacteria are bound to a plant (Fig. 1), they act as a sink for ACC ensuring that plant ethylene levels do not become elevated to the point where root growth is impaired (Glick *et al.*, 1998). This activity is important during normal plant development and also protects plants from the deleterious effects of numerous abiotic stresses (Grichko and Glick, 2001a Mayak *et al.*, 2004; Belimov *et al.*, 2001; Burd *et al.*, 1998). Moreover, ACC deaminase-containing plant growth promoting bacteria up-regulate genes involved in plant growth and protein production while down-regulating plant genes involved in ethylene stress and defence signaling pathways (Hontzeas *et al.*, 2004a). The ACC deaminase-containing plant growth promoting bacteria, in part, alleviate the need for the plant to actively defend itself against various abiotic stresses (Hontzeas *et al.*, 2004a; van Loon and Glick, 2004).

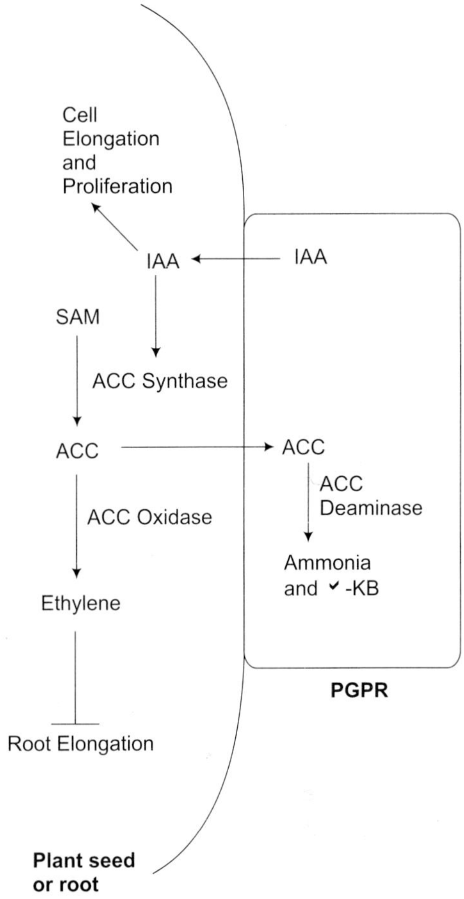


**Fig.1** Schematic model to explain how ACC deaminase containing PGPR lowers the ethylene concentration thereby preventing ethylene inhibition of root elongation (Adapted from Glick et al., 1998). Key: IAA, Indole acetic acid; ACC, 1- aminocyclo-propane-1-carboxylic acid; SAM, S-adenosyl methionine; α-KB, α-Ketobutyrate.

Very few ACC deaminase proteins have been purified and biochemically characterized, and their crystal structures have been produced for only two organisms (Table 1). For the first time Minami and co-workers (1998) reported ACC deaminase enzyme crystal structure for the yeast *Hansenula saturns*. Recently, the crystal structures for bacterial ACC deaminase enzymes were determined by Karthikeyan *et al.*, (2004); the biochemical and thermodynamic properties of the ACC deaminase from *Pseudomonas putida* UW4 have been measured (Hontzeas *et al.*, 2004b). Characteristic of all ACC deaminase enzymes is their low affinity for the substrate ACC, which is always in the millimolar range. In the past few years a large number of bacterial ACC deaminase genes have been isolated and shown to encode ACC deaminase activity (Glick, 2005). Alignment of the amino acid sequences from some of the genes indicates that there is conservation of key amino acid residues (i.e., $Lys^{51}$ $Cys^{162}$), both thought to be important for enzymatic activity.

**Table 1** Biochemical characteristics of some ACC deaminases

| *Characteristics* | *Pseudomonas putida UW4* | *Pseudomonas putida GR12-2* | *Pseudomonas sp. ACP* | *Hansenula saturnus* | *Penicillium citrinum* |
|---|---|---|---|---|---|
| Subunit molecular mass (Daltons) | 36,874 | 35,000 | 36,500 | 40,000 | 41,000 |
| Estimated subunits | 3 | 3 | 3 | 2 | 2 |
| $P^H$ optimum | 8.0 | 8.5 | 8.5 | 8.5 | 8.5 |
| $\Delta T_m$ | 60.2 °C | nd | nd | nd | nd |
| $\Delta K_m$ | 3.4mM | nd | 1.5mM | 2.6mM | 4.8mM |
| $\Delta G^{\#}$ at 298K | 69.6KJ/mol | nd | nd | nd | nd |
| $\Delta H^{\#}$ | 46.8KJ/mol | nd | nd | nd | nd |
| $\Delta S^{\#}$ | -78J/mol K | nd | nd | nd | nd |
| Crystal structure | No | No | Yes | Yes | No |
| References | Hontzeas *et al.*, (2004b) | Jacobson *et al.*, (1994) | Karthikeyan *et al.*, (2004) | Minami *et al.*, (1998) | Jia *et al.*, (2000) |

nd = not determined.

## PGPR IN MANAGEMENT OF ABIOTIC STRESSES

Most agricultural and agronomic practices are designated to optimize crop growth by avoiding or reducing abiotic stress (irrigation, fertilization, etc.). Plant breeding is designated, among other things, to develop genetically stable crop cultivars that are more resilient under stresses. However, application of plant growth promoting rhizobacteria is an alternative approach in the management of abiotic stresses in crop production.

The most successful strategy for managing the abiotic stresses in plants is lowering the level of stress ethylene production. A number of labs have reported decreased ethylene production by suppressing the expression of some of the enzymes of the ethylene biosynthesis and signalling pathways. However, since the activity of many of these enzymes affects processes other than ethylene synthesis, there may be some advantages in utilizing the microbial enzyme ACC deaminase, which cleaves ACC to ammonia and $\alpha$-ketobutyric acid to establish a sink for ACC and lowering the ethylene level in plants (Glick *et al.*,1998). This can be achieved either through the interaction of ACC deaminase-containing plant growth-promoting bacteria with plant roots or by the development of transgenic plants expressing this enzyme. In either case, ACC deaminase can significantly decrease ACC levels in plants, especially plants subjected to various abiotic stresses, thereby decreasing the amount of stress ethylene and subsequent damage to the plant that might occur as a consequence of stress ethylene (Grichko and Glick, 2001a; Mayak *et al.*, 2004; Belimov *et al.*, 2005).

## FLOODING STRESS

Deprivation of oxygen to the roots of plants is the main consequence of flooding (Jackson, 1985; Kozolowski, 1984). Under normal conditions, water dissolves about 230 m mol·m$^{-3}$ oxygen, and hypoxia occurs when the oxygen level falls below 50 m mol·m$^{-3}$. Within 1 h of flooding, the partial pressure of oxygen declines from 20.8 to 7.9 kPa, and further decreases to 1 kPa after 1 d of flooding (Nilsen and Orcutt, 1997). Shortly after the onset of flooding, soil microorganisms consume all of the available oxygen, and different toxic compounds begin to accumulate in the soil. Plants respond to flooding with a reduced root permeability, water absorption and mineral uptake; closure of stomata followed by a decrease in photosynthesis; alterations in hormone balance; inhibition of stem and root growth; hypertrophy of lenticels; development of aerenchyma and adventitious roots; epinasty, chlorosis, leaf abscission and premature fruit drop (Vartapetian and Jackson, 1997).

In flooding, roots release a high amount of ACC into the surrounding soil (Else *et al.*, 1995). Soil bacteria capable of degrading ACC are relatively common (Glick *et al.*, 1995). The establishment of microbial populations able to break down ACC may be an effective non-invasive approach to ameliorating the damage to plants caused by flooding. It utilizes the beneficial relationships of a plant and ACC deaminase-containing plant growth-promoting bacteria that can significantly decrease ACC levels in plants (Glick *et al.*, 1994; Glick *et al.*,1998). In plant growth-promoting bacteria, the expression of an ACC deaminase gene, which is increased under anaerobic conditions, helps to lower the level of endogenous ACC and hence, the concentration of ethylene in a plant (Grichko and Glick, 2001a). In flooded soils, the growth of fungi and actinomycetes is suppressed and bacteria predominate. Moreover, bacteria that are able to proliferate in the absence of oxygen and utilize the ACC that is exuded by plant roots are likely to be selectively enriched. Grichko and Glick (2001a) reported that tomato plants that were grown from the seeds bacterized with organisms expressing ACC deaminase such as *Enterobacter cloacae* UW4, *E. cloacae* CAL2, *Pseudomonas putida* ATCC17399/pRKACC or *P. putida* ATCC17399/pRK415 showed a substancial tolerance to flooding stress.

Klee *et al*., (1991) developed transgenic tomato plants expressing microbial ACC deaminase gene and showed that expression of an ACC deaminase gene under the transcriptional control of the 35*S* promoter from cauliflower mosaic virus greatly reduced ethylene production and delayed fruit ripening, and ACC treatment did not cause an increase in either ethylene production or epinasty in petioles. In fact, when transgenic plants were infiltrated with ACC, there was a low level of ethylene production but no epinastic response. Grichko and Glick (2001b) studied the response of transgenic tomato plants expressing the bacterial gene 1-aminocyclopropane-1-carboxylic acid (ACC) deaminase to flooding stress and showed some increase to flooding and hence less deleterious effects of root hypoxia on plant growth.

## Drought and Salt Stress

Drought or water stress is an environmental condition resulting due to deficiency or dearth of water, which affects the physiological processes of plant. By far it is the most important abiotic factor that adversely affects the crop production. Relatively few mechanisms have been demonstrated to explain the increased resistance of plants treated with plant growth promoting rhizobacteria to drought or water stress. The mechanisms that have been suggested include reduction of stress ethylene production via ACC deaminase activity (Glick *et al*., 1998) and increased expression of the ERD15 gene, which is responsive to drought stress (Timmusk and Wagner, 1999). Mayak *et al*., (2004a) isolated the bacterium containing ACC deaminase from soil samples collected in dry river beds and shown to increase resistance in tomato and peppers to water stress.

Natural soil forming processes in warm and dry regions frequently produce saline soils with low agricultural potential. In these areas most crops are grown under irrigation, and inadequate irrigation management leads to secondary salinization that affects 20% of irrigated land worldwide. Soil salinity is an enormous problem for agriculture under irrigation. In addition to the use of traditional breeding and plant genetic engineering with production of transgenic plants (Apse *et al*., 1999; Blumwald, 2000), the use of plant growth promoting rhizobacteria may prove useful in developing strategies to facilitate plant growth in saline soils. A bacterium *Achromobacter piechaudii* ARV8 containing ACC deaminase increases the resistance of tomato seedlings to salt. The bacterium reduces the production of ethylene by the seedlings, hence alleviating the suppression of growth. It slightly increased the uptake of phosphorus and potassium, which may contribute in part to activation of processes involved in the alleviation of the effect of salt. The bacterium did not influence the status of water parameters. It did, however, increase the water use efficiency (Mayak *et al*., 2004b).

### Heavy Metal Stress

A number of heavy metals are required by plants as micronutrients to act as cofactors as part of prosthetic group of enzymes, which are involved in a wide variety of metabolic pathways.

However, when they are present in excess, most heavy metals can act as toxicants, reducing the success of plant life (Van Assche and Clijsters, 1990). Among the metals that are most toxic to plant and animal life are those that displace essential metal ions in biological processes. These include cadmium, zinc, mercury, copper, lead and nickel (Martinez *et al.*, 1994). Nickel is one of the most abundant heavy metal contaminants of the environment due to its release during mining and smelting practices, and it has been the target of many cleanup strategies (Prasad and Strazalka, 2000). Arsenic is one of the metal contaminants of soil that requires remediation. Arsenic affects seed germination, and reduces root length and mass. Arsenite is 4-100 times more toxic than arsenate. The amounts of arsenic found in plant tissue are generally proportional to its level in soil. In plants, damage from heavy metals occurs due to both iron deficiency and the evolution of active oxygen species, both of which cause a reduction in chlorophyll and the shutdown of photosynthesis (Buchanan *et al.*, 2000). In addition to this, high metal concentrations in the soil have been shown to cause increased ethylene production (Burd *et al.*, 1998; Goren and Siegel, 1976), inhibit root and shoot development, reduce $CO_2$ fixation and limit sugar translocation (Prasad and Strazalka, 2000).

The low iron content of plants, which are grown in the presence of high levels of heavy metals, generally results in these plants becoming chlorotic, since iron deficiency inhibits both chloroplast development and chlorophyll biosynthesis. Once heavy metals bound to iron, microbial-siderophore complexes can be taken up by plants, and thereby serve as an iron source for plants (Bar-Ness *et al.*, 1991; Wang *et al.*, 1993). Therefore, the best way to prevent plants from becoming chlorotic in the presence of high levels of heavy metals is to provide them with an associated siderophore-producing PGPR that could provide a sufficient amount of iron to the plant. Burd *et al.*, (2000) demonstrated that a siderophore-over producing mutant of *Kluyvera ascorbata* SUD165 decreases the plant toxicity to three different heavy metals such as nickel, zinc and lead in three different crop plants such as tomato, canola and Indian mustard.

ACC deaminase containing plant growth-promoting bacteria can significantly increase the growth of plants in the presence of heavy metals including nickel, lead and zinc (Burd *et al.*, 1998) by lowering the level of stress-induced ethylene, thus allowing plants to develop longer roots and establish better during early stages of growth (Glick *et al.*, 1998). As part of a phytoremediation strategy, transgenic tobacco and canola plants expressing a bacterial ACC deaminase gene were constructed. These transgenic plants were tested for their resistance to growth inhibition by nickel, arsenic and high salt. Similar to what had previously been found with tomatoes, the transgenic plants, especially those in which the ACC deaminase gene was under the control of the *rollD* promoter, were significantly protected from the growth inhibition that often occurs as a consequence of these stresscs (Nic *et al.*, 2002; Stearns *et al.*, 2005).

## PGPR IN MANAGEMENT OF BIOTIC STRESSES

Biotic stresses in plants are mainly represented by pests and diseases, which constitute the single greatest threat to crop production. These include many thousands of species and types of phytopathogens (fungi, bacteria and viruses), insects, nematodes, weeds and other organisms. Modern farming practices with their reliance on agrochemical pest and disease control are responsible for considerable pollution and can have harmful effects on non-target organisms.

Exploiting naturally occurring plant growth promoting rhizobacteria as biological control agents to manage the biotic stresses represents one means of addressing the problems associated with agrochemical control.

## PGPR in Management of Phytopathogens

The beneficial free-living soil bacteria, PGPR, protect plants from damages caused by phytopathogens by a number of different indirect mechanisms such as production of antibiotics, antifungal metabolites and defence enzymes, exhibiting rhizospheric competition with phytopathogens, secretion of iron chelating siderophores and induction of systemic resistance (ISR) in plants (Glick, 1995a; Glick *et al.*,1999 ).

***PGPR producing antibiotics:*** One of the most effective mechanisms that a PGPR can employ to prevent phytopathogen proliferation is the synthesis of antibiotics. The antibiotics synthesized by biocontrol pseudomonads include, but are not limited to, agrocin 84, agrocin 434, 2, 4-diacetylphloroglucinol, herbicolin, oomycin, phenazines, pyoluteorin and pyrrolnitrin (Fig. 2). The biocontrol activity of a number of strains has been shown to be directly related to the ability of the bacterium to produce one of these antibiotics. However, an antibiotic that is effective in the

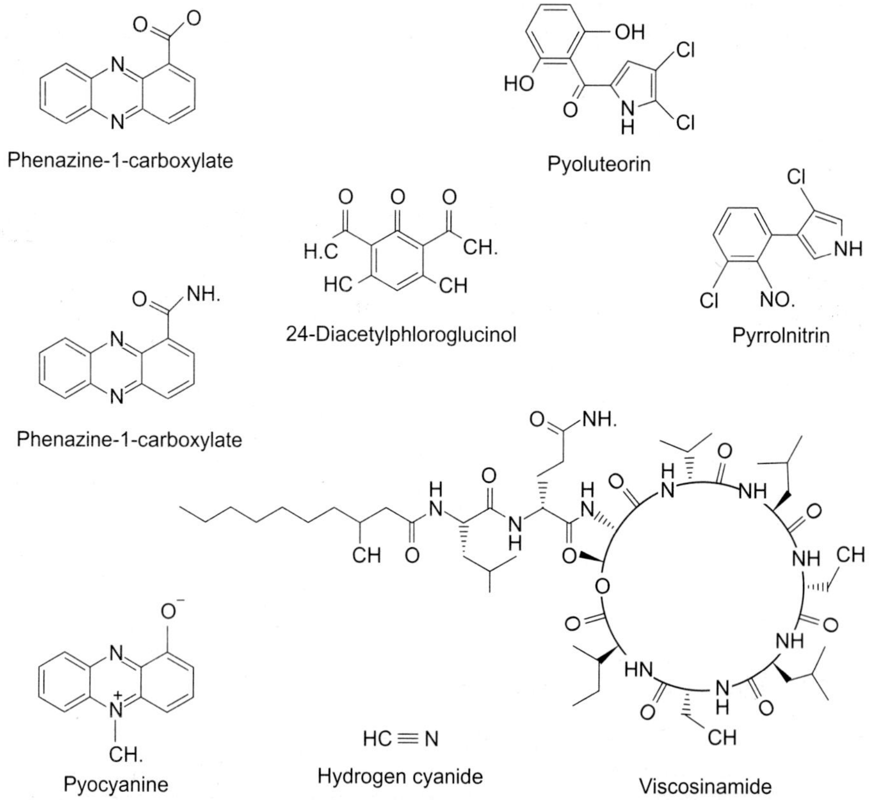


**Fig. 2** Antibiotic compounds produced by PGPR strains of fluorescent pseudomonads that are relevant for biocontrol of phytopathogens (Haas and Defago, 2005).

laboratory against one strain of a pathogenic agent may not prevent damage to the plant that occurs as a consequence of other strains of the same pathogen, and may not be as effective under more variable field conditions.

Evidence for the direct involvement of antibiotic production in PGPR-mediated disease-suppression comes from several different types of experiments:

(i) In a number of instances, the antibiotics that were isolated and purified from PGPR were shown to inhibit the same spectrum of fungal pathogens as the biocontrol strain itself (Carmi *et al*., 1994).

(ii) Non-antibiotic producing mutants of several different disease-suppressive bacterial strains either were no longer able to prevent phytopathogens (e.g., *Gaeumannomyces graminis* var. *tritici, Pythium ultimum* and *Rhizoctonia solani)* caused damage to plants or protected the plant to a much lesser extent than the wild-type bacterium (Thomashow and Weller, 1988; Howie and Suslow, 1991; Hamdan *et al*., 1991; Keel *et al*., 1992; Pierson *et al*., 1994).

(iii) In one study, it was reported that mutants of the PGPR *Pseudomonas fluorescens* BL915, which no longer produced the antibiotic pyrrolnitrin, lost the ability to prevent *Rhizoctonia solani* induced damping-off of cotton plants (Hill *et al*., 1994). Moreover, when a DNA fragment isolated from the wild-type bacterium, which restored pyrrolnitrin synthesis to these mutants, was transferred to two strains of *P. fluorescens* that did not normally synthesize this antibiotic, the transformed strains gained both the ability to synthesize pyrrolnitrin and inhibit *Rhizoctonia solani*-induced damping-off of cotton plants.

(iv) When an antibiotic-producing (wild-type) strain of *Pseudomonas fluorescens* was genetically manipulated to overproduce the antibiotics pyoluteorin and 2, 4-diacetylphloroglucinol, the resultant strain protected cucumber plants against the disease caused by *Pythium ultimum* to a greater extent than did the wild-type strain (Maurhofer *et al*., 1992; Schnider *et al*., 1994).

Since one of the major ways in which PGPR act as biocontrol agents is through the antifungal phytopathogen activity of the antibiotics that they produce, the activity, and hence the utility of a PGPR may be improved by providing it with genes that encode the biosynthesis of antibiotics normally produced by other bacteria (Gill and Warren, 1988). In this way, it should be possible to extend the range of phytopathogens that a PGPR is able to suppress. Another way in which the activity of a PGPR strain can be enhanced is by genetic manipulation to increase the amount of antibiotic that the bacterium synthesizes. While some increase in the amount of antibiotic produced by a particular bacterium might be obtained by conventional mutagenesis and selection, more extensive manipulation of antibiotic production will in all likelihood only be obtained through the use of recombinant DNA technology.

Since a number of antifungal metabolites produced by pseudomonads appear to be regulated by a global regulator (Laville *et al*., 1992), it should be possible to enhance the antibiotic production of a PGPR by modifying this global regulation. In fact, it was recently reported that amplification of the gene from *Pseudomonas fluorescens* CHAO encoding the housekeeping sigma factor $\sigma^{70}$ enhanced both antibiotic production and improved protection against *Pythium ultimum*-induced damping-off of cucumber (Maurhofer *et al.*, 1995; Schnider *et al.*, 1995a). In another study, inactivation of the *pqq* genes, which are involved in the biosynthesis of pyrroloquinoline quinone, a cofactor of different hydrogenases, in *Pseudomonas fluorescens* CHAO stimulated the production of the antibiotic pyoluteorin (Schnider *et al.*, 1995b). While the precise mechanism of this stimulation is unclear, it is possible that this mutant might enhance the flux of metabolites from other metabolic pathways into the pyoluteorin biosynthesis pathway.

***PGPR producing antifungal-metabolites:*** PGPR produces a wide range of low molecular weight metabolites with antifungal activity (Dowling and O'Gara, 1994). For example, some pseudomonads can synthesize hydrogen cyanide—to which these pseudomonads are themselves resistant—a metabolite that has been linked to the ability of those strains to inhibit some pathogenic fungi, e.g., *Thielabiopsis basicola,* the causative agent of black root rot of tobacco (Voisard *et al.*,1989). One group of researchers reported that several different microorganisms including strains of *Cladosporium werneckii, Pseudomonas cepacia* (*Burkholderia cepacia*) and *Pseudomonas solanacearum* are able to hydrolyze the compound, fusaric acid (Toyoda and Utsumi, 1991). Fusaric acid is the causative agent of the damage to plants that occurs upon *Fusarium* infection. As a consequence of the ability to hydrolyze fusaric acid, these bacterial strains can prevent the damage that is caused by various species of the fungus *Fusarium.*

***Rhizosphere competition:*** In addition to the more common antibiosis mechanisms, which include the disease-suppressive effects of siderophores and antibiotics, there are number of other ways in which PGPR can inhibit phytopathogens. For example, competition for nutrients and suitable niches on the root surface (Kloepper *et al.*, 1988; O'Sullivan and O'Gara, 1992) is a somewhat overlooked mechanism by which some PGPR may protect plants from phytopathogens. In one study, Stephens *et al.*, (1993) concluded that the major factor influencing the ability of a pseudomonad isolate to act as a biocontrol agent against *Pythium ultimum* on sugar beets in soil is their ability to metabolize the constituents of seed exudate in order to produce compounds inhibitory to *P. ultimum.* They also observed that there was not necessarily any relationship between the ability of a bacterium to inhibit a fungal pathogen when the bacterium was grown in the laboratory on media that favored the production of either antibiotics or siderophores, and the biocontrol activity of the bacterium *in vivo* (Stephens *et al.*, 1993).

An important facet of the competitiveness of a PGPR is the ability of the bacterium to persist and proliferate. However, it is often difficult to predict the behaviour in the environment of a particular PGPR, since the soil persistence of a bacterium may be influenced by a number of different factors including soil composition (Bashan *et al.*, 1995), temperature (Sun *et al.*, 1995) and the presence of recombinant plasmids (Tang *et al.*, 1995; Glick, 1995b). Since in cold and temperate climates many fungal phytopathogens are most destructive when the soil temperature is low, it is reasonable to expect that PGPR that are cold tolerant and can also function at low temperatures will be much more effective in the field than mesophilic biocontrol strains. One strategy that plants sometimes use to limit root colonization by phytopathogens is through the production of active oxygen species such as the hydroxyl radical, the superoxide anion and hydrogen peroxide that can inhibit a variety of pathogen cell processes (Doke, 1983; Sutherland, 1991). Plant roots may also respond to colonization by PGPR by producing active oxygen species (Katsuwon and Anderson, 1989; Katsuwon and Anderson, 1990). Phytopathogens that contain higher levels of enzymes such as superoxide dismutase, catalase and peroxidase that can reduce the amount of active oxygen species have been shown to be more effective pathogens presumably reflecting the ability of the fungus to survive this plant defence (Klotz and Hutcheson, 1992). It should, therefore, be possible, by genetic manipulation of PGPR, to increase the levels of one or more of the enzymes that reduce the amount of active oxygen species (e.g., Gruber *et al.*,1990), so that PGPR strains with an increased root colonizing ability and hence increased effectiveness against fungal pathogens might be created.

The soil contains a large number of different microorganisms, and those strains that are able to utilize an unusual carbon or nitrogen source such as an opine, 1-amino cyclopropane carboxylate (ACC) or a xenobiotic compound (such as a herbicide or pesticide) should be able to proliferate and then persist longer than other microorganisms in those soils that contain such unusual compounds. For example, the ability of some PGPR to hydrolyze ACC, the immediate precursor of ethylene in plants and a compound naturally found in root exudates, may provide these strains with a competitive advantage over other microorganisms in the rhizosphere because they can use ACC as a source of nitrogen (Jacobson *et al.*, 1994; Glick *et al.*, 1994a, 1994b; Glick 1995a). In an effort to engineer a more soil persistent biocontrol bacterium, another group of researchers transferred the NAH7 plasmid, which carries the genes encoding the enzymes of the naphthalene and salicylate biodegradative pathway, into an established biocontrol strain (Colbert *et al.*, 1993). The introduced plasmid was stably maintained and conferred increased persistence upon the host bacterium when salicylate was present in the soil.

Similarly, the presence of a herbicide, pesticide or other organic pollutant in soil may facilitate the proliferation of bacteria engineered to degrade these compounds (Brazil *et al.*, 1995). At the same time, these chemicals may suppress the proliferation of the other microorganisms in the same soil and possibly provide a "biodegradative PGPR" with a significant competitive advantage. This strategy has the advantage that in addition to increasing the competitiveness of a PGPR strain, it may also be a useful strategy for the biodegradation of some recalcitrant organic molecules in the soil. It has recently been postulated that an additional mechanism for plant growth promotion by PGPR could be their altering of microbial rhizosphere communities (Ramos *et al.*, 2003). Agreeing with such an indirect mechanism, it would be interesting to evaluate the actual impact of this activity in rhizosphere biology.

***PGPR producing defence enzymes***: Many plants respond to pathogen attack by synthesizing pathogenesis related (PR) proteins that can hydrolyze the cell walls of some fungal pathogens (Mauch *et al.*, 1988). Similarly, some PGPR strains have been found to produce enzymes including chitinase, β-1, 3-glucanase, protease and lipase that can lyse fungal cells (Chet and Inbar, 1994). For example, Lim *et al.*, (1991) isolated a strain of *Pseudomonas stutzeri* that produced extracellular chitinase and laminarinase, and found that these enzymes could digest and lyse *Fusarium solani* mycelia thereby preventing the fungus from causing crop loss due to root rot. Similarly, Fridlender *et al.*, (1993) were able to reduce the incidence of plant disease caused by the phytopathogenic fungi *Rhizoctonia solani, Sclerotium rolfsii* and *Pythium ultimum* by using a β-1, 3 glucanase-producing strain of *Pseudomonas cepacia*, which was able to damage fungal mycelia. Furthermore, it was recently shown (Chernin *et al.*, 1995) that three different strains of the biocontrol PGPR *Enterobacter agglomerans* that are antagonistic to fungal pathogens including *Rhizoctonia solani,* possess a complex of four separate enzymes that is responsible for the chitinolytic activity of the bacteria. These bacteria significantly decreased the damage to cotton plants following infection with *Rhizoctonia solani.* Moreover, Tn5 mutants of one of these biocontrol strains that were deficient in chitinase activity were unable to protect the plant against damage caused by the fungal pathogen. Since many of the enzymes from biocontrol PGPR that have been found to lyse fungal cells, including chitinases and β-1, 3-glucanases, are encoded by a single gene, it should be a relatively straightforward matter to isolate some of these genes and then transfer them to other PGPR, thereby constructing biocontrol PGPR that, for example, produce both antibiotics and fungus-degrading enzymes.

In one series of experiments, a chitinase gene was isolated from the bacterium *Serratia marcescens* and then transferred into *Trichoderma harzianum* and *Rhizobium meliloti* cells (Chet and Inbar, 1994). In both cases, the transformed microorganisms expressed the chitinase and subsequently displayed increased antifungal activity. When *S. marcescens* chitinase gene was introduced into a strain of *P. fluorescens* that acts as a biocontrol PGPR, the transformant stably expressed and secreted active chitinase and was an effective biocontrol strain against the pathogen *Rhizoctonia solani* (Koby *et al.*, 1994).

***PGPR producing siderophores*:** Although iron is one of the most abundant minerals on Earth, in the soil it is unavailable for direct assimilation by microorganisms because ferric ion, or $Fe^{3+}$, which is the predominant form of iron in nature, is only sparingly soluble, i.e., about 10-18M at pH 7.4 (Neilands *et al.*, 1987). Since the amount of soluble iron in the soil would be much too low to support microbial growth, soil microorganisms secrete low molecular mass (400-1000 daltons) iron-binding molecules known as siderophores, which bind $Fe^{3+}$ with a very high affinity (Kd = 10-20 to 10-50) (Castignetti and Smarrelli, 1986) and transport it back to the microbial cell where it is taken up by means of a cellular receptor, and then make it available for microbial growth (Neilands and Leong, 1986; Briat,1992). One way that PGPR can prevent the proliferation of phytopathogens (Fig. 3), and thereby facilitate plant growth, is through the production and secretion of siderophores that bind most of the $Fe^{3+}$ that is available in the rhizosphere, and as a result effectively prevent any fungal pathogens in this immediate vicinity from proliferating because of lack of iron (O'Sullivan and O'Gara, 1992). The bacteria that originally synthesized the siderophores take up the iron-siderophore complex using a receptor, which is specific for the complex and is located in the outer cell membrane of the bacterium (O'Sullivan and O'Gara, 1992). Although fungal phytopathogens also synthesize siderophores, the fungal siderophores generally have a lower affinity for iron than do the siderophores produced by PGPR (Schippers *et al.*, 1987), so that PGPR in effect out-compete fungal phytopathogens for available iron. Unlike microbial phytopathogens, plants are not generally harmed by the localized depletion of iron in the soil caused by PGPR. Most plants can grow at much lower (about 1000-fold) iron concentrations than microorganisms (O'Sullivan and O'Gara, 1992). In addition, a number of plants have mechanisms for binding the bacterial iron-siderophore complex, transporting it through the plant, and then reductively releasing the iron from the bacterial siderophore, so that it can be used by the plant (Bar-Ness *et al.*, 1992; Wang *et al.*, 1993).

The ability of siderophores to act as effective "disease-suppressive" agents is affected by the particular crop plant, the specific phytopathogen being suppressed, the soil composition, the bacterium that synthesizes the siderophore, and the affinity of the specific siderophore for iron. Thus, even though a particular PGPR is an effective disease-suppressive agent in the laboratory under controlled conditions, its behaviour in the field is extremely difficult to predict. This caveat

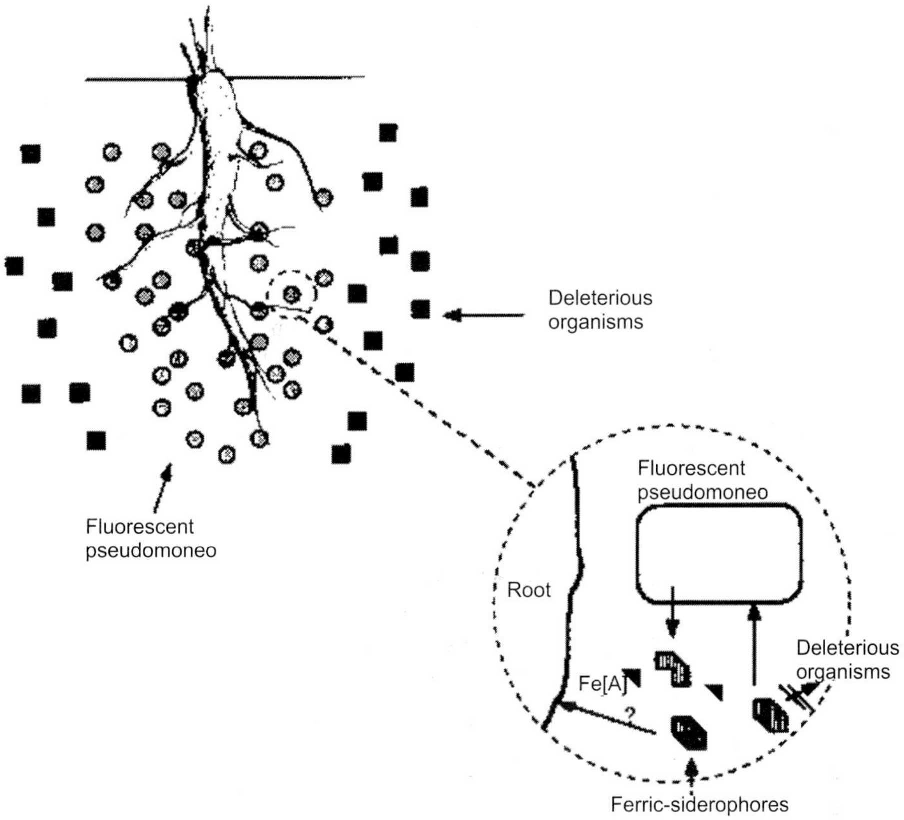


**Fig. 3** Model for suppression of root pathogens by siderophores from fluorescent pseudomonads. The growth of deleterious organisms is restricted by the unavailability of the growth-limiting ferric iron (Adapted from O'Sullivan and O'Gara, 1992).

notwithstanding, there is every reason to believe that the ability of bacterial siderophores to suppress phytopathogenic organisms is an important trait that could have a significant agronomic impact. Consistent with the involvement of siderophores in fungal pathogen-caused disease suppression, it was shown that:

(i) A mutant strain of *Pseudomonas putida* that overproduced siderophores was more effective than the wild-type bacterium in controlling a strain of *Fusarium oxysporum* that is pathogenic to tomatoes (Vandenburgh and Gonzalez, 1984).

(ii) A mutant strain of *Pseudomonas aeruginosa* that was selected for its lack of siderophore production no longer had the ability, shown by the wild-type, to protect tomato plants against *Pythium* damping-off (Buysens *et al*., 1994).

(iii) Increasing the amount of iron present in the soil to 40 mol $Fe^{+}3$/L caused a parallel decrease in both the amount of fluorescent siderophores produced and the inhibitory effect against *Gaeumannomyces graminis* var. *tritici,* a pathogen of wheat, in a collection of 70 separate isolates of fluorescent pseudomonads (Elsherif and Grossmann, 1994).

(iv) A single Tn5 insertion into the genome of *Alcaligenes* sp. strain MFA1 resulted in the simultaneous loss of the ability of the bacterium to grow in the absence of iron, and to inhibit microconidial germination and germination-tube elongation of the pathogen *Fusarium oxysporum* (Martinetti and Loper, 1992).

(v) Direct confirmation that PGPR in the rhizosphere actually synthesize siderophores in response to iron-limiting conditions comes from a study in which monoclonal antibodies were used to develop an ELISA assay to quantify the amount of siderophore from a fluorescent pseudomonad that was present in a barley rhizosphere sample (Buyer *et al.*, 1993).

Since each siderophore is encoded by a number of different genes, genetically engineering bacteria to produce modified siderophores is not a simple matter. On the other hand, it is possible to improve PGPR by extending the range of iron-siderophore complexes that a particular strain can utilize, so that a genetically altered biocontrol PGPR strain could take up and use siderophores synthesized by other soil microorganisms, thereby giving it a competitive advantage. This was done by cloning the genes for iron-siderophore receptors from one PGPR and introducing them into other strains (Marugg *et al.*, 1989).

***PGPR inducing plant systemic resistance (ISR):*** Certain bacteria trigger a phenomenon known as ISR phenotypically similar to systemic acquired resistance (SAR). As SAR, ISR is effective against different types of pathogens but differs from SAR in that the inducing PGPR does not produce visible symptoms on the host plant (Van Loon *et al.*, 1998). PGPR-elicited ISR was first observed on carnation (*Dianthus caryophillus*) with reduced susceptibility to wilt caused by *Fusarium* sp. (Van Peer *et al.*, 1991) and on cucumber (*Cucumis sativus*) with reduced susceptibility to foliar disease caused by *Colletotrichum orbiculare* (Wei *et al.*, 1991). Manifestation of ISR is dependent on the combination of host plant and bacterial strain. Most reports of PGPR-mediated ISR involve free-living rhizobacterial strains, but endophytic bacteria have also been observed to have ISR activity. Several bacterial traits (i.e. flagellation and production of siderophores and lipopolysaccharides (Table 2) have been proposed to trigger ISR (Leeman, *et al.*, 1995; Leeman *et al.*, 1996; Van Wees *et al.*, 1997;), but there is no compelling evidence for an overall ISR signal produced by bacteria (Haas *et al.*, 2002). It has recently been reported that volatile organic compounds may play a key role in this process (Ryu *et al.*, 2004). A major distinction often drawn between ISR and SAR is the dependence of the latter on the accumulation of salicylic acid (SA). Some PGPR do trigger an SA-dependent signaling pathway by producing nanogram amounts of SA in the rhizosphere (De Meyer *et al.*, 1999). However, the majority of PGPR that activate ISR appear to do so via a SA-independent pathway involving jasmonate and ethylene signals. ISR is associated with an increase in sensitivity to these hormones rather than an increase in their production, which might lead to the activation of a partially different set of defense genes (Pieterse and Van Loon, 1999).

PGPR-triggered ISR fortifies plant cell wall strength and alters host physiology and metabolic responses, leading to an enhanced synthesis of plant defense chemicals upon challenge by pathogens and/or abiotic stress factors. After inoculation of tomato with endophytic *P. fluorescens* WCS417r, a thickening of the outer tangential and outermost part of the radial side of the first layer of cortical cell walls occurred when epidermal or hypodermal cells were colonized (Duijff *et al.*, 1997). In *Burkholderia phytofirmans* PsJN-grapevine interaction, a host defence reaction coinciding with phenolic compound accumulation and strengthening of cell walls in the exodermis and in several cortical cell layers was also observed during endophytic colonization of the bacterium (Compant *et al.*, 2005). The type of bacterized plant response induced after challenge with a pathogen resulted in the formation of structural barriers, such as thickened cell wall papillae due to the deposition of callose and the accumulation of phenolic compounds at the site of pathogen attack (Benhamou *et al.*, 1996). Biochemical or physiological changes in plants include induced accumulation of pathogenesis-related proteins (PR proteins) such as PR-1, PR-2, chitinases, and some peroxidases (Park and Klopper, 2000). However, certain PGPR do not induce PR proteins (Hoffland *et al.*, 1995) but rather increase accumulation of peroxidase, phenylalanine ammonia lyase, phytoalexins, polyphenol oxidase, and/or chalcone synthase (Van Peer *et al.*, 1991). Recent evidence indicates that plants are capable of expressing SA-, JA-, and ethylene-dependent defence responses at the same time without antagonistic effects, leading to an elevated

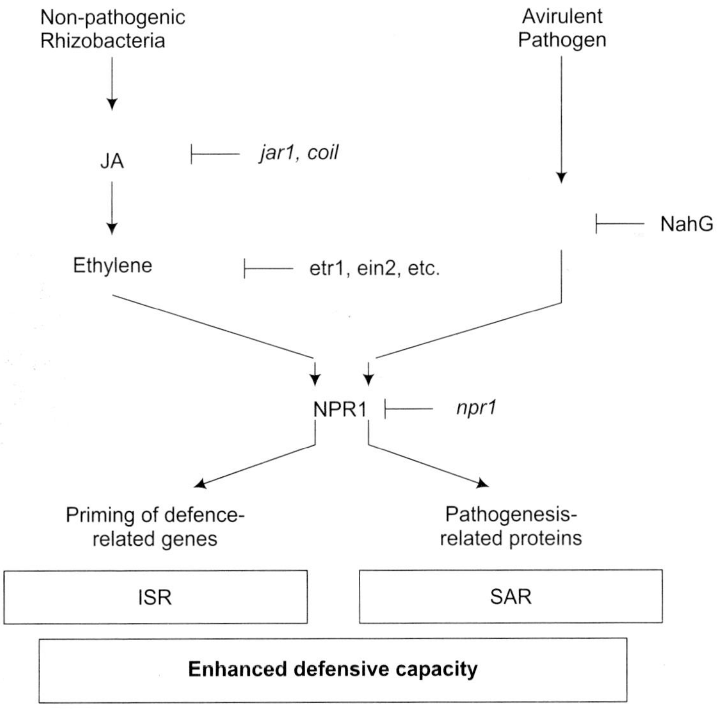


**Fig. 4** Current model of signal-transduction pathways leading to pathogen-induced systemic acquired resistance (SAR) and rhizobacteria-induced systemic resistance (ISR). Some non-pathogenic rhizobacteria may trigger a SA-dependent signalling pathway that leads to a state of induced resistance resembling SAR (Adapted from Pieterse *et al.*, 1998).

level of protection against pathogen attack. Therefore, simultaneous activation of ISR and SAR (Fig. 4) provides an attractive tool for the improvement of disease control (Van Wees *et al.*, 2000).

## PGPR IN MANAGEMENT OF INSECTS

Reports on PGPR in the control of insects are restricted to very few crops. Generally, fluorescent pseudomonads influence the growth and development of insects at all stages of their growth. *Pseudomonas maltophila* affects the growth of larval stage of *Helicoverpa zea*, the corn earworm, leading to more than 60% reduction in adult emergence, while pupae and adults that emerged from bacteria-infected larvae were smaller (Bong and Sikorowski, 1991). Similarly, the relative growth rate, consumption rate and digestibility of feed by *Helicoverpa armigera* have been affected when larvae fed on cotton plants treated with *Pseudomonas gladioli* due to an increase in their polyphenol and terpenoid content (Qingwen *et al.*, 1998). Induction of systemic resistance by PGPR strains, viz., *P. putida* strain 89B-27, *S. marcescens* strain 90-166, *Flavomonas oryzihabitans* strain INR-5 and *Bacillus pumilus* strain INR-7 have significantly reduced populations of the striped cucumber beetle, *Acalyma vittatum* and the spotted cucumber beetle,

**Table 2** Rhizobacterial determinants of ISR in different plants

| *Bacterial strain* | *Plants* | *ISR determinant* |
|---|---|---|
| *B. amyloliquefaciens* IN937a | *Arabidopsis* | 2, 3-butandiol |
| *B. subtilis* GB03 | *Arabidopsis* | 2, 3-butandiol |
| | Bean | Salicylic acid |
| *P. aeruginosa* 7NSK2 | Tobacco | Salicylic acid |
| | Tomato | Pyochelin + Pyocyanin |
| | *Arabidopsis* | 2, 4-diacetylphloroglucinol |
| *P. fluorescens* CHA0 | Tobacco | Siderophore |
| | Tomato | 2, 4-diacetylphloroglucinol |
| *P. fluorescens* Q2-87 | *Arabidopsis* | 2, 4-diacetylphloroglucinol |
| | | Lipopolysaccharide |
| *P. fluorescens* WCS374 | Radish | Siderophore |
| | | Iron-regulated compound |
| | *Arabidopsis* | Lipopolysaccharide |
| *P. fluorescens* WCS417 | Carnation | Lipopolysaccharide |
| | Radish | Lipopolysaccharide |
| | | Iron-regulated compound |
| | *Arabidopsis* | Lipopolysaccharide |
| | | Siderophore |
| | | Flagella |
| *P. putida* WCS358 | Bean | Lipopolysaccharide |
| | | Siderophore |
| Tomato | | Lipopolysaccharide |
| | | Siderophore |
| *Rhizobium etli* G12 | Potato | Lipopolysaccharide |
| *S. marcescens* 90-166 | Tobacco | Iron-regulated compound |

*Diabrotica undecimpunctata howardi* on cucumber. Among these strains, *S. marcescens* strain 90-166 was more effective in reducing the population of both the beetles and its efficacy was better than application of the insecticide Eesfenvalerate (Zehnder *et al.*, 1997). As certain fluorescent pseudomonads are effective rhizosphere colonizers and are endophytic in nature in the plant system, attempts have been made to transfer the insecticidal crystal protein from *Bacillus thuringiensis* to *P. fluorescens*. *P. fluorescens* thus genetically engineered was effective against a lepidopteran insect pest. Transgenic *P. cepacia* strain 526 with the crystal protein gene has consistently shown insecticidal activity against tobacco hornworm (Stock *et al.*, 1990). Thus, PGPR treatment of crops can be effective for insect pest management and has a great potential for future use.

## PGPR IN MANAGEMENT OF NEMATODES

PGPR also induce systemic resistance against nematode pests (Oostendorp and Sikora, 1990; Sikora, 1992; Sikora and Holmann-Hergarten, 1992). *P. fluorescens* has induced systemic resistance and inhibited early root penetration of *Heterodera schachtii*, the cyst nematode in sugar beet (Oostendorp and Sikora, 1990). Similarly, *B. subtilis* has induced protection against *Meloidogyne incognita* and *M. arenaria* in cotton (Sikora, 1992). Though attempts to use PGPR for nematode control are limited, the use of PGPR as biological control agents of plant-parasitic nematodes, especially for sugar beet and potato cyst nematode, has been reported as a successful strategy in management of these nematodes (Sikora, 1992). Treatment of rice seed with PGPR alone or in combination with chitin and neem cake has reduced the root and soil population of the rice root nematode, *Hirschmanniella oryzae* (Swarnakumari and Lakshmanan, 1999). The level of infestation of root-knot nematode *M. incognita* on tomato was reduced with fewer galls and egg masses in the soil following root dipping with *P. fluorescens* strain Pf1 (Santhi and Sivakumar, 1995). Similarly, application of the bacterium, *P. chitinolytica* reduced the root-knot nematode infection in tomato crop (Spiegel *et al.*, 1991).

## CONCLUSION

Environmental stresses which include both abiotic and biotic stresses are the major force that governs the food production in tropics. Drought, high and low temperature, flood, salinity and air pollution are most frequent abiotic stresses which are caused by various environmental factors, and phytopathogens, insect pests, nematodes and weeds act as biotic stresses which affecting the agricultural production. To achieve sustainable crop production to feed growing human population, strategic measures should be taken in management of these environmental stresses. One of the approach/strategy is the application of plant growth promoting rhizobacteria in agriculture. Large-scale application of PGPR to crops as inoculants would be attractive to increase crop yield as it would substantially reduce the use of chemical fertilizers and pesticides, which often pollute environment and contaminate the foodstuffs. Research and field trials of PGPR over decade have opened up new horizons for the agricultural bioinoculants industry. Development of superior or novel PGPR strains with improved plant growth promotion traits and development of transgenic crop plants expressing PGPR gene with increased resistance to various abiotic and biotic stresses are possible by genetic manipulations. These PGPR-biotechnologies can be exploited as a low input, sustainable and environment-friendly technology for the management of plant stresses.

## References


Abeles, F.B., Morgan, P.W. & Saltveit, M.E. Jr. (1992) *Ethylene in Plant Biology*, Academic Press, San Diego, CA, pp.105-110.

Apse, M.P., Aharon, G.S., Snedden, W.A. & Blumwald, E. (1999) *Science* **285**: 1256-1258.

Barea, J.M., Azcon, R. & Azcon-Aguilar, C. (2004) In *Plant: surface microbiology,* (Eds. A. Varma, L. Abbott, D.Werner & R.Hampp), Springer-Verlag, Heidelberg, Germany, pp. 351-371.

Barea, J.M., Gryndler, M., Lemanceau, P., Schuepp, H. & Azcon, R. (2002b) In: *Mycorrhiza technology in agriculture: from genes to bioproducts,* (Eds. S. Gianinazzi, H.Schuepp, J.M. Barea & K. Haselwandter), Birkhauser-Verlag, Basel, Switzerland, pp.1-18.

Bar-Ness, E., Hadar, Y., Shanzer, A. & Libman, J. (1992) *Plant Physiol.* **99:** 1329-1335.

Bashan, Y. & Levanony, H. (1990) *Can. J. Microbiol.***36:** 591-608.

Bashan, Y., Puente, M.E., Rodriguez-Mendoza, M.N., Toledo, G., Holguin, G., Ferrera-Cerrato, R. & Pedrin, S. (1995) *Appl. Environ. Microbiol.* **61:** 1938-1945.

Belimov, A.A., Hontzeas, N., Safronova, V.I., Demchinskaya, S.V., Piluzza,G. & Bullitta, S. (2005) *Soil Biol. Biochem.***37:** 241-50.

Belimov, A.A., Safronova, V.I., Sergeyeva, T.A., Egorova, T.N., Matveyeva V.A. & Tsyganov, V.E. (2001) *Can. J. Microbiol.* **47:** 642-52.

Benhamou, N., Kloepper, J. W., Quadt-Hallmann, A. & Tuzun. S. (1996) *Plant Physiol.* **112:**919-929.

Blumwald, E. (2000) *Curr. Opin. Cell Biol.* **12:** 431-434.

Bong, C.F.J. & Sikorowski, P.P. (1991) *J. Invertebr. Pathol.* **57:**406-412.

Bowen, G.D. & Rovira, A.D. (1999) *Advances in Agronomy* **66:** 1-102.

Brazil, G.M., Kenefick, L., Callanan, M., Haro, A., de Lorenzo, V., Dowling, D.N. & O'Gara, F. (1995) *Appl. Environ. Microbiol.* **61:** 1946-1952.

Briat, J.F. (1992) *J. Gen. Microbiol.* **138:** 2475-2483.

Buchanan, B.B., Gruissem, W. & Jones, R.L. (2000) In: *Biochemistry and molecular biology of plants*, American Society of Plant Physiologists, Rockville, MD U.S.A., pp. 575-588.

Burd, G.I., Dixon, D.G. & Glick, B.R. (1998) *Appl. Environ. Microbiol.***64:** 3663-3668.

Burd, G.I., Dixon, D.G. & Glick, B.R. (2000) *Can. J. Microbiol.* **46:** 237-245.

Buscot, F. (2005) In: *Microorganisms in soils: roles in genesis and functions,* (Eds. Buscot, F. & Varma, S.), Springer-Verlag, Heidelberg,Germany, pp.3-18.

Buyer, J.S., Kratzke, M.G. & Sikora, L.J. (1993) *Appl. Environ. Microbiol.* **59:** 677-681.

Buysens, S., Poppe, J. and Höfte, M. (1994) In: *Improving plant productivity with rhizosphere bacteria,* (Eds. M.H. Ryder, P.M. Stephens & G.D. Bowen), CSIRO, Adelaide, pp. 139-141.

Carmi, R., Carmeli, S., Levy, E. & Gough, F.J. (1994) *J. Nat. Prod.* **57:** 1200-1205.

Castignetti, D. & Smarrelli, J.,Jr. (1986) *FEBS Lett.* **209:**147-151.

Chernin, L., Ismailov, Z., Haran, S. & Chet, I. (1995) *Appl. Environ. Microbiol.* **61:** 1720-1726.

Chet, I. & Inbar, J. (1994) *Appl. Biochem. Biotechnol.* **48:** 37-43.

Ciardi, J.A., Tieman, D.M., Lund, S.T., Jones, J.B., Stall, R.E. & Klee, H.J. (2000) *Plant. Physiol.* **123:** 81-92.

Colbert, S.F., Hendson, M., Ferri, M. & Schroth, M.N. (1993) *Appl. Environ. Microbiol.* **59:** 2071-2076.

Compant, S., Reiter, B., Sessitsch, A., Nowak, J., Clement, C. & Barka. E. A. (2005) *Appl. Environ. Microbiol.*71:1685-1693.

De Meyer, G., Audenaert, K. & Hofte. M. (1999) *Eur. J. Plant Pathol.* **105:**513-517.

Doke, N. (1983) *Physiol. Plant Pathol.* **23:** 345-357.

Dowling, D.N. & O'Gara, F. (1994) *Trends Biotechnol.* **12:** 133-141.

Duijff, B. J., Gianinazzi-Pearson, V. & Lemanceau. P. (1997) *New Phytol.* **135:**325-334.

Else, M.A., Hall, K.C., Arnold, G.M., Davies, W.J. & Jackson, M.B. (1995) *Plant Physiol.* **107:** 377-384.

Elsherif, M. & Grossmann, F. (1994) *Microbiol. Res.* **149:** 371-377.

Fridlender, M., Inbar, J. & Chet, I. (1993) *Soil. Biol. Biochem.* **25:**1211-1221.

Frommel, M.I., Nowak, J. & Lazarovits, G. (1991) *Plant Physiol.* **96:**928-936.

Gill, P.R. & Warren, G.J. (1988) *J. Bacteriol.* **170:** 163-170.

Glick, B.R. (1995a) *Can. J. Microbiol.* **41:**109-117.

Glick, B.R. (1995b) *Biotechnol. Adv.* **13:**247-261.

Glick, B.R. (2005) *FEMS Microbiol. Lett.* **251:**1-7.

Glick, B.R., Jacobson, C.B., Schwarze, M.M.K. & Pasternak, J.J. (1994) *Can. J. Microbiol.* **40:** 911-915.

Glick, B.R., Karaturovic, D.M. & Newell, P. (1995) *Can. J. Microbiol.* **41:** 533-536.

Glick, B.R., Patten, C.L., Holguin, G., Penrose, D.M. (1999) *Biochemical and Genetic Mechanisms Used by Plant Growth Promoting Bacteria*, Imperial College Press, London, pp. 351-376.

Glick, B.R., Penrose, D.M. & Li, J. (1998) *J. Theor. Biol.***190:** 63-68.

Goren, R. & Siegel, S.M. (1976) *Plant Physiol.***57:** 628-631.

Grichko, V.P. & Glick, B.R. (2001a) *Plant Physiol Biochem.* **39:** 11-7.

Grichko, V.P. & Glick, B.R. (2001b) *Plant Physiol. Biochem.***39:** 19-25.

Gruber, M.Y., Glick, B.R. & Thompson, J.E. (1990) *Proc. Natl. Acad. Sci. (USA)* **87:**2608-2612.

Haas, D., Keel, C. & Reimmann. C. (2002) *Antonie Leeuwenhoek* **81:** 385-395.

Hamdan, H., Weller, D.M. & Thomashow, L.S. (1991) *Appl. Environ. Microbiol.* **57:** 3270-3277.

Hill, D.S., Stein, J.I., Torkewitz, N.R., Morse, A.M., Howell, C.R., Pachlatko, J.P., Becker, J.O. & Ligon, J.M. (1994) *Appl. Environ. Microbiol.* **60:** 78-85.

Hontzeas, N., Saleh, S. & Glick, B.R. (2004a) *Mol. Plant-Microbe Interact.***12:** 951-959.

Hontzeas, N., Zoidakis, J., Glick, B.R. & Abu-Omar, M.M. (2004b) *Biochem. Biophys. Acta.***1703:**11-9.

Howie, W.J. & Suslow, T.V. (1991) *Molec. Plant-Microbe Interact.* **4:** 393-399.

Jackson, M.B. (1985) *Annu. Rev. Plant Physiol.* **36:** 145-174.

Jacobson, C.B., Pasternak, J.J. & Glick, B.R. (1994) *Can. J. Microbiol.* **40:**1019-1025.

Karthikeyan, S., Zhou, Q., Zhao, Z., Kao, C., Tao, Z. & Robinson, H. (2004) *Biochemistry* **43:**13328-13339.

Katsuwon, J. & Anderson, A.J. (1989) *Appl. Environ. Microbiol.* **55:** 2985-2989.

Katsuwon, J. & Anderson, A.J. (1990) *Appl. Environ. Microbiol.* **56:**3576-3582.

Keel, C., Schnider, U., Maurhofer, M., Voisard, C., Laville, J., Burge r, U., Wirthner, P., Haas, D. & Défago, G. (1992) *Molec. Plant-Microbe Interact.* **5:** 4-13.

Kennedy, A.C. (1998) In: *Principles and applications of soil microbiology,* (Eds. D.M. Sylvia, J.J, Fuhrmann, P.G. Hartel & D.A. Zuberer), Prentice Hall, Upper Saddle River, New Jersey, pp. 389-407.

Klee, H.J., Hayford, M.B., Kretzmer, K.A., Barry, G.F. & Kishore, G.M. (1991) *Plant Cell* **3:** 1187-1193.

Kloepper, J.W., Lifshitz, R. & Zablotowicz, R.M. (1989) *Trends Biotechnol.* **7:** 39-43.

Kloepper, J.W., Lifshitz, R. & Schroth, M. N. (1988) *ISI Atlas Sci. Anim. Plant Sci.* **2:**60-64.

Klotz, M.G. & Hutcheson, S.W. (1992) *Appl. Environ. Microbiol.* **58:** 2468-2473.

Koby, S., Schickler, H., Chet, I. & Oppenheim, A.B. (1994) *Gene* **147:** 81-83.

Kozlowski T.T., (1984) In Flooding and Plant Growth, (Ed. T.T. Kozlowski), Academic Press, New York, pp. 1-5.

Laville, J., Voisard, C., Keel, C. Maurhofer, M., Defago, G. and Haas, D. (1992) *Proc. Natl. Acad. Sci. U.S.A.* **89:**1562-1566.

Leeman, M., Denouden, E. M., van Pelt, J. A., Dirkx, F., Steijl, H., Bakker, P. & Schippers. B. (1996) *Phytopathology* **86:**149-155.

Leeman, M., van Pelt, J. A., Denouden, F. M., Heinsbroek, M., Bakker, P. & Schippers. B. (1995) *Phytopathology* **85:**1021-1027.

Lim, H. S., Kim, Y.S. & Kim, S.D. (1991) *Appl. Environ. Microbiol.* **57:** 510-516.

Linderman, R.G. (1992) In: *Mycorrhizae in sustainable agriculture,* (Eds. Bethlenfalvay, G.J. & Linderman, R.G.), ASA Special Publication, Madison, Wisconsin, pp. 45-70.

Lynch, J. & Brown K.M. (1997) *Physiol. Plant.* **100:** 613-619.

Lynch, J.M. (1990) *The rhizosphere,* John Wiley, New York, pp. 351-371.

Martinetti, G. & Loper, J.E. (1992) *Can. J. Microbiol.* **38:** 241-247.

Martinez, V., Nunez, J.M., Ortiz, A. & Cerda, A. (1994) *J. Plant. Nutr.* **17:** 1359-1368.

Marugg, J.D., de Weger, L.A., Nielander, H.B., Oorthuizen, N., Recourt, K., Lugtenberg, B., van der Hofstad, G.A.J.M. & Weisbeek, P.J. (1989) *J. Bacteriol.* **170:** 1812-1819.

Mattoo, A.K. & Suttle, J.C. (1991) *The Plant Hormone Ethylene*, CRC Press, Boca Raton, FL, pp.124-129.

Mauch, F., Mauch-Mani, B. & Boller, T. (1988) *Plant Physiol.* **88:** 936-942.

Maurhofer, M., Keel, C., Haas, D. & Défago, G. (1995) *Plant Pathol.* **44:** 40-50.

Maurhofer, M., Keel, C., Schnider, U., Voisard, C., Haas, D. & Défago, G. (1992) *Phytopathology* **82:** 190-195.

Mayak, S., Tirosh, T. & Glick, B.R. (2004a) *Plant Sci.* **166:** 525-30.

Mayak, S., Tirosh, T. & Glick, B.R. (2004b) *Plant Physiol. Biochem.* **42:** 565-72.

Minami, R., Uchiyama, K., Murakami, T., Kawai, J., Mikami, K. & Yamada, T. (1998) *J. Biochem.* **123:** 1112-1118.

Nannipieri, P., Ascher, J., Ceccherini, M.T., Landi, L., Pietramellara,G. & Renella, G. (2003) *European Journal of Soil Science* **54**: 655-670.

Neilands, J.B. & Leong, S.A. (1986) *Ann. Rev. Plant Physiol.* **37:**187-208.

Neilands, J.B., Konopka, K., Schwyn, B., Coy, M., Francis, R.T., Paw, B.H. and Bagg, A. (1987) In: *Iron transport in microbes, plants and animals,* (Eds. Winkelmann, G., van tier Helm, D. & Neilands, J.B.), Verlagsgesellschaft-mbH, Weinheim, pp. 3-33.

Nie, L., Shah, S., Rashid, A., Burd, G.I., Dixon, D.G. & Glick, B.R. (2002) *Plant Physiol. Biochem.***40:** 355-361.

Nilsen, E.T. & Orcutt, D.M. (1997) *The Physiology of Plants Under Stress. Abiotic Factors,* John Wiley and Sons Inc., New York, NY, pp. 241-248.

Oostendorp, M. & Sikora, R.A. (1990) *Rev. Nematol.* **13:** 269-274.

O'Sullivan & O'Gara, F. (1992) *Microbiol. Rev.* **56:** 662-676.

Pace, N.R. (1997) *Science.* **276:** 734-740.

Park, K. S. & Kloepper. J. W. (2000) *Biol. Control* **18:**2-9.

Paul, E.A. & Clark, F.E. (1989) *Soil microbiology and biochemistry,* Academic Press, San Diego, Calif, pp. 10-28.

Piao, C.G., Tang, W.H. & Chen, Y.X. (1992) *Chin. J. Microecol.***4:** 55-62.

Pieterse, C. M. J., & van Loon, L. C. (1999) *Trends Plant Sci.* **4:** 52-58.

Prasad, M.N.V. & Strazalka, K. (2000) In: *Physiology and biochemistry of metal toxicity and tolerance in plants,* Kluwer Academic Publishers, Boston, pp. 153-160.

Qingwen, Z., Ping, L., Gang, W. & Qingnian, C. (1998) *Acta Phytophylacica Sinica* **25:** 209-212.

Robison, M.M., Griffith, M., Pauls, K.P. & Glick, B.R. (2001) *J. Phytopathol.* **2:** 385-388.

Ryu, C., Farag, M., M. A., Hu, C. H., Reddy, M. S., Kloepper, J. W. & Pare, P.W. (2004) *Plant Physiol.* **134:**1017-1026.

Santhi, A. & Sivakumar, V. (1995) *J. Biol. Control* **9:** 113-115.

Schippers, B., Bakker, A.W. & Bakker, A.H.M. (1987) *Ann. Rev. Phytopathol.* **25:** 339-358.

Schnider, U., Blumer, C., Troxler, J., Défago, G. & Haas, D. (1994) In: *Improving plant productivity with rhizosphere bacteria,* (Eds. Ryder, M.H., Stephens, P.M. & Bowen, G.D.), CSIRO, Adelaide, pp. 120-121.

Schnider, U., Keel, C., Blumer, C., Troxler, J., Défago, G. & Haas, D. (1995a) *J. Bacteriol.* **177:** 5387-5392.

Schnider, U., Keel, C., Voisard, C., Défago, G. & Haas, D. (1995b) *Appl. Environ. Microbiol.* **61:** 3856-3864.

Sikora, R.A. & Ho!mann-Hergarten, S. (1992) *Arab. J. Plant Prot.* **10:**53-58.

Sikora, R.A. (1992) *Annu. Rev. Phytopathol.* **30:** 245-270.

Spiegel, Y., Cohn, E., Galper, S., Sharon, E. Chet, I. (1991) *Biocontrol Sci. Technol.* **1:** 115-125.

Stearns, J.C. & Glick, B.R. (2003) *Biotechnol. Adv.* **5257:** 1-18.

Stearns, J.C., Shah, S., Greenberg, B.M., Dixon, D.G. & Glick, B.R. (2005) *Plant Phys. Biochem.***43:** 701-708.

Stephens, P.M., Crowley, J.J. & O'Connell, C. (1993) *Soil Biol. Biochem.* **25:**1283-1288.

Stock, C.A., Mcloughlin, T.J., Klein, J.A. & Adang, M. (1990) *Can. J. Microbiol.* **36:** 879-884.

Sun, X., Griffith, M., Pasternak, J.J. & Glick, B.R. (1995) *Can. J. Microbiol.* **41:** 776-784.

Sutherland, M.W. (1991) *Mol. Physiol. Plant. Pathol.* **39:** 79-93.

Swarnakumari, N. & Lakshmanan, P.L. (1999) In: *Proce.International Seminar on Integrated Pest Management,* Hyderabad, India, pp. 101.

Tang, W.H. (1994) In: *Improving plant productivity with rhizosphere bacteria,* (Eds. Ryder, M.H., Stephens, P.M. & Brown, G.D.), CSIRO, Adelaide, Australia, pp.267-278.

Tang, W., Pasternak, J.J. & Glick, B.R. (1995) *Can. J. Microbiol.* **41:** 455-451.

Thomashow, L.S. & Weller, D.M. (1988) *J. Bacteriol.* **170:** 3499-3508.

Timmusk, S. & Wagner, G.H. (1999) *Mol. Plant-microbe interact.***12:** 951-959.

Toyoda, H. & Utsumi, R. (1991) *U.S. Patent* #4,988,586.

Van Assche, F. & Clijsters, H. (1990) *Plant Cell Environ.***13:** 195-206.

Van Loon, L. C., Bakker, P. A. H. M. & Pieterse, C. M. J. (1998) *Annu. Rev. Phytopathol.* **36:** 453-483.

Van Loon, L.C. & Glick, B.R. (2004) In: *Molecular ecotoxicology of plants,* (Ed. Sandermann H.), Springer-Verlag, Berlin, pp. 177-205.

Van Peer, R., Niemann, G. J. & Schippers, B. (1991) *Phytopathology* **81:**728-734.

Van Peer, R. & Schippers, B. (1989) *Can. J. Microbiol.* **81:** 456-463.

Van Wees, S., de Swart, E., van Pelt, J., van Loon, L. C. and Pieterse, C. (2000) *Proc. Natl. Acad. Sci. (USA)* **97:** 8711-8716.

Van Wees, S., Pieterse, C., Trijssenaar, A., Van't Westende, Y., Hartog, F. & van Loon, L. C. (1997) *Mol. Plant-Microbe Interact.* **10:**716-724.

Vandenbergh, P.A. & Gonzalez, C.F. (1984) *U.S. Patent* #4,479,936.

Vartapetian, B.B. & Jackson, M.B. (1997) *Ann. Bot.* **79:** 3-20.

Voisard, C., Keel, C., Haas, D. & Defago, G. (1989). *EMBO J.* **8:** 351-358.

Wang, Y., Brown, H.N., Crowley, D.E. & Szaniszlo, P.J. (1993) *Plant, Cell and Environ.* **16:** 579-585.

Wei, L., Kloepper, J. W. & Tuzun. S. (1991) *Phytopathology* **81:** 1508-1512.

Yang, S.F. & Hoffman, N.E. (1984) *Annu. Rev. Plant. Physiol.* **35:** 155-189.

Zehnder, G., Kloepper, J., Yao, C. & Wei, G. (1997) *J. Econ. Entomol.* **90:** 391-396.

3

# Nature and Applications of *Bacillus* Species for Improving Plant Growth

**Pankaj Trivedi, Bhavesh Kumar and Anita Pandey**

*Environmental Physiology and Biotechnology, G B Pant Institute of Himalayan Environment and Development, Kosi-Katarmal, Almora, India-263 643*

## ABSTRACT

The widespread application of agrochemicals for improving plant growth is known to severely impact arable soils. An alternative is the reduction in the use of these chemicals by the development of formulations based on microbial agents, such as plant growth promoting rhizobacteria (PGPRs). There has been a large body of literature describing potential uses of *Bacillus* species as agents for stimulating plant growth and management of soil and plant health. The members of this genus are associated with most of the plant species and are commonly found in all the agricultural systems. This ubiquity is due to their physiological traits such as production of multilayered cell wall structure and formation of a stress resistant endospore. *Bacillus* spp. may promote growth directly, e.g., by fixation of atmospheric nitrogen, solubilization of minerals such as phosphorus, production of siderophores that solubilize and sequester iron, or production of plant growth regulators (hormones). Some support plant growth indirectly, by improving growth restriction conditions either via production of antagonistic substances or by inducing resistance against plant pathogens. Several methods including antibiotic markers or Tagging bacteria by reporter gene or genes products have been chosen for studying root colonization pattern. A number of strains of genus *Bacillus* have been developed as bioinoculants for use in different crops in greenhouse and/or field conditions. A successful formulation of *Bacillus* strain is potentially more readily achieved than a successful formulation of microbial biomass that does not produce a resistant life stage comparable to *Bacillus* spore. This feature along with a vast base of industrial knowledge associated with *Bacillus* species for other purposes has paved way for commercialization of PGPR based products in various countries including India.

## INTRODUCTION

The potential negative effect of chemical fertilizers on the global environment and the cost associated with the production has led to research with the objective of replacing chemicals with microbial inoculants (Lynch, 1990; Glick *et al.,* 1999; Bloemberg and Lugtenberg, 2001; Pandey *et al.,* 2004). Rhizospheric bacteria that favourably affect plant growth and yield of commercial important crops are designated as plant growth promoting rhizobacteria (PGPR) (Kloepper and Schroth, 1981; Bashan and Holguin, 1998). PGPR can improve seed germination, root development, mineral nutrition and water utilization (Phytostimulation) (Glick, 1995; Bowen and Rovira, 1999; Gerhardson and Wright, 2001; Persello-Cartieaux *et al.,* 2003; Vessey, 2003). They can also suppress bacterial, fungal and nematode pathogens (biological control) (Weller, 1988; Whipps, 2001; Zehnder, 2001). The use of PGPR has become rather common practice and is on rise in many parts of the world. Inoculation of plants with PGPRs, namely, *Pseudomonas*, *Serratia*, *Azospirillum* and *Bacillus* have been reported to enhance growth of whole plant and often limit the growth of certain soil-borne plant pathogens (Weller, 1988; Pandey and Kumar, 1989; Kloepper, 1994; Tang, 1994; Glick, 1995; Whipps, 2001; Sharma and Johri, 2003).

The Gram-positive bacteria, especially those belonging to genus *Bacillus* have received less attention than Pseudomonads in the literature on PGPR development. This is mainly because: (i) *Bacillus* spp. has been less feasible for genetic studies, and (ii) there is insufficient knowledge about their beneficial mechanisms (Halndelsman and Stabb, 1996). However, there is a substantial base of industrial experience especially with *Bacillus subtilis* and *B. thuringiensis* that can be applied to the use of other members of the genus to overcome current obstacles hindering the commercialization of PGPR technology. *Bacillus* spp. can form spores, which survive and remain metabolically active under harsh environmental conditions (Rodgers, 1989), making them appropriate for formulation of stable and viable products (Kloepper, 1997).

The genus *Bacillus* includes a variety of important species used in the fermentation industry. *Bacillus* species are good secretors of proteins and metabolites and easy to cultivate. Products currently available commercially include enzymes, antibiotics and insecticides. In this review, we seek to highlight the PGPR abilities of the genus *Bacillus* spp. with reference to their mechanism of actions, rhizosphere competence and commercialization of *Bacillus*-based bioinoculants under following heads.

## THE GENUS *BACILLUS*

In 1872, Ferdinand Cohn, a student of Robert Koch, recognized and named the bacterium *Bacillus subtilis.* The organism represented a large and diverse genus of bacteria, *Bacillus*, and was placed in the family *Bacillaceae.* The distinguishing feature of the family was the production of endospores, which are highly refractile resting structures formed within the bacterial cells. Members of the genus *Bacillus* are characterized as Gram positive, rod-shaped, aerobic or facultative, endospore-forming bacteria (Slepeckey and Hemphill, 1992).

Early attempts at classification of *Bacillus* species were based on two major characteristic features:

(i) Their aerobic growth
(ii) The endospore formation.

This resulted in placing together many bacteria possessing different physiology and occupying a variety of habitats. Hence, the heterogeneity in physiology, ecology and genetics made it difficult to categorize the genus *Bacillus* or to make generalizations about this bacteria. In view of the existing diversity within the genus *Bacillus* and related genera, numerous valid descriptions of new genera and species as well as many classifications have been emerged (Garrity, 2001). On the basis of extensive studies of the small-subunit ribosomal RNA sequence, the genus *Bacillus* contains 88 species and 2 subspecies (Fritze, 2004).

The ubiquity of *Bacillus* species in nature is attributed to the formation of resting structures called endospores (Driks, 2004). Endospores have proven to be the most durable type of cell found in nature, and in their cryptobiotic state of dormancy, they can remain viable for extremely long periods of time, perhaps millions of years (Dubnau *et al.*, 1988; Piggot and Hilbert, 2004). They are highly resistant to environmental stresses such as high temperature, irradiation, strong acids, disinfectants, etc. Endospores are formed by vegetative cells in response to environment signals that indicate a limiting factor for vegetative growth, such as exhaustion of an essential nutrient (Sonenshein, 2000). They germinate and become vegetative cells when the environmental stress is relieved. Hence, endospore-formation is a mechanism of survival rather than a mechanism of reproduction (Stephens, 1998).

## DIVERSITY AND DISTRIBUTION IN SOIL/RHIZOSPHERE

Multiple *Bacillus* and related *Paenibacillus* spp. can be readily cultured from both bulk and rhizosphere soils. Culturable counts of these bacteria generally range from 3 to log 6 cells per fresh gram weight, with soil counts typically exceeding those obtained from rhizosphere (Kilian *et al.*, 2000). Derbyshire and Greaves (1973) and Hallmann *et al.*, (1998) reported that species of *Bacillus* were among the most common types of bacteria isolated from soil samples and can account up to 36% of the total bacterial population. Morphologically, most distinctive among these *B. mycoides*, which often confound attempts to accurately enumerate cultured populations by virtue of their rapid mycelial like growth patterns on agar media (Buyer, 1995). *B. megaterium* has been reported to be one of the most abundant in some soil (Zheng and Sinclair, 2000). But like the total number of microbes, amount of *Bacillus* spp. in soil also vary depending upon the environmental factors, plantation type, cultural practices and various edaphic factors and it seems unlikely that a single species will dominate numerically in most soils (Darbyshire and Greaves, 1973; Lynch, 1983; Mahafee and Kloepper, 1996). Shoda (2000) summarized the accepted view that while *B. subtilis* is a common microbe in the soils, it is not widespread in the rhizosphere except when introduced massively. However, there is an emerging evidence that *B. subtilis* is a common inhabitant of certain type of plant roots, probably because of biofilm formation (Fall *et al.*, 2004). A number of species belonging to the genus *Bacillus* were found to be well adapted to the rhizoplane and rhizosphere of established tea bushes grown under temperate climate. Amongst the species, *B. subtilis* and *B. mycoides* appeared to be closely associated with tea roots. While these two species comprised the major part of bacterial population, these were the only species that could survive the extreme cold including snowfalls (Pandey and Palni, 1996; 1997; 2004).

Culture independent methods have confirmed the presence of easily cultured species and revealed additional unculturable diversity in *Bacillus* rRNA linkages (Brousseau *et al.*, 1993;

Jurtshuk *et al.*, 1992; Martel *et al.*, 2000). Sequencing of dominant bands excised from the rhizosphere patterns of *Verticillium dahliae*, field-grown strawberry (*Fragaria ananassa* Duch.), oilseed rape (*Brassica napus* L.), and potato (*Solanum tuberosum* L.), revealed that 6 out of 10 bands resembled gram-positive bacteria (Smalla *et al.*, 2001). DNA was extracted from soil sampled to a distance of 2.5 mm beneath a root mat of oilseed rape (*Brassica napus*) and the 16S rDNA amplified, cloned and sequenced revealed that 14% of the clones were close relatives of *B. megaterium* (Macrae *et al.,* 2000). Petersen *et al.*, (1995) developed a polymerase chain reaction amplification of 23S rDNA to identify *Bacillus* spp. recovered from roots, mycorrhizae, and rhizosphere soil of conifers. A PCR based method for screening of *Bacillus* spp. with antifungal activity in suppressive soybean rhizosphere has been described (Giacomodonato *et al.*, 2001). Recently, novel genetic markers associated with the biological control activities in *Bacillus subtilis* have been identified and characterized (Joshi and McSppadden, 2006).

## MECHANISMS OF PLANT GROWTH PROMOTION

The promotion in plant growth by *Bacillus* species can occur through at least one of the following three mechanisms:

(1) Suppression of plant disease (Bioprotectants)
(2) Improved nutrient acquisition (Biofertilizers)
(3) Phytohormone production (Biostimulants).

### Bioprotectants

Antibiotic producing *Bacillus* spp. release compounds that prevent the growth of pathogens. A considerable number of antibiotics from genus *Bacillus* have been identified, many of which are known to have antifungal activity (Berdy, 1974; Fravel, 1988; Emmart and Handelsman, 1999). *Bacillus* strains with biocontrol activity can produce the lypopeptide fengycin (Chen *et al.,* 1995) and kanosamine (δ-amino-3-deoxy-D-glucose) (Silo-Suh *et al.,* 1994). *B. cereus* UW85 produces two antibiotics, zwittermicin A and kanosamine, which contribute to biocontrol of alfalfa damping off (Silo-Suh *et al.*, 1994). The purified antibiotics directly inhibit growth and development of pathogens and suppress diseases (He *et al.,* 1994). A new antifungal antibiotic of the Iturin group, Bacilliomycin Lc secreted by *Bacillus subtilis* has been purified (Eshita *et al.*, 1995). The ability of *B. subtilis* PY-1 strain to inhibit the growth of *Fusarium oxysporum* was postulated due to the production of Iturin A (Gong *et al.,* 2006). *B. mycoides* and eight other strains of *Bacillus* isolated from tea rhizosphere exhibited antifungal activity against twelve test fungi *in vitro*, which included nine minor (saprophytes) and three major pathogens of tea (Pandey *et al.,* 1997). Antagonism of two bacilli isolates MR-11(2) and MRF, isolated from maize rhizosphere, against the pathogens *Fusarium moniliforme*, *F. graminearum* and *Macrophomina phaseolina*, was shown to be due to production of antibiotics (Pal *et al.*, 2001). The structural deformities in six pathogenic fungi under *in vitro* culture conditions by an efficient antagonistic strain of *Bacillus subtilis*, originally isolated from the rhizosphere of established tea bushes, was attributed to the production of diffusible and volatile antifungal compounds (Chaurasia *et al.*, 2005). Genetic techniques such as high frequency transformation, transposon mutagenesis, and reporter gene fusion have been used to analyze thoroughly the antibiotic biosynthesis and regulation in bacilli (Milner *et al.,* 1996; Raffel *et al.,* 1996).

Induced systemic resistance (ISR) is the process of active resistance dependent on the physical or chemical barriers of host plants, activated by biotic or abiotic agents. Evidence that some strains of *Bacillus* sp. can elicit systemic protection against foliar pathogens has been provided in comprehensive studies aimed at determining mechanisms for disease protection (Kloepper *et al.*, 2004 and reference within). Reductions in the severity or incidence of viral diseases have also been reported in greenhouse studies with some strains of *Bacillus* species. Zehnder *et al.*, (2000) conducted a greenhouse screen of PGPR for potential to elicit ISR against *Cucumber mosaic virus* (CMV) on tomato. Each of the three strains selected from 26 tested significantly reduced the mean percentage of symptomatic plants in each of five experiments, with disease incidence ranging from 88 to 98% in nonbacterized controls and 32 to 58% for PGPR treated plants. Similarly, Murphy *et al.*, (2003) used five two-strain combinations of *Bacillus* spp. incorporated into potting mix. When compared with controls, bacterial treatments significantly reduced disease severity based on symptoms.

It has been established in recent years that many bacterial pathogens have sophisticated genetic control networks to enable coordination of production of virulence factors with cell population size, thus ensuring a concerted attack to overcome host defence responses. This mechanism is widely known as quorum sensing (Fuqua *et al.*, 1996). The quorum sensing bacteria produce, detect and respond to small signal molecules known as quorum sensing signals or autoinducers. Acyl homoserine lactones (AHLs) are one family of the most characterized quorum-sensing signals found in many Gram-negative bacterial species (Xu *et al.*, 2003). A simple strategy to keep production of AHLs in check and thus suppress the expression of virulence genes is to inactivate the AHL signals produced by pathogenic bacteria. The first AHL-inactivation enzyme encoded by *aiiA* gene was identified in *Bacillus* sp. 240B1 (Dong *et al.*, 2000). *AiiA* homologue were later found in many subspecies of *B. thuringiensis* and closely related *Bacillus* species, including *B. cereus* and *B. mycoides* (Dong *et al.*, 2000; Lee *et al.*, 2002, Reimmann *et al.*, 2002). All *aiiA* protein contains a motif of several enzymes in the metallohydrolase superfamily (Dong *et al.*, 2000). For practical application all A240B1 was cloned in a plant expression vector PBI 121 and introduced into tobacco and potato by *Agrobacterium*-mediated transformation. Transgenic plant expressing AHL lactonase showed significantly enhanced resistance to *Erwinia carotovora* infection (Dong *et al.*, 2001).

## Biofertilizers

A substantial number of bacterial species, many of which belong to genus *Bacillus*, have the ability to increase the availability of phosphorus (P) to the plants, not only by mineralizing organic P compounds but also rendering inorganic phosphates more available to them (Illmer *et al.*, 1995; Richardson, 2001). Eight bacteria, each of the genus *Bacillus*, were isolated on $AlPO_4$-sucrose agar from a typical Indian lateritic soil (Banik and Dey, 1983). All the organisms produced free aliphatic organic acid in detectable amounts. *B. megaterium* var. *phosphaticum* is well-known for its ability to solubilize rock P materials (Schilling *et al.*, 1998). P-solubilization potential of several species of *Bacillus* including *B. megaterium* isolated from the rhizosphere of rice grown in acidic soils has been reported (Thakuria *et al.*, 2004). Vazquez *et al.*, (2000) isolated three potent phosphate solubilizers viz. *B. amyloliquefaciens, B. licheniformis* and *B. atrophaeus* from the rhizosphere of mangroves. A new phosphate-mobilizing *Bacillus* strain, *B. subtilis* IMV V-7023 displayed a very high ability to mobilize phosphorus from its sparingly soluble inorganic and organic compounds and the capability of synthesizing biologically active substances (Roi *et al.*, 2004). Trivedi *et al.,* (2006) described the plant growth promoting potential of *B. megaterium* and *B. subtilis* initially isolated from different Himalayan locations.

Microbial system can siphon out appreciable amounts of nitrogen (N) from the atmospheric reservoir and enrich soil with this important but scare nutrient, and many studies have described the N fixing properties of the bacteria belonging to genus *Bacillus.* Witz *et al.*, (1967) reviewed the nitrogen fixing property of the growing cells and cell-free extracts of the Bacillaceae. Cultures of *B. macerans* and *B. polymyxa* isolated from wheat roots showed significant N fixation in lab assays (Nelson *et al.*, 1976). Several *Bacillus* strains isolated from the rhizosphere of *Ammophila arenaria* appeared on 'nitrogen-free' agar plates and some of these were found positive for their ability to fix nitrogen using the acetylene reduction assay (Abdel Wahab, 1975). *Bacillus polymyxa*-like nitrogen fixers capable of reducing acetylene and assimilating $^{15}N_2$ were isolated from the rhizosheaths of xeric grasses (Wullstein *et al.,* 1979). Twenty-four *Bacillus* strains including *B. azotofixans* sp. nov from five different Brazilian soils isolated using media lacking a nitrogen source showed N fixation as determined by their acetylene-reducing ability (Seldin *et al.,* 1983). Chanway and Holl (1991) isolated a nitrogen-fixing *Bacillus* L6 strain from the rhizosphere of pine. Nitrogen fixing *Bacillus* sp. was found to be associated with Douglas-fir tuberculate ectomycorrhizae under microaerophilic conditions (Li *et al.,* 1992). Poonguzhali *et al.*, (2006) reported two strains of *B. megaterium* and one of the *B. subtilis* strain from the rhizosphere of field grown Chinese cabbage having nitrogenase activity as determined by ARA.

## Biostimulants

Many actual and putative biofertilizing PGPR belonging to genus *Bacillus* produced phytohormones that are known to be related to their ability to stimulate plant growth. In most cases these phytohormones are believed to be changing assimilate partitioning patterns in plants and affecting growth patterns of roots to result in bigger roots, more branched roots, and/or roots with greater surface area. Indole-3-acetic acid is a phytohormone, which is known to be involved in root initiation, cell division, and cell enlargement (Salisbury, 1994). This hormone is very commonly produced by *Bacillus* spp. (Barazani and Friedman, 1999). Srinivasan *et al.*, (1996) reported *Bacillus* spp., isolated from the rhizosphere of *Phaseolus vulgaris*, to produce large amounts of IAA. *Bacillus* spp. also produce high levels of gibberellins, $A_1$, $A_3$, $A_4$ and $A_{20}$, inducing positive effect on stem and shoot elongation in *Alnus glutinosa* (Gutierrez Manero *et al.*, 1996; Gutierrez Manero *et al.*, 2001). Some species of *Bacillus* can produce as yet not well-characterized phytohormones or growth regulators that cause plants to have greater amounts of fine roots, which have the effect of increasing the absorptive surface of plant roots for uptake of water and nutrients. The potent growth promoting strains, *B. subtilis* GB03 and *B. amyloliquefaciens* IN937a produced volatiles of 3-hydroxy-2-butanone (acetoin) and 2, 3 butanediol, postulated to act as plant growth hormone (Ramos *et al.*, 2000). Idris *et al.*, (2004) reported the phytohormone-like action of culture filtrates prepared from *B. amyloliquefaciens* and *B. subtilis* FZB24, FZB42, FZB45 and FZB37.

## ROOT COLONIZATION EFFICIENCY

The efficiency of beneficial bacteria to augment overall plant growth depends on the colonization ability of the inoculated strains (Bloemberg and Lugtenberg, 2001). *Bacillus* species have been considered less rhizosphere-competent compared to *Pseudomonas* species and, perhaps for this reason, most research aimed at development of biofertilizers for introduction into the rhizosphere has been with *Pseudomonas* species (Weller, 1988). However, considering the genetic diversity within the genus *Bacillus* in soil and in rhizosphere of various plant species (Mavingui *et al.*, 1992), the evidence-come-principle that rhizosphere competence is a characteristic of the strain (genotype) rather than genus or species of soil microorganisms, and the reports of plant rhizosphere colonization by *Bacillus* species (Juhnke *et al.*, 1987; Maplestone and Campbell, 1989; Milus and Rothrock, 1993), it would seem likely that rhizosphere competent genotypes occur within the genus *Bacillus*.

Plant colonization studies have revealed that *Bacillus* has various colonization patterns in plants. Studies of Juhnke *et al.*, (1987) with spring wheat in Montana, and Milus and Rothrock (1993) with winter wheat in Arkansas showed that selected strains of *Bacillus* species introduced on seed established in the rhizosphere of wheat as well as, or better than, strains of fluorescent *Pseudomonas* species. Halverson *et al.*, (1993) demonstrated in field studies that *B. cereus* UW85 was the dominant strain on roots of soybean at harvest after its introduction on the seed at planting, but they also reported that the population of this biocontrol agent, selected for its ability to protect alfalfa and soybeans against damping-off, remained confined mostly to the top 2 to 3 cm of soybean roots. Liu and Sinclair (1992; 1993) reported that *B. megaterium* B153-2-2 colonized the rhizosphere of soybean under field conditions to a depth of 20 cm in the soil when introduced into the seed furrow at planting as a dry product consisting of bacteria grown on ground soybean stems. The effects of traits of *B. megaterium* on seed and root colonization and their correlation with the suppression of *Rhizoctonia* root rot of soybean has been described (Zheng and Sinclair, 2000). Using rifampicin (Rif) resistance gene, the colonization pattern of *B. pumilus* SE34 in tomato roots, stems, and leaves at 6 weeks after inoculation (Yan *et al.,* 2003) and *B. mojavensis* AB1 in leaves and twigs of the coffee plants (Nair *et al.*, 2002) has been reported. Using intrinsic antibiotic markers, Trivedi *et al.*, (2005) reported persistence of *B. megaterium* and *B. subtilis* in the rhizosphere of different tea clones.

Recent advances in molecular techniques are also encouraging in respect that tools are becoming available to determine the mechanism by which crop performance is improved using PGPR and track survival and activity of PGPR organisms in soil and roots. The rarity of repetitive extragenic palindrome (REP) or enterobacterial repetitive intergenic consensus (ERIC) elements in Gram-positive bacteria precludes their utilization for fingerprinting by polymerase chain reaction (Fox *et al.,* 1992; Wisotzkey *et al.*, 1992). Consequently, alternative primers have been employed to distinguish closely related *Bacillus* species (Brousseau *et al.*, 1993; Jurtshuk *et al.*, 1992). A protocol from pulsed field gel electrophoresis using a restriction enzyme for *B. subtilis* B2g was obtained (Martel *et al.*, 2000). Bacterial biocontrol agents *Bacillus thuringiensis BS8, B. cereus* L, and *B. mycoides* S were tested with PCR probes for the presence of genes coding for known antifungal compounds (Zhang *et al.*, 2006). Recently, Liu *et al.*, (2006) reported the colonization pattern of *B. megaterium* C4, a nitrogen-fixing bacterium in maize and rice seedlings using green fluorescent proteins (GFP) gene. Using molecular markers such as GFP or fluorescent antibodies, it is possible to monitor the location of individual rhizobacteria on root using confocal laser scanning microscopy (CLSM).

## *BACILLUS* SPP. AS NOVEL BIOINOCULANTS

*Bacillus* spp. may be used separately or co-inoculated with other bacterial species to increase plant growth. *B. polymyxa* (H5), a phosphate solubilizing bacterium, increased available soil phosphorus levels when co-inoculated with *Rhizobium* on chickpea (Alagawadi and Gaur, 1988). Colonization and nodulation by *B. japonicum* on soybean can be enhanced with *Bacillus* spp., thereby increasing plant dry weight and seed yield (Li and Alexander, 1988). An IAA producing *Bacillus* sp. increased nodule number, nodule weight, nitrogenase activity and leghaemoglobin content of *Phaseolus vulgaris* when co-inoculated with *Rhizobium meliloti* (Srinivasan *et al.*, 1996). Nitrogen fixation and phosphate solubilization increased in case of mangroves when the mixture of slow-growing $N_2$-fixing bacterium *Phyllobacterium* sp. and the fast-growing phosphate-solubilizing bacterium *Bacillus licheniformis* was used (Adriana *et al.*, 2001). Co-inoculation of one-month-old *Pinus contorta* Dougl. seedlings with mycorrhizal fungi (*Wilcoxina mikolae* isolate R947) and nitrogen fixing *Bacillus* strain L6 resulted in increase in mycorrhizal colonization and nitrogen content of leaves (Chanway and Holl, 1991). Parmer and Dadarwal (1999) have studied the co-inoculation of *Bacillus* sp. with *Rhizobium* in field trials. *Bacillus* sp. CECT 450 increased nodulation on bean when co-inoculated with *Rhizobium tropici* CIAT 899 (Camacho *et al.*, 2001).

Application of *Bacillus* spp., alone has also shown growth promoting effects on various plants that may vary depending on soil types, illustrated with Alder and *B. licheniformis* (Ramos *et al.*, 2003). A super strain of *Bacillus firmus* (NCIM-2636) producing IAA, in addition to its high ability to solubilize insoluble inorganic phosphates increased rice production when applied in acid soils of Nagaland, India (Datta *et al.*, 1982). *B. subtilis* promoted growth of peanut seedlings resulting in yield increase of 3.5 to 37% (Turner and Blackman, 1991), as well as improved germination, seedling emergence, plant nutrition and increased root growth. Legume species vary in the magnitude of response to the plant-growth promoting bacterium *B. polymyxa* (Chanway *et al.*, 1988a), and *Bacillus* strains isolated from wheat root enhance growth of wheat in a cultivar related manner (Chanway *et al.*, 1988b). In China, plant associated *Bacillus* mainly *B. cereus* Frankland and Frankland, collectively called yield-increasing bacteria (YIB) increased by 11%, the yield of wheat and of vegetables by 25%. (Chen *et al.*, 1996). Other strains of *Bacillus*, including *B. subtilis* (Edrenberg) Cohn have been isolated and tested for biological control of plant diseases as well as growth promotion in China (Tang, 1994). These strains initially isolated from China, when tested for their ability to suppress take-all and *Rhizoctonia* root rot, and to promote seedling growth of glasshouse grown wheat in Australian soils yielded positive results (Ryder *et al.*, 1999). Inoculation with *B. licheniformis* CECT 5106 and *B. pumilus* CECT5105 enhanced the growth of *Pinus* seedlings. *B. cereus* UW85 enhanced soybean nodulation under both field and growth chamber conditions (Halverson *et al.*, 1993). In greenhouse experiments, Assis *et al.*, (1996) observed that *B. subtilis* R14, isolated from the surface of kale leaves, completely controlled the disease on cabbage when tested against three strains of *Xanthomonas campestris* pv.

**Table 1** Recent examples of *Bacillus* spp. that benefit the growth of various plant species in greenhouse and/or field trials

| *Bacillus spp.* | *Benefits* | *Reference* |
|---|---|---|
| *B. subtilis* AF1 | • Peat based formulations promoted seed germination and biomass of groundnut and pigeon pea against pathogen pressure (*Aspergillus niger* and *Fusarium oxysporum*) | Manjula and Podile (2001) |
| *B. licheniformis* | • Inoculation doubled the rate of N fixation in mangrove plants in greenhouse trials<br>• Increased leaf development | Adriana *et al.*, (2001) |
| *B. licheniformis* CECT 5106 & *B. pumilus* CECT 5105 | • Co-inoculation with mycorrhizal fungus *Pisolithus tinctorius* promoted the growth of *Pinus pinea* seedlings | Probanza *et al.*, (2001) |
| | • Inoculation increased the degree of mycorrhization in test plants<br>• Alterations in the microbial community | |
| *B. amyloliquefaciens* FZB 45 | • Diluted culture filtrates stimulated growth of maize seedlings under phosphate limitation in presence of phytate | Idriss *et al.*, (2002) |
| *B. amyloliquefaciens*, IN937 | • Statistically significant increases in plant growth of tomato and pepper in two growing years in field trials | Kokalis-Burelle *et al.*, (2002) |
| *B. pumilis* INR7,SE34, *B. subtilis* GB03, *B. cereus* C4 | • Improvement in transplant vigour and fruit yield | |
| *B. cereus* | • Stimulation of rhizobia-legume symbiosis in soybean<br>• Increase in nodule number and N accumulation | Vessey and Buss (2002) |
| *Bacillus* sp. OSU 142 | • Spraying of bacterial suspension in apricot cultivar increased the mineral nutrition content of leaves<br>• Significant increase in yield compared to control | Esitken *et al.*, (2002) |
| *Bacillus* spp. | • Increase in survival rate of tissue culture raised tea (*Camellia sinensis*) and *Picrorhiza kurrooa* | Pandey *et al.* (2002) |
| *B. subtilis* strains BS21, BS22, BS23 | • Antagonistic against *Colletotrichum lindemuthianum* (casual agent of anthracnose of cowpea) in laboratory, greenhouse and field assays<br>• In field spray application suppressed the incidence and severity of disease | Adebanjo and Bankole, (2003) |
| *B. pumilus* T4; INR7; SE34, *B. amyloliquefaciens* IN | • Increase in seedling germination and vigor of peal millet under lab/ greenhouse and field trials | Niranjan Raj *et al.*, (2003) |
| 937a, *B. subtilis* GB03; IN937b | • Suppression of downy mildew caused by *Sclerospora graminicola* (sacc.) J. Schröt) by ISR | |
| *B. pumilus* SE34r | • Incorporation into soil less media in tomato transplant enhanced plant growth<br>• Effectively colonized the plant system in seed treatment | Yan *et al.*, (2003) |
| *Bacillus* sp. | • Production of 2R, 3R-Butanediol and acetoin and induction of systemic resistance<br>• Growth promotion in *Arabidopsis thaliana* | Ryu *et al.*, (2003) |
| *B. subtilis* NEB4, NEB5 | • Inoculation resulted in growth promotion of soybean greenhouse trials | Bai *et al.*, (2003) |

*Table 1 Contd.*

*Table 1 Contd.*

| | | |
|---|---|---|
| | • Co-inoculation with *Bradyrhizobium japonicum* at various root zone temperature and in field increased nodule number, total biomass, total nitrogen and yield | |
| *B. mycoides* strain JC192 & K194 | • Reduction in influence of *Fusarium culmorum* on winter wheat in pot experiment<br>• Increase in yield in microplot experiment | Czaban *et al.*, (2004) |
| *B. subtilis* strain 6051 | • Protect Arabidopsis root against bacterial pathogen *Pseudomonas syringae* pv *tomato* DC 3000 *in vitro* and soil conditions<br>• Formation of extensive biofilm and secretion of surfactin was postulated to contribute towards the biocontrol activity | Bais *et al.*, (2004) |
| Consortium of *Bacillus* spp. | • Application during first developmental stages of two micropropogated banana varieties increased plant health and survival rates<br>• Increase in foliar mineral contents | Vega *et al.*, (2004) |
| *B. licheniformis* | • Increment in growth of pepper and tomato in seedbed and greenhouse<br>• Considerable colonization and competitive ability | Garcia *et al.*, (2004) |
| *B. megaterium, B. subtilis* | • Increased growth parameters of four different tea clones in nethouse<br>• Stimulation of native microflora<br>• High root colonization efficiency | Trivedi *et al.*, (2005) |
| *B. firmus* GRS123 & *B. megaterium* GPS55 | • Greenhouse trials increased (upto 26%) plant biomass of groundnut<br>• Actively growing cells and peat based formulations significantly increased plant growth and pod yield (upto 19%) in field | Kishore *et al.*, (2005) |
| *B. subtilis* M4 | • Reduce disease incidence caused by *Colletotrichum logenarium* and *Pythium aphanidermatum* on cucumber and tomato<br>• cDNA-AFLP analysis in gene transcription in host plant after inoculation indicated towards defence associated modification at transcriptional level | Ongena *et al.*, (2005) |
| *B. megaterium* de Bary TRS-4 | • Growth promotion and protection against brown root rot disease in tea<br>• Increase in polyphenolics and defence related enzymes in inoculated plants | Chakraborty *et al.*, (2005) |
| *B. subtilis* strain CBE4 | • Effective in controlling damping off of chilli in greenhouse trials<br>• Efficiency to control disease was at par to the standard fungicide Ridomil<br>• Talc based formulations had a shelf life of more than 4 months | Kumaresan *et al.*, (2005) |
| *B. pumilus* INR7, T4 & SE34; *B. subtilis* GB03; *B. amyloliquefaciens* IN937a | • Decrease in incidence of bacterial canker of tomato in greenhouse<br>• Increase in phenylalanine ammonia lyase (PAL) and total phenol contents<br>• Increase in seed quality parameters like seed germination and seedling vigor | Girish and Umesha, (2005) |
| *B. subtilis* | • Growth promotion of maize under greenhouse conditions due to various formulations including alginate beads<br>• High root colonization efficiency in alginate formulations<br>• Greater shelf life of inoculant in alginate based formulations | Trivedi *et al.*, (2005) |
| *B. edaphicus* | • Significant increase in root and shoot growth of wheat (*Triticum 'aestivum* L.) Yangmai-158' by inoculation of *B. edaphicus* NBT and its mutants in pot experiment | He and Fang, (2006) |

*Table 1 Contd.*

*Table 1 Contd.*

| | | |
|---|---|---|
| | • Significantly higher N, P, and K contents | |
| *B. megaterium* var *phosphaticum* & *B. mucilaginosus* | • Growth promotion of pepper and cucumber by co-inoculation with one P solubilizing (*B. megaterium*) and other K solubilizing (*B. mucilaginosus*) bacteria<br>• Synergistic solubilization of P & K by co-inoculation | Han *et al.*, (2006) |
| *B. megaterium* (c 96) | • Reduction in *Fusarium* crown and root rot of tomato in greenhouse<br>• Integrated control by bacterial isolate and chemical fungicide carbendazium gave significant control over disease | Omar *et al.*, (2006) |
| *B. cereus* & *B. circulans* | • Formulations in carrier (lignite) based retained high population of bacteria at ambient temperatures<br>• Significant increase in maize, pigeonpea and wheat in long-term field trials | Tilak and Reddy, (2006) |
| *B. lentimorbus* B-30485, *B. subtilis* 30487 & | • Inhibition of *Fusarium moniliforme* and *Colletotrichum falcatum* causing disease in sugar cane | Nautiyal *et al.*, (2006) |
| *B. lentimorbus* B-30488 | • Commercial scale formulation using molasses, pressmud and spent wash decreased disease severity and increased plant growth of two different cultivars | |
| *Bacillus* species OSU-142, RC07 and M-13 | • Increase in growth parameters of sugar beet under greenhouse and two different field soil conditions<br>• Increase in leaf, root and sugar yield for two successive years in field conditions | Cakmakci *et al.*, (2006) |
| *B. cereus* | • Isolate produced antifungal hydrolytic enzymes when grown in medium containing shrimp and crab shell powder (SCSP)<br>• Inhibited growth of various phytopathogens in greenhouse trials<br>• Increase the growth of Chinese cabbage | Chang *et al.*, (2006) |
| *B. megaterium, B. subtilis* | • Increase in yield parameters of rice at field trials in Himalayan location<br>• Increase in mineral content<br>• Stimulation of native microflora | Trivedi *et al.*, (2006) |

*campestris* in all application periods tested, suggesting its utilization as a preventive treatment for the disease. Besides, a 73% reduction in the disease in field tests corroborated the efficiency of the same isolate in controlling cabbage black rot (Assis *et al.*, 1997). Pandey *et al.*, (2000) used the biocontrol property of two *Bacillus* spp. in biological hardening of tissue culture raised tea plants. Recent examples of greenhouse and field trials of *Bacillus* spp. on the growth of various plants are presented in Table 1.

## FORMULATION DEVELOPMENT

Formulation development of PGPR poses special problems because the active ingredient is a living organism, which must be kept relatively immobile and inactive in storage, but quickly resumes its normal metabolic processes once applied to target site (Bashan, 1998).

A successful formulation of a *Bacillus* spp. for use in plant growth is potentially more readily achieved than a successful formulation of microbial biomass that does not produce a resistant life stage comparable to *Bacillus* spore (Emmart and Handelsman, 1999). The durability of *Bacillus* spore (Driks, 2004) allows a formulation scientist to consider some processes that may be chemically, physically, or environmentally more severe and would, therefore, not be feasible for formulating less resistant microbial biomass (Schisler *et al.*, 2004). Experimental formulations of *Bacillus* spp. included clay (Osburn *et al.*, 1995); peat and chitin (Manjula and Podile, 2001; Sid Ahmed *et al.*, 2003); methylcellulose (Racke and Sikora, 1992), Ca-alginate, Na-alginate,

**Table 2** Commercially available formulations of *Bacillus* spp. in some developed countries and India

| *Trade Name* | *Microorganism(s)* | *Properties/Formulation* | *Reference* |
|---|---|---|---|
| Companion | *B. subtilis* GB03, *B. lichenformis, B. megaterium* | Biocontrol/Liquid | www.growthproducts.com |
| Kodiak | *B. subtilis* GB03 | Biocontrol/Powder | www.gustafson.com |
| EcoGuard | *B. licheniformis* SB3086 | Biocontrol/Liquid spore concentrate | www.novozymes.com |
| HiStick N/T | *B. subtilis* (rhizobia also in formulation) | Biocontrol/Slurry | www.beckerunderwood.com |
| Rhizo-Plus | *B. subtilis* FZB24 | Biocontrol/Granule | www.home.fhtw-berlin.de |
| Serenade | *B. subtilis* QWT713 | Biocontrol/Powder | www.agraquest.com |
| Subtilix | *B. subtilis* MBI600 | Biocontrol/Powder | www.beckerunderwood.com |
| YieldShield | *B. subtilis* GB34 | Biocontrol/Powder | www.gustafson.com |
| Shakti | *Bacillus* sp. | Biocontrol/Powder | www.amitbiotech.com |
| Phosphobacterin | *B. megaterium* var. *phosphaticum* (*Pseudomonas* sp. Also in formulation) | P-solubilization/Powder | www.iari.res.in |
| Vitagro | *B. firmus, B. polymyxa, B. megaterium, B. licheniformis* (*P. striata, Trichoderma viride,* also in formulation) | P- solubilization, biocontrol/Powder | www.microbax.com |
| Phosfert | *B. megaterium* var. *phosphaticum* (*Azotobacter chroococcum, A. vinelandii* also in formulation) | P- solubilization and N-fixation/Powder | www.krishimitra.net |
| Teeka | *Bacillus* sp. | P-solubilization, Biocontrol/ Liquid, granular | www.sribio.com |

manicol, or carob (Schmidt *et al.,* 2001; El-Komy, 2005; Trivedi *et al.,* 2005); vermiculite, kaolin, peat moss, wheat bran (Amer and Utkhede, 2000) carboxymethyl cellulose, vegetable oil, and polyvinyl pyrrolidone (Kanjanamaneesathian *et al.*, 2000) and peptone (Ferreira *et al.*, 1991; Schmidt *et al.*, 2001) and nutrient medium (Smith *et al.*, 1993). Such formulations are then usually used as a wettable powder, water dispersible granules or dust where proxel will generally be used to prevent the microbial contamination. Some of the commercially available *Bacillus*-based formulations of plant growth promotion products are listed in Table 2.

## FUTURE PROSPECTS

The use of *Bacillus* species as plant growth promoting rhizobacteria offers an environmentally sustainable approach to increase crop production and health. The application of molecular tools is providing new insight into the diversity of *Bacillus* spp. in the rhizosphere, the heterogeneity within the root environment and the importance of environmental and biological factors in determining the community structure. Development of species and strain specific molecular markers has overcome barriers of precise identification of isolates and has contributed greatly towards understanding of PGPR involvement in the rhizosphere. These efforts will enhance our ability to understand and manage the rhizosphere and will lead to new products of *Bacillus* based formulations with improved effectiveness and better compatibility with existing agricultural practices.

## ACKNOWLEDGEMENTS

The authors thank Drs. D.K. Maheshwari and R.C. Dubey for giving the opportunity to prepare this review article. Professor A.N. Purohit and Dr. L.M.S. Palni (Former Directors) and Dr. U. Dhar (Present Director) are gratefully acknowledged for the guidance and encouragement. The Department of Biotechnology, Council of Scientific and Industrial Research, Department of Science and Technology, the Union Ministry of Environment and Forests, Govt. of India, and Uttarakhand State Council for Science and Technology, Govt. of India, Uttarakhand are thanked for supporting the research in this area.

## References

Abdel Wahab, A.M. (1975) *Plant Soil* **42:** 703-708.

Adebanjo, A. & Bankole, S.A. (2003) *J. Basic Microbiol.* **44:** 3-9.

Adriana, R., Holguin, G., Glick, B.R. & Bashan, Y. (2001) *FEMS Microbiol. Ecol.* **35:** 181-187.

Alagawadi, A.R. & Gaur, A.C. (1988) *Plant Soil* **105:** 241-246.

Amer, G.A. & Utkhede, R.S. (2000) *Can. J. Microbiol.* **46:** 809-816.

Assis, S.M.P., Mariano, R.L.R., Michereff, S.J. & Coelho, R.S.B. (1996) In: *Advances in Biological Control of Plant Diseases*, (Eds. Wenhua, T., Cook, R.J. & Rovira, A.), China Agricultural University Press, Beijing, China, pp. 347-353.

Assis, S.M.P., Mariano, R.L.R., Michereff, S.J., Silva, G. & Maranhão, E.A.A. (1997) In: *Plant Growth-Promoting Rhizobacteria œ Present Status and Future Prospects*, (Eds. Ogoshi, A., Kobayashi, Y., Homma, Y., Kodama, F., Kondo, N. & Akino, S.), Sapporo: Univ. of Hokaido/OECD, pp. 345-348.

Bai, Y., Zhou, X. & Smith, D.L. (2003) *Crop Sci.* **43:** 1774-1781.

Bais, H.P., Fall, R. & Vivanco, J.M. (2004) *Plant Physiology* **134:** 307-319.

Banik, S. & Dey, B.K. (1983) *Zentralbl Mikrobiol.* **138:** 17-23.

Barazani, O. & Friedman, J. (1999) *J. Chem. Ecol.* **25:** 2397-2406.

Bashan, Y. (1998) *Biotechnol. Adv.* **16:** 729-770.

Bashan, Y. & Holguin, G. (1998) *Soil Biol. Biochem.* **30:** 1225-1228.

Berdy, J. (1974) *Adv. Appl. Microbiol.* **18:** 309-406.

Bloemberg, G. V. & Lugtenberg, B.J.J. (2001) *Curr. Opin. Plant Biol.* **4:** 343-350.

Bowen, G. D. & Rovira, A.D. (1999) *Adv. Agron.* **66:** 1-102.

Brousseau, R., Saint-Onge, A., Préfontaine, G., Masson, L. & Cabana, J. (1993) *Appl. Environ. Microbiol.* **59(1):** 114-119.

Buyer, J.S. (1995) *Appl. Environ. Microbiol.* **61:** 1839-1842.

Cakmakci, R., Donmez, F., Aydin, A. & Sahin, F. (2006) *Soil Biol. Biochem.* **38:** 1482-1487.

Camacho, M., Santamaria, C., Temprano, F., Rodriguez-Navarro, D.N. & Daza, A. (2001) *Can. J. Microbiol.* **47:** 1058-1062.

Chakraborty, U., Chakraborty, B. & Basnet. M. (2005) *J. Basic Microbiol.* **46(3):** 186-195.

Chang, W.T., Chen, Y.C. & Jao, C.L. (2006) *Bioresource Technol.* **98:** 1224-1230.

Chanway, C.P. & Holl, F.B. (1991) *Can. J. Bot.* **69:** 507-511.

Chanway, C.P., Holl, F.B. & Turkington, R. (1988a) *Plant Soil* **106:** 281-284.

Chanway, C.P., Nelson, L.M. & Holl, F.B. (1988b) *Can. J. Microbiol.* **34:** 925-929.

Chaurasia, B., Pandey, A., Palni, L.M.S., Trivedi, P., Kumar, B. & Colvin, N. (2005) *Microbiol. Res.* **160:** 75-81.

Chen, C.L., Chang, L.K., Chang, Y.S., Liu, S.T. & Tschen, J.S.M. (1995) *Mol. Gen.Genetics.***243:** 121-125.

Chen, Y.X., Mei, R.H., Lu, S., Liu, L. & Kloepper, J.M. (1996) In: *Management of Soil Borne Diseases*, (Eds. Utkhede, R.S. & Gupta, V.K.), Kalyani Press, New Delhi, pp. 165-184.

Czaban, J., Ksiezniak, A. & Perzynski, A. (2004) *Polish J. Microbiol.* **53:** 175-182.

Datta, M., Banik, S. & Gupta, R.K. (1982) *Plant Soil* **61:** 365-373.

Derbyshire, J.F. & Greaves, M.P. (1973) *Pestic. Sci.* **4:** 349-360.

Dong, Y.H., Wang, L.H., Xu, J.L., Zhang, H.B., Zhang, X.F. & Zhang, L.H. (2001) *Nature* **411:** 813-817.

Dong, Yi-Hu., Xu, Jin-Ling., Li, Xian-Zhen. & Zhang, Lian-Hui. (2000) *Microbiol.* **97:** 3526-3531.

Driks, A. (2004) *Phytopathol.* **94:** 1249-1251.

Dubnau, E., Weir, J., Nair, G., Carter, L., Moran, C., Jr. & Smith, I. (1988) *J. Bacteriol.* **170:** 1054-1062.

El-Komy H.M.A. (2005) *Food. Technol. Biotechnol.* **43:** 19-27.

Emmart, E.A.B. & Handelsman, J. (1999) *FEMS Microbiol. Lett.* **171:** 1-9.

Eshita, S.M., Roberto, N.H., Beale, J.M., Mamiya, B.M. & Workman, R.F. (1995) *J. Antibiot. (Tokyo)* **48:** 1240-1247.

Esitken, A., Karlidag, H., Ercisli, S., Turan, M. & Fikrettin, S. (2002) *Australian J. Agri. Res.* **54:** 377-380.

Fall, R., Kinsinger, R.F. & Wheeler, K.A. (2004) *Systematic Appl. Microbiol.* **27:** 372-379.

Ferreira, J.H.S., Matthee, F.M., & Thomas, A.C. (1991) *Phytopathol.* **81:** 283-287.

Fox, G.E., Wisotzkey, J.D. & Jurtshuk, Jr, P. (1992) *International J. System. Bacteriol.* **42:** 166-170.

Fravel, D.R. (1988) *Ann. Rev. Phytopathol.* **26:** 75-91.

Fritze, D. (2004) *Phytopathol.* **94:** 1245-1248.

Fuqua, C., Winans,. S.C & Greenberg, E.P. (1996) *Annu. Rev. Microbiol.* **50:** 727-751.

Garcia, J.A.L., Probanza, A., Ramos, B., Palomino, M.R. & Manero, F.J.G. (2004) *Agronomie.* **24:** 169-176.

Garrity, G.M. (2001) *Bergey's Manual of Systematic Bacteriology*,(Ed. Garrity, G.M.) Vol. 2, pp. 2816.

Gerhardson, B. & Wright, S. (2001) In: *Microorganisms in Plant Conservation and Biodiversity*, (Eds. Sivasithamparam, K., Dixon, K.W. & Narrett, R.L.), Kluwer Academic Press, London, pp. 79-103.

Giacomodonato, M.N., Pettinari, M.J., Souto, G.I. & Mendez, B.S. (2001) *World J. Microbiol. Biotechnol.* **17:** 51-55.

Girish, N. & Umesha, S. (2005) *Archives Phytopathol. Plant Protection.* **38:** 235-243.

Glick, B.R. (1995) *Can. J. Microbiol.* **41:** 109-117.

Glick, B.R., Patten, C.L., Holguin, G. & Penrose, D.M. (1999) In: *Biochemical and Genetic Mechanisms used by Plant Growth Promoting Bacteria*, Imperial College Press, London. pp. 270.

Gong, M., Wang, J.D., Zhang, J., Yang, H., Lu, X.F., Pei, Y. & Cheng, J.O. (2006) *Acta. Biochim. Biophys. Sin.* **38:** 233-240.

Gutierrez Manero, F.J., Acero, N., Lucas, J.A. & Probanza, A. (1996) *Plant Soil* **182:** 67-74.

Gutierrez Manero, F.J., Ramos-Solano, B., Probanza, A., Mehouachi, J., Tadeo, F.R. & Talon, M. (2001) *Physiol. Plant.* **111:** 206-211.

Hallmann, J., Rodriguez-Kabana, R. & Kloepper, J.W. (1998) *Soil Biol. Biochem.* **31:** 551-560.

Halverson, L.J., Clayton, M.K. & Handelsman, J. (1993) *Soil Biol. Biochem.* **25:** 485-493.

Han, H.S., Supanjani, Lee, K.D. (2006) *Plant Soil Environ.* **52:** 130-136.

Handelsman, J & Stabb, E.V. (1996) *Plant Cell.* **8:** 1855-1869.

He, H., Silo-Suh, L.A., Handelsman, J. & Clardy, J. (1994) *Tetrahedron Lett.* **35:** 2499-2502.

He, L.Y. & Fang, S.X. (2006) *Can. J. Microbiol.* **52:** 66-72.

Idris, E.E., Bochow, H., Ross, H. & Borriss, R. (2004) *J. Plant Dis. Protection.* **111:** 583-597.

Idriss, E.E., Makarewicz, O., Farouk, A., Rosner, K., Greiner, R., Bochow, H., Richter, T. & Borriss, R. (2002) *Microbiol.* **148:** 2097-2109.

Illmer, P., Barbato, A. & Schinner, F. (1995) *Soil Biol. Biochem.* **27:** 265-270.

Joshi, R. & McSppadden, G.B.B. (2006) *Phytopathol.* **96:** 145-154.

Juhnke, M. E., Mathre, D. E. & Sands, D. C. (1987) *Appl. Environ. Microbiol.* **53:** 2793-2799.

Jurtshuk, R.J., Blick, M., Bresser, J., Fox, G.E. & Jurtshuk, P. Jr. (1992) *Appl. Environ. Microbiol.* **58(8):** 2571-2578.

Kanjanamaneesathian, M., Pengnoo, A., Jantharangsri, A., Niratana, L. & Kusonwiriyawong, C. (2000) *World J. Microbiol. Biotech.* **16:** 523-527.

Kilian, M., Steiner, U., Krebs, B., Junge, H., Schmiedeknecht, G. & Hain, R. (2000) *Pflanzenschutz-Nachrichten Bayer.* **1/00:** 72-93.

Kishore, G.K., Pande, S. & Podile, A.R. (2005) *Lett. Appl. Microbiol.* **40:** 260-266.

Kloepper, J. W., Ryu, C-M. & Zhang, S. (2004) *Phytopathol.* **94:** 1259-1266.

Kloepper, J. W. & Schroth, M.N. (1981) *Phytopathol.* **71:** 642-644.

Kloepper, J.W. (1994) In: *Azospirillum/plant Associations*, (Ed. Okon, Y.), CRC Press, Boca, Raton. pp. 137-166.

Kloepper, J.W. (1997) In: *International Symposium on Clean Agriculture*, Sapporo: OECD, pp. 49-52.

Kokalis-Burelle, N., Vavrina, C.S., Rossskopf, E.N., & Shelby, R.A. (2002) *Plant Soil.* **238:** 257-266.

Kumaresan, K., Subramanian, M., Vaithiyanathan, S., Sevagaperumal, N. & Gopal, C. (2005) *Archives Phytopathol. Plant Protection.* **38:** 19-30.

Lee, S. J., Park, S.Y., Lee, J.J., Yum, D.Y., Koo, B.T. & J. K. Lee. (2002) *Appl. Environ. Microbiol.* **68:** 3919-3924.

Li, C.Y., Massicote, H.B. & Moore, L.V.H. (1992) *Plant Soil* **140:** 35-40.

Li, D. & Alexander, M. (1988) *Plant Soil* **108:** 211-219.

Liu, X., Zhao, H. & Chen, S. (2006) *Curr. Microbiol.* **52:** 186-190.

Liu, Z. L. & Sinclair, J. B. (1992) *Phytopathology* **82:** 1297-1301.

Liu, Z. L. & Sinclair, J. B. (1993) *Soil Biol. Biochem.* **25:** 849-855.

Lynch, J.M. (1983) In: *Soil Biotechnology*, Blackwell Scientific Publication, New York.

Lynch, J.M. (1990) *Biotechnol. Adv.* **8:** 335-346.

Macrae, A., Rimmer, D.L. & O'Donnell G.A. (2000) *Antonie van Leeuwenhoek.* **78:** 13-21.

Mahafee, W.F. & Kloepper, J.W. (1996) *Microbiol. Ecol.* **34:** 210-223.

Manjula, K. & Podile, A.R. (2001) *Can. J. Microbiol.* **47:** 618-625.

Maplestone, P. A. & Campbell, R. (1989) *Soil Biol. Biochem.* **21:** 543-550.

Martel, P., Smalla, K. & Berg, G. (2000) *J. Appl. Microbiol.* **89:** 463-471.

Mavingui, P., Laguree, G., Berge, O. & Heulin, T. (1992) *Appl. Environ. Microbiol.* **58:**1984-1993.

Milner, J.L., Stohl, E.A. & Handelsman, J. (1996) *J. Bacteriol.* **178:** 4266-4272.

Milus, E. A. & Rothrock, C. S. (1993) *Can. J. Microbiol.* **39:** 335-341.

Murphy, J.F., Reddy, M.S., Ryu, C.M., Kloepper, J.W. & Li, R. (2003) *Phytopathol.* **93:** 1301-1307.

Nair, J.R., Singh, G. & Sekar, V. (2002) *J. Appl. Microbiol.* **93:** 772-780.

Nautiyal, C.S., Mehta, S. & Singh, H.B. (2006) *J. Microbiol. Biotechnol.* **16:** 184-192.

Nelson, A.D., Barber, L.E., Tjepkema, J., Russell, S.A., Powelson, R. & Evans, H.J (1976) *Can J. Microbiol.* **22(4):** 523-30.

Niranjan Raj, S., Chaluvaraju, G., Amruthesh, K.N. & Shetty, H.S. (2003) *Plant Dis.* **87:** 380-384.

Omar, I., O'Neill, T.M. & Rossall, S. (2006) *Plant Pathol.* **55:** 92-99.

Ongena, M., Jacques, P., Toure, Y., Destain, J., Jabrane, A. & Thonart, P. (2005) *Appl. Microbiol. Biotechnol.* **69:** 29-38.

Osburn, R.M., Milner, J.L., Oplinger, E.S., Smith, R.S. & Handelsman, J. (1995) *Plant Dis.* **79:** 551-556.

Pal, K.K., Tilak, K.V., Saxena, A.K., Dey, R. & Singh, C.S. (2001) *Microbiol. Res.* **156:** 209-223.

Pandey, A. & Kumar, S. (1989) *J. Sci. Ind. Res.* **48:** 134-144.

Pandey, A., Bag, N., Chandra, B., & Palni L.M.S. (2002) In: *Role of Plant Tissue Culture in Biodiversity Conservation and Development*, (Eds. Nandi, S.K, Palni, L.M.S. & Kumar, A.), Gyanodaya Prakashan, Nainital, pp. 565-577.

Pandey, A., & Palni, L.M.S. (1996) *Biol. Fertil. Soils.* **21:** 131-137.

Pandey, A. & Palni, L.M.S. (1997) *Microbiol Res.* **152(4):** 359-65.

Pandey, A., & Palni, L.M.S. (2004) *Internat. J Tea Sci.* **3(3&4):** 285-290.

Pandey, A., Palni, L.M.S. & Bag, N. (2000) *Biotechnol. Lett.* **22:** 1087-1091.

Pandey, A, Palni, L.M.S. & Coulomb, N. (1997) *Microbiol. Res.* **152:** 105-112.

Pandey, A., Trivedi, P., Kumar, B., Chaurasia, B., Singh, S. & Palni, L.M.S. (2004) In: *Biotechnological Approaches for Sustainable Development*, (Eds. Reddy, M.S. & Kumar, S.), Allied Publishers Ltd., New Delhi. pp. 13-20.

Parmar, N & Dadarwal, K.R. (1999) *J. Appl. Microbiol.* **86:** 36-44.

Persello-Cartieaux, F., Nussaunme, L. & Robaglia, C. (2003) *Plant Cell & Environ.* **26:** 189-199.

Petersen, D.J., Shishido, M., Holl, F.B. & Chanway, C.P. (1995) *FEMS Microbiology Lett* **133:** 71-76.

Piggot, P.J. & Hilbert, D.W. (2004) *Curr. Opinion Microbiol.* **7:** 579-586.

Poonguzhali, S., Madhaiyan, M. & Sa, T. (2006) *Plant Soil.* **286:** 167-180.

Probanza, A., Mateos, J.L., Garcia, J.A.L., Ramos, B., deFeliepe, M.R. & Manero, F.J.G. (2001) *Microbiol. Ecol.* **41:** 140-148.

Racke, J. & Sikora, R.A. (1992) *Soil Biol. Biochem.* **24:** 521-526.

Raffel, S., Stabb, E.V., Milner, J.L. & Handelsman, J. (1996) *Microbiol.* **142:** 3225-3236.

Ramos, B., Lucas Garcia, J.A., Probanza, A., Barrientos, M.L., Gutierrez Manero, F.J. (2003) *Environ. Exp. Bot.* **31:** 61-68.

Ramos, H.C., Hoffmann, T., Nedjari, H., Presecan-Siedel, E., Dreesen, O., Glaser, P. & Jahn, D. (2000) *J. Bacteriol.* **182:** 3072-3080.

Reimmann, C., Ginet, N., Michel, L., Keel, C., Michaux, P., Krishnapillai, V., Zala, M., Heurlier, K., Triandafillu, K., Harms, H., Defago, G. & Haas, D. (2002) *Microbiol.* **148:** 923-932.

Richardson, A.E. (2001) *Aust. J. Plant. Physiol.* **28:** 897-906.

Rodgers, P.B. (1989) *Pesticide Sci.* **27:** 155-164.

Roi, A.A., Reva, O.N., Kurdish, I.K. & Smirnov, V.V. (2004) *Prikl. Biokhim Mikrobiol.* **40(5):** 551-557.

Ryder, M.H., Yan, Z., Terrace, T.E., Rovira, A.D., Tang, W. & Correll, R.L. (1999) *Soil Biol. Biochem.* **31:** 19-29.

Ryu, C-M., Farag, M.A., Hu, C.H., Reddy, M.S., Kloepper, J.W. & Pare, P.W. (2004) *Plant Physiology* **134:** 1017-1026.

Salisbury, F.B. (1994) In: *Plant-Environment Interactions*, (Ed. Wilkinson, R.E.), Marcel Dekker, New York, USA, pp. 39-81.

Schilling, G., Gransee, A., Deubel, A., Lezovic, G. & Ruppel, S. (1998) *Z. Pfl. Ernahr. Bodenkde.* **161:** 465-478.

Schisler, D.A., Slininger, P.J., Behle, R.W. & Jackson, M.A. (2004) *Phytopathol.* **94:** 1267-1271.

Schmidt, C.S., Lorenz, D., Wolf, G.A. & Jager, J. (2001) *J. Phytopathol.* **149:** 437-445.

Seldin, L., Van Elsas, J.D. & Elisa G. C. Penido, E.G.C. (1983) *Plant Soil* **70:** 243-255.

Sharma, A. & Johri, B.N. (2003) *Microbiol. Res.* **158:** 243-248.

Shoda, M. (2000) *J. Biosci. Bioeng.* **89:**515-521.

Sid Ahmed, A., Ezziyyani, M., Perez Sanchez, C. & Candela, M.E. (2003). *Eur. J. Plant Pathol.* **109:** 633-637.

Silo-Suh, L.A., Lethbridge, B.J., Raffel, S.J., He, H., Clardy, J. & Handelsman, J. (1994) *Appl. Environ. Microbiol.* **60:** 2023-2030.

Slepeckey, R.A. & Hemphill, H.E. (1992) In: *The Prokaryotes 2* (Eds. Balows, A., Truper, H.G., Dworkin, M.D., Harder, W. & Schleifer, K.) Springer-Verlag, New York, pp. 1663-1696.

Smalla, K., Wieland, G., Buchner, A., Zock, A., Parzy, J., Kaiser, S., Roskot, N., Heuer, H. & Berg, G. (2001) *Appl. Environ. Microbiol.* **67:** 4742-4751.

Smith, K.P., Havey, M.J. & Handelsman, J. (1993) *Plant Dis.* **77:** 139-142.

Sonenshein, A.L. (2000) *Curr. Opinion Microbiol.* **3:** 561-566.

Srinivasan, M., Peterson, D.J. & Holl, F.B. (1996) *Can. J. Microbiol.* **42:** 1006-1014.

Stephens, C. (1998) *Curr. Biol.* **8:** R45-R48.

Tang, W.H. (1994) In: *Improving Plant Productivity with Rhizosphere Bacteria*, (Eds. Ryder, M.H., Stephens, P.M. & Bowen, G.D.), Division of Soils CSIRO, Adelaide, Australia, pp. 267-273.

Thakuria, D., Talukdar, N.C., Goswami, C., Hazarika, S., Boro, R.C., & Khan, M.R. (2004) *Curr. Sci.*, **86(7):** 978-985.

Tilak, K.V.B.R. & Reddy, B.S. (2006) *Curr. Sci.* **90:** 642-644.

Trivedi, P., Kumar, B., Pandey, A. & Palni, L.M.S. (2006) *Plant Soil.* (In Press).

Trivedi, P., Pandey, A. & Palni, L.M.S., (2005) *World J. Microbiol. Biotechnol.* **21:** 941-945.

Trivedi, P., Pandey, A. Palni, L.M.S., Bag, N. & Tamang, M.B. (2005) *International J. Tea Sci.* **4:** 19-25.

Trivedi, P., Pandey, A., Palni, L.M.S. (2005). *World J. Microbiol. Biotech.* **21:** 941-945.

Turner, J.T & Blackman, P.A. (1991) *Plant Dis.* **75:** 347-353.

Vazquez, P., Holguin, G., Puente, M.E., Lopez-Cortes, A. & Bashan, Y. (2000) *Biol. Fertil. Soils.* **30:** 460-468.

Vega Jaizme, M.D.C., Romero-Rodriguez, A.S. & Guerra, M.S.P. (2004) *Fruits* **59:** 83-90.

Vessey, J.K. & Buss, T.J. (2002) *Can. J. Plant. Sci.* **82:** 282-290.

Vessey, J.K. (2003) *Plant Soil.* **255:** 571-586.

Weller, D.M. (1988) *Ann. Rev. Phytopathol.* **26:** 379-407.

Whipps, J.M. (2001) *J. Experimental Bot.* **52:** 487-511.

Wisotzkey, J.D., Jurtshuk Jr P., Fox, G.E., Deinhard, G. & Poralla, K. (1992) *International J. System Bacteriol.* **42:** 263-269.

Witz, D.F., Detroy, R.W. & Wilson, P.W. (1967) *Archives Microbiol.* **55:** 369-381.

Wullstein, L.H., Bruening, M.L. & Bollen, W.B. (1979) *Physiologia Plantarum* **46:** 1- 4.

Xu, F., Byun, T., Deussen, H.J., Duke, K.R. & Dussen, H.J. (2003) *J. Biotechnol.* **101:** 89-96.

Yan, Z., Reddy, M.S. & Kloepper, J.W. (2003) *Can. J. Microbiol.* **49:**383-389.

Zehnder, G.W., Yao, C., Wei, G. & Kloepper, J.W. (2000) *Biocontrol.* **45:** 127-137.

Zhang, Y.., Fernando, W.G.D., de Kievit, T.R., Berry, C., Daayf, F. & Paulitz, T.C. (2006) *Can. J. Microbiol.* **52(5):** 476-481.

Zheng, X.Y. & Sinclair, J.B. (2000) *Biocontrol* **45:** 223-243.

4

# Biofertilizers for Sustainability, Agroresources and Technology

**Sunita Gaind and B.D. Kaushik**

*Microbiology Division, Indian Agricultural Research Institute, New Delhi-110012*

## ABSTRACT

Biofertilizers are the renewable inputs, which can maximize ecological benefits and minimize environmental hazards. The integrated use of organic and inorganic fertilizer has been found to be more effective as far as crop yield and produce quality are concerned. The isolation of endophytic diazotrophs viz. *Acetobacter diazotrophicus, A. johannae, A. azotocaptans, Herbaspirillum seropedicae, H. rubrisubalibicans* and ammonia secreting strains of *Azotobacter* and *Azospirillum* may help in extending the biofertilizer benefit to other crops like sugar cane, cereals, forage grass, oil palm trees and coffee plants. Application of carrier based nitrogen and phosphorus biofertilizers in soil for cereals, millets, pulses, legumes, oilseeds, sugar cane and cotton grown under different agroclimatic conditions have shown an increase of 15-20% in crop yield. Moreover, mixed biofertilizers application has an edge over single biofertilizer in improving crop yield. For desired effect of carrier based biofertilizer, its quality needs due attention and should be as per BIS standard. As there is wide gap between production and demand of biofertilizers based on the area under cultivation, efforts should be made to minimize the gap by setting up new biofertilizer production units, developing production technology using cheap and environment-friendly biodegradable carrier material as well as by creating awareness among farmers.

## INTRODUCTION

Green revolution though made the country self-reliant in food grain production (50.8 MT in 1950 to 213 MT in 2004) but at the cost of soil health (Kanwar, 1997). The increasing use of chemical fertilizers, pest and weed control agents has deprived the soil of its organic matter, damaged the ecology, adversely affected the surrounding environment and most importantly the human health. This coupled with irrigation has led to rise in salinity and alkalinity of land. The cumulative effect of environment degradation resulted in a decline in yields of crops in rice-wheat system in the last ten years. The food grain production has stagnated at 1.5%. To overcome the adverse effects of chemical cultivation and to keep pace with ever growing population, there is utter need to bring out some new strategies to ensure further growth in agricultural output. The production and productivity patterns should be related to social equity, environmental integrity and economic prosperity. Adopting a strategy of integrated nutrient supply by using a combination of chemical fertilizers, organic manures and biofertilizers, the soil can be protected from further impoverishment. Biofertilizers are essential components of this approach and are being promoted to harvest the naturally available biological system of nutrient mobilization.

## Reasons for Considering Biofertilizer Technology

(i) Agriculture in tropical countries like India produces a considerable amount of crop residues and other biomass, which offers opportunities for its conversion into beneficial products.

(ii) This technology makes use of methods, which are relatively simple and employ pre-existing microorganisms.

(iii) The technology lends itself to decentralized application and can therefore be used in rural industries (Venkataraman, 1984).

## BIOFERTILIZERS

Biofertilizers are live formulations of agriculturally beneficial microorganisms, which upon application to seed, root or soil can mobilize the availability of nutrients by their biological activity and help to improve the soil health. Microbes involved in these formulations not only mobilize N and P but also secrete various growth promoting and health promoting substances. The overall effect is reflected in increased productivity of crop, microbial life and stability of soil. In broad sense, the term "biofertilizer" may be used to include all organic resources for plant growth, which are rendered in an available form for absorption through microorganisms or plant association or interactions.

## CLASSIFICATION

Soil harbours large number of microorganisms. These microorganisms play a significant role in the life cycle of plant through a number of processes such as $N_2$ fixation and supply to plants, decomposition of organic matter, solubilization of insoluble phosphates, etc. Nitrogen is the most limiting nutrient for increasing crop production. Only few prokaryotic microorganisms fix nitrogen directly through biological process. Annual biologically fixed nitrogen is estimated to be 175 mt of which 79% is accounted for terrestrial fixation.

A classical discovery by Hellriegel in 1886, mentioning the role of legume in harvesting atmospheric nitrogen made the scientific basis for biological fixation of nitrogen. Since then biofertilizers came a long way. The concept of microbial inoculation started with legume-

*Rhizobium* inoculant first patented by Nobe and Hiltner in 1896 (Fred *et al.*, 1932). In developed countries like USA, UK, France, Australia, biofertilization is restricted to *Rhizobium*, whereas in Brazil, China and India it has been diversified and large number of bacteria, fungi and actinomycetes have been included in this group.

Biofertilizers may be broadly classified into nitrogen biofertilizers and phosphate biofertilizers.

**Major biofertilizers and target crops**

| *S.No.* | *Biofertilizers* | *Target crops* |
|---|---|---|
| | **(A) Nitrogenous** | |
| (i) | *Rhizobium* (symbiotic) | • Legume: Chickpea, red gram, pea, lentil, black gram<br>• Oilseed legumes: Soybean and groundnut<br>• Forage legumes: Clover and Lucerne<br>• Tree legumes: *Leucaena* (0.026-20kg N/ha/hr) |
| (ii) | *Azotobacter* (Non-symbiotic) | Wheat, maize, cotton, sorghum, sugar cane, pearl millet, rice, vegetable crops |
| (iii) | *Azospirillum* (Associative) | Rice, wheat, vegetable crops, sugar cane (10-20 kg N/ha/yr) |
| (iv) | Blue-green algae or cyanobacteria (phototrophic) | Flooded rice fix (25 kg N/ha/yr) |
| (v) | *Azolla* (symbiotic) | Only for flooded rice (30-100 kg $N_2$/ha/yr) |
| | **(B) Phosphatic Biofertilizer** | |
| (vi) | Phosphate solubilizing microorganisms | For all crops (solubilize 50-60% of fixed P in soil) |
| (vii) | Vesicular arbuscular mycorrhizae | For all crops especially agronomic, forest and horticultural crops (Solubilize 60% of fixed P in soil) |

Discovery of some endophytic diazotrophs like *Acetobacter diazotrophicus*, *Herbaspirillum seropedicae* from sugar cane, cereals, forage grass and oil palm trees (Olivares *et al.*, 1996), *Herbaspirillum rubrisubalbicans* from 13 members of graminae family including rice, sugar cane and palm trees (Baldani *et al.*, 1996 Olivares *et al.*, 1996), *Acetobacter azotocaptans* and *A. johannae* from coffee plants (Fuentes – Ramirez *et al.*, 2001) have raised the hopes for more crops. Biotechnological innovations leading to development of ammonia secreting strains of *Azotobacter* and *Azospirillum* (Narula, 2000) may lead to development of promising endophytic diazotrophic system. Formulation prepared by using these microorganisms may be more effective than the present-day biofertilizers. Discovery of *Pirimoformaspora indica* (Verma *et al.*, 2001), a laboratory cultivable VAM-like fungus, has also opened up fresh avenues in the commercial prospects of VAM inoculants. Briones *et al.*, (2003) showed that $NH_3$ oxidizing bacteria form biofilms on the root surface of rice. These bacteria can play an important role in N nutrition of plants because of their intimate contact with roots, and N mineralized by them could be directly taken up by roots, thus reducing the N competition by rhizospheric microorganisms. About 2/3 of the fixed nitrogen of the world is made available by biological nitrogen fixation. It is contributed by *Rhizobium*-legume symbiotic system, non-legume and free-living associations. These organisms can utilize molecular $N_2$ from air at the expense of solar energy.

**Estimates of biological nitrogen fixation**

| *$N_2$ fixing system symbiotic* | *$N_2$ fixed kg N/ha* | *Reference* |
|---|---|---|
| | (a) **Legume-*Rhizobium*** | |
| *Leucaena leucocephala* | 100-300/yr | |
| *Glycine max* | 0-237/crop | Keyser and Li (1992) |
| *Trifolium repens* | 13-280/crop | Ledgard and Steel (1992) |
| *Sesbania rostrata* | 320-360/crop | Ladha *et al.* (1990) |
| | **(b) Non-legume** | |
| *Frankia casuarina* | 40-60/yr | Gauthier *et al.* (1995) |
| | **(c) Free-living associative** | |
| Rice-BGA | 10-80/crop | Roger and Ladha (1992) |
| Rice-Bacterial association | 10-30/crop | Roger and Ladha (1992) |
| Sugar cane-Bacterial association | 20-160/crop | Urquiaga *et al.* (1992) |

*Source:* Motsara *et al.* (1995)\

# TYPES OF BIOFERTILIZERS

## *RHIZOBIUM*

Symbiotic $N_2$ fixation by *Rhizobium* in legumes contribute substantially to total biological $N_2$ fixation. *Rhizobium* inoculation is a well-known agronomic practice to save chemical nitrogen fertilizer. Different species of *Rhizobium* are mainly classified in two main groups:

(a) Fast growing *Rhizobium*

(b) Slow growing under genus *Bradyrhizobium*.

**Table 1** Estimates of nitrogen fixed by some legumes

| *Crop* | *Nitrogen fixed (kg per hectare)* |
|---|---|
| Alfalfa | 100-300 |
| Berseem | 120-150 |
| Chickpea | 26-63 |
| Clusterbean | 37-196 |
| Clover | 100-150 |
| Cowpea | 53-85 |
| Black gram | 38-50 |
| Green gram | 50-55 |
| Groundnut | 112-152 |
| Lentil | 35-100 |
| Peas | 59-80 |
| Pigeonpea | 60-200 |
| Soybean | 49-130 |
| Sesbania | 69-90 |

Fast growing rhizobia can produce acid on YEMA medium and are also known as acid producers. They can be grown on simple synthetic medium with most of the carbohydrates as carbon source. Glutamate is considered superior to nitrate or ammonium as N source. *R. trifolii, R. leguminosarum* and *R. meliloti* have been grouped as acid producers and fast growers, whereas *Bradyrhizobium japonicum* and cowpea rhizobia have been classified under slow grower, non-acid producing rhizobia. Slow growing rhizobia are more specific in their needs and grow in medium containing either of the carbohydrates viz. xylose, mannitol, arabinose, galactose, fructose and rarely in dextran. They need thiamine, biotin and calcium pantothenate for their growth. However, nicotinic acid, pyridoxin, folic acid, p-aminobenzoic acid, inositol, vitamin $B_{12}$ and riboflavin are inhibitory for its growth. Low concentrations of vanadium and cobalt are stimulatory. Iron concentration upto 10 ppm is also desirable.

Rhizobia are Gram-negative, rods, motile, when young, have bipolar, subpolar or peritrichous flagella. Cells contain polymerized β - hydroxybutyrate (40-50% of cell dry weight). They are non-spore formers. In addition to *Rhizobium* and *Bradyrhizobium, Azorhizobium, Sinorhizobium* and *Photorhizobium* also fix atmospheric nitrogen in symbiotic association with plant nodules. The use of *Sesbania rostrata* nodulated by *Azorhizobium caulinodans* in rice cultivation is preferred because nodules are present both on root and stem. Nodulation is insensitive to the presence of combined N in soil, and non-stem nodulating *Neptunia* could prove useful as green manure legume plant for harnessing in rice cultivation (Subba Rao *et al.*, 1995).

## *Azotobacter*

*Azotobacter chroococcum*, the most abundant species of genus *Azotobacter*, is commonly found in arable soils. Its number in Indian soil never exceeds $1 \times 10^5$/g soil due to poor organic matter content and presence of antagonistic microorganisms in soil. The organism produces a lot of slime responsible for soil aggregation. In addition, it produces antifungal antibiotic, which inhibits the growth of several plant pathogenic fungi. Inoculation with efficient strain of *Azotobacter* is known to improve the yield of brinjal, tomato, cabbage and onion by 2-40%, and in sorghum, maize and cotton in the range of 9–71% (Sundra Rao *et al.*, 1963; Shende and Apte, 1982). The organism fixes atmospheric $N_2$ in the range of 2-15 mg $N_2$/g of carbon source.

## *Azospirillum*

It is a non-symbiotic, microaerophilic nitrogen fixer, most abundantly found in association with the roots of millets and grasses. They are Gram-negative, vibrio or spiral-shaped, containing polybeta-hydroxybutyrate in their cytoplasm (Okon and Itzigsohn, 1992). *Azospirillum brasilense, A. lipoferum, A. amazonense, A. halopraeferens* and *A. irakense* are most commonly used species of this genus for inoculation. The production of plant growth promoting substances, e.g., gibberllin, cytokinin and auxin by this organism are responsible for growth improvement in plants (Prinsen *et al.*, 1993; Bar and Okon, 1993). Phytohormones produced by *Azospirillum* stimulate the density and length of root hair (Dubrovsky *et al.*, 1994). The concentration of free indole acetic acid and indole butyric acid along with enzymes involved in TCA cycle and glycolysis pathway gets affected by *Azospirillum* inoculation, resulting in improved absorption of mineral nutrients by roots (Andreeva *et al.*, 1993). *A. lipoferum* is known to produce siderophores when grown in iron deficient medium (Wani, 1990; Mori *et al.*, 1992). Siderophores provide protection against minor phytopathogens. Certain *Azospirillum* sp. offer protection to cotton plant against *Fusarium oxysporum* (Redkina, 1990).

## Blue-Green Algae

They belong to a class of prokaryotic photosynthetic microscopic plants having striking morphological and physiological similarities with bacteria. Hence, these are also known as cyanobacteria. These are most abundantly found in tropical zone, never form fibrous growth and are enveloped by a sheath. Blue-green algae need simple inorganic nutrients, moisture and diffused sunlight for their growth. Heterocysts, devoid of oxygen, evolving photosystem II, act as the sites for nitrogen fixation under aerobic environment. The trophic independence of algae for C and N makes them responsible for a natural fertility build-up of rice field soils. Non-heterocysts cyanobacteria are also known to utilize atmospheric $N_2$ under anaerobic or microaerobic conditions. However, *Gloeocapsa* fix $N_2$ under aerobic conditions. It may be due to spatial and temporal separation of processes of carbon and $N_2$ fixation. Most commonly found sp. of blue-green algae are *Anabaena, Nostoc, Cylindospermum, Calothrix, Scytonema, Tolypothrix,* etc. *Nostoc* and *Hapalosiphon* are known to produce auxin, indole-3-acetic acid, indole-3-propionic acid or 3-methyl indole in their extracts. *Tolypothrix tenuis*, *Nostoc muscorum* and *Hapalosiphon fontinalis* are known to synthesize vitamin $B_{12}$ (Mishra and Kaushik, 1989). The growth of BGA in saline-alkaline habitat reduces salinity by 25-30%, pH, EC and exchangeable Na.

### *Azolla*

*Azolla* is a freshwater fern with small leaves, inside which grows the nitrogen fixing blue-green algae *Anabaena azollae*. It grows in ditches and stagnant water and forms a green mat over there. The plant has floating branched stem, deeply bilobed leaves and true root that penetrate through water. Leaves show an alternate arrangement. Each leaf has dorsal and ventral lobe. The dorsal lobe is exposed to air and contains chlorophyll. It has an algal symbiont *Anabaena* within the central cavity. *Anabaena* contains 3-4% nitrogen and produces organic matter in soil. *Azolla* is readily decomposed to $NH_4^+$ that is used by rice plant. The most common species of *Azolla* are *A. caroliniana, A. nilotica, A. mexicana, A. filiculoides, A. microphylla* and *A. pinnata*. *Azolla* application is known to increase the rice yield by 0.5 – 2 t/ha. Nitrogen content of rice receiving *A. mexicana* is 10-35 kg/ha higher than control. The potentialities of *Azolla* have been observed for fish culture and carp culture ponds. It prefers to grow in colder regions.

## Phosphate Solubilizing Microorganisms

Phosphorus, both native in soil and applied as inorganic fertilizers, becomes mostly unavailable to crops because of its low level of solubility, mobility and its tendency to become fixed in soil. The phosphate solubilizing bacteria/fungi life can help in improving phosphate uptake of plants in different ways. The PSB also have the potential to utilize India's abundant deposits of rock phosphate. The most efficient PSB belong to genus *Pseudomonas* and *Bacillus*, though species of genus *Escherichia, Micrococcus* and *Sarcina* have also been reported. These microorganisms are known to dissolve inorganic phosphate by production of organic acid and organic phosphate by phosphatase enzyme (Gaind and Gaur, 1990; Gaur, 1972). Most of these bacteria are aerobic, both Gram positive and Gram negative, some are spore formers, and most of them prefer a temperature of 30-32°C.

## Vesicular /Arbuscular Mycorrhizae

Mycorrhizae literally stand for fungus root association. The vesicles are meant for storage of nutrients and arbuscules for funnelling those nutrients into the root system. VAM is an obligate symbiont and it is not possible to culture it on a nutrient medium. Mycorrhizae are known to improve plant growth due to improved uptake of phosphorus. Hyphae absorb phosphorus from the labile pool rather than dissolving insoluble phosphates. These fungi produce enzyme phosphatase that allows the utilization of organic phosphorus especially under humid and tropical conditions. VAM is also known to improve the uptake of micronutrients as zinc and copper and increases absorption of water. They also produce plant hormones, increase nitrogen-fixing activity of bacteria in root zone and reduce the severity of disease caused by pathogens (Bagyaraj, 1984). They are not host specific (Bagyaraj, 1992). The most common genera are *Glomus, Gigaspora*, *Acaulospora, Sclerocysts* and *Endogone* commonly used for biofertilizer production. However, the selection of strain depends on its effectiveness, extent of colonization, ability to survive in rhizosphere and ability to stimulate plant growth.

## PRODUCTION TECHNOLOGY

It involves three major steps:

(i) Selection of efficient strains

(ii) Inoculant preparations, including selection and processing of carrier material and mass cultivation of specific culture of $N_2$ fixer and P solubilizer.

(iii) Quality control

**(A) Mass Production of *Rhizobium/Azotobacter/Azospirillum*/PSB**

(i) ***Strain efficiency***: Since the rhizobial strains within a species vary in their effectiveness (Boonkerd *et al*., 1978; Kucey *et al.,* 1988), they show a great degree of host specificity. A rhizobial strain effective on one cultivated legume may not be effective on others.

**Most Promising Strains of *Rhizobium* for Biofertilizer Production**

| (a) Host Legume | Strain * | At IARI |
|---|---|---|
| Groundnut | IGR-6, IGR-40 | NC – 92 |
| Pigeon pea | CC-I, F-4, BDN-A2 | AR-4 |
| Chickpea | F-6, F-76, H-45 | SP-4 |
| Soybean | SB-119, 4 ASB-29 | SB – 103 |
| Lentil | L-6, L-I | L-288 |
| Pea | RL (P) – 114 | P-677 |
| Green gram | M-10, GMBS-I, MO-5, M-11 | M-493 |
| Black gram | K4-I, BDN-F | D-10 |
| Cowpea | D6 –28, DC-2, DC-6 | |

*Source:* Motsara *et al*., 1995

**(b) *Azotobacter* species**

*Azotobacter chroococcum*

*A. beijerinckia*

*A. vinelandii*

*A. macrocytogenes*

*A. agilis*

*A. insignis*

**(c) *Azospirillum* species**

*A. lipoferum*

*A. brasilense*

**(d) Phosphate solubilizing bacteria**

*Pseudomonas striata*

*P. putida*

*Bacillus polymyxa*

*B. megaterium*

**(e) Composting**

*Aspergillus awamori*

*Trichoderma viride*

**(f) Mycorrhiza**

*Glomus fasciculatum*

*G. mossae*

*Gigaspora nigra*

(ii) ***Culture production***: There are two methods for culture production of *Rhizobium*:

(a) By mechanical agitation

(b) By fermenter

**(a)** The efficient strain of fast growing *Rhizobium* is grown in a medium supplemented with sucrose as carbon source which is cheap and easily available (Balatti *et al.*, 1987). However, *Bradyrhizobium*, which is a slow grower, prefers mannitol that is quite expensive. Glycerol has been reported as a good substitute for mannitol. Lie *et al.*, (1992) developed procedure for utilization of sucrose as carbon source for *Bradyrhizobium*. Flasks containing sterilized medium and inoculated with specific strain are put on shaker for desired growth.

**(b) Fermenter**

In order to produce high number of viable cells of *Rhizobia* in a short time, *Rhizobia* are cultured asceptically (1% v/v) in fermenters. The fermenter is a stainless steel pressure vessel with a total volume varying from 50 L to 3000 L. The vessel is used for sterilization of culture medium and subsequently for cell multiplication. Initial sterilization of medium is achieved by heating with 98,000 BTU four ring gas burner. In medium size vessel, 100 L of culture medium can be sterilized per run and viable cell counts of $2.5 \times 10^9$/ml (Keyser *et al.*, 1992) can be achieved. An oxygen partial pressure of 0.15 atm is optimum for respiration and *Rhizobia* grows best at 28 to 30°C. Incubation period is 3-5 days for fast growers and 5-7 days for slow growers.

**(iii) Selection of carrier**

Selection of carrier material for transportation of culture from broth to rhizosphere is very important. It should meet the following requirements:

(i) High waterholding capacity
(ii) No heat of wetting
(iii) cheap
(iv) easily available
(v) non-toxic to inoculant strain
(vi) near neutral pH
(vii) nutritive
(viii) should favour the survival, both on seed inoculation and during distribution (Date and Roughley, 1977). Though, peat is a most preferred carrier in USA, Austria, Canada, New Zealand and USSR, but a number of carriers have been explored in India.

Lignite and wood charcoal have been found most satisfactory (Jauhri *et al.*, 1979). Coal, lignite, corn oil and bentonite have also been found as good as peat (PacZ Kowski and Berryhill, 1979; Kremmer and Peterson 1983). Contrarily, a problem of lump formation with coal during storage was reported (Halliday and Grahim, 1978). The quality of oil-based inoculant (freeze dried rhizobia in vegetative oil base) remained high even when stored at high temperature (Kremmer and Peterson, 1983). Gelatin and soymeal improved the moisture content of charcoal-soil (3:1) mixture (Jauhri and Gutpa 1987). Filter mud or press mud, spent compost obtained from mushroom cultivars dehydrated bagasse with traces of mahaua cake and silk cocoon waste have also been tried as carrier (Jauhri and Koshy Philip, 1984; Bahl *et al.*, 1989, Jauhri, 1989). Hegde and Brahmparkash (1992) tried alginate and perlite. Bagasse compost and mineral microgranules were used as carrier by Philpotts (1976) and Foullieux *et al.*, (1994), respectively. Virulence of *Rhizobium* was lost in groundnut husk. Viveganandan and Jauhri (2002) reported the success of rock phosphate-based inoculant. Talcum powder wettable powder, polyethylene glycol paste have shown the shelf life of *Bacillus subtilis* beyond one year (Jayaraj *et al.*, 2005).

## ISI STANDARDS FOR CARRIER BASED BIOFERTILIZERS

| *Biofertilizer* | *Viable cell count* | *pH* | *Moisture %* | *Shelf life (month)* | *Carrier material/N fixation/P solubilization* |
|---|---|---|---|---|---|
| *Rhizobium* | x$10^7$ cells/g of carrier before expiry | 6.0-7.5 | 35-40 | 6 | Peat, lignite, peat soil, humus shall be in powdered form should pass through 150–212 micron IS sieve |
| *Azotobacter* | x$10^8$ cells/g of carrier before expiry | 6.5-7.0 | 35-40 | 6 | Able to fix minimum of 10 mg N/g of sucrose utilized |
| *Azospirillum* | x$10^7$ cells/g | 7.0-8.0 | 35-40 | 6 | Peat lignite, charcoal capable of passing through 100 micron IS sieve |
| PSB | x$10^7$ cells/g | 6.5-7.5 | 35-40 | 6 | 35-50% phosphate solubilizing efficiency, 10 mm solubilization zone |

**Quality Standards for *Rhizobium* Inoculants**

| *Country* | *Rhizobium standards* | *Contaminant standard* |
|---|---|---|
| Australia | $1 \times 10^9$ rhizobia/g *Lotononis*; gram –ve, serologically reactive, nodules on test host at $10^7$ and $10^8$ dilution | At $10^{-6}$ free of contaminants |
| Canada | $10^3$ (alfalfa, clover, birdsfoot trefoil), $10^4$ (Sainfoin) $10^5$ (beans, pea, soybean), viable rhizobia cells/seeds | None |
| Czechoslovakia | $3 \times 10^8$ viable rhizobia/g | None |
| France | Soybean $10^6$, alfalfa $10^3$ rhizobia/seed | None |
| India | $10^8$ cfu at manufacturing; $10^7$ cfu at expiry | None |
| Netherland | 4-25 $\times 10^9$ rhizobia/g | None |
| New Zealand | $1 \times 10^8$ rhizobia/g | None |
| South Africa | $1 \times 10^8$ rhizobia/g | None |
| Thailand | $1 \times 10^7$ rhizobia/g | None |
| U.K. | Manufacturer discretion | None |
| U.S.A. | Manufacturer discretion | None |
| U.S.S.R. | 5-10 $\times 10^7$ rhizobia/g | None |

*Source:* Brahmaprakash and Hegde 2002

New carrier materials with the concept of protection from environmental stress replacing the old concept of carrier as source of nutrient should be tried.

**(iv) *Quality Standards***

The presence of sufficient number of viable rhizobia is a prerequisite for any biofertilizer used for legumes. The standards may vary from $1 \times 10^7$ to $4 \times 10^9$ viable cells/g carrier. In India, standards set by BIS is $1 \times 10^8$ cfu/g at the time of manufacturing, and $1 \times 10^7$ cfu/g carrier before expiry. Use of MPN method for counting rhizobia is most preferable (Rupela and Hegde, 1995). ELISA is used as a strain specific serological technique (Olsen *et al.*, 1983). The immunoblot analysis (Olsen and rice, 1989) allows accurate strain specific identification and viable quantification even when number of contaminants equal the number of *Rhizobia*. There are new biotechnological approaches for strain identity, though not confirming their infectivity and efficiency (Smith, 1992).

**(B) Mass Production of Algal Biofertilizer**

Depending upon the requirement, algal biofertilizer can be produced by individual farmers and also on commercial basis.

***At individual level***:

1. Prepare shallow troughs, at least 2 or multiple of two, approximately 2 m × 1 m × 23 cm of galvanized iron sheet, permanent cement and mortar tanks or polythene lined pit on ground. All these structures should be in an open area so as to receive sunlight for the entire day. A light intensity of 1000 to 4500 lux is good enough for the growth of BGA.
2. Spread about 3 kg soil (1.4 kg $m^2$) per multiplication unit and add 200 g single super phosphate. For best growth of BGA, soil pH should be between 7.0 and 7.5, however, growth does occur at higher or lower soil pH. To bring the pH to 7.5, soils can be amended with gypsum or lime depending upon initial soil pH.

3. Add water up to a height of 7-10 cm to the multiplication units and mix the contents.
4. As soon as the suspended soil settles and water becomes clear, sprinkle/broadcast the starter culture containing (i) *Anabaena* and *Nostoc* (mixture 1), (ii) *Aulosira* and *Tolypothrix* (mixture 2) in two different troughs/multiplication units. This is essential to give equal opportunity to all the forms to grow and multiply equally.
5. To prevent insect breeding, add carbofuran (3% granules, 20 g/tray or unit) or malathion, BHC or any other suitable insecticide.
6. In hot summer months (temperature more than 30°C) BGA grows fast, and in about seven to ten days a thick mat is formed. During this period if rate of evaporation is high, add water intermittently. Once the algal growth is thick, stop watering. Allow the algal mat to dry. Dried algae form flakes along with soil.
7. Collect the dried algal flakes, mix from both the multiplication unit to make composite mixture and store in polythene line bags in cool dry place.
8. Repeat the process till the requirement is met. From a single harvest about 3 kg dried material per unit is obtained.

**Storage of soil-based BGA**

- The sun-dried BGA can be stored for a long time (more than 2 years) in a dry state at normal room temperature in shade.
- Storage bags for BGA biofertilizers should be fresh, and bags earlier containing urea or other chemical fertilizer and other agrochemicals should not be used.

**Field application**:

- Broadcast the dry algal inoculum over the standing water in the rice field at a rate of 10 kg/ha. In clay soils, water remain turbid for 2-3 days, hence BGA inoculum should be applied 2-3 days after transplantation of rice seedlings. Whereas, in loamy soils, application should be done immediately after transplantation is completed.
- If no chemical fertilizers are used, BGA inoculum should be uniformly used to get a benefit of 25-30 kg N/ha.
- When chemical fertilizers are used to the extent of only 50-60 kg N/ha, BGA inoculum should be used to obtain 10-15 per cent increase in crop yield.
- Where chemical fertilizers are used at high doses (100 kg N or more), the use of chemical fertilizers should be reduced by 25-30 kg N/ha and BGA inoculum should be used to get the same yield as at the full dose of chemical fertilizers.
- Use of BGA fertilizers is mainly recommended for rice. However, it has given good results with crop, which require large amount of water like sugar cane.

***Production technology for commercial purposes***

For the large-scale production of BGA biofertilizer at the commercial dimensions, the technology has gone for several changes:

(a) It involves indoor production to maintain round the year production schedule.

(b) Soil as carrier is to be avoided since it is heavy and differ in its physico-chemical properties from area to area as well as liable to carry contamination.

(c) A suitable growth medium is to be developed which supports fast growth rate, when soil is not used as the source of nutrients.

(d) A suitable and cheap carrier is to be selected to carry the BGA inoculum. Taking all these and other related parameters into consideration, a technology was developed. Following given schedule can be employed.

2. **Production unit**: Indoor production can be undertaken in poly house or glasshouse. The dimensions and number of such production units are variable depending upon the turnover. A smallest recommended size is 12 m × 5 m × 3.5 m (l × b × h). Covering material (polythene/glass) quality depends upon the cost input capacity.
3. **BGA multiplication units**: These units can either be of RCC brick and mortar or polythene lined pits on ground as suggested for soil base production units. Possible size of polythene line pits is 4 m × 0.9 m × 0.23 m (l × b × d). The smallest recommended pits of RCC or brick and mortar is 1.7 m × 0.9 m × 0.23 m (l × b × d).
4. A special formulated medium involving 3 major constituents is to be filled in each pit/ trough.
5. Only one form of BGA is to be inoculated per pit to avoid species competition and dominance during multiplication.
6. BGA is allowed to grow for 3-8 days depending upon the ambient temperature, light intensity and humidity till final biomass of 14-15 g is attained.
7. Once fully grown, culture is mixed with rice or wheat straw carrier and sun dried.
8. Once culture is sun dried, packed in suitable size packets (polythene), sealed and stored for future use.
9. Every time fresh medium is added and pits are freshly inoculated.

**Field application of straw based BGA biofertilizers**

- BGA biofertilizer produced in a formulated medium in indoor culture facility contains 100,000 to 1,000,000 units (propagules) per gram carrier, hence 1 kg/ha (400 g/acre) straw based BGA inoculum is enough.
- Broadcast 1 kg/ha in the fields immediately after transplantation of rice seedlings.

**Benefits**

- Benefits of using straw based biofertilizers are alike of soil-based biofertilizer and are even better at many locations.

## (C) Mass Production of Azolla

The flooded rice ecosystem is favourable for BNF and *Azolla* – *Anabaena* are ideal biological system under tropical conditions for better conservation of $N_2$ in paddy soil. As the growth of *Azolla* is too fast so much so that it doubles its weight in 2-3 days. Its production can be carried out in (i) Nursery plots (ii) Ponds/ditches/canal (iii) Concrete tanks/polythene sheets lined pits or simple earthen ditches. However, the basic production method remains the same. The selected field is thoroughly prepared and levelled uniformly and divided into plots of size 20 m × 2 m with suitable bunds and irrigation channel. Water is maintained at a depth of 10 cm. Ten kg of fresh cattle dung mixed in 20 litre of water is added in each plot. 8-10 kg fresh *Azolla* is inoculated in each plot. 100 g super phosphate is applied in 3 split doses at an interval of four days. For insect/ pest control, 100 g furadon/plot is applied a week after inoculation. 100-150 kg *Azolla* is harvested after every 15 days from each plot. Some *Azolla* is left in the plot and serves as inoculum for next crop. *Azolla* can also be produced in bigger plots by adjusting the inputs accordingly. When the production is carried out in ponds/canal, no such addition of fertilizer or insecticide/pesticide is required. For mass production, generally *A. microphylla* is used as inoculant.

**(D) VAM Production**

Being an obligate symbiont, commercial production of VAM is not so simple. Till now there are three major methods for inoculum production of VAM:

(i) Pot culture

(ii) Solution culture

(iii) Root organ culture

Traditionally, AMF are cultivated by sand-based pot cultures. However, it is a labour intensive process and not economical for mass production of inoculum. Solution culture technique such as nutrient film and aeroponics (Mosse and Thompson, 1984; Sylvia and Hubbell, 1986) are mainly adopted for the production of clean soil less inoculum of AMF. Nevertheless, these methods have been found successful in cultivating only *Glomus* species. (Hung and Sylvia, 1988; Sylvia and Jarstfer, 1994).

**(a) Isolation of VAM**: The most common technique to recover VAM propagules from soil is wet sieving and decanting to remove clay and sand fractions of the soil while retaining spores and other similar sized soil and organic matter particles on sieves of various sizes (Table 4).

(i) Wet sieving and decanting: 250 g of soil is suspended in 1 litre of water. Heavier particles are allowed to settle for few minutes and liquid is passed through the sieve of size that can retain the large particles of organic matter, but coarse enough to allow the desired spores to pass through. The suspension that passes through this sieve is saved and stored to resuspend all particles. The heavier particles are allowed to settle for few minutes and liquid decanted through sieve. The sievings retained on different sieves are washed into separate Petri dishes for further observations or purification by sucrose centrifugation.

(ii) Sucrose centrifugation: Spores and minimal amount of organic particles are further purified by suspending the sieving in 40% sucrose solution and centrifuging at 2000 rpm for 1 min. Supernatant is poured through the sieve of 400 mesh and rinsed with distilled water to remove sucrose.

(iii) Soil inoculum is produced by pot culture technique containing all VAM fungal structures and is highly infective. But the success of this method depends on selection of host plant and ambient conditions for the mass multiplication of VAM fungus. Srinivasa and Bagyaraj (1990) observed Rhodes grass as the best host, perilite-soilrite mixture as the best substrate, and calcium ammonium nitrate and rock phosphate as the best N and P source for mass production of *Glomus fasciculatum*.

(iv) Storage: The plants over pruned to soil level and compacted soil adhering to roots could be shaken and fragmented. It is mixed with soil from pots and air dried. The cultures are packed in polythene bags, sealed and stored at low temperature upto 4 years.

## CROP RESPONSE TO BIOFERTILIZERS

The response of *Azospirillum* and *Azotobacter* has been studied in both cereals and non-cereals crops to know their effectiveness in yield improvement, nitrogen economics, soil quality, etc. The response may vary with crop cultivar, location, bacterial strain inoculated, its compatibility with the native soil flora, agronomic practices, nitrogen fertilizer dose applied, season, soil pH, etc. A summary of *Azospirillum* and *Azotobacter* response to different crops is given in Table 2 and 3.

**Table 2** Average increase in yield of crops in response to Azospirillum inoculation

| *Crop* | *Inoculant* | *% increase in yield* | *Reference* |
|---|---|---|---|
| Wheat, rice, barley, oat, sorghum | *A. brasilense* | 0 – 81% grain yield | Subba Rao, 1982 |
| Pearlmillet, finger millet | *A. brasilense* | 10-17, 13-31% in grain yield respectively | Subba Rao *et al.*, 1982, 1983, 1985. |
| Sorghum | *A. brasilense* | 7-13% grain yield | Subba Rao *et al.* 1986 |
| Rice | *A. lipoferum* | 11% Grain yield | Rao *et al.* (1986) |
| Pearlmillet, Corn, Wheat, Sorghum, *Setaria italica, Pannicum millaceum* | *A. lipoferum* | 11,16,9,17, 43 and 13% grain yield respectively | Okon *et al.* (1988) |
| Barley | *A. brasilense* | 10-20 % grain yield | Tilak and Dwivedi (1989) |
| Pearlmillet, Sorghum; maize, wheat and rice | *A. brasilense* | 11-15, 15-20 % respectively | Wani (1990) |
| Pearlmillet | | 11% | Wani and Lee; 1992 |
| Paddy | *Azospirillum sp.* | 4.16 – 16.6% | Mowade *et al.* 2005 |

**Table 3** Crop yield as affected by *Azotobacter* biofertilizer inoculation

| *Crop* | *% increase in yield* | *Reference* |
|---|---|---|
| Finger millet | 37.0 | Sundara Rao *et al.*, 1963 |
| Wheat | 7.18 – 7.54 | Shende, 1972 |
| | 26.4 – 34.5 | Meshram and Shende 1982 |
| Sorghum | 4.5 – 24.17 | Shende *et al.*, 1984 |
| Potato | 3.4 – 11.5 | Shende *et al.*, 1986 |
| Cotton | 7.1 – 31.6 | Pandey *et al.*, 1989 |
| | 29.1 – 38.5 | Malik *et al.*, 1994 |
| Onion | 4.33 – 16.97 | Paul *et al.*, 2002 |
| Okra | 8.3 | Mowade *et al.*, 2005 |
| Brinjal | 5.78 | Mowade *et al.*, 2005 |
| Chilli | 10.30 | Mowade *et al.*, 2005 |
| Cauliflower | 9.52 | Mowade *et al.*, 2005 |
| Jowar | 10.29 | Mowade *et al.*, 2005 |

**Table 4** Crop response to VAM inoculation

| *Crop* | *Endosymbiont* | *% in grain yield or other parameter* | *Reference* |
|---|---|---|---|
| Soybean | *Glomus sp.* | 30-40 % increase in yield & higher content of P, N, Ca, Cu, Mn | Ross and Harper, 1970 |
| Lucerne | *Glomus mossae* | Satisfactory nodulation dependent on VAM fungi | Azcon-G de Anguilor and Barea, 1978 |
| Cowpea | *G. fasciculatum* | Crop benefitted by VAM | Islam *et al.*, 1980 |
| Fingermillet | *G. fasciculatum* | Response to VAM inoculation reduced with increasing rate of P application | Govindrao *et al.*, 1983. |
| Pigeon pea | *G. fasciculatum* | Significant increase in yield in 1st year | Subba Rao, 1988 |
| Chickpea | *G. versiforme* | Under P deficient soil 10% increase in yield | Singh and Tilak, 1989 |
| Pearlmillet | *G. fasciculatum* | Improved performance with 50 kg $P_2O_5$/ha as superphos than rockphos. | Singh and Rana, 2005 |
| Wheat | *G. fasciculatum* | Improved grain yield | Singh and Rana, 2005 |

## Crop Response to Application of BGA and *Azolla*

Nitrogen fixing cyanobacteria and *Azolla* have proved their importance in increasing productivity of rice fields under flooded conditions compared to dry land conditions (Watanabe and Roger, 1984). Both BGA and *Azolla* neither contaminate the environment nor consume photosynthates of rice plants. The release of extracellular organic compound and $O_2$ due to photosynthesis as well as biomass after death are responsible for positive changes in physical, chemical and biological properties of soil (Kaushik, 2003) as given in Table 5.

**Table 5** Changes in available N ($\mu g\ N\ g^{-1}$ soil) and available P ($\mu g\ g^{-1}$ soil) due to incorporation of blue green algae, *Azolla* and urea.

| *Days of flooding* | *Treatment ($\mu g\ N\ g^{-1}$ soil)* | | | | *Treatment ($\mu g\ P\ g^{-1}$ soil)* | | | |
|---|---|---|---|---|---|---|---|---|
| | *Control* | *BGA* | *Azolla* | *Urea* | *Control* | *BGA* | *Azolla* | *Urea* |
| 7 | 5.06 | 13.28 | 17.06 | 23.86 | 7.50 | 7.92 | 9.37 | 9.08 |
| 14 | 8.46 | 23.59 | 14.85 | 16.33 | 3.58 | 3.94 | 4.89 | 4.48 |
| 21 | 7.99 | 23.58 | 21.78 | 15.38 | 5.32 | 13.55 | 13.97 | 6.29 |
| 28 | 14.10 | 29.79 | 29.70 | 20.12 | 6.97 | 14.06 | 18.45 | 7.90 |
| 35 | 10.10 | 33.02 | 29.70 | 13.49 | 7.45 | 10.89 | 11.62 | 7.85 |

*Source:* Saha *et al.*, 1982

| | *Treatment Mean* | *Days Mean* | *Treatment Mean* | *Days Mean* |
|---|---|---|---|---|
| LSD at 5% | 2.52 | 2.82 | 1.19 | 1.33 |

*Source:* Singh and Singh 1987

## CURRENT STATUS OF BIOFERTILIZERS

Presently, there is a wide gap of about 10 MT of plant nutrients between removal by crops and replenishment by fertilizers. Application of organic manure could not fill this gap. Moreover, the nutrient supply is unlikely to improve due to competitive demand for the use of agricultural waste as fuel and feed. So, biofertilizers are the only alternative left for supplementing the need of plant nutrients.

India is one of the important producers of biofertilizers. Its capacity of biofertilizer production during 2003-2004 was 10892.8 MT spread in 23 states and Union territories (Yadav and Ray Chaudhari, 2004). The maximum production capacity is in Agro Industries Corporation, State Agriculture Department, National Centre on Biofertilizer, Organic farming Centre, State Agriculture Universities and Private Sector. Among different states, maximum production capacity is in Tamil Nadu followed by Maharashtra, Andhra Pradesh and closely followed by M.P. (Table 6).

The approximate share of different biofertilizers produced is as under:

Phosphate solubilizing microorganisms - 48 %

*Rhizobium* – 16 %

*Azotobacter* – 18 %

*Azospirillum* – 16 %

Others - 2 %

It is observed that most of the production units use unsterilized carrier for mixing inoculum. Technically untrained labourers do the packing, which results in multiplication of contaminants.

- Though recommended shelf life for inoculum packet is six months, but most of the products do not fulfil the requirements.
- Only 6-8 firms have acquired the BIS registration, as the specifications set up by BIS for all the four categories of biofertilizers are purely voluntary in nature.
- Of the 1865 samples tested by NBDC and its 6 regional centres during 1990-91 to 1997-98 for quality standards, only 864 were found to be fully conforming to BIS standards (Singh and Dixit, 1999). Annual percentage has improved from 72.11 in 1999-2000 to 86.04 % in 2002-2003 (Annonymous, 2003). However, Dwivedi *et al.*, (2004) reported pass % of 70.9 % in 86 samples collected from M.P., Chhattisgarh, Rajasthan and Gujarat.

**Table 6** Pass % of different biofertilizer samples during 2002-2003 and 2003-04

| *Biofertilizers* | *RBDC Jabalpur 2002-2003* | | *RBDC Nagpur 2003-2004* | |
|---|---|---|---|---|
| | *Total sample* | *Pass %* | *Total Sample* | *Pass %* |
| *Rhizobium* | 18 | 44.4 | 14 | 14.28 |
| *Azotobacter* | 17 | 76.4 | 29 | 44.8 |
| *Azospirillum* | 07 | 85.7 | 04 | 100 |
| PSM | 39 | 74.3 | 87 | 77 |
| BGA | - | - | 08 | 25.0 |
| Others | 05 | 100 | 16 | 56.25 |
| Total | 86 | 70.93 | 158 | 61.39 |

RBDC – Regional Biofertilizer Development Centre

Estimates for total requirement of biofertilizers (*Rhizobium, Azotobacter*, *Azospirillum* and blue-green algae) based on area under different crops and biofertilizers demand as given by NBDC and BCIL to be about 5.07 and 3.44 lakh tonnes, respectively (Table 7). However, according to Tamil Nadu Govt. only demand for *Rhizobium* is 35000 tonnes, *Azospirillum* 482,000 tonnes, *Azotobacter* 162,610 tonnes, Blue green algae 267.72 thousand tonnes, *Azolla* 20,380 tonnes and PSB 275,510 tonnes. Total of this comes out to be 12.44 lakh tonnes as given by Govt. of Tamil Nadu. It is significantly higher than NBDC and BCIL, as they did not include PSM and their estimates for *Azospirillum* are also low.

**Table 7** Biofertilizer production during 2003-2004 based on information by production units to NBDC.

| *States* | *Number of units* | *Installed production capacity* | *Actual production* | *Capacity utilization %* |
|---|---|---|---|---|
| Andhra Pradesh | 10 | 1500 | 1120.58 | 74.7 |
| Arunachal Pradesh | 1 | 75 | - | 0 |
| Assam | 5 | 375 | 248.52 | 66.27 |
| Bihar | 3 | 300 | NA | NA |
| Delhi | 3 | 151 | - | 0 |
| Gujarat | 5 | 900 | 473.38 | 52.65 |
| Haryana | 2 | 99 | 7.10 | 7.17 |
| Himachal Pradesh | 1 | 75 | 5.40 | 7.02 |
| Jharkhand | 2 | 75 | 5.40 | 7.2 |
| Karnataka | 19 | 2210 | 848.58 | 38.39 |
| Kerala | 6 | 425 | 370.68 | 87.21 |
| Madhya Pradesh | 12 | 1475 | 632.86 | 42.9 |
| Maharashtra | 37 | 3218 | 2189.73 | 68.04 |
| Meghalaya | 1 | 75 | - | - |
| Mizoram | 1 | 75 | - | - |
| Nagaland | 1 | 150 | 8.12 | 5.41 |
| Orissa | 10 | 600 | 61.36 | 10.22 |
| Pondicherry | 3 | 150 | 80.32 | 53.54 |
| Punjab | 1 | 2 | - | - |
| Rajasthan | 3 | 750 | 381.79 | 50.90 |
| Tamil Nadu | 30 | 4575 | 4074.02 | 89.04 |
| U.P. | 8 | 375 | 149.48 | 39.86 |
| West Bengal | 14 | 715 | 235.61 | 32.95 |
| Total | 178 | 18,345 | 10,892.8 | 59.38 |

Response of biofertilizers in cereals, millets, pulses, legumes, oilseeds, sugar cane and cotton grown under different agroclimatic conditions showed the increase in yield of crops and this increase is of the order of 15-20 %. Biofertilizer *Rhizobium* can supply 20-25 kg N/ha. However, complex biofertilizers have been found to be more effective in increasing crop yield. Technology for bulk production of BGA inoculum has been developed at IARI using cheap and environment-friendly biodegradable carrier material. Montmorillonite clay and wheat/rice straw has showed good results for storing BGA for longer duration. Technology to biodegrade coir pith and its use as carrier for BGA has been developed at Bharathidasan University, Trichi.

## MAJOR AREAS OF THRUST

In order to have sustainable agriculture system, dependence on use of chemical fertilizers needs to be reduced. One possible way to achieve this is the integrated use of biofertilizers in combination with chemical fertilizers and other organic manures in cropping system prevalent in various agroclimatic zones of country. Though a lot of research is going on biofertilizer development and their awareness, but still these areas need to be emphasized.

1. Development of transgenic biofertilizers
   i. Competitive ability of strain to ward off inefficient strains and to form efficient nodules in large number is the key factor in *Rhizobium* biofertilizer technology.
   ii. A possible approach is the construction of strains capable of oxidizing $H_2$ evolved during nitrogen fixation by an uptake of hydrogenase system, so that electron produced may be used for $N_2$ reduction.
2. Shift from carrier-based inoculant to liquid inoculants, as they have high cell count, longer shelf life, and minimum contamination.
3. Use of mixed inoculants, which can provide both N and P.
4. Development of formulations, which can be stored at room temperature for 1-2 years.
5. Viability of propagules in these carriers should not be less than 80 %.
6. Quality of the product should be given due attention by opting for BIS standards and employing technically skilled personnel.
7. Exploration of some cheap and easily available nutritive carrier materials.
8. Proper marketing and awareness among farmers regarding positive effects of biofertilizers on soil and plant health.
9. Developing technology packages for integrated plant health and production management for crop plants through the use of diverse bioinoculants in combination.
10. Studying molecular biology of biofertilizer microorganisms and soil-plant biofertilizer microorganisms relation.

## References

Andreeva, I.N., Red'kina, T. V. & Izamailov, S.F. (1993) *Russian J. Pl. Physiol.* **40(6)**: 780-784.

Annonymons (2003) Annual Report 2002-2003, National Biofertilizer Development Centre, Deptt. of Agriculture and Co-operation, Govt. of India, Ghaziabad, pp. 1-58.

Azcon-Gdeangnlar C. & Barea, J.M. (1978) *Can. J. Microbiol.* **24**: 520-524.

Baldani, J.J., Pot, B., Kitchhof, G., Falson, E., Baldani, V.L.D., Olivares, F.L., Hoste, B., Kersters, K., Hartman, A., Gillis, M. & Dobereiner, J. (1996) *Int. J. Syst. Bacteriol.* **46:** 802-810.

Ballati, A.P., Mazza, L.A. & Moretti, E. (1987) *MIRCEN J.* **3**: 227-234.

Bagyaraj, D.J. (1984) In: *VA mycorrhiza* (Eds. Powell, C.L. & Bagyrarj, D.J.) CRC Press Boca Raton, pp. 131-153.

Bagyaraj, D.J. (1992) In: *Methods in Microbiology* (Eds. Norris, J.R., Read, J.J. & Varma, A.K.) Academic Press, London, pp. 359-374.

Bar, T. & Okon, Y. (1993) *Can. J. Microbiol.* **29**: 924-929.

Briones, A.M., Okabe, S., Umemiya, Y., Ram Singh, N.B., Reichardt, W. & Okuyama, H. (2003) *Plant Soil* **250**: 335-348.

Date, R.A. & Roughley, R.J. (1977) In: *A Treatise on Dinitrogen Fixation.* Section-IV (Eds. Hardy, R.F & Gibson, A.H) John Wiley. 253-275.

Dubrovsky, J.G., Puente, M.E. & Bashan, Y. (1994) *Soil Biol. Biochem.* **26(12)**: 1657-1664.

Dwivedi, A., Ghosh, Vandana, Singh, T.K. & Rajput, A.S. (2004) In: *SOUVENIER, National Conference on Quality Control of Biofertilizers.* 14-15 Jan. 2004. National Biofertilizer Development Centre, Ghaziabad, India, pp.33-41.

Fouilleux, G., Revellin, C. & Catroux, G. (1994) *Can. J. Microbiol.* **40**: 322-325.

Fred, E.D., Baldwin, I.L. & McCoy, E. (1932) Leguminous Plants Univ. Root Nodule Bacteria and Wisconsin, Madison, Wis, USA. pp. 343

Fuentes-Ramirez, L.E., Bustillos-Cristales, R., Tapia – Hernandez, A., Jimenez Salgado, T., Wang, E.T., Martinez-Romero, E. & Caballero-Mellado, J. (2001) *Int. J. Syst. Evo. Microbiol.* **51**: 1305-1314.

Gaind Sunita & Gaur, A.C. (1990) *Indian J. Microbiol.* **30**: 305-310.

Gaur, A.C. (1972) In: *Symposium on Soil Productivity*, National Academy of Sciences. 259-268.

Gauthier, D., Diem, H.G., Dommergues, Y.R. & Ganry, D. (1995) *Soil Biol. Biochem.* **17**: 375-379.

Govinda Rao, Y.S., Bagyaraj, D.J. & Rai, P.V. (1983) *Zbl. Microbiol.* **138**: 413-419.

Hung, L.L. & Sylvia, D.M. (1988) *Appl. Environ. Microbiol.* **54**: 353-357.

Islam, R., Ayanaba, A. & Sanders, F.E. (1990) *Plant Soil* **54**: 107-117.

Jauhri, K.S. & Gupta, M. (1987) *Curr. Sci.* **56**: 1061-1062.

Jauhri, K.S., Bhatnagar, R.S. & Iswaran, V. (1979) *Curr. Sci.* **48**: 170-171.

Jayaraj, J., Radhakrishnan, N.V. Kanan, R., Sakthivel, K., Suganya, D., Venkatesan, S. & Velazhahan, R. (2005) *Biocontrol Sci. Technol.* **15(1)**: 55-66.

Kanwar, J.S. (1997) In: *Plant Nutrient Needs, Supply, Efficiency and Policy Issues 2000-2025.* (Eds. Kanwars J.S. and Katyal, J.C.) National Academy of Agricultural Sciences, New Delhi, pp. III-VIII.

Kaushik, B.D. (2003) In: *Biotechnology in Sustainable biodiversity and food security* (Ed. Prasad B.N.) Sci. Publishers Inc. Enfield (NH) USA.

Keyser, H.H. & Li, F. (1992) *Plant and Soil.* **141**: 119-135.

Kremner, R.J. & Peterson, H.L. (1983) *Appl. Environ. Microbiol.* **45**: 1790-1794.

Ladha, J.K., Pareek, R.P., So, R. & Becker, M. (1990) In: *Nitrogen fixation. Achievements and objectives* (Eds. Gresshoff, P.M., Roth, L.E., Staley, G. & Newton, W.L.), Chapman and Hall, New York, London, pp. 633-640.

Ledgard, S.F. & Steel, K.V. (1992) *Plant and Soil.* **141**: 137-153.

Lie, T.A., Muilanberg, M.I., Rep. N.H. & Ayhan, K. (1992) *Can J. Microbiol.* **38**: 569-572.

Malik, R.K. Shende, S.T. & Singh, M. (1994) *Ind. J. Agric. Sci.* **64**: 552-554.

Meshram, S.U. & Shende, S.T. (1982) *Plant Soil.* **69**: 265-273.

Mishra, S. & Kaushik, B.D. (1989) *Proc. Indian National Sci. Acad.* **B55**: 295-300.

Mori, E., Fani, R. Gallori, E., Fantappie, O. & Bazzicalupo (1992) *Symbiosis (Rehrot)* **13(1-3):**115-122.

Motsara, M.R., Bhattacharya, P. & Srivastava, B. (1995) In: *Biofertilizer Technology, Marketing and Usage – A source book-cum glossary.* Fertilizer Development and Consultation Organisation, New Delhi, India. pp.184 + viii.

Motsara, M.R., Bhattacharyya, P. & Srivastava, B. (1995) In: *Biofertilizer Technology, Marketing and usage* (Ed. Tandon, H.LS.) Fertilizer Development and consultation organization 204-204A. Bhanot Corner, 1-2, Pamposh Enclave, New Delhi. 48.

Mowade, Sarita, Shukla, A.L., Deoghare, V.Y. & Yadav, A.K. (2005) In: *National Conference-cum-workshop on upgradation of production technology for quality production of "Biological inputs in agriculture"* March 3-4,2005 at Nagpur, India.

Narula, Neeru (2000) In: *Azotobacter in sustainable agriculture.* (Eds. Narula, Neeru) CBS Publishers and Distributors, New Delhi, pp. 1-12.

Okon,Y., Kapulnik, Y. & Sarig, S. (1988) In: *Biological Nitrogen Fixation : Recent Developments* (Ed. N.S. Subba Rao), Oxford and IBH Publishing Company Pvt. Ltd. New Delhi. India. pp. 175-196.

Okon, Y. and Itzigsohn, R. (1992) *FEMS Microbiol. Rev.,* **103(2-4)**: 131-140.

Olivares, F.L., Baldani, V.L.D., Reis, V.M. Baldani, J.J. & Dobereiner, J. (1996) *Biol. Fertil. Soils.* **2**: 197-200.

Olsen, P.E. and Rice, W.A. (1989) *Appl. Environ. Microbiol.* **55**: 520-522.

Olsen, P.E., Rice, W.A. Stemke, G.W. & Page, W.J. (1983) *Can J. Microbiol.* **29**: 225-230.

Paczkowski, M..W. & Brerryhill, B.L. (1979) *Appl. Environ. Microbiol.* **38**: 612-615.

Pandey, A., Shende, S.T. & Apte, R.G. (1989) *Zentralbl. Microbiol.* **144**: 596-604.

Paul, S., Verma, O.P., Rathi, M.S. & Tyagi, S.P. (2002) *Ann. Agric. Res.* **28**: 297-299.

Philpotts, H. (1976) *J. Appl. Bacteriol.* **41**:277-281.

Prinsen, E., Costacurta, A., Michiels, K., Vanderleyden, J. & Van H. Onckelen (1993) *Mol. Plant Microbe Interact.* **6(5)**: 609-615.

Redkina, T.V. (1990) *Agrokemiae's Tolajtan*, **39(3-4)**: 465-468.

Roger, P.A. & Ladha, J.K. (1992) *Plant and Soil.* **141**:41-55.

Ross, J.P. & Harper, J.A. (1970) *Phytopathol.* **60**: 1552-1556.

Saha, K.C., Panigrahi, B.C. & Singh, P.K. (1982) *Soil Biol. Biochem.* **14**: 23-26.

Shende, S.T. Rudraksha, G.B. Apte, R. & Raut, R.S. (1984) In: *Proceeding of working group meeting* 9-12 Oct. 1984. ICRISAT, pp. 75-76.

Shende, S.T., Verma, O.P. & Bisht, N.S. (1986) *Seed and Farm.* **12**: 41-43.

Shende, S.T. (1972) In: *Proceedings of symposium on soil productivity.* Nat. Acad. Sci. India 191-192.

Shende, S.T. & Apte, R. (1982) In: *Biological Nitrogen Fixation.* Proc. Natl. Symp. IARI, New Delhi, pp. 532-543.

Singh, A.L. & Singh, P.K. (1987) *Plant Soil.* **102**: 49-54.

Singh, C.S. & Rana, J.P.S. (2005) In: *Advances in Microbiology at IARI. 1961-2004* (Ed. Kaushik, B.D.), Publisher, Director, IARI, New Delhi-12.

Singh, T. & Dixit, S.K. (1999) In: *Use and Development of Microbial Inoculants*, (Eds. Singh, T., Yadav, A.K. & Raychaudhari, S), Publ. North Eastern Regional Conference on Biofertilizers 22-23 Jan. 1999 (Second Print), pp. 19-52.

Singh, M. & Tilak, K.V.B.R. (1989) *Plant & Soil.* **119**: 281-284.

Smith, R.S. (1992) *Can. J. Microbiol.* **38:** 485-492.

Sreenivasa, M.N. & Bagyaraj, D.J. (1990) *J. Soil Biol Ecol.* **10**: 1-4.

Subba Rao, N.S. (1988) In: *Proc. Mycorrhiza. Round Table IDRC. CRDI, CIID*, New Delhi, 211-225.

Subba Rao, N.S., Mateos, P.F., Baker, D., Pankratz, H.S. Palma, J., Dazzo, F.D. & Sprent, J.I. (1995) *Planta.* **196**: 311-320.

Subba Rao, N.S. Tilak, K.V.B.R. & Singh, C.S. (1985) *Soil Biol. Biochem.* **17**: 119-121.

Subba Rao, N.S., Tilak, K.V.B.R., Singh, C.S. & Gautam, R.C. (1982) In: *Proceeding of National Symposium on Biological Nitrogen Fixation.* IARI. New Delhi, Feb. 25-27.

Subba Rao, N.S., Tilak, K.V.B.R., Singh, C.S. & Gowda, B.K.L. (1983) *Curr. Sci.* **52:** 439-440.

Subba Rao, N.S., Tilak, K.V.B.R., Singh, C.S. & Nair, P.V. (1986) In: *Cereal nitrogen fixation* (Eds. Wani, S.P), International Crop Research Institute for Semi Arid Tropics, Patancheru, India, pp. 69-74.

Sundra Rao, W.V.B., Mann, H.S., Pal, N.B. & Mathur, R.S. (1963) *Indian J. Agric. Sci.* **33**: 279-290.

Sylvia, D.M. & Hubbel, D.H. (1986) *Symbiosis* **1**: 259-267.

Sylvia, D.M. & Jarstfer, A.G. (1994) In: *Management of mycorrhiza in agriculture, horticulture and forestry* (Eds. Robson, A.D., Abbott, L.L. and Malajczuk, N), Kluwer Academic Publishers, Dordrecht Netherland, pp. 231-238.

Tilak, K.V.B.R. and Dwivedi, A. (1989) *Indian J. Microbiol.* **29:** 61-64.

Urquiaga, S., Cruz, K.H.S. & Boddy, R.M. (1992) *Soil Sci. Soc. of Am.* **156**:105-114.

Venkataraman, K. (1984) Biotechnology: *The role of UNIDO. Development.* **4**: 71-76.

Verma. A., Singh, A., Sudha Sahay, N., Sharma, J., Roy, A., Kumari, M., Rana, D., Thakaran, S., Deka, D., Bharti, K. Franken, P., Hurek, T., Blechert, O., Rexer, K.H., Khost, G., Hahn, A., Hock, B., Maier, W. Walter, M., Satrk, D & Kranner, I. (2001) *Pirimoformaspora indica* : A cultivable mycorrhiza like endosymbiotic fungus. In: Mycota, IX chapter 8. Springer Series, Germany, pp. 123-150.

Viveganandan, G. & Jauhri, K.S. (2002) *Indian J. Biotechnol.* **1**: 180-187.

Wani, S.P. (1990) *Indian J. Microbiol.* **30**: 363-393.

Wani, S.P. & Lee, K.K. (1992) In: *Fertilizers of Organic Manures Recycle Wastes and Biofertilizers* (Ed. Tandon, H.L.S), Fertilizer Development and Consultation Organization. 204-204A. Bhanot Corner, 1-2, Pamposh Enclave, New Delhi 110 048, India, pp. 91-112.

Watanabe, I. & Roger, P.A. (1984) In: *Current development in biological nitrogen fixation* (Ed. Subba Rao, N.S.), Oxford IBH New Delhi, pp. 237-276.

Yadav, A.K. & Raychudhari, S. (2004) In: *Biotechnology in Sustainable and Organic Farming.* Shree Publishers and Distributors. New Delhi, pp. 1-22.

5

# Biodiversity of Plant Pathogenic *Fusarium* Species

**Vishwa Dhar, Subhojit Datta and Rita Rai**

*Indian Institute of Pulses Research, Kanpur-208 024, Uttar Pradesh*

## ABSTRACT

***Fusarium*** species are common soil-borne plant pathogens with worldwide distribution. Some of the common species like *F. oxysporum, F.udum and F.solani* incite wilt disease of several important agricultural, horticultural and plantation crops causing huge production losses. *Fusarium* species have been identified using combinations of morphological and colony characters. Further characterization usually requires time-consuming procedures such as pathogenicity tests on a range of hosts, the determination of cross fertility groups, or examination of the mycotoxin profile. Although these attributes are important for characterization and diversity studies, very often these alone are not enough. The need of the hour is to substantiate morphological, cultural and pathaogenic variability with the DNA polymorphism based studies for a rational understanding of the diversity prevailing in this group of pathogens. The current international scenario is ripe with work progressing on molecular characterization of a range of fungal species exploiting PCR based techniques and utilizing different markers.

Some of the important techniques suitable for diversity studies in *Fusarium* include use of RAPD, RFLP, AFLP, SSR, ISSR and rDNA markers. The understanding of the pathogen diversity can effectively contribute in disease management strategies by allowing resistant genes and genotypes to be identified and characterized relative to the spectrum of pathogen diversity and in streamlining the resistant breeding programme.

## INTRODUCTION

For understanding the evolutionary history and to design sustainable crop management strategies, characterization of pathogen biodiversity is very important. The conventional methods for identifying the variability in the pathogens have been the study of virulence reactions using disease rating scales on sets of host differentials. Characterization of the pathogen populations based on pathogenicity data has often been difficult and time-consuming. Biotechnological methods are being increasingly used to characterize pathogen populations and assess the genetic variability much more precisely. These methods do not necessarily replace the virulence tests that are carried out in the field and it is even more important to validate these molecular markers by correlating with the pathogenicity data.

*Fusarium* is a group of filamentous fungi widely distributed on plants and in the soil. This is a large cosmopoliton genus of imperfect fungi and is of primary interest because numerous species are important plant pathogens (Nelson et al., 1981). *Fusarium* spp. occur frequently among the fungal microflora associated with seedling disease, and *Fusarium* wilt is a major cause of plant mortality in many important crops (Minton and Garber, 1983). While most *Fusarium* species are prevalent in tropical and subtropical areas, some inhabit soil in cold climates. Traditional classification of *Fusarium* has been based exclusively on morphology. However, this system of classification has been controversial for many years because it has resulted in the description of markedly different numbers of sections and species by taxonomists (Matuo 1980; Aoki 1998). The genus *Fusarium* currently contains over 20 species. The most common of these are *Fusarium oxysporum*, *Fusarium graminearum*, *Fusarium chlamydosporum* and *Fusarium solani*. The high host specificity found among plant pathogenic strains of *Fusarium* led Snyder and Hansen (1940) to create new taxa of species based on this specificity. Isolates that cause disease in members of a particular family, genus or species are designated as *forma specialis*. Race characterizes isolates by resistance genes found in particular cultivars of specific host species. Identifying *Fusarium* at the species, *forma specials*, and race levels is difficult because many variables must be considered. Different methods have been used to assess genetic variability in *Fusarium* spp. and to identify markers at the species, *forma specialis*, and race levels. Molecular phylogenetic analysis have helped to clarify ambiguities in traditional classification system of *Fusarium*. Genome organization and molecular mechanisms of pathogenicity are still not well understood in many *Fusarium* species.

The criteria used for diversity study and comparative analysis based on morphological, pathological and molecular characters are presented here.

## MORPHOLOGICAL CHARACTERIZATION

Identification at species level could be possible in several cases based on important cultural and morphological characters (Booth, 1971; Gerlach and Nirenberg, 1982, Nelson *et al.*,1983). Pigmentation produced by the fungus in culture media, kind of mycelium and their growth, characters of the sporodochia, nature of conidiogenous cells, micro and macroconidia and presence, absence and type of chlamydospores are among the major cultural and morphological characters used to differentiate *Fusarium* at species level. Isolates of *Fusarium* produce colony growth with fluffy, appressed or intermediate types of mycelium. Key macroconidial characters are shape, size and septation that distinguish the isolates from each other. Shape of conidiogenous cells and kind of phialides on which microconidia are produced are also among the primary characters for distinction of the isolates.

Although these cultural and morphological attributes are important for diversity studies leading to characterization at species level often, these alone are not sufficient in several instances. Additional information on cross fertility has been effectively used in some cases, for example in *Gibberella fujikuroi* species complex (Kerry O' Donnell *et al.* 1998; Leslie, 1999). The unknown strain used as a male parent is crossed with a tester strain of known species and mating types as female parent. Standard methods are now available for cross fertility studies in *Fusarium* spp. which have been widely used (Klittich and Leslie, 1988; Summerell *et al.,* 2003).

## PATHOGENIC CHARACTERIZATION

In addition to a wide range of morphological and cultural diversity, pathogenic variability has been used to characterize the fungus at species, subspecies and intra-subspecies level. The concept of *forma specialis* was given to accommodate host specific pathogenic forms of *Fusarium oxysporum* group (Snyder & Hansen, 1940). Till date, over 100 *forma specialis* in *F. oxysporum* have been described from identical number of hosts. Isolates within one *forma specialis* have also been reported to differ in their virulence and characterized by assigning pathogenic races. Races are defined by their differential reaction on a set of host differential genotypes, which may include cultivars known to carry one or more genes for resistance.

Quick induction of the disease symptoms and a standard set of host differential genotypes with known genes of resistance are the two major requirements for characterization of pathogenic diversity and identification of races. Therefore, the time-consuming procedure of testing pathogenicity must be reduced by developing reliable and quick artifical inoculation and screening techniques. Also developing a uniform and standard differential set of host genotypes based on genetic information is essential. In several cases, these are either inadequate or completely lacking, and therefore precise information on these aspects need to be generated for elucidation of the extent of pathogenic diversity present in the pathogens.

## MOLECULAR DIVERSITY

Molecular techniques are most precise tools for differentiation between species, and identification of new strain/isolates collected from infected samples. The molecular methods vary with respect to discriminatory power, reproducibility, ease of use and interpretation (Lasker, 2002). DNA fingerprinting has been successfully used for *Fusarium* in characterization of individual isolates and grouping them into standard racial classes. This is also particularly useful when any unknown fungal sample is to be identified. For breeding of resistant crop varieties, knowledge about the pathogen races in that particular crop area is very important. This is more so when the breeding objective is to provide resistance against multiple races or to pyramid several resistance genes in an elite genotype. Comparison at the DNA sequences level provides accurate classification of fungal species; they are beginning to elucidate the evolutionary and ecological relationships among diverse species.

## THE APPLICATIONS OF MARKER TECHNOLOGY

The development of molecular techniques for genetic analysis has led to a great increase in our knowledge of fungus genetics and our understanding of the structure and behaviour of fungal genome. Molecular biology has brought many powerful new tools to fungal taxonomists including the potential for rapid identification of isolates, methods for rapid determination of virulence or toxicity of strains. Molecular methods have also been used to distinguish between closely related species with few morphological differences and to distinguish strains (or even specific isolates) within a species.

Molecular markers monitor the variations in DNA sequences within and between the species and provide accurate identification. In recent years, different marker system such as Restriction Fragment Length Polymorphisms (RFLP), Random Amplified Polymorphic DNA (RAPD), Sequence Tagged Sites (STS), Amplified Fragment Length Polymorphisms (AFLP), Simple Sequence Repeats (SSR) or microsatellites, Single Nucleotide Polymorphism (SNPs) and others have been developed and applied to different fungus species. The ribosomal DNA (rDNA) based classification is also the method of choice especially when classifying the related species. The nucleotide sequence analysis of rDNA region has been widely accepted to have phylogenetic significance, and is therefore useful in taxonomy and the study of phylogenetic relationships (Hibbett, 1992). RAPD, AFLP, SSR and ITS are currently the most popular markers in fungus. An increasing amount of sequence information and determination of gene functions in fungus is likely to lead in the near future to the preferred use of new marker types. The rRNA genes commonly used in identification and taxonomic studies, particularly appropriate for the purpose of providing target sequences for molecular detection. Differences in the nucleotide composition of the variable ITS region have been successfully employed to design specific primer sets that amplify DNA selectively among and within species of plant pathogens (Nazar *et al.*,1991; Moukhamedov *et al.*, 1994; Schilling *et al.*, 1996; Moricca *et al.*, 1998).

PCR (Polymerase Chain Reaction) has revolutionized the molecular marker technology and being increasingly used to generate molecular markers, which are useful for taxonomy and for characterizing fungal populations. Once the primers are designed and conditions for a robust assay are optimized, PCR assay is very sensitive, rapid and relatively easy to handle. Ever since the PCR technique became available (Mullis and Faloona, 1987), it has tremendously influenced research in diverse areas of biological sciences leading to an unprecedented understanding of pathogen. Using PCR, it is now possible to make virtually unlimited copies of a fragment of DNA (Field and Wills, 1998). The organism of interest can be detected directly through PCR assays in a much shorter time than conventional culture takes. As the primary factors in developing a PCR detection protocol are specificity and sensitivity, it is essential that the parameters used in the PCR are optimal, resulting in a maximum outcome. Some of the molecular methods that have been successfully used to characterize the pathogen populations of the crops are described below.

## RANDOM AMPLIFIED POLYMORPHIC DNA (RAPD)

This is one of the simplest PCR based molecular methods available for the characterization of pathogen population. It uses random primers (Williams *et al.*, 1990) and can be applied to any species without requiring any information about the nucleotide sequence. The amplification products from this analysis exhibit polymorphism and thus can be used as genetic markers. The presence of a RAPD band, however, does not allow distinction between hetero- and homozygous states. Welsh and McClelland (1990) described RAPD-PCR technique for detecting genetic variation among different organisms. Genetic variability is assessed by employing short single primer of arbitrary nucleotide sequences. Specific sequence information of the organism under investigation is not required and amplification of genomic DNA is initiated at target sites which are distributed throughout the genome. Polymorphic fragments are the results of variation in the number of appropriate primer-matching sites of different DNA.

RAPD offers several advantages that are useful in studying *formae specialis* and races of *Fusarium*. RAPD reduces the time needed for race identification in disease plants, and provides genetic information on isolates, allowing for their fingerprinting. Nijs *et al.* (1997) studied variation in RAPD patterns within *Fusarium* species of cereals from various parts of the Netherlands. Gherbawy (2002) used RAPD technique to analyze different *formae specials* of *Fusarium oxysporum*. Möller *et al.* (1999) studied fungal populations of *F. moniliforme* and *F. subglutinans* using RAPD technique. Gherbawy *et al.* (2002) used RAPD technique for identifying *Fusarium subglutinans*, *F. proliferatum* and *F. verticillioides* strains isolated from maize in Austria. Pasquali *et al.* (2003) characterized isolates of *Fusarium oxysporum* pathogenic on *Argyranthemum frutescens* L. using RAPD technique. Genetic similarity between isolates of *F. oxysporum* f. sp. *ciceri* was studied using 40 RAPD and 2 IGS primers and results indicate that there was little genetic variability among the isolates collected from the different locations in India (Singh *et al.* 2006). Honnareddy and Dubey (2006) have generated significant information in terms of pathogenic and genetic variability of *F. oxysporum* f. sp. *ciceri*, which could be used further for development of area-specific resistant varieties of chickpea. Cramer *et al.*, (2003) reported specific RAPD banding which distinguish among races *F. oxysporum* f.sp.*phaseoli* and *F.oxysporum* f.sp.*betae*. RAPD analysis using only three primers provide differentiating between isolate of the yellowing and wilt-inducing pathotypes of *F.o.ciceri,* non-pathogenic isolates, different *formae speciales*, and other fungi (Kelly *et al.*, 1994). Genetic variability within four races of *Fusarium oxysporum* f.sp.*pisi.*was assessed by 40 RAPD Primers by Grajal-Martin (1993). Identification of pathogenic races 0,1B/C,5, and 6 of *Fusarium oxysporum* f.sp.*ciceri* has been reported using 40 RAPD primers (Jimenez-Gasco *et al.*, 2001). They subsequently (2003), generated SCAR primers for *F.oxysporum* f.sp.*ciceri*. The SCAR primer pairs amplified a single diagnostic PCR product from all isolates of the target *forma specialis* and races.

## RESTRICTION FRAGMENT LENGTH POLYMORPHISMS (RFLP)

The procedures involve isolation of DNA, digestion of DNA with restriction endonucleases, size fractionation of the resulting DNA fragments by electrophoresis, DNA transfer from electrophoresis gel matrix to membrane, preparation of radio-labelled and chemiluminiscent probes, and hybridization to membrane-bound DNA. RFLP fingerprinting technique is regarded as the most sensitive method for strain identification. Direct analysis of DNA polymorphism is a more general approach to establishing genetic variation in organism. RFLP is comparatively more time-consuming than PCR based methods in analyzing a large number of strains. It requires large quantities of pure DNA sample, probe preparation and fastidious procedures of Southern-blotting and hybridization.RFLP analysis of nuclear or mitochondrial DNA have been used to estimate the genetic diversity of *Fusarium oxysporum* within (Jacobson 1990 and Elias 1993) and between *formae speciales* (Kistler 1991) and among non-pathogenic strains (Gordon 1992). Edel *et al.* (1995) compared three methods –RFLP, PCR with primers matching ERIC sequences, REP elements and PCR-RFLP analysis of IGS for diversity studies in *Fusarium oxysporum.* The two PCR based procedures find widespread application for large-scale characterization of *F.oxysporum* strains. PCR/RFLP analysis of the IGS region appears suitable to categorize groups of closely related strains at the intraspecific level.

## AMPLIFIED FRAGMENT LENGTH POLYMORPHISMS (AFLP)

Amplified fragment length polymorphism (AFLP) is a variation of RAPD, able to detect restriction site polymorphisms without prior sequence knowledge using PCR amplification (Vos, 1993). AFLP analysis is one of the robust multiple-locus fingerprinting techniques among genetic marker techniques that have been evaluated for genotypic characterization. It is technically similar to restriction fragment length polymorphism analysis, except that only a subset of the fragments are displayed and the number of fragments generated can be controlled by primer extensions. The advantage of AFLP over other techniques is that multiple bands are derived from all over the genome. This prevents over-interpretation or misinterpretation due to point mutations or single-locus recombination, which may affect other genotypic characteristics. The main disadvantage of AFLP markers is that alleles are not easily recognized (Majer *et al.*, 1998).

The utility, repeatability, and efficiency of the AFLP technique are leading to broader application of this technique to the analysis of *Fusarium* populations (Abd-Elsalam *et al.*, 2002; Kiprop *et al.*, 2002; Sivaramakrishnan *et al.*, 2002). AFLP has been more effective in bringing out the polymorphism among the pathogen isolates and placing them in different groups in contrast to RAPD. The genetic diversity existing within the pathogen isolates of *Fusarium udum* in the different pigeon pea growing states of India has been clearly demonstrated using AFLP. Sivaramakrishnan (2002) used both RAPD and AFLP analysis to classify 36 isolates of *Fusarium udum* from India into 3 major groups suggesting the existence of a minimum of 3 specific races of the pathogen prevailing in the pigeon pea growing areas of India. AFLP and RAPD banding patterns separated the Algerian isolates of *F.oxysporum* f.sp.*lentis* in two distinct groups (Belabid *et al.*, 2004).

## SIMPLE SEQUENCE REPEATS (SSR)

Simple sequence repeats (SSR) provide a powerful tool for taxonomic and population genetic studies. This method facilitates DNA fingerprinting by the use of mini- and microsatellites, which are hyper variable and dispersed DNA sequences in the form of long arrays of short tandem repeat units throughout the genome.The fragments generated by flanking primers differ in length based on the number of repeats in the amplified fragments.This length polymorphism is revealed by Agarose / metaphor or PAGE. In the RFLP based DNA fingerprinting method di-, tri-, and tetra nucleotide repeats have been radio-labelled and used as probes to characterize the pathogen populations. However, it is different from conventional RFLP in terms of the probe used and the way the reaction is carried out. Though the method is very precise, because of the DNA hybridization it can be laborious and time-consuming. With the use of simple sequence repeats (SSR or microsatellites), $(GAA)_6$, $(GACA)_4$ and especially $(GATA)_4$ high levels of polymorphism have been detected among the pathotypes. SSR offers the greatest potential for studies of comparative fitness, as multiple combinations of alleles are possible at each specific locus, thus increasing the likelihood of identifying unique test isolates for any given experiment. For tracking particular strains, or monitoring inoculum movement on a larger scale, SSR again has the greatest potential to uniquely discriminate each strain. However, further work is needed to investigate whether the resolution offered by SSRs will be sufficient in populations with limited genetic diversity.

DNA fingerprinting using microsatellite markers has been carried out in several plant pathogens, including the downy mildew pathogen *Sclerospora graminicola* (Sastry *et al.*, 1995), the chickpea blight pathogen *Ascochyta rabei* (Kaemmer *et al.*, 1992). SSR marker distinguished the four races of *Fusarium oxysporum ciceri* causing varied levels of wilting with differential host cultivars (Barve *et al.*, 2001). Bogale *et al.* (2005) has shown that, the polymorphism revealed with 8 SSR markers was sufficient for study of genetic diversity in *Fusarium oxysporum* complex. VNTR (Variable Number of Tandem Repeats) markers have been successfully applied to a preliminary population study of *F. graminearum* and *F.asiaticum* isolates (Suga *et al.*, 2004). Molecular variability based Universal Rice Primers (URPs) designed from the rice microsatellites has indicated clearly that *F.oxysporum* is heterogeneous and most of the *formae speciales* have close evolutionary relationship (Prasad *et al.*, 2004). The microsatellite loci characterized by 8 SSR markers proved to be highly polymorphic among *F.culmorum, F. graminearum*. This opens opportunities to study the *Fusarium* species complex (Giraud *et al.*, 2002). After testing thirteen oligonucleotide probes complementary to microsatellite loci, in combination with 11 restriction enzymes, Barve *et al.* (2001) identified that $(AGT)_5$ $(ATC)_5$ and $(GATA)_4$ are the best fingerprinting probes for *F.oxysporum.*

## rDNA MARKERS

The ribosomal RNA genes (rDNA) possess characteristics that are suitable for the detection of pathogens at the species level. The cluster of ribosomal DNA consists of tandem repeat of three coding (18S, 5.8S, and the 28S), and two non-coding internal transcribed spacer-ITS and intergenic spacer region- (IGS). These rDNA are highly stable and exhibit a mosaic of conserved and diverse regions within the genome (Hibbett, 1992). This approach, designing primers from the rDNA region has far superior reliability compared to the use of random non-defined probes or primers. These markers are most effective in detecting polymorphism. They also occur in multiple copies with up to 200 copies per haploid genome arranged in tandem repeats (Bruns *et al.*, 1991; Yao *et al.*, 1992). ITS regions have been used successfully to generate specific primers capable of differentiating closely related fungal species (Bryan *et al.*, 1995). In the broader context, taxon-selective amplification of ITS regions is likely to become a common approach in molecular identification strategies.

Taxon-selective ITS amplification has already been used for detection of the fungal pathogens such as *Fusarium* (ODonnell, 1992) and *Verticillium* spp. (Nazar *et al.*, 1991). O'Donnell (1992) found a surprising level of divergence for ITS sequences within the species of *F. sambucinum*. Amplification of the IGS region and digestion with restriction enzymes is also useful tool for fungal fingerprinting. Chakrabarti *et al*., (2000) have shown digestion of amplified IGS region with EcoR1 produced similar bands for both race1 and race 4 of *F.oxysporum* f.sp.*ciceri*. But individual and distinctive banding patterns were observed for race 2 and race 3.

## INTER SIMPLE SEQUENCE REPEATS (ISSR)

This method is a robust, PCR based technique that produces dominant molecular-markers by DNA amplification of putative microsatellite regions (Zietkiewicz *et al.,* 1994). ISSR markers show a higher level of polymorphism than RAPD markers (Esselman *et al.,* 1999) and have been used extensively in other fungal population analysis (Muller *et al.,* 1997; Hantula *et al.,* 2000; Tooley *et al.,* 2000). Intra-specific variation of ISSR products has been shown to be high in some fungal species (Hantula & Muller, 1997), while low in others (Longato & Bonfante, 1997).The ISSR fingerprinting with four primers generates highly polymorphic markers for *F. culmorum* and proved to be authentic and reliable molecular markers for inferring the genetic relationships within and between *Fusarium* species (Mishra *et al.*, 2003).

## References

Abd-Elsalam, K.A., Schnieder, F. & Verreet, J.A. (2002) *Phytomedizin* **3**: 18-19.

Aoki, T. (1998) *Soil Microorganisms* **52**: 73-83.

Barve, M. P., Haware, M. P., Sainani, M. N., Ranjekar, P, K. & Gupta, V. S. (2001) *Theoretical and Applied Genetics* **102:**138-147.

Bogale, M., Wingfield, B.D., Wingfield, M.J. Steenkamp E.J. (2005). *Ecology Notes* **5**: 622-624.

Booth, C. (1977) The genus *Fusarium*. Common Wealth Mycological Institute, Kew, England.

Bruns, T.D., White, T.J. & Talyor, J.W. (1991) *Annual Review of Ecology and Systematic* **22**: 525-564.

Bryan, G.T., Daniels, M.J. & Osbourn, A.E. (1995) *Applied and Environmental Microbiology* **61**: 681-689.

Chakrabarti, A., Mukherjee, P.K., Sherkhane, P.D., Bhagwat, A.S. & Murthy, N.B.K. (2000) *Current Science* **80**(4): 571-575.

Cramer, R.A., Byrne, P.F., Brick, M.A., Panella, L., Wickliffe, E. & Schwartz, H.F. (2003) *Phytopathology* **151**: 352-360.

Edel, V., Steinberg, C., Avelange, I., Laguerre, G. & Alabouvette, C. (1995) *Phytopathology* **85**(5): 579-585.

Elias, K. S., Zamir, D., Lichtman-Pleban, T. & Katan, T. (1993) *Molecular Plant - Microbe Interaction* **6**: 565-572.

Esselman, E. J., Jianqiang, L., Crawford, D. J., Windus, J. L. & Wolfe, A. D. (1999) *MolecularEcology* **8**: 443-451.

Field, D. & Wills, C. (1998) Abundant microsatellite polymorphism in *Saccharomyces cerevisiae*, and the different distributions of microsatellites in eight prokaryotes and *S. cerevisiae*, result from strong mutation pressures and a variety of selective forces. Protocol. National. Academy of Science U. S. A. **95**: 1647-1652.

Gerlach, W. & Nirenberg, H.I. (1982) *Land-Forst (Berlin-Dahlem)* **209** : 1-406.

Gherbawy, Y.A.M.H., Adler, A. & Prillinger, H. (2002) *Mycobiology* **30**: 139-145.

Giraud, T., Fournier, E., Vautrin, D. Solignac, M., Vercken, E., Bakans & Brygoo. (2002) *Molecular Ecology Notes* **2**: 121-123.

Gordon, T.R. & Okamoto, D. (1992) *Phytopathology*.**82**: 73-77.

Grajal-Martin, M.J., Simon, C.J. & Muehlbauer, F.J. (1993) *Phytopahology* **83**: 612-614.

Hantula, J., Lilja, A., Nuorteva, H., Parikka, P. & Werres, S. (2000) *Mycological Research* **104**:1062-1068.

Hantula, J. & Muller, M. (1997) *Mycological Research* **101**:169-175.

Hibbett, D.S. (1992) *Trans. Mycol. Soc*. **33**: 533-556.

Honnareddy, N. &. Dubey, S.C. (2006) *Current Science* **91**(5): 661-666.

Jacobson, D.J. & Gordon, T.R. (1990) *Mycological Research* **94** :734-744.

Jimenez-Gasco, M. M. & Jimenez-Diaz, R.M. (2003) *Phytopathology* **93**: 200-209.

Jimenez-Gasco, M. M. Perez-Artes, E. Jimenez-Diaz, R.M. (2001) *European Journal of Plant Pathology* **107**: 237-248.

Kaemmer, D., Ramser, J., Schon, M., Weigand, F., Saxena, M.C., Driesel, A.J., Khal, G. & Weising, K. (1992) *Advances in Molcular Genetics* **5:** 225-270.

Kelly, A., Alcala-Jimenez, R. Bainbridge, W., Heale, J.B., Perez, Artes, E. & Jimenez-Diaz, R.M. (1994) *Phytopathology* **84**: 1293-1298.

Kiprop, E.K., Baudoin, J.P., Mwangómbe, A.W., Kimani, P.M., Mergeai, G. & Maquet, A. (2002) *Journal of Phytopathology* **150**: 517-527.

Kistler, H.C., Momol, E.A. & Benny, U. (1991) *Phytopathology* **81:** 331-336.

Klittich, C.J.R. & Leslie, J.F. (1988) *Genetics* **118**: 417-423.

Lakhdar, B., Michael, B., Zohra, F., Zouaoui, B. & Imad, E. **(**2004) *African Journal of Biotechnology* **3**:25-31.

Lasker, B.A. (2002) *Clinical. Microbiology* **40**: 2886-2892.

Leslie, J.F. (1999) *Plant Pathology Journal*. **15**: 259-269.

Longato, S. & Bonfante, P. (1997) *Mycological Research* **101**: 425-432.

Majer, D., Lewis, B.G. & Mithen, R. (1998) *Plant Pathology* **47**: 22-28.

Majer, D., Mithen, R., Lewis, B., Vos, P. & Oliver, R.P. (1996) *Mycological Research* **100**: 1107-1111.

Minton, B.F. & Garber, R.H. (1983) *Plant Disease* **67**: 115-118

Mishra, P. K., Fox, R. T. V. & Chulham, A. (2003). *Annals of Applied Biology* **143**: 291-301.

Mishra, P. K., Roland, T.V.F. & Culham, A. (2003) *Annals of Applied Biology* **143**: 302-307.

Möller, E.M., Chelkowski, J. & Geiger, H.H. (1999) *Journal of Phytopathology* **147**: 497-508.

Moricca, S., Ragazzi, A., Kasuga, T. & Mitchelson, K.R. (1998) *Plant Pathology* 47: 486-494

Moukahmedov, R., Hu, X, Nazar, R.N. & Robb, J. (1994) *Phytopathology* **84**: 256-259.

Muller, R., Pasberg-Gauhl, C., Gauhl, F., Ramser, J. & Kahl, G. (1997). *Journal of Phytopathology* **145**: 25-30.

Mullis, K.B. & Faloona, F. (1987). Specific synthesis of DNA in vitro via a polymerase catalysed chain reaction. *Methods in Enzymology* **155**:335-350.

Nazar, R.N., Hu, X., Schmidt, J., Culham, D. & Robb, J. (1991). Potential use of PCR-amplified ribosomal intergenic sequences in the detection and differentiation of *Verticillium* wilt pathogens. *Physiological and Molecular Plant Pathology* **39**: 1-11.

Nelson, P.E., Toussoun, T.A. & Marasas, W.F.O. (1983). *Fusarium* species : An Illustrated Manual for Identification. Pennsylvania State University, University Park.

Nelson, P.E. 1991. History of *Fusarium* systematics. *Phytopathology,* **81**: 1045-1048.

Nijs, M, Larsen, J.S., Gams W, Rombous, F.M., Wernars, K., Thrane, U. & Notermans, S.H.W. (1997). Variations in random amplified polymorphic DNA patterns and secondary metabolite profiles within *Fusarium* species from cereals from various parts of the Netherlands. *Food Microbiology* **14:** 449-457.

O'Donnell, K. (1992). Ribosomal DNA internal transcribed spacers are highly divergent in the phytopathogenic ascomycete *Fusarium sambucinum* (*Gibberella pulicaris*). *Current Genetics* **22**: 213-220.

O'Donnell, K., Cigelnik, E. and Nirenberg, H.I. (1998). Molecular Systematics and Phylogeography of the *Gibberella fujikuroi* species complex. *Mycologia* **90**: 465-493.

Pasquali, M., Acquadro, A., Balmas, V., Migheli, Q., Garibaldi, A. & Gullino, M.L. (2003). RAPD characterization of *Fusarium oxysporum* isolates pathogenic on *Argyranthemum* Frutescens L. *Journal of Phytopathology* **151**: 30-35.

Prasad, R.D., Sharma, T.R., Jana, T.K., Prameela, D., Singh, N.K. & Koundal, K.R. (2004). Molecular analysis of genetic variability in *Fusarium* species using microsatellite markers. *Indian Phytopathology* **57**: 272-279.

Restrepo, S., Duque, M., Tohme, J. & Verdier, V., (1999). AFLP fingerprinting: an efficient technique for detecting genetic variation of Xanthomonasaxonopodis pv. Manihotis. *Microbiology* **145**: 107-114.

Saiki, R.H., Scharf, S., Faloona, F., Mullis, K.B., Horn, G.T., Ehrlich, H.A. & Arnheim, N. (1985). Enzymatic amplification of beta-globulin genomic sequences and restriction site analysis for diagnosis of sickle cell anemia. *Science* **230**: 1350-1354.

Sastry, J.G., Ramakrishna, W., Siwaramakrishnan, S., Thakur, R.P., Gupta, V.S. & Ranjekar, P.K. (1995). DNA fingerprinting detects genetic variability in the pearl millet downy mildew pathogen *(Sclerospora graminicola)*. *Theoretical and Applied Genetics* **91**: 856-861.

Schilling, A.G. (1996). Characterization and differentiation of the cereal pathogens *Fusarium culmorum* and *F. graminearum* by PCR-based molecular markers. Ph.D. Thesis. University of Hohenheim, Germany.

Singh, B.P., Saikia, R., Yadav, M., Singh, R., Chauhan, V.S. & Arora, D.K. (2006). Molecular characterization of *Fusarium oxysporum* f. sp. *ciceri* causing wilt of chickpea. *African Journal of Biotechnology*. **5(6):** 497-502.

Sivaramakrishna, S., Kannan, S. & Singh, S.D. (2002). Detection of genetic variability in *Fusarium udam* using DNA markers. *Indian Phytopathology*.**55**(3):258-263.

Sivaramakrishnan, S., Kannan, S. & Singh, S.D. (2002). Genetic variability of *Fusarium* wilt pathogen isolates of chickpea (*Cicer arietinum* L.) assessed by molecular markers. *Mycopathologia* **155**: 171-178.

Snyder, W.C. & Hansen, H.N. (1940). The species concept in *Fusarium. American Journal of Botany* **27**: 64-67.

Snyder, W.C. & Hansen, H.N. (1940). The species concept in *Fusarium* with reference to section Martiella. *American Journal of Botany* **28**: 738-742.

Suga, H., Gale, L.R. & Kistler, H.C. (2004). Development of VNTR markers for two *Fusarium graminearum* clade species. *Molecular Ecology Notes* **4**: 468-470.

Summerell, Brett, A., Salleh, B. & Leslie, J.F. (2003). A Utilitarian Approach to *Fusarium* Identification. *Plant Disease* **87**: 117-128.

Tooley, P.W.O., Neill, N.R., Goley, E.D. & Carras, M.M. (2000). Assessment of diversity in *Claviceps africana* and other *Claviceps* species by RAM and AFLP analysis. *Phytopathology* **90**: 1126-1130.

Vos, P., Hogers, R., Bleeker, M., Reijans, M., van de Lee, T., Hornes, M., Frijters, A., Pot, J., Peleman, J., Kuiper, M. & Zabeau, M. (1995). AFLP: A new technique for DNA fingerprinting. *Nucleic Acids Research* **23**: 4407-4414.

Welsh, J. & McClelland, M. (1990). Fingerprinting genomes using PCR with arbitrary primers. *Nucleic Acid Research* **18**: 7213-7218.

Williams, J.G.K., Kubelik, A.R., Livak, K.J., Rafalski, J.A. & Tingey, S.V. (1990). DNA polymorphisms amplified by arbitrary primers are useful as genetic markers. *Nucleic Acids Research* **18**: 6631-6635.

Yao, C., Frederiksen, R.A. & Magill, C.W. (1992). Length heterogeneity in ITS2 and the methylation status of CCGG and GCGC sites in the rRNA genes of the genus *Peronosclerospora*. *Current Genetics* **22**: 415-420.

Zietkiewicz, E., Rafalski, A. & Labuda, D. (1994). Genome fingerprinting by simple sequence repeat (SSR)-anchored polymerase chain reaction amplification. *Genome* **20**:176-183.

# 6

# Biocontrol of Soil-Borne Phytopathogens by Fluorescent Pseudomonads as a Sustainable Alternative to Agrochemicals

**N.K. Arora and Ekta Khare**

*Department of Microbiology, Institute of Biosciences and Biotechnology, CSJM University, Kanpur-208024, India*

## ABSTRACT

Bacterial strains in certain natural environments prevent infectious diseases of plant roots. During root colonization, these bacteria produce allelochemicals, elicit induced systemic resistance in the host plant or interfere specifically with fungal pathogenicity factors, representing a real alternative to the application of chemical fungicides. This review tries to evaluate what is available and what should be explored in the future to uplift the stature of the bioinoculants. For being competent to chemicals, the inoculants apart from being ecofriendly should also be market-friendly and above all true to the work assigned to them. The review looks from the history to future of the bioinoculants with special reference to fluorescent pseudomonads.

## INTRODUCTION

Pathogenic microorganisms affecting plant health are a major and chronic threat to food production and ecosystem stability worldwide. As agricultural production intensified over the past few decades, producers became more and more dependent on agrochemicals as a relatively reliable method of crop protection. However, increasing use of chemical inputs causes several negative effects, development of pathogen resistance to the applied agents and their non-target environmental impacts (Gerhardson, 2002). Further, the growing cost of pesticides, particulary in less-affluent regions of world, and consumer demand for pesticide-free food has led to a search for substitutes for these products. There are also a number of fastidious diseases for which chemical solutions are few, ineffective, or non-existent (Gerhardson, 2002). However, it is known that there are certain places on earth where plants do not suffer from certain diseases and where disease severity is highly reduced, although a pathogen might be present and the host plant is susceptible to the disease. Such exceptional places exist and are known as natural suppressive soils (Weller *et al.*, 2002). They occur, for instance, in the Salinas Valley (California, United States), the Chateaurenard region near Cavaillon (France), Canary Islands and the Broye Valley (Switzerland).

Certain suppressive soils when pasteurized (for example, by wet heat at 60°C for 30 min.) lose their suppressiveness, and other harsher antimicrobial treatments (like gamma radiation or autoclaving) have same effect (Stutz *et al.*, 1986). Suppressiveness can be transferable; an inoculation of 0.1-10% of a suppressive soil introduced into a conductive soil can establish disease suppression (Stutz *et al.*, 1986; Weller *et al.*, 2002). Sensitivity to antimicrobial treatments and transferability indicates that disease suppression results from the activities of soil microorganisms that act as pathogen antagonists.

The zone of intense bacterial activity around plant roots is termed rhizosphere. It is assumed to extend from the root surface into the soil upto a few millimetres. The rhizosphere can be divided into the rhizoplane (i.e., the root surface), the endorhizosphere (i.e., the internal root parts) and the ectorhizosphere (the thin layer of sand or soil adhering to roots). The bacterial level in the rhizosphere is 10-1000 folds higher than the surrounding soil. This effect is ascribed to the secretion of nutrients by roots and seeds.

Within the rhizosphere microbial populations are a group of plant beneficial bacteria referred to as plant growth promoting rhizobacteria (PGPR). PGPR are beneficial to plants via nutrient acquisition (Rodriguez and Fraga, 1999; Dobbelaere *et al.*, 2003), plant hormone production, lowering of plant ethylene level (Glick, 1995; Steenhoudt and Vanderleyden, 2000), induction of systemic resistance (van Loon *et al.*, 1998) and biocontrol of phytopathogens (Walsh *et al.*, 2001; Chin-A-Woeng *et al.*, 2003). The direct mechanism of plant growth promotion may involve the synthesis of substances by the bacterium or facilitation of the uptake of nutrients from the environment (Glick *et al.*, 1999). The indirect promotion of plant growth, i.e., reduction of the incidence of plant disease occurs when PGPR lessen or prevent the deleterious effects of plant pathogens on plant by production of inhibitory substances or by increasing the natural resistance of the host (Cartieaux *et al.*, 2003). The mechanisms that are involved in plant growth promotion process include: nitrogen fixation, phosphate solubilization, production of phytohormones (such as auxins, cytokinins), sequestering of iron through production of siderophores, lowering of ethylene concentration and volatile growth stimulants (such as ethylene and 2, 3-butanediol (Vessey, 2003). Biofertilizers are not discussed further in this review. In other PGPR, which are sometimes called biopesticides, the biocontrol aspect is most conspicuous. The mechanism of biocontrol by PGPR includes antibiotic production, depletion of iron from the rhizosphere, production of fungal cell wall lysing enzymes, competition for sites on roots, detoxification and degradation of virulence factors and induced systemic resistance (Weller and Cook, 1986; Glick, 1995). PGPR competitively colonize plant roots and the closely adhering soil interface, the rhizosphere (Kloepper and Schroth, 1978). Some of them can also enter root interior and establish endophytic populations (Gray and Smith, 2005). Many of them are able to transcend the endodermis barrier, crossing from the root cortex to the vascular system and subsequently thrive as endophytes in stem, leaves, tubers and other organs (Compant *et al.*, 2005a; Gray and Smith, 2005). Despite their different ecological niches, free-living rhizobacteria and endophytic bacteria use some of the same mechanisms to promote plant growth and control phytopathogens (Glick, 1995; Bloemberg and Lugtenberg, 2001; Dobbelare *et al.*, 2003). The observation indicates that the bacterial community in the rhizosphere soil is likely to be more sensitive to modifications in the agricultural practices rhizoplane than in the endorhizosphere (Roesti *et al.*, 2005). Because of their catabolic versatility, their excellent root-colonizing abilities and their capacity to produce a wide range of antifungal metabolites, the soil-borne fluorescent pseudomonads have received particular attention. Biological control is thus being considered as an alternative or a supplemental way of reducing the use of chemicals in agriculture. In this review, we attempt to explain principles and mechanisms by which fluorescent pseudomonads control plant diseases and some of the scientific challenges that still exist in biocontrol research.

## BIOCONTROL ABILITY OF FLUORESCENT PSEUDOMONAS

Members of the genus *Pseudomonas* are rod-shaped Gram-negative bacteria that are characterized by metabolic versatility, aerobic respiration (some strains also have anaerobic respiration with nitrate as a terminal electron acceptor and for arginine fermentation), motility owing to several polar flagella, and high genomic GC content (59.68%) (Holt *et al.*, 1994). The term "pseudomonads" (*Pseudomonas*-like bacteria) is often used to describe strains for which the taxonomic affiliation has not been established in detail. Fluorescent pseudomonads produce the fluorescent pigment pyoverdin (Pvd) (also known as pseudobactin). This large and heterogenous group comprises most notably, *P. aeruginosa*, *P. putida*, *P. fluorescens* and *P. syringae* (Kumar *et al.*, 2005).

*Pseudomonas* spp. are particularly suitable for application as agricultural biocontrol agents since they can use many exudates compounds as a nutrient source, which is indicative of their adaptive potential (Lugtenberg *et al.*, 1999). *Pseudomonas* possess diverse mechanisms by which they can exert inhibitory activity towards phytopathogens and thereby mediate crop protection (Dunlap *et al.*, 1996). They have a high growth rate relative to many other rhizosphere bacteria, are easy to grow *in vitro,* and can subsequently be reintroduced into the rhizosphere by seed bacterization (Rhodes and Powell, 1994).

## BIOCONTROL ACTIVITY MEDIATED BY COMPETITIVE ROOT COLONIZATION

Despite their potential as low-input practical agents of plant protection, application of PGPR has been hampered by inconsistent performance in field tests (Thomashow, 1996). This is usually attributed to their poor rhizosphere competence (Weller, 1988). Rhizosphere competence of biocontrol agents (BCA) comprises effective root colonization combined with the ability to survive and proliferate along growing plant roots over a considerable time period in the presence of the indigenous microflora (Whipps, 1997; Lugtenberg and Dekkers, 1999). Competition between

BCAs and phytopathogens for nutrients and infection sites at the root surface is one of the known mechanisms of biocontrol. Among the most rhizosphere competent bacteria are several *Pseudomonas* sp. (Lugtenberg *et al.*, 2001). Given the importance of rhizosphere competence as a prerequisite for effective biological control, understanding root-microbe communication (Bais *et al.*, 2004; Ping and Boland, 2004), as affected by genetic and environmental (Petterson and Baath, 2004) determinants in spatial (Bais *et al.*, 2004) and temporal (Chatterton *et al.*, 2004) contexts, will significantly contribute to improve the efficacy of these biocontrol agents.

Using molecular markers such as green fluorescent protein (GFP), it is possible to monitor the location of individual rhizobacteria on the root using confocal laser scanning microscopy (Sørensen *et al.*, 2001; Liu *et al.*, 2006). Tombolini *et al.*, (1997) investigated the colonization of GFP-tagged *P. fluorescens* on the roots of Japanese lotus plant. *P. fluorescens* was chromosomally tagged with gfp fused to strong constitutive promoter present in a Tn5 delivery vector. This approach has also been combined with an rRNA targeting probe to monitor the metabolic activity of a rhizobacterial strain in the rhizosphere and showed that bacteria located at the root tip were most active (Sørensen *et al.*, 2001).

## EFFECT OF PLANTS ON COLONIZATION ABILITY OF FLUORESCENT PSEUDOMONADS

The root surface and surrounding rhizosphere are significant carbon sinks (Rovira, 1965). Photosynthate allocation to this zone can be as high as 40% (Degenhardt *et al.*, 2003). The roots of the plant release soluble exudates components, mucigel and sloughed off cells. Compounds detected in plant root exudates include amino acids, proteins, fatty acids, flavonoids, hormones, organic acids, polysaccharides, organic phosphorus compounds, purines, pyrimidines, sterols, sugars including oligosaccharides, vitamins and unidentified compounds which inhibit or stimulate fungi, bacteria and nematodes. The amount and composition of exudates is largely affected by multiple factors such as plant species, root region, plant age, pH, temperature and surrounding microbes (Meharg and Killham, 1995). The mutants unable to use the major exudate components are poor competitive colonizers and indicate that efficient utilization of major exudates carbon sources is the nutritional basis for competitive root tip colonization. Despite the fact that sugar has often been reported as the major carbon source in exudates, the ability to use specific sugar does not play a major role in tomato root colonization (Lugtenberg *et al.*, 1999). Similarly, although amino acids are present in root exudates, the bioavailability of amino acids alone is considered insufficient to support root tip colonization by auxotrophic mutants of *P. fluorescens*, indicating that amino acid prototrophy is involved in rhizosphere competence. Root mucilage also offers an utilizable carbon source for PGPR (Knee *et al.*, 2001) to use for the competitive colonization. Due to these exudates along root surfaces, there are various suitable nutrient rich niches attracting a great diversity of microorganisms, including phytopathogens. Competition for these nutrients and niches is a fundamental mechanism by which PGPR protects plants from phytopathogens (Duffy, 2001). In addition, PGPR regulate the rate of uptake of polyamines such as putrescine, spermine, and spermidine, since their high titer could retard bacterial growth and reduce their ability to competitively colonize roots (Kuiper *et al.*, 2001). A potential multidrug efflux pump could be involved in protection against such toxins in exudates.

The effect of plant species on microbial communities (Wieland *et al.*, 2001; Kuske *et al.*, 2002) and populations of specific bacterial genera has been well-documented (Berg *et al.*, 2002; Gu and Mazzola, 2003). For example, the composition of fluorescent *Pseudomonas* sp. may vary significantly within the rhizosphere of different plant species cultivated in the same soil (Lemanceau *et al.*, 1995), and may change dramatically over time in response to the age-related physiological changes in plant species (Mazzola and Gu, 2000), i.e., modification in the amount, release site, and chemical composition of root exudates available for microbial growth and activity. In contrast, relatively few studies have addressed the effect of plant species on the dynamics, composition and activity of indigenous bacterial populations that share a specific antagonistic trait. Recent studies support the adverse effect of oats and lily on populations of 2, 4-diacetylphloroglucinol (Phl or DAPG) producing pseudomonads. HPLC analysis showed that plant species had a significant effect on Phl production by the indigenous *phl* $D^+$ population: the wheat and potato rhizospheres supported significantly higher amount of Phl produced per cell basis than the rhizospheres of sugar beet and lily. Interestingly in this context is the fact that both oats and members of liliaceae are well-known for their production of steroid saponins in the rhizosphere (Mimaki *et al.*, 1999; Park *et al.*, 2003). Saponins including venacoside A and B from oats, which exhibit activity against fungi, Gram-negative and Gram-positive bacteria (reviewed in Mahato *et al.*, 1982). Studies on the chemical constituents of root exudates of lily, including saponins and other allelochemical compounds (Bertin *et al.*, 2003), are being initiated to further unravel their role, if any, in the decline of indigenous Phl producing pseudomonads. Another plant trait is its root glycoprotein complex known as agglutinin can be involved in the short-term adherence of pseudomonads (Glandorf, 1992).

## BACTERIAL TRAITS INVOLVED IN COMPETITIVE COLONIZATION

A variety of bacterial traits and specific genes contribute to this process, but only a few have been identified (Benizri *et al.*, 2001; Lugtenberg *et al.*, 2001). These include motility, chemotaxis to seed and roots exudates, production of pili or fimbriae, production of specific cell surface components, ability to use specific components of root exudates and root mucilage, ability to synthesize vitamins and building blocks, phase variation, secretion and quorum sensing (Lugtenberg *et al.*, 2001). The generation of mutants altered in expression of these traits is aiding our understanding of the precise role each one plays in the colonization process (Cartieaux *et al.*, 2003).

**Motility and Chemotaxis:** Non-motile and flagella-less mutants lose competition for root tip colonization from their wild type (Delany *et al.*, 2000). Using *che* $A^-$ chemotaxis mutants of various *P. fluorescens* strains, De Weert *et al.*, (2002) showed that the presence of flagella is not sufficient for competitive colonization and that chemotaxis is also essential. Based on concentrations estimated to be present in rhizosphere, the results indicated that malic acid and citric acid are major chemo-attractants in the tomato rhizosphere (De Weert *et al.*, 2002).

**Type IV Pili:** Another determinant of root colonization ability by bacteria is type IV pili. Using mutants in *pil A*, encoding prepilin, and *pil T*, involved in pilus retraction, Hahn (1997) showed slight defects in competitive root tip colonization for both mutants.

**Cell Surface Components:** Bacterial lipopolysaccharides (LPS), in particular the O-antigenic chain, can also contribute to root colonization (Dekkers *et al.*, 1998b). However, the importance of

LPS in this colonization might be strain dependent, since the O-antigenic side chain of *P. fluorescens* WCS374 does not contribute to potato root adhesion, whereas the O-antigen chain of *P. fluorescens* PCL1205 is involved in tomato root colonization (Dekkers *et al.*, 1998b).

**Ability to Synthesize Vitamins and Building Blocks for Macromolecules:** A study using amino acid auxotrophic mutants of *P. fluorescens* and mutants with reduced pyrimidine synthesis indicates that ability to synthesize building blocks for macromolecules is required for competition in the rhizosphere (Camacho-Carvajal, 2001). A vitamin B1 auxotroph of *P. fluorescens* appeared to be a competitive tomato root tip colonizer (Simons *et al.*, 1997). This and other observations on different PGPR strains that concentrations of vitamins present in exudates are not sufficient to allow growth of these vitamin auxotrophs.

**Phase Variation:** Bacterial traits required for effective root colonization are subject to phase variation, a regulatory process for DNA rearrangements orches traced by site-specific recombinase (Dekkers *et al.*, 1998a; Van der Brock *et al.*, 2003). Transfer of site-specific recombinase gene from a rhizosphere-competent *P. fluorescens* into a rhizosphere-incompetent *Pseudomonas* strain enhances its ability to colonize root tips (Dekkers *et al.*, 2000). *Pseudomonas* isolates from maize roots foım opaque phase I and translucent phase II colonies. It appeared that important biocontrol traits are correlated with only phase I colony morphology.

**Quorum Sensing:** Fluorescent *Pseudomonas* species colonizing plant roots were reported by Elasri *et al.*, (2001) to produce N-acyl-L-homoserine lactone (AHL) molecules, serving as bacterial cell-cell communication signal. Production of antifungal metabolites is cell density dependent and is regulated in certain strains by quorum sensing (Wood *et al.*, 1997).

Specific plant root exudates or fungal signals exerting a direct control on gene expression of *Pseudomonas* sp. are currently a popular research topic. The most fascinating observation was that of Steidle *et al.*, (2001), reporting that root signals of molecules may actually mimic and thus interact with bacterial (including *Pseudomonas* species) cell-cell communication in the rhizosphere (Tepletski *et al.*, 2000; Juhas *et al.*, 2005).

**Secretion:** Antagonism between microorganisms is mediated by antibiotics, lytic enzymes, exotoxins and bacteriocins. Bacteriocin producing isolates occur frequently among plant associated fluorescent *Pseudomonas* sp. Bacteriocin is a proteinaceous compound that is produced by certain strains and kill closely related species (Riley and Wertz, 2002). It has been proposed that such strains may have a competitive advantage in the rhizosphere (Parrel and De Mot, 2002; Validov *et al.*, 2005).

**Other Traits:** It has been recently demonstrated that NADH dehydrogenase contribute to plant colonization by PGPR (Dekkers *et al.*, 1998b). NADH dehydrogenase I (encoded by a 14 gene nuo operon), identified in *P. fluorescens* WCS365 (Camacho-Carjaval *et al.*, 2002), is involved in the generation of proton motive force, which can be used for the uptake of various nutrients, generation of ATP and ATP dependent rotation of flagella. Since efficient nutrient uptake and chemotaxis are very important processes for rhizosphere competence, it is clear why expression of *nuo* in the rhizosphere is important.

Interestingly, a gene identified in the *in vivo* expression technology (IVET) study showed homology to type III secretion system genes—a finding which may suggest a more intimate and specific interaction between plants and associated beneficial bacteria (Rainey, 1999).

## DIRECT BIOCONTROL INTERACTION WITH PHYTOPATHOGENS VIA SYNTHESIS OF ALLELOCHEMICALS

Offensive PGPR colonization and defensive retention of rhizosphere niches are enabled by production of bacterial allelochemicals. Biochemical and molecular approaches are providing new insight into the genetic basis of biocontrol traits, the biosynthetic pathways involved, their importance for biocontrol in laboratory and field studies (Glick, 1995; Cartieaux *et al.*, 2003).

### Siderophores: Competition for Iron

Fluorescent pseudomonads owe their fluorescence to an extracellular pigment called pyoverdin (Pvd) or pseudobactin. This pigment has high affinity for $Fe^{3+}$ ions (the association constant of the interaction (Kass) is ~ $10^{24}$ at pH 7 and is a siderophore (iron carrier) of the producer strain (Meyer and Abdallah, 1978). Ferripyoverdin (Pvd complexed with $Fe^{3+}$) interacts with a specific outer membrane receptor, which is present in the producer but might also occur in some non-producers. Subsequently, $Fe^{3+}$ is transported into the cytoplasm and reduced to $Fe^{2+}$ (Visca, 2004).

The siderophores hypothesis postulates that PGPR exert their biocontrol activity by depriving pathogens of iron (Schroth and Hancock, 1982). Under greenhouse conditions, *Pseudomonas putida* suppressed *Fusarium* wilt and take-all, but this suppression was lost when the soil was amended with iron, which repress siderophore production. (Kloepper *et al.*, 1980). Though Pvd-type siderophores might contribute to disease suppression in some situations, alone they are not sufficient to account for suppression. If they were, it would be difficult to explain why most fluorescent pseudomonads do not have biocontrol activity.

Another pseudomonad siderophore, pyochelin, has been identified as an antifungal antibiotic in a screening program (Phoebe *et al.*, 2001). As pyochelin is a relatively weak $Fe^{3+}$ chelator, but a good $Cu^{2+}$ and $Zn^{2+}$ chelator (Cuppeis *et al.*, 1987), it might be able to deprive some fungi of copper and/or zinc.

***Regulation of siderophore biosynthesis:*** Siderophore biosynthesis is generally tightly regulated by iron-sensitive Fur proteins, the global regulators Gac S and Gac A, the sigma factors Rpo S, Pvd S and Fpv I, quorum sensing autoinducers such as N-acyl homoserine lactone, and site-specific recombinase (Cornelis and Matthijs, 2002; Ravel and Cornelis, 2003). However, some data demonstrate that none of these global regulators is involved in siderophore production. A myriad of environmental factors can also modulate siderophore synthesis, including the pH, the level of iron and the form of iron ions, the presence of other trace elements, and an adequate supply of carbon, nitrogen, and phosphorus (Duffy and Défago, 1999).

### Antibiosis

A variety of antibiotics have been identified including compounds such as amphisin, 2, 4-diacetylphloroglucinol (Phl or DAPG), hydrogen cyanide, oomycin A, phenazine, pyoluteorin, tensin, tropolone and cyclic lipopeptides produced by pseudomonads (Nielsen *et al.*, 2002; Nielsen and Sørensen, 2003). The distinction between siderophores and typical antibiotics is blurred. Although siderophores are part of primary metabolism (because iron is an essential element), occasionally they also behave as antibiotics (which are commonly considered as secondary metabolites) (Haas and Défago, 2005). On low-iron agar plates, pyoverdin (Pvd) that is deposited on a filter disc can produce a halo of inhibition on a susceptible microorganism. Therefore, under certain conditions, Pvd functions as a diffusible, biostatic or fungistatic antibiotic, whereas ferripyoverdin does not (Scher and Baker, 1982). In this review, we focus on the six classes of antibiotic compounds for which the experimental evidence most clearly supports a function in the biocontrol of root disease: phenazines, phloroglucinols, pyoluteorin, pyrrolnitrin, cyclic lipopeptides (all of which are diffusible) and hydrogen cyanide (HCN; which is volatile) (Haas and Keel, 2003).

***Mode of action:*** In the presence of ferripyochelin, phenazines catalyze the formation of hydroxyl radicals, which damage lipids and other macromolecules. Interestingly, reduced phenazine-1-carboxamide can release soluble $Fe^{2+}$ $(OH^-)_3$ at neutral pH, which raises the possibility that phenazines might contribute to iron mobilization in soils (Hernandez *et al.*, 2004). 2, 4-diacetylphloroglucinol is the best known phloroglucinol compound in a family of related molecules that includes monoacetylphloroglucinol and uncharacterized condensation products of Phl and monoacetylphloroglucinol (Mavrodi *et al.*, 2001). Phl causes membrane damage to *Pythium* sp. and is particularly inhibitory to zoospores of this oomycete (de Souza *et al.*, 2003). In addition to its direct antimicrobial activity, Phl may also function as a signal inducing systemic plant resistance against phytopathogens (Iavicoli *et al.*, 2003; Bottiglieri and Keel, 2006). To our knowledge, no mode of action has been published for pyoluteorin, whereas pyrrolnitrin has been described as an inhibitor of fungal respiratory chains (Tripathi and Gottlieb, 1969). Synthetic analogues of pyrrolnitrin have been developed for use as agricultural fungicides. Cyclic lipopeptides with antibiotic activity include amphisin (Koch *et al.*, 2002), tensis (Nielsen *et al.*, 2000). These are able to insert into membranes and perturb their function, which results in broad antibacterial and antifungal activities. Finally, the cyanide ion derived from HCN is a potent inhibitor of many metalloenzymes, especially copper containing cytochrome C oxidase (Blumer and Haas, 2000). In summary, with the exception of pyrrolnitrin, these antibiotic compounds show little selectivity towards fungi.

***Molecular regulation:*** Regulatory cascades of these antibiotics involve Gac A/Gac S or Grr A/ Grr S, Rpo D, and Rpo S, N-acylhomoserine lactone derivatives (Bloemberg and Lugtenberg, 2001; Haas and Keel, 2003) and positive autoregulation (Brodhagen *et al.*, 2004). Antibiotic synthesis is tightly linked to the overall metabolic status of the cell, which in turn is dictated by nutrient availability and other environment stimuli (Thomashow, 1996), such as major and minor minerals, type of carbon source and supply, pH, temperature, and other parameters (Milner *et al.*, 1996; Ownley *et al.*, 2003). It is important to note that conditions favouring one compound may not favour another (Duffy and Défago, 1999). Biotic conditions can also influence antibiotic biosynthesis (Haas and Keel, 2003; Petterson and Bååth, 2004). For example, bacterial metabolites, salicylates and pyoluteorin can effect Phl production by *P.fluorescens* CHA0 (Schnider-Keel *et al.*, 2000). Furthermore, plant growth and development also influence antibiotic production, since biological activity of Phl producers is not induced by the exudates of young plant roots but is induced by the exudates of older plants, which results in selective pressure against other rhizosphere microorganisms.

## FUNGAL CELL WALL LYSING ENZYMES

Antagonistic effects on phytopathogens can also be mediated by extracellular lytic enzymes, and many biocontrol pseudomonads produce such enzymes (Chernin and Chet, 2002). It has been demonstrated that extracellular chitinase and laminarinase synthesized by *Pseudomonas stutzeri* digest and lyse mycelia of *Fusarium solani* (Lim *et al.*, 1991). Relatively good growth of the bacterial strains noticed in chitin amended media also suggests their ability to degrade the complex chitin polymer for their growth (Ajit *et al.*, 2006). Antagonistic studies with the bacterial strains and their cell-free culture filtrate grown on chitin-amended media showing maximum chitinase production also exhibited direct inhibitory activity on mycelial growth of the pathogen. The variation in the level of antagonism might be due to the variation in producing the extracellular enzymes. Chitinase and cellulose produced by *Pseudomonas* strains showed mycelial growth inhibition of different fungal pathogens on PDA medium (Sindhu and Dadarwal, 2001).

Production of extracellular β-1,3 glucanases, chitinases, and proteinase increases significantly when *Pseudomonas* sp. are grown in media supplemented with either autoclaved mycelium or isolated purified host fungal cell walls (Viswanathan and Samiyappan, 2000). Such induction of chitinases and their antifungal activity together with the fact that chitin, β-1,3 glucan and protein are the main structural components of most fungal cell walls suggested that hydrolytic enzymes produced by some fluorescent pseudomonads play an important role in destruction of plant pathogens (Peberdy, 1990).

Similar to siderophores and antibiotics, regulation of lytic enzyme production (protease and chitinase in particular) involves the Gac A/Gac S (Sanchez-Contreras *et al.*, 2002) or Grr A/ Grr S regulatory systems (Ovadis *et al.*, 2004) and colony phase variation (Lugtenberg *et al.*, 2001).

## DETOXIFICATION AND DEGRADATION OF VIRULENCE FACTORS

Pathogen toxins (virulence or pathogenicity factors) display board-spectrum activity and can suppress the growth of microbial competitors, or detoxify antibiotics produced by some biocontrol microorganisms as a self-defence mechanism against biocontrol agents (Schouten *et al.*, 2004; Siddiqui and Shaukat, 2005). It has been observed that lytic enzymes produced by biocontrol agents can also act by interfering with pathogenicity factors of the pathogens or with the signal molecules that trigger production of secondary metabolites involved in pathogenicity. For instance, potato plant can be protected against wilt caused by *Fusarium oxysporium* f. sp. *lycopersici* by inoculation with a *Pseudomonas* strain able to degrade a virulence factor (fusaric acid) produced by the fungus (Toyoda *et al.*, 1988). Recently, it has been found that several nonpathogenic bacteria (including pseudomonads) can degrade N-acylhomoserine lactones quorum sensing signals that turn on gene cascades for their key virulence factors like cell degrading enzymes and phytotoxins (Molina *et al.*, 2003; Dong *et al.*, 2004; Newton and Fray, 2004).

## BIOCONTROL INTERACTIONS VIA THE PLANTS

In addition to direct effect on phytopathogens, some *Pseudomonas* strains can contribute to disease suppression via a positive, direct effect on the plant. This systemic resistance is long-lasting and protects plants against a broad range of pathogens and parasites attacking leaves or roots, and in some case against nematodes and insects (Zehnder *et al.*, 1997; Siddiqui and Shaukat, 2003).

## Induced Systemic Resistance

Certain bacteria trigger a phenomenon known as induced systemic resistance (ISR) phenotypically similar to systemic acquired resistance (SAR). SAR develops when plants successfully activate their defence mechanism in response to primary infection by a pathogen, notably when the latter induces a hypersensitive reaction through which it becomes limited in a local necrotic lesion of brown, desiccated tissue (van Loon *et al.*, 1998). As SAR, ISR is effective against different types of pathogens but differs from SAR in that the inducing PGPR does not cause visible symptom on the host plant and bacterial strain (van Loon *et al.*, 1998; Kilic-Ekici and Yuen, 2004). Studies with *Arabidopsis thalina* mutants indicate that the jasmonate/ethylene inducible defence pathway is important for ISR, whereas the salicylate-inducible pathway mediating SAR seems to be less important (Bakker *et al.*, 2003). ISR is associated with an increase in sensitivity to these hormones rather than an increase in their production, which might need for the activation of a partially different set of defence gene (Pieterse and van Loon, 1999; Hase *et al.*, 2003).

***Bacterial signals for ISR:*** $Phl^-$ mutants of *P. fluorescens* are less effective than the wild type bacteria in protecting *Arabidiopsis* from the leaf pathogen (Lavicoli *et al.*, 2003). A salicylate overproducing recombinant of *P. fluorescens* strain P3 affords enhanced protection to tobacco against tobacco necrosis virus compared with the wild type P3, which indicates that salicylate might also stimulate defence. In another *Pseudomonas* biocontrol strain, a combination of pyocyanin and pyochelin seems to be most effective for inducing resistance in tomato (Audenaert *et al.*, 2002). Recent evidence indicates that induction of some of these plant defence compounds (e.g., chalcone synthase) may be triggered by the N-acyl homoserine lactones that bacteria use for intraspecific signaling (Mathesius *et al.*, 2003) in several ISR-competent strains of fluorescent pseudomonads, thus been difficult to isolate specific ISR elicitors; it has therefore been proposed that a combination of siderophores, O-antigens and flagella might account for the ISR effect (Bakker *et al.*, 2003; van Loon and Bakker, 2003).

***Mechanisms of ISR:*** The type of bacterized plant response induced after challenge with a pathogen resulted in the formation of structural barriers, such as thickened cell wall papillae due to the deposition of callose and the accumulation of phenolic compounds at the site of plant attack (Benhamou *et al.*, 1998; Mishra *et al.*, 2006). Biochemical and physiological changes in plants induced accumulation of pathogenesis related proteins (PR proteins) such as PR-1. PR-2, chitinases, and some peroxidases (Ramamoorthy *et al.*, 2001; Jeun *et al.*, 2004). However, certain PGPR does not include PR proteins (Hoffland *et al.*, 1995) but rather increase accumulation of peroxidase, phenylalanine ammonia lyase, phytoalexins, polyphenol oxidase, and/or chalcone synthase (Chen *et al.*, 2000). The revelation that some PGPR genes involved in antibiotic biosynthesis (e.g. *phl D*) are highly homologous with some plant genes involved in defence (e.g., chalcone synthase) (Bangera and Thomashow, 1999; Austin and Noel, 2003) raises the intriguing but as yet unexplored possibility that the products of these genes may have interspecies activity benefiting plant protection, in addition to their currently known functions (Compant *et al.*, 2005b).

## Future Prospects to Make Better use of *Pseudomonas* as Biocontrol Agent

At present, there are many scientific challenges for research in the field of biocontrol pseudomonads. By contrast with natural suppressive soils whose characteristic is consistent over years and seems to be pathogen specific, when biocontrol PGPR are added to conductive soils, the degree of suppression can vary considerably over time and the spectrum of disease that is suppressed can be broad. One possible explanation is that natural suppressive soils might contain several PGPR strains with different biocontrol properties, and suppressive ness could be brought about by a consortium of PGPR strains that are specifically adapted to the roots of the host plant and to the stress caused by a specific root pathogen. Another explanation might be that some as yet-unrecognized soil properties, in particular the bioavailability of minerals, are important in the success of biocontrol (Stutz *et al.*, 1989). In addition, many specific questions remain to be addressed:

Signal molecules of microbial and plant origin turn on the expression of biocontrol traits but the mechanism of action is still unclear.

- Knowledge about microbial signals which are most effective in eliciting pathogen resistance in plant through ISR, SAR or alternative pathways is uncertain
- Mechanism of plant-specific root colonization of microbes
- The use of multi strain inocula with different biocontrol properties can increase disease suppressiveness in comparison to their monoinoculation
- The antagonistic activities between biocontrol agents are still unchecked
- Role of type III secretion system of certain biocontrol PGPR in their biocontrol activity (root colonization, direct action on pathogens)
- Influence of soil characteristics and other abiotic factors on biocontrol mechanisms
- The apparent specificity of natural suppressive soils
- Regulation of pathogenicity factors of fungal pathogens in the rhizosphere
- Types of interactions amongst PGPRs.

As our understanding of the complex environment of the rhizosphere, of the mechanisms of action of biocontrol agents and of the practical aspects of inoculant formulation and delivery increases, we can expect to see new PGPR products being available. The success of these products will depend on our ability to manage the rhizosphere to enhance survival and competitiveness of these beneficial microorganisms (Bowen and Rovira, 1999). Rhizosphere management will require consideration of soil and crop cultural practices as well as inoculant formulation and delivery (Mc Spadden Gardner and Fraver, 2002). The application of molecular tools in enhancing our ability to understand and manage the rhizosphere will lead to new products with improved effectiveness. Genetic enhancement of PGPR strains to enhance colonization and effectiveness may involve addition of one or more traits associated with plant growth promotion (Bloemberg and Lugtenberg, 2001). Genetic manipulation of host crops for root-associated traits to enhance establishment and proliferation of beneficial microorganisms (Mansouri *et al.*, 2002) is being pursued. However, regulatory issues and public acceptance of genetically engineered organisms may delay their commercialization. Researchers are now directing their studies toward the discovery and engineering of novel enzymes that are more robust with respect to their kinetics and the identification of active site residues. In the future, protein engineering will offer the possibility of generating chitinases with entirely new functions.

Prior to registration and commercialization of PGPR products, a number of hurdles must be overcome (Mc Spadden Gardner and Fraver, 2002). These include scale up and production of the organism under commercial fermentation conditions while maintaining quality, stability and efficacy of the product. Formulation development must consider factors such as shelf life, compatibility with current application practices, cost and ease of application. Health and safety testing may be required to address such issues as non-target effects on other organisms including toxigenicity, allergenicity and pathogenicity, persistence in the environment and potential for horizontal gene transfer. Capitalization costs and potential markets must be considered in the decision to commercialize. However, the future marketing of *Pseudomonas* inoculant products as environmental-friendly alternatives to chemical fungicides will depend on the generation of biosafety data required for the registration of biocontrol agents.

## Acknowledgements

Thanks are due to Department of Science and Technology, New Delhi and Prof. S.S. Katiyar Vice Chancellor, CSJM University, Kanpur, India for their support. Authors also acknowledge Dr. Ram Naraian for helping in compiling the study.

## References

Ajit, N.S., Verma, R. & Shanmugam, V. (2006) *Curr. Microbiol.* **52:** 310-316.

Audenaert, K., Pattery, T., Cornelis, P. & Hoffe, M. (2002) *Mol. Plant Microbes Interact.* **15:** 1147-1156.

Austin, M.B. & Noel, A.J.P. (2003) *Nat. Prod. Rep.* **20:** 79-110.

Bais, H.P., Park, S.W., Weir, T.L., Callaway, R.M. & Vivaneo, J.M. (2004) *Trends Plants Sci.* **9:** 26-32.

Bakker, P.A.H.M., Ran, L.X., Pieterse, C.M.J. & van Loon, L.C. (2003) *Can. J. Plant Phytol.* **25:** 5-9.

Bangera, M.G. & Thomshow, L.S. (1999) *J. Bacteriol.* **181:** 3155-3163.

Benhamou, N., Kloepper, J.W. & Tuzun, S. (1998) *Planta* **204:** 153-168.

Benizri, E., Baudoin, E. & Guckert, A. (2001) *Biocontrol Sci. Technol.* **11:** 557-574.

Berg, G., Roskot, N., Steidle, A., Ebert, L., Zock, A. & Smalla, K. (2002) *Appl. Environ. Microbiol.* **68:** 3328-3338.

Bertin, C., Yang, X. & Westen, L.A. (2003) *Plant Soil* **256:** 67-83.

Blumer, C. & Haas, D. (2000) *Arch. Microbiol.* **173:** 170-177.

Bottiglieri, M. & Keel, C. (2006) *Appl. And Environ. Microbiol.* **72:** 418-427.

Bowen, G.D. & Rovira, A.D. (1999) *Adv. Agron.* **66:** 1-102.

Brodhagen, M., Henkels & Loper, J.E. (2004) *Appl. Environ. Microbiol.* **70:** 1758-1766.

Camacho Carjaval, M.M., Lugtenberg, B.J.J. & Bloemberg, G.V. (2002) *Mol. Plant Microbe Interact.* **15:** 662-671.

Camacho Carvajal, M.M. (2001) Molecular characterization of the roles of type 4 pili, NDH-I and *Pyr* R in rhizosphere colonization of *Pseudomonas fluorescens* WCS35. PhD Thesis, Univ. Leiden. The Netherlands.

Cartieaux, F.P., Nussaume, L. & Robaglia, C. (2003) *Plant Cell and Environ.* **2:** 189-199.

Chatterton, S., Sutton, J.C. & Bolarid, G.J. (2004) *Biol. Control* **30:** 360-370.

Chen, C., Bélanger, R.R., Benhamou, N. & Paulitz, T.C. (2000) *Physiol. Mol. Plant Pathol.* **56:** 13-23.

Chernin, L. & Chet, I. (2002) In: *Enzymes in the environment: activity, ecology, and applications,* (Eds. Burns, R.G. & Dick, R.P.), Marcel Dekker, New York, N.Y.

Chin-A-Woeng, T.E.C., Bloemberg, G.V. & Lugtenberg, B.J.J. (2003) *New Phytol.* **157:** 503-523.

Compant, S., Duffy, B., Nowak, J., Clement, C. & Barka, E.A. (2005b) *Appl. Environ. Microbiol.* **71(9):** 4951-4959.

Compant, S., Reiter, B., Sessitsch, A., Nowak, J., Clement, C. & Ait Barka, E. (2005a) *Appl. Environ. Microbiol.* **71:**1685-1693.

Cornelis, P. & Malthijs, S. (2002) *Environ. Microbiol.* **4:** 787-798.

Cuppeis, D.A., Stipanovie, R.D., Stoessl, A. & Stothers, J.B. (1987) *Can. J. Chem.* **65:** 2126-2130.

Degenhardt, J., Gershenzon. J., Baldwin, I.T. & Kessler, A. (2003) *Curr. Opin. Biotechnol.* **14:** 169-176.

Dekkers, L.C., Phoelich, C.C., Vander Fits, L. & Lugtenberg, B.J.J. (1998a) *Proc. Natl. Acad. Sci.* USA **95:** 7051-7056.

Dekkers, L.C., Mulders, J.H., Phoelich, C.C., Chin-A-Woeng, T.F.C., Wijfjes, A.H. & Lugtenberg, B.J.J. (2000) *Mol. Plant Microbe Interact.* **13:** 1177-1183.

Dekkers, L.C., Vander Bij, A.J., Mulders, I.H.M., Phoelich, C.C., Wentwoord, R.A.R., Glandorf, D.C.M., Wijffelman, C,.A. & Lugtenberg, B.J.J. (1998b) *Mol. Plant Microbe Interact.* **11:** 763-771.

Delany, I., Sheehan, M.M., Fenton, A., Bardin, S., Aarons, S. & O' Gara, F. (2000) *Microbiology* **146:** 537-546.

De Souza, J.T., Arnould, C., Deulvot, C., Lemanceau, P., Gianinazzi-Pearson, V. & Raaijmakers, J.M. (2003) *Phytopathology* **93:** 966-975.

De Weert, S., Vermeiren, H.,Mulders, I.H.M., Kuiper, I., Hendricky, N. Bloemberg, G.V., Van derleyden, J., De Mot, R. & Lugtenberg, B.J.J. (2002) *Mol. Plant Microbe Interact.* **15:** 1173-1180.

Dobbelare, S., van derleyden, J. & Okon, Y. (2003) *Crit. Rev. Plant Sci.* **22:** 107-149.

Dong, Y.H., Zhang, X.F., Xu, J.L. & Zhang, L.H. (2004) *Appl. Environ. Microbiol.* **70:** 954-960.

Duffy, B.K. (2001) In: *Encyclopedia of plant pathology*, (Eds. Maloy, O.C. & Murray, T.D.), John Wiley & Sons, Inc., New York, N. Y.

Duffy, B.K. & Défago, G. (1999) *Appl. Environ. Microbiol.* **65:** 2429-2438.

Dunlap, C., Delaney, I., Fenton, A., Lohrke, S., Moënne-Loccoz, Y. & O' Gara, F. (1996) In: *Biology of Plant microbe interactions*, (Eds. Stacey, G., Mullin, B., Gresshoff, P.M.), St Paul, MN, USA: International Society for Plant-Microbe Interactions, pp. 441-448.

Elasri, M., delorme, S., Lemanceau, P., Stewart, G., Lava, B., Glickmann, E., Oger, P.M. & Dessaux, Y. (2001) *Appl. Environ. Microbiol.* **67:** 1198-1209.

Gerhardson, B. (2002) *Trends Biotechnol.* **20:** 338-343.

Glandorf, D.C.M. (1992) Ph. D. thesis. Univ. Utrecht, The Netherlands.

Glick, B. (1995) **41:** 109-117.

Glick, B.R. Patten, C.L., Holguin, G. & Penrose, D.M. (1999) In: Biochemical and genetic mechanisms used by plant growth promoting bacteria. Imperical college Press. London, United Kingdom, pp. 267.

Gray, E.J. & Smith, D.L. (2005) *Soil Biol. Biochem.* **37:** 395-412.

Gu, Y.H. & Mazzala, M. (2003) *Appl. Soil Ecol.* **24:** 57-72.

Haas, D. & Défago, G. (2005) In: Biological control of soil-borne pathogens by fluorescent pseudomonads. *Nature Rev. Microbiol. / AOP, doi: 10. 10. 1038/nrmicrol 129.*

Haas, D. & Keel, C. (2003) *Annu. Rev. Phytopathol.* **41:** 117-153.

Hahn, H.P. (1997) *Gene* **192:** 99-108.

Hase, S., van Pelt, J.A., van Loon, L.C. & Pieterse, C.M.J. (2003) *Physiol. Mol. Plant Pathol.* **62:** 219-226.

Hernandez, M.E., Kappler, A. & Newman, D.K. (2004) *Appl. Environ. Microbiol.* **70:** 921-928.

Hoffland, E., Pieterse, C., Bik, L. & van Pelt, J.A. (1995) *Physiol. Mol. Plant Pathol.* **46:** 309-320.

Holt, J. H., Kreig, N.R., Sneath, J.J. & Williams, S.T. (1994) In: *Bergey's Manual of Determinative Bacteriology*, 9th edition, Williams and Wilkins Publication, Baltimore, Maryland, USA.

Iavicoli, A., Boutet, E., Buchala, A. & Métraux, I.P. (2003) *Mol. Plant Microbe Interact.* **16:** 851-858.

Jeun, Y.C., Park, K.S., Kim. C.H., Fowler, W.D. & Kloepper, J.W. (2004) *Biol. Control.* **29:** 34-42.

Juhas, M., Eberl, L. & Tümmler, B. (2005) *Environ. Microbiol.* **7(4):** 459-471.

Kloepper, J.W., Loeng, J., Teintze, M. & Schroth, M.N. (1980) *Curr. Microbiol.* **4:** 317-320.

Kloepper, J.W. & Schroth, M.N. (1978) In: *Station de pathologie vegetable et phyto-bacteriologie*, (Eds. Gilbert- Clarey, Tours), Proceeding of the 4th International Conference on Plant Pathogenic Bacteria. Tours, France, Vol. II, pp. 879-882.

Knee, E.M., Gong, F.C., Gao, M.,Teplitski, M., Jones, A.R., Foxworthy, A., Mort, A.J. & Bauer, W.D. (2001) *Mol. Plant- Microbe Interact.* **14:** 775-784.

Koch, B., Nielsen, T.H., Sørensen, D., Andersen, J.B., Christophersen, C., Molin, S., Givskov, M., Sørensen, J. & Nybroe, O. (2002) *Appl. Environ. Microbiol.* **68:** 4509-4516.

Kuiper, I., Bloemberg, G.V., Noreen, S., Thomas-Oates, J.E. & Lugtenberg, B.J.J. (2001) *Mol. Plant Microbe Interact.* **14:** 1096-1104.

Kumar, N.R., Arasu, V.T. & Gunasekaran, P. (2005) *Indian J. Experimental Biology* **43:** 84-89.

Kuske, C.R., Ticknor, L.O., Miller, M.E., Dunbar, J.M., Davis, J.A., Barns, S.M. & Belnap, J. (2002) *Appl. Environ. Microbiol.* **68:** 1854-1863.

Lavicoli, A., Boutet, E., Buchala, A. & Metraux, J.P. (2003) *Mol. Plant Microbe Interact.* **16:** 851-858.

Lim, H.S., Kim, Y.S. & Kim, S.D. (1991) *Appl. Environ. Microbiol.* **57:** 510-516.

Liu, X., Zhao, H. & Chen, S. (2006) *Curr. Microbiol.*, Published online 10.10.07/s 00284-005-0162-3.

Lugtenberg, B.J.J., Dekkers, L.C. & Bloemberg (2001) *Annu. Rev. Phytopathol.* **39:** 461-490.

Lugtenberg, B.J.J. & Dekkers, L.C. (1999) *Environ. Microbiol.* **1:** 439-446.

Lugtenberg, B.J.J., Kravchenko, L.V. & Simons, M. (1999) *Environ. Microbiol.* **1:** 439-446.

Mahato, S.B., Ganguly, A.N. & Sahu, N.P. (1982) *Phytochemistry* **21:** 959-978.

Mansouri, H., Petit, A., Oger, P. & Dessaux, Y. (2002) *Appl. Environ. Microbiol.* **68:** 2562-2566.

Mavrodi, B.B.M., Gardener, D.V., Mavrodi, R.F. & Bonsall, D.M. (2001) *Phytopathology* **91:** 35-43.

Mathesius, U., Mulders, S., Gao, M.S., Teplitski, M., Caetanoanolles, G., Rolfe, B.G. & Baver, W.D. (2003) *Proc. Natl. Acad. Sci.* **100:** 1444-1449.

Mazzola, M. & Gu, Y.H. (2000) *Phytopathology* **90:** 114: 119.

Meharg, A.A. & Killham, K. (1995) *Plant soil* **170:** 345-349.

Meyer, J.M. & Abdallah, M.A. (1978) *J. Gen. Microbiol.* **107:** 319-328.

Mc Spadden Gardner, B.B. & Fravel, D.R. (2002) *Plant Health Progress* doi: 10.1094/PHP-2002-0510-01-RV.

Milner, J.L., Silo-Sun, L., Lee, J.C., He, H., Clardy, J. & Handelsman, J. (1996) *Appl. Environ. Microbiol.* **62:** 3061-3065.

Mimaki, Y., Satiu, T., Kuroda, M., Saahida, Y. & Hatakeyama, Y. (1999) *Phytochemistry* **51:** 567-573.

Mishra, R.P.N., Singh, R.K., Jaiswal, H.K., Kumar, V. & Maurya, S. (2006) *Curr. Microbiol.* **52:** 383-389.

Molina, L., Constantinescu, F., Reimmann, C., Duffy, B. & Défago, G. (2003) *FEMS Microbiol. Ecol.* **45:** 71-81.

Newton, J.A. & Fray, R.G. (2004) *Cell Microbiol.* **6:** 213-224.

Nielsen, T.H. & Sørensen, J. (2003) *Appl. Environ. Microbiol.* **69:** 861-868.

Nielsen, T.H., Sørensen, D., Tobiasen, C., Andersen, J.B., Christeophersen, C., Givskov, M. & Sørensen, J. (2002) *Appl. Environ. Microbiol.* **68:** 3416-3423.

Nielsen, T.H., Thrane, C., Christophersen, C., Anthoni, U. & Sørensen. J. (2000) *J. Appl. Microbiol.* **89:** 992-1001.

Okubara, P.A., Kornoel, J.P. & Landa, B.B. (2004) *Biocontrol.* **30:** 392-403.

Ovadis, M., Liu, X., Graviel, S., Ismailov, Z., Chet, J. & Chernin, L. (2004) *J. Bacteriol.* **186:** 4986-4993.

Ownley, B.H., Duffy, B.K. & Weller, D.M., (2003) *Appl. Environ. Microbiol.* **69:** 3333-3343.

Park, H.J., Lee, J.Y., Moon, S.S. & Hwang, B.K. (2003) *Phytochemistry.* **64:** 997-1001.

Parrel, A.H.A. & De Mot, R. (2002) *Trends Microbiol.* **10.** 107-112.

Peberdy, J.F. (1990) In: *Biochemistry of cell walls and membranes in fungi*, (Eds. P.J., Kuhn, A.P.J., Trinchi, M.J., Jung, M.W. Goosey, & L.G., Copping), Berlin: Springer-Verlag, pp. 5-30.

Petterson, M. & Bååth, E. (2004) *Soil Biol. Biochem.* **36:** 1383-1388.

Phoebe, C.H., Cambia, J., Aldert, F.G., Van Tran, K., Cabrera, J., Correira, H.J., Guo, Y., Lindermuth, J. *et al.* (2001) *J. Antibiol.* **54:** 56-65.

Pieterse, C.M.J. & van Loon. L.C. (1999) *Trends Plant Sci.* **4:** 52-58.

Ping, L. & Boland, W. (2004) *Trends Plant Sci.* **4:** 263-269.

Rainey, P.B. (1999) *Environ. Microbiol.* **1:** 243-257.

Ramamoorthy, V., Viswanathan, R., Raguchander, T., Prakasam, V. & Smaiyappan, R. (2001) *Crop Prot.* **20:** 1-11.

Ravel, J. & Cornelis, P. (2003) *Trends Microbiol.* **11:** 195-200.

Rhodes, D.J. & Powell, K.A. (1994) *BCPC Monograph* 57: 303-310.

Riley, M.A. & Wertz, J.E. (2002) *Annu. Rev. Microbiol.* **56:** 117-137.

Rodriguez, H. & Fraga, R. (1999) *Biotechnol. Adv.* **17:** 319-339.

Roesti, D., Gaur, R., Lohri, B.N., Imfeld, G., Sharma, S., Kawaljeet, K. & Aragho, M. (2005) *Soil Biol. Biochem.* **20:** 1-10.

Rovira, A.D. (1965) *Annu. Rev. Microbiol.* **19:** 241-266.

Sanchez- Contreras, M., Martin, M., Villacieros, M., O' Gara, F., Bonilla, J. & Rivilla, R. (2002) *Appl. Environ. Microbiol.* **184:** 1587-1596.

Scher, F.M. & Baker. R. (1982) *Phytopathology.* **72:** 1567-1573.

Schnider-Keel, U., Seematter, H., Maurhofer, M., Blumer, C., Duffy, B., Gigot-Bonnefoj, C., Reimmann, C., Notz, R., Défago, G., Haas, D. & Keel, C. (2000) *J. Bacteriol.* **182:** 1215-1225.

Schouten, A., van der Berg, G., Edel-Hermann, V., Steinberg, C., Gautheron, N., Alabouvette, de Vos, C.H., Lemanceau, P. & Raaijmakers, J.M. (2004) *Mol. Plant Microbe Interact.* **17:** 1201-1211.

Schroth, M.N. & Hancock, J.G. (1982) *Science* **216:** 1376-1381.

Siddiqui, I.A. & Shaukat, J.S. (2003) *Soil Biol. Biochem.* **35:** 1359-1368.

Siddiqui, I.A. & Shaukat, S.S. (2005) *J. Appl. Microbiol.* **98(1):** 43-55.

Simons, M., Permentier, H.P., de Weger, L.A., Wijffelman,C.A. & Lugtenberg, B.J.J. (1997) *Mol. Plant Microbe Interact.* **10:** 102-106.

Sindhu, S.S. & Dadarwal, K.R. (2001) *Microbiol. Res.* **156:** 353-358.

Sørensen, J., Jensen, L.E. & Nybroe, O. (2001) *Plant soil* **232:** 97-108.

Steenhoudt, O. & Vanderleyden, J. (2000) *FEMS Microbiol. Rev.* **24:** 487-506.

Steidle, A., Sigl, K., Schuhegger, R., Ihring, A., Schmid, M., Gartner, S., Stoffels, M., Riedel, K., Givskov, M., Hartmann, A., Lagebartels, C. & Eberl, L. (2001) *Appl. Environ. Microbiol.* **67:** 5761-5770.

Stutz, E.W., Défago, G. & Kern, H. (1986) *Phytopathology* **76:** 181-185.

Stutz, E.W., Kahr, G. & Défago, G. (1989) *Soil Biol. Biochem.* **21:** 361-366.

Tepletski, M., Robinson, J.B. & Baver, W.D. (2000) *Mol. Plant Microbe Interact.* **13:** 637-648.

Thomashow, L.S. (1996) *Curr. Opin. Biotechnol.* **7:** 343-347.

Tombolini, R., Enge, A., Darey, M.E., de Bruijn, F.J. & Jansson, J.K. (1997) *FEMS Microbiol. Ecol.* **22:** 17-28.

Toyoda, H., Hashimoto, H., Utsumi, R., Kobayashi, H. & Ouchi, S. (1988) *Phytopathology.* **78:** 1307-1371.

Tripathi, R.K. & Gottieb, D. (1969) *J. Bacteriol.* **100:** 310-318.

Validov, S., Mavrodi, O., Fuente, L.D.L., Boronin, A., Weller, D., Thomashow, L. & Mavrodi, D. (2005) *FEMS Microbiol. Lett.* **224:** 249-256.

Van der Brock, D., Chin-A-Woeng, T.F.C., Eijkemans, K., Mulders, I.H.M., Bloemberg, G.V. & Lugtenberg, B.J.J. (2003) *Mol. Plant-Microbe Interact.* **16:** 1003-1012.

Van Loon, L.C., Bakker, P.A. H.M. & Pieterse, C.M.J. (1998) *Annu. Rev. Phytopathol.* **36:** 453-483.

Van Loon, L.C. & Bakker, P.A.H.M. (2003) In: *Root Ecology*, (Eds. de Kroon, H., & Visser, W.J.W.), Springer-Verlag Berlin, pp. 297-330.

Vessey, K.J. (2003) *Plant soil* **255:** 571-586.

Visca, P. (2004) In: *Pseudomonas* (Ed. Ramos, J.L.), Vol.2, Kluwer. Academic/Plenum Publishers, New York.

Viswanathan, R. & Samiyappan, R. (2000) *Microbiol. Res.* **155:** 1-6.

Walsh, U.F., Morrissey, J.P. & O'Gara, F. (2001) *Curr. Opin. Biotechnol.* **12:** 289-295.

Weller, D.M. (1988) *Annu. Rev. Phytopathol.* **26:** 379-407.

Weller, D.M. & Cook, R.J. (1986) *Can. J. Plant Pathol.* **8:** 328-334.

Weller, D.M., Raaijmakers, J.M., Gardner, B.B.M. & Thornashow, L.S. (2002) *Annu. Rev. Phytopathol.* **40:** 309-348.

Whipps, J.M. (1997) *Adv. Bot. Res.* **26:** 1-133.

Wieland, G., Neumann, R. & Backhaus, H. (2001) *Appl. Environ. Microbial.* **67:** 5849-5854.

Wood, D.W., Gong, F., Daykin, M.M., Williams, P. & Pierson III, L.S. (1997) *J. Bacterial.* **179:** 7663-7670.

Zehnder, G., Kloepper, J., Tuzun, S., Yao, C.B., Wei, G., Chambliss, O. & Shelby, R. (1997) *Entomol. Exp. Appl.* **83:** 81-85.

# 7

# Biological Control: Alternative Paradigms for Commercialization

**Swati[2] and Alok Adholeya[1*]**

*[1]Centre for Mycorrhizal Research, Bioresources and Bioresources Management, The Energy and Resources Institute, Darbari Seth Block, Habitat Place, Lodhi Road, New Delhi-110003, India*

*[2]TERI School of Advanced Studies, Bioresources and Biotechnology, The Energy and Resources Institute, Darbari Seth Block, Habitat Place, Lodhi Road, New Delhi-110003, India*

## ABSTRACT

The dramatic worldwide increase in agricultural and industrial productivity has created severe environmental problems. Soil and groundwater reservoirs have been polluted with pesticides, xenobiotics and heavy metals. Biological agents such as bacteria with beneficial properties (as plant growth promoting strains) could in some cases substitute for pesticides, or serve to remediate contaminated environments like soil and sediment. In the near future, improved management of insect pests will rely largely on novel biologically based agents. Recent advances in this field of study include natural products from plants or microbes, as well as living pathogens and transgenic crops.

## INTRODUCTION

Agriculture in developed countries undergoes continuous change often based on new introductions, some based on customer preferences and some based on ethical considerations. This is particularly true of crop protection. The global consensus to reduce inputs of chemical pesticide, which are perceived as being hazardous by some consumers, has provided opportunities for the development of novel, benign, sustainable crop protection strategies. A great many chemical pesticides have been or are being phased out (e.g., organochlorine insecticides, methyl bromide) either because of potential human health risks, environmental pollution, effects on non-target organisms or the development of pest resistance.

The identification and the biological and molecular characterization of microorganisms are useful as biocontrol agents or as producers of bioactive compounds, is of great interest for the modern and echo-compatible agriculture. In this respect the role of these secondary metabolites could play in the induction of disease symptoms (phytotoxins, entomotoxins, antibiotics and phytohormones) or of defence response (elicitors).

Genes involved in the pathogenicity of agriculturally important microorganisms are also examined for a better comprehension of the molecular and physiological mechanisms of plant-pathogen interaction. Finally, researches on genes and environmental factors involved in the production of growth compounds (auxins, cytokinins) and biopolymers (exopolysaccharides and lypopolysaccharides) from *Pseudomonas* are carried out with the aim of identifying hyper-producer strains of these metabolites for possible biotechnological purpose.

In this context, isolation and chemical/biological characterization of new bioactive compounds produced by fungi and bacteria, such as visoltricin (with potential therapeutical use for its miotic and anticholinesterasic activity), chlamydosporol, fusapyrones, fusaproliferin, cytochalasyns, trichothecenes, enniatins, *Alternaria* toxins, phomenonea, have been carried out.

## USE OF BIOLOGICAL CONTROL AGENTS IN AGRICULTURE

Many microorganisms and insects attack several crops, thereby causing significant damage. These include bacteria, fungi and several insects of different species. Fungi belonging to about 50 genera cause most plant diseases. For this reason, fungicides are the main means to control fungal pathogens on plants (Eckert and Ogawa, 1988). Current legislation in many countries restrict their use in order to protect public safety and the environment (Norman 1988; Wisniewski and Wilson, 1992). In recent years, the need to develop disease control measures as alternatives to chemicals has become a priority for scientists worldwide.

Biological control, especially using microbial antagonists against different plant pathogens, has gained considerable attention and appears to be promising as a viable supplement or alternative to chemical control (Woodhead *et al.*, 1990). Biological control agent is any agent that is non-chemical and natural and acts to control the incidence of disease or pest. Biological control agents could be antagonists. These agents help to conduct studies on developing artificial diets for host insects and natural enemies, behavioural studies on natural enemies and insect viruses, identifying and developing fungal and bacterial antagonists.

All over the world several researchers are trying to develop effective biocontrol agents for better crop production. The development of microbial antagonists effective against fungal pathogens of pome (Janisiewicz 1987; Janisiewicz and Marchi 1992; Usall *et al.* 2001; Nunes *et al.* 2001), stone (Pusey and Wilson 1984), citrus (Chalutz *et al.* 1988; Teixido′ *et al.* 2001) and various subtropical fruits (Koomen and Jeffries 1993; Korsten *et al.* 1993) has been achieved.

In order to establish a real alternative to the use of chemical pesticides and fungicides, it is necessary to develop new products from biological control agents. A number of commercial products available in the USA, Europe, India and many other parts of the world markets includes

Decco I-182 (*Candida oleophila* strain 182, Ecogen Inc., Langhorne, PA, USA), Bio-Save 100 and 110 (*Pseudomonas syringae* strains ESC10 and ESC11, EcoScience Corp., Orlando, FC, USA) and Yield Plus (Anchor Bio-Technologies, Cape Town, South Africa). Brief details regarding available fungal biocontrol agents are provided in Tables 1, 2 and 3.

## MICROBIAL BIOCONTROL AGENTS

### Bacterial Biocontrol Agents

Several researchers have been continuously working on this area for many years and developed numerous fungal, bacterial biological control agents for disease suppression and for better crop-protection. Several bacterial species and sub-species have been developed as biopesticides, primarily for insect and plant disease control.

#### *Bacillus thuringiensis (Bt)*

Most prominent among the bacterial biocontrol agents are insecticides based on various sub-species of *Bacillus thuringiensis* Berliner (*Bt*). These include *B. thuringiensis kurstaki*, with activity against lepidopteran larval species; *B. thuringiensis aizawai*, also with activity against lepidopteran larvae; *B. thuringiensis israelensis*, with activity against mosquito larvae, black fly (simuliid) and fungus gnats; *B. thuringiensis tenebrionis*, with activity against coleopteran adults and larvae, most notably, the Colorado potato beetle; *Leptinotarsa decemlineata*; and *B. thuringiensis* japonensis strain *buibui*, with activity against soil-inhabiting beetles.

Among the sporeforming bacilli, there are major groups (*Bacillus thuringiensis and B. sphaericus*), which produce protoxins during sporulation. These proteins are deposited as parasporal inclusions and, in some cases, are also found on the surface of the spore. When the inclusion (or spore) is ingested by susceptible larvae (i.e., Lepidoptera and Diptera), the protoxin is solubilised in the alkaline, reducing environment of the midgut. The midguts of these larvae also contain proteases necessary to convert the protoxins to toxins and perhaps receptors on the surface of midgut epithelial cells to which the toxins (or protoxins) bind to initiate their action.

*B. thuringiensis* is a widely occurring Gram-positive, spore-forming soil bacterium that produces parasporal, proteinaceous, crystal inclusion-bodies during sporulation. The insecticidal properties of *Bt* are largely a function of the presence of extra-chromosomal plasmids in the cell. These carry genes, termed *cry* genes, which encode a diverse array of these protein crystalline inclusion bodies that are toxic to insects. Upon ingestion by an insect, the crystal proteins are solubilised and the insect gut proteases convert the original pro-toxin into a combination of up to four smaller toxins. These hydrolyzed toxins bind to the insect's midgut cells at high-affinity specific receptor binding sites where they interfere with the potassium ion dependent, active amino acid symport mechanism, which leads to formation of large cation selective pores that increase the water permeability of the cell membrane. A large uptake of water causes cell swelling and eventual rupture, disintegrating the midgut lining. Different toxins have different spectra of activity and different strains and serotypes have been developed by different companies. Many *Bt* genes are themselves potent and many workers transformed them into many crops, and expressed *Bt* gene crops (GM crops) have been introduced by many companies. So far, over 25 genes, their products, insect host range, and crystal size have been reported (Carlton, 1993).

The *cry* genes and cry proteins are grouped into five groups that differ in insecticidal activity or crystal protein size and sequence 74. *cry I* and *cry II* genes produce proteins that are toxic to lepidopterous larvae, *cry III* genes produce proteins toxic to coleopterans, *cry IV* genes produce proteins toxic to dipteran larvae (mosquitoes, black flies) and *cry V* genes produce proteins toxic to both Lepidoptera and Coleoptera.

The worldwide market for *Bt* products is still very small, estimated at US$ 100 million, approximately 1.5% of the total insecticide market worldwide (Woodburn, 1998). *Bt* nowadays has getting more commercial success as the *Bt* genes have been successfully transformed into different crops by many companies. For example, Monsanto has used *cryIA(c)* genes from *B. thuringiensis* subsp *kurstaki* in cotton and tomatoes and *cryIIIA* genes from *B. thuringiensis* subsp *tenebrionis* in potatoes and Novartis and Mycogen used *cryIA (b)* genes from *B. thuringiensis* subsp *kurstaki* and Monsanto *cryIIA* and *cryIA (b)* genes, both from *B. thuringiensis* subsp *kurstaki* in corn. AgrEvo (Plant Genetic Systems) is also using *cry9C* toxin genes from *B. thuringiensis* subsp *tolworthi* in transgenic corn.

## *Bt* commercial formulations

Currently, 26 Bt-based products are registered by the US-EPA and marketed worldwide and also in many countries, specially in Europe and other parts of the world, exists different commercial products. Of these, 15 are derived from naturally occurring strains, e.g., Able (Thermo Trilogy), Bactospeine (Valent BioSciences), Dipel (Valent BioSciences), Javelin (Thermo Trilogy), Thuricide (Thermo Trilogy) and Xentari (Valent BioSciences). Three products are derived from plasmid conjugation, resulting in *cry* gene exchanges producing new toxin arrays. Examples of these products pioneered by Ecogen are Condor (Ecogen), Design (Thermo Trilogy), Foil (Ecogen), and Cutlass (Ecogen). Few more commercial products by plasmid conjugation result in *cry* gene exchanges producing new toxin arrays (Copping, 1993; Baum *et al.*, 1998). These products manufactured by Ecogen are Condor (Ecogen), Design (Thermo Trilogy), Foil (Ecogen), and Cutlass (Ecogen).

## *Bacillus sphaericus*

*Bacillus sphaericus* Neide, a closely related species of *Bt* that occurs widely in nature, has been used to control mosquito larvae, notably *Culex spp*, primarily in Europe and Africa. *B. sphaericus* is similar in activity to *B. thuringiensis* subs *israelensis* (*Bti*), which is used commercially for control of mosquito larvae and black fly (*Simulium sp*), primarily in Africa where black flies are the vector of river blindness disease. There are few registered products of *B. sphaericus*, including Vectolex CG (Valent BioSciences). Several bacterial species, including *Bacillus subtilis* (Ehrenberg) Cohn, *Agrobacterium radiobacter* (Beijerink and Van Delden) Conn, *Burkholderia cepacia* (Yabuchi *et al.)* ex Burkholder, *Pseudomonas fluorescens* (Trevisan) Migula, *P. syringae* Van Hall and *Streptomyces griseoviridis* (Anderson *et al.)* have been found highly efficacious in controlling soil-borne seedling diseases, and fruit and vegetable pathogens.

*Bacillus subtilis* has achieved a modicum of commercial success in North America and Europe as an arrestant or antagonist of soil pathogens. Kodiak and Quantum 4000 by Gustafson and as System 3 by Uniroyal are applied to many field crops, primarily cotton. *B. subtilis* also controls brown rot of peaches. It produces an antibiotic, iturin that inhibits spore germination (Gueldner *et al.*, 1988 and Pusey, 1996). *Agrobacterium radiobacter* strain K84 registered by the US-EPA in 1979 was already utilised on a commercial scale in 1973 to control gall infections caused by *Agrobacterium tumefaciens* Conn. Strain K1026 was developed in Australia by Bio-Care Technology. Deny (CCT) and Intercept (Soil Technologies) a trade name for *Burkholderia cepacia*, Wisconsin strain (formally *Pseudomonas cepacia)* is used as seed treatment or as a powder treatment for transplanted seedlings in the control of a wide range of soil-borne fungal and nematode pathogens and is being sold under the trade names Deny (CCT) and Intercept (Soil Technologies). Nissan's *Erwinia carotovora* Holl, under the trade name Bio-Keeper, to prevent the invasion of pathogenic strains of the bacterium from becoming established on Chinese cabbage is one more example. *Pseudomonas gladioli* pv *gladioli* is the only one bacterial BCA commercialized and sold by Japan Tobacco as AM 301 and Microgen as Camperico for the control of annual meadow grass (*Poa annua* L) in turf.

### *Pseudomonas* spp.

Many active biocontrol microbes are bacteria of the genera *Pseudomonas* and *Burkholderia* (formerly included in *Pseudomonas*). This group of bacteria is known to produce a diverse array of antifungal compounds, and in many cases the production of these compounds has been directly correlated with biocontrol activity. Among the organisms considered to have the greatest potential as commercially viable biocontrol products are also the bacterial strains of the soil-borne fluorescent *pseudomonad* group (Dowling and O'Gara, 1994).

## Fungal Biocontrol Agents

A few fungal agents have been registered in the USA, Europe, India and many parts of the world for control of plant pathogenic fungi (Table 1, 2 and 3). *Ampelomyces quisqualis* Ces (AQ10) was developed by Ecogen for control of powdery mildews in vegetable and fruit crops, but primarily in viticulture. It is available in water-dispersible granule under the trade name AQ10. The product works very effectively in a spray programme to control a wide range of powdery mildews, although high humidity is essential for spore germination of the hyperparasite and for powdery mildew hyphal invasion, and this can be enhanced with the addition of mineral oil adjuvants.

### *Trichoderma* spp.

One of the most successful special biofungicides is *Trichoderma* spp., which are present in substantial numbers in nearly all agricultural soils and in other environments such as decaying wood. Among their other activities, they grow tropically towards hyphae of other fungi, coil about them in a lectin-mediated reaction, and degrade cell walls of the target fungi. This process (mycoparasitism) limits the growth and activity of plant pathogenic fungi. In addition to, or sometimes in conjunction with mycoparasitism, individual strains may produce antibiotics. However, numbers and the physiological attributes of wild strains are not sufficient for highly effective control of plant diseases. The antifungal abilities of these beneficial microbes have been known since the 1930s, and there have been extensive efforts to use them for plant disease control since then. However, they are only now beginning to be used commercially.

Of all the fungal biocontrol agents, *Trichoderma* spp., particularly strain T22, are extensively studied organisms. These are 5highly diverse and ecologically successful fungi. Members of the

**Table 1** Fungicides-bactericides

| *Name Fungicide Bactericide* | *Organism* | *Trade name and available formulation* | *Mode of Action* | *Crop(s)* | *Reference* |
|---|---|---|---|---|---|
| Blasticidin (Blasticidin S) | *Streptomyces griseochromogenes* | Bla-S ((Kaken et al) Inhibition of protein Dustable powder (DP), Emulsifiable concentrate (EC) and Wettable Powder (WP) | Contact fungicide<br>biosynthesis by Binding to the 50S ribosome in prokaryotes<br>Inhibits spore germination and mycelial growth of Pyricularia oryzae | Rice blast disease *et al.,*1958 | Isolated by Takeuchi<br>Fungicidal properties were First described by T Misato *et al.,* 1959 |
| Kasugamycin | *Streptomyces kasugaensis* | Kasugamin and Kasumin (Hokko). Wettable Powder (WP), Dustable Powder (DP), Ultra-low-Volume (UL), Soluble Concentrate (SL) and Granular (GR) | Inhibits protein biosynthesis, interfering with the binding of aminoacyl-tRNA to both the mRNA-30S and the mRNA-70S ribosomal sub-unit complexes, thereby preventing the incorporation of amino acids into proteins. | Used in rice, tree fruit and vegetables for the control of rice blast *(P oryzae)*, leaf spot in sugar beet and celery (*Cercospora spp*), bacterial disease in rice and vegetables, and scab (*Venturia spp*) in apples and pears. | Isolated by Umezawa and first described by H Umezawa *et al.,*1965 |
| Mildiomycin | *Streptoverticillium rimofaciens* strain B-98891, | Mildiomycin<br>Wettable Powder (WP) | Active against the pathogens that cause powdery mildew.<br>Believed to be fungicidal through an inhibition of protein biosynthesis in fungi by blocking peptidyl -transferase (Om *et al.*). | Used for ornamentals for the control of powdery mildews *(Erysiphe sp, Uncinula necator Burr, Podosphaera sp and Sphaerotheca sp*) | Harada and Krishi, 1978 |
| Natamycin | *Streptomyces natalensis S chattanoogensis,* | Delvolan (Gist-Brocades), | Controls fungal diseases, in particular basal rots caused by Fusarium oxysporum Schlecht, in bulbs such as daffodils. | Daffodils | Struyk *et al.,* 1958 |
| Oxytetracycline | | Mycoshield and Terramycin (Novartis) Water-soluble powder | Inhibitor of protein Biosynthesis in bacteria, it binds to the 30S and 50S Bacterial ribosomal subunits | Stone tree fruit, pome fruit and turf grass | Finlay *et al.,* 1950 |

Table 1 Contd.

Table 1 Contd.

| | | | | | |
|---|---|---|---|---|---|
| | | | and inhibits the binding of aminoacyl-tRNA and the termination factors RF1 and RF2 to the A site of the bacterial ribosome. Diseases such as Fireblight (*Erwinia amylovora* Winsl) and diseases caused by *Pseudomonas* and *Xanthomonas* Species and diseases caused by mycoplasma -like organisms. | | |
| Polyoxin | *Streptomyces cacaoi var asoensis* | Polyoxin AL (Kaken, Kumiai, Nihon Nohyaku and Hokko)-Wettable powder (WP), Emulsifiable concentrate (EC) and water-soluble granule (SG) | Exert their effects through an inhibition of cell wall (chitin) biosynthesis. | Pear black spot and apple cork spot (Alternaria spp), grey moulds (*Botrytis cinerea* Pers) and other sclerotia-forming plant pathogens. | Isono *et al.,* 1965 |
| Polyoxorim | | Kakengel (zinc salt) and Polyoxin Z (zinc salt) (Kaken), and as Stopit (Kaken)-Wettable powder (WP) and paste (PA) | | | |
| Streptomycin | *Streptomyces griseus* (Krainsky) | Agrimycin 17 and AS-50 (Novartis), Plantomycin (Aries Agro-Vet Industries) and Paushamycin (Paushak)-Wettable powder (WP) and liquid concentrate | Inhibits protein biosynthesis by binding to the 30S ribosomal subunit and causing a mis-reading of the genetic code in protein biosynthesis. | Tree fruit orchards, vegetables, tobacco, Cotton and ornamentals, diseases caused by *Xanthomonas oryzae* Dows, *X citri* Dows, *Pseudomonas tabaci* Stevens and *P lachrymans* Carsner | A Shatz *et al.,* 1944 |
| Validamycin | *Streptomyces hygroscopicus* (Jensen) | Validacina and Valimun (Takeda), Solacol (Takeda and Aventis), Mycin (Sanonda) and Vivadamy (Vietnam Pesticide). Dispersible powder (DP), soluble concentrate (SL), powder seed treatment (DS) and liquid formulations | Inhibitor of Trehalase-Abnormal branching of the tips of the pathogen followed by a cessation of further development | Excellent control of R solani in rice, potatoes, vegetables, strawberries, tobacco, ginger, Cotton, sugar beet and several other crops. | Waksman & Henrici var *limoneus* by Horii *et al* |

Table 1 Contd.

Table 1 Contd.

| | | | | | |
|---|---|---|---|---|---|
| Bilanafos. | Streptomyces *hygroscopicus* (Jensen) Waksman & Henrici *Streptomyces viridochromeogenes* (Krainsky) | Meiji Herbiace (Sodium salt) (Meiji Seika) Soluble powder (SP) and liquid formulations | Irreversible inhibitor of glutamine synthetase, causing ammonia Accumulation and inhibition of photophosphorylation in photosynthesis. | Vines, apples, brassicas, cucurbits, mulberries, azaleas, rubber and many other crops and on uncultivated land for the control of annual weeds in crop situations and control of annual and perennial weeds in uncultivated land. | Waksman & Henrici |

genera have long been known to act as biocontrol agents of plant pathogens. Recently, these fungi have been used in quite significant amounts in commercial agriculture (e.g., about 30% of the total soil fungicides used in the greenhouse industry in the USA is products based on *T. harzianum* strain T22) (Harman, 2000). There are many successful formulations available in the market in many parts of the world including USA, Europe, Australia and India.

*Fusarium oxysporum* Schlechtendal strain Fo 47 is a naturally occurring mutant strain of the fungus that was found in suppressive soil at Chateaurenard, South-east France by INRA researchers, and was selected for commercialisation as it is not phytopathogenic and competes with pathogenic strains of the fungus. Natural Plant Protection (NPP) is producing commercial formulations, which are used for the control of vascular wilts caused by the fungal pathogens *F. oxysporum* and *F. moniliforme* Sheldon (*Gibberella fujikuroi* Wr) as Fusaclean L for treating rock-wool blocks and Fusaclean G for use on field and glasshouse-grown crops.

*Verticillium lecanii* (Zimmerman) Viegas is sold as two different strains: either as Mycotal for control of whitefly with a side-effect on thrips, or as Vertalec for aphid control. *Beauveria brongniartii* Var Bb 96 (Betel) was isolated from a mycosed white grub (*Hoplochelus marginalis* (Fairmaine) larva found in Madagascar by CIRAD/IRAT, France. The isolated strain was particularly virulent against white grubs and cockchafers (*Melolantha melolantha* (L)). *B. bassiana* entomopathogenic fungi that can be formulated as conidia have the best shelf-life. Formulations of blastospores and mycelia loose viability within a few months at room temperature. A brief description of different microbial biocontrol formulations for crop protection has been given in Tables 2 and 3.

**Table 2** Fungi being developed or commercially available for the biological control of weeds (Butt *et al.*, 1999, 2001; Templeton and Heiny, 1989).

| Product | Commercial Name | Supplier or country where registered | Target Weed |
|---|---|---|---|
| *Alternaria cassiae* | Casst | USA | Sicklepod (*Cassia obtusifolia*) and coffee senna (*C. occidentalis*) in soybeans and peanuts |
| *Cercospora rodmanii* | 'ABG 5003' | Abbott Labs, USA | Water hyacinth (*Eichhornia crassipes*) |
| *Colletotrichum coccodes* | Velgo | USA, Canada | Velvetleaf (*Abutilon theophrasti*) in corn and soybeans |
| *Colletotrichum gloeosporioides* f. sp *cuscutae* | Luboa 2 | PR China | *Cuscuta chinensis, C. australis* in soybeans |
| *Colletotrichum gloeosporioides f. sp. Malvae* | Biomal | Canada | Mallow (*Malva pusilla*) in wheat and lentils |
| *Colletotrichum gloeosporioides f. sp aeschynomene* | Collego | Encore Technologies, USA | Northern jointvetch (*Aeschynomene virginica*) in rice |
| *Chondrosterium purpureum* | BioChon | Koppert, The Netherlands | Black cherry (*Prunus serotina*) in forestry in the Netherlands |
| *Phytophthora palmivora* | Devine | Sumitomo, Valent, USA | Milkweed vine (*Morrenia odorata*) in Florida citrus |

**Table 3** Fungi developed or being developed for the biological control of diseases (Burges, 1998; Butt *et al.*, 1999, 2001).

| *Product* | *Fungus* | *Target* | *Producer* |
|---|---|---|---|
| Mycoparasites | | | |
| Rotstop | *Phlebiopsis (Peniophora) gigantea* | Heterobasidium annosus | Kemira Agro Oy, Finland |
| Primastop | *Gliocladium catenulatum* | Several plant diseases | Agro Oy, Finland |
| SoilGard (= GlioGard) | *Gliocladium virens* | Several plant diseases Damping off & root pathogens | ThermoTrilogy, USA |
| Cotans WG | *Coniothyrium minitans* | *Sclerotinia* species | Prophyta, Germany. KONI, Germany |
| AQ10 Biofungicide | *Ampelomyces quisqualis* | Powdery mildews | Ecogen Inc.USA |
| YIELDPLUS | *Cryptococcus albidus* | *Botrytis spp., Penicillium spp* | Anchor Yeast, S. Africa |
| Aspire | *Candida oleophila* | *Botrytis spp., Penicillium spp.* | Ecogen Inc.USA |
| Fusaclean | *Fusarium oxysporum* | *Fusarium oxysporum* | Natural Plant Protection, France |
| Biofox C | *Fusarium oxysporum* | *Fusarium oxysporium, F. moniliforme* | SIAPA, Italy |
| Polygandron | *Pythium oligandrum* | *Pythium ultimum* | Plant Protection Institute, Slovak |
| Polyversum | | | Republic |
| Trichoderma 2000 | *Trichoderma harzianum* | *Rhizoctonia solani, Sclerotium rolfsii, Pythium* | Mycontrol (EfA1)Ltd, Israel |
| Trichopel | *Trichoderma harzianum* | Wide range of fungal diseases | Agrimm Technologies Ltd, New Zealand |
| T-22 and T-22HB Bio-Trek, RootShield | *Trichoderma harzianum* | *Pythium, Rhizoctonia, Fusarium, Sclerotina* | BioWorks (TGT Inc) Geneva, USA |
| Trichodowels, | *Trichoderma harzianum* | *Trichoderma harzianum* and other | Agrimms Biologicals, New |
| Trichoject, Trichoseal and others | *and T. viride* | soil and foliar pathogens | Zealand ands others |
| Binab T | *Trichoderma harzianum, T. polysporum* | Fungi causing Wilt, wood decay and take all | Bio-Innovation, Sweden |
| Trichodex | *Trichoderma harzianum* | Fungal diseases; e.g. *Botrytis cinerea* | Makhteshim-Agan, Several European companies e.g. DeCeuster, Belgium. |

## *AM Fungi and their role in disease control*

The study and interactions between mycorrhizal fungi and plant pathogens began in the 1970s. Mycorrhizal fungi have inhibitory effect on some fungal pathogens. These also reduce the disease severity in majority of the studies with some reports on their effect as neutral (Baath and Hayman, 1983; 1984). Reduction of fungal disease has been reported such as *Phytophthora*, *Gaeumannomyces*, *Fusarium*, *Chalara* (thielaviopsis), *Phythium*, *Rhizoctonia*, *Sclerotium*, *Verticillium*, *Aphanomyces*, *Pyrenocheta*, *Ganoderma*, *Macrophomina*, *Bipolaris*, *Phoma*,

*Urocystis*, *Microcyclous*, *Olpidium*, *Cylindrocorpon mycelium* and bacterial pathogens like *Erwinia* (Garcia-Garrido and Ocampo, 1989), *Pseudomonas* (Garcia-Garrido and Ocampo, 1989) for nematodes *Pratylenchys*, *Radopholus*, *Heterodera*, *Tylenchorynchus*, *Aphelenchus* and *Meliodogyne*. There are also few reports in which it has been said mycorrhiza will increase disease severity (Jayaram and Kumar, 1995; Nemec and Myhre, 1984).

The management of crop diseases caused by root pathogens has become a challenging research area in plant pathology. Increasing knowledge and concern about the environmental consequences of pesticide application has aroused interest in alternative methods of plant protection. This interaction is performed between the roots of over 80% of terrestrial plant species and *Zygomycetes* fungi in the order Glomales. These fungi are termed AM fungi and are obligate symbionts, which form endomycorrhizal symbioses. The fungus biotrophically colonizes the root cortex and develops an extramatrical mycelium that helps the plant to acquire phosphorus and mineral nutrients from soil (Harley and Smith, 1983). In addition, AM fungi may also enhance the resistance of the plant to biotic and abiotic stresses.

Generally, AM fungi cause few changes in root morphology, but the physiology of the host plant changes significantly. The improved potential for mineral uptake from the soil accounts for changes in the nutritional status of the host tissue, which in turn changes the structural and biochemical aspects of the cells. This can alter the membrane permeability and thus the quality and quantity of root exudation. Altered exudation induces changes in the composition of the microorganisms in the rhizospheric soil. The net effect of these alterations is a tolerant plant, better able to withstand environmental stress and plant diseases (Linderman, 1988). The influence of AM fungi is not direct, but indirect and evaluate the interactions among plant, pathogen and the symbiont. Generally, host mediates the interaction between the pathogen and symbiont. These interactions are complex and it is difficult to generalize the effects of AM fungi on disease (Tanu *et al*., 2003).

## Nematodes as Biological Control Agents of Insects

Nematodes are one of the most abundant groups of living animals, and although morphologically they are very simple, they have exploited a wide range of diverse habitats including invertebrates. Nematodes are usually considered pests because of the diseases they cause in humans and animals and the economic impact they have on many agricultural products. There are, however, a small but significant number of beneficial entomogenous nematodes, i.e., nematodes associated (often parasitically) with insects.

Some of these entomopathogenic (insect-parasitic) nematodes are of considerable interest because of their potential as biological control agents of pest insects. If the nematode attacks an insect pest; kills or hampers the development of the insect host; and is capable of mass production it can be used as an effective biological control agent. Nine families of nematodes (Allantonematidae, Diplogasteridae, Heterorhabditidae, Mermithidae, Neotylenchidae, Rhabditidae, Sphaerulariidae, Steinernematidae, and Tetradonematidae) include species that attack insects and kill or sterilize them, or alter their development. An increasing number of nematode products are entering the marketplace and increased communications among producers and users on quality control procedures are desired to facilitate the production and use of viable and effective nematode products.

## Viral Biocontrol Agents

Insects are attacked by many different types of virus, but baculoviruses show the most promise in insect control because of their specificity to insects. Interest in baculoviruses as bioinsecticides has been rising and waning over the past 50 years. The family *Baculoviridae*, comprising baculoviruses, contains two genera of occluded viruses, nuclear polyhedroviruses (NPVs) and granuloviruses (GVs), and one group of nonoccluded viruses (NOVs).

### *Baculoviruses*

Baculoviruses are rod-shaped DNA viruses, many of which begin their life cycle reproducing inside cells. In the nuclei of caterpillar cells infected with baculoviruses, viral progeny multiply and are incorporated into protective polyhedron-shaped protein structures called occlusion bodies. Infected caterpillars die and contaminate the leaf surfaces with the occlusion bodies. Then, healthy caterpillars ingest the occlusion bodies and release the virus when feeding on contaminated leaves, thus continuing the life cycle of infection and replication.

Baculovirus as a complementary tool for insect management is apparent as the insect resistance for many chemical and biocontrol agents developed. It is more evident from the fiasco of stopping *Bt* endotoxins for the control of Diamondback moth (development of more resistance to *Bt* endotoxins in transgenic crops). The NRC report which was concluded in 2000, arguing that Integrated Pest Management is the better management tool for crop protection (Berenbaum *et al.*, 2000)

Baculoviruses expressing insect-selective neurotoxins are showing great potential as integrated pest management tools for the selective control of major insect pest complexes. Natural baculovirus epidemics nearly wipe out populations of some caterpillar pests, such as corn earworm, cotton bollworm, tobacco budworm, and cabbage's nemesis—the diamondback moth," and many other crop pests. Because of the economics of present biopesticide-versus-chemical control technology in most of the countries including US is still in nascent stage where much research innovation is required. In some countries, like Thailand, subsistence farmers spray suspensions of last year's diseased pest-insect larvae on crops, giving the biological control agent a jump-start. In Brazil, velvetbean caterpillars are reared to produce the baculovirus known as AgMNPV, which is marketed to control the caterpillars on 2 million acres of soybean fields.

## Nuclear Polyhedro Viruses (NPVs)

NPVs are most prevalent in lepidopterous insects, with over 500 isolates presently known. Most NPVs have a narrow host range, but some, such as the NPV of the alfalfa looper *Autographa californica* (Speyer) have a broader host range, attacking many pest species in the genus Noctuidae. Commercial introduction of an NPV to control the cotton bollworm, *Helicoverpa zea* (Boddie) (HzNPV) was originally made by Sandoz (now Novartis) under the name Elcar, in the early 1970s.

The economics of production of NPVs is another reason for their lack of commercial success. At present it is not possible to produce NPVs economically by deep fermentation methods in insect cell cultures and all products are currently produced *in vivo*, being harvested from infested insect cadavers and processed as powders. These concentrated powders can be formulated as liquids, Suspension concentrates (SC) or Wettable powders (WP) but problems of microbial contamination remain. Lots of researches are being done in the development of large-scale tissue culture production techniques and hopefully this will get success.

Because of the economics of present biopesticide-versus-chemical control technology in most of the countries including US, it is still in nascent stage where much research innovation is required. In some countries, like Thailand, subsistence farmers spray suspensions of last year's diseased pest-insect larvae on crops, giving the biological control agent a jump-start. In Brazil, velvetbean caterpillars are reared to produce the baculovirus known as AgMNPV, which is marketed to control the caterpillars on 2 million acres of soybean fields. Several NPV products are registered for control of insect pests and they are listed in the Table 4.

### *Granuloviruses* (GVs)

Two granuloviruses have also been commercialized, the most important of which is that of the codling moth, *Laspeyrsia pomonella* (L) (CpGV). The codling moth is a universal, major pest of pome fruits, plums and walnuts, and the granulovirus has been developed by several organizations in Europe and USA for the control of this important pest species, under the trade names Madex 3 (Andermatt Biocontrol), Granupom (AgrEvo), Carpovirusine (NPP-Calliope), Carposin (Agrichem), Virin-Gyap (NPO Vector) and CYD-X (Thermo Trilogy). The other granulovirus that has been commercialized is *Adoxophyes orana* Fischer GV (AoGV) for the control of the summer fruit tortrix (Capex 2; Andermatt Biocontrol).

Even though viral biopesticides show considerable potential in insect control, many obstacles have yet to be overcome to make them economic and effective in the field. Amongst these, the most important elements requiring improvement are: the ability for large-scale continuous production of NPVs in deep fermentation; increased photostability in the field; increased storage stability; and broadened biological activity spectrum (Inceoglu *et al.,* 2001).

## Plant Derivatives used in Biocontrol Agents

### *Azadirachtin*

Neem grows wild in more than 50 countries worldwide, including India and on much of the African continent. So, many medicinal and agricultural uses have been attributed to neem that it is often called "nature's pharmacy" by those who have studied its uses. The important compound found in neem is azadirachtin composed only of carbon, hydrogen and oxygen, works as a broad spectrum insect growth regulator by disrupting molting during an insect's larval, or juvenile, stage. By altering insect metamorphosis, azadirachtin prevents larvae from developing into adults and producing a subsequent generation of insect pests.

In addition, many insects are strongly repelled by the anti-feedant properties of azadirachtin, so much so that dozens of leaf-eating species will starve to death rather than eat leaves treated with azadirachtin. These combined modes of action are unique among currently available synthetic chemical and microbial insecticides. Proprietary methods have been developed for extracting the insecticidal components from neem oil. The resulting "neem extract" contains the essential natural insecticide azadirachtin. A large number of companies formulate and sell products based on azadirachtin. A brief description with formulations available, names of the manufacturers, crops to which it can be applied and mode of the action for each biocontrol agent has been given in Table 5.

**Table 4** Viral biocontrol agents currently in use for insect pest control in various countries

| *Virus* | *Trade name* | *Target Host(s)* | *Crop(s)* | *Registration status* | *country* |
|---|---|---|---|---|---|
| *Spoptera exigura* NPV | SPOD-X | Beet army worm, *S exigura* (Hubner) | Glasshouses | Yes | USA,The Netherlands, Thailand |
| *S littoralis* NPV | Spodopterin | Egyptian cotton leaf worm, *S littoralis* (Boisd) | Cotton | Yes | France |
| *Anagrapha falcifera* (Kirby) NPV Celery looper | AfNPV | Broad spectrum control of lepidopterous larvae | Many | Yes | USA |
| *Lymantria disper* NPV | Disparvirus | Gypsy math larvae, *L disper* (L) | Forests | Yes | Canada, USA |
| NPV | Mamestrin | *Mamistra brassicae, Helicoverpa, Plutella, Lebisia sp* | Vegetable crops | Yes | France |
| *Helicoverpa Zea* NPV | GemStar | Cotton bollworm, *H Zea* Tabacco budworm, *Heliothis virescens* Fabricius | Corn, Cotton | Yes | USA |
| *Anticarsia gemmatalis* NPV | Polygen Multigen | Velvet bean caterpillar, *Anticarsia gemmantalis* Hubner | Soybean | Yes | Brazil |
| Genetically Engineered NPV | – | Various species of Lepidoptera | Various crops | Yes | USA |
| *Adoxyphyes crana* GV | Capexa | *Adoxyphyes crana* | Various Crops | Yes | Germany |
| AcMNPV | Gusano | Various species of Lepidoptera | Various crops | Yes | USA |
| CpGV | Cyd-X, Madex, Granupom | *Cydia pomanella* | Various crops | Yes | USA, Germany, Spain, France and Switzerland |
| Oryctes NPV | – | Oryctes beetles | Various crops | Not necessary | Maldives, Philippines, Samoa, Tonga, Fiji, Indonesia |

Data compiled from Copping and Menn and Society of Invertebrate Pathology web page (www.sipweb.org) and Inceoglu *et al.*, 2001).

Abbreviations: NPV: Nucleopolyhdrosis Virus; MNPV, Multiple Nucleocapsid NPV; SNPV, Single Nucleocapsid NPV; GV Granulovirus; Af, *Autographa falcifera*, AC-*Autographa californica*; Hz *Heliothis Zea*, Se, *Spodoptera exigua*

**Table 5** Plant derived biocontrol agents

| *Name of Biocontrol agent* | *Plant* | *Trade name and available formulation* | *Mode of action* | *Crop(s)* |
|---|---|---|---|---|
| Azadirachtin. | *Azadirachta indica* | Neemix 90 EC (90g litreÿ1), Neemazid, Trilogy 90 EC (Neem oil for disease control), Triact 90 EC (Neem oil for disease control), Bio-neem, Margosan-O, Azatin, Align, Turplex and Bollwhip (all Thermo Trilogy), Fortune Aza and Fortune Biotech (Fortune), Azatin (Agridyne), Neem Suraksha, Proneem, Neem Wave and Aza Technical (all Karapur Agro), NeemAzal (Trifolio-M), Kayneem (Krishi Rasayan), Neemolin (Rallis), SureFire and Neemachtin (Consep) and Nimbecidine (T Stanes). | Disrupt insect molting by antagonising ecdysone, leading to morphological defects in insects and also reduce the reproductive capabilities of phytophagous insects by disrupting normal mating behaviour and thereby reducing fecundity | Vegetable, fruit and many horticultural crops, cotton, etc. |
| Nicotine. | *Nicotiana rustica* L | Nico Soap (United Phosphorus Ltd), No-Fid (Hortichem), XLAll Nicotine (Vitax) and Nicotine 40% Shreds (Dow AgroSciences). Dispersible powder (DP), Soluble concentrate (SL) and fumigant formulation | Binds to the cholinergic acetylcholine nicotinic receptor in the nerve cells of insects, leading to a continuous ring of the neuroreceptor. | Many crops, specially for sucking insects, and whitefly |
| Plant pelargonic and related fatty acids. | — | M-Pede, Scythe, De-Moss Thinex, Neo-Fat<br><br>Liquid concentrate (SL) | Fatty acid extracts interfere with the cell membrane constituents of the target organism, leading to a breakdown of the integrity of the membrane and death. | Vegetables, fruit and ornamentals; fungicidal uses include disease control in grapes, roses and other crops and herbicide uses include total weed control and moss control in lawns. |
| Pyrethrins | *Chrysanthemum cinerariaefolium* | Alfadex (Novartis), Pyrocide and Evergreen (MGK), Pyronyl (mixture), Excite R and Prentox Pyrethrum Extract (all Prentiss), Milon (Delicia), Pycon (for concentrated mixture with piperonyl butoxide) (Agropharm) and CheckOut (Consep)-Aerosol dispensers (AE), dispersible powders (DP), Emulsifiable concentrates (EC), fogging concentrates, | Pyrethrins have been shown to bind to the sodium channels in insects, prolonging their opening and thereby causing knockdown and death. | Many horticultural crops |

Table 5 Contd.

Table 5 Contd.

| | | | | |
|---|---|---|---|---|
| | | wettable powders (WP) and ultralow-volume liquids (UL) | | |
| Rotenone | *Derris*, *Lonchocarpus* and *Tephrosia* species. | Chem Sect, Cube Root and Rotenone Extract (all Tifa), Noxfire and Rotenone FK-11 (AgrEvo Environmental Health) and Prenfish (mixture), Prentiss, Synpren Fish (mixture) and Prentox (all Prentiss). Dispersible powder (DP), Emulsifiable concentrate (EC) and Wettable powder (WP) | Inhibitor of Site I respiration within the electron transport chain and is used as a selective, non-systemic insecticide with contact and stomach action, with secondary acaricidal activity. | Many crops and organic farming systems. |
| Ryania extracts. | *Ryania speciosa* Vahl | Natur-Gro R-50 and Natur-Gro Triple Plus (AgriSystems International) and Ryan 50 (Dunhill Chemical). Wettable powder (WP) | Affect muscles by binding to the calcium channels in the sarcoplastic reticulum. This causes calcium ion flow into the cells and death follows very rapidly. | Many crops—good for boring insects. |

*Source:* Copping and Menn, 2000

## Other Plant Derived Compounds used in Biocontrol of Pests

Even though there are many compounds derived from plants are identified as pest controlling agents, but none of them reached to the stage of commercialization (Ikawa *et al*, 1945 and Crosby 1971). Quassia, a weak insecticide that is derived from the wood of *Quassia amara* L and related genera such as *Aeschrion excelsa*. Aqueous extracts of Quassia chips were used as an insecticide from the late eighteenth to the mid-twentieth centuries. Ceveratrum alkaloids derived from *Sabadilla* and *Veratrum* species are effective by blocking the sodium channel in insects (Ujvary, 1999) but they also have rather high mammalian toxicity. *Sabodilla* extracts were extensively used in USA and Europe and in California for the control of thrips in citrus (Hare, 1997).

Natural plant derivatives, which are synthesized unsaturated isobutylamides, which are not commercialized, include pipercide from *Piper nigrum* L, affinin from *Heliopsis longipes* Pers and pellitorine from *Anacyclus pyrethrum* Link. The compounds cause rapid knockdown and kill flying insects, but are too unstable for use as products themselves. They act as voltage-dependent blockers of the sodium channel (Elliott and More, 1989). Some naphthoquinones (NQs), have been isolated and characterized from the Chilean plant, *Calceolaria andina* (Scrophulariaceae). Ubiquinone and plastoquinone that are structurally similar to NQs are also known to mimic their action in the respiration and photosynthesis pathways. The NQs which having low mammalian toxicity are BTG 504 and BTG 505.

Plant essential oils has known to mankind as food products is often cited as sufficient evidence of safety and there is no doubt that, as far as ingestion by humans is concerned, are also used for many insect pests appear to be the result of effects on the insect nervous system, either by inhibition of acetylcholinesterase or by antagonism of the octopamine receptors. Often the amounts of plant oil needed to kill insects are quite high compared with synthetic insecticides and this will lead to higher residues in the environment. Although the effects of these products on the environment and humans is not well studied. Nevertheless, these essential oils are used for many crops from several years. Pulegone, which is found in pennyroyal, has a somewhat higher mammalian toxicity than other edible plant oils. Although there are many reports in the literature of the insecticidal effects of essential oils on stored product pests (Dales, 1996) it is unlikely that such aromatic compounds would be acceptable as contact insecticides for application to foodstuffs in the developed world.

Other than crop protection, citronella oil is used as a mosquito repellent and limonene for the control of pet fleas in some countries. A number of products are coming on stream in the near future, including a mixture of essential oils for control of the Varroa mite, cinnamon oil aphicidesa and acaricides, and eugenol-based products for urban insect control. An American company intends to market a range of insecticide products which will be a blend of plant oils, primarily clove and peanut oil, but with thyme, lemon grass, cinnamon and pennyroyal as other components.

## Genetic Engineering in Enhancing Efficacy of Biocontrol Agents and its Role in Commercialization

Genetic modification of biocontrol agents can play a great role in improving the biocontrol potential of antagonists. It has been visualized that the next decade will see wide utilization of genetic engineering and biotechnology in the service of biological control. Biocontrol agent has been manipulated to improve production of antifungal metabolites and antagonistic potential of biocontrol agents to control a broad spectrum of phytopathogens (Rey *et al.,* 2000).

The modification of *Trichoderma viride* was done by a chemical mutagenesis and the resulted mutant sporulate heavily and had shown better biocontrol ability than the parent strain when used against the wilt disease of pigeon-pea caused of *Fusarium udum.* Given the key role of antimicrobial metabolite production in biocontrol, strategies for increasing the biocontrol efficacy of BCAs based on genetic modification for enhanced antimicrobial metabolite production have been investigated. Maurhofer *et al.*, (1995) demonstrated that increased antifungal metabolite 2,4-diacetylphloroglucinol (Phl) and pyoluteorin production by *Pseudomonas fluorescens* CHA0, via introducing the recombinant cosmid bearing a insert of CHA0 DNA, resulted in increased protection of cucumber from *Fusarium oxysporum* f. sp. *cucumerinum* and *Phomopsis sclerotioides*. Significantly, the antimicrobial metabolite overproducing strain, proved to be phytotoxic to both tobacco and sweet corn. A genetic approach was used to further enhance the biocontrol ability of *Pseudomonas fluorescens* F113 by developing two genetically modified (GM) strains for enhanced *Phl* production and assessed for biocontrol efficacy (Delany *et al.,* 2001).

The success of any biocontrol agents depends on their virulence when applied against insect pests. Hypervirulence has been obtained in many organisms by engineering genes into a biocontrol agent that will each incrementally enhance the extent or rapidity of its establishment, lower the level of inoculum needed, or increase its speed of debilitating the host. Enhanced virulence can be gained through the overexpression of known functional genes from the biocontrol agent itself, or from other organisms.

At a molecular genetic level, attempts to increase the biocontrol ability of many organisms have been focussed in order to improve the strain for better traits. For example, a lot of research being done on *Trichoderma* on increasing chitinase or proteinase activity either by increasing the number of copies of the appropriate genes or by fusing them with strong promoters (eg. p*cbh1*::*ech42*). These strategies have not always resulted in the expected results in the desired organism with a better biocontrol control activity (Limon *et al.,* 1999 and Margolles-Clark *et al.,* 1996). The availability of a versatile expression system for *T. atroviride* based on application of recently characterized biocontrol-related promoters (Mach *et al*., 1999 and Zeilinger *et al.,* 1999) suggested that it might be possible to improve the disease control effect of this biocontrol agent.

Pesticide companies have genetically engineered a number of crops to contain the naturally-occurring bacterial insecticide, *Bacillus thuringiensis,* or *Bt*. Genetic engineering on *Bacillus thuringiensiss (Bt)* is noteworthy to mention here. Several of *Bt* strains naturally are good biocontrol agents and also their genes are even more toxic to several pests. Many *Bt* genes have been isolated and used to transform crops thereby rendering them insecticidal. *cry* genes and protein which are identified as the toxic to insect pests. Many more cry proteins and genes will be identified and introduced into new crops by genetic engineering techniques. Growth of *Bt* is reached to stage in the world market that almost $500 million dollars business world wide and is expected to rise steadily. The real growth is mainly depended on the expression of *Bt* toxins in genetically modified crops. There are number of prominent companies which have commercialised many *Bt* transgenic crops as Monsanto, Novartis, Mycogen, AgrEvo (Plant Genetic Systems) are the major players. During the past decade, genetically engineered or modified (GE or GM) crops have been widely planted by farmers in many countries around the world. The agricultural biotechnology industry claims that GM crops will benefit developing countries by reducing pesticide use and increasing yields, promising to make farming more profitable and less environmentally harmful. *Bt* cotton, has been widely planted in India and it's initial reports are favourable and in many other countries these crops had created their mark.

## Inoculum Preparation and Formulation of a Biocontrol Agent

Standard method of inoculum preparation varies between microorganisms to microorganisms and also depends on the type of the crop for which it is being applied. Most of the times adding a concentrated cell suspension to the formulation material with a proper mixing and keeping for some time for incubation with adequate drying at ambient temperatures are the commonly followed procedure. The formulation material can be anything like talcum powder, lignite, vermiculite, wheat gluten, carboxymethyl cellulose, dextrin, PEG are few of the formulation raw materials which normally used for. In order to obtain a commercial biocontrol product, it should be produced in an economical feasible medium. One such good example is the biocontrol agent *Candida sake* CPA-1 which is being produced in a medium from a cheap industrial waste material molasses (Abadias *et al.,* 2000; 2001).

A procedure for the formulation of biocontrol fungi involving the immobilization of wet biomass within gluten was developed. The release of active fungi from formulated preparations to soil was governed by the soil pH, the moisture content in soil, and the entrapped biomass content. The use of formulated preparations reduced the amount of biomass required, compared with unformulated fungi as biocontrol agents (Chen-Fu and Wen-Chien, 1999).

## Constraints for Commercialization of Biocontrol Agents

The major bottleneck to the commercial use of microbial biocontrol agents is their inconsistent performance in the field. The use of inappropriate formulations for strain delivery is sometimes responsible for this lack of consistency (Paau, 1988). Indeed, carefully designed formulations enabled the effective control of some diseases, for example, *Pseudomonas fluorescens* enabled good control of chickpea wilt and resulted in yields similar to, higher than those obtained when using chemical control (Vidhyasekaran and Muthamilan, 1995).

The greatest reason in lack of adoption of biological control is the lack of consistency of biological control performance based on the application of a strain. In addition to filed studies integrating biological control into commercial production systems, a more through understanding of the genetics, biology, and ecology of biocontrol agents, and their mode of action will enable optimal exploitation of these biocontrol agents for disease control.

Temperature plays an important role on the viability of the stored biocontrol agents (Torres *et al.,* 2003), lower temperature always favours and the viability of the stored biocontrol agents will be better than the one which are stored at ambient or higher temperatures. This factor in many cases is a big constraint in commercializing the biocontrol agents since many traders and sellers are not able to maintain the required temperatures for longer duration as is required.

## Indian Agriculture, Present Trends and Future Prospects

India is an agriculturally dependent nation and more than 70% are in the rural agriculture sector. According to Directorate of Economics and Statistics, Deptt. of Agriculture and Cooperation, Ministry of Agriculture, Govt. of India's estimates for total foodgrains in 2001-2002 as 204.069 million tonnes to 224.207 million tonnes for the year 2006-2007 (Table 6).

**Table 6** Demand Projections of Cereals, Pulses and Foodgrains (2001-02 to 2006-07)

| Year | Population (in millions) | Total Consumption in Million Tonne | | |
|---|---|---|---|---|
| | | Cereals | Pulses | Foodgrains |
| 2001-02 | 1046.513 | 187.744 | 16.326 | 204.069 |
| 2002-03 | 1066.397 | 191.311 | 16.636 | 207.947 |
| 2003-04 | 1086.658 | 194.946 | 16.952 | 211.898 |
| 2004-05 | 1107.305 | 198.65 | 17.274 | 215.924 |
| 2005-06 | 1128.344 | 202.424 | 17.602 | 220.026 |
| 2006-07 | 1149.782 | 206.27 | 17.937 | 224.207 |

Source: Bulletin on Food Statistics 1998-2000, Directorate of Economics and Statistics, Dept. of Agriculture and Cooperation, Ministry of Agriculture, Govt. of India.

Data source: www.indiastat.com, accessed on 15. 08. 06

Note: According to the provisional results compiled quickly for the Census of India 2001, the Population of India stood at 1027 millions on 1st March 2001. The population is projected to rise by 1.9 per cent during the next five years.

In India, much attention is not paid to reduce post harvest losses of what is produced. Timely crop management steps and storage conditions will certainly make a better in form produce. Following IPM practices and reducing chemical inputs will lead to a better export demand for cash crops there by the economic status of the Indian former will improve. According to Ministry of Agriculture, Government of India, the total land covered under the IPM programme in 1994-95 to 2001-2002 is 42.63 million hectares against an estimate of 38.50 million hectares. Even though the achievement is more than the target set for the period being a large agriculturally dependent nation this has to be extended to much more land coverage in order to maximize crop production with good quality produce (Table 7).

**Table 7** Total IPM accomplishments in India during 1994-95 to 2001-2002

| (Area in Lakh Ha.; Release in Million) State/UTs | Pest Monitoring | Biocontrol Release | Area Coverage | IPM Training & Demonstration | | |
|---|---|---|---|---|---|---|
| | | | | No. FFSs | AEOs Trained | Farmers Trained |
| Total (Achievement) | 58.89 | 14925.70 | 42.63 | 7257 | 30381 | 219141 |
| Total (Targets) | 54.00 | 14000.00 | 38.50 | 7620 | 37560 | 224960 |

Abbr: IPM- Integrated Pest Management, FFSs- Farmer's Field Schools

AEOs- Agriculture Extension Officers

Source: National Conference of Agriculture for Rabi Campaign 2002-03, Ministry of Agriculture, Govt. of India.

Data source: www.indiastat.com, accessed on 15. 08. 06

## Government Policy Support

Crop failure due to natural calamities and very low prices for agricultural produce due to market gluts are not adequately addressed by the State and Central Governments through comprehensive minimum support price and crop insurance scheme. Hence, for Indian farmers, farm sector is very much insecure. There is no strong institutionalized credit support for farmers.

In India a large number of farmers are with high lend of illiteracy. Farmers have limited access to market information and intelligence, lack of market infrastructure, inadequate storage and transport facilities with weak farm extension services which has lead to limited scope for Indian farmers to compete in the free global market.

Vast agro-climatic diversity, production potential, cheap farm labour availability and domestic and overseas market potential of India (Shaji, 2001) provides greater scope for private sectors' participation. Interestingly, in the recent past, private agri-business firms and Multi National Corporations (MNCs) have also received offers from State Governments such as Gujarat, Uttar Pradesh, Maharashtra, Karnataka and Tamil Nadu for contract farming (Balaji, 1999; Srinivas, 1998). National Agricultural Policy of India also favours private sector participation through contract farming arrangements (Anonymous, 2000). In this existing scenario, Multi National Corporations (MNCs), agri-business firms and processing industries are entering into agricultural sector in a big way for contract farming in India. The advantage of contract farming is that there will an adequate and timely crop protection and pest management practices with a monocrop culturing practices in large acres of lands where there is better scope for application of biocontrol agents and IPM practices.

## Effective Technology Transfer and Crop Management

Contract farming with effective technology transfer and better crop management practices have given more stress on IPM practices which is having considerable reduction of chemicals inputs and encouragement for biocontrol agents application. Case study on tomato contract farming in Punjab reveals that, it provides new and better farming skills and better soil management practices. Continuous extension work ensured new varieties which have been adopted soon and grew successfully (Sukhpal Singh, 2000). Pepsi Ltd. extension staffs frequently inspect and contract farmers and farmers are educated at training camps with a variety of teaching methods (Sukhpal Singh, 2000). The case study on gherkin contract farmers (Hindustan Lever Limited (Kissan)) in Karnataka state, gherkin contracts in Tamil Nadu and tomato contracts in Punjab state shows that, technical guidance regarding plant protection and harvesting is given by field officers (Subramanyam, 2000; Ramesh, 1999a; Sukhpal Singh, 2000a) which had given better output to the farmers and there by increasing their economic status with a better lifestyle and of course a better and unpolluted environmental condition made it possible with IPM practices.

## Investment on Agricultural Research and Development

India's public investment in agricultural research has been stagnating at 0.32 per cent of total agricultural GDP for the last two decades (Manmohan Attavar, 1998). Moreover, government has to support agricultural research projects, extension work etc. by funding liberally in order to give boosting to the farming community. Giving tax breaks for the industries who are engaged in production of organic products, biocontrol agents. Participation of private sector in Agriculture Research and Development will complement the public sector research. Considering the Indian agriculture scenario and issues such as disappointing technology transfer, problems in supplying timely and quality farm inputs, market gluts and lack of infrastructure development, lack of appropriate government policy support and challenges of WTO era calls for increased private sector participation with a much better crop protection measures like biocontrol agents and organic chemicals.

## Role of Different Stakeholders in Commercialization

The All India Biotech Association (AIBA) a non-profit society that aims to promote and safeguard the overall interests of biotechnology in India, suggested that biopesticides and other biocontrol agents should be exempted from levy of excise duty as well as customs duty (in respect of imported items). This is important for penetrating these biocontrol agents in Indian agriculture. This will go a long way in making these environmentally safe products available to the Indian farmers at cheaper prices. At the same time biopesticides and biocontrol agents should be exempted from levy of VAT (Value Added Tax). Beside this, Indian regulatory system should be made to take faster decision in order to get the product clearances in a much faster way.

The researchers in India have to develop different formulations of the biocontrol agents at cheaper and affordable price to the farming community. An intense research and development strategy is needed for this to achieve with a specific policy from government. Government has to come forward in funding research in biotechnology and biocontrol agents and also should establish specific centres of excellence. In the Union Budget for 2005-06 an announcement was made for biotechnology sector regarding extension of terminal date, from March 31, 2005 to March 31, 2007, for a weighted deduction of 150 per cent of expenditure on in house R&D facilities of companies engaged in the business of biotechnology.

## CONCLUSION

Even though synthetic pesticide gives faster and effective control to pest problems, its excessive use for almost four decades caused several problems such as resistance, pest resurgence and environmental problems has triggered the need for alternative methods and ways of crop protection. Biocontrol agents are the best alternative for the increasing crop protection and production with less or almost no environmental hazards. The replacement of the pesticides and hazardous crop protection chemicals by controlled use of microorganisms can be considered as a key component in the development of a sustainable agriculture (Barea *et al.*, 2005; Barea *et al.*, 1998) to protect crops antagonist pathogens in an economically profitable manner with a minimal environmental pollution (Bethlenfalvay and Linderman, 1992).

The best advantages of biocontrol agent are its cheaper and easier way of production which is a big push for many developing countries that lack the economic resources to produce or purchase synthetic chemical pesticides. Most of the disadvantages of biocontrol agents have been or can be addressed by molecular biology and well designed ecological studies. Biocontrol agents are currently sold in various market segments with various brand names, of which some are most effective and some are good for supplementing the present synthetic pesticides in IPM programmes.

On the other side biocontrol agents market penetration was possible due to competitive prices, reliable formulations and efficacy comparable to standards. Advances in IPM programmes will increase the usage of biocontrol agents, especially in situations where the performance of biological insecticides are below the standard insecticides due to high insect pressure. Advancements in genetic engineering and manufacturing processes (fermentation, synthesis, formulation) will strengthen the position of biocontrol agents in the marketplace. This will be achieved largely with a strong policy and encouragement from government to the industry as well as end user by means of liberal tax benefits, incentives, etc. Finally, creating awareness among the farmer community about the beneficial effects of biocontrol agents and harmful effects of chemical protectants will certainly creates a congenial and longstanding effect which can lead to commercial success for the biocontrol agents produced.

## References

Abadias, M., Teixido, N., Usall, J., Vin˜ as I. & Magan, N. (2000) *J. Applied Microbiolog.* **89:** 1009-1017.

Abadias, M., Teixido, N., Usall, J., Vin˜as, I. & Magan, N. (2001) *Can J. Microbiol.* **47:** 123-129.

Akpa, E., Jacques, P.,Wathelet, B., Paquot, M., Fuchs, R., Budzikiewicz, H. & Thonart, P. (2001) *Appl Biochem Biotechnol* **91:** 551-561.

Anonymous. (2000) Agri Business: Where corporates falter? Agriculture Today, June issue, pp. 28-29.

Ando, K., Oishi, H., Hirano, S., Okutomi, T., Suzuki, K., Okazaki, H., Sawanda, M. & Sagawa, T. (1971) *J. Antibiot.* **24:** 347.

Ando, K., Sagawa, T., Oishi, H., Suzuki, K. & Nawata, T. (1974) Tetranactin, a pesticidal antibiotic, Proc 1st Intersect Congr IAMS (Sci Counc Jpn) **3:** 630.

Asaka, O. & Shoda, M. (1996) *Appl Environ Microbiol.* **62:** 4081-4085.

Backman, P.A., Wilson, M. & Murphy, J.F. (1997) In: Environmentally safe approaches to crop disease control (Eds. Rechcigl, N.A.& Rechcigl, J.E.) Lewis, Boca Raton, pp. 95-109.

Balaji, R. (1999) In: Agricultural News, Vol. V (5), National Academy of Agricultural Sciences, New Delhi, India, pp. 59.

Barea, J.M., Werner, D., Azcón-Aguilar, C. & Azcón, R. (2005) In: Agriculture, forestry, ecology and the environment (Eds.Werner, D, & Newton W.E.). The Netherlands: Kluwer Academic Publishers.

Barea, J.M., Andrade, G., Bianciotto, V., Dowling, D., Lohrke, S., Bonfante, P., O'Gara, F. & Azcón-Aguilar, C. (1998) *Appl Environ Microbiol.* **64:** 2304-2307.

Baum, J.A., Johnson, T.B. & Carlton, B.C. (1999) In: Biopesticides Use and Delivery, (Eds. Hall, F.R. & Menn, J.J.) Humana Press, Totowa, New Jersey, pp. 189-209.

Benhamou, N., Kloepper, J., Quadt-Hallman, A. & Tuzun, S. (1996) *Plant Physiol.* **112:** 919-929.

Berenbaum, M., Brusseau, M., Dipietro, J., Good, R., Gould, F., Gunsolus, J., Hammock, B.D., Hartung, R., Marrone, P., Maxwell, B., Raffa, K., Ryals, J., Shaner, D., Seiber, J. & Zilberman, D., (2000) The future role of pesticides in US agriculture, National Research Council, Washington, D.C.

Bethlenfavay, G. J. & Linderman, R. G. (1992) Mycorrhizae in Sustainable Agriculture, ASA Special publication No. 54, Madison, Wisconsin.

Bonning, B.C. & Hammock, B.D. (1996) *Annu. Rev. Entomol.* **41:** 191-210.

Boyette, C.D. (1994) *Weed Technol.* **8:** 526-528.

Boyette, C.D. & Abbas, H. K. (1994) *Weed Sci.* **42:** 487-491.

Boyette, C.D., Walker, H.L. & Abbas, H.K. (1999) *Weed Sci. Soc.* **52:** 94.

Boyette, C.D. (1986) *Plant Sci.* **45:** 223-228

Brannen, P.M. & Kenney, D.S. (1997) *J. Ind. Microbiol. Biotechnol.* **19:** 169-171.

Burges, H.D. (1998) In: *Formulation of microbial pesticides*. Kluwer Academic Publishers. Dordrecht, The Netherlands.

Butt, T.M., Goettel, M.S. & Papierok, B. (1999) *Directory of specialists involved in the development of fungi as biocontrol agents.* Colin Butt Design & Print, Warley, West Midlands.

Butt, T.M., Jackson, C.W. & Magan, N. (2001) *Fungal biological control agents: Progress, Problems and Potential.*(Eds. Butt, T.M., Jackson, C.W. & Magan, N.) CABI International, Wallingford, Oxon, U.K.

Carlton, B.C. (1993) *Genetics of Bt Insecticidal Crystal Proteins and Strategies for the Construction of Improved Strains, in Pest Control with Enhanced Environmental safety,* (Eds. Duke, S.O., Menn, J.J. & Plimmer, J.R.) ACS Symposium Series 524, American Chemical Society, Washington, D.C., pp. 326-337.

Chen, T.W. & Wu, W.S. (1999) *J. Phytopathol* **147:** 99-104.

Chen-Fu, C. & Wen-Chien, L. (1999) *J. Biosci. Bioeng.* **87:** 822-4.

Cook, R. J. (2000) *Phytopathol.* **38:** 95-116.

Cook, R. J. & Baker, K. F. (1983) *The Nature and Practice of Biological Control of Plant Pathogens.* American Phytopathol Society, St. Paul, MN.

Copping, L.G. & Menn, J.J. (2000) *Pest Manag. Sci.* **56:** 651-676.

Copping, L.G. (1998) *The BioPesticide Manual,* (Ed. Copping, L.G.) First Edition, British Crop Protection Council, Farnham, UK, pp. 1-333.

Crosby, D.G. (1971) *Minor insecticides of plant origin, in Naturally Occurring Insecticides,* (Eds. Jacobson, M. & Crosby, D.G.) Marcel Dekker Inc, New York, USA, pp. 127-239.

Dales, M.J. (1996) A review of plant materials used for controlling insect pests of stored products, NRI Bulletin 65.

Delany, I.R., Walsh, U.F., Ross, I., Fenton, A.M., Corkery, D.M. & O'Gara1, F. (2001) *Plant Soil* **232:** 195-205.

Dowling, D.N. & O'Gara, F. (1994) *Trends Biotechnol.* **12:** 133-141.

Eckert, J.W. & Ogawa, J.M. (1988) *Annu. Rev. of Phytopathol.* **26:** 433-469.

Elliott, M., Farnham, A.W., Janes, N.F., Johnson, D.M. & Pulman, D.A. (1989) *Pestica Sci* **26:** 199-208.

Federici, B.A. (1999) *In: Biopesticides Use and Delivery,* (Eds. Hall, F.R. & Menn, J.J.) Humana Press, Totowa, New Jersey, pp 301- 320.

Finlay, A.C., Hobby, G.L., Pan, S.Y., Regna, P.P., Routier, J.B., Seeley, D.B., Shull, G.M., Sobin, B.A., Solomons, I.A., Vinson, J.W. & Kane, J.H. (1950) *Science* **111:** 519.

Ferreira, J.H.S., Matthee, F.N. & Thomas, A.C. (1991) *Phytopathol* **81:** 283-287.

Granados, R.R. & Federici, B.A. (1986) *The Biology of Baculoviruses,* vols 1 and 2, Ed by CRC Press, Boca Raton, Florida, USA.

Gressel, J. (2001) *Trends Biotechnol.* **19:** 149-154.

Gueldner, R.C., Reilly, C.C., Pusey, P.L., Costello, C.E., Arrendale, R.F., Cox, R.H., Himmelsbach, H.S., Crumley, F.G. & Cutler, G. (1988) *J. Agric. Food Chem.* **36:** 366-370.

Handelsman, J. & Stabb, E.V. (1996) *Plant Cell* **8:** 1855-1869.

Harada, S. & Kishi, T. (1978) *J. Antibiot* **31:** 519.

Hare, J.D. & Morse, J.G. (1997) *J. Econ. Entomol.* **90:** 326-332.

Harley, J.L. & Smith. S.E. (1983) *In: Mycorrhizal Symbiosis.* New York: Academic Press.

Harman, G.E. (2000) *Plant Dis.* **84:** 377-393.

Harris, A.R. & Adkins, P.G. (1999) *Biol. Control.* **15:** 10-18

Ikawa, M., Dicke, R.J., Allen, T.C. & Link K.P. (1945) *J. Biol Chem.* **159:** 517-524

Inceoglu, A.B., Kamita, S.G, Hinton, A.C., Huang, Q., Severson, T. F., Knag, K. & Hammock, B.D. (2001) *Pest Manag Sci.* **57:** 981-987.

Isono, K., Nagatsu, J., Kawashima, Y. & Suzuki, S. (1965) *J. Antibiot Ser. A.* **18:** 115.

Jacques, P., Hbid, C., Destain, J., Razafindralambo, H., Paquot, M., De Pauw, E. & Thonart, P. (1999) *Appl. Biochem. Biotechnol.* **77:** 223-233

Janisiewicz, W.J. (1987) *Phytopathol.* **77:** 481-485.

Janisiewicz, W.J. & Marchi, A. (1992) *Plant Dis.* **76:** 555-560.

Kloepper, J.W., Tuzun, S. & Kùc, J.A. (1992) *Biocontrol Sci. Technol.* **2:** 349- 351.

Koomen, I. & Jeffries, P. (1993) *Plant Pathol.* **42:** 230-237.

Korsten, L., De Villiers, E.E., Rowell, A. & Kotze, J.M. (1993) *Postharvest biological control of avocado fruit diseases.* South Africa Avocado Growers Association Yearbook 16: 65-69.

Leifert, C., Li, H., Chidburee, S., Hampson, S., Workman, S., Sigee, D., Epton, H.A. & Harbour, A. (1995) *J. Appl Bacteriol.* **78:** 97-108.

Leisy, D.J. & Fuxa, J.R. (1996) In: Crop Protection Agents from Nature: Natural Products and Analogues, (Ed. Copping, L.G.) Royal Society of Chemistry, Cambridge, UK.

Likover, T.E. & Kurland, C.G. (1967) *The contribution of DNA to translation errors induced by streptomycin in vitro.* Proc Natl Acad Sci USA. 58: 2385.

Limon, M.C., Pintor-Toro, J. A. & Benitez, T.(1999) *Phytopathology* **89:** 254-261.

Linderman, R.G. (1988) *Phytopathol.***78:** 366-371.

Mach, R. L., Peterbauer, C. K., Payer, K., Jaksits, S., Woo, S. L., Zeilinger, S., Kullnig, C. M., Lorito, M., & Kubicek, C. P., (1999) *Appl. Environ. Microbiol.* **65:** 1858-1863.

Margolles-Clark,E., Hayes, C.K., Harman, G.E. & Penttila, M. (1996) *Appl. Environ. Microbiol.* **62:** 2145-2151.

Manmohan, A. (1998) *Corporatisation of Agriculture- for No active Agricultural Policy,* Agriculture Today, Nov-Dec Issue, pp. 60-62.

Maurhofer, M., Keel, C., Haas, D. & Défago, G. (1995) *Plant Pathol.* **44:** 40-50.

McSpadden Gardener, B. B., & Fravel, D. R. (2002) *Biological control of plant pathogens: Research, commercialization, and application in the USA.* Online. Plant Health Progress *doi*: 10.1094/PHP-2002-0510-01-RV.

Misato, T., Ishii, I., Asakawa, M., Okimoto, Y. & Fukunaga, K. (1959) *Ann. Phytopathol. Soc. Jpn.* **24:** 302.

Norman, C. (1988) *Science* **242:** 366-367.

Nunes, C.,Usall, J., Teixido, N. & Vin˜ as I. (2001) *International J. Food Microbiol.* **70:** 53-61.

Ogawa, Y., Yoshida, H., Inouye, S. & Niida, T. (1973) *Studies on a new antibiotic SF-1293, III. Synthesis of a new phosphorus containing amino acid, a component of antibiotic SF-1293.* Meiji Seika Kenkyu Nempo **13:** 49.

Om, Y., Yamaguchi, I. & Misato, T. (1984) *J. Pestic. Sci.* **9:**317.

Omura, S., Murata, M., Hanaki, H., Hinotozawa, K., Oiwa, R. & Tanaka, H., (1984) *J. Antibiot.* **37:** 829.

Paau, A.S. (1988) *Trends Biotechnol.* **6:** 276-279.

Pusey, P.L. & Wilson, C.L. (1984) *Plant Dis.* **68:** 753-756.

Pusey, P.L. (1996) In: *Crop Protection Agents from Nature: Natural Products and Analogues,* (Eds. Copping, L.G.), Royal Society of Chemistry, Cambridge, UK, pp. 426-443.

Ramesh, T. (1999) *M.Sc. thesis, submitted to Department of Agricultural Economics,* UAS, GKVK, Bangalore, India.

Raupach, G.S. & Kloepper, J.W. (1998) *Phytopathol.* **88:** 1158-1164.

Rey, M., Delgado, J.J., Rincon, A.M., Limon, C.M., Benitez, T., Perez, E.A. & Cantoral, F.J. (2000) *Micopatotogia* **17**: 531-536.

Roush, R.T. (1997) *Pestic. Sci.* **51**: 328-334.

Schatz, A., Bugie, E. & Waksman, S.A. (1944) *Proc. Soc. Exp. Biol. Med.* **55**: 66.

Schatz, A. (1944) *Proc. Soc. Exp. Biol. Med.* **55**: 66.

Shaji, T.K. (2001) India crucially awaits agripreneurship, Agriculture today, May issue 2001, New Delhi, India, pp.10-12.

Sinha, I. & Upadhyay, R.S. (1993) *Acta. Phytopathol. Et. Entomol. Hung.* **28**: 209-214.

Srinivas, N.N. (1998) In: *Agricultural News*, Vol. VI (6), National Academy of Agricultural Sciences, New Delhi, India, pp. 56.

Struyk, A.P., Hoette, I., Drost, G., Waisvisz, J.M., Van Eek, J. & Hoogerheide, J.C. (1958) In: *Antibiotics Annual 1957-1958* (Eds. Welch, H. & Marti-Ibanez, F.) Medical Encyclopedia Inc, New York, USA, pp. 878.

Subramanyam, K.V. (2000) Linkages between farm and non-farm sector: Role of processing of horticultural products. Occasional paper 16, NABARD, Mumbai, pp. 29-42.

Sukhpal, Singh. (2000) *Ind. J. Agric. Econ.* vol.55 (3), New Delhi. Pp. 283-294.

Sukhpal, Singh. (2000) Multi National Corporations and Agriculture technology transfer: A case study of contract farming in the Indian Punjab, Paper presented at National Seminar on Private Extension Approaches and Challenges in the Millennium (July 28-29, 2000), National Institute of Agricultural Extension Management (MANAGE), Hyderabad, India.

Suzuki, S., Isono, K., Nagatsu, J., Mizutani,T., Kawashima, Y. & Mizuno, T., (1965) *J. Antibiot. Sen. A.* **20:**109.

Takeuchi, S., Hirayama, K., Ueda, K., Sasaki, H. & Yonehara, H., Blasticidin, S. (1958) *J. Antibiot. Ser. A.* **11:**1.

Tanaka, N., Yamaguchi, H. & Umezawa, H. (1966) *J. Biochem.* **60:** 429.

Tanu, A. P. & Adholeya, A.(2006) In: Hand book of Microbial Biofertilizers ( Ed. Rai, M. K.) . Haworth Press, New York. 223-239.

TeBeest, D.O. (1991) Microbial Control of Weeds. Chapman and Hall, New York.

TeBeest, D.O. & Dickman, M.B. (1989) *Phytopathol.* **79:** 1173.

TeBeest, D.O. (1982) *Plant Dis.* **66:** 469-472.

TeBeest, D.O. (1988) *Plant Dis.* **72:** 16-18.

TeBeest, D.O, Yang, X..B.& Cisar, C.R. (1992) *Phytopathol.* **30:** 637-657.

TeBeest, D.O. & Templeton, G.E. (1985) *Plant Dis.* **69:** 6-10.

Teixido, N., Usall, J., Palou, L., Asensio, A., Nunes, C. & Vin˜as, I. (2001) *Eur. J. Plant Pathol.* **107:** 685-694.

Templeton, G.E. (1982) In: Biological Control of Weeds with Plant Pathogens. (Eds. Charudattan, R., Walker, L. & John, H.) Wiley and Sons, New York.

Templeton, G.E., Smith, R..J. Jr. & TeBeest, D.O. (1986) *Rev. Weed Sci.* **2:** 1-14.

Templeton, G. E. & Heiny, D. K. (1989) In: Biotechnology of fungi for improving plant growth (Eds.Whipps, J. M. & Lumsden, R. D.) Cambridge University Press, UK, pp. 127-151.

Torres, R.J., Usall, J., Teixido, N., Abadias, M. & Vin˜as, I. (2003) *J. Appl. Microbiol.* **94:** 330-339.

Treacy, M.F. (1999) In: Biopesticides Use and Delivery, (Eds. Hall, F.R. & Menn, J.J.) Humana Press, Totowa, New Jersey, pp 321- 340.

Ujvary, I. (1999) In: Nicotinoid Insecticides and the Nicotinic Acetylcholine Receptor, (Eds. Yamamoto, I. & Casida, J.E.) Springer-Verlag, Tokyo, pp 29-69.

Umezawa, H., Okami, Y., Hashimoto, H., Suhara,Y. & Otake, N. (1965) *J. Antibiot Ser A.* **18:** 101.

Usall, J., Teixido, N., Torres. R., Ochoa de Eribe, X. & Vin˜, as I. (2001) *Postharvest Biol. Biotechnol.* **21:** 147-156.

USDA-ARS. (2001) U.S. Department of Agriculture, ARS National Programs, Crop Protection and Quarantine. http://www.nps.ars.usda.gov/programs.htm?npnumber=304.

USDA-ERS (1988) Agricultural Resources - Inputs, Situation, and Outlook Report. U.S. Dept. Agriculture, Economic Research Service. Publ. AR- 9. U.S. Govt. Print. Office, Washington, D.C.

Van Loon, L.C., Bakker, P.A.H.M., Pieterse, C..M..J. (1998) *Annu. Rev. Phytopathol.* **36:** 453-483.

Vidhyasekaran, P. & Muthamilan, M. (1995) *Plant Disease* **79:** 782-786.

Walker H.L. & Connick, W.J. (1983) *Weed Science.* **31:** 333-338.

Walker, H.L. & Tilley, A.M. (1997) *Biol Control.* **10:** 104-112.

Wei, G., Kloepper, J.W., Tuzun, S. (1996) *Phytopathol.* **86:** 221-224.

Whipps, J..M. (2001) *J. Exp. Bot* **52:** 487-511.

Wisniewski, M.E. & Wilson, C.L. (1992) *Hortscience* **27:** 94-98.

Woodhead, S.H., O'Leary, A.L., O'Leary, D.J. & Rabatin, S.C. (1990) *Can. J. Plant Pathol.* **12:**328-331.

Woodburn, A., Agrochemical sales- 1998, (1999). Allan Woodburn Associates, Edinburgh, UK. www.indiastat.com, Bulletin on Food Statistics 1998 - 2000, Directorate of Economics & Statistics, Dept. of Agriculture & Cooperation, Ministry of Agriculture, Govt. of India. Accessed on 15. 08. 06

Yoshida, S., Hiradate, S., Tsukamoto, T., Hatakeda, K., Shirata, A. (2001) *Phytopathol.* **91:** 181-187.

Yu, G.Y., Sinclair, J.B., Hartman, G.L., Bertagnolli, B.L. (2002) *Soil Biol. Biochem* **34:** 955-963.

Zehnder, G.W., Murphy, J.F., Sikora, E.J., Kloepper, J.W. (2001) *Eur. J. Plant Pathol.* **107:** 39-50.

Zehnder, G.W., Yao, C.B., Murphy, J.F., Sikora, E.R., Kloepper, J.W. (2000) *Biocontrol* **45:** 127-137.

Zeilinger, S., Galhaup, C., Payer, K., Woo, S. L., Mach, R. L., Fekete, C., Lorito, M., & Kubicek, C. P. (1999) *Fungal Genet. Biol.* **26:** 131-140.

# 8

# Screening, Evaluation and Formulation of Delivery Systems for Biocontrol Agents

**R. Saravanamuthu**

*Department of Botany, A.V.C. College (Autonomous),* Mannampandal, Mayiladuthurai-609 305
Tamil Nadu, India

**ABSTRACT**

Biological control is a way of thinking about the plant disease problems and its management, which seek solution through manipulation of associated microorganisms and in terms of restoring the biological equilibrium. *Trichoderma* as an ideal antagonist exhibited various mechanisms of biological control such as production of antibiotic substances such as viridian, plant cell wall degrading enzymes such as cellulases, mycoparasitism and fungal cell wall degrading enzymes such as chitinases. The rDNA technology may optimistically be extended to develop strains which could exhibit extended range of effectiveness against multiple pathogens. It is true that the biological control as a research field still remains young, perhaps just entering its exponential growth phase and in some ways it is maturing and taking new dimensions as stated by Cook and Baker some two decades ago. So, the main goal of this article is to make those, who are interested, look into the organisms with a critical eye to elucidate them as biocontrol agents.

## INTRODUCTION

Direct application of antagonistic fungi to control damping off of pine seedlings caused by *Pythium debaryanum* was first demonstrated by Harley in 1921. Since then several attempts were made and are still being made to understand the mechanisms of biological control and to develop delivery systems for biocontrol agents. Weindling's interesting discovery of antagonism of *Trichoderma viride* against the soil-borne plant pathogen, *Rhizoctonia solani,* during 1931-1941 has made the attention of scientists diverted towards various strains of *Trichoderma*. Species of *Trichoderma* including *T. harzianum* and *T. viride* have been well exploited for understanding the mechanisms of biological control, and used extensively as biocontrol agents in glasshouse and field trials in different formulations. Intensive studies on *Trichoderma* have made us to understand the molecular mechanisms of antagonism, which involve different kinds of enzyme systems such as chitinolytic enzymes and the genes encoded for the synthesis of such enzymes. This helps us to isolate such gene sequences, construction of plasmids with desired genes and to develop transgenic strains of antagonists with desired traits.

The propagules, when directly released into the soil, generally do not readily establish as there is the wide spread phenomenon of fungistasis operating. Hence, food based inoculum formulations of *Trichoderma* were attempted and registered for commercial exploitation in countries such as US, Israel and Britain. Some of such formulations have also been patented. Thus, development of feasible, efficient and appropriate delivery system for appropriate microorganism and application to appropriate crop is an important component of biocontrol research. Delivery systems incorporating various organic and inert carriers in different formulations such as dusts, alginate pellets, starch granules, extruded granules, seed coatings, etc., have been made mostly with reference to *Trichoderma.*

Most of the above mentioned researches on the biological control were commonly centered around the strains of *Trichoderma,* especially *T. harzianum* and *T. viride*. But there are thousands of microorganisms, many of them still remain unidentified, in soil environment with antagonistic potentialities remaining unexplored and unutilized. There are many candidates of fungi for mycologists whose insight and perseverance into such organisms would definitely introduce many more biocontrol organisms to the world.

## BIOCONTROL MECHANISMS

### Antibiotic Interactions

In soil, probably more than any other habitat, growth and development of individual species is dependent upon the nutrient status and the various types of interactions between diverse groups of organisms. These interactions may be either between different species of the same group or between different groups of organisms of the soil. The interaction between fungi is a matter of interest for the mycologists and plant pathologists as it provides a mechanism for biological control (Baker, 1968; Baker and Cook, 1974; Mangenot and Diem, 1979; Papavizas and Lumsden, 1980; Cook and Baker, 1093). The different types of inter-fungal interactions mainly include antagonism and symbiosis (Park, 1960). The term "antagonism" is one of the major subdivisions of various possible types of interactions between pairs of species and includes all those associations in which at least one of the interacting species is harmed. It has been frequently demonstrated *in vitro*, and accepted that antagonism is not uncommon in soil too (Park, 1960, 1968; Jackson, 1965; Garrett, 1979; Griffin, 1972). The antagonistic interactions include antibiosis, competition, and exploitation.

## Antibiosis

The term 'antibiosis' was first applied by Vuillemin (1889) to situations where one organism destroyed another to sustain its own life. The phenomenon of antibiosis has been accepted as the principal mechanism of interference competition by which fungi may exclude other organisms from resource potentialities available to each of the competitor (Brian,1951, Park, 1967). Waksman (1947) defined antibiotic as 'a chemical substance produced by microorganism which has the capacity to inhibit the growth (of) and even destroy bacteria and other microorganisms'. Thus, antibiosis is defined as the condition in which one or more metabolites excreted by an organism have a harmful effect on one or more organisms. This definition has generally been accepted in the field of ecology of fungi (Park, 1960, 1968; Garrett, 1970; Ikediugwu and Webster, 1970; Griffin, 1972; Wicklow, 1981; Dwivedi and Saravanamuthu, 1985). The idea of possible involvement of 'staling products', now called 'antibiotic substances' or 'metabolites', in antagonism was first proposed by Sanford and Broadfoot (1931). Antibiotics or staling products are produced by fungi in pure culture on a sufficiently rich nutrient medium when mycelial growth has been halted by some limiting factor, such as nutrient deficiency. In this situation, the mycelium is unable to stop absorbing sugar from the culture medium, and the excess is excreted as a shunt metabolite, which may have antibiotic properties (Garrett, 1970). Thus, staling is a complex phenomenon and is usually detected initially by a reduction in growth rate and ultimately by a cessation of growth (Park, 1961; Robinson and Park, 1966; Robinson, 1969).

Robinson (1969) stated that it is difficult to define the exact time that a culture became stale and it is more useful to think of staling as a complex process resulting in a gradually increasing inhibition of growth. The studies on the factors influencing staling have been reviewed (Prat, 1924; Robinson, 1969).

Change in the pH of the medium, oxygen tension, exhaustion of the available nutrients and the volatile and non-volatile substances are considered as factors of staling. The production of stable, labile, and volatile substances in liquid cultures, which are inhibitory to spore germination and mycelial growth, has been reported (Brian, 1957, 1960; Park, 1960, 1968; park and Robinson, 1964; Robinson and Park, 1966; Gottlieb and Shaw, 1970; Dwivedi and Singh, 1971; Shukla et al., 1977; Arora and Upadhyay, 1978; Gandy, 1978; Upadhyay and Arora, 1980). It has been strongly believes that these antibiotic substances are active in soil too (Brian, 1960; Park, 1960, 1968; Baker, 1968; Garrett, 1970; Griffin, 1972; Baker and Cook, 1974; Dwivedi and Saravanamuth, 1985). It has been demonstrated that the antibiotics are produced by *Penicillium, Aspergillus, Trichoderma* and *Streptomyces* spp. in soil. It has also been proved that antibiotics are produced in direct contact between microorganism, which may function at continuously low concentration and are subjected to biotic decomposition (Grossbard, 1954). But all the attempts to extract specific antibiotic substances from soil proved futile. However, these substances have been extracted from the organic substrates buried in soil. Some of such substances include griseofulvin, chloramphenicol, gliotoxin and actinomycin (Wright, 1954,1955,1956; Brian, 1957). The antibiotic substances may reduce or prevent the germination of fungal propagules, involve lysis or inhibit growth after germination (Robinson, 1969; Papavizas and Lumsden, 1980). It is also believed that the widespread soil fungistasis is 'most probably a summation of production of substances by the activity of microbes generally in the soil' (Park, 1968). This conclusion is generally based on the strongest evidences as summarized by Brock (1986), which include the following:

(a) Antibiotics can be extracted from non-sterile soil.

(b) Antibiotic production will occur in non-sterile soil inoculated with antibiotic producing organism and supplemented with organic enrichments such as straw.

(c) Antibiotic producing microbes can be isolated easily from the natural soil.

(d) Antibiotics producing organisms are able under natural conditions, to antagonized other organism sensitive to antibiotics.

(e) Conditions unfavourable for the accumulation of antibiotics reduce the antagonistic activities of the organisms producing the antibiotics.

In recent years, it has been emphasized and demonstrated that antibiosis can provide a weapon for biological control of plant diseases (Garrett, 1965; Baker,1968; Wilhelm, 1973; Baker and Cook, 1974; Cook and Rovira, 1976; Cook, 1977; Mangenot and Diem, 1979; Papavizas and Lumsden, 1980; Papavizas and Lewis, 1981; Kraft and Roquebert, 1981; Marois *et al.*, 1981, 1982; Bell *et al.*, 1982; Utkhede, 1982; Utkhede and Rahe, 1983; Cook and Baker, 1983; Hadar *et al.*, 1984; Upadhya and Rai, 1985).

## Competition

The most frequently used definition of competition, as defined by Clark (1965), is 'the endeavour of two or more organisms to gain the measure each wants from the supply of a substrate in the specific form and under specific conditions in which the substrate is presented when the supply is not sufficient for both' (Baker and Cook, 1974; Lockwood, 1981; Cook and Baker, 1983). The competition between fungi, particularly in soil, is mainly for specific 'substrates' and also may be for space (Park, 1960; Garrett, 1963, 1970; Lockwood, 1981). The success of a fungal species in the competition is determined by its inherent competitive saprophytic ability (Garrett, 1950). Garrett (1956) defined it as 'the summation of physiological characteristics that make for success in competitive colonization of dead organic substrates'. He attributed it to the following characters of the fungus: (a) growth rate; (b) antibiotic production; (c) tolerance to antibiotics produced by competitors; and (d) enzyme producing ability. The competitive saprophytic ability of certain pathogenic fungi received the attention of many workers (Butler, 1953; Rao, 1959; Macek 1961; Wastie, 1961; Park, 1965; Burges and Griffin, 1967; Dingra *et al.*, 1976; Upadhyay and Rai, 1983).

The antibiotic production and the tolerance of the fungus to the antibiotics produced by antagonists determine the saprophytic ability. Rao (1959) and Wasties (1961) investigated the interspecific fungal competition for substrate colonization but, however, failed to correlate the saprophytic success with the degree of tolerance to fungistatic growth substances of competing microorganism. The ingenious attempts made by Dwivedi and Garrett (1968) bridged this gap. According to them, the competitive success of fungi on the staled agar is dependent on the tolerance of fungi to the antibiotic staling growth products produced by the associated microorganisms.

The success in the competition is not only determined by the competitive saprophytic ability of the fungus but also by the ability of the fungus to survive saprophytically in the pre colonized host substrates, as it imposes resistance for colonization by other microorganisms (Garrett, 1970).

The saprophytic survival of the fungus, as reviewed by Garrett (1970), is influenced by the conditions of moisture, temperature, and soil aeration. Nitrogen content of the soil and cellulolysis must be adequate enough to satisfy the requirement of nutrients (Garrett, 1966, 1967, 1970, 1975, 1978,1980,1983a, 1983b). The greater tolerance to antibiotics has been considered as an important condition of saprophytic survival (Anwar, 1949; Butler, 1953; Park, 1956; Barton, 1961). The production of antifungal antibiotics by *Cephalosporium* has also been demonstrated *gramineum* and it has been suggested that the continued production of the wide spectrum antibiotic substances helped the fungus to prolong its saprophytic survival (Breuhl and Lai, 1966; Lai and Bruehl, 1966). Garrett (1970), while critically analyzing the saprophytic survival of root infecting fungi, as influenced by various factors, concluded that the production of antifungal antibiotics is not essential for saprophytic survival. Mehrotra and Claudius (1973) studied the competitive saprophytic ability of *Fusarium oxysporum* and *Sclerotium rolfsii* and reported that tolerance to antibiotic substance was not a deciding factor of the saprophytic status of these fungi in soil. Nevertheless, Upadhyay and Rai (1983) reported that the colonization of pigeonpea substrates by *F.udum* was highly suppressed by antagonism from *Penicillium citrinum, A. niger, Micromonospora globosa, A. flavus, A.terreus* and *Trichoderma viride.*

## Exploitation

Exploitation involves the direct interactions between the fungal species and includes mycoparasitism and lysis. The interaction between fungi involving the parasitism of one fungus on another (Barnett, 1964; Boosalis, 1964) is not uncommon in soil (Boosalis and Mankau, 1965; Barnett and Binder, 1973; Ayers and Adams, 1981; Lumsden, 1981). This can be classified into biotrophic and necrotrophic mycoparasitism (Barnett and Binder, 1973). The biotrophic mycoparasites obtain their nutrients directly from the living cells of its fungus host, either by growing in intimate contact with the host or by penetrating and growing inside the host cell. The necrotrophic mycoparasites, on the other hand, obtain their nutrients from the host structures by killing the cells, sometimes without infecting the host cells. The method of killing the host structures is not understood, however, it has been attributed to the release of toxic substances, cell wall degrading enzymes, or other effects (Boosalis and Mankau, 1965; Barnett and Binder, 1973; Ayers and Adams, 1981). Thus, all those fungi coming in contact with each other in necrotrophic mycoparasitic interactions need not necessarily penetrate.

The cytological studies on the mycoparasitism showed that it involves morphological changes such as coiling, penetration, vacuolization, granulation, hyphal bursting and lysis. (Ikediugwu and Webster, 1970a, b; Skidmore and Dickinson, 1976). The necrotrophic mycoparasitic interactions between different taxonomic groups of fungi have been reported and are reviewed (De Vay, 1956; Sanford, 1959; Weindling, 1959; Darpoux, 1960; Barnett, 1963; Boosalis, 1964; Boosais and Mankau, 1965; Barnett and Binder, 1973; Ayers and Adams, 1981; Lumsden, 1981; Upadhyay and Rai 1982). *In vitro* mycoparasitic interactions of some soil-borne pathogens, namely *Cephalosporium roseo-grisueum* (Chaturvedi and Dwivedi, 1982), *Drechslera australiensis* (Dwivedib and Saravanamuthu, 1983*), F.oxysporum f. sp. lini* (Arora *et al.*, 1979*), F. oxysporum f. sp. lycopersici* (Pathak *et al.*, 1981). *F. udum* (Upadhyay *et al.*, 1981) and *Rhizoctonia solani* (Arora and Dwivedi, 1980) with different taxonomic groups of fungi have been reported. It has also been reported that most of the mycoparasitic interactions that affect the persistent structures of soil-borne plant pathogens are necrotrophic (Lumsden, 1981). However, little is known about the occurrence of mycoparasitism in soils. Nevertheless, it has been reported that several persistent structures of soil-borne plant pathogens were invaded by a variety of mycoparasitic fungi in natural soil, which suggests that the mycoparasitic interactions may be of some importance in the regulation of the population of soil-borne plant pathogens. Thus, the possible mycoparasitic suppression of fungal pathogens towards biological control has been emphasized in recent years (Baker and Cook, 1974; Mangenot and Diem, 1979; Papavizas and Lumsden, 1980; Ayers and Adams, 1981; Cook and Baker, 1983; Upadhyay and Rai, 1985; Lockwood, 1988, Saravanamuthu, 2001).

Thus, it is obvious that there are various mechanisms operating synergistically or in multitude in the biocontrol process. So, the organisms are to be screened for the possession of the following characteristics:

- The organism should have a short generation time. It should germinate more rapidly than the pathogenic organism.
- The growth rate of the organism should be more than the pathogenic organism.
- The organism should possess the ability to competitively colonize the root zone even in the presence of the root pathogen. Thus, it should exhibit 'rhizosphere competence'.
- It should competitively and rapidly utilize the available nutrients so as to deprive of the nutrients for the pathogen.
- It should produce antibiotic substances, which could inhibit the germination of the propagules and the growth of the germinated germ tube.
- It should be a potential mycoparasite.
- It should produce the plant cell wall degrading enzymes such as cellulases and pectinases, which enable the organism to dislodge the pathogen from the corpse of the host tissue.
- It should produce chitinolytic enzymes, which are involved in the destruction of fungal cell wall.

## Primary Screening

Soil is the most complex and heterogeneous environment, which harbours a diverse array of microorganisms. These include saprophytes, potential antagonists and plant pathogens. These microorganisms, for most part of their life, remain as dormant propagules, and become active only at the vicinity of the available energy rich substrates. When they become active, they delimit their niche space as multidimensional configuration and also by means of biological interactions called realized niche. The biological interactions that restrict one organism at the vicinity of the other can be comprehensively called antagonism. It is also true with soil saprophytes and soil-borne plant pathogens. Hence, the antagonistic interactions between the soil fungi and the soil-borne plant pathogens should be primarily screened.

## Interfungal Interactions

***Collection of soil samples:*** Soil samples are collected within 15 cm from the surface of the soil. The collected soil samples are brought to the laboratory, handpicked, air dried and stored for future use.

## Isolation of Fungi from Soil

***Serial Dilution Technique:*** Weigh 10 g of the soil sample and transfer it to a 250 ml conical flask containing 100 ml of sterilized distilled water. Shake the flask on an electric shaker (150 throws $min^{-1}$ and 1.5 cm displacement) in order to get a homogeneous suspension of soil solution. Then prepare 1:100 and 1:1000 dilutions by serially transferring to sterile distilled water in conical flasks. One ml of the soil solution from the flask containing 1:1000 dilution is to be plated in each Petri dish containing Martin's agar (dextrose 10/ g, peptone 5 g, $KH_2 PO_4$ 1g, Mg $SO_4$. $7H_2O$ 0.5 g, rose Bengal 0.066 g, agar-agar 18 g, distilled water 1000 ml and pH 5), and Czapek-Dox + 0.5% yeast extract agar ($MgSO_4$. $7H_2O$ 0.5 g, KCl 1 g, Na $NO_3$ 2 g, $KH_2PO_4$ 1 g, FeSo4 0.066 g, yeast extract 0.5 g, sucrose 30 g, agar-agar 18 g, distilled water 1000 ml and pH5. 6) separately. Incubate the plates at 25 ± 2°C for five days. Record the fungi appearing over the surface of the media. Identify the individual colonies and subculture them for future use.

The test organisms, the pathogens against which the soil fungi are to be screened for their antagonistic potentialities, can be either obtained (procured) or isolated from the infected plant tissues after surface sterilization with chemicals such as mercuric chloride (0.01%) solution or sodium hypochlorite solution and culturing on the suitable media.

Colony interactions between the soil fungi and the individual species of test fungi (pathogens) can be studied by adopting 'Dual Culture Experiment' (Ikediugwu and Webster, 1970) and with cell free culture filtrates.

## DUAL CULTURE EXPERIMENT

Individual species of soil fungi, isolated from the soil sample, and the test pathogens are to be grown in pure cultures in Petri dishes containing solidified Czapek-Dox + 0.05% yeast extract agar medium. Agar blocks (5 mm diameter) cut from the actively growing margins of the test pathogen and the soil fungus are to be inoculated juxtaposed to other 3 cm apart on solidified Czapek-Dox + 0.05% yeast extract agar medium in 90 mm diameter Petri dishes. At least five replicates are to be maintained for each isolate of soil fungi. Controls can be maintained by single and dual inoculated cultures of the fungus. The position of the colony margins on the reverse side of disc is to be measured and recorded daily. The growth of the soil fungus and the pathogen in each plate will be affected as and when they approach each other. Hence, the assessment is to be made for both soil fungus and the pathogen. It should be made only after the soil fungus (antagonist) and the pathogen have achieved state of equilibrium, i.e., after which there is no alteration in the pattern of growth. The per cent inhibition of growth can be calculated as follows:

$$\text{Inhibition of growth (\%)} = \frac{r - r'}{r} \times 100$$

where, 'r' is the growth of the fungus (pathogen) measured from the centre of the colony towards the centre of the plate in the absence of the antagonistic fungus, and r' is the growth of the fungus (pathogen from the centre of the colony towards the antagonistic fungus).

***Assessment:*** Assessment of the colony interactions can be done following the model proposed by Dickinson and Broadman in (1971). There are five types of interaction grades proposed (Fig.1) (Skidmore and Dickinson, 1976). They include.

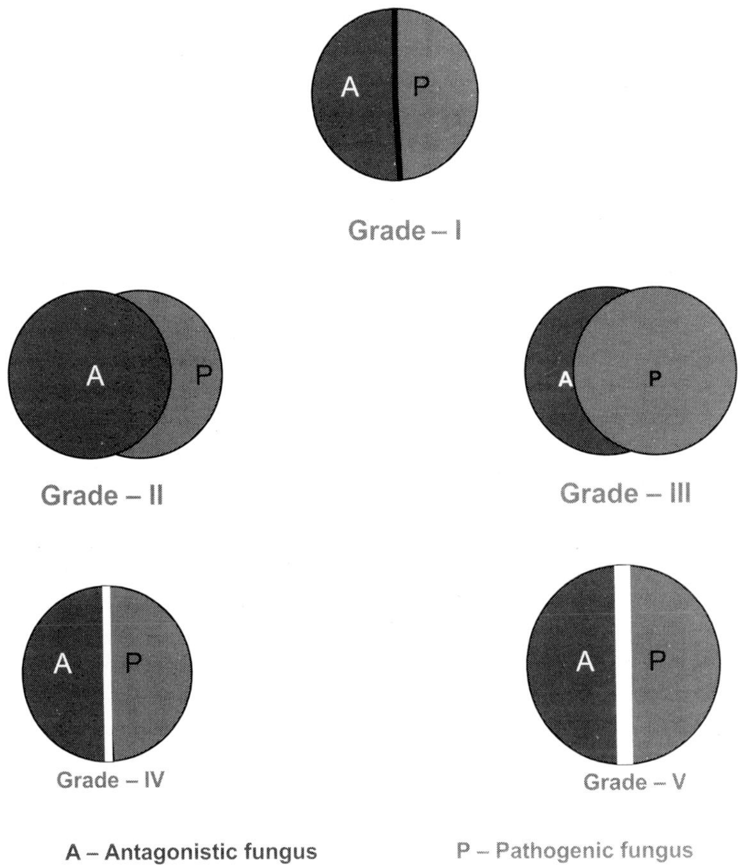


1. Mutual intermingling growth without any macroscopic sights of interaction — GRADE I
2. Intermingling growth where the growth of the fungus (pathogen) is ceased, and is being overgrown by the opposed fungus — GRADE II
3. Intermingling growth where the fungus under observation (pathogen) is growing into the opposed fungus either above or below — GRADE III
4. Slight inhibition of both the interacting fungi with a narrow line of zone of inhibition (1 – 2 mm) — GRADE IV
5. Mutual inhibition of growth at a distance with a zone of inhibition greater than 2 mm — GRADE V

The dual culture experiment gives an idea about the type of interfungal interactions or hyphal interference and the possible production of antibiotic substances by the soil fungi and the test pathogens or vice versa.

## EFFECT OF STALING GROWTH PRODUCTS OF SOIL FUNGI

Staling growth products of fungi most likely include the antibiotic substances. The effect of staling growth products of the soil fungi can be studied by cellophane agar diffusion technique and by using cell free culture filtrates of the soil fungi against the test organisms (pathogens).

### Cellophane Agar Diffusion Technique

The effect of staling growth products diffused through the cellophane paper can be studied by the method described by Gibbs (1967). The soil fungi and the pathogens are in pure cultures. De-plasticized and sterilized cellophane discs of 9 cm diameter are laid aseptically over the solidified Czapek-Dox + 0.05% yeast extract agar medium in 90 mm diameter Petri dishes containing approximately 20 ml of the medium. All the plates are to be left as such over night so as to allow the excess moisture to get evaporated. Each of the thus prepared Petri plates is inoculated centrally with 5 mm diameter agar block cut from the actively growing margin of the individual species of soil fungi. At least five replicates are maintained. The plates are then incubated at 25 ± 2°C. The cellophane paper along with the fungus is to be removed after 2, 3, 4 and 5 days of incubation. Control is to be maintained without inoculation of fungal disc over the cellophane paper. The cellophane paper from the control plate is also removed. Now, agar bocks (5 mm diameter) cut from the actively growing margin of the pathogen are to be inoculated individually in each treated plate at the centre and incubated at 25 ± 2°C. The radial growth of the pathogen is measured after 2, 3, 4 and five days of incubation. The percentage of inhibition of the growth of the pathogen in the treated plates is assessed as follows.

$$\text{Inhibition of growth (\%)} = \frac{\text{Radial growth in control} - \text{Radial growth in treatment}}{\text{Radial growth in control}} \times 100$$

## EFFECT OF CELL FREE CULTURE FILTRATES

### Preparation of Culture Filtrates

The isolated soil fungi are grown in pure cultures in Petri plates. Czapek-Dox + 0.05% yeast extract liquid medium is prepared and sterilized. 100 ml of the liquid medium is then distributed in each 250 ml conical flask. Each flask is then inoculated with agar blocks (5 mm) cut from the actively growing margin of the soil fungi separately. The flasks are incubated at 25 ± 2°C for about 15 days. After 15 days, filter the staling growth products through Whatman number 1 filter paper and then through seitz filter (G5). The filtrates are then transferred to conical flasks, condensed and stored at 4°C for further use. The efficacy of the culture filtrates can be assayed for conidial germination and for the hyphal growth.

**Assay:**

### Radial Growth

Prepare Czapek-Dox + 0.05% yeast agar medium. Distribute this medium into the conical flasks and sterilize them. Add the cell free culture filtrate of each fungus into the flasks separately so as to give the concentrations of 10, 15 and 20%. Then disperse the amended media into Petri plates under aseptic condition and allow the medium in the Petri plates to solidify. Cut agar blocks (5 mm) from the actively growing margin of the pathogen and inoculate a block at the centre of the solidified amended medium. Incubate the plates at 25 ± 2°C for five days. Measure the radial growth of the fungus at periodic interval of 24 h for about five days. Calculate the mean growth rate. Maintain control without inoculation. Assess the per cent inhibition of growth of the pathogen due to cell free culture filtrate of individual species of fungi. It is calculated as follows.

$$\text{Growth Inhibition (\%)} = \frac{\text{Growth in control} - \text{Growth in treatment}}{\text{Growth in control}} \times 100$$

## Conidial Germination

Amend the sterilized Czapek-Dox + yeast extract medium with cell free culture filtrates of the individual species of fungi at the concentrations of 10 and 20%. Distribute the amended liquid medium into the cavity slides placed in the sterilized Petri plates. Harvest the conidia from the pure cultures of the pathogen by splashing sterilized distilled water over the cultures and collect the conidial suspension. Wash the conidial suspension with repeated changes of sterilized distilled water in a centrifuge at 5000 rpm. Determine the conidial density in the suspension. Add a standard drop of the conidial suspension in each cavity slide containing the amended liquid medium and incubate the cavity slides at 25 ± 2°C. Control can be maintained by adding conidial suspension into the cavity slides containing only distilled water. After 24 h, mount the conidia from the amended liquid medium on clean slides and count the number of conidia germinated and calculate the per cent inhibition of conidia germination. The length of the germ tube is measured by using a calibrated ocular micrometer, and the per cent inhibition of growth in the hyphal length can be calculated. The determinations can be made as follows:

$$\text{Inhibition in conidia germination} = \frac{\begin{matrix}\text{No. of conidia germi-} \\ \text{nated in control}\end{matrix} - \begin{matrix}\text{No. of conidia germi-} \\ \text{nated in treatment}\end{matrix}}{\text{No. of conidia germinated in control}} \times 100$$

$$\text{\% inhibition of germ tube length} = \frac{\begin{matrix}\text{Length of the germ} \\ \text{tube in control}\end{matrix} - \begin{matrix}\text{Length of the germ} \\ \text{tube in treatment}\end{matrix}}{\text{Length of the germ tube in control}} \times 100$$

## MYCOPARASITIC INTERACTIONS

The mycoparasitic interactions between fungi can be studied by using the methods of Ikediugwu and Websster (1970) modified by Skidmore and Dickinson (1976). The fungi isolated from the soil by the conventional dilution plate technique and the pathogenic fungus used as the test organism can be used for this study. The individual species of soil fungi and the pathogenic fungi are grown

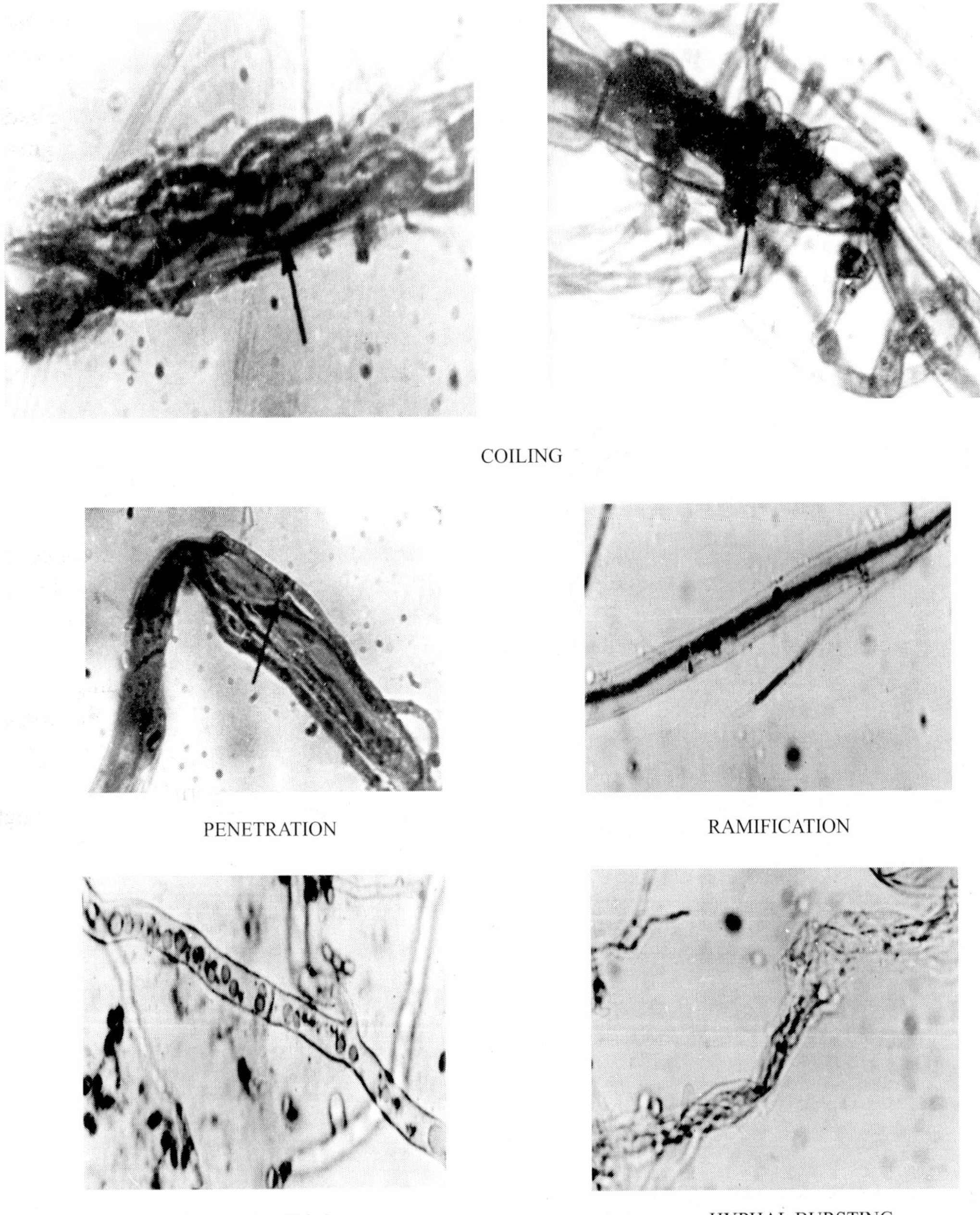


**Fig. 2** Mycoparasitic Interactions

individually in pure cultures. Potato-dextrose-agar (PDA) medium is prepared, sterilized and distributed into the 90 mm diameter sterilized Petri plates and the medium in the Petri plates is allowed to solidify. After solidification, a de-plasticized and sterilized cellophane disc of the size of the inner circumference of the plate is placed over the solidified medium in each plate. Then inoculate the agar blocks (5 mm diameter) cut from the actively growing margin of the individual species of fungi and the pathogen juxtaposed to each other approximately 3 cm apart. Incubate the plates at 25 ± 2°C for about a week. After incubation, cut small squares of cellophane paper from the intermingling growth region of the colony. The squares of the cellophane papers are to be stained with cotton blue and observed under the microscope. The parasitism of one fungus over the other can be observed. It can be observed as coiling, penetration, granulation, vacuolization, lysis and cell wall bursting. All these symptoms ultimately lead to the death of the opposed fungus. Such observations are to be made for each individual species of soil fungi dual inoculated with the pathogen.

## Electron Microscopic Studies

The mycoparasitic interaction can also be best studied by using staining as well as electron microscopic technique studies. The antagonist and the test organism are grown in pure cultures. Agar discs (5 mm diameter) are to be cut from the actively growing margin of the colonies and placed 3 cm apart on the surface of the Petri dishes containing freshly prepared, sterilized and solidified potato dextrose agar medium. Incubated Petri dishes at 25°C with continuous light after incubation of 4 days collect mycelial sample from the region of interaction samples are fixed with 3%(vol/vol) glutaraldehyde in 0.1 m sodium cacodylate buffer (pH 7.2) overnight at 4°C and post fixed with 1% (wt/vol) osmium tetroxide in the same buffer for one hour 4°C then dehydrate the sample in a graded serious of methanol and embedded in Epon 812. Cut ultra thin section (70 mm) thin section with diamond knife. Collect the sections on fornvar-coated with nickel grid. Then make them contrasted either with uranium acetate and lead citrate for direct examination with transmission electron microscope at 80 KV and are processed for cytochemical labelling to detect enzyme activities.

## Secondary Screening

In the foregoing paragraphs, the possible one to one interactions at the region of contact between the saprophytic soil fungi and the pathogenic fungi during their active phase were highlighted. But the types of interactions that occur are not as simple as explained. A battery of complexity of interactions could occur in nature. It is well-known the pathogens not only survive in the soil as propagules but also as saprophytes in the pre-colonized dead host tissues. Hence, disinfestations of the pathogen from the corpus of the host need a system that is involved in the production of cell wall degrading enzymes. Generally, such substrates are mainly composed of cellulose, hemicellulose and pectin. So, it is expected that the organism that is screened as potential antagonist and could be used as biocontrol agent should also possess the operating system, which facilitates the cell wall degrading enzymes. The antagonist during its mycoparasitic phase succumbs its host also by means of complete exploitation of the protoplast and cell wall bursting. It is also well known that the cell wall of most of the fungi is made up of chitin. The efficiency of the antagonist in the production of chitinolytic enzyme can be considered as an additional power with the biocontrol agents. Hence, the potential antagonists should also be screened for the production of plant and fungal cell wall degrading enzymes.

## CELL WALL DEGRADING ENZYMES

The cell wall of the plant tissues is mainly made up of cellulose, and the primary walls of the two adjacent cells are cemented together by pectic substances. Cellulose is an important carbohydrate that constitutes 40 to 60 per cent of plant materials and is also present in the plant cell wall (Norkans, 1963). It has been reported that the pathogen produces the cellulolytic enzymes during pathogenesis.

Cellulolytic enzymes also produced when the pathogen survives as saprophyte in the pre-colonized host tissues or in the dead host tissues. Thus, make the pathogen to possess the substrate in its hold. Hence, the disinfestations of the substrate with an predefined antagonist needs the antagonist to produce these enzymes. Thus, it becomes essential that the cellulolytic enzyme producing potentialities of the proposed antagonist is to be determined. It would be a value addition to the antagonist. The cellulolytic activity of an organism can be determined by filter paper disc colonization method or by viscometric method.

### Filter Paper Colonization and Loss in Dry Weight of the Filter Paper Discs

In general, the growth and activity of fungi in or over the surface plant substrates could be attributed to the production of cell wall degrading enzymes. Garrett (1966) mainly attributed it to the 'cellulolysis rate' of the fungus. It is nothing but the measure of the cellulolytic activity of the fungus. He determined it by estimation of the loss dry weight of the filter paper due to colonization by the fungus. The technique used by Garrett (1980, 1983) is described below:

### Determination of 'Cellulolysis Rate'

The 'cellulolysis rate', a measure of cellulolytic activity, of the organism can be assessed by determining the percentage of loss in dry weight of filter paper disc. In the first, trial two circular discs of Whatman No.1 filter paper (85 mm diameter) are placed in Petri plates (90 mm dia.) and sterilized. Another set is maintained without sterilization. The discs are then saturated with 2 ml of mineral solution containing $NaNO_3$ 660 mg, $KH_2PO_4$ 70 mg, $MgSO_4$. $7H_2O$ 35 mg and $FeSO_4$. $7H_2O$ 0.07 mg in one litre of distilled water. Then each plate is inoculated with agar blocks (5 mm diameter each) cut from the actively growing margin of the fungus grown on PDA medium. They are placed equidistantly round the margin of the disc, but 1 cm inside from the rim, so as to give more room for the development of fungal colonies. Three replicates are maintained and the plates are incubated at 25 ± 2°C. The growth rate of the organism is recorded after 7, 15 and 22 days of incubation. Mean loss in dry weight of each filter paper is calculated after 22 days. The weight that is lost due to colonization of the fungus indicates its cellulolytic activity.

### Viscometric Method

The cellulolytic activity can also be determined by the viscometric method (Alexander, 1954). The activity expressed has percentage of reduction in viscosity of one per cent carboxy-methyl cellulose (CMC) solution in Ostvald's viscometer pipettes. The test organisms are grown in 250 ml conical flasks containing 100 ml of Czepek's medium with one gram of CMC instead of glucose as carbon source. The pH is adjusted to 5.5. The sterilized medium is inoculated with agar blocks

(5mm diameter) cut from the actively growing margin of the pure cultures of the test organism individually and separately. The flasks are then incubated at 25 ± 2°C. At least three replicates are maintained for each organism. The contents are filtered through Whatman No.32 filtered paper, the cell free culture filtrates are taken in viscometer and a flex time is recorded at 12-hour interval. The procedure is repeated at least 5 times and a mean value is calculated. The per cent loss in viscosity is calculated as follows:

$$\% \text{ loss in viscosity} = \frac{\text{ETo} - \text{ET1}}{\text{ETo} - \text{ETw}} \times 100$$

where,

To = Flow time of reaction mixture at 0 minute

T1 = Flow time of reaction mixture at a particular interval

Tw = Flow time of distilled water

Control was maintained without inoculation of the organism.

## Determination of Pectinolytic Activity

The pectinolytic activity of the antagonistic organism can also be determined by the viscometric method. Czapek's medium containing $NaNO_3$ – 3.0 g; $K_2HPO_4$ – 1.0 g; $MgSO_4$ – 0.5 g; $FeSO_4$ – 10 mg and pectin 10.0 g in one litre distilled water was prepared. The pH adjusted to 5.5 and 50 ml of the medium was distributed in each 100 ml conical flask and sterilized in an autoclave at 15 lbs pressure and for 15 minutes. The flasks are then inoculated with agar blocks cut from the actively growing margin of test organism and some 5 replicates were maintained for each organism. Percentage loss in viscosity can be calculated as described in the previous paragraph.

## Chitinases

Chitin is the most abundant biopolymer, next to cellulose, found in nature. It is a linear polymer of β-1-4 linked N-acetyl glucosamine units. It is commonly found in the cell wall of most of the fungi, exoskeleton of crustaceans and cuticle of insects. The breakdown chitin means the deterioration of fungal cell wall and thus the destruction of pathogens by the antagonists, which potentially secrete chitinases.

Chitinases are enzymes that catalyze the hydrolysis of beta-1-4 bonds of the polymer into N-acetyl-chitooligosaccharides. They are broadly classified into exochitinases and endochitinases. The exochitinase activity is defined as the progressive action starting at the non-reducing end of the chitin with release of chitobiose or N-acetyl-beta-glucoaminidase, and chitobiosidases are considered as exochitinases. The endochitinase activity is defined as the random cleavage at the internal points in the chitin chain. The exo- and endochitinases synergistically increase the chitinolytic activity. The chitinolytic activities can be described as two reactions. In Reaction I, the chitinases catalyse the random cleavage of the internal points into oligomers, which are very rapidly hydrolysed into the dimer, chitobiose, and in the Reaction II, the chitobiase has a progressive action, starting at the non-reducing ends of chitin chains with release of successive chitobiose units and finally into N-acetyl-glucosamine.

$$\text{chitin} \xrightarrow{\text{ChitinasE}} \text{oligomer} \longrightarrow \text{dimer (reaction I)}$$

$$\text{chitin} \xrightarrow{\text{Chitobiase}} \text{dimer} \longrightarrow \text{n} - \text{acetyl} - \text{glucosamine (r}$$

It has been reported that not only the production of antibiotic substances, but the production of chitinolytic enzymes is also responsible for the antagonistic properties in the case of the strains of *Trichoderma*. The production of chitinolytic enzymes has been demonstrated in *T. harzianum* and *T. viride*. Chitinases of mycoparasites have been suggested to play a role in host infection since they can be localized on infection structures (Chet *et al.*, 1997; Manocha and Zhanghua, 1997).

Thus, in the foregoing paragraphs, chitins, chitinases and their activities were highlighted. It is obvious that the potential antagonist, which could be used as a potential biocontrol agent, should possess the ability to produce chitinases. Hence, screening of the selected organisms for their chitinolytic activity becomes essential.

## Determination of Production of Chitinases

Colloidal chitin is prepared from practical grade crab shell chitin with 85% $H_3PO_4$ at 6°C for 48 h. Aqueous colloidal chitin containing 10 mg of dried chitin/ml, after sterilization at 120°C for 30 minutes and kept at 6°C, can also be used.

The fungi, which are to be assessed for their chitinase production, are cultured in corn meal agar medium. Erlenmeyer flask (11) containing 150 ml of growth medium (0.2% dry weight colloidal chitin, 0.03% NaCl, 0.03% $MgSO_4$ $7H_2O$. 0.03%$K_2HPO_4$. 0.02% yeast extract and 1 (w/v glass wool) in distilled water and sterilized at 120°C in an autoclave) in each flask is inoculated with seven agar blocks (5 mm diameter each), cut from the actively growing margin of the individual species of fungi. The flasks are then incubated at 25 ± 2°C for about 37 days. After incubation the culture medium is first filtered through Whatman No. 3 filter paper and then through 0.2 μm Millipore polydifluoropropylene membrane. The filtrate is then analyzed for its chitinolytic activity, especially for beta-D-glucosaminidase activity. The filtrate can also be used for the analysis of proteolytic activity. The filtrate thus collected is lyophilized and kept at –20°C.

The chitinolytic activity is assayed by measuring the release of reducing saccharides from colloidal chitin. It is done as described below:

A reaction mixture containing 1 ml of filtrate, 0.3 ml of 1 M sodium acetate buffer (SA buffer), pH 4.7, and 0.2 ml of colloidal chitin is incubated at 40°C for 6-24 h, and then centrifuged at 12,225 g for 5 minutes at 6°C. After centrifugation, an aliquot of 0.75 ml of the supernatant is added with 0.25 ml of 1% solution of dinitrosalicylic acid in 7 M NaOH and 0.1 ml of 10 M NaOH in 1.5 ml of Eppendorf tubes, mixed and heated at 100°C for 5 minutes. The reaction mixture is then cooled at room temperature and its absorbance is measured at 582 nm (A582), as described by Miller (1959). A standard calibration curve is prepared with N-acetyl-D-glucosamine (NAGA) to determine the concentration of reducing saccharide. A linear correlation between A582 and the concentration of N-acetyl-D-glucosamine (NAGA) has to be found at an interval of 40-800 μg/ml. The NAGA chitinolytic activity is expressed as NAGA units. One NAGA unit is defined as 1 μ mol of NAGA released i.e.,

$$1 \text{ NAGA unit} = 1\ \mu \text{ mol of NAGA}, \text{ under the described conditions.}$$

Specific chitinolytic activity is defined as the quantity of BAGA units released by the activity of 1 mg of enzyme per one hour under said condition i.e.,

Specific chitinolytic activity = A582, NAGA-U. $h^{-1}$ $ml^{-1}$or NAGA-U. $h^{-1}$.$mg^{-1}$

The exochitinase, N-acetyl-beta-D-glucominidase, activity can be measured and monitored spectrophotometrically. It is measured by determining the p-nitrophenyl-N-acetyl phenol (pNP) from p-nitrophenyl-N-acetyl-D glucosaminide.

A mixture containing 1-25 μl of testing solution, 0.2 ml of PNP solution and 0.1 ml of 0.1 M SA buffer at pH 4.7 is heated at 40°C for 6-24 h and centrifuged. After centrifugation, the supernatant is added with an aliquot of 0.3 ml of 0.125 m sodium tetraborate – NaOH buffer (pH 10.7). Immediately after mixing, the absorbance is measured at 400 nm. The PNP concentration in the solution is calculated by using the NP molar extinction coefficient (18,500 $M^{-1}$). So, evaluation and optimization of conditions under which maximum expression of characters become essential.

## Cytochemical Labelling

The chitinase activity of fungi can also be studied in the case of mycoparasites by a technique called cytochemical labelling of the ultra thin sections prepared for the electron microscopic studies. The technique of cytochemical labelling is described below.

Colloidal gold (average particle diameter 12 nm) is prepared, which is then used for studying the distribution of chitin. Ovomucoid, a high molecular weight glycoprotein, is used as a second step reagent, because it has specific affinity for wheat germ agglutinin (WGA). Then a glycoprotein is complexed with a colloidal role at pH 5.4. The sections are then incubated with one drop of phosphate buffered saline solution (PBS-pH 7.2) for 5 minutes and then with one drop of WGA (12.5 μg/ml in PBS) for 30 to 60 minutes at room temperature in moist chamber. Now the sections are washed thoroughly with PBS. Then the sections are incubate with the ovomucoid gold complex (1:60) in PBS-polyethlene glycol 20000). Now, the sections are washed with PBS, rinsed with distilled water and contrasted with uranil acetate and lead citrate. This specificity of labelling can be assessed by (i) Incubation with WGA to which N-N'-N"- tri acetylchitotriose(1mg/ml) in PBS , which is added previously, (ii) Incubation with WGA, then with uncomplexed ovomucoid and finally with ovomucoid-gold complex, and (iii) direct incubation with gold complex ovomucoid.

## Quantification of Labelling

The density of labelling obtained with WGA-ovomucoid-gold complex can be determined by counting the number of gold particles per square micrometer. Areas can be determined by the point counting method by using negatives of electron micrographs projected onto a lattice. The amount of labelling in a specified wall area (Sa) is estimated by counting the number of gold particles (Ni) on a photographic enlarged plate. The density of labelling (Ns) is calculated as follows:

Ns = Ni/Sa where Ns is the number of gold particles per unit surface.

## IMPROVEMENT OF ANTAGONISTIC ORGANISMS

The genetic make of the antagonistic organisms can be modified and the maximum expression of the interactive forces can be improved. They can be achieved through mutation and recombination which in turn through protoplast fusion and genetic recombination with rDNA technology.

Genetic manipulation in fungi and other microorganism that produce antibiotics have led to substantial increases in the yields of the secondary metabolites. The most widely used method is mutation through mutagenic treatment of individually species of *Trichoderma, Gliocladium and Pythium* have been extensively studies. United States and Hebrew University team produced new useful mutant stains of *Pythium* which attaches its pathogenic relatives. They have been patented by both US and Israeli partner.

The American and Israelian team has produced new 'super strains' of *Trichoderma* by combining cells from different variation of *Trichoderma* effective on wide range of crops and pests. The super strains were produced by suing protoplast fusion technique. Protoplast fusion is a technique of combining useful derivatives from two strain into single more broad by effective strain through fusion of the protoplasts of the two strains. Osmotic stabilizers such as KCl (0.6M), sugar, sugar alcohol, other mineral salts can be used for obtaining protoplast of the strains. The protoplast fusion can be induced with polyethylene glycol (PEG) and $Ca^{+2}$. It has been reported that protoplast fusion introduced great genetic diversity into progeny strain and very small percentage were improved biocontrol ability over the wild strain of *T. harzianum* and *T.viride*.

### Gene Manipulation

It has been well established that the antibiotic compounds are secondary metabolites. They are small molecules and many genes are involved in their biosynthesis. Hence, it has been suggested that it would be realistic to investigate the genetics of secondary metabolite production and ways of optimizing yield rather than attempt to transfer all the genes concerned to another organism. On the other hand production of compounds such as eukaryotic enzymes is controlled by single genes, and hence they can be manipulated and the enzymes can be manufactured in prokaryotes.

Chitinases are the important enzyme involved in biocontrol mechanisms, which have already been dealt with. A complete survey of *Trichoderma* chitinases reveals that there are three subgroups in the family of eighteen chitinases (Seidl *et al*, 2005). The chitinolytic system of *T. harzinam* has been studied (Haran *et al.,* 1995, Carolina Carsolio *et al.*, 1999, Steyaert *et al.*, 2004). The system apparently has two B-(1,4) – N – acetyl glucosaminidase (102 and 73 K Da) and four endochitinases (52, 42, 33 and 31 K Da). Although the entire system is required for the maxium efficacy, the most interesting individual enzyme of the complex is the 42 K Da (Ech 42). The corresponding gene has been indentified as 'ech 42'. It has been experimentally proved that it strongly induced during fungus-fungus interactions and when the fungus is grown in the media containing autoclaved fungal mycelia of several fungi or in the media with colloidal chitin as the only carbon source. Thus, the production of 'ECh 42' endochitinase plays a key role in the degradation of fungal cell wall and essentially in biocontrol.

Carolina Carsolio *et al.,* (1999) studied the role of *Trichoderma harzianum* endochitinase (ECh 42) in mycoparasitism by genetically manipulating the gene (*ech 42*) that encodes Ech 42. They constructed several transgenic *T. harzianum* strains carrying multiple copies of ech 42 the corresponding gene disruptants. In the following paragraphs, the methods as described by them are provided in order to facilitate the readers to understand the back ground technique of gene manipulation. Carolina carsolio *et al.*,1999 used the organisms *T. harzianum* (MI 2060-40) and *Escherichia coli* JM 103 and MC 1061 for DNA manipulation, pBlueScript (Stratagene), P sport (Gibco – BRL, Madison, Wis) and pHA T alpha, which carries the *E.coli* hygromycin phosphotransferase (*hph*) gene as a dominant selectable marker, were used as plasmid.

## Plasmid Construction

The DNA fragment Ag-kb BamHI from the '*ech* 42' genomic clone in lambda EMBL3 was subcloned into the plasmid m pBlue script and the plasmid pQui49 was generated. The entire coding region for 'Ech 42' and approximately 2 kb of the 3' region and 5 kb of 5' region are included in the inserted fragment. Plasmid pQuiA2d is generating plasmid pquiA 2.3. In this way 228 pp from the gene 'Ech42' was replaced by the hygromycin resistance cassette.

## PCR Amplification

Amplification of genomic DNA from *T. harzianum* transformant with plasmid pQui 2.3 was carried out through PCR technique. They designed primers in order to make homologous reconstitution or ectopic integration in different band patterns. According to them the forward primer (5' GGACCAGGTGGTGTT3') hybridized on the 5' side of the gene 'each 42' but outside the fragment contained in p Qui 2.3 Δ, and the reverse primer (5'TAGTTGAGACCGCTTCG3') hybridized in the coding region down stream of the insertion site of the cassette. They used 'Taq' polymerase (Gibco – BRL) for all the amplification processes. They also used PCR programme as a hot start consisting of 5 minutes at 95°C, 25 cycles consisting of 71 minutes at 94°C and final extension period consisting of 7 minutes at 72°C. It is expected that there should be a 4.2 kb band for disruptants, where as the transformants in which integration had occurred ectobically should have a 3.0 kb fragment.

## DNA and RNA Manipulation

Isolation of fungal chromosomal DNA was made as described by Raeder and Broda (1985). Sambrook *et al.*, (1989) have described the molecular cloning techniques which were followed in DNA manipulation of *T. harzianum*. Fungal RNA was isolated following the protocol described by Jones *et al.* (1985). Standard procedures of Southern blotting and Northern blotting were followed.

## Chitinase Assays

Determination of chitinase activities was made by using the culture fultrates. Carolina Carsolio *et al.*, (1999) used the culture filtrates of transgenic *T. harzianum* strain. The culture filtrates were freeze dried and resuspended in reaction buffer containing 50 mM sodium acetate, 0.15 mg of phenyl methyl sulfonyl fluoride per ml, 10 mM EDTA and pH 5.6; and the activity was determined by SDS PAGE.

Protein samples (20 ml) from *T. harzianum* were freeze dried, suspended in 3 ml of deionized water dialyzed against $H_2O$ for 36 h at 4°C frozen, and lyophilized again. The samples were resuspended in 300 to 500 ml of $H_2O$ divided into aliquots, frozen and kept at 70°C, until they were used one to 10 microgram of each sample was resuspended in Laemmli buffer without beta-mercapto ehanol and subjected to sodium dodecyl sulfate (SDS) polyacrylamide gel electrophoresis (PAGE) without boiling in 0.7 mm thick 4 per cent acrylamide stacking gel and 10 per cent acrylamide separating gels. After electrophoresis the SDS was removed by using the casein EDTA procedure as described by Mc Grew *et al.* (1990) and as modified by Haran *et al.* (1995). The gels were overheated in 2 ml of 1% low melting agarose in acetate buffer containing 300 μg of 4-methyl umbelliferyl beta-D-N,N',N"- tri acetylchito triose and incubated at 37°C for 10 minutes, and the bands were visualized under short wavelength uv light.

## Western Blot Analysis

Proteins were electrophorased in polyacrylamide SDS gels as described previously. The enzyme activity analysis was also made in the same way except that the samples were boiled in the precence of beta mercarpto ethanol. The gels were electrotransfered to Immbilon membranes. The blots were sequencing treated with rabbit anti – ECH 42 antibodies and alkaline phosphatase – conjugated goat anti rabbit immunoglobulin G. Phosphatase colour was developed by incubating the preparation with BCIP (5-bromo-4chloro-3-indolyl phosphate) and nitriblue tetrazolium.

## EVALUATION

The foregoing paragraphs clearly illustrate the various mechanisms involved in biological control process and various methods of determining them. It is obvious that the various antagonistic interactions are either acting independently or complementing to each other or there is multitude of interactions. Various forms of *in vitro* studies indicate the genetic potentialities of the organisms. Nevertheless, the expression of the interactive forces varies in relation to organisms, strain, ecosystem and environmental conditions.

It has been suggested that the accuracy of the presumptive tests in agar media could undoubtedly be influenced by the composition of nutrients supplied, alteration of the salt concentration above or below the level of water potential, the type of agar, the pH and temperature (Baker and Cook, 1974). The influence of pH and nitrogen content of the medium and of soil has often been emphasized. Generally, fungi have a wide range of tolerance to pH. However, the expression of the genetic make up of the individual species of fungi for the inhibitory effect is influenced by specific levels of pH.

Temperature as a factor influencing the activities of fungi has received the attention of many workers in temperate countries. But in tropical countries such as India the fluctuation of temperature is very wide. It ranges between 20 and 42°C in most of the regions. Generally, it falls within the tolerance range of the mesophilic fungi.

## FORMULATION OF DELIVERY SYSTEM FOR THE ANTAGONIST

Once when the organism is screened evaluated and improved, it becomes essential to formulate and to develop a suitable delivery system. The design of the delivery systems varies depending on the organism, crop and mode of application. Hence, the development of feasible and efficient delivery system for the application of appropriate organism is an important component of the biocontrol research. Dusts, alginate pellets, starch granules, extruded granules, seed coatings, complex cellulose granules, etc., are some of the commonly used delivery systems for the biocontrol agents. Various organic and inert carriers such as diatomaceous earth, lignite/stillage, wheat bran and formulated semi solid fermentation products have also been practiced. In the subsequent paragraphs, a brief review of the various formulations is provided.

The mass introduction of antagonistic fungi into the soil has been suggested. As Garrett (1956) pointed out, at different times, various workers have proposed inoculation of soil with such antagonistic organisms. If an alien organism with such desirable characteristics be introduced into a soil population from which it is absent, there is always a chance, albeit somewhat remote, that it will succeed in establishing itself, and that beneficial result will follow. Most attempts at soil inoculation have had a still smaller chances of success, because the inoculant organism have been isolated in the first place from the soils in which it was desired to establish than at a higher population level. Such attempts to boost the population of an antagonistic organism by inoculation alone have been doomed to failure from their inception, because they are in flagrant contradiction to the ecological axiom that the population is a reflection of the habitat, and that any change due to plant introduction without change of the habitat much be a transient one. Early experiments in direct introduction of the antagonistic microorganisms into the soil were effective against certain plant pathogens in previously sterilized soil (Garrett, 1956). Most of the experiments when extended to the natural soil provided futile. Attempts to control soil-borne plant pathogens in natural soil with direct soil microbial amendment did not always fail. Some limited success has been reported (Weinding and Fawcett, 1936; Wood and Tveit, 1955; and Stevenson, 1956).

The demonstration of suppression of the damping off of caused seedlings of Citrus by *Rhizoctonia solani* in the acidified soil (pH 4) introduced with *Trichoderma lignorum* is the only practical success with *Trichoderma*. The registration in France of *Trichoderma* preparation in a nutrient base for the control of dry bubble disease of cultivated mushrooms caused by *Verticillium* sp. has been reported. The formulated products of *T. harizianum* have been developed for application to wounds of *Acer rubrum*. Wells *et al.* (1972) first reported the field control of *Sclerotium rolfsii*, a soil borne pathogen on tomato transplants, with *T. harzianum*. Elad *et al.*, (1980) grew *T. harzianum* in a wheat bran + saw dust medium and introduced into the soil infested with pathogens, *S.rolfsii* and *R. solani,* which delayed the progress and incidence of damping off of beans caused by the pathogens in the field and decreased the severity of disease. However, Elad *et al.*, (1980) found that *T. harzianum* was less effective against *S. rolfsii* and *R. solani* in the field than in the glasshouse.

Papavizas (1981) grew *Corticium sp., T. hamatum, I* and *Gliocladium roseum* in sterilized sand-corn meal, and when amended in the soil controlled the cucumber fruit rot disease caused by *R. solani*. The various media used for the growth and development of antagonistic fungi have been reviewed by Papavizas and Lewis (1981) which included milled rice (Ahmed and Tribe, 1977) and barley, rye and sunflower seed (Huang, 1976) for *Coniothyrium minitans*, corn leaf meal (Hoch and Abawi 1979) sand-cornmeal (Lewis and Papavizas, 1980) for *Corticium* sp. peat soil, and nutrients (Kommedahl and Windels, 1978) for *Gliocladium virens*: and grain bran (wells *et al.*, 1972) celatom and molasses (Backman and Rodriquez Kabana, 1975), bark pellet (Sundheim, 1977), wheat straw (Akhtar, 1977), wheat bran (Henis *et al.*, 1978; Hadar *et al.*, 1979), Corn meal

– sand (Mangenot and Diem, 1979), wheat bran and saw dust (Elad *et al.*, 1980). The biological control aspects of *Trichoderma* spp. and *Gliocladium spp.* were elaborately discussed by Papavizas (1985). Other systems include incorporating antagonists into a fluid drilling liquid (Conway, 1986); adding chemicals to enhance antagonist germination (Nelson *et al.*, 1988); and most recently, hydration of seed before sowing, coupled with addition of biological control agents (Harman and Taylor, 1988; Osburn and schroth 1988; Callan *et al.,* 1990). Alginate wheat bran formulation of *Gliocladium virens* has been found to control damping off of seedlings of cabbage, cotton and zinnia caused by *Pythium* sp and *R. solani* (Lumsden and Locke, 1989; Lumsden *et al.*, 1990).

The induction of *Rhizoctonia* suppressiveness due to increase in activity of *Trichoderma* spp. was not easily accomplished in alkaline soils was reported by Chet and Baker (1980, 1981). Many investigations also reported (Henis *et al.*, 1978: Liu and Baker, 1980; Chet and Baker 1980, 1981: Elad *et al.*, 1981, 1982, 1987: Marshal, 1982: Papavizas *et al.*, 1982; Hadar *et al.*, 1983; Lim Tong Kwee and The Boon Keng, 1990) that seed treatment or soil treatments with isolates of *Trichoderma* spp. reduced damping off of induced by *R. solani*. Several other diseases have been reduced in severity following the application of antagonists in soil. For example, damping off caused by *R. solani* on bean, cucumber, egg plant, potato and tomato was reduced by *T. harzianum* (Hadar *et al.*, 1979; Beagle Restalino and Papavizas, 1985; Lewis and Papavizsa, 1987 a, 1989; Oyarzum *et al.*, 1993; Yuen, *et al.*, 1994) and reduction of damping off and crown rot caused by *R. solani* on cotton, radish, and wheat through introduction of non pathogenic isolates of *Rhizoctonia* (Ichielevich-Auste *et al*, 1985). These reports suggested that a considerable increment of control of *R. solani* was achieved through mycoparasitism. Many antagonists have been described as having some potential as biological disease control agents. They come mainly from screening or 'mode of action', studies *in vitro* and commonly produces metabolites or parasitizes pathogens in the absence of a plant. Many reports have indicated reproducible biocontrol in pot tests or seedlings bio assay under controlled conditions or in the glass house for less than a month.

Despite many examples of biocontrol in both glasshouse and field, there are relatively few commercial successes in current practices (Whipps, 1992). There is often lack of reproducibility between trials in the laboratory or glasshouse and the field, which is because and antagonist is taken from one environment and expected to act in another to which it is not adapted. Whipps and Lumsden (1991) suggested that the degree of control is often varying from time to time and from place to place. Similarly, it has been suggested that the soil conditions such as temperature, water availability, pH and nutrients which are sub optimal for good growth and development of antagonistic fungi can produce effective biological control (Kommedahl and Windels, 1978; Whipps, 1992).

Schneider (1954) and Dube (1971) have reported that *R. solani* cause greatest disease incidence in wetter soils. Eastburn and Butler (1991) reported that the colonization of alfalfa pieces was very low under flooded condition. Per cent colonization increased as metric potential of the soil decreased to 1.0 bar. Thus, the optimum requirements of moisture vary among different species of fungi. Huber and Christmas (1992) reported that the crown rot was generally more prevalent in wet and poorly drained areas of fields but also was severe on early seeded canola on sand soils. They have also suggested that seeding date; nitrogen use and cultivar selection may be important consideration in reducing severity of this disease.

The soil pH plays an important role in biological control as well the disease caused by the soil-borne plant pathogens. Chet and Baker (1980) have reported that the induction of *Rhizoctonia* suppressiveness due to increase in activity of *Trichoderma* spp. was not easily accomplished in alkaline soils whereas *Trichoderma* spp. germinate best in acid soils. Scher and Baker (1980) have reported that the soil pH 8.1 was suppressive to *Fusarium* and the suppressiveness was not when the soil pH was lowered to 7.0. Thus, the soil suppressiveness and the activity of the antagonistic fungi against the pathogen vary with the soil pH. Huber (1981) has reported that the mineral nutrients, especially nitrogen affect the expression of disease. He also emphasized that the form of nitrogen frequently influence disease expression more than its rate. It has been reported (Mac Nish and Speijers 1982) that ammonium sulphate reduced severity of take all of wheat in field experiments while sodium nitrate failed to reduce the disease.

## Preparation of Conidial Suspension as Inoculum

The repeated *in vitro* demonstrations of the antagonistic interaction between the antagonistic fungi and plant pathogens have made to attempt to find out their possible application in the natural soil. The antagonistic fungi are grown as pure cultures on PDA medium in Petri dishes. The conidia are harvested from the pure culture by flushing with sterile distilled water and collected in culture tubes. The conidia are then washed at least thrice with sterile distilled water in a centrifuge at 5000 rpm. The pelletted condia are then diluted to a standard volume. The density of conidia is determined using haemocytometer. Then this conidial suspension is used for direct inoculations in to the soil.

## Preparation of Sand Maize Meal Inoculum

It has been reported that the addition of individual antagonistic fungi in the form spore/conidial suspensions it not have any effect in declining the disease. There are reports of early experiments with direct introduction of antagonistic fungi in the soil proved effective against soil-borne plant pathogens in previously sterilized soil. Nevertheless, test in natural soil were discouraging and the antagonistically active fungi against pathogens in pure cultures failed to exert any effect of biological control on pathogens when mixed in natural soils. The success in sterile soil may be due to the abundant nutrients and freedom from competition by other microorganisms in contrast to lack or infrequency of success in natural soils that oppose the establishment of the introduced antagonistic fungi. The conidia when directly introduced into the soil do not germinate because the phenomenon of soil fungistasis does not permit them to do so. Soil fungistasis, a phenomenon wide spread in soil, does not allow the conidia to germinate which are not under the influence of either constitutive are exogenous dormancy. Annulment of fungistasis can be made by amending the soil with energy rich substrates. So, food bases have often been emphasized has the essential requirement for the mass production of inocula. The first attempt was made in Cambridge with sand-maize meal inocula.

The sand-maize meal inoculum of the antagonist fungi can be prepared following the method described by Garrett (1963). 200 g of acid washed sand is taken in 500 ml conical flask. Maize meal is added to the acid washed sand at 3% level then the flask is sterilized in an autoclave. The flasks are then inoculated with 3 agar blocks (10 mm diam) of pure culture of the individual species of the fungi. The moisture is maintained at 45% MHC. The pH is adjusted to 6.5. These flasks are incubated for a period of one month. They are shaken at regular interval in order to maintain a homogenous continuum of propagules. The inocula, prepared in such a way are used for the biocontrol experiments.

## Preparation of Rice Bran Medium

Apparently no attempts have been made to standardize the growth media for fungal antagonist against soil-borne plant pathogens. There are various method suggested for the mass production of the inoculum of the antagonist fungi using different substrate. The rice bran inoculum can be prepared with exogenous carbon and nitrogen. Empty saline bottles (500 ml capacity) are collected, are washed, rinsed with distilled water, dried in hot air oven at 60°C and stored. The bottles then filled with rice bran (300 g). Mix the bran with sucrose (3%) and Na NO3 (1%). The bottles are shaken thoroughly in order to get uniform distribution and suitable moisture is adjusted. Care is to be taken to see that there is no excessive are insufficient moisture. The mouth of the conical flasks is closed with cotton plug and covered with rapper. They are then sterilized in an autoclave at 15 lbs pressure for 15 minutes at 110°C at least twice in alternate days. The bottle containing rice bran medium is then stored for further use. Individual species of selected antagonistic fungi are grown in pure culture on PDA medium, distributed in 90 mm diameter Petri dishes. Agar blocks (5 mm) cut from the actively growing margin of the individual species of the fungi are inoculated in to each conical flask containing the rice bran medium, at least 5 blocks in each flasks. The inoculated conical flasks are then incubated at 25 ± 2°C for a month. The flasks are periodically observed for the confirmation of the successful establishment of the inoculated fungus, and to ensure that there is no contamination. This rice bran medium is used for the bio control experiments.

## Preparation of 'BIODAC' Formulation

The Biocontrol of Plant Disease Laboratory (BPDL) in US has recently developed a complex cellulose granule with a trade name 'Biodac' and got the patent. The 'Biodac' is a product of consistent particle. It is composed of recycled cellulose based paper waste. The granules can be prepared in a variety of sized particles. They are used as a carrier. The 'Biodac' granule (size 16/30 mesh) formulations are prepared and activated by heat treatment in a convection oven for three days at 80°C. This treatment reduces the microbial contaminations. Sodium alginate (1% w/v) is prepared, autoclaved and used as sticker. The 'Biodac' formulation is mixed with 45 ml of the alginate solution with spatula and the biomass of the biocontrol agent (*T. harzianum, T. viride* and *Gliocldium virens*) is added up to 2.25 g. The mixture is spread over aluminium foil and dried over night with fowing air. The dried formulation stored at 4°C and used.

In the foregoing paragraphs the significance of food-bases for the preparation of delivery systems of biocontrol agents are highlighted. The delivery systems thus developed should satisfy the following requirements.

- They should be cost-effective.
- They should permit the production in large quantities.
- They should remain viable with out any specialized storage systems.

- They should possess the ability of tolerance to desiccation.
- They should have the mechanism prolonged shelf life.
- The designed formulation could be extended to as many crops as possible.
- They should have effectiveness against multiple pathogens.
- They should be contamination free.
- They could be potentially applied to diverse environmental conditions.
- They should not cause any environmental problem.
- They should not cause any health problem such allergy or infection on handling.
- They should get readily regenerated and established on application so as to make the pathogen ineffective in the pathozone.
- The food bases used should not alter the inherent potentialities of the biocontrol agent and at the same time they should cause any adverse effect on the crop.

## CONCLUSION

Biocontrol strategies take the advantage of the environmentally safe microorganisms. Many antagonists have been described as having some potential biological disease control agents. They have come mainly from screening or 'mode of action' or *in vitro* studies for the production of antibiotic metabolites or *in vitro* mycoparasitism. *Trichoderma* strains had been the most potential candidate for such evaluations. There are several strain of *Trichoderma* discovered as potential biocontrol agents. Emerging modern technologies such as protoplast fusion and rDNA technology have been used for the development of 'super strains' of *Trichoderma.*

Most of such organisms were successfully demonstrated as biocontrol agents in both glass houses and pot culture experiments but failed when extended to the nature. Improved strains of biocontrol organisms have been developed employing modern technologies and they have been registered, patented and even approved by the environmental agencies for commercialization. Nevertheless, the commercial uptake of biocontrol agents for control of plant pathogens and crop diseases has been limited due to variable efficacy under field conditions. Considerable performance of such candidates in nature is often elusive. Hence, the impact of the environmental conditions on the expression of biocontrol potentialities needs to be addressed. Moreover, the available evidences clearly indicate that most of the biocontrol researches have been dominatingly centred on different strains of *Trichoderma*. Therefore, it is high time to give attention towards the indigenous resident microorganisms for the screening which could definitely introduce new candidate of biocontrol agents.

## References

Ahamed, A.H.M. & Tribe, H.T. (1977) *Plant Pathology*, **26**: 75-78.

Akhtar, C.M. (1977) Ann. N.Y. *Acad.sci.*, **287**: 45-56.

Alexander, M. (1971) *Introduction to Soil Microbiology* 2nd Ed. John Wiley, New York.

Backman, P.A. & Rodriqez Kabana (1975) *Phytopathology*, **65**:819- 821.

Benhamou, N. (1989) In: *Colloidal Gold. Principles, Methods and application*, Vol. I (Ed. M.A. Haejat) pp 95-143, Acad. Press, New York.

Bragle-Ristaino, J.E. & Papavizas, G.C. (1985) *Phytopathology,* **75**: 560- 564.

Brian, P.W. (1951) In: *Ecology of Soil Fungi* (Eds Parkinson, D. & Waid, J.S.) pp 115-129, Liverpool Univ. Press.

Brock, T.D. (1966) Principles of Microbial Ecology, Prentice-Hall, Englewood, Clifts,N.J

Callan, N.W., Mathre, D.E. & Miller, J.B. 1990 *Plant Disease*, **74**: 368-372.

Cherif, M. & Benhamou, N. (1990) *Phytopathology*, **90**: 1406-1414.

Chet, I. (1987) *Trichoderma- application, mode of action and potential as biocontrol agent of soil-borne plant pathogenic fungi*, Wiley and sons, New York.

Chet, I. (1987) *Innovative Approaches to Plant Disease Control*, Wiley, New York.

Chet, I. & Baker, R. (1980) *Phytopathology,* **70**: 994-998.

Chet, I. & Baker, R. (1981) *Phytopathology*, **71**: 286-290.

Chet, I., Inbar, J. & Hadar, Y. (1997) In:*The Mycology IV Environment and Microbial Relationship* (eds D.T. Wicklow and B. Soderstorm) pp 168-184 Springer Verlog, Berlin.

Conway, K.E. (1986) *Plant Diseas*, **70**: 835-839.

Carolina Carsolio, Nicole Benhamou, Shoshan Haran, Carles Cartes, Ana Gutierrez, Ilan Chet, & Alfredo Herrera- Estrella (1999) *Appl. Environ. Microbial.*, **65**: 929-935.

Cook, R.J. (1977) In: *Plant Disease. An Advanced Treatise* vol 1., (Eds. Horsfall, J.G. & Cowling, E.B.) pp 145-166, Acad. Press, New York.

Cooper, R.M. 1983 In: *Biochemical Plant Pathology* (Ed. Callow, J.A.), John Wiley, New York.

De la Cruz, J., Hidalgo-Gallego, A. Lora, J.M., Benitez, T, Pintor-Toro, J.A. and Llobell *1992 Eur. J. Biochem*. **206**: 859-867.

Eastburn, D.M. & Butler, E.E. (1991) *Mycologia,* **83**: 257-263.

Elad, Y., Chet, I. & Katan, J. (1980) *Phytopathology*, **70**: 119-121.

Elad, Y., Barak, R. Chet, I. & Henis, H. (1983) *Phytopathology* Z **107**: 168-175.

Elad, Y. Hadar, Y., Hadar, E., Chet, I. & Henis,Y. (1981) *Plant Disease* **65**: 675-677.

Elad, Y., Chet, I. & Henis, Y. (1982) *Canadian J microbiology* **28**: 719-725.

Elad, Y., Sadowsky, Z. & Chet, I. (1987) *Trans. British Mycol soc*. **88**: 259-263.

Fravel, D.R. (1988) *Ann. Rev. Phytopathol.*, **26**: 75-91.

Gracia, I., Lora, J.M., De la Cruz, Benetez, T., Llobel, A. & Pintor- Tora, J.A. (1994) *Curr. Genetics* **27**: 83-89.

Hadar, Y., Chet, I. & henis, Y. (1979) *Phytopathology*, **69**: 64-68.

Hadar, Y., Harman, G.E., Taylor, A.G. & Norton, J.M. (1983) *Phytopathology*, **73**: 1322-1325.

Haran, S. Schickler, H., Oppenheim,A. & Chet, I. (1995) *Mycol. Res*. **99**: 441-446.

Harman, G.E. & Taylor, A.G. (1988) *Phytopathology*, **78**: 520-525.

Henis, Y. Ghaffar, A. & Baker, R. (1978) *Phytopathology*, **68**: 900-907.

Hoch, H.C. & Abawi, G.S. (1979) *Phytopathology*, **69**: 417-419.

Huang, H.C. (1977) *Canadian j. Botany*, **55**: 289-295.

Huber, D.M. (1981) In: *Biology and Control of Take-All*, (Eds. Acer, M.J.C. & Shipton, P.J.) pp 317-341.

Huber, D.M. & Christmas, E.P. (1992) *Plant Disease*, **76**: 1251-1253.

Ichielvich-Aster, M., Sneh, B., Barash, I. & Koltin, Y. 1985 *Phytopathology*, **75**: 1080-1084.

Jones, J.D.G., Dunsmuir, P. & Bedbrook, J. (1985) *EMBO J.* **4**: 2411-2418.

Kommedahl, T. & Windels, C.E. (1978) *Phytopathology*, **68**: 1087-1085.

Kumar, T., & Saravanamuthu, R. (2004) *Geobios* **31**: 137-139.

Lewis, J.A. & Papavizas, G.C. (1980) *Phytopathology*, **70**: 85-89.

Lewis, J.A. & Papavizas, G.C. (1987) *Plant Pathology*, **36**: 438-446.

Lewis, J.A. & Papavizas, G.C. (1987a) *Soil Biol. and Biochem.* **19**: 195-201.

Lewis, J.A., Papavizas, G.C. & Lumsden, R.D. (1989) *Phytopathology,* **79**: 1160.

Lewis, J.A., Papavizas, G.C. & Lumsden, R.D. (1991) *Biocontrol Science and Technol.*, **1**: 59-69.

LimTong-Kwee & The Boon Keng (1990) *J. pl. dis. prot.*, **97**: 33-41.

Liu, S. & Baker, R. (1980) *Phytopathology*, **70**: 404-417.

Lockwood, J.L. (1988) Ann. Rev. *Phytopathol.*, **26**: 93-121.

Lumsden, R.D., Lewis, J.A. & Papavizas, G.C. (1989) *Phytopathology,* **79**: 361-366.

Lumsden, R.D., Carter, J.P.. Whipps, J.M. & Lynch, J.M. (1990) *Soil boil. and biochem.*, **33**: 187-194.

Lumsden, R.D., Locke, J.C., Adkins, S.t. Walter, J.F. & Ridout, C.J., (1992) *Phytopathology*, **82**: 230-235.

Lumsden, R.D., Ridout, C.J., Vendemia, M.E., Harrison, D.J. Waters, R.M. & Walter, J.F. (1992) *Canadian j. microbial.*, **38**:1274-1280.

Mc Grew, B.R. & Green, D.M. (1990) *Anal Biochem.*, **189**: 68-74.

Mac Nish, G.C. & Speijers, J. (1982) *Annals appl.biol.*, **100**: 83-90.

Manocha, M.S. & Zhonghua, Z. (1997) *Mycologia*, **89**: 189-194.

Marshal, D.S. (1982) *Plant Disease*, **66**: 788-789.

Miller, (1959) *Anal Chem.,* **31**: 426-428.

Muthukumar, C., Kumar, T. & Saravanamuthu, R. (2004) *Geobios* **31**: 13-16.

Nelson, E.B., Harman, G.E. & Nash, G.T. (1988) *Soil boil. and biochem.*, **20**: 145-150.

Norkans, B. (1963) *Ann. Rev. Phytopathol.*, **1**: 325.

Osburn, R.M. & Schroth, M.N. (1988) *Phytopathology,* **78**: 1246-1250.

Oyarzun, P.J., Postma, J., Lultikholt, A.T.G. & Hoagland, A.E. *Canadian J botany*, **72**: 843-852.

Papavizas, G.C. (1981) *Biological Control in Crop Production*, Beltsville Symposia in Agricultural Res. 5, Allanheld, Osmun pub. Inc. N.J.

Panneerselvam, A. & Saravanamuthu, R. (1999) *J. Indian Bot Soc.* **78**: 95-97.

Papavizas,G.C. (1982) *Phytopathology,* **72**: 121-125.

Papavizas, G.C. (1985) *Ann. Rev. Phytopathol.,* **23**: 23-54.

Papavizas, G.C. & Lewis, J.A. (1981) In: *Biological Control in Crop Production* Beltsville Symposia in Agricultural Res. 5, Allanheld, Osmun pub. Inc., N.J.

Papavizas, G.C. & Lewis, J.A. (1989) *Plant Pathology*, **38**: 277-286.

Papavizas, G.C., Lewis, J.A. & Al Elmoity, T.H. (1982) *Phytopathology*,**72**:126-132.

Park, D. (1967) In: *Soil Biology* (Eds. Burges, A. & Raw, F.) pp 435- 447, Acad. Press New York.

Sambrook, J., Fritsch, E.F. & Miniatis, T. (1989) *Molecular Cloning: A Laboratory Manual* 2nd ed. Vol.I, Cold Spring Harbour Laboratory Press, Cold Spring Harbour, New York.

Saravanamuthu, R. (2004) In: *Innovative Approaches in Microbiology* (Eds. Maheshwari, D.K. & Dubey, R.C.) pp 23 -52, Bishen Singh and Mahendra Paul Singh Pub., India.

Scher, F.M. & Baker, R. (1980) *Phytopathology*, **70**:412-417

Schneider, R. (1954) *Phytopathology Z* **21**: 63-78.

Seidl, V. Huemer, B., Seiboth, B. & Kubicek, C. P. (2005) *FEBS J.* **272**: 5923-5939.

Stevensen, I. L. (1956) *J. Gen. Microbiol.* **15**: 372 -380.

Styaert, J.M., Stewart, A>, Jaspers, M. Carpenter, M. & Ridgway, H.J. (2004) *Mycologia,* **96**: 1245-1252.

Sundheim, L. (1977) *Netherland J. plant pathol.* **83**: 439-442.

Ten Vave, A. Mulder, W., Visser, J. & Kan, J.A.L. (1998) *Mol. Pl Microbe. Interact.*, **11**: 1009-1016.

Walton, J.D. (1994) *Pl. Physiol.* **104**: 1113-1118.

Weindling, R. & Fawcett, H.S. (1936) *Hilgardia*, **10**: 1-16.

Wells, H.D., Bell, D.K. & Jaworski, C.A. (1972) *Phytopathology*, **62**: 442-447.

Whipps, J.M. & Lumsden, R.D. (1991) *Biocontrol Sci. and Technol.* **1**: 75-90.

Whipps, J.M. (1992) *Biocontrol Sci. and Technol.* **2**:3-24.

Wood, R.K.S. & Tveit, M. (1955) *Botanical Rev.*, **21**:441-492.

9

# Molecular Basis of Plant Growth Promotion and Biocontrol Activities of Rhizobacteria

**Yogesh Kumar Negi**[1], **S.K. Garg**[2] **and J. Kumar**[3]

[1]*Sardar Bhagwan Singh (PG) Institute of Biomedical Sciences and Research, Balawala, Dehradun;*

[2]*R.M.L. Avadh University, Faizabad, U.P.;*

[3]*University of Horticulture and Forestry, Ranichauri, Tehri Garhwal*

## ABSTRACT

In the plant rhizosphere, bacteria, defined as rhizobacteria, are abundantly present more often organized in microcolonies and feed on the nutrients secreted by the plant root. Some of these rhizobacteria beneficially influence the host plant in a direct or indirect way, resulting either in to direct stimulation of its growth or through suppression of phytopathogens and diseases caused by them. A vast body of research evidences exist in literature on many soil-borne non-pathogenic bacteria with biocontrol and plant growth promotion activities. To commemorate their spectrum of action, several terms designate these as plant growth promoters, biocontrol agents, biopesticides, biofertilizers, etc. The plant growth promoting rhizobacteria (PGPR) consist of a wide range of root colonizing beneficial bacteria including, *Acenatobacter* spp., *Azotobacter* spp., *Bacillus* spp., fluorescent *Pseudomonas* spp., *Rhizobium* spp., etc., and are known to enhance the growth and health of the host plant.

Besides direct plant growth promotion (PGP) and biocontrol of plant pathogens (BC), some of the rhizobacteria exhibit induction of systemic resistance (ISR) in the host plants against a wide range of phytopathogens. Rhizobacteria with such abilities qualify to be the potential candidates for plant growth promotion and biocontrol of different pathogens. Thus, use of rhizobacteria having better PGP, BC, and ISR activities is a qualitative shift towards adopting non-chemical means of plant protection against resident pathogen populations. The interactions of PGPRs with their biotic environment, for example plants and other microorganisms, are often complex. Nevertheless, considerable improvement in finding the genetic basis of the beneficial effects of PGPRs on plants has been made. This progress will lead to a more efficient use of these microbes and possibly to their improvement by genetic modification.

In response to environment and health concerns about extended use of pesticides, there is increasing interest in finding alternative approaches for plant growth promotion and management of crop diseases. It seems inevitable that fewer pesticides will be used in future and that greater reliance will be placed on biological technologies including the use of different microbes. A growing number of PGPRs are already being marketed. These strains can only be used and improved genetically if the molecular basis of their beneficial traits is understood. This chapter focuses on the recent advancements in the molecular basis of plant growth promotion and biocontrol of several dangerous phytopathogens by rhizobacteria.

## INTRODUCTION

### Rhizosphere Colonization

Plant growth-promoting rhizobacteria were first defined by Kloepper and Schroth (1980) as the soil bacteria that colonize the roots of plants and enhance plant growth. Inoculant bacteria are often applied to seeds or roots of the plant for rapid colonization. After sowing or transplanting, the inoculant bacteria must be able to establish in the rhizosphere at population densities sufficient to produce a beneficial effect. Therefore, efficient inoculant bacteria should survive in the rhizosphere, make use of nutrients exuded by the plant root, proliferate, be able to efficiently colonize the entire root system and compete with indigenous microorganisms. Following are implicit in the colonization process: ability to survive inoculation onto seed, multiply in the spermosphere (region surrounding the seed) in response to seed exudates, attach to the root surface, and colonize the developing root system. A variety of bacterial traits and specific genes contribute to this process, but only a few have been identified. These include motility, chemotaxis towards seed and root exudates, and production of specific cell surface components, ability to use specific components of root exudates, protein secretion, quorum sensing and production of pili or fimbriae. Production of type IV pili is one of the abilities that can enhance root colonization ability of bacteria, better known for its involvement in the adhesion of animal and human pathogenic bacteria to eukaryotic cells (Steenhoudt and Venderleyden 2000, Strom and Lory 1993). The type IV pili also play a role in plant colonization by endophytic bacteria such as *Azoarcus* sp (Steenhoudt and Venderleyden 2000). The generation of mutants altered in expression of these traits is aiding our understanding of the precise role each one plays in the colonization process.

Bacterial lipopolysaccharides (LPS), in particular the O-antigen chain, can also contribute to root colonization. However, the importance of LPS in this colonization might be strain dependent since the O-antigenic side chain of *Pseudomonas fluorescens* WCS374 does not contribute to potato root adhesion, whereas the O-antigen chain of *P. fluorescens* PCL1205 is involved in tomato root colonization. Furthermore, the O-antigenic aspect of LPS does not contribute to rhizoplane colonization of tomato by the plant beneficial endophytic bacterium *P. fluorescens* WCS417r but, interestingly, this bacterial determinant was involved in endophytic colonization of roots (Dekkers *et al.*, 1998a, De Werger et al., 1987). It has also been recently demonstrated that the high bacterial growth rate and ability to synthesize vitamin $B_1$ and exude NADH dehydrogenases contribute to plant colonization by PGPR (Dekkers *et al.*, 1998a). Another determinant of root colonization ability of PGPRs is type IV pili (described above).

An important aspect of colonization is the ability to compete with indigenous microorganisms already present in the soil and rhizosphere of the developing plant. Our understanding of the factors involved in these interactions has been hindered by our inability to culture and characterize diverse members of the rhizosphere community and to determine how that community varies with plant species, plant age, location on the root, and soil properties. Phenotypic and genotypic approaches are now available to characterize rhizobacterial community structure. Phenotypic methods that rely on the ability to culture microorganisms include standard plating methods on selective media, community level physiological profiles (CLPP) using the BIOLOG system (Garland 1996), phospholipid fatty acid (PLFA) (Tunlid and White 1992), and fatty acid methyl ester (FAME) profiling (Germida *et al.*, 1998). Culture-independent molecular techniques are based on direct extraction of DNA from soil and 16S-rRNA gene sequence analysis, bacterial artificial chromosome or expression cloning systems. These are providing new insight into the diversity of rhizosphere microbial communities, the heterogeneity of the root environment, and the importance of environmental and biological factors in determining community structure. These approaches can also be used to determine the impact of inoculation of plant growth-promoting rhizobacteria on the rhizosphere community (Ciccillo *et al.*, 2002).

## Root Colonization and Site-specific Recombinase

Bacterial traits required for effective root colonization are subjected to phase variation, a regulatory process for DNA rearrangements orchestrated by site-specific recombinase. In certain PGPRs, efficient root colonization is linked to their ability to secrete a site-specific recombinase (Dekkers *et al.*, 1998b). Transfer of the site-specific recombinase gene from a rhizosphere-competent *P. fluorescens* into a rhizosphere-incompetent *Pseudomonas* strain enhanced its ability to colonize root tips (Dekkers *et al.*, 2000).

### *Bacterial determinants directly involved in root colonization*

Progress in the identification of new, previously uncharacterized genes is being made using non-biased screening strategies that rely on gene fusion technologies. These strategies employ reporter transposons and *in vitro* expression technology (IVET) to detect genes expressed during colonization (Rainey 1999). Using molecular markers such as green fluorescent protein or fluorescent antibodies, it is possible to monitor the location of individual rhizobacteria on the root using confocal laser scanning microscopy (Sorensen *et al.*, 2001). Identification of the genes and traits involved in the processes of inoculation and root colonization will give a more detailed insight into plant-bacterial interactions and lead to the more efficient application of inoculant strains. The first step in the inoculation process is the attachment of the bacterial cells to the seed. A screen for mutants of the rhizobacterial strain *Pseudomonas putida* KT2440 resulted in the identification of a set of putative surface and membrane proteins involved in attachment to corn seeds. Among these proteins are homologous of a calcium-binding protein, of hemolysin and of a potential multi-drug efflux pump (Espinosa *et al.*, 2002).

*P. fluorescens* genes that are specifically expressed in the rhizosphere (i.e., *rhi* genes) have been identified using *in vivo* expression technology (IVET) (Rainey 1999). More than 20 *rhi* genes have been identified, of which 14 showed significant homology to genes that are involved in nutrient acquisition, stress response, or secretion. Another six *rhi* genes showed no homology to genes with identified functions. Many root colonization genes and traits from *Pseudomonas* biocontrol species have been identified (Lugtenberg and Dekkers 1999), and even been used to improve colonization by *Pseudomonas* wild-type strains (Dekkers *et al.*, 2000). Recent studies have shown that organic acids form the nutritional basis of rhizosphere colonization. A defect in the ability to utilize sugars does not result in a colonization defect (Lugtenberg *et al.*, 1999). Efficient scavenging for iron using siderophores makes *Pseudomonas* strains more competitive. On the basis of identified colonization genes and traits, mutants of *P. chlororaphis* strain PCL1391 with defective colonization ability have been constructed. These mutants have lost their ability to control tomato foot and root rot, showing for the first time that root colonization is an essential trait for biocontrol (Smith *et al.*, 1999).

For better understanding of how bacteria function in the rhizosphere, the plant should also be taken into account. Therefore, the effect of the rhizosphere of different recombinant inbred tomato lines on the ability of *B. cereus* to control *Pythium torulosum* was analyzed. The results indicated a genetic basis from the plant side for the efficient growth and performance of biocontrol agents in the rhizosphere (Smith *et al.*, 1999). Interestingly, a follow-up study showed a clear and consistent growth difference for *B. cereus* in the spermosphere of the different inbreed tomato lines, whereas this difference was not observed for the two *Pseudomonas* biocontrol species (Simon *et al.*, 2001).

## MAJOR TRAITS FOR PLANT GROWTH PROMOTION

PGPR affect plant growth by direct and indirect means. The indirect promotion of plant growth occurs when PGPR diminish or prevent the deleterious effects of one or more phytopathogenic organisms. Whereas, the direct plant growth stimulation by PGPR entails providing the plant with a compound(s) that is synthesized by the bacterium or facilitating the increased uptake of certain nutrients from the soil. Several mechanisms (Table 1) for direct improvement of plant growth by PGPRs have been reported including fixation of atmospheric $N_2$ that is transferred to host plant, production of siderophore (iron chelaters) that increase the iron availability to host plant, solubilization of minerals such as phosphorus, and synthesis of phytohormones. PGPR strains possessing two or more of these mechanisms can show better plant growth in contrast to the strains possessing only one trait. Combination of the strains possessing varied mode of activities may again result in better health in contrast to single strain application. The different traits involved in plant growth promotion are described below.

### Biofertilization

Biofertilization by rhizobacteria mainly accounts for phosphorus and nitrogen supply. Though the molecular basis for the interactions is not always well-known, several basic principles of molecular interplay between the PGPRs and plants have been successfully unravelled. In this section we have described in brief the different determinants of rhizobacteria for increased availability of N and P.

**Table 1** Selected bacterial factors involved in plant growth regulation

| *S. No.* | *Bacterial exudates* | *Microorganism* | *Effect on plant* | *Plant* |
|---|---|---|---|---|
| 1. | Phytohormones | | | |
| | Auxin (analogs) | *Pseudomonas* | Morphogenesis, growth regulation | e.g. *Arabidopsis* |
| | Gibberellin | *Rhizobium* | Growth promotion | Rice and berseem clover (*Trifolium alexandrinum*) |
| | Cytokinin | e.g. *Rhizobium* | Growth promotion | e.g. Canola (*Brassica napus*) |
| | Ethylene | *Pseudomonas* | Growth regulation | e.g. *Pueraria lobata* |
| | Salicylic acid | e.g. *Serratia* | SAR | e.g. Tobacco |
| 2. | Extracellular enzymes | | | |
| | ACC deaminase | e.g. *Enterobacter* | Reduction of ethylene | e.g. Canola production |
| | Phosphatase | e.g. *Rhizobium* | Mineralization of organic | e.g. Maize |
| | Organic acids, complexants e.g. Gluconic acid, 2-ketogluconic acid | e.g. *Pseudomonas, Rhizobium* and *Bacillus* | phosphorus solubilization of inorganic phosphates | General |
| | Surface factors | | | |
| | Flagellins | *Pseudomonas* | ISR | *Arabidopsis* |
| | Lipopolysaccharides | *Pseudomonas* | ISR | Radish and carnation |
| | Antibiotics | | | |
| | e.g. Pyoluteorin, phenazines and phoroglucinols | e.g. *Pseudomonas* | Host plant protection | General |
| | Volatile signals | | | |
| | 2R,3R-Butanediol, acetoin | *Bacillus* sp. | Growth promotion | *Arabidopsis thaliana* |

Source: Whipps, J. M. 2001. Microbial interactions and biocontrol in the rhizosphere. J. Expl. Bot. 52: 487-511.

## *$N_2$ fixation*

Approximately 65% of biofertilization by rhizobacteria accounts for the increased nitrogen supply to crops worldwide. The most efficient nitrogen fixers are symbiotic bacterial strains belonging to the genera *Rhizobium*, *Sinorhizobium*, *Mesorhizobium*, *Bradyrhizobium*, *Azorhizobium* and *Allorhizobium*, and have been studied in most detail. All of these bacteria form a host-specific symbiosis with different leguminous plants. The symbiosis, as is known well, is exhibited by the formation of root or stem nodules in response to the presence of the bacterium (Long 2001, Perret *et al.*, 2000). Lipooligosacharide signal molecules that are secreted by the bacterium play a major role in this process. The bacteria penetrate the cortex, induce root nodules, multiply and subsequently differentiate into bacteroids, which produce the nitrogenase enzyme complex. Within the root nodules, the plant creates a low oxygen concentration, which allows bacterial nitrogenase to convert atmospheric nitrogen into ammonia. In return, the plant supplies a carbon source to the bacteria. A major challenge in rhizobial research is to understand how the bacterial signals, which trigger the formation of the nodule, are perceived by the plant. Genome analyses of model legumes such as *Lotus* and *Medicago* are in progress, and comparative genomics with other non-legumes may indicate which genes make leguminous plants susceptible to the establishment of the symbiosis.

In most of the above cases, symbiotic bacterial strains are identified. Additionally, some free-living nitrogen fixers are also known which have been studied extensively. Free-living nitrogen-fixing rhizobacteria such as *Azospirillum*, *Herbaspirillum*, *Acetobacter*, *Azotobacter* and *Azoarcus* are also able to fix atmospheric nitrogen (Steenhoudt *et al.*, 2000). They use a nitrogenase complex that functions under low oxygen conditions and that is not as specific in its interaction with the plant as are rhizobia. *Azospirillum* predominantly colonizes the rhizosphere, whereas the other bacteria are predominantly found as endophytes inside roots, stems and leaves. The genes involved in nitrogen fixation, nitrogen assimilation and nitrogen regulation have been described for *Azospirillum*. Several of the *nif* genes have also been described for the other free-living nitrogen fixers, which all have similar nitrogenase complexes, except for *Azoarcus* which possesses three differently encoded nitrogenase complexes. A practical challenge is to widen the host range of the symbiosis towards major non-leguminous food crops such as rice. An exciting new finding is that rhizobial bacteria contain genes on the *Sym* (Symbiosis) plasmid (Freiberg *et al.*, 1997) or chromosomally (Gottfert *et al.*, 2001) in a symbiotic island that are homologous to those encoding type-III secretion systems, which are used by pathogenic bacteria to deliver virulence factors into host cells. Recent results show that rhizobial type-III secretion systems secrete specific proteins, and are involved in the establishment of the symbiosis suggesting that type-III secretion systems also function in symbiosis (Viprey *et al.*, 1998). Remarkably, genes encoding type-III secretion systems have also been identified in a plant beneficial *Pseudomonas fluorescens* strain (Rainey 1999). Different strains of *Pseudomonas* sp. have also been found effective for increased nitrogen availability. Many reports exist where *Pseudomonas* isolates have been found effective to increase effective nodulation in legume crops. This effect was found better if combined with rhizobium or any other symbiotic nitrogen fixer.

## Phosphate solubilization

Phosphorus is one of the key plant nutrients but less (about 0.05% w/w) available in the soil. However, only 0.1% of this soil phosphorus is available to plants. Large amount of phosphate fertilizers (chemical) added to enhance agro productivity on a regular basis but remain immobilized in large proportion, and hence not available to plants. Maintenance of available phosphorus remains a major challenge. The biogeochemical cycle of phosphorus is substantially influenced by microorganisms, which influence the phosphorus metabolization through their metabolic process in soil. The precipitated inorganic phosphate is solubilized by the action of mineral and organic acids produced by bacteria (Kucey, 1983) and fungi (Mattey, 1992). An effective interaction between P-solubilizers and plants depends on (i) a high population of P solubilizers maintained in the rhizosphere over a long period, (ii) exudation of carboxylates and protons into the rhizosphere by roots and microbes, (iii) low P uptake by these microbes, and (iv) positive interaction with mycorrhizal fungi or other beneficial microbes. Many microbial species are known to possess the ability to solubilize sparingly soluble P *in vitro* (Table 2). Inoculation with effective P solubilizers may increase P uptake and growth of plants. However, it is often not clear whether the improved P uptake by plants is directly attributable to P solubilization by the introduced microbe. Combination

**Table 2** Phosphate mineralization from P-substrates by some soil bacterial species

| S. No. | Bacterial strain | Substrate | Enzyme type |
|---|---|---|---|
| 1. | *Pseudomonas fluorescence* | Non-specific | Phosphatase |
| 2. | *Pseudomonas* sp. | Non-specific | Phosphatase |
| 3. | *Burkholderia cepacia* | Non-specific | Phosphatase |
| 4. | *Enterobacter aerogenes* | Non-specific | Phosphatase |
| 5. | *Enterobacter cloacae* | Non-specific | Phosphatase |
| 6. | *Proteus mirabilis* | Non-specific | Phosphatase |
| 7. | *Serratia marcenscens* | Non-specific | Phosphatase |
| 8. | *Bacillus subtilis* | Inositol phosphate | Phytase |
| 9. | *Pseudomonas mendocina* | Inositol phosphate | Phytase |
| 10. | *Pseudomonas fluorescens* | Phosphonoacetate | Phosphonoacetate hydrolase |
| 11. | *Bacillus licheniformis* | D-$\alpha$-glycerophosphate | D-$\alpha$-glycerophosphatase |
| 12. | Klebsiella aerogenes | Phosphate | C-P Lyase |

Source: Rodriguez and Fraga, 1999. Phosphate solubilizing bacteria and their role in plant growth promotion. Biotechnol. Advs.17: 319-339.

of P solubilizers with $N_2$ fixers or mycorrhizal fungi is superior for inoculation in comparison to inoculation with P solubilizers alone.

Large-scale inoculation with P solubilizers in farming practice is hampered by several factors that could diminish the effectiveness of the introduced microorganism. These are (i) most soils already contain P solubilizers, so the effect of inoculation may be small; (ii) introduced strains may have poor survival in the rhizosphere because of low competitiveness against indigenous, well adopted strains; (iii) microbes are selected based on their P solubilization under *in vitro* conditions that are always ideal for their growth and P solubilization by them, whereas, conditions in the rhizosphere may be far from favourable, and (iv) P solubilized by microbes may be unavailable to plants because microbe can also take it up for their growth and reproduction. It is, therefore, of great importance that the possible contribution of P solubilizing microbes to increase P uptake by host plant must be evaluated in natural soil conditions, i.e. in fields.

## Genetics of P-solubilizing bacteria

The genetic basis of mineral phosphate solubilization is not extensively understood. Two major factors have been found to be responsible for P solubilization by rhizobacteria:

(i) Production of organic acids, considered to be the principal mechanism of mineral solubilization.

(ii) Production of P hydrolyzing enzymes (Phosphatase and phytase). It could be assumed that any gene involved in organic acid synthesis might have an effect on mineral phosphate solubilization ability of these organisms.

Goldstien and Liu (1987) cloned a gene from *Erwinia herbicola*, which was involved in mineral phosphate solubilization, by screening the antibiotic resistant recombinants from the genomic library in a medium containing hydroxypeptite as the source of P. The expression of this gene allowed production of gluconic acid and thereby mineral phosphate solubilization in *E. coli* HB 101. Sequence analysis of this gene suggested its possible involvement in the synthesis of enzyme pyrroquinoline quinone (PQQ) synthase, which directs the synthesis of PQQ, a co-factor necessary for the formation of the holoenzyme glucose dehydrogenase (GDH)-PQQ. This enzyme catalyzes the formation of gluconic acid from glucose by direct oxidation pathway. Similarly, a mineral phosphate solubilization gene (*gab Y*) from *Pseudomonas cepacia* has been isolated (Babu-Khan *et al.*, 1995). Expression of this gene in *E. coli* JM109 induced the character and showed P-solubilization via gluconic acid production. No apparent homology of the gene was seen with previous PQQ synthase gene, but it did with a permease system membrane protein. The *gab Y* gene could play an alternative role in the expression and/ or regulation of direct oxidation pathway in rhizobacteria, thereby acting as functional mineral phosphate solubilization gene *in vivo*.

Very little is known regarding the genetic regulation governing the mineral phosphate solubilization trait. In fact, the information about the genetic or biochemical mechanisms involved in the synthesis of the GDH-PQQ hydrolase is little, and variations between constitutive and inducible phenotypes are observed among several bacterial species (Goldstein 1994). Glucose, gluconate, mannitol and glycerol are among the possible inducers of the holoenzyme activity (van Schie *et al.*, 1987).

Concerning the possible effect of soluble P on expression of the phosphate solubilizing activity, Goldstien and Liu (1987) found that mineral phosphate solubilization trait in *E. herbicola* is induced by P starvation and repressed by elevated exogenous P levels (complete repression was achieved at P concentration >20 mM).

## Secretion of Plant Growth Hormones

PGPRs that are able to affect plant physiology directly synthesize and exudate plant hormones (Table 1). PGPRs are reported to produce most of the plant growth hormones like auxin, cytokinin(s), gibberellin(s) (Steenhoudt and Vanderleyden 2000), ethylene, octadecanoids and jasmonate. Some of them produce salicylate (Persello *et al.*, 2003, Parker 2003) or control the biosynthesis of ethylene via an ACC deaminase (1-amino-cyclopropylcarboxylic acid) that reduces the available pool of ACC for ethylene biosynthesis by the plant (Persello *et al.*, 2003). Since ethylene often acts either synergistically or antagonistically with other plant hormones, manipulation of this pathway has an impact on many response complexes for example, induced systemic resistance, systemic acquired resistance or salicylic acid and jasmonic acid-linked gene expressions (Walling 2000). The phenomenon of phytostimulation profits particularly from manipulating the complex and balanced network of plant hormones or hormone-like compounds that directly influence growth or stimulate root formation (Figure). For example, *Azospirillum* spp. produce auxins, cytokinins and gibberellins that stimulate root development, which can lead to improved crop yield (Steenhoudt and Vanderleyden, 2000). Auxins are quantitatively the most abundant phytohormones secreted by *Azospirillum*, and it is generally agreed that auxin production, rather than nitrogen fixation, is the major factor responsible for the stimulation of rooting and, hence, enhanced plant growth. The construction of *Azospirillum* mutants that completely lack auxin production is complicated because three pathways for indole acetic acid (IAA) biosynthesis are present in *Azospirillum*. Recently, it was indicated that the pathways are differently regulated by catabolite repression, and that IAA synthesis is regulated by the autoinduction of IAA. *Azospirillum* also enhances the growth of the freshwater microalga *Chlorella vulgaris*, and therefore could potentially be used to improve water cleaning.

**Table 3** Recent examples of bacteria applied to seeds or roots providing biocontrol of fungal plant pathogens

| S.No. | Bacteria | Fungal pathogen | Plant | Environment |
|---|---|---|---|---|
| 1 | *Actinoplanes* spp. | *Pythium ultimum* | Table beet | Soil |
| 2 | *Bacillus* spp. | *Rhizoctonia solani; Gaeumannaomyces graminis* var. *tritici* | Wheat | Soil |
| 3 | *Bacillus subtilis* GB03 | *Fusarium oxysporum* f. sp. *ciceris* | Chickpea | Sterile soil |
| 4 | *B. subtilis* BACT-D | *Pythium aphanidermatum* | Tomato | Soil |
| 5 | *Burkholderia cepacia* A3R | *Fusarium graminearum* | Wheat | Soil |
| 6 | *B. cepacia* PHQM 100 | *Fusarium* spp.<br>*Pythium* spp. | Maize<br>Maize | Soil<br>Soil |
| 7 | *Comamonas acidovorans* HF42 | *Magnaporthe poae* | Kentucky bluegrass | Soil |
| 8 | *Enterobacter* sp. BF14 | *Magnaporthe poae* | Kentucky bluegrass | Soil |
| 9 | *Paenibacillus* sp. 300 *cucumerinum* | *Fusarium oxysporum* f. sp. | Cucumber | Soil-less mix |
| 10 | *Pseudomonas* spp. | *Fusarium oxysporum* f. sp. *radicis-lycopersici* | Tomato | Rockwool |
| 11 | *Pseudomomas aureofaciens* AB244 | *Pythium ultimum* | Tomato | Soil-less mix |
| 12 | *P. aureofaciens* 63-28 | *P. aphanidermatum* | Cucumber | Vermiculite |
| 13 | *Pseudomonas chlororaphis* MA342 | *Drechslera graminea*<br>D. teres<br>D. avenae<br>Ustilago avenae<br>U. hordei<br>*Tilletia caries* | Barley<br>Barley<br>Oats<br>Oats<br>Barley<br>Wheat | Soil<br>Soil<br>Soil<br>Soil<br>Soil<br>Soil |
| 14 | *P. chlororaphis* PCL 1391 | *Fusarium oxysporum* f. sp. *radicis-lycopersici* | Tomato | Soil |
| 15 | *P. chlororaphis* RD31-3A | *Fusarium* spp. | Douglas fir | Seed |
| 16 | *Pseudomonas corrugata* 13 | *Pythium aphanidermatum* | Cucumber | Vermiculite |
| 17 | *Pseudomonas fluorescens* | *Fusarium oxysporum* f. sp. *raphani* | Radish | Soil/sand |
| 18 | *P. fluorescens* WCS417 | *F. oxysporum* f. sp. *raphani* | Radish | Rockwool |
| 19 | *P. fluorescens* WCS358 | *F. oxysporum* f. sp. *lini* | Flax | Nutrient solution |
| 20 | *P. fluorescens* Q8r1-96 | *Gaeumannomyces graminis* | Wheat | Soil |
| 21 | *P. fluorescens* BTP7 | *Pythium aphanidermatum* | Cucumber | Vermiculite |
| 22 | *P. fluorescens* V061 | *Pythium ultimum*<br>*Rhizoctonia solani* | *Lotus corniculatus*<br>Rice | Soil mix<br>Soil |
| 23 | *Pseudomonas putida* | *Fusarium oxysporum* f. sp. *raphani* | Radish | Soil/sand |
| 24 | *P. putida* BTP1 | *Pythium aphanidermatum* | Cucumber | Vermiculite |
| 25 | *Serratia plymuthica* | *Pythium ultimum* | Cucumber | Peat-perlite-vermiculite |
| 26 | *Stenotrophomonas maltophilia* C3 | *Rhizoctonia solani* | Tall fescue | Soil mix |
| 27 | *Streptomyces* sp. 385 *cucumerinum* | *Fusarium oxysporum* f. sp. | Cucumber | Soil-less mix |

Source: Whipps, J. M. 2001. Microbial interactions and biocontrol in the rhizosphere. J. Expl. Bot. 52: 487-511.

## Siderophore production

Iron, one of the most abundant minerals on earth, is an essential requirement for almost all organisms including microbes. To date, the only known exception is lactobacilli, which are devoid of haem proteins, and hence have no iron requirement (Archibald, 1983). However, iron in the soil is unavailable for direct assimilation by microbes because of its unavailable form i.e., ferric iron or $Fe^{3+}$, which is prominent in the soil and sparingly soluble about $10^{-18}$ M at pH 7.4 (Neilands *et al.*, 1987). This amount of soluble iron is too low to support microbial growth in soil, which generally needs concentrations approaching $10^{-6}$ M for normal growth. Consequently, to survive in such environments, organisms secrete iron-binding legends (siderophores), which can bind the ferric iron and make it available to the host microbe. These compounds have been identified as "low molecular weight, ferric specific legends, the biosynthesis of which is carefully regulated by iron and the function of which is to supply iron to the cell" (Neilands 1981).

Siderophore secreted by PGPRs have been shown to have a very high affinity for iron and bind most of the $Fe^{3+}$ that is available in the rhizosphere, and prevent the pathogens present in immediate vicinity from proliferation because of lack of iron (O' Sullivan and O' Gara, 1992). Although various bacterial siderophores differ in their abilities to sequester iron, in general, they deprive pathogenic fungi of this essential element as the fungal siderophores have lower affinity. Some PGPR strains go one step forward and draw iron from heterologous siderophores produced by cohabitating microbes.

The basic features of high-affinity, siderophore-mediated iron (III) assimilation are illustrated in Fig. 1. Since ferric-siderophores are usually too large to diffuse through the small, water-filled porin channels in the cell envelope of Gram-negative bacteria, active iron transport is initiated by outer-membrane receptor proteins for ferric siderophores. The synthesis of these receptor proteins is usually also induced by iron-limiting conditions. Inner-membrane and periplasmic proteins are also required for ferric-siderophore transport. Release of iron from the ferric-siderophore complex probably proceeds via reduction since siderophores have little affinity for iron (II). The released iron is then incorporated into iron-containing cellular constituents, where it may interact as iron (II) with repressor protein to regulate the biosynthesis of the siderophore and the receptor protein for the ferric siderophore.

Siderophore biosynthesis is generally tightly regulated by iron-sensitive Fur proteins, the global regulators GacS and GacA, the sigma factors RpoS, PvdS, and FpvI, quorum-sensing autoinducers such as N-acyl homoserine lactone, and site-specific recombinases (Cornelis and Metthijs 2002, Ravel and Cornelis 2003). However, some data demonstrate that none of these global regulators is involved in siderophore production. Neither GacS nor RpoS significantly affected the level of siderophores synthesized by *Enterobacter cloacae* CAL2 and UW 4 (Saleh and Glick 2001). RpoS is not involved in the regulation of siderophore production by *Pseudomonas putida* strain WCS358 (Kojic *et al.*, 1999). In addition, GrrA/GrrS, but not GacS/GacA, are involved in siderophore synthesis regulation in *Serratia plymuthica* strain IC1270, suggesting that gene evolution occurred in the siderophore-producing bacteria (Ovadis et al., 2003). A myriad of environmental factors can also modulate siderophore synthesis, including pH, the level of iron and the form of iron ions, the presence of other trace elements, and an adequate supply of carbon, nitrogen, and phosphorus (Duffy and Defago 1997).

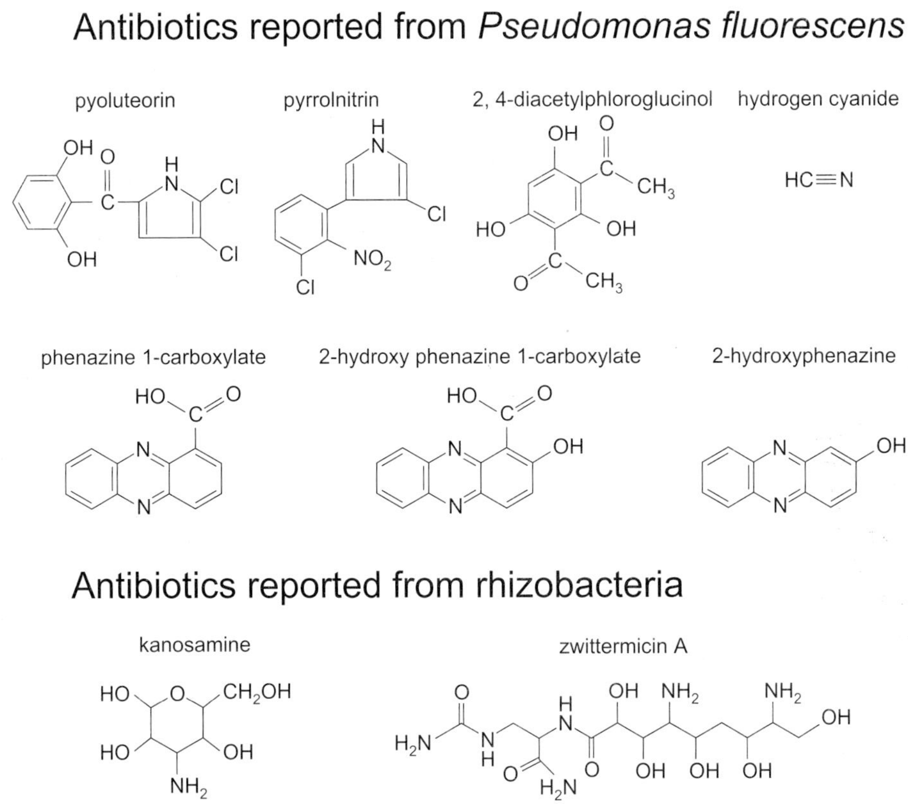


**Fig. 1** Chemical structures of some antibiotics produced by bacterial biocontrol agents (*Pseudomonas fluorescens* and *Bacillus cereus*). Antibiotics were found antagonistic to a wide range of phytopathogens. (Source: Handelsman and Stabb, 1996. Biocontrol of soil borne pathogens. Plant Cell 8: 1855-1869)

## BIOCONTROL ACTIVITY OF RHIZOBACTERIA

Throughout the history of agriculture, humans have struggled to reduce the crop losses due to different plant disease on their crops. Early agriculturists realized the benefits of cultural practices such as crop rotation and the use of organic soil amendments in promoting plant productivity. It is now well established that many of these effects are achieved by promoting the natural microbiological processes that keep plant disease in control. Cultural practices were mainstays of traditional agricultural systems, and still provide the primary approaches for management of many soil-borne diseases. Recently, a vast range of rhizobacteria that produce specific secondary metabolites, have been associated convincingly with soil suppressiveness. This compelling evidence for the role of these rhizobacteria in natural processes of biological control has built even greater enthusiasm for the development of these bacteria as biological control agents for different crop diseases worldwide. Isolation of effective biocontrol agent has been reported with 'suppressive soils'. Suppressive soils are defined as those soils where soil-borne diseases or pathogens are naturally in control. This nature of such soils is supposed to be due to presence of good biocontrol agents.

The mechanisms responsible for this biocontrol activity include competition for nutrients, niche exclusion, induced systemic resistance (ISR), and the production of antifungal metabolites (AFMs). The biocontrol agents that are best characterized at the molecular level belong to the genus *Pseudomonas*.

There is no doubt that strains representing diverse genera of Gram-negative and Gram-positive bacteria can suppress plant diseases. Fluorescent pseudomonads are typically among the most effective biocontrol agents selected for suppression of both soil-borne and aerial diseases of plants (Loper *et al.,* 1997). Most of the identified *Pseudomonas* biocontrol strains produce antifungal metabolites (AFMs), of which phenazines (Thomashow and Weller, 1988), pyoluteorin (Howell and Stipanovic, 1980), pyrrolnitrin (Howell and Stipanovic, 1979) and 2, 4-diacetylphloroglucinol (DAPG) (Keel *et al.,* 1990) are the most frequently detected classes. Chemical structures of some of the antibiotics are illustrated in Fig. 1. However, new AFMs belonging to the class of cyclic lipopeptides, such as viscosinamide and tensin have been discovered (Nielsen *et al*., 2002). Viscosinamide prevents the infection of sugarbeet by *Pythium ultimum.* The genetic basis of the biosynthesis of the more frequently detected AFMs in *Pseudomonas* has been elucidated. More recently, new information has become available on the biosynthesis of pyoluteorin in *P. fluorescens* Pf-5 and of 2,4-diacetylphloroglucinol (DAPG) in *P. fluorescens* Q2-87. Individual genes have been discovered that are responsible for the presence of functional groups on phenazine compounds, such as *phzO,* which is required for the 2-hydroxy group and *phzH*, and which is responsible for the 1-carboxamide group. On the basis of homology and mutant studies, functions have been proposed for many of the identified biosynthetic genes. Biochemical studies are required, however, to prove these functions (Walsh *et al*., 2001, ). *Streptomyces* and *Bacillus* species can also exert biocontrol, which is being characterized at the molecular level. For example, the biosynthetic gene cluster responsible for the production of the antibiotic zwittermicin A in *Bacillus cereus* has been identified (Milner 1995).

## Biocontrol Activity Mediated by the Synthesis of Secondary Metabolites of Rhizobacteria

Several secondary metabolites have been identified and tested for effective biocontrol of different phytopathogens by several rhizobacterial strains, including siderophores, antibiotics, biocidal volatiles, lytic enzymes, and detoxification enzymes. All these components are explained as under.

### *Competition for iron*

Iron is an essential growth element for all living organisms. The scarcity of bio-available iron in soil habitats results in a furious competition for this element. Under iron-limiting conditions, certain rhizobacteria produce low-molecular-weight compounds called siderophores to competitively acquire ferric ion. Although various bacterial siderophores differ in their abilities to sequester iron, in general, they deprive pathogenic fungi of this essential element since the fungal siderophores have lower affinity. Siderophore mediated biocontrol of many soil-borne fungal plant pathogens has been shown by many researchers (Fig. 2). Siderophore producing rhizobacteria have been found more promising as compared to non-siderophore producing strains or mutant strains. Seong and Shin (1996) reported siderophore-mediated inhibition of *Pythium ultimum*, *Pyricularia oryzae, Rhizoctonia solani* and *Xanthomonas oryzae* by *Pseudomonas fluorescens* ps88. Additionally, cucumber seeds treated with ps88 grown in greenhouse soil showed increased seed emergence and reduced disease incidence (up to 50 %) caused by *Fusarium oxysporum*. Regulation of iron-siderophores can be improved by the addition of $Zn^{2+}$ in the medium. Increased level of siderophores in the presence of $Zn^{2+}$ have been reported by different workers and concluded inhibition of ferric-reductase by $Zn^{2+}$, which is involved in low iron uptake (Duffy and Defego 1997).

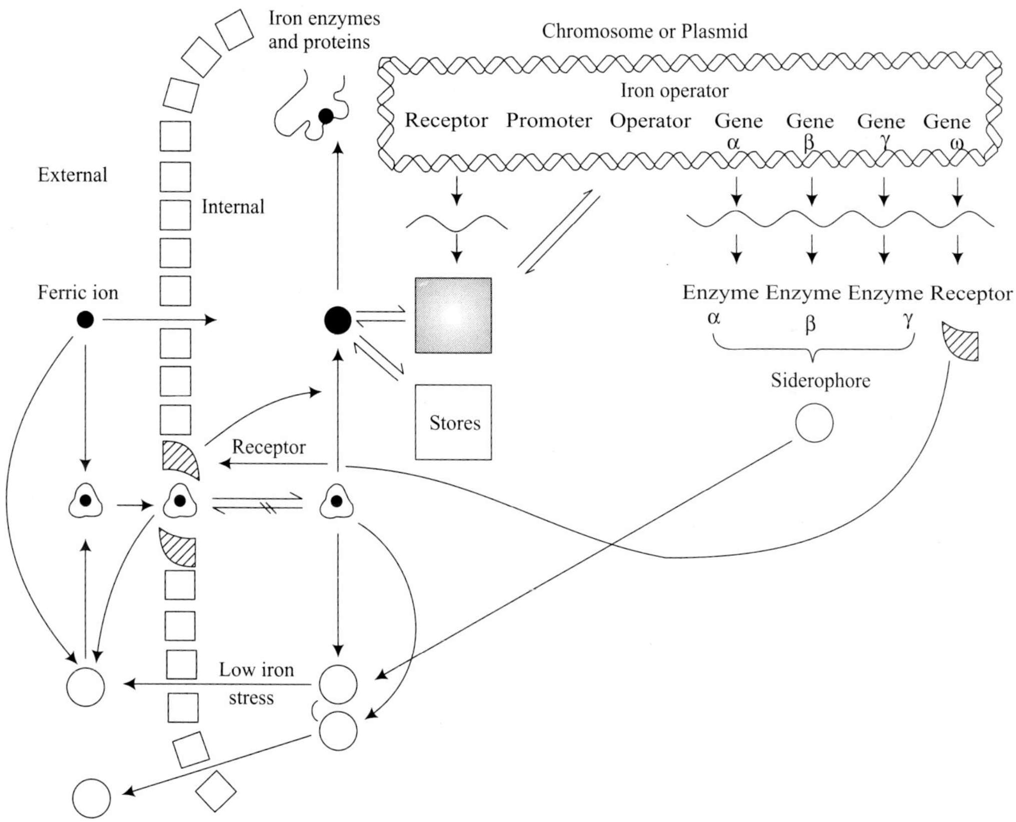


**Fig. 2** Schematic representation showing the pathway of iron chelation by siderophore producing rhizobacteria. (Source: Leong, J, 1986. Siderophores: Their biochemistry and possible role in biocontrol of plant pathogens. *Ann. Rev. Phytopathol.* 24: 187-209)

## *Antibiosis*

The basis of antibiosis as a biocontrol mechanism of PGPR has become increasingly better understood over the past two decades (Fig. 3). A variety of antibiotics have been identified, including compounds such as amphisin, 2, 4-diacetylphloroglucinol (DAPG), hydrogen cyanide, oomycin A, phenazine, pyoluteorin, pyrrolnitrin, tensin, tropolone, and cyclic lipopeptides produced by pseudomonads and oligomycin A, kanosamine, zwittermicin A, and xanthobaccin produced by *Bacillus*, *Pseudomonas*, *Streptomyces*, and *Stenotrophomonas* spp (de Souza *et al.*, 2003, Nielsen *et al.*, 2002, 2003, Milner *et al.*, 1995, 1996). Interestingly, some antibiotics produced by PGPR are finding new uses as experimental pharmaceuticals, and this group of bacteria may offer an untapped resource for compounds to deal with the alarming ascent of multidrug-resistant human pathogenic bacteria.

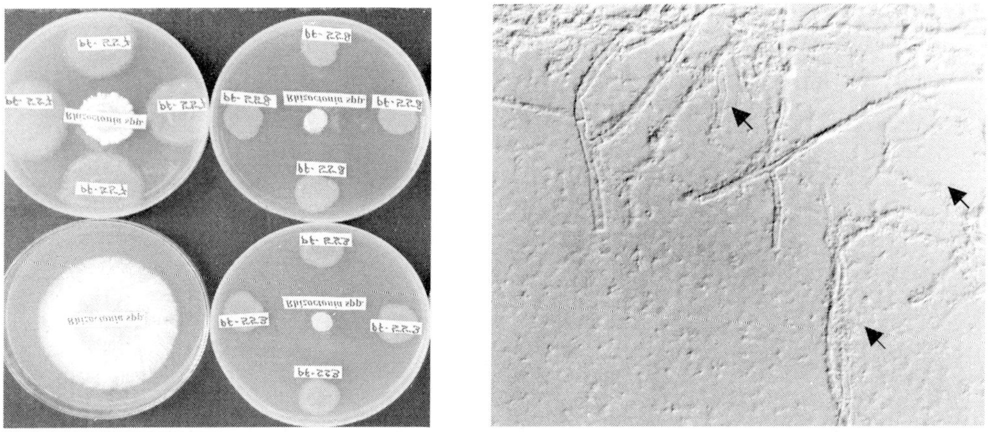

**Fig. 3** a: *In vitro* inhibition of *Rhizoctonia solani* by *Pseudomonas* isolates. b: lysis of *R. solani* mycelium by antifungal metabolites of *Pseudomonas* isolates. [Source: Negi, Y. K. 2006. Strain improvement of fluorescent *Pseudomonas* spp. With respect to their PGPR activity using molecular approaches (Ph.D. thesis)]

Regulatory cascades of these antibiotics involve GacA/GacS or GrrA/GrrS, RpoD, and RpoS, *N*-acyl homoserine lactone derivatives and positive autoregulation (Chancey *et al.*, 1999, Hass and Keel 2003, Pierson *et al.*, 1998). Antibiotic synthesis is tightly linked to the overall metabolic status of the cell, which in turn is dictated by nutrient availability and other environmental stimuli, such as major and minor minerals, type of carbon source and supply, pH, temperature, and other parameters (Milner *et al.*, 1995, 1996). Trace elements, particularly zinc, and carbon source levels influence the genetic stability/instability of bacteria, affecting their ability to produce secondary metabolites. It is important to note that many strains produce pellet of secondary antimicrobial metabolites and that conditions favouring one compound may not favour another (Duffy and Defago 1999). Thus, the varied arsenal of biocontrol strains may enable antagonists to perform their ultimate objective of pathogen suppression under the widest range of environmental conditions. For example, in *P. fluorescens* CHA0 biosynthesis of DAPG is stimulated and pyoluteorin is repressed in the presence of glucose as a carbon source. As glucose is depleted, however, pyoluteorin becomes the more abundantly antimicrobial compound produced by this strain (Duffy and Defago 1999). This ensures a degree of flexibility for the antagonist when confronted with a different or a changeable environment. Biotic conditions can also influence antibiotic biosynthesis. For example, bacterial metabolites, salicylates and pyoluteorin can affect DAPG production by *P. fluorescens* CHA0 (Schinder *et al.*, 2000). Furthermore, plant growth and development also influence antibiotic production, since biological activity of DAPG producers is not induced by the exudates of young plant roots but is induced by the exudates of older plants, which results in selective pressure against other rhizosphere microorganisms (Picard *et al.*, 2000).

## *Lytic enzyme production*

A variety of microorganisms also exhibit hyperparasitic activity, attacking pathogens by excreting cell wall hydrolases (Chernin and Chet. 2002). Chitinase produced by *S. plymuthica* C48 inhibited spore germination and germtube elongation in *Botrytis cinerea* (Frankowski *et al.*, 2001). The ability to produce extracellular chitinases is considered crucial for *Serratia marcescens* to act as antagonist against *Sclerotium rolfsii* (Ordentlich *et al.*, 1988), and for *Paenibacillus* sp. strain 300 and *Streptomyces* sp. strain 385 to suppress *Fusarium oxysporum* f. sp. *cucumerinum*. It has also been demonstrated that extracellular chitinase and laminarinase synthesized by *Pseudomonas stutzeri* digest and lyse mycelium of *F. solani* (Lim *et al.*, 1991). Although, chitinolytic activity appears less essential for PGPR such as *S. plymutica* IC14 when used to suppress *S. sclerotiorum* and *B. cinerea*, synthesis of proteases and other biocontrol traits are involved. The ß-1,3-glucanase synthesized by *Paenibacillus* sp. strain 300 and *Streptomyces* sp. strain 385 lyse fungal cell walls of *F. oxysporum* f. sp. *cucumerinum* (Singh *et al.*, 1999). *B. cepacia* synthesizes ß-1,3-glucanase that destroys the integrity of *R. solani*, *S. rolfsii* and *Pythium ultimum* cell walls. Similar to siderophores and antibiotics, regulation of lytic enzyme production (proteases and chitinases in particular) involves the GacA/GacS or GrrA/GrrS regulatory systems (Ovadis *et al.*, 2004, Sacherer *et al.*, 1994) and colony phase variation (Lugtenberg 2001).

## *Suppression of virulence factors*

Another mechanism of biological control is the suppression of pathogen virulence factors. For example, certain biocontrol agents are able to detoxify albicidin toxin produced by *Xanthomonas albilineans* (Zhang and Birch, 1996, 1997). The mechanisms here include production of a protein that reversibly binds the toxin in both *Klebsiella oxytoca* (Walker *et al.*, 1988) and *Alcaligenes denitrificans* (Basnayake and Brich 1995), as well as an irreversible suppression of albicidin mediated by an esterase that occurs in *Pantoea dispersa* (Zhang *et al.*, 1996, 1997). Several different microorganisms, including strains of *B. cepacia* and *P. fluorescens*, can hydrolyze fusaric acid, a phytotoxin produced by various *Fusarium* species (Toyoda *et al.*, 1988). $Zn^{+2}$ again plays a vital role in the inactivation of fusaric acid. Addition of $Zn^{+2}$ with biocontrol agent may give better effects to suppress the action of fusaric acid (Duffy and Defego 1997). More often though, pathogen toxins display a broad-spectrum activity and can suppress growth of microbial competitors, or detoxify antiobiotics produced by some biocontrol microorganisms as a self-defence mechanism against biocontrol agents (Schouten *et al.*, 2004).

Recently, it has been discovered that certain PGPR quench pathogen quorum-sensing capacity by degrading autoinducer signals, thereby blocking expression of numerous virulence genes (Dong *et al.*, 2000, 2004, Morello *et al.*, 2004). Since most, if not all, bacterial plant pathogens rely upon autoinducer-mediated quorum-sensing to turn on gene cascades for their key virulence factors (e.g., cell-degrading enzymes and phytotoxins) (von Bodman 2003), this approach holds tremendous potential for alleviating disease, even after the onset of infection, in a curative manner.

## GENETIC DIVERSITY AMONG PGPRS

Recent trend in most plant growth promoting and biocontrol studies is first to make a repository of large number of isolates, and then to screen them for different traits required for plant growth promotion and biocontrol activities. Generally, greater diversity of introduced bacterial inoculants results in a diverse but potentially more stable rhizosphere community to colonize the root system and survive against biological, physical and chemical changes occurring in rhizosphere throughout the plant growth. These changes are very likely to occur in rhizosphere of host plant in all growing seasons. Further, diverse populations of PGPRs increase the spectrum of action, as different strains possess different modes of action. Therefore, it is quite reasonable to assume that performance of the isolates of the same genotype will be more or less same depending upon the environment provided during characterization. Nevertheless, genetic characterization tends to be necessary to provide a clearer picture. Different techniques have been employed to assess the genetic diversity among different rhizobacterial PGPRs including biochemical, immunological and molecular tools. A number of PCR based techniques have been devised and are strengthened over last decade. These include RAPD-PCR, BOX and ERIC PCR, Rep PCR, etc., which have been used extensively to assess the genetic diversity among PGPRs. Higher diversity level among the pseudomonads was also reported by Raaijmakers *et al*., (2001) and Ramesh Kumar *et al*., (2002) Specific groups could be identified using different molecular tools including RAPD-PCR. Mavrodi *et al*., (2001) estimated genetic diversity among 123 *P. fluorescens* using RAPD-PCR, BOX-PCR and correlated identification of 2, 4-DAPG producing strains on the basis of *phlD* gene by RFLP analysis. Similarly, McSpadden Gardener *et al*., (2000) developed a rapid and reliable PCR-based assay for rapid characterization of 2, 4-DAPG producing *Pseudomonas* population based on the amplification of *phlD* gene sequence. Recently, Negi (2006) showed a tremendous diversity in rhizospheric *Pseudomonas* population isolated from Uttarakhand hills using RAPD PCR. In this study, RAPD-PCR profile could group the different pseudomonads in different lineages on the basis of their host of origin and habitat. The study remarks first large scale characterization of *Pseudomonas* spp. from Uttarakhand region.

The complexity of the interactions involved in biocontrol and the wide range of environmental conditions found globally in agriculture make it unlikely that any one strain will suppress even a single disease in all settings. The genetic diversity of microorganisms with disease-suppressive potential remains a powerful yet largely untapped resource for biocontrol of plant disease (Kerr and Htay, 1974). There is a need to seek new biocontrol strains, particularly strains adapted to the site where they will be used (Cook, 1993; Stabb *et al*., 1994), but if random bacterial isolates from each site are screened for disease suppressiveness, as has been done for years, the effort will continue to be labor intensive and will not make use of existing knowledge of biocontrol mechanisms. Recently, there has been some success in identifying diverse biocontrol strains that suppress plant pathogens through common mechanisms (Stabb et al., 1994). This approach has been employed successfully in insect biocontrol programs, most notably in the development of *B. thuringiensis*-based strategies for the control of insects. Initial application of *B. thuringiensis* resulted in limited and variable insect control. Biocontrol by *B. thuringiensis* is due to accumulation of a large protein toxin, which forms a crystal in the bacterial cell. Identification of the crystal toxin as the basis for insect control led to searches for genetically diverse *B. thuringiensis* strains with related but unique crystal toxins. *B. thuringiensis* based biocontrol strategies are now employed, and their development has generated a tremendous knowledge base of fundamental genetics and protein chemistry (Hoefte and Whitely, 1989; Feitelson *et al.*, 1992; Carlton, 1993). The limitation of this approach is that widespread use of organisms that share a mechanism of biocontrol will increase the selection for resistance; therefore, this approach must be coupled with an understanding of the mechanism and frequency of resistance and of strategies to avoid it.

Recently, a comparable approach has been applied to biocontrol of plant disease, and there have been some notable successes in finding diverse strains that suppress disease based on a common mechanism. Genes for antibiotic production or antibiotic self-resistance are conserved among antibiotic producers and therefore form the basis of molecular probes for detecting new antibiotic-producing strains. Zwittermicin A-producing *B. cereus* strains and 2, 4-diacetylphloroglucinol producing *P. fluorescens* strains have been identified from geographically and chemically diverse soils (Stabb *et al.*, 1994; Keel *et al.*, 1996). Strains that produce the same antibiotic can be phenotypically diverse and may express traits useful under particular conditions (Fig. 1). Some of these strains may provide effective control in certain soils in specific geographic regions or on particular crops (Fig. 1). In addition, the genetic diversity of these strains may be tapped by combining them in mixed inoculants. Certain mixtures of fluorescent pseudomonads, or fluorescent pseudomonads and fungi suppressed disease more effectively than did single strain inoculants (Park *et al.*, 1988; Pierson and Weller, 1994; Duffy and Weller, 1995; Duffy *et al.*, 1996).

## GENETIC IMPROVEMENT OF PGPRS USING MOLECULAR APPROACHES

The identification of genes involved in the ability of rhizobacterial strains to improve plant growth creates the potential to improve the performance of biocontrol strains or to construct novel biocontrol strains by genetic modification. Complete operons, as well as single genes under the control of their own regulatory genes or regulated by the constitutive expression of the *tac* or *lac* promoters, have been transferred to rhizobacterial strains. *P. fluorescens* strains carrying a mini-Tn5 vector that included the complete biosynthetic operon for the antifungal metabolite phenazine-1-carboxylic acid (PCA) were enhanced in their rhizosphere competence, as well as in their ability to suppress fungal diseases (Timms *et al.*, 2000). Similarly, the biocontrol ability of *Pseudomonas* strains producing PCA was extended after the introduction of *phzH* gene from *P. chlororaphis* PCL1391. The introduction of this gene resulted in the production of phenazine-1-carboxamide (PCN) by these strains and in their ability to biocontrol tomato root and foot rot. There have been other studies in which the introduction of genes into rhizobacterial strains has enhanced biocontrol and/or plant growth promotion. These genes include the Cry-toxin-encoding *cry1Ac7* gene of *Bacillus thuringiensis* (Downing *and* Thompson, 2000), and the chitinase-encoding *chiA* gene of *Serratia marcescens* (Downing *et al.*, 2000). In addition, Dekkers *et al.*, (2000) showed that the transfer of the *sss* gene of *P. fluorescens* WCS365 could enhance the competitive colonization ability of other *P. fluorescens* strains.

Genetic studies on important PGPR traits of fluorescent *Pseudomonas* spp. have also shown that there existed a latent potential to improve the performance of the strains. Siderophore receptors have been cloned and introduced to new strains, allowing them to use a wider range of siderophores for better iron uptake (Marugg *et al.*, 1989; O' Sullivan *et al.*, 1990). Studies on antibiotic production have shown that transfer of particular segment from one strain to another could increase the range of pathogens inhibited by the modified strain (Gill and Warren, 1988). As certain *fluorescent pseudomonads* are effective rhizosphere colonizers and are endophytic in nature in the plant system, attempts have been made to transfer the insecticidal crystal protein from *Bacillus thuringiensis* to *P. fluorescens*. Thus, genetically engineered *P. fluorescens* was found effective against a lepidopteran insect pest. Transgenic *P. cepacia* strain 526 with the crystal protein gene has consistently shown insecticidal activity against tobacco hornworm (Stock *et al.*, 1990). Recently, a simplified and rapid protocol for strain improvement of *Pseudomonas fluorescens* isolates has been developed by Negi (2006). The technique i.e., bacterial transformation, involves the transfer of necked DNA fragments directly to competent cells and has been found effective to transfer phosphate solubilization and siderophore production activities to the recipient isolates lacking these activities. The strain improvement through genetic modification can introduce or improve the desired trait in suitable organism and this therefore offers to be an attractive option for the commercialization of these improved strains.

## VISUALIZATION OF BACTERIA AND GENE EXPRESSION IN THE RIZOSPHERE

After inoculation, bacteria must become established in the rhizosphere where they interact with the root and the endogenous population of microorganisms, which includes phytopathogenic as well as non-pathogenic microbes. Progress has been made during the past years in using confocal laser scanning microscopy (CLSM) in combination with various fluorescent markers to visualize and monitor bacterial populations in the rhizosphere (Fig. 4). Using a combination of immunofluorescence and an rRNA-targeting probe that monitors the presence and metabolic activity of *P. fluorescens* DR54 inoculant cells in the sugar beet rhizosphere, Lubeck *et al.*, (2000) showed that bacteria at the root tip are metabolically most active and that indigenous bacteria enter the rhizosphere two days after inoculation. Visualization of interactions among the carrot roots,

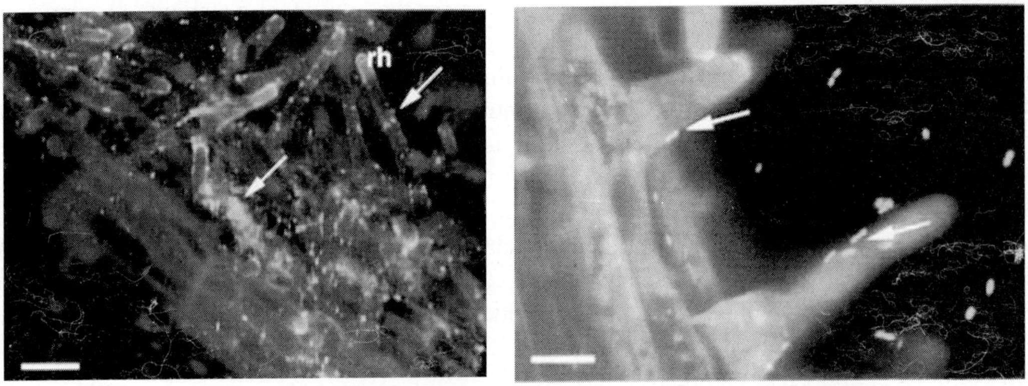


A B

**Fig. 4** Confocal laser scanning micrograph carrot roots A-B: Bacteria are visible as small green spots on the root surface. Plant nuclei are stained in red. Bars = 30 mm.

(Source: Valeria Bianciotto et al., 2001. MPMI Vol. 14, No. 2, 2001, pp. 255–260)

mycorrhizal mycelium and *P. fluorescens* CHA0 bacteria showed that mucoid mutant strains of CHA0 adhere much better to the root, indicating that acidic extracellular polysaccharides can enhance root colonization (Bianciotto *et al.*, 2000).

The study of microbial communities has been facilitated by the use of combinations of the green fluorescent protein (GFP), its color variants cyan fluorescent protein (CFP) and yellow fluorescent protein (YFP), and DsRed as markers (Fig. 4). Using rhizosphere stable vectors, broad-host-range plasmids expressing the *e-gfp*, *e-yfp*, *e-cfp* or the *rfp* (*red fluorescent protein*) have been constructed (Bloemberg *et al.*, 2000, Stuurman *et al.*, 2000). Using these plasmids, up to three differently marked bacterial populations can be studied on the root. Such experiments have indicated that *Pseudomonas* microcolonies on the root surface are initiated by one bacterial cell, and bacteria from outside the growing colony can join, as indicated by the frequent presence of mixed-color colonies on the older part of the root (Dekkers *et al.*, 2000, Bloemberg *et al.*, 2000). The construction of unstable variants of the GFP makes it possible to study transient gene expression in the rhizosphere, as has been shown for ribosomal activity in *P. putida* cells (Ramos *et al.*, 2000). A gfp-based system for the detection of AHLs should allow the visualization of quorum sensing and cross-talk among bacteria in the rhizosphere (Andersen *et al.*, 2001). Visualization of gene expression in the rhizosphere will provide detailed information on the functioning of bacterial cells in a specific environment. Spatio-temporal analysis of gene expression in the rhizosphere will be made possible by visualizing bacterial cells that harbor an unstable *gfp* variant under control of the promoter to be analyzed and the constitutive expression of *rfp*.

## CHALLENGES IN SELECTION AND CHARACTERIZATION OF PGPR

One of the challenges in developing PGPR for commercial application is ensuring that an effective selection and screening procedure is in place, so that the most promising organisms are identified and brought forward (Fig. 5). In the agricultural chemical industry, several prospective compounds are screened in efficient high-throughput assays to select the best one or two compounds for further development. Similar approaches are not yet in place for PGPR. Effective strategies for initial selection and screening of rhizobacterial isolates are required. It may be important to consider host plant specificity or adaptation to a particular soil, climatic conditions or pathogen in selecting the isolation conditions and screening assays (Bowen and Rovira 1999, Chanway *et al.*, 1989). The spermosphere model, an enrichment technique that relies on seed exudates as the nutrient source, has been used for selection and isolation of promising $N_2$-fixing rhizosphere bacteria from rice (Thomas *et al.*, 1982). One approach for selection of organisms with the potential to control soil-borne phytopathogens is to isolate from soils that are suppressive to that pathogen (Weller *et al.*, 2002). Other approaches involve selection based on traits known to be associated with PGPR such as root colonization, 1-aminocyclopropane-1-carboxylate (ACC) deaminase activity, and antibiotic and siderophore production (Cattelan *et al.*, 1999, Silva *et al.*, 2003). The development of high throughput assay systems and effective bioassays will facilitate selection of superior strains.

### Challenges in Field Application of PGPR

The application of PGPR for control of fungal pathogens in greenhouse systems shows considerable promise (Paulitz and Belanger 2001), due in part to the consistent environmental

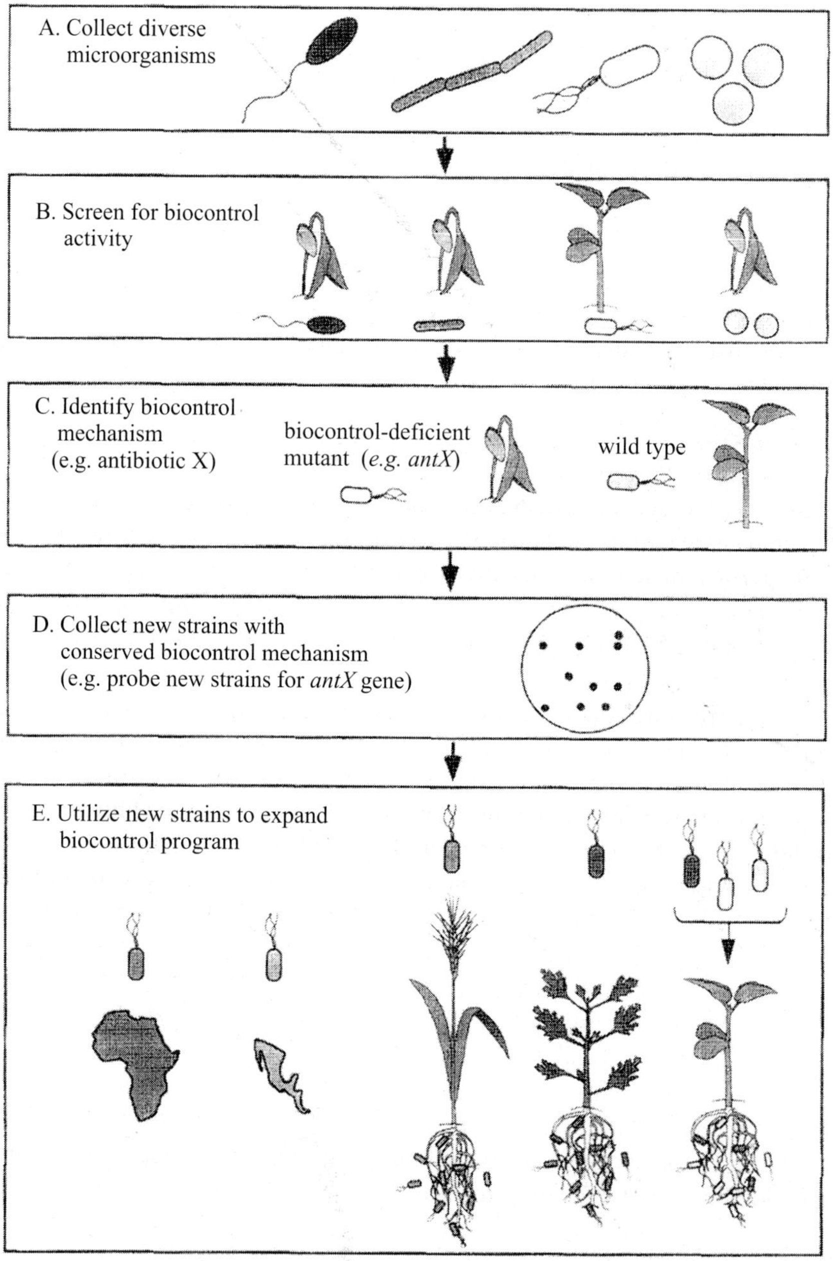


**Fig. 5** Schematic representation for select and utilize as biocontrol agent. (Source: Handelsman and Stabb, 1996. Biocontrol of soil borne pathogens. Plant Cell 8: 1855-1869)

conditions and high incidence of fungal disease in greenhouses. Achieving consistent performance in the field where there is heterogeneity of abiotic and biotic factors and competition with indigenous organisms is more difficult. Knowledge of these factors can aid in determination of optimal concentration, timing and placement of inoculant, and of soil and crop management strategies to enhance survival and proliferation of the inoculant (Bowen and Rovira 1999). The concept of engineering or managing the rhizosphere to enhance PGPR function by manipulation of the host plant, substrates for PGPR, or through agronomic practices, is gaining attention (Bowen and Rovira 1999, Mansauri *et al.*, 2002). Development of better formulations to ensure survival and activity in the field and compatibility with chemical and biological seed treatments is another area of focus; approaches include optimization of growth conditions prior to formulation and development of improved carriers and application technology (Bashan 1998, Bowen and Rovira 1999, Mathre *et al.*, 1999).

## CHALLENGES IN COMMERCIALIZATION OF PGPR

Prior to registration and commercialization of PGPR products, a number of hurdles must be overcome (Mathre *et al.*, 1999). These include scale up and production of the organism under commercial fermentation conditions while maintaining quality, stability, and efficacy of the product. Formulation development must consider factors such as shelf life, compatibility with current application practices, cost, and ease of application (Fig. 6). Health and safety testing may be required to address such issues as non-target effects on other organisms including toxigenicity, allergenicity and pathogenicity, persistence in the environment and potential for horizontal gene transfer. Capitalization costs and potential markets must be considered in the decision to commercialize the product (Walsh *et al.*, 2001).

## CONCLUSION

Since PGPRs are an environmental-friendly alternative to chemical fertilizers and pesticides, the use of which is regulated and sometimes forbidden, the market for bioinoculants is still expanding. *Rhizobium* and *Bradyrhizobium* inoculants have been marketed with success for over a century. Approximately twenty bacterial biocontrol products based on *Pseudomonas*, *Bacillus*, *Streptomyces* and *Agrobacterium* strains have been marketed (for an inventory list see URL http://www.barc.usda.gov/psi/bpdl/bpdlprod/bioprod.html). There is still a need to optimize the efficacy of these products. The better understanding of many traits and genes that are involved in the beneficial effects of PGPRs can result in a better understanding of the performance of bioinoculants in the field, and provide the opportunity to enhance the beneficial effects of PGPR strains by genetic modification. The different techniques discussed in the chapter may help the beginners to choose the right area and strengthen the rhizobacterial research, thereby contributing in raising agriculture and productivity in an agriculture-based country, like ours.

## FUTURE PROSPECTS

As our understanding of the complex environment of the rhizosphere, of the mechanisms of action of PGPR, and of the practical aspects of inoculant formulation and delivery increases, we can

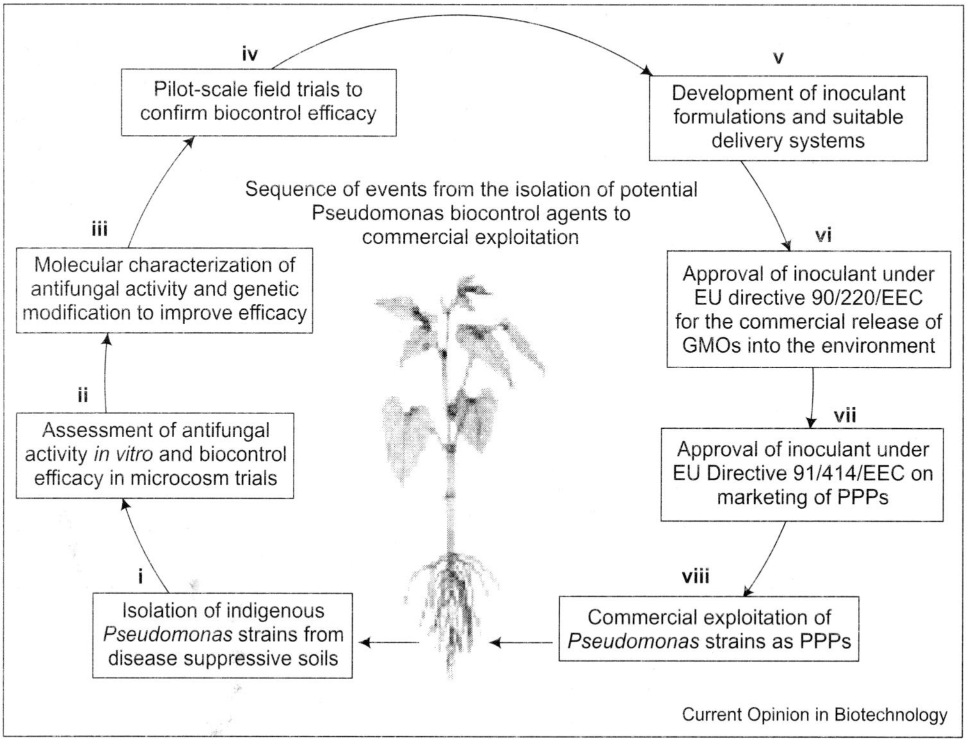


**Fig. 6** Schematic representation representing different steps for isolation, select and development of a biocontrol agent. (Source: Walsh *et al.*, 2001.*Pseudomonas* for biocontrol of phytopathogens: from functional genomics to commercial exploitation. Currt. Oppn. Biotechnol. 12: 289-295)

expect to see new PGPR products becoming available. The success of these products will depend on our ability to manage the rhizosphere to enhance survival and competitiveness of these beneficial microorganisms (Bowen and Rovira 1999). Rhizosphere management will require consideration of soil and crop cultural practices as well as inoculant formulation and delivery (Bowen and Rovira 1999). Genetic enhancement of PGPR strains to enhance colonization and effectiveness may involve addition of one or more traits associated with plant growth promotion (Glick 1995, Lubeck *et al.*, 2000). Genetic manipulation of host crops for root-associated traits to enhance establishment and proliferation of beneficial microorganisms is being pursued. However, regulatory issues and public acceptance of genetically engineered organisms may delay their commercialization. The use of multi-strain inocula of PGPR with known functions is of interest as these formulations may increase consistency in the field. They offer the potential to address multiple modes of action, multiple pathogens, and temporal or spatial variability (Guetsky *et al.*, 2001; De Boer *et al.*, 2003, Duffy *et al.*, 1995, Pierson and Weller 1994).

PGPR offer an environmentally sustainable approach to increase crop production and health. The application of molecular tools is enhancing our ability to understand and manage the rhizosphere and will lead to new products with improved effectiveness.

## References

Andersen, J. B., Heydorn, A., Hentzer, M., Eberl, L., Geisenberg, O., Christensen, B. B., Molin, S. & Givskov, M. (2001) **67**: 575-585.

Archibald, F. (1983) *FEMS Microbiol. Lett.* **19**: 29-32.

Babu-Khan, S., Yeo, T. C., Martin, W. L., Duron, M. R., Rogers, R. D. & Goldstein, A. H.(1995) *Appl. Environ. Microbiol.* **5**: 72-74.

Bashan, Y. (1998) *Biotechnol. Adv.* **16**: 729-770.

Basnayake, W. V. S. & Brich, R. G. (1995) *Microbiol.* **141**: 551-560.

Bianciotto, V., Lumini, E., Lanfranco, L., Minerdi, D., Bonfante, P. & Perotto, S. (2000) *Appl. Environ. Microbiol.* **66**: 4503-4509.

Bloemberg, G. V., Wijfjes, A. H. M., Lamers, G. E. M., Stuurman, N. & Lugtenberg, B. J. J. (2000) *Mol. Plant Microbe Interact.* **13**:1170-1176.

Bowen, G. D. & Rovira, A. D. (1999) *Adv. Agron.* **66**: 1-102.

Carlton, B. C. (1993) In: Pest control with enhanced environmental safety, (Eds. Duke, S. O., Menn, J. J.& Plimmer, J. R.) American chemical society, Washington, D.C. pp. 258-266.

Cattelan,A. J., Hartel, P. G. & Fuhrmann. (1999) *Soil Sci. Soc. Am. J.* **63**: 1670-1680.

Chancey, S. T., Wood, D. W. & Pierson, S. L. (1999) *Appl. Environ. Microbiol.* **65**: 2294-2299.

Chanway, C. P. Nelson, L. M. & Holl, F. B. (1989) *Can. J. Microbiol.* **34**: 925-929.

Chernin, L. & Chet, I. (2002) In: (Eds. Burns, R. G. & Dick, R. P.) Enzymes in environment: activity, ecology and applications. Marcel Dekker, New York. pp. 171-225.

Chin,A.,Woeng, T. F. C., Bloemberg, G., V., Mulders, I., Dekkers, L. & Lugtenberg, B. J. J. (2000) *Mol. Plant Microbe Interact.* **13**:1340-1345.

Ciccilo, F., Fiore, A, Bavivino, A., Dalmastri, C., Tabbachioni, S. & Chiarini, L. (2002) *Environ. Microbiol.* **4**: 238-245.

Cook, R. J. (1993) *Annu. Rev. Phytopathol.* **31**: 53-80.

Cornelis, P. & Matthijs, S. (2002) *Environ. Microbiol.* **4**: 787-798.

De Boer, M., Bom, P., Kindt, F., Keurentjes, J. J. B., van der Sluis, I., van Loon, L.C. & Bakker, P. A. H. M. (2003) *Phytopathol.* **93**: 626-632.

De Souza, J. T., de Boer, M., de Ward, P., van Beek T. A. & Raaijmakers, J. M. (2003) *Appl. Environ. Microbiol.* **69**: 7161-7172.

De Werger, L. A., van der Vlugt, C. I. M., Wijffelman, C. A., Bakker, P. A. H. M., Schippers, B. & Lugtenberg, B. (1987) *J. Bacteriol.* **169**: 2769-2773.

Dekkers, L. C., Mulder, I. H., Phoelich, C. C., Chin,A.,Woeng, T. F. C., Wijfjes, A. H. & Lugtenberg, B. J. J. (2000) *Mol. Plant-Microbe Interact.* **13**: 1177-1183.

Dekkers, L. C., Phoelich, C. C., van der Fits, L. & Lugtenberg, B. J. J. (1998b) *Proc. Natl. Acad. Sci.* USA. **95**: 7051-7056.

Dekkers, L. C., van der Bij, A. J., Mulders, I. H. M., Phoelich, C. C., Wentwoord, R. A. R., Glendorf, D. C. M., Wijffelman, C. A. & Lugtenberg, B. J. J. (1998) *Mol. Plant Microbe Interact.* **11**: 763-771.

Dong, Y. H., Xu, J. L., Li, X. Z. & Zhang, L. H. (2000) *Proc. Natl. Acad. Sci.* USA **97**: 3526-3531.

Dong, Y. H., Zhang, X. F., Xu, J. L. & Zhang, L. H. (2004) *Appl. Environ. Microbiol.* **70**: 954-960.

Downing, K. J. & Thomson, J. A. (2000) *Can. J. Microbiol.* **46**:363-369.

Downing, K. J., Leslie, G. & Thomson, J. A. (2000) *Appl. Environ. Microbiol.* **66**: 2804-2810.

Duffy, B. K. & Defago, G. (1997) *Phytopathol.* **87**: 1250-1257.

Duffy, B. K. & Defago, G. (1999) *Appl. Environ. Microbiol.* **65**: 2429-2438.

Duffy, B. K. & Weller, D. M. (1995) *Plant Dis.* **79**: 907-911.

Duffy, B. K., Simon, A. & Weller, D. M. (1996) *Phytopathol.* **86**: 188-194.

Espinosa-Urgel, M., Salido, A. & Ramos, J. L. (2000) *J. Bacteriol.* **182**: 2363-2369.

Feitelson, J. S., Pyne, J & Kim, L. (1992) *Bacillus thuringiensis*: Insect and beyond. Bio/Technology.

Frankowsky, J., Lorito, M., Scala, F., Schmidt, R., Berg, G. & Bahl, H. (2001) *Arch. Microbiol.* **176**: 421-426.

Freiberg, C., Fellay, R., Bairoch, A., Broughton, W. J., Rosenthal, A. & Perret, X. (1997) *Nature*, **387**: 394-401.

Garlend, J. L. (1996) *Soil Biol. Biochem.* **28**: 223-230.

Germida, J. J., Siciliano, S. D. de Freitas, J. R., & Seib, A. M. (1998) *FEMS Microbiol. Ecol.* **26**: 43-50.

Gill, P. R. & Warren, G. J. (1988) *J. Bacteriol.* **170**: 163-170.

Goldstein, A. H. (1994) In: Phosphate in Microorganisms: Cellular and Molecular Biology. (Eds. Torriani-Gorini, A., Yagil, E. & Silver, S.) Washington, DC: ASM Prss, pp. 197-203.

Goldstein, A. H. & Liu, S. T. (1987) *Bio Technology*, **5**: 72-74.

Gottfert, M., Rothlisberger, S., Kundig, C., Marty, R. & Hennecke, H. (2001) *J. Bacteriol.* **183**:1405-1412.

Guetsky, R., Shtienberg, D., Elad, Y. & Dinoor, A. (2001) *Phytopathol.* **91**: 621-627.

Hass, D. & Keel, C.(2003*) Annu. Rev. Phytopathol.* **41**: 117-153.

Hoefte, H. & Whitely, H. R. (1989) *Microbiol. Rev.* **53**: 242-255.

Howell, C. R. & Stipanovic, R. D. (1979) *Phytopathol.* **69**: 480-482.

Howell, C. R. & Stipanovic, R. D. (1980). *Phytopathol.* **70**: 712-715.

Keel, C., Weller, D. M., Defago, G., Cook, R. J. & Thomashow, L. S. (1996) *Appl. Environ. Microbiol.* **62**: 552-563.

Keel, C., Wirthner, P. H., Oberhansli, T. H., Voisard, C., Burger, D., Hass, D. & Defago, G. (1990) *Symbiosis.* **9**: 327-341.

Kerr, A. & Htay, K. (1974) *Physiol. Plant Pathol.* **4**: 37-44.

Kloepper, J. W. & Scroth, M. N. (1978) In: *Proc.* of 4$^{th}$ Int. Conf. on plant pathogenic bacteria (Vol. 2), Station de Pathologie Vegetaleet Phytobacteriologie, INRA, Angers, France. pp. 879-882.

Kojic, M., Degrassi, G. & Venturi, V. (1999) *Biochem. Biophys. Acta.* **1489**:413-420.

Kucey, R. M. N. (1983) *Can. J. Soil. Sci.* **63**: 671-678.

Lim, H. S., Kim, Y. S. & Kim, S. D. (1991) *Appl. Environ. Microbiol.* **57**: 510-516.

Long, S. R. (2001) *Plant Physiol.* **125**: 69-72.

Lubeck, P. S., Hansen, M. & Sørensen, J. (2000) *FEMS Microbiol. Ecol.* **33**:11-19.

Lugtenberg, B. J. J. & Dekkers, L. C. (1999) *Environ. Microbiol.* **1**: 9-13.

Lugtenberg, B. J. J., Dekkers, L. & Bloemberg, G. V. (1999) *Annu. Rev. Phytopathol.* **39**: 461-490.

Lugtenberg, B. J. J., Kravchenko, L. V. & Simons, M. (1999) *Environ. Microbiol.* **1**: 439-446.

Mansauri, H., Petit, A., Oger, P. & Dessaux, Y. (2002) *Appl. Environ. Microbiol.* **68**: 2562-2566.

Marugg, J. D., De Werger, L. A., Nielander, H. B., Oorthuizen, N. & Weisbeek, P. J. (1989) *J. Bacteriol.* **171**: 2819-2826.

Mathre, D. E. Cook., R. J. & Callan, N. W. (1999)*Rev. Biotechnol.* **12**: 87-132.

Mavrodi, Olga. V., Gardener, B. B. M., Mavrodi, D. V., Bonsal, R. F., Weller, D. M. & Linda, S. T. (2001) *Phytopathol.* **91**: 35-43.

McSpadden Gardener, B. B., Mavrodi, D. V., Linda S. T. & Weller, D. M. (2000) *Phytopathol.* **91**: 44-54.

Milner, J. L., Raffel, S. J., Lethbridge, B. J. & Handelsman. (1995) *Appl. Microbiol. Biotechnol.* **43**: 685-691.

Milner, J. L., Silo-Suh, L., Lee, J. C., He, H., Clardy, J. & Handelsman. (1996) *Appl. Environ. Microbiol.* **62**: 3061-6065.

Morello, J. E., Pierson, E. A. & Pierson, L. S. (2004) *Appl. Environ. Microbiol.* **70**: 3103-3109.

Negi, Y. K. (2006) Ph. D. thesis submitted to Dr. R. M. L. Avadh University, Faizabad U. P. (India).

Neilands, J. B. (1981) *Ann. Rev. Biochem.* **50**: 715-731.

Neilands, J. B., Konopka, K., Schwyn, B., Coy, M., Francis, R. T., Paw, B. H. & Bagg, A. (1987). In: Iron transport in microbes, plants and animals. (Eds. Winkelmann, G., Van Der Helm, D. & Neilands, J. B.). Verlagsellschaft MBH, Weinheim. pp. 3-33.

Nielsen, T. H., Sorensen, D. Tobiasen, C., Andersen, J. B., Christophersen, C., Givskov, M. & Sorensen, J. (2002) *Appl. Environ. Microbiol.* **69**: 861-868.

O'Sullivan, D. J & O'Gara, F. (1992) *Microbiol. Rev.* **56**: 662-672.

O'Sullivan, D. J., Moris, J. & O'Gara F. (1990) *Appl. Environ. Microbiol.* **56**: 2056-2064.

Ordentlich, A., Elad, Y. & Chet, I. (1988). *Phytopathol.* **78**: 84-88.

Ovadis, M., Liu, X., Gavriel, S., Ismailov, Z., Chet, I. & Chernin, L. (2004.) *J. Bacteriol.* **186**: 4986-4993.

Park, C., Paulitz, T. C. & Baker, R. (1988) *Phytopathol.* **78**: 190-194.

Parker, J. E. (2003) *Trends Plant Sci.* **8**: 245-247.

Paulitz, T. C. & Belanger, R. B. (2001). *Annu. Rev. Phytopathol.* **39**: 103-133.

Perret, X., Staehelin, C. & Broughton, W. J. (2000) *Microbiol. Mol. Biol. Rev.* **64**:180-201.

Persello-Cartieaux, F., Nussaume, L. & Robaglia, C. (2003). *Plant Cell Environ.* **26**: 189-199.

Picard, C., Cello, D., Ventura, M. F., Fani, R. & Guckert, A. (2000) *Appl. Environ. Microbiol.* **66**: 948-955.

Pierson, E. A. & Weller, D. M. (1994) *Phytopathol.* **84**: 940-947.

Pierson, S. L., Wood, D. W. & Pierson, E. A. (1998) *Annu. Rev. Phytopathol.* **36**: 207-225.

Raaijmakers, J. M. & Weller, D. M. (2001)*Appl. Environ. Microbiol.* **67**: 2545-2554.

Rainey, P. B. (1999) *Environ. Microbiol.* **1**: 243-257.

Ramesh Kumar, N., Arasu, V. T. & Gunasekaran, P. (2002) *Curr. Sci.* **82**: 1463-166.

Ramos, C., Molbak, L. & Molin, S. (2000) *Appl Environ Microbiol* **66**: 801-809.

Ravel, J. & Cornelis, P. (2003) Genomics of pyoverdin mediated iron uptake in pseudomonads. *Trends Microbiol.* **11**: 195-200.

Saleh, S. S. & Glick, B. R. (2001) *Can. J. Microbiol.* **47**: 698-705.

Schinder-Keel, U., Seemater, A., Maurhofer, M., Blumer, C., Duffy, B., Gigot-Bionnefoy, C., Reimmann, C., Notz, R., Defago, G., Hass,. D. & Keel, C. (2000) *J. Bacteriol.* **182**: 1215-1225.

Schouten, A., van der Berg, G., Edel-Harmann,V., Steinberg, C., Gautheron, N., Alabbouvette, C., de Vos, C. H., Lemanceau, P. & Raaijmakers, J. M. (2004). *Mol. Plant-Microbe Intarct.* **17**: 1201-1211.

Seong, Ki-Young & Shin, P. G. (1996) *Agric. Chem. Biotechnol.* **39**: 20-24.

Silva, H. S. A., Romeiro, R. S. & Mounteer, A. (2003) *J. Phytopathol.* **151**: 42-46.

Simon, H., Smith, K. P., Dodsworth, J. E., Guenthner, B., Handelsman, J. & Goodman, R. M. (2001)*Appl. Environ. Microbiol.* **67**:514-520.

Singh, P. P., Shin, Y. C., Park, C. S. & Chung, Y. R. (1999) *Phytopahol.* **89**: 92-99.

Smith, L. M., Tola, E., de Boer, P. & O'Gara, F. (1999) *Environ. Microbiol.* **1**: 495-502.

Sorensen, J., Jensen, L. E. & Nybroe, O. (2001) *Plant Soil.* **232**: 97-108.

Stabb, E. V. (1994) *Science*. **264**: 388-393.

Stabb, E. V. Jcobson, L. M. & Handelsman, J. (1994) *Appl. Environ. Microbiol.* **60**: 4404-4412.

Steenhoudt, O. & Venderleyden, J. (2000) *FEMS Microbiol. Rev.* **24**: 487-506.

Storm, M. S. & Lory, S. (1993) *Rev. Microbiol.* **30**: 565-596.

Stuurman, N., Bras, C. P., Schlaman, H. R., Wijfjes, A. H., Bloemberg, G. V. & Spaink, H. P (2000) *Mol. Plant Microbe Interact.* **13**:1163-1169.

Thomas-Bauzon, D., Wheinhard, P., Villecourt, P. & Balandreau, J. (1982) *Can. J. Microbiol.* **28**: 922-928.

Thomashow, L. S. & Weller, D. M. (1988) *J. Bacteriol.* **170**: 3499-3508.

Timms-Wilson, T. M., Ellis, R.J., Renwick, A., Rhodes, D. J,, Mavrodi, D. V., Weller, D. M., Thomashow, L. S. & Bailey, M. J. (2000) *Mol Plant Microbe Interact,* **13**:1293-1300.

Toyoda, H., Hashimoto, H., Utsumi, R., Kobayashi, H. & Ouchi, S. (1988) *Phytopathol.* **78**: 1307-1311.

Tunlid, A. & White, D. (1992) In: *Soil Biochemistry* Vol. 7. (Eds. Stotzky, G & Bollag J. M.) pp. 229-262. Marcel Dekker Inc., New York USA.

Van Schie, B. J., Helingwerf, K. J., van Dijken, J. P. Elferink. M. G. L., van Dijl, J. M., Kuenen, J. G. & Konigns, W. N.(1987) *J. Bacteriol.* **163**: 493-499.

Viprey, V., Del Greco, A., Golinowski, W., Broughton, W. J. & Perret, X. (1998) *Mol. Microbiol.* **28**:1381-1389.

Von Bodman, S. B., Baur, W. D. & Coplin, D. L. (2003) *Annu. Rev. Phytopathol.* **41**: 455-482.

Walker, M. J., Birch, R. G. & Pemberton, J. M. (1988) *Mol. Microbiol.* **2**: 443-454.

Walling, L. L. (2000) *J. Plant Growth Regul.* **19**: 195-216.

Walsh, U. F., Morrissey, J. P. & O'Gara, F. (2001) *Currt. Oppn. Biotechnol.* **12**: 289-295.

Weller, D. M., Raaijmakers, J. M., McSpadden-Gardener, B. B. & Thomshow, L. S. (2002) *Annu. Rev. Phytopath.* **40**: 309-348.

Zhang, L. & Birch, R. G. (1996) *Lett. Appl. Microbiol.* **22**: 132-136.

Zhang, L. & Birch, R. G. (1997) *Proc. Natl. Acad. Sci.* USA. **94**: 9984-9989.

10

# Rhizospheric Competent Microorganisms in Biological Control of Plant-Parasitic Nematodes

**T. Kumar [1], S. C. Kang [2] and D. K. Maheshwari [1]**

*[1]Department of Botany and Microbiology, Gurukul Kangri University, Haridwar-249404, Uttarakhand, India*

*[2]Department of Food, Biological and Chemical Engineering, Daegu University, Gyungsan 712-714, Korea*

## ABSTRACT

A range of microorganisms in the rhizosphere attack disease-causing plant-parasitic nematodes. The use of such microorganisms for the management of plant-parasitic nematodes has great significance. These microorganisms are beneficial in comparison to synthetic nematicides because they are easy to apply are free from environmental pollution and restore soil fertility. Hence, there proved as alternative to chemicals that fulfil all requirements of farmers and bring sustainability to agricultural crops. Efforts should be made to find out the toxic metabolites secreted by microorganisms as nematicides and evaluate their mode of action to make their use more economic, feasible and environmentally safe.

## INTRODUCTION

Biological control of plant pathogens including plant-parasitic nematode has gained considerable attention and is quite promising as a viable alternative to chemical control strategies (Rebafka *et al.*, 1993; Trigalet *et al.*, 1994). Bio-agents reduce the pathogen attack and its disease producing activities. They induce resistance in host and create environment for unsuccessful disease development. Different microorganisms including both beneficial and deleterious one, heavily populate the rhizosphere region. Therefore, there is an opportunity to find rhizosphere competent microorganisms, which may act as potential biocontrol agents. Inefficient and overuse of synthetic chemicals (nematicides) leads to economic loss to the nation and the farmers. Loss of soil fertility resulted due to continuous nutrient mining without adequate replenishment that also led in deterioration not less dangerous than other forms of environmental degradation. An important factor on this aspect is to enhance soil fertility through appropriate sustainable technology in order to replenish the nutrients so as to build up the nutrient status of soils (Heera, 1996). Although, synthetic nematicides are recognized to be a quick and sure way of boosting crop production, but their cost and other constraints frequently deter farmers (Finck, 1994). Excessive application of nematicides for enhancement of crops yield is not only undesirable from the economic point of view, but also exerts an adverse effect on the crop quality (Kenney, 1982).

The interest in biological control agents for the control of plant-parasitic nematode has arisen due to number of reasons. The high cost of conventional nematicides; the withdrawal of nematicides from the market due to detrimental side effects, in particular their residual effect; ground water contamination; undesirable harmful effects on non-target species; development of nematode-resistant host cultivars, etc. further aggravated the situation of chemical control of plant-parasitic nematodes (Jatala, 1986; Kerry, 1995). Interest in biological control started in the years 1920s and 1930s, when some plant pathogens were suppressed by introducing some antibiotic producing microbes to the natural habitats. De Bach (1964) has defined biological control as the "action of parasites, predators and other pathogens in maintaining the population density of other organisms at a lower average than would occur in their absence". The definition of biological control was further elaborated by Garret (1965), who stated that "Any condition under which, or practice where the survival or activity of a pathogen is reduced through the agency of any other living organisms, results in the reduction of disease incidence". Beirner (1967) proposed "Biological control is the induced or natural, direct or indirect limitation of harmful organism or its effect on another organism or group of organisms". But the most comprehensive definition of biological control is "The reduction of inoculum density or disease producing activities of pathogen or parasite in its active or dormant state, by one or more organisms, host or antagonists or by mass interaction of one or more antagonists" (Baker and Cook, 1974).

## CROP LOSSES DUE TO ATTACK OF PLANT-PARASITIC NEMATODES

Plant-parasitic nematodes are involved in huge loss of revenue in agricultural crops. They constitute an important group of crop pests wherein crop yield is considerably reduced both at nursery stage and at farm level. The magnitude of the loss depends upon the virulent nature of nematode species, degree of invasion, the host plant genotype and the abiotic and edaphic factors during cultivation. In fact, quantitative losses are estimated to the extent of 50% in some horticultural and vegetable crops and 5-10% loss occurs due to nematode attack. In India, over all crop losses due to nematode attacks contributed to about 10.6% (Sheshadri, 1970), which has now increased due to development of nematicides resistant races of the pests. Generally, in normal conditions, 15-20 eggs and larvae in a gram of soil has been reported to be critical threshold level for economic loss in wheat and barley as reported by Gill and Swarup (1971). Similar to that,

Handa *et al.*, (1975) have observed populations as high as 50 eggs and larvae per gram soil in some states of our country. The ear cockle disease due to seed gall nematode, *Anguina tritici* may result in 10,000 tones of annual loss of wheat crop. Similarly, the cereal cyst nematode, *Heterodera avenae* causes molya disease of wheat and barley in north-western India. The root lesion nematode, *Pratylenchulus coffeae* is a serious pest of coffee infesting hundreds of hectares of land with a huge loss of revenue in coffee cultivation industry (Reddy, 1983). The reniform nematode, *Rotylechulus reniformis* has been reported to cause 5-15% of crop losses in cash crop (Palaniswamy and Balasubramanian, 1981). The burrowing nematode, *Rhadopholus similes* causes slow wilt in pipper whereas golden nematode of potato *Globodera rostochiensis* is a serious pest of nilgiri hills. Sometimes, complete loss has been recorded due to nematode attack on the crops. The citrus nematode, *Tylenchulus semipenetrans,* the causal organism of 'slow decline' in citrus, incurred a loss of 15% in crop yield (Anon, 1971).

One of the major pests is root knot nematode *Meloidogyne* spp. The nematodes attack at seedling stage and cause tremendous losses. More than 3000 host plants are attacked by root knot nematode (Parvatha Reddy and Singh, 1981) and *Meloidogyne incognita* alone causes about 45% yield loss in brinjal (Krishnappa *et al.*, 1981), 90.6% in okra, 42.4% in tomato (Bhatii and Jani, 1977) and 33.4% in chilli (Sudar Babu, 2000) as shown in Table 1. The small size and hidden way of life beside lack of definite information on their occurrence and pathogenicity allows them to attack the crops.

## ATTRIBUTES OF SUCCESSFUL BIOLOGICAL CONTROL AGENTS

The effectiveness and success of any such organisms, which is antagonistic to plant-parasitic nematodes, depends upon certain desired specific characteristics as given below:

(i) Biological competitiveness: The biocontrol agent should be biologically competitive with other existing or introduced parasites and predators in the host ecosystem.

(ii) Easy culture conditions: The biocontrol agent should be produced easily on economically viable raw materials for its mass production for field application.

(iii) Non-pathogenic: They should not be pathogenic to the agricultural crops, human beings and should not impart mammalian toxicity.

(iv) High nematode reducing potential: A successful biocontrol agent must suppress nematode population in large quantities, so that after a given period, the nematode population is reduced to a level below the damage threshold.

(v) Capability to produce deleterious compounds: Biocontrol agents should be capable of producing antibiotics or other deleterious compounds.

(vi) Long term storage: The biocontrol agent should be genetically stable so as not to loose the beneficial characteristics during storage.

(vii) High reproductive potentials: The biocontrol agents must have high reproductive rate for keeping desirable population density. However, some times high reproductivity adversely affects its efficiency because more energy is spent for such activity. There must be a judicious balance between reproductive and infective potentials, besides these should aggressively colonize the soil.

**Table 1** Crop losses due to attack of plant parasitic nematodes.

| *Nematode species* | *Crops affected* | *Annual loss or % loss estimated/ determined* | *References* |
|---|---|---|---|
| *Heterodera avenae* | Wheat and Barley | 8M rupees, AL 8-49%, 29.8-38.7% grain and 34.03-42.9% fodder | Bhatii *et al.* (1990) |
| *H. zeae* | Maize | 11-29% in N., W. and north-western belt 12.3% | Gill (1994) |
| *Meloidogyne incognita* | Cowpea | 12.3% | Patel *et al.* (1993) Gupta and Verma (1990) |
| *M. incognita* | Mungbean | AL 42.1-93.4, 42.1-93.4%, 12.7%, 7-48%, 50-61% | Mohanty and Mohapatra (1994) |
| *M. incognita* | Pointed gourd | 43.8% (at 1-2 J/g) | Verma and Anwar (1996) |
| *M. javanica* | Okra, Brinjal, Tomato | 30-47% avoidable loss | Jain *et al.* (1994) |
| *M. hapla* | Carrot | Germination loss 14.9%, yield loss 50.95%, marketability loss 86.2% | Sivakumar and Vadivelu (1994) |
| *Meloidogyne spp.* | Vegetables | 20-50.6% (AL) | Parvatha Reddy (1994) |
| *Rotylenchulus* | Tomato, | 42.25% (AL) | Palaniswami and |
| *reniformis* | Cotton | 10-15% | Balasubramanian (1981), |
| | Black gram | 13% | Parvatha Reddy (1994) |
| | Cowpea | 9%, 13.2% (AL) | |
| | Maize and millets | 6% | |
| *Globodera pallida, G. rostochiensis* | Potato | 2800 ha affected, total crop loss under severe infestation 99.5-99.8% (AL) | Parvatha Reddy (1994) |
| *Hirschmanniella spp.* | Rice | 0.5-20% loss in eastern and north-eastern coastal and riverine plains | Gill (1994) |
| *Radopholus similes* | Banana, Coconut, Arecanut, Black pepper and Betelvine may be Coffeae and Sugar cane | 30-60%, 41% Extensively affected | Parvatha Reddy (1993) Krishna Murthy (1993) |

(viii) Efficient dispersal: Biological agents should have a high dispersal capability.

(ix) Searching capability: Searching capability of any organism is an important phenomenon that always affects its mobility as well as other activities including predation. The organisms having a good searching capacity would appear more effective against plant-parasitic nematodes.

(x) Environmental adaptability: The biocontrol agents must be able to adapt and tolerate existing and changing environmental conditions, both biotic and abiotic.

(xi) Resistance and survival capacity: Shelf life and resistance to adverse condition are prerequisite for their existence in extreme soil environment, even in the absence of its host.

(xii) Last, but certainly not least, the application of a successful biocontrol agent to field soils should result in an increased crop yield.

Although, an individual organism may have several desirable attributes, it is doubtful that any single organism fulfils all the attributes of a successful biocontrol agent. Therefore, the selection of an organism as a possible candidate should be based on its ability to possess a large number of beneficial attributes of plant growth enhancement.

## FUNGI

Nematode parasitic fungi consist of a great variety of organisms attacking various developmental stages of nematodes. Spores of some of these organisms are often adhesive and presumably remain dormant in soil until active nematodes make contact. Others produce mycelium and colonize the soil or the surface of plant roots. Some organisms, although basically parasitic on specific stages of nematodes, may overlap in their parasitic activities, attacking more than one developmental stage of nematodes. Parasitic categories of these fungi are nematode trapping fungi, mycorrhizal fungi and opportunistic fungi.

### Nematode-Trapping Fungi

Various fungi, particularly members of the *Arthrobotrys* group, produce different trapping structures that may or may not produce adhesions (Barron 1977). In general, fungi such as *A. oligospora* that produce adhesive trap networks are better saprophytic competitors than *A. dactyloides* because the former contains more specialized constricting ring (Jansson and Nordbring, 1980). The constriction ring helps in trapping the nematodes. The colonization of nematode-trapping fungi varies from host to host. Even certain groups bear different structures, which also affects the extent of colonization. The morphology of the structures used in capturing nematodes varies from simple, undifferentiated hyphae to very highly specialized trapping devices (Sayre, 1971).

Different isolates within the species of trapping fungi affect the extent of nematode colonization in various host plants (Persson and Jansson, 1999). Gaspard and Mankau (1986) isolated some of the species of nematode-trapping fungi from citrus orchards. These fungi belong to different genera, namely *A. arthrobotryoides, A. dectyloides, Monoacrosporium endermatum, M. gephyropagum* etc. In general, nematode-trapping fungi fall into two groups:

(a) Those that have adhesive hyphal networks are good saprophytes and grow rapidly.

(b) Those that capture nematodes by adhesive knobs, adhesive branches, or constricting rings. These are more predacious and grow slowly. Apparently, a correlation exists between the intensity of nematode attraction to these fungi and the increasing dependence of fungi on nematodes for nutrients (Jansson and Nordbring-Hertz, 1980). The rapidity with which nematodes induce trap formation is dependent both on their mobility and the concentration of the captured in equal numbers in the presence of trap (Jansson and Nordbring-Hertz, 1980). Besides morphological structures such as adhesive products or traps, these fungi may also produce certain toxins that paralyze or kill the nematodes prior to penetration in host plants (Clthof and Estey, 1963). While working on sterile substrates, the presence of saprophytic fungi increases competition for nutrients and the activity of *A. oligospora* and *Monoacrosporium cionopagum*; nematode population densities become smaller than where the trapping fungi are present alone (Persmark and Jansson, 1997). Earlier, Jaffee *et al.* (1992) proposed that there is little growth of *A. dactyloides, Monoacrosporium ellipsosporum*, and *M. cionopagum* in soil and little need to compete for nutrients that are mostly supplied by their hosts.

## Mycorrhizal Fungi

The vesicular-arbuscular mycorrhizal (VAM) fungi in the plant root system are well known to improve plant health and growth. A plant with a well-established mycorrhizal symbiont is better off because of (a) increased resistance to fungal pathogens and nematode parasites, (b) increase tolerance to drought, salt and other abiotic stress factors, (c) improved phosphate uptake (d) and less root injury followed by transplantation (Sikora, 1988; Schroth and Hancock, 1982; Hussey and Roncadori, 1982; Dehne, 1987). Many workers have demonstrated a reduction in nematode population or disease incidence due to application of mycorrhizal fungi (Sikora and Schonbeck; 1975; Bagyaraj *et al.*, 1979; Kassab and Taha, 1990; Sharma, 1992; Sharma and Trivedi, 1994). In some cases, nematode population remains unaffected (O'Bannon *et al.*, 1974; Cason *et al.*, 1983) or even increase under the influence of mycorrhiza (Atilano *et al.*, 1981; Kassab and Taha, 1990).

Endomycorrhizal fungi limit the densities of plant-parasitic nematodes on a broad range of host plants (Sikora, 1981; Sitaramaiah and Sikora, 1981; 1982). In such cases, mechanism of action is related to induce resistance/tolerance or competition for nutrients in the root (Sikora, 1978; Smith, 1987). The survey of literature reveals that the *Glomus* sp. is most predominant genera of VAM fungi in India, especially *Glomus fasciculatum* has shown great promise in the management of nematodes (Jalali and Chand, 1990; Trivedi, 1995).

## Opportunistic Fungi

In recent years, a group of fungi, which are saprophytic deriving their nutrition through decomposition of organic matter termed as "opportunistic fungi" has emerged as promising biocontrol agents for management of endoparasitic nematodes. These fungi parasitize developmental stages of nematodes whenever they get an opportunity to come in their contact. The most frequently occurring fungal genera belong to *Cylindrocarpan, Enophiala, Fusarium, Gliocladium, Paecilomyces, Phoma* and *Verticillium* species. These genera are not nematode feeders by adaptation but infect, colonize and consume reproductive structures of various genera of nematode viz. *Globodera*, *Heterodera*, *Meloidogyne* and *Rotylenchulus* at sedentary stages of their life cycle. When the females of *Globodera* and *Heterodera* break in through the root cortex and mature into the cysts, due to their exposure on the root surface, they become infected and colonized by these fungi. Eggs of *Meloidogyne* and *Rotylenchulus* released into the soil are rapidly colonized and destroyed. Fungi survive in host free periods and can be grown axenically.

*Paecilomyces lilacinus* is found to be highly parasitic on eggs of *Meloidogyne incognita* and females and cysts of *Globodera pallida* (Jatala *et al.,* 1979). Jatala (1986) gave an account of the development of this fungus, which first grows in the gelatinous matrix of the species of *Meloidogyne*, *Tylenchulus* and *Nacobbus* and later the infection spreads to the eggs that are completely surrounded by the mycelial network. This organism has given good control of various species of *Meloidogyne* (Gaspard, 1986), *M. incognita* (Adiko, 1984; Guevara *et al.*, 1985; Ibrahim *et al.*, 1987; Khan and Hussain, 1988; Shahzad and Ghaffar, 1989), *M. arenaria* (Rodriguez-Kabana *et al.*, 1984; Culbreath *et al.*, 1986), *Tylenchulus semipenetrans* (Herrera *et al.*, 1984; 1985), *Rotylenchulus reniformis* (Khan and Hussain, 1988; Reddy and Khan, 1988; 1989), *Globodera pallida* (Franco *et al.*, 1981), *G. rostochiensis* (Saifullah *et al.*, 1988) and *Nacobbus aberrens* (de Sisler *et al.*, 1985). Cabanillas *et al.*, (1988) studied the histology of the interaction of *P. lilacinus* with *Meloidogyne incognita* on tomato. Root galling and giant cell formation did not form in roots inoculated with fungus-infected eggs of *M. incognita*. Almost similar condition prevailed when roots dipped in spore suspension of *P. lilacinus* and inoculated with *M. incognita* raised for the cultivation of crop in case of tomato.

Morgan-Jones *et al.*, (1984) observed *P. lilacinus*, which penetrates the eggs of *M. arenaria* through minute pores, which are dissolved in the vitalline layer. There is a change in the permeability of shell of eggs infected with *P. lilacinus* resulting in the swelling-up. The infection also causes splitting of the vitalline layer into three distinct layers, where the chitin layer becomes vacuolated with the disappearance of the lipid layer (Morgan-Jones and Rodriguez-Kabana, 1985). It has also been reported antagonistic to pathogenic fungi such as *Macrophomina phaseolina* and *Rhizoctonia solani* (Shahzad and Ghaffar, 1988; 1989). This behaviour of *P. lilacinus* helped in the control of disease complexes due to *M. incognita-R. solani* and *R. reniformis-R. solani* in cowpea (Khan and Hussain, 1988), *M. incognita-R. Solani* and *M. incognita-M. phaseolina* on many beans and okra (Shahzad and Gaffur, 1989). Because of many good qualities, *P. lilacinus* has been the subject of many investigations. In Philippines, it has been produced on commercial scale under the trade name of "BIOCON" as reported by Jatala (1986). *Verticillium chlamydosporium* parasitizes females of *M. arenaria* (Morgan-jones *et al.*, 1981), *H. glycines* (Ginitis *et al.*, 1983), *H avenae* (Sharma and Trivedi, 1995; Bhardwaj and Trivedi, 2000) and *H. cajani* (Bhardwaj and Trivedi, 1998; 2000).

## BACTERIA

Majority of rhizosphere bacteria are abundant in soil in free-living form and some of them have shown great potential as biocontrol agents for the control of nematodes and their diseases (Sayre and Walter 1991; Stirling, 1991; Siddiqui and Mahmood, 1993; 1995). Bacteria are numerically the most abundant organisms in the field soil but their biomass is somewhat less than that of the fungi although they exceed combined form of algae, protozoa and nematodes (Clark, 1967). The bacteria used in the biocontrol of nematodes may be broadly grouped into parasitic bacteria and non-parasitic rhizobacteria.

## Parasitic Bacteria

Various bacterial genera have been reported as parasitic to nematodes, of which *Pasteuria* spp. has been widely studied (Sayre, 1980; Stirling, 1991). Three species namely *P. penetrans*, *P. thornei* and *P. nishizawe* are widely reported as parasitic to nematodes (Sturhan *et al.*, 1994). *P. penetrans* is the most widely prevalent species and has been reported from more than 200 host nematodes belonging to diverse genera (Sayre and Starr, 1988) from all over the world.

*Pasteuria penetrans*, a prokaryotic obligate parasite of nematodes, has shown great promise as potentially significant biocontrol agent. Initially, it was described as a microsporidian, *Duboscqia penetrans* by Thorne (1940), and then renamed as *Bacillus penetrans* (Mankau, 1975) as evidenced by electron microscopic studies. This bacterium has a very wide host range (Table 2). Obligate parasites of nematodes grow little outside their nematode hosts and are less likely to be affected by abiotic and biotic factors in the rhizosphere than those pathogens and antagonists of nematodes that are able to feed saprophytically in soil (Kerry and Hominick, 2000), However, there are subtle interactions between obligate parasites of nematodes in the rhizosphere, for example, second stage juveniles of root knot nematodes, *Meloidogyne* species encumbered with spores of the bacterium *P. penetrans* are less able to invade the roots of hosts than unencumbered nematodes. This effect is influenced by the species of plant host, age and density of nematodes (Stirling, 1984, Brown and Smart, 1985; Davies *et al.*, 1991).

*P. penetrans* is multiplied for experimental purposes using the host *Meloidogyne* sp. on a suitable host plant. The nematode-infected roots containing the endospores of the bacterium are powdered sieved through fine mesh and used as spore powder (Stirling and Wachtel, 1980; Sharma and Stirling, 1991). Attempts to multiply *P. penetrans in vitro* on synthetic media have so far not been successful. Even though the endospore could germinate in some media, but failed to complete the life cycle (Bishop and Ellar, 1991). The biocontrol potential of *P. penetrans* has been established against *Meloidogyne* spp. by using pot culture experiments (Mankau, 1973; Sayre *et al.*, 1984; Bhattacharya and Swarup, 1988; Stirling *et al.*, 1990; Walia and Dalal, 1994; Melki *et al.*, 1995; Tzortzakakis *et al.*, 1997; Sosamma and Koshy 1997).

## Rhizospheric Bacteria

Rhizospheric microorganisms are ideal for use as biocontrol agents against plant-parasitic nematodes (Table 1), because the rhizosphere is the front line defence for roots against attack by plant-parasitic nematodes (Sikora, 1988). Kloepper and Schorth (1978) and Suslow (1982) coined the term 'rhizobacteria' for those bacteria in the rhizosphere that aggressively colonize the root system. The premier example of rhizobacterial biocontrol agents occurs in many genera including *Actinoplanes, Agrobacterium, Alcaligens, Amorphosporangium, Arthrobacter, Azotobacter, Bacillus, Burkholderia, Cellulomonas, Enterobacter, Erwinia, Flavobacterium, Gluconacetobacter, Micromonospora, Pseudomonas,* Rhizobia, *Serratia, Streptomyces, Xanthomonas*, etc. as stated by large number of microbiologists (Kloepper *et al.*, 1989; Tang,

**Table 2** Host specificity of Pasteuria to nematodes.

| *Bacteria* | *Nematode* | *Effect of bacteria on nematodes* | *References* |
|---|---|---|---|
| *P. penetrans* | *M. incognita, Pratylenchus brachurus* | Spore population of *M. incognita* and *P. brachurus* attached only to *Meloidogyne* spp. and *P. brachurus* respectively and not to the range of other nematodes tested. | Dutky and Sayre (1978) |
| *P. penetrans* | 13 nematode species | Large spores (4.3-6.6 ìm) were associated with the endoparasitic nematodes | Spaull (1981) |
| *P. penetrans* | *Heterodera elastica* | This isolate has distinct specificity within cyst nematodes. | Nishizawa (1984) |
| *P. penetrans* | *Meloidogyne* spp. | Spores of *P. penetrans* derived from *M. incognita* infected a much larger proportion of larvae of this species than the larvae of *M. incognita*. In contrast spores from *M. incognita* were equally effective. | Spaull (1984) |
| *P. penetrans* | *Meloidogyne* spp. | Specificity of *P. penetrans* isolate may be a response to nematode population rather than to nematode species. | Stirling (1985) |
| *P. penetrans* | *M. incognita, M. javanica* | J2 were readily infested with spores of bacteria but varietal differences in susceptibility were observed. | Sell and Hansen (1987) |
| *P. penetrans* | *Meloidogyne* spp. | Spores of *P. penetrans* from six populations of *Meloidogyne* only adhered to *Meloidogyne* and they adhered in greatest numbers to the species from which they are isolated. | Davies *et al.* (1988) |
| *P. penetrans* | *Meloidogyne* spp. | The isolate varied greatly in their attachment to different nematode species and genera. | Oostendorp *et al.* (1990) |
| *P. penetrans* | *Helicotylenchus pseudorobustus, Rotylenchus capensis* | Endospores of *P. penetrans* released from infected nematodes or adhering to their cuticle did not reveal significant morphological differences apart from endospores diameters. | Volvas *et al.* (1993) |
| *P. penetrans* | *M. javanica, M. arenaria* | *P. penetrans* spores attached to *M. javanica* more readily than *M. arenaria* irrespective with the species of the host on which they had been produced and had no adaptation to the poorer hosts like *M. arenaria.* | De Silva and Gowen (1994) |
| *P. penetrans* | *H. cajani* | The spores of this isolate had a mean diameter of 2.36 μm and were characterized using host range spore morphometrics are not adequate for characterization of a species. | Sharma and Davies (1996) |

**Table 3** Effect of rhizobacteria on nematode control and plant growth.

| *Bacteria* | *Nematode* | *Effect of bacteria on nematode* | *References* |
|---|---|---|---|
| (A) *Pseudomonas* sp. | | | |
| *P. aureofaciens* | *Criconemella xenoplax* | One strain inhibited nematode multiplication in green house test. | Westcott and Kluepfel (1992) |
| *P. fluorescens* | *Panagrellus* sp. | Bacteria cultivated in plant count broth for 24 h at 30°C reduce nematode up to 57.4%. | Weidenborner and Kunz (1993) |
| *P. solanacearum* | *R. reniformis* | Resulted in slight inhibition of nematode activity on aubergine roots. | Kermarrec *et al.* (1994) |
| *P. fluorescens* | *M. javanica* | Reduced nematode multiplication and morphometrics of root-knot females on tomato in different soil types. | Siddiqui and Mahmood (1997) |
| (B) *Bacillus* sp. | | | |
| *Bacillus thuringiensis* | *Meloidogyne* | Prevented *M. incognita* from forming galls on tomato. | Ignoffo and Dropkin (1977) |
| *Bacillus subtilis, B. cereus, B. pumilus* | *M. incognita, Heterodera cajani, H. zeae, H. avenae* | The most effective isolates were *Bacillus subtilis, B. pumilus* against all tested species. The non-cellular extract exhibited a high degree of larvicidal properties. | Gokte and Swarup (1988) |
| *Bacillus subtilis* | *Meloidogyne* spp. *Rotylenchulus reniformis* | Reduced nematode reproduction and galling on cotton, tomato and peanut. | Sikora (1988) |
| *B. subtilis* | *M. incognita* | Nematode multiplication was reduced on tomato in pot test. | Gautam *et al.* (1995) |
| *B. subtilis* | *M. incognita* race 3 | Greater growth in chickpea plants and reduces nematode multiplication when seeds are treated with bacteria. | Siddiqui and Mahmood (1995a) |
| *B. subtilis* | *H. cajani* | Reduced multiplication of pigeon pea cyst nematode in the presence of bacteria. | Siddiqui and Mahmood (1995b) |
| (C) *Clostridium* spp. | | | |
| *Clostridium butyricum* | *Tylenchorhynchus martini* | Nematicidal acids produced by the bacteria reduced nematode population. | Johnston (1958) |
| *Clostridium* sp. | *T. martini* | Nematode concentrations of n-butyric acid and lesser amount of propionic acid were quickly formed in treated pots 4 days after flooding which resulted in rapid killing of nematodes. | Hollis and Rodriguez-Kabana (1966) |
| (D) *Serratia* spp. | | | |
| *Serratia marcescens* | *M. incognita* | Bacteria produced a volatile metabolite and were nematoxic. | Zavaleta-Mejia (1985) |
| (E) *Agrobacterium* sp. | | | |
| *Agrobacterium radiobacter* | *Globodera pallida* | Resulted in reduced nematode infection by 40% when sprayed on seed pieces of potato. | Sikora *et al.* (1989) |

1994; Weller and Thomashow, 1994; Glick *et al.*, 1995, 1998, 1999; Lucy *et al.*, 2004; Pandey *et al.*, 2005). A number of other rhizobia are being discovered day by day. Recently, a novel PGPR, *Delftia tsuruhatensis* HR4 has been reported both as diazotroph as well as potential biocontrol against various plant pathogens (Han *et al.*, 2005). The bacterial species inhibits phytoparasitic nematodes of many genera such as *Pseudomonas* (Oostendorp and Sikora, 1989; Spiegel *et al.*, 1991), *Bacillus subtilis, B. cereus* (Oka *et al.*, 1993), *B. sphaericus* (Racke and Sikora, 1992), *Arthrobacter* (Kloepper *et al.*, 1988), *Serratia* (Kloepper *et al.*, 1988) and *Agrobacterium* (Racke and Sikora, 1992).

Among the rhizobacteria, *Pseudomonas* spp. has been investigated for antagonism to nematodes more extensively than other group of bacteria. Both fluorescent pseudomonads and other *Pseudomonas* spp. are quite effective against *Heterodera cajani, H. avenae* and *H. zeae* (Gokte and Swarup, 1988; Kumar *et al.,* 2005b), *H. schachtii* (Oostendorp and Sikora, 1989) and *Meloidogyne* spp. (Gokte and Swarup, 1988; Kumar *et al.,* 2005a).

The antagonistic effects of naturally occurring rhizobacteria on plant-parasitic nematodes have been found significant. About 7-10% of the rhizobacteria isolated from the root system of potato, sugar beet, and tomato exhibited antagonistic activity toward different cyst and root knot nematodes (Oostendorp and Sikora, 1989; 1990; Racke and Sikora, 1986; 1992; Sikora, 1988). The antagonistic potential within rhizobacteria population is probably larger than indicated by this percentage considering that over 90% of soil microorganisms have not been cultured or studied (Macdonald, 1986).

Specific group of PGPR have the ability to improve plant growth and/or root health (Kloepper *et al.,* 1980; Schroth and Hancock, 1982; Suslow and Schroth, 1982; Schippers *et al.,* 1987; Sikora, 1988; Weller, 1988). A key factor of all PGPR is that they all colonize seed, root and behave as endophytes. Such traits are desirable for considering them suitable for biocontrol activity (Lugtenberg and Bloemberg, 2004; Compant *et al.*, 2005). Further, the phenomenon of chemotaxis, flagellar mobility, lipopolysaccharides (LPS) secretion, the outer membrane protein OprF and to a lesser extent, presence of pili are all important for competitive root colonization (Lugtenberg and Bloemberg, 2004). The Rhizo-competence is one of the important traits of PGPR. However, in field soil, environmental conditions and competition or displacement by the myriad of microorganisms present in the rhizosphere limit colonization. Certainly, use of mutants and promoter probe techniques are the beginning to identify genes in bacteria that are important to root colonization and these are often related to nutrient uptake (Roberts *et al.*, 2000; Rediers *et al.*, 2005).

Plant-growth promoting bacteria (PGPR) are distinguished from other groups of bacteria that are unable to promote plant growth (Kloepper *et al.,* 1991). According to their mode of action, rhizobacteria have been divided into two groups: (a) plant-growth promoting rhizobacteria (PGPR) and (b) plant-health promoting rhizobacteria (PHPR) (Sikora, 1988; 1991). Earlier workers (Becker *et al.,* 1988; Sikora, 1988; Oostendorp and Sikora, 1989; Racke and Sikora, 1992) have applied PHPR to seed or planting material in attempts to establish them in the rhizosphere to control cyst nematodes and root knot nematodes. The often-short duration of PHPR activity in the rhizosphere, as a seed treatment, may limit their use to control of nematodes that damage the plant in the seedling stage. Nematodes with multiple generations, e.g., species of *Meloidogyne* and *Heterodera*, cause damage throughout the season, may be more difficult targets

(Sikora, 1991). Thus, management techniques are needed to improve in terms of both, the intensity and duration of root colonization by PHPR and simultaneously increase the effectiveness of biological control across a broader spectrum of nematode species. Comments on the advantages and limitations of the major groups of organisms with potential as biological control agents are summarized in Table 4.

## VIRUSES

Our knowledge of viral disease of nematodes is limited. Significant discoveries have been made regarding the virus-vector relationship between some plant viruses and their nematode vectors. Loewenberg *et al.*, (1959) reported sluggishness of *M. incognita* presumably caused by virus that had passed through the bacteriological filter. Juveniles of this nematode were incapable of forming galls. However, no virus particles were found in the diseased nematodes. The phenomenon of swarming of viruses particles was first observed by Hollis (1958) and is known to occur in at least 13 species of plant-parasitic nematodes (Mc Bride and Hollis, 1966).

Since viruses affect a large number of organisms, it is conceivable that nematode pathogenic viruses might be operative but have escaped our detection. However, this assumption has its drawbacks in that plants nematodes ingest nutrients through the orifice of the stylet after it perforates and penetrates the plant cells. Viruses then have to be acquired from the cytoplasm or nuclei of the infected plant. It is very unlikely that plant viruses can also be pathogenic to nematodes.

## RICKETTSIAS

Rickettsias are intracellular, bacterium-like microorganisms that have been reported in *Heterodera goettingiana* Liebscher, *Globodera rostochiensis* woll. and *H. glycines* Ichinohe (Endo, 1979; Shepherd *et al.*, 1973; Walsh, 1979; Walsh *et al.,* 1983 a,b). Rickettsias are minute rod-shaped organisms, which divide by binary fission. Such organisms have been recorded in *Xiphinema index* associated with yellows disease of grapevines (Rumbos *et al.*, 1977). They have been reported to be transmitted transversally from generation to generation (Walsh, 1979). Despite their presence in the sperms cells, they cannot be transmitted from one generation to the next generation (Walsh *et al.*, 1983). Some rickettsias are pathogenic to insects and multiply in specialized cellular structures called mycetomes, which are not found in nematodes.

## ACTINOMYCETES

Some members of actinomycetes are known as antagonist to plant-parasitic nematodes (Burg *et al.*, 1979; Garabedian and Van Gundy, 1983). A new class of macrocyclic lactones, designated "avermactins" was isolated from *Streptomyces avermitilis* and proved to have potent broad-spectrum nematicidal activities (Burg *et al.*, 1979; Miller *et al.*, 1979). Granular and liquid formulations of avermactins have been reported to be 10 times more effective than oxamyl or aldicarb in controlling *M. incognita* on tomatoes (Garabedian and Van Gundy, 1983). Avermectins are compounds having antijuvenile hormone activity. They retard the sequential developmental

**Table 4** Effect of rhizobacteria in combination with other biocontrol agent on nematode reproduction and plant growth.

| *Management combination* | *Nematodes* | *Effect of nematodes and plant growth* | *References* |
|---|---|---|---|
| *Bacillus subtilis* + *Bradyrhizobium japonicum*+ | *Heterodera cajani* | Application of all the three management agents resulted in greater plant growth and greater reduction in nematode multiplication. | Siddiqui and Mahmood (1995c) |
| *Glomus fasciculatum* *Pseudomonas fluorescens*+ *Glomus mosseae* in different soil types | *M. javanica* | Reduced nematode multiplication and morphometrics of root knot females in different soil types. | Siddiqui and Mahmood (1997) |
| *Bacillus subtilis*+*P. lilacinus* | *M. incognita* race 3 | Combined application improved plant growth and reduced nematode multiplication more than individual inoculation on chickpea. | Siddiqui and Mahmood (1993) |
| *Bacillus subtilis*+*P. lilacinus*, *Eicchornia crassipes* | *M. incognita* | Application of all the three together resulted in greater reduction in nematode multiplication. | Gautam *et al.*, (1995) |
| *Bacillus subtilis*+culture filtrate of fungi | *M. incognita* race 3 | Increase in plant growth and reduction in nematode multiplication was greater when *B. subtilis* was used with *A. niger* or these two combined with other filtrates. | Siddiqui and Mahmood, (1995a) |
| *Bacillus licheniformis*+ *Pseudomonas mindocina*+ *Acrophialophora fusispora*+ *Aspergillus flavus* | *M. incognita* | Combination of all four biocontrol agents was best for reducing root knot index. *P. mindocina* had adverse effect on plant growth while others increased plant growth. | Siddiqui and Hussain (1991) |

presses, both in insects and nematodes, resulting in disruption of molting subsequent to incomplete adult formation.

## PROTOZOANS

Many protozoans have been known to attack plant-parasitic nematodes (Doncaster and Hooper, 1961; Webster, 1972). An amoeboid organism, *Theratomyxa weberi*, attacking *Heterodera trifolii, M. incognita, Aphelenchus avenae, Aphelenchoides rutgerri* and *Rotylenchulus reniformis* was observed (Sayre, 1973). Doncaster and Hooer (1961) noted that *Urostyla* sp. can catch nematodes but sometimes the prey escaped after destroying itself. The extent to which these organisms can be efficient and economically useful is not known.

## MITES

Mites have also been observed to damage the cysts of *Heterodera* sp. (Murphy and Ishibashi, 1957). Apparently, mites exhibit density-dependent traits of effective and specific predators. There was an increase in mite population corresponding with the increase of nematode prey (Imbriani and Mankau, 1983). The ability of mites to use fungi as alternative food sources in the absence of nematodes is an advantageous characteristic. Some genera namely, *Lasioseius salpatus* and *Tyrophagus putrescentiae* fed voraciously on plant and soil nematodes.

## PLANTS AND GREEN MANURE

Several plants are inhibitory to nematodes (Rodriguez-Kabana *et al.*, 1991). Number of juveniles of *Meloidogyne arenaria* lowered in soil at peanut harvest time in plots rotated two years with castor bean and then rotated two years with *Pensacola bahiagrass*. The differences are mainly attributed due to the nematicidal or nematostatic effects of compounds secreted by root and other plant tissues (Rodriguez-Kabana *et al.*, 1991).

Alsayed *et al.*, (1992) found that leaf powder of *Canna indica, Ricinus communis, Clerodendrum inerme, Lantana camara* and *Tagetes erecta* suppressed reproduction in *Rotylenchus reniformis* and improved soybean shoot and root weight. El-Nagar *et al.*, (1993) reported that addition of dry ground leaves of the Egyptian neem, *Melia azadirach, Asparagus sprengeri* (*Asparagus densiflorus*), *Eucalyptus* spp., *Psidium guajava* significantly reduced the number of egg masses of *Rotylenchus reniformis* on Sunflower. Sobita Devi (2000) reported a significant reduction in *Heterodera cajani* population with neem bark, seed, leaf cake, powder, flowers and ash treatment as compared to control. Similarly, Pandey and Trivedi (2000) observed significant reduction in *Heterodera cajani* population in cowpea plant treated with neem leaf powder, neem bark and nemboli. Earlier, Mojumdar (1999) studied the effect of neem seed, seed kernel, seed coat and cake for checking root knot nematode infestation in many beans and their residual effect on *Meloidygyne incognita* in chickpea and reported minimum number of galls in neem seed kernel treated plants.

Nematicidal values of certain botanicals have also been tried in the form of aqua extract of their various parts. Bhatti and Nandal (1994) tested the extracts of several plants/plant materials against plant-parasitic nematodes.

# MECHANISM OF BIOCONTROL AGENTS

## Antimicrobial Metabolites

Some bacterial metabolites, such as valinomycin (Mishra *et al.,* 1987) and 2, 4-diacetylphloroglucinol (Cronin *et al.,* 1997) have adverse effects on nematodes (Stiring, 1991). Fluorescent *Pseudomonas* spp. have been reported to produce 2, 4-diacetylphloroglucinol (2, 4-DAPG) by which they provide biological control of soil-borne pathogens on a wide range of crops and have a key role in the suppressiveness of some plant pathogens (Raaijmakers and Weller, 2001; Weller *et al.*, 2002; Validov, 2005). 2, 4-diacetyphloroglucinol has also shown to reduce juvenile mobility (Fravel, 1988; Thomashow and Weller, 1991; Keel, 1992). It proved a potential source of biological control due to its production by a wide range of pseudomonads including rhizobacteria used for the biological control of root diseases (Keel *et al.,* 1996; Sharifi-Tehrani *et al.,* 1998). It is a phenolic compound with broad-spectrum nematicidal, antifungal, antibacterial, antihelminthic and phytotoxic activity (Keel *et al.,* 1992; Dowling and O'Gara, 1994; Thomashow and Weller, 1995; Cronin *et al.,* 1997).

Bacterial symbionts namely *Xenorhabdus* and *Photorhabdus* spp. produce antimicrobial metabolites *in vitro* (McInerney *et al.,* 1991; Li *et al.,* 1997). Grewal *et al.*, (1999) have also shown the nematicidal properties of culture filtrate from *Xenorhabdus nematophilus*. Similarly, Hu *et al.*, (1999) have reported the indole derivative from the culture filtrate of an isolate of *Photorhabdus luminescens* having nematicidal properties *in vitro* against the root knot nematode *Meloidogyne incognita*.

## Induced Systemic Resistance

Systemic resistance induced by plant growth promoting bacteria has also been considered as a mechanism for biocontrol of pathogens (Wei *et al.*, 1996). Mechanisms for induced systemic resistance may be due to increase in activity of chitinases, β-1, 3 glucanases, peroxidases and other pathogenesis related proteins (Lawton and Lamb, 1987; Bakker *et al.*, 1991). The type of bacterized plant response induced after challenge with a pathogen resulted in the formation of structural barriers, such as thickened cell wall papillae due to the deposition of callose and the accumulation of phenolic compounds at the site of plant attack (Benhamon *et al.*, 1998; Mishra *et al.*, 2006). Biochemical and physiological changes in plants induced accumulation of pathogenesis related proteins (PR proteins) such as PR-1, PR-2, chitinases and some peroxidases (Ramamoorthy *et al.*, 2001; Jeun *et al.*, 2004).

## Hydrolytic Enzymes

Recently, the hydrolytic enzymes have received considerable attention due to their role in controlling diseases caused by plant-parasitic nematodes. The enzymatic digestion or deformation of cell wall components of nematodes occurs by chitinase and protease mediated enzymatic reactions involving PGPR (Tikhonov *et al.*, 2002). Tikhonov *et al.*, (2002) observed that the combined action of protease and chitinase resulted in more severe egg shell damage. Chitin gets depolymerized by chitinase and releases ammonia, which has nematicidal activity (Spiegel *et al.*, 1987). Chitinase killed *Tylenchorhynchus dubius* by producing structural changes in the nematode.

cuticle (Miller and Sands, 1977). Purified chitinase inhibited egg hatching of *Globodera rostochiensis* up to 70% *in vitro*, and in soil (Cronin *et al.,* 1997). *Pseudomonas chitinolytica* with strong chitinolytic activity reduced *Meloidogyne incognita* infection (Spiegel *et al.,* 1991). The chitinolytic fungus, *Paecilomyces lilacinus*, destroyed nematode eggs and efficiently controlled *M. incognita* (Morgan-Jones *et al.,* 1984). Furthermore, application of both *Paecilomyces lilacinus* and chitin to sterilized soil suppressed *M. incognita* population levels more than that of using alone (Mittal *et al.,* 1995).

## Toxins

*Bacillus thuringiensis* producing exotoxins have been demonstrated to be antagonistic to nematodes. *Bacillus thuringiensis* suppressed root knot nematodes in soil under natural conditions (Prasad and Tilak, 1972). *In vitro* studies established that the exotoxin produced by bacterium during growth and sporulation had high nematicidal activity against juveniles and eggs of *Meloidogyne* spp. (Prasad *et al.,* 1972).

Similarly, endophytic fungi such as *Fusarium oxysporum* proved toxic to *M. incognita* (Hallmann and Sikora, 1996). The toxins from various *Fusarium* spp. reduced nematode viability (Anke and Sterner, 1997; Nitao *et al.,* 2001). The wood rotting basidiomycetes *Pleurotus ostreatus* produced nematoxin tran-2-decenedioic acid inhibited nematodes (Kwok *et al.,* 1992). The culture filtrate of *P. lilacinus* and *Trichoderma longibrachiatum* contain acetic acid (Djian *et al.,* 1991) and *Arthrobotrys conoides* and *Arthrobotrys oligospora* produce nematoxic effect due to the presence of linoleic acid in their culture filtrate (Anke *et al.,* 1995). On the other hand, Khambay *et al.*, (2000) observed nematicidal activity in *P. chlamydosporia* due to presence of phomalactone, which acted against *M. incognita*.

## Root Exudates

Root exudates of many plants have been found to be highly deleterious to plant-parasitic nematodes. Rohde and Jenkins (1958) reported that the roots of *Asparagus* species produce a nematicidal glycoside. Suatmadji (1969) reported nematicidal nature of root exudates *Tagetes patula* and suggested that the toxic principle is made up of several components in addition to thiophenes. Nematicidal and nematostatic nature of root exudates of *Azadirachta indica* has been observed by Alam *et al*., (1975) and Siddiqui and Alam (1989) along with that of *Persian lilac, Melia azedarach.*

Many plants based chemical compounds like polythienyls acetylenes, phenols, fatty acids, alkaloids, phenolic glucosidine, tannis and terpenoides are known to be toxic to a variety of nematodes affecting various crops (Valette *et al*., 1953; Tarjan and Cheo, 1956; Hussain and Masood, 1975). The roots of *Tagetes patula* and *Tagetes erecta* contain terthienyl and derivative of bithienyl *Pratylenchus* species (Oostenbrink *et al*., 1957; Winoto, 1969). Davis *et al*., (1989) reported that phytolecteins, particularly concanavalin A from jack bean *Canavalia eusiformis* reduced *Meloidogyne incognita* infestation of host plants in *in vivo* experiments. Das *et al*., (1989) found that lectins may play some role in the natural defence mechanism of plants and are accumulated in the galled region of the root of okra infested by *Meloidogyne incognita*.

**Table 5** The advantages and limitations of potential biological control agents with different modes of action against plant parasitic nematodes.

| *Type of agent* | *Mode of action* | *Comments* |
|---|---|---|
| Rhizosphere bacteria | Toxins or modification of root exudates | Advantages: easy to culture *in vitro*; can be applied as seed treatments; reduce plant damage.<br>Limitations: effective for a relatively short period; activity affected by crop cultivar and nematode species; little effect on nematode multiplication. |
| Obligate parasites *Pasteuria* spp. | Adhesive spores | Advantages: most isolates highly virulent; infective spores resistant to drying; good shelf life; reduce infectivity of nematodes as well as fecundity.<br>Limitations: very difficult to culture *in vitro*; isolates are very specific; no proliferation in soil in absence of nematodes. |
| *Aecilomyces lilacinus* | Hyphal penetration | Advantages: easily produced *in vitro*; rhizosphere competent attacks the eggs of several nematodes species; treatment of planting material (e.g. seed tubers) can be effective.<br>Limitations: requires high soil temperature; has given variable control in range of conditions, large number of propagules ($10^6$/g soil) required for nematode control; some isolates are pathogenic to humans. |
| *Verticillium chlamydosporium* | Hyphal penetration | Advantages: easily produced *in vitro*; some isolates rhizosphere competent, and virulent ($10^3$ propagules/g soil required for nematode control); resistant resting spores produced; survives throughout growing season in soil.<br>Limitations: seed treatments ineffective; efficacy dependent on nematode species, density and plant host. |
| Endophytic fungi (non-pathogenic root infecting fungi and mycorrhizae) | Competition in roots and modification of root exudates | Advantages: include agents with potential to control migratory endoparasitic nematodes in roots; may improve plant growth even in absence of nematodes; reduce damage caused by wide range of nematodes and limit their multiplication; can be mass produced and formulated; could be applied to seeds or transplant material; may reduce fungal root rots.<br>Limitations: non-mycorrhizal fungi may be difficult to register as closely related to plant pathogens; efficacy affected by plant cultivar and other crops in rotation such as crucifers can reduce survival. |

## Others

Soil bacteria like *Clostridium butyricum* (Hollis and Rodrigues-Kabana, 1966; 1967) and *Disulfovibrio desulficans* (Rodrigues-Kabana *et al.*, 1965) producing butyric acid and hydrogen sulphide, respectively, have been demonstrated to be antagonistic to nematodes. The effect of rhizobacteria on penetration of nematodes is possibly due to bacteria binding to lectins on the root surface and therefore, interfering with host recognition by the nematodes. In aerobic soil, ammonia is usually produced by ammonifying bacteria during decomposition of nitrogenous organic materials (Rodriguez-Kabana, 1986) and this has resulted in reduced nematode population.

## CONCLUSION

Large numbers of studies have indicated that microorganisms have great potential in the management of nematodes. Understanding these interactions is essential for the successful development of biological control, and at a molecular could lead to the identification of new nematicidal compounds. Use of *Pasteuria* for nematode control is very limited. The major hurdle is lack of techniques for its mass production *in vitro*. It is important to device synthetic medium for culturing of *Pasteuria* to expand the use of this organism as a commercially viable biocontrol agent of phytonematodes. The rhizobacteria hold a good promise in reducing the damage caused by phytonematodes. These bacteria are easy to culture and produce in large quantity. However, research on use of rhizobacteria for nematode control has a long way to go to decide whether or not these bacteria provide a practical option for nematode management. We need to identify and select the most effective strains, study their mode of action, and demonstrate their efficiency in farmers' fields.

In the case of predaceous and opportunistic parasitic fungi, the most important disadvantage is the difficulty in their mass production and the establishment in the soil. Whereas those fungi that rapidly colonize eggs and cysts are free from these inhibitions and thus provide a guarded optimism about their possible role in future control strategies of nematodes. Botanicals for the management of nematodes are also observed to be easy in application, free from environmental pollution, and improved soil health structurally and nutritionally. Nematicidal compounds from plants would be economical only when such compounds are available in large quantity, either through their isolation from plants or through their synthesis on a commercial scale.

## References


Adiko, A. (1984) M.S. Thesis, Dept. Plant Pathol., North Carolina St. Univ., Raleigh. pp. 14.

Alam, M.M., Masood, A. & Husain, S.I. (1975) *Indian J. Exp. Biol.* **13:** 412-414.

Al-sayed, A.A., Ahmad, S.S. & Abdel-Hameed, S.H. (1992) *Annals of Agril. Sci. Moshtohor.* **30:** 615-620.

Atilano, R.A., Menge, J.A. & Vangundy, S.D. (1981) *J. Nematol.* **213:** 52-57.

Bagyaraj, D.J., Manjunath, A. & Reddy, D.D.R. (1979) *Plant Soil.* **51:** 397-403.

Baker, K.F. & Cook, R.J. (1974) In: Biological Control of Plant Pathogens, (Eds. Freeman, W.H. & Co), San Francisco, California. Amer. Phytopathol. Soc., St. Paul, Minnesota. pp. 433.

Barron, G.L. (1977) The nematode destroying fungi. Canadian Biological Publications, Guelph, Ontario. pp. 140.

Becker, J.O., Zavaleta-Majia, E., Colbert, S.F., Schroth, M.N., Weinhold, A.R., Hancock, J.G. & Vangundy, S.D. (1988) *Phytopathol.* **78:** 1466-1469.

Beirner, B.P. (1967) *World Rev. Pest Control.* **6:** 7-20.

Benhamon, N., Kloepper, J.W. & Tuzun, S. (1998) *Planta.* **204:** 153-168.

Bhardwaj, P. & Trivedi, P.C. (1998) *J. Ind. Bot. Soc.* **77:** 5-8.

Bhardwaj, P. & Trivedi, P.C. (2000) In: Microbe: Agriculture industry and environment (Eds. Maheshwari, D.K., Dubey, R.C., Gupta, G.P. & Navneet, Bishen Singh Mahendra Pal Singh, Dehradun). pp. 125-135.

Bhattacharya, D. & Swarup, G. (1988) *Indian J. Nematol.* **18:** 61-70.

Bhatti, D.S. & Nandal, S.N. (1994) In: Nematode pest management in crops (Eds. Bhatt, D.S. & Walia, R.K.), CBS Publishers, Delhi. pp. 148-164.

Bishop, A.H. & Eller, D.J. (1991) *Biocontrol Sci. Technol.* **1:** 101-114.

Brown, S.M. & Smart, G.C. (1985) *J. Nematol.* **17:** 123-126.

Brown, S.M., Kepner, J.L. & Smart, G.C. (1985) *Soil Biol. Biochem.* **17:** 483-486.

Burg, R.W., Miller, B.M., Baker, E.E., Birnbaum, S.A. & Hartman, C.R. (1979) *Antimicrob. Agents Chemother.* **15:** 361-367.

Cabanillas, E., Barker, K.R. & Daykin, M.E. (1988) *J. Nematol.* **20:** 362-365.

Caron, T.K.M., Hussey, R.S. & Roncadori, R.W. (1983) *J. Nematol.* **15:** 410-417.

Clark, F.E. (1967) In: Soil Biology, (Eds. Eurges, A. & Raw, F.) Academic Press, London. pp. 15-19.

Compant, S., Duffy, B., Nowak, J., Clement, C. & Barka, E.A. (2005) *Appl. Environ. Microbiol.* **71(9):** 4951-4959.

Culbreath, A.K., Rodriguez-kabana, R. & Morgan-Jones, G. (1986) *Nematropica.* **16:** 153-166.

Das, S., Sukul, N.C., Mitra, D. & Sarkar, M. (1989) *Nematol. Meditt.* **17:** 123-125.

Davies, K.G., Laird, V. & Kerry, B.R. (1991) *Rev. Nematol.* **14:** 611-618.

Davis, E.L., Kaplen, D.T., Dickson, D.W. & Mitchless, D.J. (1989) *Rev. Nematol.* **12:** 257-263.

De Bach, P. (1964) Biological control of insect, pests and weeds. London: Chapman & Hall.

De Sisler, G.M., Silvestri, L. & Acita, J.O. (1985) *Fitopatologia.* **20:** 17-20.

Dehne, H.W. (1987) Zur Nutzang der VA Mycorrhiza als Antistressfaktor. Z. Angew.

Doncaster, C.C. & Hooper, D.J. (1961) *Nematologica.* **6:** 333-335.

El-Nagar, H.I., Hendy, H.H., Abdel Hameed, S.H., Farhat, A.A. & Osman, A.A. (1993) *Bull. Faculty Agric. Univ. Cairo.* **44:** 205-216.

Endo, V.Y. (1979) *J. Ultrastru. Res.* **67:** 124.

Eren, J. & Pramer, D. (1965) *Soil Sci.* **99:** 285.

Finck, A. (1994) From the fertilization of crops to the management of plant nutrients in crop rotation and farming systems. Expert consultation on integrated plant nutrition systems, Rome.

Franco, J., Jatala, P. & Bocangel, M. (1981) *J. Nematol.* **13:** 438-439.

Garabedian, S. & Van Gundy, S.D. (1983) *J. Nematol.* **15:** 503-510.

Garret, S.D. (1965) Toward biological control of soil-borne plant pathogens Prelude to Biological control. (Eds. Baker, K.F. & Synder, W.C.) California Univ. Press, Berkeley. pp 4-17.

Gaspard, J.T. (1985) Strategies of biocontrol of *Meloidogyne* spp. using the nemotophagous fungi *Monacrosporium ellipsosporum, Paecilomyces lilacinus* and *Verticillium chlamydosporium*, Thesis, Univ. Calif., Riverside, U.S.A.

Gaspard, J.T. & Mankau, R. (1986) *Nematologica.* **32:** 359-363.

Gautum, A., Siddiqui, Z.A. & Mahmood, I. (1995) *Nematol. Medit.* **23:** 245-272.

Gintis, B.O., Morgan-Jones, G. & Rodriguez-Kabana, R. (1983) *Nematropica.* **13:** 181-200.

Glick, B.R., Karatuprovic, D.M. & Newwell, P.C. (1995) *Can. J. Microbiol.* **41:** 533-536.

Glick, B.R., Patten, C.L., Holguin, G. & Penrose, D.M. (1999) Biochemical and genetic mechanism used by plant growth-promoting bacteria. Imperial College Press, London, UK.

Glick, B.R., Penrose, D.M. & Li, J. (1998) *J. Theor. Bio.* **190:** 63-68.

Gokte, N. & Swarup, G. (1988) *Indian J. Nematol.* **18:** 152-153.

Guevara, E., Jatala, P., Salas, R. & Herrera, E. (1985) Eficiencia de algunos nematicidas y *Paecilomyces lilacinus* en el control de *Meloidogyne* spp. en olivos. Resumenes IX Congr. Peru. Fitopatol. Huanuco, Peru. pp. 63.

Han, J.S., Cheng, J.H., Yoon, T.M., Song, J., Rajkarnikar, A., Kim, W.G., Yoo, I.D., Yang, Y.Y. & Suh, J.W. (2005) *J. Appl. Microbiol.* **99:** 213.

Hasky-Gunther, K., Hoffmann-Hergarten, S. & Sikora, R.A. (1998) *Fund. Appl. Nematol.* **21:** 511-517.

Heera, C. (1996) *Fertil. Environmen.* 131-149.

Herrera, E., Jatala, P. and Canicoba, R. (1984) Estudio de la dinamica poblacional de *Tylenchulus semipenetrans* Cobb1913, mediante aplicaciones de nematicidas y *Paecilomyces lilacinus* (thom.) Samson. 33: 22.

Herrera, E., Jatala, P. & Canicoba, R. (1985) Respuesta al control quimico y biologico del nematodo de los citricos. Inst. Nac. Invest. Promoc. Agropecu. (INIPA). Informe Especial, Year **4:** 2: 16.

Hollis, J.P. (1958) *Nature.* **182:** 956.

Hollis, J.P. & Rodriguez-Kabana, R. (1966) *Phytopathol.* **6:** 1015-1019.

Husain, S.I. & Mahmood, A. (1975) *Acta Botanica Indica.* **3:** 142-146.

Hussey, R.A. & Roncadori, R.W. (1982) *Plant Dis.* **66:** 9-14.

Ibrahim, I.K.A., Rezk, M.A., El-Saedy, M.A. & Ibrahim, A.A.M. (1987) *Nematol. Medit.* **15:** 265-268.

Ignoffo. C.M. & Dropkin, V.H. (1977) *J. Krans Entomol. Soc.* **50:** 394-395.

Imbriani, J.L. & Mankau, R. (1983) *J. Nematol.* **15:** 523-528.

Jaffee, B.A., Phillips, R., Muldoon, A. & Mangel, M. (1992) *Ecology.* **73:** 495-506.

Jalali, B.L. & Chand, H. (1990) Current trends in mycorrhizal research. Proc. of the national conference on mycorrhiza, HAU, Hisar, Hariyana, India. pp. 210.

Jansson, H.B. & Nordbring-Hertz, B. (1980) *Nematologica.* **26:** 383-389.

Jansson, H.G. & Nordbring-Hertz, B. (1985) *Nematologica.* **26:** 383-389.

Jatala, P. (1986) *Annu. Rev. Phytopathol.* **24:** 453-489.

Jatala, P., Kaltenbach, R. & Bocangel, M. (1979) *J. Nematol.* **11:** 303.

Jeun, Y.C., Park, K.S., Kim, C.H., Fowler, W.D. & Kloepper, J.W. (2004) *Biol. Control.* **29:** 34-42.

Johnston, T.M. (1958) Ph.D Thesis Louisiana State University.

Kassab, A.S. & Taha, A.H.Y. (1990) *Annu. Agric. Sci.* **35:** 509-520.

Kenny, D.R. (1982) Nitrogen management of maximum efficiency and minimum pollution. In: Nitrogen in agricultural soils. (Ed. Stevenson, F.J.). Am. Soc. Agron. Madison Wisconsin.

Kermarrec, A., Jacqua, G. & Anais, J. (1994) *Nematologica.* **40:** 152-154.

Kerry, B. (1995) *Arab. J. Plant Protection.* **13:** 47-52.

Kerry, B.R. & Jaffee, B.A. (1997) In: The Mycota IV: Environmental and Microbial Relationships (Eds. Wicklow, D.T. & Soderstrom, B.), Berlin, Springer. pp. 203-218.

Khan, T.A. & Husain, S.I. (1988) *Nematol. Meditt.* **16:** 229-231.

Kloepper, J.W. & Schroth, M.N. (1978) Plant growth promoting rhizobacteria on radish. Proc. 4$^{th}$ Int. Conf. Plant Pathol. Bacteria, Vol. II: 879-882. Tours: Gilbert-Clarey. pp. 979.

Kloepper, J.W., Hume, D.J, Scher, F.M., Singeleton, C., Tipping, B., Laliberte, M., Frauley, K., Kutchaw, T., Simonson, C., Lifshitzr, R., Zeleska, I. & Lee, L. (1988). *Plant Dis.* **72:** 42-46.

Kloepper, J.W., Leong, J., Teintze, M. & Schroth, M.N. (1980) *Curr. Microbiol.* **4:** 317-320.

Kloepper, J.W., Lifshitz, R. & Zablotowich, R.K. (1989) *Trends. Biotechnol.* **7:** 39-43.

Kloepper, J.W., Rodriguez-Kabana, R., McInroy, J.A. & Collins, D.J. (1991) *Plant Soil.* **136:** 95-102.

Kumar, T., Kang, S.C. & Maheshwari, D.K. (2005b) *Agric. Chem. Biotechnol.* **48(4):** 161-166.

Loewenberg, J.R., Sullivan, T. & Schuster, M.L. (1959) *Nature.* **184:** 186.

Lucy, M., Reed, K. & Glick, R. (2004) *Antonie van Leeuwenhoek.* **86:** 1-25.

Lugtenberg, B.J.J. & Bloemberg, G.B. (2004) In: *Pseudomonas* Vol. I, (Ed. Ramos, J.L.) Kluwer, Academic Publishers, New York. pp. 403-430.

Macdonald, R.M. (1986) *Biol. Agric. Hort.* **3:** 361-365.

Mankau, R. (1975) *J. Invertebr. Pathol.* **26:** 333-339.

McBridge, J.M. & Hollis, J.P. (1966) *Nature.* **211:** 545-546.

Melki, T. Tzorzakakis, E.A. & Gowar, S.R. (1995) *Nematologica.* **41:** 321.

Miller, T.W., Charet, L., Cole, D.J. & Flor, J.E. (1979) *Antimicrob. Agents Chemother.* **15:** 368-371.

Mishra, R.P.N., Singh, R.K., Jaiswal, H.K., Kumar, V. & Maurya, S. (2006) *Curr. Sci.* **52:** 383-389.

Mojumdar, V. (1999) *Indian J. Nematol.* **29:** 202-203.

Morgan-Jones, G. & Rodriguez-Kabana, R. (1984) *Phytopathol.* **74:** 831.

Morgan-Jones, G., Godoy, G. & Rodriguez-Kabana, R. (1981) *Nematropica.* **11:** 115-120.

Murphy, P.W. & Doncaster, C.C. (1957) *Nematologica.* **2:** 202-204.

Obannon, J.H., Inserra, R.N., Nemic, S. & Vovlas, N. (1979) *J. Nematol.* **11:** 247-250.

Oka, K., Chal, I. & Spiegal, Y. (1993) *Biol. Sci. Technol.* **3:** 115-126.

Oostenbrink, M., Kuiper, K. & S'Jacob, J.J. (1957) *Nematologica.* **2:** 424-433.

Oostendorp, M. & Sikora, R. A. (1989) *Review de Nematologie.* **12:** 77-83.

Oostendorp, M. & Sikora, R.A. (1990) *Rev. Nematol.* **13:** 269-274.

Pandey, S. & Trivedi, P.C. (2000) Effects of various neem products on *Heterodera cajani* infecting cowpea. $87^{th}$ Indian Science Congress, 3-7 January 2000, Pune, Botany Section. pp. 25-26.

Persmark, L. & Jansson, H.B. (1997) *FEMS Microbiol. Ecol.* **22:** 303-312.

Persson, C. & Jansson, H.B. (1999) *J. Invertebr. Pathol.* **49:** 170-174.

Racke, J. & Sikora, R.A. (1986) *Rev. Nematol.* **9:** 305.

Racke, J. & Sikora, R.A. (1992) *Soil Biol. Biochem.* **24**: 531-536.

Ramamoorthy, V., Viswanathan, R., Raguchandra, T., Prakasam, V. & Smaiyappan, R. (2001) *Crop Prot.* **20**: 1-11.

Rebafka, F.P., Ndungunu, B.J. & Marschner, H. (1993) *Plant Soil.* **150**: 213-222.

Reddy, P.P. & Khan, R.M. (1988) *Nematol. Medit.* **16**: 113-116.

Reddy, P.P. & Khan, R.M. (1989) *Pak. J. Nematol.* **7:** 55-59.

Rediers, H., Rainey, P., Vanderlayden, B.J. & De Mot, R. (2005) *Microbiol. Mol. Biol. Rev.* **69(2):** 217-261.

Rodriguez-Kabana, R., Morgan-Jones, G., Godoy, G. & Gintis, B.O. (1984) *Nematropica.* **14:** 155-170.

Rodriguez-Kabana, R., Robertson, D.G., King, P.S. & Wells, L. (1991) *Nematropica.* **21:** 97-103.

Rohde, R.A. & Jenkins, W.R. (1958) *Allen. Md. Agric. Exp. Stn. Bull.* **A-97:** 1-19.

Rumbos, I., Sikora, R.A. & Nienhaus, F. (1977) *Pflanzenschutz.* **84:** 240-243.

Saifullah, H., Gul, A. & Saeed, M. (1988) *Int. Nematol. Network Newsl.* **5:** 20-22.

Sayre, R.M. (1971) In: Plant parasitic nematodes. (Eds. Zuckarman, B.N., Mai, W.F. & Rohde, I.) New York, Academic Press. pp. 235-266.

Sayre, R.M. (1973) *J. Nematol.* **5:** 258-264.

Sayre, R.M. & Starr, M.P. (1988) In: Diseases of Nematodes (Eds. Poinar, G.O. and Jansson, H.B.), Florida, CRC Press. pp. 69-101.

Sayre, R.M. & Walter, D.E. (1991) *Ann. Rev. Phytopathol.* **29:** 149-166.

Sayre, S.M. (1980) *J. Nematol.* **12:** 260-270.

Schippers, B., Bakker, A.W. & Bakker, P.A.H.M. (1987) *Ann. Rev. Phytopathol.* **25:** 339-358.

Schroth, M.N. & Hancock, J.G. (1992) *Science.* **216:** 1376-1381.

Shahzad, S. & Ghaffar, A. (1988) In: Advances in Plant Nematology (Eds. Maqbool, M.A., Golden, A.M., Ghaffar, A. & Krusberg, L.R.). Nat. Nematol. Res. Centre, Univ. of Karachi, Pakistan. pp. 249-251.

Shahzad, S. & Ghaffar, A. (1989) *Pak. J. Nematol.* **7:** 47-53.

Sharma, R.D. & Stirling, G.R. (1991) *Nematol.* **37:** 483-484.

Shepherd, A.M., Clark, S.A. & Kempton, A. (1973) *Nematologica.* **19:** 31-34.

Siddiqui, M.A. & Alam, M.M. (1989) *Pflanzenschutz Umweltschutz.* **62:** 33-35.

Siddiqui, Z.A. & Hussain, S.I. (1991) *Curr. Nematol.* **2:** 5-6.

Siddiqui, Z.A. & Mahmood, I. (1993) *Fund. Appl. Nematol.* **16:** 215-218.

Siddiqui, Z.A. & Mahmood, I. (1995a) *Fund. Appl. Nematol.* **18:** 71-76.

Siddiqui, Z.A. & Mahmood, I. (1995b) *Biores. Technol.* **54:** 217-226.

Siddiqui, Z.A. & Mahmood, I. (1995c) *Fund. Appl. Nematol.* **18:** 556-559.

Siddiqui, Z.A. & Mahmood, I. (1997) *Appl. Soil Ecology.*

Sikora, R.A. (1978) *Pflanzenkr. Pflanzenschutz.* **85:** 197-202.

Sikora, R.A. (1981) In: biological and chemical interaction in the rhizosphere, Symp. Proc. Stockholm. Swedish Nat. Scie. Res. Counc. Press. pp. 115-136.

Sikora, R.A. (1988) *Med. Fac. LandbouwwetRijksuniv. Gent.* **53:** 867-78.

Sikora, R.A. (1991) The concept of using plant health-promoting rhizobacteria for the biological control of plant parasitic nematodes. Proc. Int. Work-shop Plant Growth-Promoting Rhizobacteria, 2nd Interlaken, Switzerland, WRRS Bull. XIV/ **8**: 3-10.

Sikora, R.A., Racke, J. & Bodenstein, F. (1989) *J. Nematol.* **21:** 588.

Sitaramaiah, K. & Sikora, R.A. (1981) *Med. Fac. LandbouwwetRijksuniv. Gent.* **46:** 695-702.

Sitaramaiah, K. & Sikora, R.A. (1982) *Nematologica.* **28:** 412-419.

Smith, G.S. (1987) Interaction of nematodes with mycorrhizal fungi. pp. 292-300.

Sobita Devi, L. (2000) *Indian J. Nematol.* **30:** 250-252.

Sosamma, V.K. & Koshy, P.K. (1997) *J. Plantation crop.* **25:** 75-76.

Spiegel, Y., Chon, E., Galper, S., Sharon, E., & Chet, I. (1991) *Biocontrol Sci. Technol.* **1:** 115-125.

Stirling, G.R. (1984) *Phytopathol.* **74:** 55-60.

Stirling, G.R. (1991) Biological control of plant parasitic nematodes: Progress, Problems and Prospects. UK: CAB Int. pp. 282.

Stirling, G.R. & Wachtel, M.F. (1980) *Nematropica.* **26:** 308-312.

Stirling, G.R. & Wachtel, M.F. (1980) *Nematol.* **26:** 308-312.

Stirling, G.R., Sharma, R.D. & Perry, J. (1990) *Nematropica.* **36:** 246-252.

Sturban, D., Winkelheide, R., Sayre, R.M. & Wergin, W.P. (1994) *Fund. Appl. Nematol.* **17:** 29-42.

Sturhan, D. (1985) Studies on distribution and host of the nematode parasite *Bacillus penetrans*. Biologischen Bund Fur Land-Und Fonstwirtschaft Berlin-Dalhem. **226:** 75-93.

Suatmadji, R.W. (1969) Studies on the effect of Tagetes spp. on plant nematodes. Diss. Agric. Univ., Wageningen. pp. 132.

Suslow, R.V. & Schroth, M.N. (1982) *Phytopathol.* **72:** 199-206.

Suslow, T.V. (1982) In: Phytopathogenic Prokaryotes, (Eds. Mount, M.S. & Lacy, G.H.), London: Academic. **1:** 187-223.

Tang, W.H. (1994) In: improving plant productivity with rhizosphere bacteria. (Eds. Ryder, M.H., Stephens, P.M. & Bowen, G.D.), Commonwealth scientific and industrial research organization, Adelaide, Australia. pp. 267-278.

Tarjan, A.C. & Cheo, P.A. (1956) Nematicidal value of some fatty acids. Bull. Rhod. Island. Agr. Exp. Stn. **112:** 41.

Thorne, G. (1940) Duboscqia penetrans sp. (Sporozoa: Microsporidia, Nosematidae) a parasite of the nematode *Pratylenchus pratensis* Filipjev. Proc. Helminthol. Soc. Wash. **7:** 51-53.

Trigalet, A., Frey, P. & Demmerry-Trigalet, D. (1994) In: Bacterial wilt (Eds. Hartman, G.L. & Hayward, A.C.): The disease and its causative agent, *Pseudomonas solanacearum.* pp. 225-234.

Trivedi, P.C. (1995) Platinum Jubilee Vol I. *Ind. Bot. Soc.* **74:** 165-171.

Tzorzakakis, E.A., Channer, A.G.R. De & Gowen, S.R. (1997) *Plant Pathol.* **46**: 44-45.

Valette, G., Cavier, R. & Debtmas, I. (1953) *Am. Pharm. France.* **11:** 649-653.

Walia, R.K. & Dalal, M.R. (1994) *Pest Management and Economic Zoology.* **2:** 19-21.

Walsh, J.A. (1979) *J. Nematol.* **11:** 317.

Walsh, J.A., Lee, D.L. & Shepherd, A.M. (1983) *Nematologica.* **29:** 227-239.

Walsh, J.A., Shephered, A.M. & Lee, D.L. (1983) *J. Zool.* **199:** 395-419.

Webster, J.M. (1972) In: Economic nematology (Ed. Webster, J.M.), Academic Press, London, NY. pp. 469-496.

Weidenborner, M. & Kunz, B. (1993) *Pfleanzenkrankheiten und Pflanzenschuts.* **100:** 90.

Weller, D.M. (1988) *Annu. Rev. Phytopathol.* **26:** 279-407.

Weller, D.M. & Thomashow, L.S. (1994) In: Molecular ecology of rhizosphere microorganisms: Biotechnology and the release of GMO's. (Eds. O'Gara, F., Dowling, D.N. & Boesten, B.), Vch, Weinheim, Germany. pp. 1-18.

Wescott, S.W. III & Kluepfel, D.A. (1992) *J. Nematol.* **24:** 626.

Winoto, S.R. (1969) Studies on the effect of sargeted species on plant parasitic nematodes sticht founds hand bouw Exporrt. But. Pub. Weemanns Zonen. N.V.

Zavaleta-Mejia, E. (1985) *Science and Engineering.* **46(4):** 108.

11

# Recent Advances in Biotechnology of *Beauveria Bassiana*: A Mycopesticide Against Insect Pests

**S.R. Niranjana, N. Haraprasad and H.S. Shetty**

*Department of Applied Botany & Biotechnology, Biocontrol Research Laboratory, Manasagangotri, University of Mysore, Mysore- 570006.*

*Department of Biotechnology, Sri Jayachamarajendra College of Engineering, Manasagangotri, Mysore- 570006*

## ABSTRACT

Biological control—the use of an organism to reduce the population density of another organism—is the most successful, most cost effective and environmentally safest way of pest management. It is nature's own way to keep numbers of pest organisms at low levels. Biological control is present in all ecosystems, both natural and man made, and is always active. The result of natural biological control is that the earth is green and that plants can produce sufficient biomass to sustain other forms of life. Without biological control, the production of energy by plants would be a tiny fraction of what is produced currently. In such an approach, pest management will be a guiding theme instead of being the marginal issue it was during the past 60 years. Modern pest management will strongly depend on biological control, because it is the most sustainable, cheapest and environmentally safest pest management method. In additon, it has important benefits for farmers and consumers.

## INTRODUCTION

In addition to natural forms of biological control, man started to use arthropod biological control around the year 300 by using predatory ants for control of pests in citrus orchards. Large-scale use of biological control started in 1888 with the release of *Rodolia* ladybird beetles to control scale insect in citrus in California. Many permanent successes have been obtained since, resulting in annual profits of millions of dollars, and these profits are accumulating continuously as biological control is permanent in contrast with chemical control where resistance against the pesticide develops.

Due to the facts that: (1) earth will have to feed about 11 billion human beings in the near future, (2) fossil energy is running out, and thus are conventional synthetic pesticides, (3) man cannot continue to pollute the environment and reduce biodiversity at the same dramatic rate as during the past 100 years, agricultural research needs to be redirected to a systems approach. Biological control is expected to make-up 35-40% of all crop protection methods in the year 2050.

## DISCOVERY OF PATHOGENS OF INSECTS

Diseases of silkworms were recognized as early as 18th century, although diseases of bees were known to the Greeks and the Romans. Many publications in the sixteenth, seventeenth and eighteenth century deal with diseases of silkworm, a very important industry at that time. Vallisnieri was the first to mention the muscardine disease of silkworm. De Reamur was the first to described and illustrate a fungus, *Cordyceps*, infecting a noctuid larva in 1726. The microbial nature of these diseases was not yet realized. From William Kirby's chapter on "Diseases of Insects" (Vol. 4 (1826) of An Introduction to Entomology by Kirby and Spence), it was recognized that true fungi grew in the bodies of insects as saprophytes and possibly as parasites (Kirby and Spence, 1826). Agustino Bassi was the first to experimentally demonstrate in 1837 that a microorganism, *Beauvaria bassiana*, caused an animal disease, namely the muscardine disease of silkworms. It was also Bassi who published the idea to use microorganimsm for insect pest control in 1836. Later in 1874, Pasteur suggested the use of microorganisms against the grape phylloxera in France. These suggestions did not result in practical application. Metchnikoff tried to develop biological control for the wheat cockchafer *(Anisopilia austriaca*), a serious pest of cereal crops in the area of Odessa, Russia. In 1879 he published a paper on *Metarrhizium anisopliae*, and his experiments led to the conclusion that the fungus, when mass produced, and properly introduced in the field, might result in effective control. Based on Metchnikoffs work, *Metarrhizium* was mass produced in 1884 in the Ukraine, and the spores were tested in the field against a curculionid in sugar beet *(Cleonus punctiventris*).

Colombia is well-known for its research & application of entomopathogenic fungi such as *Beauveria bassiana*, *Verticillium lecanii*, *Metarrhizium anisopliae* and *Paecilomyces fumosoroseus*. The largest applications concern: (1) the spraying of *Beauveria bassiana* and *Metarhizium anisopliae* on 550,000 ha of coffee against the coffee berry borer (*Hypothenemus hampei*) and (2) the application of *Beauveria bassiana* against *Opsiphanes cassina* on 130,000 ha of oil palm, but the entomopathogens are also used for control of *Anthonomus* in cotton, thrips in ornamentals, whiteflies in beans and tomatoes, grasshoppers in pastures and insect pests in rice and citrus. Currently, Colombia has 5 producers of entomopathogenic fungi.

### *Beauveria bassiana*

*Beauveria* spp., belongs to a class of Euascomycetes and Phylum Ascomycota. This hyaline, filamentous fungus, first recognized as the etiologic agent of the devastating muscardine disease of silkworm, is ubiquitous in plant debris and soil. *Beauveria* is also isolated from foodstuff, infected insects and indoor air environment. The genus *Beauveria* contains several species. The most common ones are *B. basisana* and *B. alba*. The hyphae are hyaline, septate and narrow. The conidiogenous cells on the hyphae are typically flask-shaped with inflation at the base and narrow zigzagging filaments at the apex. Laterally, from the filament, conidia are produced from each bending point. This type of conidium production is called sympodial geniculate growth.

*B. bassiana* is present throughout the world in various climatic zones and has been found to attack 200 or more species of insect with a predominance for Coleoptera, Lepidoptera, Auchenorrhyacha, Sternorrhyncha and Heteroptera (Sponagel, 1994). The entomopathogenic fungus *Beauveria bassiana* Vuillemin is one of the several fungi that are of particular research interest because of their potential as commercial bioinsecticides. Some studies have focused on identifying nutrient substrates that *B. bassiana* can utilize with application to industrial production, while others have focused on the pathogenic processes of *B. bassiana* and interactions with insect cuticle. *B. bassiana* can utilize a wide variety of substrates including insect cuticle, chitin and its monomers, complex proteins, amino acids and several carbohydrates. In particular carbohydrate supplementation in *B. bassiana* growth media affects several physiological factors, for instance, biomass production, extracellular enzyme synthesis, blastospore yield and culture pH dependent on the amount and types of carbohydrates provided (Bidochka *et al.,* 1990).

Entomopathogenic fungi can provide safe and effective control of many important insect pests. Currently, economic interest lies in focusing upon fungi from three genera within the class Hyphomycetes—*Beauveria, Verticillium* and *Metarhizum*. Research has attempted to meet the challenges of ensuring the stability efficacy and mass production of inexpensive yet high quality fungal formulations (Feng *et al*., 1994).

The driving force for studying entomopathogenic fungal spores is that they initiate insect pathogenesis and are the active ingredient of mycoinsecticide formulations. Environmental conditions, especially temperature and humidity have long been recognized to play a significant role in the incidence of epizootics of insect pathogens. It has been shown that temperature and humidity affect both survival and germination rates in *B. bassiana.* If temperature affects germination percentage in conidia, then it also affects the concentrations of infective units in the field. Increase in infections was observed with increase in humidity. Increasing the period of high ambient humidity allowed a large number of conidia to germinate (Feng *et al*., 1994).

Temperature is one of the principal factors influencing entomopathogen efficacy. It is well documented that ambient temperature influences the rate of infection and time to death of insects treated with entomopathogenic hyphomycetous fungi (Inglis *et al.,* 2001).

Ideally, the thermal characteristics of isolates should be matched to the microhabitats in which they will be deployed for optimal performance. The optimum temperature for most entomopathogenic hyphomycetes is between 20-25°C, but infection and diseases can occur at temperatures ranging between 15-30°C. Above 30°C, the vegetative growth of most taxa is inhibited and growth usually ceases at –37°C. Considerable variability exists among the genotypes in their thermal characteristics. Some studies have reported no or weak correlations between geographical origin and thermal characteristics (Fargues *et al.,* 1997).

Others have observed a stronger relationship between temperature characteristics *in vitro* and place of origin. Vidal *et al*., (1997) observed considerable variability among isolates in their ability to grow at various temperatures. The thermal biology of insects in field settings is extremely complex and it still remains to be determined whether isolate possessing slightly higher optimal growth rates *in vitro* will translate into increased efficacy in field conditions. Inglis *et al*., (2001) found that growth rates for both fungi decreased as the amplitude of temperature increased under field settings. The cardinal temperature of *B. bassiana* conidial germination and hyphal development is ~35°C and the prevalence of mycosis decreased by ~ 45°C and 90% in nymphus allowed to optimize their body temperature by basking for only 1 and 2 h/ day respectively (Inglis *et al.,* 1997).

Relative humidity (RH) can influence fungal efficacy in several ways. Humidity in combination with temperature influences evaporation of spray droplets, which can result in the loss of small particles and thereby adversely affects targeting, Furthermore, moisture can also have very significant effects on the persistence of fungal inocula. Fungal conidia usually exhibit greatest stability under cool and dry conditions. In all instances, conidiogenesis on insect surfaces requires high moisture and the production of conidia may influence horizontal transmission of the pathogen. Fargues and Luz (2000) observed that a RH of at least 97% was required for conidial production of *B. bassiana* and suggested that the occurrence of high humidity is the most crucial climatic constraint limiting the recycling ability of *B. bassiana* against *Rhodnius prolinus*. A number of studies indicate that dry conditions during or immediately following application of fungal propagules are less detrimental than previously thought. The ability of *B. bassiana* and *M. anisopliae* to germinate and infect the host under conditions of low ambient humidity is attributed to sufficient moisture within microhabitats.

In addition to increasing humidity, rainfall can serve to dislodge and disperse conidia from substrate as well as air and contributes to dispersion of propagules. However, evidence indicates that rain does not cause significant removal of *B. bassiana* and *M. anisopliae* conidia from foliage of monocotyledonous and dicotyledenous plants (Inyang *et al.,* 2000). Many hyphomycetous fungal entomopathogens are considered to be soil-borne microorganisms and have demonstrated considerable potential against soil pests (Inglis *et al.,* 2001).

The influence of other parameters such as pH and ionic conductivity is less understood. A number of studies have demonstrated no or minimal effects of soil pH on the distribution of entomopathogenic hyphomycetes (Inglis *et al*., 2001). In contrast, Groden and Lockwood (1991) observed that fungistasis against *B. bassiana* was correlated with pH. In a later study, Groden and Dunn (1996) found that soil pH and nitrogen fertilizers had an impact on the germination of *B. bassiana* conidia.

Lipase is one of the fungal enzymes, which hydrolyzes cuticular components for penetration of the insect cuticle, playing a role in the degradation for fungal penetration of the cuticular lipid barrier of some insects (St-Leger *et al.,* 1986).

Distinction between *Beauveria* species has always been problematic possibly due to the comparatively larger heterogeneity of spherical/globose-spored *B. bassiana*. Mugnai *et al*., (1989) found that spore form was the most useful single character of the 64 morphological and biochemical tests they compared for separating *Beauveria* spp.

Production of spore types depends on the type of culture; conidia are produced in semi-solid culture and blastospores are formed in submerged liquid culture. In submerged shake flask culture, fungi grow as dispersed filamentous mycelia, as dense mycelial aggregates known as pellets or as some mixtures of two. Manipulation of growth medium can alter fungal morphology (Humphreys *et al.,* 1989). Daoust and Roberts (1983) studied the effect-prolonged storage of *Metarhizium anisopliae* conidia grown on different growth substrates on its conidial viability and virulence. Compounds such as glucose may exert their effect on fungal morphology by changing water activity of the medium (Humphreys *et al.*, 1989).

Biochemical characterization of *Beauveria* isolates carried out by esterase analysis showed difference in esterase bands but still cluster analysis showed that all isolates were closely related (Mugnai *et al.,* 1989). Molecular techniques using DNA have been useful in identifying distinct intraspecific groups when a large number of fungal isolates are examined. Techniques involving polymerase chain reaction amplification of DNA, e.g., RAPD and RFLP analyses of nuclear or mitochondrial DNA have been particularly useful in resolving taxonomic and evolutionary problems in fungi (Chew *et al.,* 1998). Analyses of anonymous loci with the RAPD technique have previously detected genetic differences within and between fungal species (Hegedus & Khachatourians, 1996).

Isozymes have been used as genetic markers to analyze the variability among the isolates of *B. bassiana* (Berretta *et al.,* 1998). However, systems based on DNA markers provide more detailed genetic information suitable for identifying individuals and determining their relatedness to others (Castrillo & Brooks, 1998). Since its development, the technique called RAPD or AP-PCR has been extensively used in fingerprinting of several organisms including entomopathogenic fungi (Viaud *et al.*, 1998). Bidochka *et al.,* (1994) assessed RAPD marker patterns from three isolates of *B. bassiana*. It was found that RAPD analysis was a very efficient method to assess genetic variability in *B. bassiana* and was compatible with automated fragment analysis using a simultaneous four color fluorescence laser detection system (Beretta, *et al.,* 1998). Reproducibility of RAPD patterns of up to 1000 bp was not affected by using template DNA over a four-fold range. However, the level of amplification of each fragment was most constant through replicates of a reaction with 5 ng of DNA.

## CHARACTERIZATION OF *Beauveria bassiana* FOR EFFECTIVE USE AS BIOLOGICAL CONTROL AGENT

Microorganisms that are used in biological control are of value only if they can be stored for future use in such a fashion that their growth and productive capacities remain unaltered. The preservation method also depends on the priorities for use of the stored cultures, including long-term maintenance of virulence or viability. Thus, maintenance of stock culture plays an important role in research and production. There are two types of stock cultures—"working stocks" and "primary stocks". Working stock cultures are used frequently, hence maintained as agar slants, agar stabs, spore preparations or broth cultures, and will be held under refrigeration. They must be checked constantly for possible changes in growth characteristics, nutrition, productive capacity, and contamination.

Primary stocks are cultures that are held in reserve for biological assays, or for possible later screening programs. These cultures are not maintained in a state of high physiological activity and they are delved into only rarely. Transfers for these cultures are made only when a new working-stock culture is required or when the primary-stock culture must be sub cultured to avoid death of cells. Thus, primary-stock cultures are stored in such a manner as to require the least possible number of transfers over a period of time.

Loss of virulence is a reversible process and can be regained by passing the entomopathogenic fungus through a living insect host.

There are many methods available for preserving fungal cultures, each having its own advantages and shortcomings. The problems include changes in morphology or virulence after improper or prolonged storage. Variability in fungal morphology and physiology after repeated subculture on artificial media is a well-known phenomenon and in case of pathogenic fungi, it may result in altered (usually reduced) virulence levels. In biocontrol projects, periods of saprophytic growth may induce modifications of virulence. Subsequent subculture may select for greater saprophytic growth rates, which may pleiotropically affect virulence. (Hajek *et al.*, 1995).

Proper storage and regular maintenance of cultures is of prior importance in any biocontrol project. Cryopreservation techniques offer one of the most reliable means to assure the long-term storage of entomopathogens. Storage at low temperatures (–80°C) has minor effects on cell viability and phenotypic characters of entomophthoralean fungi (Lastra *et al.,* 2001).

As early as 900 AD it was known in the orient that fungus could grow in insects (Steinhaus, 1975). From 1880s through the early 1900s the spectacular epizootics caused by entomopathogenic fungi led to studies of their potential use for pest control. More recently, owing to the myriad difficulties that have been gradually encountered in the development and use of chemical insecticides, the field of biological control has been extensively studied and experimented upon as a promising alternative (Roberts & Hajek, 1992).

*Beauveria bassiana* (Bals.) Vuillemin, which belongs to the family Deuteromycetes (Fungi imperfectii), is a non-specific entomogenous fungus with a broad geographical distribution and has been reported to be a common fungus isolated from dead and moribund insects in nature. There is some confusion in the literature, however, over the taxonomic placement of this fungus. According to Ainsworth, 1956, Agostino Bassi described the causative agent of "muscardino" a disease of silk worms, to be a fungus in 1835. In 1838, the organism was named *Botrytis bassiana* by Giuseppe Balsamo Crivelli. Since the mid-1800s this fungus has had several names, including *Sporotrichum densum, Beauveria densa, Sporotrichum globuliferum, Beauveria globulifera* and *Beauveria bassiana*, subsp. *Tenella* (Domsch *et al.,* 1980). Since then the genus *Beauveria* has been the subject of considerable taxonomic investigation (Mugnai *et al*., 1989). Petch (1926) and Macleod (1954) distinguished two species, *B. bassiana* Vuill and *B. tenella* (Delacr). The entomopathogenic fungus *B. bassiana* has been extensively used for the control of many important pests of various crops and has been tested on different target insects (De La Rosa *et al.*, 1997; Bustillo *et al.*, 1999; Jackson *et al.*, 2000; Damon, 2000; Wagner & Lewis, 2000; Lecuona *et al.*, 2001; Mannion, *et al.*, 2001).

The importance of *Beauveria* species as a biological control agent of insect pests requires the use of complementary methods for estimating the diversity and polymorphism between species.

The use of biochemical characters in microbial taxonomy is well-established, particularly in bacteria (Cowan, 1974). In the taxonomy of the filamentous fungi, the presence of an abundance of diverse morphological features in most groups has obviated the need to consider biochemical characters (Grodona *et al.*, 1997). In other groups, lack of suitable methods has prevented their inclusion in taxonomic schemes. However, it has become apparent that fungal classifications based almost exclusively on morphological features may not always result in defined data.

Attempts have been made to introduce physiological and biochemical characters having distinction between different isolates of *B. bassiana.* Moreover, hydrocarbon assimilation and specific enzyme activities have been investigated for its use revealing the polymorphisms existing among the isolates of the entomopathogen *B. bassiana* (Bridge *et al.*, 1989). A small number of physiological tests have also been included to differentiate the isolates of *B. bassiana.*

As part of a major multidisciplinary approach to the heterogeneity within the species, *B. bassiana*, a range of physiological and biochemical tests that may be of value in understanding and developing *B. bassiana* for efficient control of CBB have been evaluated and the results of the study are presented in this chapter.

## TOXICITY OF *Beauveria bassiana*

Entomopathogens have been developed as microbial insecticides for inundative augmentation, although there remains a greater potential in using them with inoculative augmentation, classical and conservation approaches. Ecological effects of microbial control agents of non-target organisms or through indirect effects, principally depletion of the target population. Direct effects can be avoided or minimized by careful selection of highly host specific pathogens. In inundative biological control, both direct and indirect effects are usually reversible even when pathogens with a broader host range are used, effects that have documented are minimal and pathogen levels eventually return to background levels following application. Permanent indirect effects could be possible if an exotic or genetically engineered pathogen were to become established and provide a long-term control of the target host. To date there is no evidence of detrimental ecological effects from deliberate introductions of entomopathogens for use in classical biological control (Goettel & Hajek, 2001). Insights into possible side effects can be obtained from studies on natural epozootics of indigenous strains.

The safety of entomogenous fungi to non-target invertebrates and vertebrates present in the surrounding environments supposed to be brought under the ambit of biocontrol agents has to be determined before their widespread use as biocontrol agents. For its use as a bioinsecticide it is important to identify a range of characteristics favourable to survival and virulence against the insect. Difference in virulence to a host among isolates is reported in all the entomopathogenic fungi assayed (Glare & Milner, 1991).

The term non-target organisms have been adopted to indicate all living things aside from pests being treated for control. However, studies on the effects of entomopathogens on non-target organisms became more focused because of pathogen specificity. There has always been interest in the effect of entomopathogens on commercially important non-target organisms such as bees and silk worms. The effects of entomopathogens on non-target organisms are not only determined by infectivity, but can also be influenced by the ecological niches occupied by the pathogen and potential hosts. The range of species that an entomopathogen is able to infect usually under optimized conditions, often referred to as its physiological host range manifested by pathogens under field conditions, is of primary concern regarding non-target impacts. In essence, the ecological host range of a pathogen can only be tested after a pathogen species are present in the environment (Hajek & Goettel, 2000). For some organisms, pathogens do not kill but instead cause sub lethal effects such as reduced food consumption, reduced adult longevity, and decreased fecundity. The occurrences in the ecosystems life cycle depends on the target hosts (Flexner *et al.*, 1986). However, non-targets exposed to a pathogen could themselves be affected sub-lethally. In such cases, detailed observations are necessary to detect these effects.

There have always been primary concerns about the direct impact of entomopathogens on insects being reared for a product. This would principally include honey bees and silk worms although other insects are reared for food, dyes, etc., according to regional practices with the increased use of parasitoids and predators for biological control or beneficial agents is critical to integrated pest management (IPM) programs. Direct effects are more easily documented than indirect effects although both can occur. As a direct effect one can readily understand how entomopathogens applied in the field would contact free ranging bees or parasitoids causing an effect if these non-targets were susceptible. In contrast, indirect effects of pathogens can be highly variable. Studies of effects on non-targets can be focused depending on the control method and type of pathogen used.

Classical biological control is aimed at permanent establishment of an exotic pathogen. If introduced, it has the potential to persist in the environment and spread from the sites of release. Consequently, before release of a pathogen for classical biological control, knowledge of its potential effects on non-targets is critical. Emphasis should be placed on those organisms that live in the same habitats within an ecosystem as the target pests (Maddox *et al.*, 1992). In addition to learning about the effects on non-target organisms, it is important to evaluate the potential impact of depletion of the target host population non-target population showing an indirect effect. For inundative applications of pathogens it is only the non-target organisms coming in direct contact with the microbial insecticide that are at risk and therefore risk is essentially very similar to chemical insecticide application (Hajek & Goettel, 2000).

In developing protocols for evaluating effects of pathogens on non-target organisms, regardless of whether the testing is to take place in the laboratory, semi-field and field, the inclusion of both positive and negative controls is critical for comparisons. Typically, number of hosts killed is compared with negative controls. However, other parameters that might directly have negative effect on non-target populations should be considered. Studies of potential effects on non-target organisms should be carried out on species potentially present in the field rather than readily available organisms (Hajek & Goettel, 2000).

One major difficulty in testing the specificity in the laboratory as a preparation for understanding non-target impacts in the field is determining an appropriate dose. Attempts should be made to mimic as much as possible the parameters expected in the field. Consequently, parameters such as dose, inoculation method and temperature and humidity should be as close as possible to those that the non-target organisms are expected to encounter under field conditions. Nevertheless, in many instances, semi-field assays can provide information that is far superior to the results obtained from laboratory conditions as they mimic many of the complex environmental parameters that are difficult or next to impossible to duplicate in the laboratory. A popular method for assessment of target as well as non-target effects of field applications of pathogens is to periodically collect insects from the experimental plots to incubate them under laboratory conditions and then assess mortality. Such assays can measure exposure of the non-target organisms to the inoculum under field conditions and not necessarily infection and disease progression that would occur in the field. As another type of semi field-testing strategy instead of rearing field collected insects, the pathogen can be collected in the field and used in the laboratory to challenge insects. The susceptibility of non-target organisms to the levels of inoculum present in the field can be tested by exposing test insects to field collected substrates, preferably in the field or under stimulated field conditions in the greenhouse or laboratory. Virtually, all trials of the effects of microbial control agents on non-targets require tests of honeybees – *Apis mellifera.*

Cantwell *et al.*, (1986) administered pathogens by feeding. Bromenshenk *et al.*, (1996) argued that spraying adult honey bees is more realistic than feeding since the most direct exposure route of commercially applied microbial control agents to honeybees would be direct contact and not through the ingestion of pollen or nectar. A third method of bee exposure would be to apply the pathogen directly to a crop field as in an operational application against the target host. Very few detrimental effects of biocontrol fungi have been studied among which *Beauveriosis* in silkworm was a notable disaster. With time, strict hygienic practices and legislation were adopted to bring this disease under control. In China and Japan, the application of *B. bassiana* and other microbial control agents was restricted to areas where no silkworms are bred, to avoid contamination of the mulberry leaves used to feed the silkworms (Goettel *et al.*, 1990) and through careful strain selection, it may even be possible to use *B. bassiana* to control a pest in the immediate vicinity of a silkworm rearing facility. In a pilot test in a Chinese 110 ha pine plantation, good control of pine caterpillars was achieved through six applications of 2100 kg of *B. bassiana* with no significant effect on silkworm within a rearing shed built within the plantation (Anon. 1981). The strains of *B. bassiana* used against the pone caterpillars were 100 times less virulent to silkworms than strains isolated from silkworms themselves.

Determining the effects of entomopathogens on the target host population density is usually an arduous task in itself. Consequently, it is next to impossible to include every non-target organism exposed to the pathogen in such evaluations. However, at times it may be warranted to investigate population densities of specific non-target organisms. In some situations, it may be prudent to monitor population densities of either certain sentinel beneficial organisms known to be essential in Integrated Pest Management programmes or non-targets that are important members of the invertebrate community. Experimental design is essentially the same as that used to evaluate the effects of the pathogen as target host population. Methods for monitoring non-target organisms in particular must be carefully chosen to ensure that population density assessments are valid. It is imperative that the population densities of target organisms be assessed concurrently. If changes in target host population densities are not apparent, then information on lack of effects on the non-target populations are not of much use. Replicated treatments should consist of pathogen and non-treated controls.

Effects of microbial control agents on non-targets require tests of honeybees, *Apis Mellifera.* Bromenshenk *et al.*, (1996) argues that spraying adult honeybees is more realistic than feeding, since the most direct exposure route of commercially applied microbial control agents to honeybees would be direct contact and not through the ingestion of pollen or nectar. But Cantwell *et al.*, (1986) administered pathogens by feeding. Regardless of the inoculation method, at the time of administration, the propagules are enumerated and the viability assessed according to established protocols for the pathogen in question (Lacey *et al.,* 1999).

The safety of fungi being contemplated for use in biological control must be considered at many levels, with primary concern for direct effects on vertebrates especially humans. In considering safety towards all organisms, vertebrates, invertebrates and plants, not intentionally being affected by the biological control fungus, are referred to as non-target organisms. Following potentially safety issues have been considered:

1. Competitive displacement of non-target organisms
2. Allergenicity
3. Toxigenicity to non-target organisms, and
4. Pathogenicity to non-target organisms (Cook *et al.*, 1996).

In addition, indirect effects such as those that could come about through the depletion of the target host itself must be considered as a potential safety issue (Goettel & Hajek, 2001). The potential unintended effects will differ depending on the potential target and non-target organisms and the ecosystem they inhabit.

Fungi secrete a wide variety of compounds, many of which are toxic to plants, invertebrates and vertebrates. *M. anisopliae* produces destruxins and cytochalasins; *B. bassiana* produces oosporein, beauvericin, bassianolide and beauveriolide, *Hirsutella Thomposonii* produces hirsutellin and *Trichoderma harzianum peptaibols* to name a few. However, toxin production varies according to isolate and does not necessarily play an important role in virulence in every case. The production of toxins by a candidate fungal biocontrol agent should not necessarily preclude its use in biocontrol. Considerations must be made as to possible effects from the presence of the toxin in the formulated product, fate of the toxin after application and possible bioaccumulation of the toxin within the host or environment.

Pathogenicity towards the target host is usually the desired effect. However, pathogenicity towards non-target organisms could be an unintended effect. Several cases of human infection by *M. anisopliae* have been reported in both immunocompetent and immunoincompetent individuals, with one fatality in an immunocompetent child (Revanker, *et al.*, 1999). There are numerous case reports that the nematode destroying fungus *Paecilomyces lilacinus* is a causative agent of human infections in both immunocompetent and immunoimcompetent individuals (Gutierrez-Rodero *et al.*, 1999). Another entomopathogenic fungus *Conidiobolus coronatus* is commonly associated with lesions in both humans and horses (Saik *et al.*, 1990).

*B. bassiana* is an entomopathogenic fungus known to be pathogenic to the insects. Ability of *B. bassiana* to colonize mammalian tissues and cause disease has not been clearly demonstrated, but there are a few conflicting reports implicating fungi of this genus as a cause of natural and experimental disease in mammals. No animal mortality or overt toxic effects were noted in rats dosed orally with $1.9 \times 10^8$ CFU/animal of *B. bassiana* strain ATCC 74040. Red foci were noted on the lungs of three of the treated animals indicating possible pulmonary toxicity. However, the acute pulmonary toxicity/pathogenicity tests showed clearance within 15 days (EPA fact sheet, 2000). *B. bassiana* was not pathogenic infective or toxic in rabbits dosed dermally at 2 gm per animal containing $4.2 \times 10^7$ cfu/ml. Though low toxicity has been exhibited by the *B. bassiana* cultures when tested on rats/mice, it is essential to test the mammalian toxicity of the culture. Based on the low toxicity potential, this microbial pesticide can be used as an active ingredient in the commercialized formulation.

## MASS PRODUCTION OF *Beauveria bassiana*

*B. bassiana* gains its importance due to its ease in mass production. Both liquid and solid fermentation systems are used to obtain a desired quantity of biomass and the former is the preferred approach. Liquid fermentation has obvious advantages of controlled processing and the organism can be recovered easily (Deshpande, 1999).

The success of any microbial control programme depends on a multitude of factors including the pathogen virulence, host specificity, persistence and our ability to disseminate it effectively in the environment. In classical biological control, a pathogen or natural enemy becomes established and disperses once it has been introduced into the environment. Therefore, mass production of the pathogen is not an absolute requirement although in the initial stages it enables widespread introduction of the pathogen through numerous releases. However, the ability to mass-produce the pathogen becomes crucial is augmentative or inundative microbial control where the pathogen in effect is used as a biological insecticide. In addition to the ability to produce the infective propagules in a suitably large scale, a successful bioinsecticide product must remain virulent during storage and be compatible with formulations and application techniques. Insect pathogenic fungi, particularly conidia forming Deuteromycetes, have the potential to be a new class of bioinsectide suitable for widespread application in agriculture.

The mode of action and penetration through the insect cuticle makes fungi effective against wide range of insect pests that cannot be controlled effectively by bacterial, viral or protozoan pathogens that act through ingestion. Fungi can be delivered in a variety of formulations that act as direct contact sprays, foliage sprays, traits and granules. Further advantages are their mammalian safety (Siegel & Shadduck, 1990) and minimal impacts on non-target insects due to the relative host specificity of different isolates of the same species (Goettel *et al.*, 1990).

The ability to select host specific isolates of the same fungus species also has significant practical importance in bioinsecticide development from one fungus species using the same production processes and equipment and having similar requirements for formulations storage application and regulatory compliance. Commercial production of mycoinsecticide development is often assumed as easy because these organisms generally conidiate profusely on agar plates and can be grown using a wide variety of techniques and media. Unfortunately, this assumption does not necessarily hold up in an analysis of the practical consideration of commercial development. The most obvious aspect of production is economics. A company must deliver a mycoinsecticide with an effective concentration of infective propagules at prices that are either competitive with other insecticides or cost-effective in reducing the crop damage. Other aspects affecting production technologies in development of mycoinsecticide are scale, virulence, stability/ formulations and regulatory aspects (Deshpande, 1999).

*Beauveria bassiana* along with other hyphomycetous entomopathogens such as *Metarhizium anisopliae* has been the focus of commercial scale efforts to use fungi to manage insect pests in the past 20-30 years especially in the former USSR, France, eastern Europe and the Republic of China and also in the United States and Canada. *B. bassiana* has been used with limited success against Colarado Potato beetle, *Leptinotarsa decemliacata*. In the former USSR and Ukraine, typical rates exceed $2 \times 10^{13}$ conidia/hectare in conjuction with chemical insecticides. Since its perfect stage is presently unknown, *B. bassiana* is placed in the Deuteromycotina (Fungi imperfectii) within the class hyphomycetes. It is characterized by the sympodial development of conidia from ampilliform conidiogenous cells. *B. bassiana* is a ubiquitous fungus that is infectious to a wide variety of insects from most orders (Goettel *et al.*, 1990). However, different genotypes exhibit difference in virulence to different insect host species.

When aerial conidia contact the cuticle of susceptible insect host, they initiate germination. Although it is generally believed that high humidity is a requirement for infection, this is not necessarily so with several hosts including grasshoppers (Marcandier & Kachatourians, 1987).

A short hypha emerging from the conidium penetrates the cuticle by mechanical pressure and enzyme action. The fungus rapidly proliferates within the insect's haemocoel either by multiplication of yeast like blastospores or by hyphal growth. The host insect dies within a few days. Time of death depends upon the dose as well as strain specific pathogenicity. Soon after the death of the insect and when relative humidity is high, the fungus produces mycelium that quickly covers the surface of the cadaver and gives rise to conidiogenous cells and conidia often appearing like a white powder covering the insect. A number of secondary metabolisms are thought to repress saprophytic microorganisms in the insect cadaver allowing *B. bassiana* to complete its development and sporulate (Jaronski & Goettel, 1997).

*B. bassiana* can produce three spore types in culture: aerial conidia, blastospores and submerged conidia. Aerial conidia are produced on insect cadavers and solid substrates. Thin walled, single cell hyphal bodies, termed blastospores along with varying amounts of mycelium are produced in submerged culture by most isolates, especially when amino nitrogen exceeds a critical level (Bidochka *et al.*, 1987). These propagules are smooth-walled, larger than conidia and contain different wall surface carbohydrates. Blastospores usually germinate much faster than either of the two other forms, 50% germination being reached within 5-6 h. Submerged conidia are produced singly at the tips of short germ tubes arising directly from blastospores (Hegedus *et al.*, 1990). There has been considerable speculation that submerged conidia could serve as an active ingredient in a commercial mycoinsectide.

A wide variety of production techniques for entomopathogenic fungi have been reported. Production processes can be described as three general types: submerged (liquid) culture, surface culture and two-step process.

The advantage of submerged culture is that fungal insecticides can be produced using conventional deep-tank fermenters. The disadvantage is that these fungi typically produce blastospores. Blastospores are unstable with limited shelf life. As a result more recent research on submerged culture production has focused on developing culture media in which the fungi produce conidia rather than blastospores. Russian patents describe media and a selected strain of *B. bassiana* with claims of $2.5 - 5.0 \times 10^{12}$ conidia per litre of culture broth (Primak *et al.*, 1976). They observed conidia produced singly from sporogenous cells that occur as short branches or directly on mycelia. Submerged conidia appear different from aerial conidia. In liquid fermentation or submerged culture, microorganisms are grown in a liquid medium either in shake flasks or on a larger scale in deep tank fermenters.

The hyphomycetes fungi are often easy to grow on synthetic media. However, they rarely produce conidia in submerged culture. More commonly they produce blastospores or hyphal bodies, which can be formed by hyphal constriction, separation at the septa or by yeast like budding. The formation of blastospores occurs naturally during infection of the insect host and enables the fungus to spread and develop within the haemocoel. Nevertheless, blastospore may have potential for use in biological control as they are infective and could potentially be produced economically in deep tank fermenters (Kleespies & Zimmermann, 1992).

Bidochka *et al.*, (1987) studied the production of blastospores by *B. bassiana* in liquid media containing varying proportions of peptone, glucose and yeast extract. They found that a nitrogen source was essential to blastospore production, but the addition of yeast extract reduced yields of blastospores and promoted production of mycelium. Manipulation of culture conditions can also increase blastospore longevity and virulence (Keller, 1992). Production of *B. bassiana* was similar in both batch and fed cultures. Macromolecular composition of the blastospores can be altered during the fed batch culture to enhance virulence.

Liquid fermentation can be used successfully for mass production of certain fungal pathogens, but culture method, physical parameters and nutritional requirements must be developed not only for each genus or species but also for each fungal isolate of interest. Blastospores lend themselves to formulating in water but are less amenable to formulation in oils for ultra low volume application (Moore & Caudwell, 1997).

Specialized sporogenous cells produce submerged conidia in the same manner as their aerial counterparts. Some isolates of *B. bassiana* produce conidia as opposed to blastospores under certain culture conditions (Hegedus *et al.*, 1990; Rombach *et al.*, 1988). A comparative study on the differences between virulence, stability and cell wall structure of aerial conidia, submerged conidia and blastospores of *B. bassiana* was carried out by Hegedus *et al.* (1992). Virulence studies on the three spore types indicated that although blastospores were slightly more virulent in comparison to aerial and submerged conidia, both conidia types retained greater viability during storage than blastospores.

Liquid fermentation conditions, which are optimum for mycelial growth, are used and the mycelium is harvested during the active growth phase by vacuum extracting. This produces a thin mycelial mat, which is washed and then treated with a solution of maltose and sucrose, which acts as protective agent during air-drying. The dried mycelium is ground and stored at 4°C. Upon rehydration, the mycelium forms conidiophores, which subsequently discharge conidia.

A simple system for production of *B. bassiana* for control of coffee berry borer in Columbia uses a reciprocal shaker and shallow trays. The final product is obtained by peeling the surface layer of fungus and spores from the trays, removing the excess water and producing a dry powder formulation using a fluidized bed drier. A similar system has been used in the Soviet Union to produce Boverin, a *B. bassiana* product. One pilot plant was reported to have produced 22 tonnes of active products in 12 months (Ferron, 1981). In surface culture the fungus is grown and conidiates on the surface of a liquid or solid medium. Conidia produced by these processes are indistinguishable from conidia produced on dead insects. Perhaps the best example of a liquid surface culture processes is that of Samsinakova *et al.*, (1981). Shallow (2-cm) layers of sterilized liquid media in polypropylene bags are inoculated with *B. bassiana*, which grows as a mat on the liquid surface. After 12 days, cultures are harvested by filtering off the cell mat. Yields of up to $10^{13}$ conidia per 0.1 $m^2$ of surface area on 2 Lt of media are reported. This technology has been patented in Canada and used commercially in Russia and Czechoslovakia for producing Boverin, a preparation of *B. bassiana* for control of Colarado potato beetle.

The solid substrate can be either nutritive such as rice bran or wheat grains or non nutritive such as vermiculite, sponge and cloth. The use of solid substrates for conidial production gives the opportunity to increase the yield by providing the larger surface area. However, parameters such as pH, temperature, nutrient status and aeration are more difficult to monitor and control on solid substrates than in liquid fermentation. Substrates, which provide a large surface area to volume ratio such as rice grains, are often used for conidial production. Rice has been shown to be superior to other substrates such as sweet potato, tapioca, papaya and coconut (Ibrahim & Low, 1993). Production of *B. bassiana* for control of rice and forest pests had been a common practice in the people's Republic of China (Husey & Tinsley, 1980). Production of conidia on simple substrates such as rice husks has been preformed from a village scale in dugout pits to highly technical and automated industrial procedures (Feng *et al.*, 1994).

Solid substrate fermentation (SSF) technology has been known for centuries; from approximately 2600 BC. Koji process is one of the solid substrate fermentation processes, which is used in the preparation of enzymes. The koji process is of great historical importance to modern SSF. SSF systems exhibit several advantages over submerged fermentation processes. This process of koji fermentation is more efficient and produces more number of conidia (Ooijkaas *et al.*, 2000).

Goettel (1984) reported production using flat pans containing moistened wheat bran covered with cellophane placed directly on the bran. Cultures grew on the cellophane, conidiated and were harvested by simply scraping the mycelia mat off the cellophane. This proved to be efficient process for producing small amounts of conidia from a number of different fungi from a number of different fungi with yields equivalent to $1.1 \times 10^{12}$ conidia/kg of bran for *B. bassiana* and $3.3 \times 10^{11}$/kg for *M. anisopliae*.

Diphasic fermentations systems combine the benefits of high biomass production in liquid fermentation and production of stable hydrophobic (lipophilic) aerial conidia on a solid substrate. Fungal biomass is produced in a simple liquid medium of waste brewers yeast and sucrose in shake flasks. After 2-3 days of fermentation, the liquid biomass is transferred to par-boiled autoclaved rice. Conidia are extracted from the rice using a cyclone extractor and further dried to 5 % moisture content in a dessicator. Conidia may be stored as a dry powder for 1-2 years.

Alves *et al.*, (1995) obtained a yield of $2 \times 10^{11}$ *B. bassiana* conidia/g of powdery preparation using rice as a basic growth substrate. Rombach *et al.*, (1988) produced conidia of *B. bassiana* on wheat bran in a small mass production unit using a diphasic fermentation process. Chinese entomologists and microbiologists have successfully improved the traditional diphasic fermentation technology for mass production of pure *B. bassiana* conidial powder for large-scale control of pine caterpillars and other insect pests (Li & Yung, 1988, Zhou, 1988, Chen *et al.*, 1990).

The improvement involved the introduction of machinery equipment into the production process and the use of cheap large source materials for growth substrates. This modified diphasic technology made mass production of high quality *B. bassiana* conidial powder simple, easy, cheap and safe (Feng *et al.*, 1994).

Production of fungi to use as biocontrol agents aims at the maximum possible number of living propagules that can infect pathogenic fungi and insects. This is the distinguishing feature between production of fungi as biopesticides and fungi for other products. The use of microbial control agents is becoming a common practice in many countries and success of such control programmes often depends on an adequate mass production method for the agent used (Deshpande 1999). The success of any microbial control programme depends on a multitude of factors including the pathogen's virulence, host specificity, persistence and our ability to disseminate it effectively in the environment. Therefore mass production is very essential in any biocontrol programme. At present mass production of *B. bassiana* is generally based on diphasic and submerged fermentation techniques. As pointed out by Soper and Ward (1981), diphasic fermentation combines the advantages of both solid and liquid media. The fungus is allowed to grow in fermenters to the end of log phase for maximal production of mycelial biomass, which is subsequently transferred onto nutritious or inert substrates for the production of aerial conidia in the form of natural inocula. However, diphasic method is though simple, considered to be the most expensive (Soper & Ward, 1981) and labour-intensive (Rombach, 1989). However, with large-scale production, precautions must be taken to properly control several factors (such as aeration, pH, temperature, etc.), which may significantly affect the quality and quantity of a fungal preparation. Furthermore, submerged fermentation often yields blastospores or a mixture of blastospores and condiospores (Belova, 1978; Thomas *et al.*, 1987; Yin *et al.*, 1988; Rombach 1989).

Many factors have to be considered when screening for a potential mycoinsecticide. The fungus must produce spores readily, the propagules must be highly pathogenic towards target insects and they must be capable of initiating epizootics. Most entomogenous fungi produce two types of spores— conidia, which are produce from phlialides and spores, generally formed in semi solid culture. Large-scale production of spores under strictly controlled conditions for use as a mycoinsecticide only be achieved only by liquid cultures. The onset or enhancement of spore production in fungi is normally associated with unfavorable growth conditions, e.g., growth on poor carbon sources (Mckoy & Trinci, 1987), nutrient limitation and nutrient exhaustion (Morton, 1961; Righelato *et al.*, 1968). In submerged shake flask culture, fungi grow as dispersed filamentous mycelia (Trinci, 1972), dense mycelial aggregates known as pellets (Burkholder & Sinnot, 1945), or as some mixture of the two. Compounds that affect fungal morphology may also affect blastospore, the production of which in submerged culture offers advantages over production of conidia on semi-solid media, since large numbers can be produced in relatively low volumes under precisely defined conditions (Humphreys *et al.*, 1989). Blastospores are potentially and easily produced source of inoculum for biological control. Their major weakness is rapid viability loss in storage and this has been a problem in several control programmes using *Beauveria* spp., (Lane *et al.*, 1991). Hence, mass production technologies have to concentrate on more stable conidiospore production.

## FORMULATION OF *BEAUVERIA BASSIANA*

Hyphomycetes are filamentous fungi that reproduce by conidia generally formed aerially on conidiophores arising from the substrate. Many genera of entomogenous fungi occur in the hyphomycetes and are described on the basis of their morphological features. Therefore, phylogenetic relationship cannot be implied based on these genera. The most common route of host invasion is through the external integument although infection through the digestive tract is possible. Conidia attach to the cuticle germinate and penetrate the cuticle. Once in the haemocoel, the mycelium ramifies through out the host, forming yeast like hyphal bodies often referred as blastospores. Host death is often due to a combination of the action of a fungal toxin, physical obstruction of blood circulation, nutrient depletion and invasion of organs. After host death, hyphae usually emerge from the cadaver and under appropriate conditions of temperature and humidity produce conidia to the exterior of the host. Conidia are then dispersed by wind or water. Under dry and cool conditions insects often remain intact as fungus filled mummies and the fungus only emerges and sporulates once the cadaver is brought under appropriate conditions for fungal growth and conidiation.

Consequently, entomopathogenic hyphomycetes are among the most commonly encountered insect pathogens as cadavers often remain intact for long periods and the external mycelium is conspicuous. Most of the entomopathogenic hyphomycetes are facultative pathogens and are relatively easily grown in pure culture on defined or semi-defined media. Hyphomycetes quickly colonize insect cadavers especially if they are not already colonized by a pathogen. Consequently, it is often a challenge to determine if a hyphomycete isolated from an insect cadaver was responsible for the insect's death (Goettel & Inglis, 1997).

Entomopathogenic hyphomycetes may be harvested directly from insect cadavers on which the fungus has already sporulated. If external sporulation has not occurred, cadavers may be placed in suitable environments and the fungus allowed producing hyphae or conidia externally. Alternatively, cadavers may be homogenized and the homogenate plated on an appropriate agar medium. If the identity of the fungus is known, it is preferable to use a selective medium developed for the particular fungus in question. However, if no selective medium exists or if the identity of the fungus is unknown, virtually any medium used for propagation of entomopathogenic hyphomycetes can be used. It is necessary to first induce external growth of entomopathogenic hyphomycetes by placing the cadaver in an environment with high relative humidity.

Once insects that have died of mycosis are placed under conditions of high humidity vegetative growth and/or sporulation on the surface of the integument usually occurs within a few days at 20-25°C. Once adequate growth is observed on cadavers, the fungus can be transferred to an appropriate agar medium with a needle. Selective media are frequently used for isolation of entomopathogenic hyphomycetes. Media based on the differential activity of the fungicide have been successfully used to isolate *Beauveria* and *Metarhizium* spp. from soil and insect cadavers. An oat meal agar medium amended with 650 mg/ml fungicide dodine suppressed the growth of *Pencillium* spp, *Trichoderma* spp., and *Mucorales* spp., but supported the growth of *B. bassiana* and *Metarhizium* spp. Subsequently Chase *et al.*, (1986) confirmed that an oatmeal medium amended with 600 mg/ml of dodine resulted in good isolation of *B. bassiana*. However, this concentration of dodine was inhibitory to *M. anisopliae*.

Detailed studies on fungal pathogenesis, virulence, physiology and evaluation, as microbial control agents require methods for the propagation and storage of mycelia and infection propagules. Most entomopathogenic fungi are easily propagated on defined or semi-defined media containing suitable nitrogen and carbon sources. Submerged culture is used for production of mycelia and blastospores. Continuous serial subculturing on artificial media is to be avoided as this can result in a change in the phenotype. In many cases virulence can be restored after a single or serial passages through the insect host (Goettel & Inglis, 1997). Surface culture is used for the routine maintenance of isolates and for production of conidia. At the laboratory level, surface culture is usually accomplished on agar media within glass bottles or disposable plastic Petri plates. For larger scale production of conidia, cheaper nutritive substrates such as rice, bran or cereal grains are used. Use of inert substrates as a base for the aerial growth and sporulation of a fungus enables the complete separation of the pathogen from the nutritive medium.

Goettel, (1984) used cellophane as a semi-permeable membrane, which allows harvest of the fungus virtually free from nutritive substrates contamination. Inert substrates can also be used as a base for the aerial growth and sporulation of entomopathogenic hyphomycetes. For instance, Bailey & Rath, (1994) used a nutrient-impregnated membrane for production of *M. anisopliae* conidia. Aerial conidia can also be produced on the surface of still liquid medium. Kybal & Vlcek, (1976) used a polyethylene hose as an inert medium. Smasinakova *et al.* (1981) obtained $10^8$ conidia of *B. bassiana* $cm^2$ liquid surface (Peptone (0.8%) and sorbitol (1%)). Bidockha *et al.* (1987) obtained highest yields of blastospores of *B. bassiana* in a peptone (1%)-glucose (2%) medium, whereas more mycelia mass was produced in a glucose (2%)- peptone (1%)-yeast extract (0.2%) medium. Under proper conditions, certain species and/or strains produce conidia in submerged culture. Van Winkelhof & McCoy, (1984) found corn steep liquor (1%) and tween 80 (0.2%) capable of producing submerged conidia in 14 isolates of *Hirsutella thompsonii*. Similarly, submerged conidia of a *B. bassiana* isolate were produced in a sucrose (2%)-yeast extract (0.5%)-basal salts medium ($1.7 \times 10^8$ conidia/ml), whereas mostly blastospores were produced in a sucrose (2.5%)-yeast extract (2.5%) medium ($7.4 \times 10^8$ blastospores/ml) (Rombach *et al.*, 1986).

It is usually necessary to quantify propagules in both laboratory and field conditions. Propagules are usually formulated in a water carrier but any number of carriers may be used (solid substrates). Mechanical suspension of propagules using micro pestles provides excellent suspension of *Beauveria bassiana* and *M. flavoviride* conidia in water without causing damage to cells. Some reports suggest that oil is a more efficacious carrier than water for propagules (Bateman *et al.*, 1993; Inglis *et al.*, 1996) and conidia with hydrophobic cell walls are readily suspended in oil using micro pestles. A haemocytometer is commonly used to quantify number of propagules per unit volume or weight. In contrast to haemocytometer counts and the turbidometric methods, plating techniques provide a measure of viable propagules per unit volume (Goettel & Inglis, 1997). Spore viability is most often determined by germinating propagules on a translucent agar medium while direct enumeration techniques may increase the sensitivity of enumeration and provide information on the spatial distribution of propagules these methods are considerably more labor intensive than indirect methods. Furthermore direct enumeration cannot be used in situations where physical and microbial contaminants obscure the propagules being enumerated. For indirect enumeration the propagules must first be recovered.

Recovery methods differ according to the substrate on which the propagules are present. Propagules must be either washed from the surface of the insect or obtained through homogenization of the host. The homogenization method may be more efficacious than the wash method for recovery of propagules that are strongly attached to the cuticle or that are located in relatively inaccessible locations like cuticular folds. Disadvantages to the homogenization method include the inability to distinguish between propagules on the integument surface with those in the hemolymph and/or alimentary canal following the homogenization of whole insects. Propagules can be recovered from number substrates that possess a defined or measurable area. Two examples of substrates used to assess propagules densities include glass cover slips.

The most popular methods for quantifying inoculum of entomopathogenic fungi or/in insects or soil is based on dilution plating.

A number of formulations have been shown to influence the persistence of propagules in soil. The application of conidia in oil or oil emulsion sprays onto the surfaces of soil have generally not substantially enhanced their persistence relative to aqueous preparation of entomopathogenic fungal conidia (Inglis *et al.,* 1997). In contrast, other formulations have been shown to increase the survival of propagules in soil for eg: coating *B. bassiana* conidia in clay increased their persistence in various soils and with various water activities under controlled environment conditions (Studdert *et al.,* 1990). Considerable interest has focused on use of granular formulations of entomopathogenic hyphomycetes. A variety of different materials have been used to prepare granules (alginate for e.g.) and conidia or mycelia are incorporated into the granule matrix. The ability of the fungus to grow and sporulate in and/or on the granule is usually desirable and a number of materials that facilitate growth and sporulation have been tested. Increased persistence and efficacy of fungi applied in granules in soils have been observed (Kruger *et al.,* 1992) and propagules applied in granule formulations have provided satisfactory field activity against a number of economically important soil pests.

The greatest potential for managing pests with entomopathogenic hyphomycetes is in green-house, where environment can be manipulated in favour of pathogen. Among the several taxa, which have been, demonstrated excellent suppression of insect pests in green house includes *B. bassiana*. *B. bassiana* with other fungi *P. fumosoroseus* (PRF-97 R) have been recently registered against an array of greenhouse pests including aphids, thrips, whiteflies & spider mites (Shah & Goettel, 1999)

The Colorado potato beetle is an economically important pests of potatoes and other solanaceous crops through out the world. The development of resistance to insecticides has prompted considerable research into alternative methods including the application of *B. bassiana* for managing this pest. Some studies have shown no or limited efficacy, while others have observed significant suppression of Colorado potato beetle (Drummond & Groden, 1996; Lacey *et al.,* 1999). Several insecticides have been tested in combination with *B. bassiana* against Colorado potato beetle in North America and some results have demonstrated an additive effect (Anderson & Gibson, 1998). A number of studies have observed that foliar application of *B. bassiana* to potatoes in North America provided satisfactory suppression of Colorado potato beetle populations. Poprawski *et al.*, (1997) observed that application of *B. bassiana* conidia at 3-4 days intervals early in the season effectively reduced densities of old larvae and provided substantial foliar protection. Significant effects of *B. bassiana* against Colorado potato beetle were observed when five weekly applications of conidia were made (Lacey *et al.,* 1999). In addition to foliar application of *B. bassiana* against larvae adults, the amendment of soils with conidia against over wintering adults has also been tested. Over wintering adults are often killed by indigenous *B. bassiana* in soil (Weber & Ferro, 1993) and application of fungus onto foliage or directly onto the soil surface have significantly reduced populations of emerging adults (Watt & LeBrun, 1984; Cantwell *et al.,* 1986; Anderson *et al*., 1989).

*B. bassiana* applied to maize foliage can effectively control European corn borer (Lewis *et al.,* 1998). As *B. bassiana* forms an endophytic relationship with maize exciting possibilities for the management of insect pest in maize have been realized.

Initial studies demonstrated that maize plants treated at a whorl stage of development with liquid or granules formulations of *B. bassiana* conidia are internally colonized by the fungus (Lewis & Papavizas, 1985). Subsequent studies confirmed that *B. bassiana* entered maize tissues and once in the plant, the fungus remained there through out the growing season and provided season long suppression of ECB larvae (Bing & Lewis, 1991, 1992). The timing of *B. bassiana* application was shown to influence the efficacy of endophytic *B. bassiana*. The adhesion subsequent growth and penetration of maize leaves are very similar morphologically to the germination and penetration of insect cuticles by *B. bassiana* (Wagner & Lewis, 2000). Lewis *et al.* (2000) observed no difference in rates of endophytism by *B. bassiana* between *B. thurigeiensis*-transgenic and near isogenic lines of maize.

Application of *B. bassiana* as a biocontrol agent requires an easily deliverable formulation, which can maintain the viability and pathogenicity of spores on storage. The study aims at identification of suitable delivery system for the most virulent isolate of *B. bassiana.*

There are specific requirements for successful, commercial production and use of fungal insect control agents (Roberts & Hajek, 1992). Firstly, a fungal isolate for mass production must be selected with rapid growth, abundant sporulation and sufficiently high pathogenicity to the target pests, secondly production costs must be minimal that is simple in composition, cheap in price and available in large quantities and a production procedure that is easy to run and minimizes labour. Thirdly, microbial products must be formulated to control different target pests with distinct biological aspects. Finally, formulated products must be suitable for long-term storage under natural conditions without significantly losing their viability and infectivity.

Shelf life is considered a pivotal factor that determines the commercial success of a biocontrol agent as well as its field efficacy (Feng *et al.,* 1994). An 18-month shelf life is recommended for the agriculture market (Couch & Ignoffo, 1981). *B. bassiana* is easily cultured on solid and liquid media. Rombach *et al*., (1988) produced conidia of *B. bassiana* on wheat bran by a small mass production unit. Commercial mycoinsecticide formulation must be formulated with two objectives in mind.

1. Ease of field application to target insects within their habitats.
2. Enhancement of shelf life and environmental persistence after application. A formulation attempting to provide the correct combination of the active and final components in crucial for agricultural application.

The proper biological aspects of fungal agents required an understanding of the life cycle and particularly the biological aspects of not only the fungus but also the target insects are concerned. The physical and chemical characteristics of the infective fungal propagules to be formulated affect the final qualities and properties of resultant formulation.

Formulation ingredients selected should enhance fungal viability, virulence, disease transmission and field persistence. The compatibility of ingredients used for formulations means that the infective stages of the fungus formulated have longer shelf life, higher viability and greater virulence than those unformulated or formulated otherwise. All these developmental stages of *B. bassiana* ie., conidia, blastospores and mycelia have been successfully formulated for small-scale field trials or large-scale application (Feng *et al.,* 1994).

Conidia of *B. bassiana* are smooth walled, globose and only 2-3um in diameter. They can be suspended in aqueous liquid or mixed with a powder carrier and sprayed as mist or dust with conventional equipment used for application of synthetic chemical insecticides. Dry formulations are the forms, which so far have been considered for *B. bassiana* conidia although oil-based formulations are being tried in experimental trials.

Biological control of insect pests with microbial agents is an appealing prospect since it would reduce inputs of chemical pesticides into the environment and can be more cost-effective hence, can be made available even to the poorest of poor farmers. However, development of improved methods to formulate and apply such agents is a prerequisite to their economical use (Papavizas *et al.*, 1984). The effectiveness of biocontrol fungi partially depends on their ability to take advantage of 'a short window of opportunity' where in environmental conditions is conducive to their growth or sporulation. Fluctuations in environmental factors may adversely affect the persistence of the biocontrol agent thus making it an unsuitable one to be delivered to the field and lack of adequate formulation of most microorganisms for insect control is a serious draw back in the field of microbial insecticides. Thus, more suitable prolonged viability and/or sporulation of fungi from a biocontrol formulation significantly enhance their efficacy in the field.

## References

Alves, S.B., Almeida, J.E.M., Moino, A., Stimac, J.L. & Pereira, R.M. (1995) *Ecossistema.* **20:** 50-57.

Anderson, D.L. & Gibson, N.L. (1998) *Austral. System. Bot.* **11:** 53-72.

Anderson, T.E., Hajek, A.E., Roberts, D.W., Preisler, H.K. & Robertson, J.L. (1989) *J. of Econ. Entomol.* **82:** 83-89.

Anon. (1981) Hawaii Pest Report. **1(5)** November:1-10.

Bateman, R. P., Carey, M, Moore, D. & Prior, C. (1993) *Annal.* in *Appl. Biol.* **122:** 145-152.

Belova, R.N. (1978) In: Proceedings of the First Joint US/USSR Conference on the Production, Selection & St&ardization of Entomopathogenic, Fungi of the US/USSR Joint Working Group on Production of Substances by Microbiological Means (Ed. Ignoffo, C.M.) National Science Foundation (USA), pp 102-119.

Berretta, F. M., Lecuona, E. R., Zandomeni, O. R. & Grau, O. (1998) *J. of Invert. Pathol.* **71:**145-150.

Bidochka, M. J., Pfeifer, T. A. & Khachatourians, G. G. (1987) *Mycopathol.* **99:** 77-83.

Bidochka, M.J., McDonald, M.A., St.Leger, R.J., & Roberts D.W. (1994) *Curr. Genet.* **25:**107-113.

Bing, L. A. & Lewis, L. C. (1991) *Environ. Entomol.* **20:** 1207-1211

Bing, L.A. & Lewis, L.C. (1992) *Entomophaga*. **37:** 525-536.

Bridge, P. D., Hawksworth, D. L., Kozakiewicz, Z., Onions, A. H. S., Paterson, R. R. M., Sackin, M. J. & Sneath, P. H. A. (1989) *J. of Gen. Microbiol.* **153:** 2941-2966.

Bromenshenk, J.J., Jaronski, S.T. & Loeser, M.R. (1996) Multiple endpoint, holistic assessment of the effects of Mycotrol WP (*Beauveria bassiana* Strain GHA) on outdoor Apis mellifera L. colonies, 69 pp. Tech. Rep. Mycotech 95-105.

Burkholder, P. R. & Sinnott, E. E. (1945) *Ameri. J. of Bot.* **32:** 424-431.

Bustillo, A. E., Bernal, M. G., Benavides, P. & Chaves. (1999) *Flor. Entomol.* **82:** 491-498.

Cantwell, G.E., Cantelo, W.W. & Schroder, R.F.W. (1986) *Great Lake. Entomol.* **19:** 81-84.

Castrillo, J. A. & Brooks, W. M. (1998) *J. of Invert. Pathol.* **72(3):** 190-196.

Chase, A. R., Osborne, L. S. & Ferguson, V. M. (1986) *Flor. Eentomol.* **69:** 285-292.

Chen, C. J., Wu, J. W., Li, Z. Z., Wang, Z. X., Li, Y. W., Chang, S. H., Yin, F. M., Wang, X. P., Dai, L. Y., Tao, L., Zhang, Y. A., Tang, J., Ding, S., Ding, G. G., Gao, Z. H. & Tan, Y. C. (1990) Application of microbial pesticides in IPM, in Integrated Management of Pine Caterpillars in China (Ed. Chen C. J.) China Forestry publishing House, Beijing, pp 214-308 (in Chinese).

Cook, R. J., Bruckart, W. L., Coulson, J. R., Goettle, M. S., Humber, R. A., Lumsden, R. D., Maddox, J. V., McManus, M. L., Moore, L., Mever, S. F., Quimby, P. C., Stack, J. P. & Vaughn, J. L. (1996) *Biol. Cont.* **7:** 333-351.

Couch, T. L. & Ignoffo, C. M. (1981) In: Microbial control of pests & plants diseases. 1970-1980 (Ed, Burges, H. D.) Academic Press, London, pp 621-634.

Cowan, S.T. (1974) Cowan & Steel's Manual for the Identification of Medical Bacteria, 2nd edn. Cambridge University Press. Cambridge.

Damon, A. (2000) *Bulletin of Entomol. Resea.* **90:** 453-465.

Daoust, R. A. & Roberts, D. W. (1983) *J. of Invert. Pathol.* **41:** 161-170.

De La Rosa, W., R. Alatorre, J. Trujillo, & J. F. Barrera. (1997) *J. Econ. Entomol.* **90:** 1534-1538.

Deshp, M.V. (1999) *Critical Reviews in Microbiology* **25:** 229-243.

Domsch, K. H., Gams, W. & Erson, T.H. (1980) Compendium of soil fungi. 1980. Academic press. London & New York.

Drummond, F.C. & Groden, E. (1996) In: The Ecology, Economics, & Management of Potato Cropping Systems: A report of the First Four Year of the Maine Potato Ecosystem Project. Maine Agricultural & Forest Experiment Station Bulletin. 843: Orono, Maine. pp 80-118.

Fargues, J. & Luz, C. (2000) *Biocon. Sci. & Technol.* **8:** 323-334.

Fargues, J., Goettel, M.S., Smits, N., Ouedraogo, A. & Rougier, M. (1997) *Mycologia*. **89(3):** 383-392.

Feng, M. G., Poprawski, T. J. & Kachatourians, G. G. (1994) *Biocont. Sci. & Technol.* **4:** 3-34.

Ferron, P. (1981) In: Microbial control of Pests & Diseases 1970-1980. (Ed. Burges, H.D.). Academic Press, London.

Flexner, J.L., Lighthart, B. & Croft, B.A. (1986) *Agricul., Ecosy. & Environ.* **16:** 203-254.

Glare, T. R., & Milner, R. J. (1991) In: Hand book of Applied Mycology (Eds. Arora, D. K., Ajello, L., & Mukerji, L. G.) Vol 2: Humans, Animals & Insects. Marcel Dekker, New York, pp. 547-612.

Goettel, M. S. & Hajek, A. E. (2001) Evalulation of non-target effects of pathogens used for management of Arthropods. Evaluating indirect ecological effects of Biological control (Eds. Wajnberg, E., Scott, J. K., & Quimby, P. C.). CAB International.

Goettel, M.S. & Inglis, G.D. (1997) In: (Ed: Lacey, L.A.) Manual of Techniques in Insect Pathology. Academic Press. London. pp. 213-24.

Goettel, M. S., Poprawaski, T. J., Vandenberg, J. D., Li, Z. & Roberts, W. D. (1990) In: (Eds Laird, M., Lacey, L. A., & Davidson, E. W.) Safety of Microbial insecticides. CRC press, Boca Raton, Florida, pp. 209-232.

Goettel, M. S. (1984) *J. of Mic. bio. Method.* **3:** 15-20.

Groden, E. & Dunn, T. (1996) In: Microbial Control of Soil Dwelling pests. (Eds. T.A. Jackson, & Glare, T.R.). The Microbial Control Group, Ag. Research, Lincoln, New Zealand. pp 137-145.

Groden, E. & Lockwood, J.L. (1991) *J. of Inverte. Pathol.* **57:** 7-16.

Grodona, I., Hermosa, R., Tejada, M., Gomis, M.D., Moteos, P.F., Bridge, P.D., Monte, E. & Garcia-Acha, I. (1997) *Appl. & Environ. Microbiol.* **63:** 3189-3198.

Gutierrz-Rodero, F., Moragon, M., Ortiz de la Tabla, V., Mayol, M. J. & Martin, C. (1999) *European Journal of Clini. Microbiol. & Infect. diseas.* **18:** 814-818.

Hajek, A. E., Butler, L. & Wheeler, M. (1995) *Biol. contr.* **5:** 530-544.

Hajek, A.E. & Goettel, M.S. (2000) In: Field Manual of Techniques in Invertebrate Pathology. (Eds. Lacey L.A., & Kaya, H.K.), Kluwer Academic Publishers, Netherlands. pp. 847-868.

Hegedus, D. D. & Khachatourians, G. G. (1996) *J. of Invert. Pathol.* **67(3)**: 289-299.

Hegedus, D. D., Bidochka, M. J., Miranpuri, G. A. & Khachatourians, G. G. (1992) *Appl. Microbiol. Biotechnol.* **36**: 785-789.

Hegedus, D.D., Bidochka, M.J. & Khachatourians, G.G. (1990) *Appl. Microbiol. Biotechnol.* **33**: 641-647.

Humphreys, A. M., Matewell, P., Trinci, A. P. J. & Gillespie, A. T. (1989) *Mycological Res.* **92(3)**: 257-264

Husey, N.W. & Tinsley, T.W. (1980) In: *Microbial Control of Pests & Plant Diseases. 1970-1980.* (Ed. Burges, H.D.) Academic press, London, pp. 785-795.

Ibrahim, Y.B. & Low, W. (1993) *Interna. Jour. of Pest Manage.* **39:**288-292.

Inglis, D. G., Johnson, L. D. & Goettel, S. M. (1996) *Biocont. Scie. & Tech.* **6:** 35-50.

Inglis, G.D., Duke, G.M., Kanagaratnam, P., Johnson, D.L. & Goettel, M.S. (1997) In: Microbial control of Grasshoppers & Locusts. (Eds. Goettel, M.S. & Johnson. D.L.) pp. 253-263.

Inglis, G.D., Goettel, M.S. Butt, T.M. & Hermann Strasser. (2001) In: Fungi as Biocontrol Agents. (Eds. Butt, T.M., Jackson C. & Magan, N.). CAB International.

Inyang, E., Butt, T.M., McCartney, A., Oyejola, B., Ibrahim, L., Pye, B.J. & Archer, S. (2000) *Mycol. Resea.* **104:** 653-6

Jackson, T.A., Alves, S.B. & Pereira, R.M. (2000) In: "Measures of success in Biological control" (Eds. Gurr, G.M. & Wratten, S.D.), pp. 297-330. Kluwer. Dordrecht.

Jaronski, S. T. & Goettel, M. S. (1997) *Mem. of Entomopathol. Soc. of Cana.* **171:** 225-237.

Kirby, W. & W. Spence. (1815) An Introduction to Entomology. Longman, Brown, Green & Longmans, London. 285 pp.

Keller, S. (1992) In: Biological Control of Locusts & Grasshoppers. (Eds. Lomer, C.J. & Prior, C.) CAB International, UK. pp 394.

Kleespies, R.G. & G. Zimmermann. (1992) *Biocon. Sci. & Technol.* **2**: 127-135.

Kruger, S.R., Villani, M.G., Martins, A.S. & Roberts, D.W. (1992) *J. of Invert. Pathol.* **59:** 54-60.

Kybal, J. & Vlcek, V. (1976) *Biotechnol. & Bioengin.* **18:** 1713-1718.

Lacey, L.A. Horton, D.R., Chauvin, R.L. & Stocker, J.M. (1999) *Entomol. Experim. et Applicata.* **93:** 189-200.

Lane, B. S., Trinci, A. P. & Gillespie, A. T. (1991) *Mycol. Res.* **95 (7):** 821-828.

Lastra, C. C., Hajek, A. E. & Humber, R. A. (2001) *Canad. J. of Bot.* **79:** 861-864.

Lecuona, R.E., Edlstein, J.D., Berretta, M.F., La Rossa, F.R. & Areas, J.A. (2001) *J. of Medi. Entomol.* **38(2):** 172-179.

Lewis, J. A. & Papavizas, G. C. (1985) *Plan. Pathol.* **34:** 571-577.

Lewis, E. E., Campbell, J. F. & Gaugler, R. (1998) In: Conservation Biological Control.(Ed. Barbosa, P.) Academic Press, San Diego, pp. 235-254.

Lewis, L. C., Bruck, D. J., Gunnarson, R. D. & Bidne, K. G. (2000) *Crop Sci.* (in press).

Li, Y.W. & Yang, J.H. (1988) In: Study & Application of Entomogenous Fungi in China. Vol. I. (Eds. Li, Y.W., Li, Z.Z., Liang, Z.Q., Wu, J.W., Wu, Z.K. & Xu, Q.F.). Academic Periodical Press, Beijing. pp 10-14.

Macleod, D.M. (1954) *Canad. J. of Bot.* **32:** 818-890.

Maddox, J. V., McManus, M. L., Jeffords, M. R. & Webb, R. E. (1992) In: Selection criteria & ecological consequences of importing natural enemies, (Eds Kauffman, W. C., & Nechols, J. E.). Thoman Say publications in entomology, Entomological Society of America, Lanham, Maryland, pp. 27-39.

Mannion, C.M., McLane, W., Klein, M.G., Moyseenko, J., Oliver, J.B. & Cowan, D. (2001) *Canad. Entomol.* **199:** 901-907.

McKoy, J. F. & Trinci, A. P. J. (1987) *Transac. of the Brit. Mycolog. Soc.* **88:** 299-307.

Moore, D. & Caudwell. (1997) In: Microbial Control of Grasshoppers & Locusts (Eds Goettel,. M.S. & Johnson. D.L.). Memoirs of the Entomological Society of Canada 171: 400 pp.

Morton, A. G. (1961) *Proceed. of the Royal Soc. B,* **153:** 548-569.

Mugnai, L., Bridge, P.D. & Evans, H.C. (1989) *Mycol. Res.* **92:** 199-209.

Ooijkaas, L.P, Weber, F.J., Reinetta M. Buitelaar, Tramper, J, & Rinzema, A., (2000) *Tren. in Biotech.* **18:** 356-360.

Papavizas, G. C., Dunn, M. T., Lewis, J. A., & Beagle-Ristaino, J. (1984) *Phytopathol.* **74:** 1171-1175.

Petch, T. (1926) *Trans Brit Mycol Soc* **10:**244-271.

Poprawski, T. J., Carruthers, R. I., Speese, J., Vacek, D. C. & Wendel, L. E. (1997) *Biol. cont.* **10:** 48-57.

Primak, T. A., Aleshina, O. A., Ilicheva, S. N., Kononova, Ye, V., Kozachenko, V. I. & Kroglova, N. P. (1976) Strain of Beauveria bassiana (Bals) Vuill 113H-278, USSR patent 507645.

Réaumur, M. de. (1726) Remarques sur la plante appellée a la Chine Hia Tsao Tom Tchom, ou plante ver. Mem. Acad. Roy. Sci. pp. 302-305.

Revankar, S.G., Sutton, D.A., Sanche, S.E., Rao, J., Zervos, M., Dashti, F. & Rinald, M.G. (1999) *J. of Clin. Microbiol.* **37:** 195-198.

Righelato, R. C., Trinci, A. P. J., Pirt, S. J. & Peat, A. (1968) *J. of Gen. Microbiol.* **50:** 399-412.

Roberts, D.W. & Hajek, A.E. (1992) In: "Frontiers in Industrial Mycology" (Ed. Leatham, G.F.), Chapman & Hall. New York, London. pp. 144-159.

Rombach, M. C. (1989) *Entomophaga* **43(1):** 45-52.

Rombach, M. C., Aguda, R. M., Shepard, B. M. & Roberts, D. W. (1986) *Environ. Entomol.* **15(5):** 1070-1073.

Rombach, M, C., Aguda, R.M. & Roberts, D.W. (1988) *Entomophaga* **33:** 315-324.

Rombach, M. C. (1989) *Entomophaga* **43(1):** 45-52.

Saik, J. E., Lacey, L. A. & Lacey, C. M. (1990) In: Safety of Microbial Insecticides. (Eds. Lairo, M. L., Lacey, L. A., & Davidson, E. W.). CRC press, Bocaraton, FL. pp. 115-132,

Samsinakova, A., Kalalova, S. Vleek, V. & Kybal, J. (1981) *J. of Invert. Pathol.* **38:** 169-174.

Shah, P. A., & Goettel, M. S. (1999) Directory of Microbial control products & services. Microbial Control division, society of Invertebrate Pathology, Gainesville, Florida.

Siegel, J. P. & Shadduck, J. A., (1990) The infectivity & toxicity of entomopathogens for mammals: Implications for human health (Eds. Laird, M., Lacey, L. A. & Davidson, E. W.), Safety of microbial insecticides. CRC press, Boca Raton, FL.

Soper, R. S. & Ward, M.G. (1981) In: Biological Control in Crop Protection (Ed. Papavizas, C. G.). Osmum Montclair, New Jersey.

Steinhaus, E.A. (1975) Disease in a Minor chord. pp. 488. Ohio State University Press, Columbus.

St. Leger, R. J., Cooper, R. M. & Charnley, A. K. (1986) *J. of Gen. Microbiol.* **312:** 1509.

Studdert, J.P., Kaya, H.K. & Duniway, J.M. (1990) *J. of Invert. Pathol.* **55:**417-427.

Thomas, K.C. Khachatourians,G.G. & Ingledew, W.M. (1987) *Canad. J. of Microbiol.* **33:** 12-20.

Trinci, A. P. J. (1972) *Transac. of Brit. Mycol. Soc.* **81:** 408-412.

Van Winkelhof, A.J. & Mc Coy, C.W. (1984) *J. of Invert. Pathol.* **43:** 59-68.

Viaud, M., Couteaudier, Y. & Riba, G. (1998) *Appl. Environ. Microbiol.* 88-93.

Vidal, C., Fargues, J. & Lacey, L.A. (1997) *J. of Invert. Pathol.* **70:** 18-26.

Wagner, B.L. & Lewis, L.C. (2000) *Environ. Microbiol.* **66:**3468-3473.

Watt, B.A. & LeBrun, R.A. (1984) *Environ. Microbiol.* **13:** 15-18.

Weber, D.C. & Ferro, D.N. (1993) *Entomol. Experimen. et Appli.* **66:**191-196.

Yin, F.M., Chen, Q.C., Chen, Y.G., Guo, G.L. & Li, Z.W. (1988) In: Study & Application of Entomogenous Fungi in China, Vol 1. (Eds. Y.W. Li, Z.Z. Li, Z.Q. Liang, J.W. Wu, Z.K. Wu, & Q.F. Xu). Academic Periodical Press Beijing. pp 105-110.

Zhou, W.F. (1988) Use of *Beauveria bassiana* against pine caterpillars *Dendorlimus superans* & *D. Tobulae formis* in study & application of entomogenous fungi in China. Vol. 1 (Eds. Li, Y.W., Li, Z.Z., Liang, Z.W., Wu, J.W., Wu, Z.K. & Zu, Q.F.) Academic Periodical Press, Beijing, pp. 83-84.

# 12

# Endophytic Fungi and Their Role in Biological Control and Plant Growth Promotion

**Jyoti Saxena, Anamika and Vinay Sharma**

*Department of Bioscience and Biotechnology, Banasthali Vidyapith-304022, Rajasthan, India.*

## ABSTRACT

The need for new and useful compounds to provide assistance and relief in all aspects of the human life is ever growing. There are enormous difficulties in raising enough food on certain areas of the earth to support local human populations. Environmental degradation, loss of biodiversity, and spoilage of land and water also add to the problems faced by mankind. In the continual search by both pharmaceutical and agricultural industries for new products, natural selection has been found superior to the combinatorial chemistry for discovering novel substances that have the potential to be developed into new industrial products. The plant is an extraordinarily common source of organic energy. It is thus likely that a huge array of microbes known as endophytes reside inside the plant tissues and interact with them. Of these, a range can be isolated from apparently healthy tissues, many of which have never been documented to be associated with disease. Others may cause disease when environmental conditions change. The intent of this review is to provide insights into the presence of endophytes in nature and their diversity, their interaction with the host plant, role in enhancing nitrogen availability, plant growth promotion, imparting resistance to host plants against insect pests, herbivores, pathogens and various abiotic stresses, and their potential in agricultural industry.

## INTRODUCTION

Endophytic microorganisms are found in virtually every plant on earth. These organisms are involved in a variety of relationships, ranging from symbiotic to slightly pathogenic. Their contribution to the host plant is that they produce a plethora of substances of potential use to modern medicine, agriculture, and industry. The potential prospects of finding new drugs that may be effective candidates for treating newly developing diseases in humans, plants, and animals are great (Strobel and Daisy, 2003). Novel antibiotics, antimycotics, immunosuppressants, and anticancer compounds are only a few examples of what has been found after the isolation, culture, purification, and characterization of some selected endophytes in the recent past. Endophytic fungi have capacity to protect their plant hosts against insect pests, pathogens and domestic herbivores, drought and flooding stress, and enhance competitive abilities of the host by producing several secondary metabolites and enzymes. Many of them are also associated in the initial stages of resource cycling.

In addition to the economical aspects, the study of endophytic microorganisms has strong academic interests concerning the discovery of new microbial species. The fact that some endophytes produce certain phytochemicals, originally characteristic of the host, might be related to a genetic recombination of the endophyte with the host that occurs in evolutionary time (Tan and Zou, 2001). This is a concept that was originally proposed as a mechanism to explain why the endophytic fungus *Taxomyces andreanae* may be producing paclitaxel (Stierle *et al.*, 1993). Thus, if endophytes can produce the same rare and important bioactive compounds as their host plants, this would not only reduce the need to harvest slow-growing and possibly rare plants but also preserve the world's ever-diminishing biodiversity. Furthermore, it is recognized that a microbial source of a valued product may be easier and more economical to produce, effectively reducing its market price.

In the past two decades, a great deal of information on the role of endophytic microorganisms in nature has been collected, but only a few plants have ever been completely studied in relation to their endophytic biology. The capability of colonizing internal host tissues has made endophytes valuable for agriculture as a tool to improve crop performance. Consequently, enormous efforts are being made to find new and interesting endophytic microorganisms among myriads of plants in different settings and ecosystems. The intent of this review is to provide insights into the presence of endophytes in nature and their diversity, their interaction with the host plant, role in enhancing nitrogen availability, plant growth promotion, imparting resistance to host plants against insect pests, herbivores, pathogens and various abiotic stresses, and their potential in agricultural industry.

## DEFINITION

Endophytic fungi were first noticed in 1940s, but only at the turn of the 21$^{st}$ century the ubiquity of these fungi was fully recognised. The meaning of "endophyte" has undergone various transformations in the last decade, and there is still considerable disagreement as what constitutes an "endophyte". Broadly, the term is referred to fungi that live asymptomatically within plant parts (intercellularly or intracellularly) for at least part of its life cycle (Saikkonen *et al.*, 1998). Thus, a wide range of fungi, from plant pathogens and saprophytes that have extended latency periods before disease to the specialized fungi that are considered obligate mutualists, have been included in this group.

Endophytic microorganisms are those that inhabit the interior of plants, especially leaves, branches and stems, showing apparently no harm to the host (Azevedo, 1998). They usually occur in above-ground plant tissues but are also found in roots. Unlike mycorrhizal fungi, fungal endophytes of roots lack extraradical (outside the root) hyphal networks and mantles (sheaths around the roots). Likewise, nitrogen-fixing bacteria such as *Rhizobium*, which form external structures called nodules, are different from other root endophytic bacteria. Nonetheless, mycorrhizae and nitrogen-fixing bacteria also live in an intimate relationship with their hosts and could be in broader sense considered as endophytic microorganisms. The distinction among endophytes, epiphytes (those that live on the surface of plants) and phytopathogens (those that cause diseases to plants) is of pure didactic meanings. There is a gradient separating them, and thus, there comes a situation where it becomes very difficult to draw limits to discriminate each category. While much information is available on pathogens and mycorrhizal fungi, comparatively little is known about symptomless (possibly neutral or mutualistic) endophytes of plants.

## PLANT ENDOPHYTE INTERACTION

Endophytic fungi have highly complex associations with their plant hosts, resulting in physiological responses of applied value in terms of plant vigour, water ratios, nutrient acquisition and use, and protection from pathogens and insects. Endophytes appear to have direct and induced effects on plant responses to biotic agents. The interaction with abiotic agents remains largely unexplored.

Endophytic fungi live like fungi imperfecti, much of the time producing mostly conidial spores or simply cloning themselves. They are between benign parasites and true symbionts in their behaviour, although many endophytic fungi become rotting agents upon the death of their host plant. Sometimes they may produce toxins in response to pests, which feed on their host's tissues. These toxins benefit the host plant in several ways. For example, there are endophytes which make grass poisonous to grazing animals, whereas other endophytes in oak leaves ward off gall midges. In the sense that these 'endophytes' can protect their host, they are symbionts. Endophytes are low-key examples of symbiosis (Schardl and Phillips, 1997; Meijer and Leuchtmann, 1999).

Endophytes belong to several fungal taxonomic groups, but most of them are members of ascomycetes. Many of the needle cast fungi (e.g., *Phomopsis* sp. and *Lophoderma* sp.) have endophytic relatives. Even some of the leaf spot diseases (e.g., *Rhytisma* sp.) have such relations. Most of the endophytes, which infect pines, are members of the Rhytismataceae family. Rhytismataceous parasites have made mycologists wonder if endophytes are not just the old familiar parasites in dormancy mode. However, the genetic studies show that these endophytes are not the same as their pathogenic cousins. They are members of the same genera but not identical species (Ganley *et al.,* 2004). Thus, endophyticity seems to be a niche different from parasitism. Possibly, this mode of life developed from true parasitism.

In theory, parasites can evolve to become even more benign to their host. If a parasite does not harm its host, it has a more secure food supply than if it weakens the host. In the long run, both host and parasite may co-evolve into an asymptomatic relationship. Probably, a fungus would evolve quicker than a long-lived tree. Severe plant diseases may often develop due to parasites in novel hosts due to lack of resistance to the 'new' parasite in them. Although highly virulent strains of a parasite would have an initial advantage in its new frontier, but the advantage would become a detriment after the contagion's food supply is consumed. In such cases the natural selection would favour the less virulent strains of the parasite. The co-evolution seems to have happened in many cases. Dutch elm disease caused by *Ceratocystis ulmi* was relatively benign in Asian elms. But when it reached Europe and USA, it encountered hosts which were not equipped to deal with the fungus and hence caused the epidemic. Similarly, *Phytophthora infestans* causing potato blight disease is relatively harmless to the potatoes of the Andes. When potatoes were cultivated in Europe, they lost their tolerance to the fungus and as soon as the fungus was introduced in Ireland in the 1840s, the results were devastating. There are many other examples of diseases which only become problem for host species that have not co-evolved mutual tolerance.

The interaction between endophytic bacteria and fungi has effect on their secretions of plant growth promoters and acidity. Many studies have clearly demonstrated that bacterial-fungal interactions give rise to enhanced effects of biodegradation/biosynthesis compared to their monocultures (Bandara *et al.,* 2006). This indicates the importance of this interaction in efficient microbial action. Such biofilms may have their potential applications as biofilmed inocula in bio-fertilizers and bio-controllers in agriculture. A rather typical example of an endophytic symbiont is a fungus *Gloeotinia temulenta,* which grows in the seeds of darnel grasses. This fungus is apparently a symbiont and appears to be a helpful endophyte on wild species, *Lolium temulentum.* Grazing animals learn to recognise the darnel, probably by taste, and avoid eating it. In the early days wheat and other grains were often contaminated by the seeds of wild grasses. Now it is known that darnel's toxicity actually hails from the fungus, not the grass itself. The darnel fungus can be a pathogen if it occurs in a different host species.

Another interesting example of a symbiotic endophytic fungus is *Taxomyces andreanna.* The fungus lives in several species of yew (*Taxus*). It produces a chemical called *paclitaxel* in the inner bark of the yew trees. This compound has been found to slow down the replication of cells. Probably, its symbiotic role is to inhibit competing parasites in its host. *Paclitaxel* is also useful to humans as a chemotherapy agent. It can be used, with a regimen of other drugs, for fighting certain kinds of tumours. Originally, the drug was obtained by harvesting wild yews but this exploitation is no longer necessary. Similarly, certain species of the *Epichloe* can harm plants, while other species in the genus are asymptomatic. However, the mutualistic versions of these *Epichloe* species do not harm their hosts, at least noticeably, but they do make the host plant distasteful to herbivores. This protective role could be considered a mild form of symbiosis.

The interrelatedness of endophytes with their respective hosts is a vastly underinvestigated but exciting field. Thus, more background information on a given plant species and its microorganismal biology would be exceedingly helpful in directing the search for bioactive products. Currently, no one is quite certain of the role of endophytes in nature and their relationship to various host plant species. While some endophytic fungi appear to be ubiquitous (e.g., *Fusarium* spp., *Pestalotiopsis* spp. and *Xylaria* spp.), one cannot definitively state that endophytes are truly host specific or even systemic within plants any more than one can assume that their associations with plants are chance encounters. Frequently, many biotypes of the same species (of endophytic fungi) are isolated from the same plant and only one of the endophytes will produce a highly biologically active compound in culture (Li *et al*., 1996). A great deal of uncertainty also exists between what an endophyte produces in culture and what it may produce in nature. It does seem apparent that the production of certain bioactive compounds by the endophyte *in situ* may facilitate the domination of its biological niche within the plant or even provide protection to the plant from harmful invading pathogens. This may be especially true if the bioactive product of the endophyte is unique to it and is not produced by the host. Seemingly, this would more easily facilitate the study of the role of the endophyte in the plant. Furthermore, little information exists relative to the biochemistry and physiology of the interactions of the endophyte with its host plant. It would seem that many factors changing in the host as related to the season and age, environment, and location may influence the biology of the endophyte. Indeed, further research at the molecular level must be conducted in this field to study endophyte interactions and ecology. These interactions are probably all chemically mediated for some purpose in nature. An ecological awareness of the role these organisms play in nature will provide the best clues for targeting particular types of endophytic bioactivity with the greatest potential for bioprospecting.

The reaction of the plant to endophytes suggests that the interaction is one of confinement by the plant. Colonisation by endophytes ranges from single cells (*Rhabdocline parkeri*) to patchy distribution through leaves and stems (*Chaetomium globosum*). The presence of specific endophytes, the lack of signs of disease and the long-lived association indicate that further study of the association may yield interesting information of the precise benefits to plants associated with these colonists. As endophytes can affect virtually every type of plant-plant, plant-pathogen, and plant-herbivore interactions, any human activities (agriculture, deforestation, pollution, etc.), which alter diversity of endophyte-plant interactions, may have unpredictable, indirect effects on population change and community structure of plants, pathogens and herbivores in terrestrial ecosystems. Nutrients are cycled between the host and fungus. The fungus clearly gains a predictable environment in which nutrients are readily available. Thus, the benefits are clear for fungi establishing endophytic associations. The loss of plant resources to the fungus, and the potential of some fungi to grow rapidly, indicates that the host regulates development of colonies.

## ECOLOGY AND CLASSIFICATION

### Occurrence, Distribution and Diversity

Of the uncountable ecosystems on earth, those having the greatest biodiversity seem to be the ones also having endophytes with the greatest number and the most biodiverse microorganisms. Tropical and temperate rainforests are the most biologically diverse terrestrial ecosystems on earth. The most threatened of these spots cover only 1.44% of the land's surface, yet they harbour more than 60% of the world's terrestrial biodiversity (Mittermeier *et al.*, 1999). As such, one would expect that areas of high plant endemicity also possess specific endophytes that may have evolved with the endemic plant species.

Ultimately, biological diversity implies chemical diversity because of the constant chemical innovation that exists in ecosystems where the evolutionary race to survive is the most active one. Tropical rainforests are a remarkable example of this type of environment in which competition is great, resources are limited, and selection pressure is at its peak. This gives rise to a high probability that rainforests are a source of novel molecular structures and biologically active compounds (Redell and Gordon, 2000). Bills *et al.*, (2002) described a metabolic distinction between tropical and temperate endophytes through statistical data, which compared the number of bioactive natural products isolated from endophytes of tropical regions to the number of those isolated from endophytes of temperate origin. They found that tropical endophytes produced more active natural products than the temperate ones. This observation suggests the importance of the host plant in influencing the general metabolism of endophytic microbes.

A study was conducted by Blodgett *et al.*, (2000) to identify and quantify fungi in asymptomatic, 5-month-old *Amaranthus hybridus* plants grown in cultivated plots and a greenhouse, and to determine the species composition in different plant parts. The most common species isolated from leaves and petioles was *Alternaria tenuissima*. An enormous number of different fungi can be isolated from plants growing in their native habitat. Most of the fungi are uncommon and are narrowly distributed, taxonomically and geographically. However, a few fungi are widely distributed with the host, suggesting a long standing, close and mutually beneficial interaction. Indeed, some fungi are found in many different terrestrial hosts, especially as endophytes of crop plants. While most information has been gathered from terrestrial ecosystems, comparatively little is known about the fungi found in aquatic ecosystems.

## Propagation and Dispersal

Dispersal of the endophytic fungi remains puzzling. Apart from *Neothyphodium* and related species, endophytes are transmitted horizontally, i.e., each plant is colonised by fungal propagules that arrive from the environment. The source of transmission has been determined in only a few cases. Propagules of some endophytes have been found in the body of insect pests of the host. Intriguingly, at least two entomopathogens have been documented as endophytic fungi. Thus, insects may disperse the fungi from host to host. Aerial dispersal either thorough wind or vectors is probably the most common mechanism for fungal dispersal. Endophytic fungi colonise various parts of the plant. Many of the fungi sporulate in culture indicating the potential to release spores in the air. Indeed, sporulation is seen after senescence of plant tissues. However, few cases of dispersal have been documented in the wild hosts and the various mechanisms remain unexplored.

## Classification

Endophytic fungi are often divided into two groups: the "clavicipitaceous system" (family Clavicipitaceae of class Ascomycetes) which contains endophytes of grasses that often confer resistance to herbivores and pathogens, and endophytes representing all other fungal groups on diverse herbaceous and woody hosts. Fungal endophytes are found in all woody plants studied to date. Trees and shrubs usually contain numerous fungal species, however, there has been little attempt to correlate ecological and evolutionary aspects of fungal endophytes of grasses versus woody plants (Saikkonen *et al.*, 1998).

***Tree endophytes:*** Most of the fungi belong to the genera *Fusicladium, Melanconium* and *Gnomonia*. These fungi were found in majority of studied birch leaves, but the extent of fungal infection and the composition of fungi appear to vary among host trees, birch species and study sites. The fungal performance can also be host genotype specific. The results of molecular biology study clearly indicate that genetic diversity of host plant population at concert with environmental factors may affect patterns of genetic variation of endophytic fungus, and thereby interactively directing speciation of the fungus. This may have unpredictable cumulative effects also on other trophic interactions particularly in managed systems.

***Grass endophytes***: Although endophyte infections are widespread in native grass species, overall occurrence and frequencies of vertically transmitted endophytes are lower. However, these endophytes may show selective advantage of infected grasses to their conspecifics in some habitats. If so, competitive advantages for endophyte infected and uninfected plants also appear to vary among grass species. Alternatively, occasional horizontal transmission of fungi by spores can also cause high infection frequencies in some grass populations. However, it is essential to know, how frequent horizontal transmission is, and whether sporulation is induced by some environmental factors characteristic of the high infection habitats. In grass cultivars, endophyte infected seeds were detected only in *Festuca pratensis* and *Lolium perenne*, and endophyte frequencies were either very high or very low. In short, these results challenge conventional notions of generally high occurrence of seed-borne endophyte infections based for decades on agronomic grasses, and support the idea that plant-endophyte interactions in native grass populations are much more complex and variable than in the agronomic arena of more genetically homogenous grasses and more uniform abiotic conditions. Although plant-fungus interactions can be antagonistic or neutral in native communities, endophytes can also be beneficial for its host plants in certain environments, and thus, direct the distribution of grass species, frequencies of endophyte infected plants in populations or alter grassland communites.

## COLLECTION AND ISOLATION TECHNIQUE

After a plant is selected for study, it is identified and its location is plotted using a global positioning device. Small stem pieces are cut from the plant and placed in sealed plastic bags after excess moisture is removed. Every attempt is made to store the materials at 4°C until isolation procedures can begin (Shrestha *et al.*, 2001; Stinson *et al.*, 2003).

In the laboratory, plant materials are thoroughly surface treated with 70% ethanol and air dried under a laminar-flow hood. This is done in order to eliminate surface-contaminating microbes. Then, with a sterile knife blade, outer tissues are removed from the samples and the inner tissues are carefully excised and placed on water agar plates. After several days of incubation, hyphal tips of the fungi are removed and transferred to potato dextrose agar. Bacterial forms also emerge from the plant tissues, including, on rare occasions, certain *Streptomyces* spp. The endophytes are encouraged to sporulate on specific plant materials and are eventually identified via standard morphological and molecular biological techniques and methods. Eventually, when an endophyte is acquired in pure culture, it is tested for its ability to be grown in shake or still culture by the use of various media and growth conditions. It is also immediately placed in storage under various conditions, including 15% glycerol at –70°C. Ultimately, once appropriate growth conditions are found, the microbe is fermented and extracted and the bioactive compound(s) is isolated and characterized. Virtually all of the common and advanced procedures for product isolation and characterization are utilized in order to acquire the product(s) of interest. Central to the processes of isolation is the establishment of one or more bioassays that will guide the compound purification processes. One cannot put too much emphasis on this point since the ultimate success of any natural-product isolation activity is directly related to the development or selection of appropriate bioassay procedures. These can involve target organisms, enzymes, tissues, or model chemical systems that relate to the purpose for which the new compound is needed.

## BIOCONTROL

The natural and biological control of pests and diseases affecting cultivated plants has gained much attention in the past decades as a way of reducing the use of chemical products in agriculture. Several reviews indicate that the biological control has been frequently used in Brazil, South America and other countries (Lecuona, 1996; Melo and Azevedo, 1998). Although the use of agrochemicals decreases the attack of insects and phytopathogenic microorganisms, it still represents a high risk to field workers and consumers. In addition, their use in certain cases is economically unviable. The biocontrol of pests and diseases by using entomopathogenic microorganisms or those that inhibit/antagonise other microorganisms pathogenic to plants is an alternative that may contribute to reduce or eliminate the use of chemical products in agriculture.

### Insecticidal Activities of Endophytes

The role of endophytic fungi in the control of insects has been the subject of several reviews (Carroll, 1991; Breen, 1994; Saikkonen *et al.*, 1998). In the early 80s, the specialized literature published the first reports showing that endophytic microorganisms could play an important role inside plants. It was demonstrated that the presence of these microorganisms in their respective hosts could result in the reduction of insect attacks. Landmark reports on the subject started in 1981, just two decades ago. The nature of the protection and the variables involved in the process were also initially addressed.

Webber (1981) was probably the first researcher to report an example of plant protection given by an endophytic fungus, in which the endophyte *Phomopsis oblonga* protected elm trees against the beetle *Physocnemum brevilineum*. It was suggested that the endophytic fungus was responsible for reducing the spread of the Dutch elm disease by controlling its vector, the beetle *P. brevilineum*. The author associated the repellent effect observed towards the insect to toxic compounds produced by the fungi. This was confirmed four years later by Claydon *et al.*, (1985), who showed that endophytic fungi belonging to the *Xylariacea* family synthesize secondary metabolites in hosts of the genus *Fagus* and that these substances affect the beetle larvae.

Other earlier observations on the control of insects-pests by endophytic fungi are those of Funk *et al.*, (1983), showing protection of the perennial ryegrass *Lolium perenne* against the sod webworm. Gaynor and Hunt (1983) observed in several ryegrasses that high fungal infection is correlated with a decrease in the attack frequency of the Argentine steem weevil, *Listronotus bonariensis*. With the same insect-plant interaction, in New Zealand Barker *et al.*, (1984) observed that plants free of endophytic fungi of the genus *Acremonium*, were severely attacked. Lasota *et al.*, (1983) had also correlated the death rate of the Homoptera pest, *Adelges abietis* with the endophytic fungus *Cladosporium sphaerospermum* in *Picea glauca*. In *L. perenne* and *Cyperus* spp., weight gain and survival of the insect-pest, *Spodoptera frugiperda,* were affected by endophytic fungi like *Balansia cyperi* (Hardy *et al.*, 1985). Studies related to this insect-pest affecting different species of *Lolium* were also conduced by Latch *et al.*, (1985) with two endophytic fungi viz., *Acremonium lolii* and *Gliocadium* sp. The authors discussed the benefits and risks resulting from the increase in toxin production to insect control and resulting harm to domestic animals that would feed on these plants. They also verified similar effects of the same fungi over the grasshopper *Acheta domesticus*. Johnson *et al*., (1985) observed that insects fed on endophyte-free *Festuca* plants rather than on infected samples.

The basidiomycetous fungus, *Crinipellis perniciosa* is the causal agent of witches' broom disease of cacao (*Theobroma cacao*), which is the main factor limiting its production in the USA. As this fungus colonizes growing plant tissues, the fungicide application becomes ineffective, leaving only biological control a potential alternative strategy. In order to evaluate the potential of this phytopathogen as a biological control agent, the endophytic fungal community of resistant and susceptible cacao plants as well as affected branches was studied and identified by morphological traits and rDNA sequencing. Out of the many isolated endophytic fungi, only *Gliocladium catenulatum* reduced the incidence of witches' broom disease in cacao seedlings to 70%.

Mutualistic interactions between host and vertically inherited symbionts such as endophytes are easily reconciled with existing theory of species interactions (Arnold *et al.,* 2003). However, these interactions depend on a complex web relation within plant tissues, including different fungal and bacterial species, as well as plant genotype. By understanding which microbial species are involved, how and when they occur and what are the advantages of these plant interactions, it is possible to use this approach to control several plant diseases in cacao. Today, an improved protection program implies the availability of highly efficient biofungicides, so due to its remarkable biological properties, the *G. catenulatum* isolate offers very good prospects for integrated management of Witches' Broom Disease. Therefore, further studies based on this endophytic fungi should focus on the analysis and development of an efficient strategy to control *C. perniciosa* in plant.

In cucumber, nonpathogenic *F. oxysporum* may induce host resistance against *Pythium ultimum* through a combination of antibiosis and mycoparasitism, as well as inducing plant defence reactions (Benhamou, 2002). In addition to the earlier reports mentioned above that correlate endophytic fungi with toxic effects towards insects, other reports kept accumulating in the following years. In fact, Clay (1989) stressed that although many endophytic fungi have already been isolated from around 80 grass species of temperate countries, the most effective one is *Acremonium.* Ahmad *et al.* (1986) showed that the blue grass billbug *Sphenophorus parvulus* is less frequent on *Lolium perenne* plants infected with endophytic fungi. This study was verified by several researchers in many grasses and their insect pests ( Kindler *et al*., 1990; Mathias *et al*., 1990; Clement *et al*., 1992).

Endophytic fungi belonging to Clavicipitaceae synthesize alkaloids during plant infection, reducing the survival and the development of *S. frugiperda* in Graminae and Cyperaceae. In the flour beetle *Tribolium castaneum*, diets composed of grinded seeds of *F. arundinaceae* and *L. perenne* infected with endophytic fungi resulted in a reduction of infested seeds and insect growth, notably in *F. arundinacea* (Cheplick and Clay, 1988). The authors concluded that the insect resistance and consequential improved performance of plants in nature is a general characteristic of infected grasses. Alcoholic extracts from *Poa ampla*, uninfected and infected with *Neotyphodium typhnium* were assayed against mosquito larvae. Only the extracts from the infected plants were effective against the insect, whereas the extract obtained directly from the fungi was not effective at all (Ju *et al.,* 1988).

Although plants other than grasses can also be protected by endophytes, the data on these cases are neither so conclusive nor numerous. Sherwood *et al.*, (1986) described in oak (*Quercus* sp.) the role of the endophytic fungus *Rhabdocline parkeri* against the insect *Contarinia* sp. Petrini *et al.,* (1989) found that the endophytic fungi *Phyllosticta* sp. and *Hormonema dematioides* were dominant in the balsam fir *Abies balsamea*. The first fungus colonized preferentially galled needles attacked by *Paradiplosis tumifex*, being a candidate for biological control. Out of 100, twenty-one hyphal extracts of the black spruce *Picea mariana*, were found toxic to *Choristoneura fumiferana*, and extracts of fungi isolated from older leaves were more effective than young ones in cell culture (Johnson and Whitney, 1994). The dry weight and development of larvae fed on hyphae extracts were reduced.

The results described above show the efficiency of some endophytic fungi in reducing host damage provoked by insect attack. The examples given so far include researches performed in field and lab conditions, as well as comparisons between field and lab experimental data. The hyphomycete *Heteroconium chaetospira* has been isolated from various habitats (Petrini *et al.* 1992; Domsch *et al.* 1993) and was found to colonize the root tissue of 19 plant species without visible symptoms such as wilt or necrosis (Narisawa *et al.*, 1998, 2000), suggesting its affinity with dark septate endophytic fungi. It has all the qualities of potential biocontrol agent for soil-borne diseases. It was used to control clubroot of *Brassica* crops and *Verticillium* yellows (Narisawa *et al.*, 2000) for which effective control measures were otherwise not available (Djatnika, 1991; Narisawa *et al.*, 1998).

## Products of Endophytes with Insecticidal Activities

Bioinsecticides are only a small part of the insecticide field right now, but their market is increasing (Demain, 2000). As the world becomes wary of ecological damage done by synthetic insecticides, endophytic research continues for the discovery of powerful, selective, and safe alternatives. Endophytic fungi colonize their hosts without causing any external disease symptoms, except when the host is under stress conditions. Studies on microorganisms from tropical plant species are recently becoming more frequent, since these fungi and bacteria have been studied for biological control and production of compounds with pharmacological properties (Bashyal *et al.*, 1999; Daisy *et al.*, 2002).

Several endophytes are known to have anti-insect properties. Nodulisporic acids, novel indole diterpenes that exhibit potent insecticidal properties against the larvae of the blowfly, work by activating insect glutamate-gated chloride channels. The first nodulisporic compounds were isolated from an endophyte, *Nodulisporium* sp., from the plant *Bontia daphnoides*. This discovery has resulted in an intensive search for more *Nodulisporium* spp. or other producers of more-potent nodulisporic acid analogues (Demain, 2000). Insect toxins have also been isolated from an unidentified endophytic fungus from wintergreen (*Gaultheria procumbens*). The two new compounds viz., 5-hydroxy-2-(1′-hydroxy-5′-methyl-4′-hexenyl) benzofuran and 5-hydroxy-2-(1′-oxo-5′-methyl-4′-hexenyl) benzofuran, are responsible for the toxicity to spruce budworm (Findlay *et al.*, 1997). Another endophytic fungus, *Muscodor vitigenus*, isolated from a liana

(*Paullina paullinioides*), yields naphthalene as its major product which is the active ingredient of a widely exploited insect repellant (common mothballs). *M. vitigenus* showed promising preliminary results as an insect deterrent and has exhibited potent insect repellency against the wheat stem sawfly, *Cephus cinctus* (Daisy *et al.*, 2002).

## Mechanisms of Insect Control

### *Toxin production*

In certain cases, the production of toxin by the endophytic fungi was a plausible explanation for interactions resulting in natural insect control. The pioneer works in the field correlated the capacity of endophytic fungus to repel insects, induce weight loss, and reduce growth and development and even to increase pest death rate with toxin production.

It is now a fact that several toxins such as ergot and neurotoxins are produced by endophytic fungi, thus they confer host protection against various pests. In contrast to the ergot toxins that are isolated directly from the endophytic fungi, the neurotoxins are produced only as precursors, e.g., paxiline, produced by *Acremonium lolii.* It is not known if the precursor paxiline is converted to lolitrems by the plant or the fungus itself is able to do so while inside the plant.

Examples of toxin production by endophytic fungi, notably those colonizing grasses, are considerably abundant in the literature. For instance, in 1986, Miller showed that the protection of Canadian fir against the spruce budworms resulted from the production of toxic secondary metabolites by endophytic fungi. In 1988, Prestidge and Gallagher established a relationship between the presence of the fungus *A. lolii* in *Lolium perenne* and the growth, survival and feeding behaviour of *Listronotus bonariensis* larvae. In this case, the reduction in insect attacks towards infected plants was due to a strong toxin, lolitrem B, which is also toxic to mammals. This toxin once added to insect diets, reduced insect growth and survival. Its assimilation occurs by ingestion but not by absorption through the insect integument. Several other grasses infected with *Acremonium* spp. and *Epichloe typhina* were also analysed. These fungi generally produce alkaloids, mainly peramine and ergovaline. Individuals of *Rhopalosiphum padi* and *Schizaphis graminum* did not survive in grasses containing the alkaloid loline. On the other hand, ergovaline did not affect both insect species.

Clark *et al.*, (1989) showed in *Abies balsamea* and red spruce *Picea rubens,* that from 900 samples of fungal isolates, five produced toxic substances and three of those produced powerful toxins that once extracted and given to insects, caused death and decreased development rate of *Choristoneura fumiferana*. Many other cases of insect control involving toxins produced by endophytic microorganisms can be found in the literature. Siegel *et al.*, (1990) verified the presence of the alkaloids N-formil, N-acetil loline, peramine, lolitrem B and, ergovaline during plant attack by aphids.

Surveys have been carried out aiming to the discovery of new toxins useful to insect control. Two new active toxins against the spruce budworm, *C. fumiferana* were found in an unidentified endophytic fungus infecting the wintergreen *Gaultheria procumben* (Findlay *et al.*, 1997). Alkaloids from *Neotyphodium lolii* and *Lolium perenne* are capable of altering insect behaviour. Several of these alkaloids were added to the diet of adult individuals of *Heteronychus aratur* belonging to Coleoptera. Peramine, lolitrem B, lysergol-type alkaloids, festuclavine and lisergic acid showed no effe:ts on the insect, ergovine showed moderate effects, whereas ergotamine, ergovaline from the ergot-type alkaloid family seem to be responsible for the plant resistance (Ball *et al.,* 1997).

The majority of works related to toxin production was performed in grasses. However, Calhoun *et al.*, (1992), for the first time, identified toxic products synthesized by endophytic fungi in woody plants and that were able to modify growth and death rates in larvae of the spruce budworm *C. fumiferana* feeding on balsam fir. The endophytes in this case were identified as *Phyllosticta* and *Hormonema dematioides* and the toxic compounds were mainly heptelidic acid and rugulosine. Bills *et al.*, (1992) also detected the existence of tremorgenic toxins in tropical woody plant infected with an endophytic fungus from the genus *Phomopsis*.

### Host –endophyte interaction leads to toxin production

In certain plants, toxin production is not solely a fungal function. There are fungus-plant interaction factors acting on this production and toxins may be produced and dissipated throughout the host tissues. Evidently, it may result in insect control even if the endophyte is not present in the attacked sites. The development and nutrition of the moth *Plutella xylostella* larvae were studied in *Brassica*. Larvae were affected by the ingestion of leaves, before the endophytic fungus *Acremonium alternatum* reached these leaves. There were also differences in feed efficiency of males and females. The results may be related with the phytoesterol metabolism in the plants and indicate that endophytic fungi in the soil, even through a tenuous interaction, alter insect-plant interactions (Raps and Vidal, 1998). In addition, levels of alkaloids and other toxins may be altered not only quantitatively but also qualitatively, depending on the plants' physiological state.

### Making the host plant unpalatable for insect

In several cases, besides toxin production, the mode of action of certain fungi was shown to be based on their capability to render the plant unpalatable to several types of pests like aphids, grasshoppers, beetles, etc. (Carroll, 1988; Clay, 1988a; 1988b).

### Indirect mode of pest control

The first work showing the importance of endophytic fungi in controlling agriculture pests was curiously an indirect demonstration that the endophytes, by blocking insect gall formation, were also able to control a disease caused by a fungus (Webber, 1981; Carroll, 1986). Several other examples were analysed in which endophytic fungi blocked or altered insect's larvae installation and, by doing so, repress potential vectors to disseminate diseases.

Other interesting effect of endophytes over insect galls was seen in oak, where apparently dormant endophytic fungi become active with gall formation and by destroying the leaves, also kill the insect, therefore, protecting their hosts. This process was described in three associations between endophytes and insects, one of them involving the fungus *Gloeosporium quercinum* and the insect *Neuroterus numismalis* (Butin, 1992).

### Influence of genetic and environmental modifications on insect control

Many workers have shown that ambient variations may affect toxin production by endophytic fungi. Breen (1992) verified that changes in temperature and seasons modify toxin levels in the plant. This is the case of the peramine toxin, produced in *L. perenne* by *A. lolii*. The antixenosis towards the aphid *Schizaphis graminum* is dependent on peramine and the endophyte concentration, so a natural consequence is that the effect on the insect-pest will also be modified.

The specificity between certain endophytic fungi and host plants led Leuchtmann (1992) to suggest a further study on the occurrence of physiological races in endophytic fungi that could be important to the development of new biological controls. The expression of insect resistance may be affected by several factors, i.e., active amounts of allelochemicals, plant genotype (Breen, 1993a; 1993b), endophyte concentration (Breen, 1992), soil fertility and endophyte genotype (Christensen *et al.*, 1991). Hydric stress, temperature, soil pH, insect-pest resistance and other factors may also affect the endophyte concentration and toxin production. Breen (1994) idealized a model with all of these interactions affecting the increase or decrease in plant resistance to insects-pests in the presence of endophytes.

There are important studies showing the existence of this interaction by alterations due to nutrients in the soil, especially nitrogen fertilizers. Arechavaleta *et al.*, (1989) had already verified that the production of ergopeptide alkaloids increased in soil amended with nitrogen fertilizers and during a moderate hydric stress, which consecutively altered insect-pest control by *A. coenophialum* in *F. arundinacea*. Other types of interactions, taking into account a main host grass (*Festuca* or *Lolium*) and others grasses living in a common space, were shown by Clay *et al.*, (1993), who analyzed changes resulting from these plant-plant interactions in insect control and presence of endophytic fungi. They noticed the effects of a member of the genus *Acremonium* in relation to attacks of *S. frugiperda* larvae to the tall fescue *F. arundinacea* and the red fescue *F. rubra* grown separately or mixed with other grasses. In general, insect attacks were reduced in hosts colonized by endophytes, because the plants were less chosen by the pests. However, different degrees of insect attacks were observed according to the plant associations analysed. These studies demonstrated that plant interactions would change the effects of endophytic fungi in insect control.

## Artificial Inoculation of Endophytic Fungi

The passage of a fungus from generation to generation through seeds is of great importance to the control of insects-pests (Clay, 1989). An endophytic fungal inoculation, as it is done for nitrogen fixing bacteria and mycorrhizial fungi, may open new perspectives to apply endophytic microorganisms to control agriculture pests. There are already studies in this field concerning the use of the recombinant DNA technology in endophytes as a way of improving pest control.

The existence of fungi and bacteria able to cause disease in insects is widely known. Fungal species like *Metarhizium anisopliae*, *Beauveria bassiana* and others are often used in the biological control of agriculture insects-pests (Melo and Azevedo, 1998). If these microorganisms inhabit plant tissues as endophytes, they will obviously be able to act in insect control. Even in situations where they are not present in the plant, artificial inoculation techniques may circumvent this obstacle. Unfortunately, works concerning this promising line of research are scarce and, when conduced, involve a temperate climate system, as usual where the pest control problem is not so urging as in tropical cultures. Koga *et al.* (1997) inoculated *F. arundinacea* and *L. perenne* seeds with the endophytic fungus *Acremonium* and, as a result, obtained plants resistant to the bluegrass webworm *Parapediasia teterrella*. In the same manner, Pereira *et al*., (1999) artificially inoculated an endophytic fungus tagged with a genetic marker for resistance to fungicides in a tropical host, the banana tree. It was observed that not only the inoculation worked well, but the mutant fungus was also kept in competition with wild type ones. These results demonstrated that inoculation of valuable strains for biological control cannot only be used with success but modified microorganisms, once inoculated into a host, will be able to coexist and even overcome wild type strains of the same species.

An important point, at least in grasses, resides in the fact that several fungal species may be transmitted by seeds, where the biocontrol can be passed to the next generation. There are works showing the importance of endophytic fungi colonizing seeds in insect control. Knoch *et al*., (1993) observed in *F. arundinacea*, in which the endophytic fungus *A. coenophialum* is transmitted by seeds, that ants often prefer to collect endophyte-free seeds, avoiding those infected. The seeds infected with the fungus, if collected, are later discarded, favouring seed dispersion and germination.

Entomopathogenic fungi can also be placed inside plants. This procedure was done in corn (Bing and Lewis, 1991; 1992a; 1992b) to verify the potential of these fungi as endophytes to control the European corn borer *Ostrinia nubilalis*. *Beauveria bassiana* in this case was inoculated by injection and aspersion and was detected endophytically, colonizing several corn tissues and blocking the insect attack in some of its developmental stages. Injection resulted in 95% colonized plants and aspersion in 98.3%. An interesting finding was that in 33.3% of the control plants the fungus was found, indicating that the fungus is a natural endophyte. Although the process by which the fungus invades the plant was not determined but the movement was shown to be passive, occurring in the vascular system. It was also noticed that the infected insects do not transmit the fungus because plants free of attacks could contain the fungus in an endophytic state.

## Recombinant DNA Technology for Developing Improved Endophytes

As is seen in the extensive literature cited above, the presence of endophytic microorganisms in plants may be of extreme importance to host protection against insect attacks and in the control of plant diseases caused by bacteria and fungi. It is possible to inoculate the endophytes and to monitor the existence of toxins inside plants by observing the presence of insects, especially aphids (Eichenser and Dahlman, 1992). Finally, improvement programs have been conduced, where there is an association between improved plant varieties with endophytic microorganisms that act efficiently in pest control but do not cause any harm to cattle fed on these plants (Saha *et al*., 1987). These applied aspects, as in the past, are being timidly investigated by research groups and are concentrated only on hosts from temperate countries. Evidently, more promising results should be expected when applied to plants from tropical regions that show much more diversity and are affected by several pests. More recently, the recombinant DNA technology has been applied as a tool to improve endophytic microorganisms, aiming to the introduction of new characteristics of agronomic interests as biological control of pests.

The reports concerning engineered endophytes for biocontrol are concentrated mainly on bacteria, particularly *C. xyli* subsp. *cynodontis*. In endophytic fungi, this type of study is restricted to the development of transformation systems, introduction of reporter genes and inoculation into plants for colonization studies. Already at the beginning of the 90's, in a review or endophytic fungi from the genus *Acremonium*, Van-Heeswijck and McDonald (1992) launched the idea of using engineered endophytic fungi in the control of insects and diseases affecting *L. perenne*. Murray *et al*., (1992) introduced a α -glucuronidase (*Gus*) gene into a lolitrem B-deficient strain of *Acremonium* (designated as 187B). The transformation protocol yielded 700-800 transformants/ mg DNA and transformants were successively introduced into *Lolium* where *Gus* activity was observed. In a similar manner, Tsai *et al*., (1992) developed a transformation system for *A. coenophialum*.

Several genes from endophytic fungi mostly related to toxin production are now being cloned and studied in depth. Wang *et al*., (1999) cloned a gene coding for a DMATS (dimetilail tryptophane synthase) responsible for the first step in the synthesis of ergot by the fungus *Claviceps purpurea*. Genes related to the latter were identified in the endophytic fungi *Balansia obtecta* and *Neotyphodium* spp., known to produce ergobalansine and ergovaline. Young *et al*., (1999) found a cluster containing genes for the synthesis of mycotoxins like paxiline and lolitrem B, in *Epichloe*. Efficient transformation systems have been established for some of these endophytic microorganisms as well as protocols for reintroducing these microorganisms into plants. These studies are among the first involving genetic manipulation of endophytic microorganisms from tropical plants (Longo, 1996; Glienke-Blanco, 1999; Pereira *et al*., 1999).

## RESISTANCE TO HERBIVORY

Fungi residing in leaf and stem have been associated with resistance to herbivory in plants ranging from a green alga, *Enteromorpha vexata* (Cubit, 1974) to Douglas fir *Pseudotsuga rnenziesii* (Carroll, 1988). The relationship between endophyte and the host plant is generally considered as mutualistic one. The infected host plant supply the fungus with photosynthate and receive in turn, increased resistance to mammalian herbivory associated with alkaloid production by the endophytes. In addition, even in the absence of herbivory, under controlled environmental conditions, endophyte-infected grasses generally outperform uninfected ones.

Most of the evidence for the defensive mutualism concept is derived from studies of agronomic grass cultivars, which may be atypical of many endophyte-host interactions. In contrast to domesticated grasses where infection frequencies of highly toxic plants often approach 100%, natural grass populations are usually mosaics of uninfected and infected plants. The latter, however, usually vary enormously in alkaloid levels, from none to levels that may affect herbivores. This variation may result from diverse endophyte and host genotypic combinations that are maintained by changing selective pressures, such as competition, herbivory and abiotic factors. Other processes, such as spatial structuring of host populations and endophytes that act as reproductive parasites of their hosts, may maintain infection levels of seed-borne endophytes in natural populations, without the endophyte acting as a mutualist.

Many endophytes form internal localized infections in foliage, root, stem and bark and are horizontally transmitted via spores. A much smaller fraction, mostly found in grasses, form systemic infections in above-ground tissues. Some of these are vertically transmitted via hyphae growing into seeds (Saikkonen *et al*., 1998). Endophytes may produce mycotoxins, or otherwise alter host physiology and morphology. Endophytic mycotoxins are thought to benefit their woody plant hosts as 'inducible defenses' against insect herbivores (Carroll, 1988; 1991) and their grass hosts as 'acquired plant defences' (Cheplick and Clay, 1988) against both vertebrate and invertebrate herbivores. Endophytes may also alter other physiological, developmental or morphological properties of host plants such that competitive abilities are enhanced, especially in stressful environments (Clay, 1988; Malinowski and Belesky, 1999). This wide array of purported benefits has led to the conclusion that many, if not most, endophytes in host plants are strong plant mutualists. The 'defensive mutualism' concept (hereafter termed DMC) (Cheplick and Clay 1988; White *et al.,* 2001) has been generally accepted as the foundation for widespread occurrence of systemic endophyte infections among pooid grass species. The *Acremonium* endophytes are a particularly important group because they are associated with both resistance to insects and toxicosis in livestock. In tuffgrasses, *Acremonium* endophytes are a positive attribute because they provide resistance to some important turfgrass pests. On the other hand, in pastures *Acremonium* endophytes can cause fescue toxicity in animals grazing on infected tall fescue as well as ryegrass staggers in livestock grazing on infected perennial ryegrass (Bacon *et al.*, 1977).

The view of endophytes as defensive mutualists has stemmed mainly from studies of systemic endophytes in domesticated grasses, especially asexual, vertically transmitted *Neotyphodium*, which are well known for causing toxicosis, staggers, intoxication, narcosis, gangrene and other severe negative effects on vertebrate grazers (Clay, 1990; Schardl and Phillips, 1997). These toxic and detrimental effects are attributable to endophyte production of alkaloids, especially ergot and indole diterpene-type (e.g., lolitrem B) in infected perennial ryegrass (*Lolium perenne*) and pyrrolizidine (lolines) and ergot alkaloids in tall fescue (*Festuca arundinacea*) (Bush *et al.* 1997).

## Defensive Mutualism Against Invertebrates and Insect Herbivores, Nematodes and Pathogens

Horizontally transmitted endophytes in woody plants were originally proposed as plant mutualists by providing increased resistance to insect herbivores of the host (Carroll, 1988). However, many of these endophytes (Faeth and Bultman, 2002) do not generally deter or reduce performance of invertebrate herbivores, except for a few cases involving sedentary insects. Some woody plant endophytes decrease resistance to herbivores, as herbivores may promote horizontal transmission and infection via damage or gut passage, one expects selection for tolerance or facilitation of herbivory rather than resistance (Faeth and Hammon, 1997). Very few horizontally transmitted endophytes have been studied relative to their vast diversity (Arnold *et al.*, 2000). On the other hand, vertically transmitted endophytes in grasses are predicted by the DMC to strongly increase resistance to invertebrate herbivores (Clay, 1990). This prediction is generally supported for *Neotyphodium*-infected, agronomic grasses, tall fescue and perennial ryegrass (Clay *et al.*, 1993; Breen, 1994), and *Epichloe* -infected *Festuca rubra*, a European forage grass (Bazely *et al.*, 1997). However, increased resistance to invertebrates even in the agronomic grasses is by no means universal, and there are counter examples of either no, or even positive, effects of infection (Saikkonen *et al.*, 1998). Similarly, infections in agronomic grasses increase resistance to root-feeding nematodes and to some of the plant pathogens but not to others (Elmi *et al.,* 2001). Moreover, the proportion of studies showing increased resistance to either invertebrate herbivores or pathogens in these agronomic species may be skewed because neutral or negative results are less likely to be published (Csada *et al.,* 1996).

## Roles of Alkaloids

The DMC has generally focused on endophytes as protective mutualists of host plants at the adult stage via alkaloids. However, endophyte-related alkaloids in adult plants of native grass populations and species, except for relatively rare cases, are often absent, or at levels (Siegel *et al.*, 1990; Leuchtmann *et al.*, 2000; Faeth and Fagan, 2002) that are ineffective against most herbivores, especially specialists. This wide variation in alkaloid levels suggests that selective pressures in addition to, or exclusive of, herbivory on adult host plants have maintained this variability. Reduction of seed and seedling predation is one such selective force. Faeth and Bultman (2002) predicted that an endophyte-associated defence against natural enemies of the host grass is more likely to occur at the seed and seedling stage, where predation, pathogen attack and herbivory have more direct effects on plant fitness. Documentation of higher levels of alkaloids in some seeds and seedlings than adult plants (Siegel *et al.*, 1990; Leuchtmann *et al.*, 2000) is consistent with this prediction. Another intriguing possibility is that alkaloids function not only as deterrents to natural enemies of the host as envisioned by the DMC, but also, and perhaps in some cases, primarily, as allelopathic agents. Alkaloids from infected domesticated grasses may be leached into the soil and inhibit growth of competing grasses (Matthews and Clay, 2001). Siemens *et al.* (2002) proposed that constitutive alkaloids of *Brassica rapa* have dual functions as herbivore defences and as allelopathic agents that increase performance during plant competition.

## ENDOPHYTES AND PLANT GROWTH PROMOTION

Endophytes inhabit majority of healthy and symptomless plants, in various tissues, seeds, roots, stems and leaves (Johri, 2006). Plants are benefitted extensively by harbouring these endophytic microbes, as they promote the plant growth (Compant *et al.*, 2005) and confer enhanced resistance to various pathogens (Arnold *et al.*, 2003) through antibiotic production (Ezra *et al.*, 2004). It has been suggested that the presence of a mutualistic endophyte acts as a "biological trigger" to activate the stress response system more rapidly and strongly than non-mutualistic plants (Redman *et al.*, 2002). Many reports found in literature strongly suggest that these endophytes have an excellent potential to be used as plant growth promoters with legumes and non-legumes (Antoun *et al.*, 1998; Bai *et al.*, 2002).

Endophytes promote the growth of plants in various ways, for example through secretion of plant growth regulators such as indole-acetic acid (Lee *et al.*, 2004), phosphate solubilizing activity (Wakelin *et al.*, 2004), by enhancing hyphal growth and mycorrhizal colonization (Will and Sylvia, 1990), production of siderophores (Costa and Loper, 1994) and by supplying biologically fixed nitrogen (James *et al.*, 1994). The production of auxin-like compounds increases seed formation and germination (Clay, 1987) along with increased shoot growth and tillering (Kevin, 2003). Other effects of endophyte infection on the host plant include osmotic adjustment, stomatal regulation, modification of root morphology, enhanced uptake of minerals and alteration of nitrogen accumulation and metabolism (Malinowski and Belesky, 1999; Belesky and Malinowski, 2000). Thus, the published studies have focused on plant endophytic interactions and their benefits to plant growth. Nonetheless, the way the interactions among endophytes influence the plant productivity has not been explained.

## Nitrogen Cycling

Often nitrogen is the limiting source for plant growth and development. It is recruited by plants either as nitrate or ammonium or for a few species by nitrogen fixation with the help of rhizobia. Mycorrhizal fungi also play an important role in delivering either nitrate or ammonium to the root cells. *Piriformospora indica*, an endophytic fungus of the Sebacinaceae family, colonizes the roots of a wide variety of plant species. Host plants for *P. indica* include the cereal crops viz., rice, wheat, and barley as well as many dicots, including *Arabidopsis* (Varma *et al.*, 1999; Peskan-Berghöfer *et al.*, 2004). The interaction of the endophytic fungus with plant roots is accompanied by an enormous requisition of nitrogen from the environment as it stimulates the expression of the genes for nitrate reductase in roots and in turn promotes plant growth (Sherameti *et al.*, 2005). By analyzing the interaction of *P. indica* with *Arabidopsis* and tobacco roots they found that in contrast to mycorrhizal associations, nitrate reduction in the roots is stimulated by *P. indica*. A homeodomain transcription factor responds to the fungus and binds to promoter regions of the *P. indica*-responsive *Nia2*, *SEX1*, and 2-nitropropane dioxygenase genes. These results suggest that the expression of *P. indica*-responsive target genes may be controlled by common regulatory elements and *trans*-factors.

Since the middle of the nineteenth century it has been known that the plants belonging to the family Ericaceae form mycorrhiza of a characteristic kind. Further knowledge of the relations between plant and endophyte in this group has been forthcoming time to time. In 1915, Rayner showed that the relationship in *Calluna vulgaris* is of a remarkable character, involving obligate symbiosis between the two organisms and a much more extensive distribution of the fungus throughout the green plants than had been suspected. As in Orchidaceae, root formation by seedlings is dependent upon early infection by the endophyte, failing which development ceases and the plant perishes at the seedling stage. Unlike the condition in Orchidaceae, infection at the appropriate moment is provided for by the presence of mycelium on the seed coat, a condition ensured by the distribution of the endophyte throughout the vegetative tissues and eventually within the ovary chamber. These facts have been demonstrated with certainty in *Calluna*, and the evidence points to a similar condition throughout the family. Thus, ovarian infection has been reported for many species in all the sub-orders of Ericaceae, and the inability of the seedlings to complete their development without the infection has already been proved for a number of plants. In such remarkable association between flowering plants and fungi, as are found in the Orchidaceae and Ericaceae, it is of great importance to learn the exact nutritive relations between the symbionts. In Orchids there is an ocular evidence of digestion of the mycelium by the cells of the roots, and it is clear that by this means the plant can draw indirectly upon organic compounds of carbon and nitrogen from the soil (Rayner, 1922).

## Other Plant Growth Promotion Activities

The widespread association between Clavicipitaceous endophytic fungi and grasses has prompted numerous investigations into the ecological significance of this symbiosis. While physiological mechanisms for the increased growth of endophyte-infected grasses are not known, the beneficial effects of the fungi presumably counterbalance any potential costs to the host of supporting a heterotrophic symbiont. However, previous studies were performed on plants growing under environmental conditions that were not limiting to plant growth. Potential metabolic costs to hosts may only appear under environmentally stressed conditions.

A preliminary experiment compared perennial ryegrass (*Lolium perenne*) infected by *Acremonium lolii* with uninfected plants grown at three soil moistures and three nutrient concentrations. Total biomass, tiller number, and relative tillering rate were significantly affected by nutrient concentration and moisture. Although major differences between infected and uninfected plants were not apparent, relative growth rates of infected plants were significantly greater than uninfected plants in 3 out of 9 treatment combinations. Interactions between infection and nutrient limitation were examined in seedlings and adults of perennial ryegrass and tall fescue (*Festuca arundinacea*, infected by *A. coenophialum*). A significant interaction between nutrient and infection for tall fescue seedlings indicated that, compared to uninfected seedlings, infected seedlings had greater biomass at the high nutrient level but a significantly lower biomass at low nutrient level, suggesting a metabolic cost to the host due to competition with the fungus for nutrients or photosynthate. Infected perennial ryegrass seedlings had significantly more biomass at high and intermediate nutrient level, but not at low nutrient level. Although infection did not significantly affect total biomass in perennial ryegrass adults, but in tall fescue adults the benefits of endophyte infection to host plants (recorded as greater biomass) became increasingly greater with increasing nutrient availability (Cheplick *et al.*, 1989).

In Japan, tall fescue (*Fectuca arundinacea* Schreb.) is a popular pasture grass used for forage, turf and soil conservation purposes. Its widespread use is attributable in part to tolerance of poor soils, temperature extremes, drought and varying management practices. Tall fescue is often infected by *Neotyphodium coenophialum, a* fungal endophyte. Host plants express increased tolerance to biotic and abiotic stresses, which result in better persistence. Plants adapted to acidic mineral soils possess several mechanisms to overcome stresses related to mineral imbalance. They exclude or detoxify nutrients in plants or in the rhizosphere. Plant species differed in their ability to mobilize and acquire P from soils and from sparingly soluble phosphate compounds. Different species and strains of endophytes also differed in their ability to produce many compounds, which influence the growth and survival of their host grass. The evidence suggests that interactions between host plants and endophytes can be quite variable depending both on environmental conditions and host genotypes and this interaction vary from mutualistic to antagonistic (Rahman and Saiga, 2005).

Soil nutrient status is a critical factor affecting tall fescue growth and mineral status responses to endophyte infection. Several processes including the participation of apical, intercalary and axillary meristems influenced the shoot growth of grasses through leaf growth and tillering (Arredondo and Schnyder, 2003). Malinowski *et al.* (2000) observed that leaf-located endophyte could influence root activity in tall fescue by enhancing mineral uptake rate and transport to shoots. Furthermore, slow growing grasses generally contain more sclerenchyma tissue and consequently have a high concentration of cell wall material with a low concentration of cytoplasmic and vacuolar compounds than fast-growing grasses. On the other hand, fast-growing grasses exhibit higher concentrations of organic N-compounds, organic acids and minerals (Poorter and Bergkotte, 1992). This role of endophyte in enhancing host plant performance implies that ample study is needed on growth hormones and organic acids in endophyte-infected plants.

The genus *Fusarium* that is described as endophyte and/or phytopathogen in many plant species and occurs as the dominant group in cacao plants, suggests that endophytes could play a role in the plant development. *In vitro* experiments have shown a broad host spectrum of *Piriformospora indica* (Varma *et al.*, 1999), including members of the Brassicaceae, like *Arabidopsis*, which are not colonized by arbuscular mycorrhizal fungi (AMF). Major causes for crop losses are abiotic and biotic stresses due to unfavorable climate and plant diseases and pests. Increased plant productivity, therefore, relies on a high chemical input and is achieved at the expense of detrimental effects on the environment (Chapin *et al.*, 2000).

*P. indica* induces root resistance against *Fusarium culmorum*, one of the fungal species causing head blight, and systemic resistance to barley powdery mildew *Blumeria graminis* f. sp. *hordei* via an unknown mechanism probably independent of salicylate or jasmonate accumulation. Moreover, *P. indica* protects barley from abiotic stress, such as high salt concentrations (Waller *et al.*, 2005). The beneficial effect on the defence status is detected in distal leaves, demonstrating a systemic induction of resistance by a root-endophytic fungus. The systemically altered "defence readiness" is associated with an elevated antioxidative capacity due to an activation of the glutathione-ascorbate cycle and results in an overall increase in grain yield. The fungus confers improved growth to many cereal crops. *P. indica* and barley form a mutualistic symbiosis in which the endophyte colonizes the plant root, proliferates by inter- and intracellular growth and produces chlamydospores in dead root tissue. After establishment of the symbiosis the fungus confers improved growth, disease resistance and abiotic stress tolerance to the host plant.

The mutualistic symbiosis of crop plants and Sebacinales has a great potential for sustainable agriculture. In contrast to AMF, *P. indica* and other members from the same order mediate resistance to root pathogens and systemic resistance to biotrophic leaf pathogens. From an agronomical point of view, it is most promising that *P. indica* can enhance crop yield in cereals (Waller *et al.*, 2005). Exploitation of endophytic fungi like *P. indica* may, however, not only complement crop production strategies, which presently rely on a high input of fungicides, but additionally may be an eminent source of molecular traits affecting both disease resistance and grain yield in cereals. For future utilization, it is important to gain additional information on effective application strategies (e.g., spore formulation), growth conditions, and the influence of environmental factors.

## CONCLUSION AND FUTURE PROSPECTS

Endophytes are a poorly investigated group of microorganisms that represent an abundant and dependable source of bioactive and chemically novel compounds. Although work on the utilization of this vast resource of microorganisms has just begun, it has already become obvious that an enormous potential for organism, product, and utilitarian discovery in this field holds exciting promise. This is witnessed by the discovery of a wide range of products and microorganisms that already hold inkling for future prospects. The mechanisms through which endophytes exist and respond to their surroundings must be better understood in order to be more predictive about which higher plants to seek, study, and spend time isolating microfloral components. This may facilitate the product discovery processes.

Certainly, one of the major problems facing the future of endophyte biology and natural-product discovery is the rapid diminishment of rainforests, which hold the greatest possible resource for acquiring novel microorganisms and their products. The total landmass of the world that currently supports rainforests is about equal to the area of the United States (Mittermeier *et al.*, 1999). Each year, a large area is lost due to clearing, harvesting, fire, agricultural development, mining, or other human-oriented activities. It can only be guessed that the potential loss of microbial diversity is also taking place perhaps with the same frequency as the loss of mega-life forms, especially since certain microorganisms may have developed unique specific symbiotic relationships with their plant hosts. Thus, when a plant species disappears, so too does its entire suite of associated endophytes. Multistep processes are needed now to secure information and life forms before they continue to be lost. Certain areas of the planet that represent unique places housing biodiversity need to be preserved immediately. Countries are required to establish information bases of their biodiversity and at the same time begin to make national collections of microorganisms that live in these areas.

Direct antiherbivore properties of endophytes (particularly in grasses) may be exploited, for instance, in biocontrol through developing natural pesticides or improvement of herbivore resistant cultivars by introducing biologically active (high mycotoxin production, etc.) fungal strains into cultivars. Alternatively, fungal strains, which do not produce mycotoxins but increase plant growth, seed production, seed germination rate and/or stress tolerance, can be used to increase forage productivity when introduced to the cultivars used as forage. Similarly, many fungi are able to control insect pests. This property can be exploited to transmit to the host the capacity to resist certain pests by their artificial inoculation in endophyte free plants.

## References


Ahmad, S., Johnson-Cicalese, J. M., Dickson, W. K. & Furik, C. R. (1986) *Entomol. Exp. Appl.* **41:** 3-10.

Antoun, H., Beauchamp, C. J., Goussard, N., Chabot, R. & Lalande, R. (1998) *Plant Soil* **204:** 57-67.

Arechavaleta, M., Bacon, C.W., Hoveland, C.S. & Radclife, D.E. (1989) *Agronomie* **81:** 83- 90.

Arnold, A. E., Meffa, L. C., Kyollo, D., Rojas, E. I., Maynard, Z., Robbins, N. & Herre, E. A. (2003) *Proc. Natl. Acad. Sci. USA,* **100:** 15649-15654.

Arnold, A. E., Maynard, Z., Gilbert, G. S. (2000) *Ecol. Lett.* **3:** 267-274.

Arredondo, J. T. & Schnyder, H. (2003) *New Phytol.* **158:** 305-314.

Azevedo, J.L. (1998) In: *Microorganismos endofiticos* in *Ecologia Microbiana*, (Eds. Melo, I. S. & Azevedo, J. L.), Jaguariuna, Sao Paulo, Brazil, pp. 117-137.

Bacon, C. W., Porter, J. K., Robbins, J. D. & Luttrell, E. S. (1977) *Appl. Environ. Microbiol.* **34:** 576-581.

Bai, Y., D'Aoust, F., Smith, D. L. & Driscoll, B. T. (2002) *Can. J. Microbiol.* **48:** 230-238.

Ball, O.J.P., Miles, C.O. & Prestidge, R.A. (1997) *J. Econ. Entomol.* **90:** 1382-1391.

Bandara, W. M. M. S., Seneviratne, G. & Kulasooriya, S. A. (2006) *J. Biosci.* **31:** 645-650.

Barker, G.M., Pottinger, R.P., Addison, P.J. & Prestidge, R.A. (1984) *NZ J Agric. Res.* **27:** 271-277.

Bashyal, B., Li, J. Y., Strobel, G. A. & Hess, W. M. (1999) *Mycotaxon* **72:** 33-42.

Bazely, D. R., Vicari, M., Emmerich, S. (1997) *J. Appl. Ecol.* **34:** 847-860.

Belesky, D. P. & Malinowski, D. P. (2000) In: *Microbial endophytes,* (Eds. Bacon, C. W. & White, J. F.), Marcel Dekker, New York, pp. 455-484.

Benhamou, N. (2002) *Appl Environ. Microbiol.* **68:** 4044 -4060.

Bills, G., Dombrowski, A., Pelaez, F. & Polishook, J. (2002) In: *Tropical mycology: micromycetes,* vol. 2 (Eds. Watling, R., Frankland, J. C., Ainsworth, A. M., Issac, S. & Robinson, C. H.) CABI Publishing, New York, N.Y. pp. 165-194.

Bills, G. F., Giacobbe, R. A., Lee, S. H., Pelaez, F. & Tkacz, J.S. (1992) *Mycol. Res.* **96:** 977-983.

Bing, L.A. & Lewis, L.C. (1991) *Environ. Entomol.* **20:** 1207-1211.

Bing, L.A. & Lewis, L.C. (1992a) *Entomophaga* **37:** 525-536.

Bing, L.A. & Lewis, L.C. (1992b) *Biocontrol Sc. Technol.* **2:** 39-47.

Blodgett, J. T., Swart, W. J., Louw, S. M. & Weeks, W. J. (2000) *Mycologia* **92(5):** 853-859.

Breen, J.P. (1992) *Environ. Entomol.* **21:** 68-74.

Breen, J.P. (1993a) *J. Econ. Entomol.* **86:** 621-629.

Breen, J.P. (1993b) *J. Econ. Entomol.* **86:** 1279-1286.

Breen, J.P. (1994) *Annual Review of Entomology* **39:** 401-423.

Bush, L. P., Wilkinson, H. W. & Schardl, C. L. (1997) *Plant Physiol.* **114:** 1-7.

Butin, H. (1992) *European J Forest Pathol.* **22:** 237-246.

Calhoun, L. A., Findrlay, J. A., Miller, J. D. & Whitney, N. J. (1992) *Mycol. Res.* **96:** 281-286.

Carroll, G. (1986) In: *Microbiology of the Phylloplane*, (Eds. Fokkema, C. H. & Van der, H. J.), Cambridge University Press, London, UK, pp. 205-222.

Carroll, G. (1988) *Ecology* **69:** 2-9.

Carroll, G. (1991) In: *Microbial mediation of plant-herbivore interactions*, (Eds. Barbosa, P., Krischik, V. A., & Jones, C.G.), John Wiley and Sons. New York, pp. 253-271.

Chapin, F. S., Zavaleta, E. S., Eviner, V. T., Naylor, R. L., Vitousek, P. M., Reynolds, H. L., Hooper, D. U., Lavorel, S., Sala, O. E. & Hobbie, S. E. (2000) *Nature,* **405:** 234-242.

Cheplick, G. P. & Clay, K. (1988) *Oikos* **52:** 309-19.

Cheplick, G. P., Clay, K & Marks, S. (1989) *New Phytol.* **111(1):** 89-97.

Christensen, M. J., Latch, G. C. M. & Tapper, B. A. (1991) *Mycol. Res.* **95:** 988-923.

Clark, C. L., Miller, J. D. & Whitney, N. L. (1989) *Mycol. Res.* **93:** 508-512.

Clay, K. (1987) *Oecologia* **73:** 358-362.

Clay, K. (1988a) *Ecology* **69:** 10-16.

Clay, K. (1988b) *Mycol. Res.* **92:** 1-12.

Clay, K. (1989) *Mycol. Res.* **92:** 1-12.

Clay, K. (1990) *Ann. Rev. Ecol. Syst.* **21:** 275-297.

Clay, K., Marks, S. & Cheplick, G.P. (1993) *Ecology* **74:** 1767-1777.

Claydon, N., Grove, J. F. & Pople, M. (1985) *Phytochemistry* **24:** 937-943.

Clement, S. L., Lester, D. G., Wilson, A. D. & Pike, K. S. (1992) *J. Econ. Entomol.* **85:** 583-588.

Compant, S., Reiter, B., Sessitsch, A., Nowak, J., Clément, C. & Barka, E. A. (2005) *Appl. Environ. Microbiol.* **71:** 1685-1693.

Costa, J. M. & Loper, J. E. (1994) *Mol. Plant-Microbe Interact.* **7:** 440-448.

Csada, R. D., James, P. C. & Espie, R. H. (1996) *Oikos* **76:** 591-593.

Cubit, J. D. (1974) *Interactions of seasonally changing physical factors and grazing affecting high intertidal communities on a rocky shore,* Ph.D. thesis. Univ. Oregon, Eugene, Oreg.

Daisy, B. H., Strobel, G. A., Castillo, U., Ezra, D., Sears, J., Weaver, D. & Runyon, J. B. (2002) *Microbiology* **148:** 3737-3741.

Demain, A. L. (2000) In: *Biodiversity: new leads for pharmaceutical and agrochemical industries,* (Eds. Wrigley, S. K., Hayes, M. A., Thomas, R. Chrystal, E. J. T.& Nicholson, N.), The Royal Society of Chemistry, Cambridge, United Kingdom. pp. 3-16.

Djatnika, I. (1991) *Cruciferae Newsletter* **142:** 14-15.

Domsch, K. H., Gams, W. & Anderson, T.H. (1993) *Compendium of Soil Fungi*, London, UK: Academic Press.

Eisencher, H. & Dahlman, D. L. (1992) *Env. Entomol.* **21:** 1046-1051.

Elmi, A. A., West, C. P., Robbins, R. T. & Kirkpatrick, T. L. (2001) *Grass Forage Sci.* **55:** 166-172.

Ezra, D., Hess, W. M. & Strobel, G. A. (2004) *Microbiology* **150:** 4023-403.

Faeth, S. H. & Bultman, T. L. (2002) In: *Multitrophic level interactions* (Eds. Tscharntke, T. & Hawkins, B. A.), Cambridge Univ. Press, pp. 89-123.

Faeth, S. H. & Fagan, W. F. (2002) *Integr. Compar. Biol.* pp. 199-226.

Faeth, S. H. & Hammon, K. E. (1997) *Ecology* **78:** 810-819.

Findlay, J. A., Bethelezi, S., Li, G. & Sevek, M. (1997) *J. Nat. Prod.* **60:** 1214-1215.

Funk, C.R., Halisky, P.M., Johnson, M.C., Siegel, M.R., Stewart, A.V., Ahmad, S., Hurley, R.H. & Harvey, I.C. (1983) *BioTechnol.* **1:** 189-191.

Ganley, R. J., Brunsfeld, S. J. & Newcombe, G. (2004) *PNAS* **101 (27):** 10107-10112.

Gaynor, D. L. & Hunt, W. F. (1983) In: *Proceedings of the New Zealand Grassland Association* **44:** 257-263.

Glienke-Blanco, C. (1999) PhD. Thesis, Escola Superior de Agricultura "Luiz de Queiroz", Universidade de São Paulo, Piracicaba, São Paulo, Brazil, pp. 200.

Hardy, T. N., Clay, K. & Hammond, A. M. (1985) *J Econ. Entomol.* **78:** 571-575.

James, E. K., Reis, V. M., Olivares, F. L., Baldani, J. I. & Dobereiner, J. (1994) *J. Exp. Bot.* **45:** 757-766.

Johnson, M. C., Dahlman, L. D., Siegel, M. R., Bush, L. P., Latch, G. C. M., Potter, D. A. & Varney, D. R. (1985) *Appl. Env.Microbiol.* **49:** 568-571.

Johri, B. N. (2006) *Curr. Sci.* **90:** 1315-1316.

Ju, Y., Sacalis, J. N. & Still, C. C. (1988) *J Agric. Food Chem.* **46:** 3785-3788.

Kevin, V. J. (2003) *Plant Soil* **255:** 571-586.

Kindler, D., Breen, J. P. & Springer, T. (1990) *J Econ. Entomol.* **84:** 685-692.

Knoch, T. R., Faeth, S. H. & Arnott, D. L. (1993) *Oecologia* **95:** 470-473.

Koga, H., Hirai, Y., Kanda, K., Tsukiboshi, T. & Uematsu, T. (1997) *Japan Agric. Res.* **31:** 109-115.

Lasota, J. A., Waldvogel, M. G. & Shetlar, D. J. (1983) *Environ. Entomol.* **12:** 245-246.

Latch, G. C. M., Hunt, W. F. & Musgrave, D. R. (1985) *N Z J Agric. Res.* **28:** 165-168.

Lecuona, R.E. (1996)In: *Microorganismos patógenos empleados en el control microbiano de insectos plaga* (Ed. Lecuona, C.), pp. 338.

Lee, S., Flores,E. M., Contreras Z. M., Garcia, F. L., Escamilla, J. E. & Kennedy, C. (2004) *J. Bacteriol.* **186:** 5384-5391.

Leuchtmann, A., Schmidt, D. & Bush, L. P. (2000) *J. Chem. Ecol.* **26:** 1025-1036.

Li, J. Y., Strobel, G. A., Sidhu, R., Hess, W. M. & Ford, E. (1996) *Microbiology* **142:** 2223-2226.

Longo, A.C. (1996) PhD Thesis, Escola Superior de Agricultura "Luiz de Queiroz", Universidade de São Paulo, Piracicaba, São Paulo, Brazil, pp. 101.

Malinowski, D. P., Alloush, G. A. & Belesky, D. P. (2000) *Plant Soil* **277:** 115-126.

Malinowski, D. & Belesky, D. P. (1999) *J. Plant Nutr.* **22:** 835-853.

Mathias, J. K., Ratcliffe, R. I., Hellman, J. L. (1990) *J. Econ. Entomol.* **83:** 1640-46.

Matthews, J. W. & Clay, K. (2001) *Ecology* **82:** 500-509.

Meijer, G. & Leuchtmann, A. (1999) *New Phytol.* **141:** 355-368.

Melo, I. S. & Azevedo, J. L. (1998) In: *Controle biológico I.* (Ed. Jaguariuna,E.), São Paulo, Brazil. pp. 262.

Miller, J.D. (1986) In: *Microbiology of the Phyllosphere,* (Eds. Fokkema, N. J. & van den Heuvel), Cambridge University Press, Cambridge, pp. 221-231.

Mittermeier, R. A., Meyers, N., Gil, P. R. & Mittermeier, C. G. (1999) In: *Hotspots: Earth's biologically richest and most endangered eco-regions,* Toppan Printing Co., Tokyo, Japan.

Murray, F. R., Latch, G. C. M. & Scott, D. B. (1992) *J Mol. G. Gen.* **233:** 1-9.

Narisawa, K., Ohki, T. & Hashiba, T. (2000) *Plant Pathol.* **49:** 141-146.

Narisawa, K., Tokumasu, S., Hashiba, T. (1998) *Plant Pathol.* **47:** 206-210.

Pereira, J.O., Carneiro-Vieira, M.L. & Azevedo, J.L. (1999) *World J. Microbiol. Biotechnol.* **15:** 37-40.

Peskan-Berghöfer, T., Shahollari, B., Giong, P. H., Hehl, S., Markert, C., Blanke, V., Kost, G., Varma, A. & Oelmüller, R. (2004) *Plant Physiol.* **122:** 465-477.

Petrini, O., Petrini, L. E., & Dryfuss M. M. (1992) *Mycol. Helv.* **5:** 9-20.

Petrini, L. E., Petrini, O. & Laflamme, G. (1989) *Phytoprotection* **70:** 97-103.

Poorter, H. & Bergkotte, M. (1992) *Plant Cell Environ.* **15:** 221-229.

Prestidge, R. A. & Gallagher, R.T. (1988) *Ecol. Entomol.* **13:** 429-435.

Rahman, M. H. & Saiga, S. (2005) *Plant Soil* **272:** 163-171.

Raps, A. & Vidal, S. (1998) *Oecologia* **114:** 541-547.

Rayner M. C. (1922) *Botanical Gazette* **73 (3):** 226-235.

Rayner, M. C. (1915) *Ann. Botany* **29:** 97-153.

Redell, P. & Gordon V. (2000) In: *Biodiversity: new leads for pharmaceutical and agrochemical industries* (Eds. Wrigley, S. K., Hayes, M. A., Thomas, R., Chrystal, E. J. T. & Nicholson, N.), The Royal Society of Chemistry, Cambridge, United Kingdom, pp. 205-212.

Redman, R. S., Sheehan, K. B., Stout, R. G., Rodriguez, R. J. & Henson, J. M. (2002) *Science* **298:** 1581.

Saha, D. C., Johnson, C. J. M., Halisky, P. M., Heemstra, M. I. & Funk, C. R. (1987) *Plant Disease* **71:** 1021-1024.

Saikkonen, K., Faeth, S. H., Helander, M. L. & Sullivan, T. J. (1998). *Ann. Rev. Ecol. Syst.* **29:** 319-343.

Schardl, C. L. & Phillips, T. D. (1997) *Plant Disease* **81:** 430-438.

Sherameti, I., Shahollari, B., Venus, Y., Altschmied, L., Varma A. & Oelmuller, R. (2005) *J. Biological Chem.* **280 (28):** 26241-26247.

Sherwood, P. M., Stone, J. K. & Carroll, G. C. (1986) *Can. J. Bot.* **64:** 1849-1855.

Shrestha, K., Strobel, G. A., Prakash S. & Gewali, M. (2001) *Planta Med.* **67:** 374-376.

Siegel, M. R., Latch, G. C. M., Bush, L. P., Fannin, F. F., Rowan, D. D., Tapper, B. A., Bacon, C. W. & Johnson, M. C. (1990) *J. Chem. Ecol.* **16:** 3301-3316.

Siemens, D. H., Garner, S. H., Mitchell-Olds, T. & Callaway, R. M. (2002) *Ecology* **82:** 505-517.

Stierle, A., Strobel, G. A. & Stierle, D. (1993) *Science* **260:** 214-216.

Stinson, M., Ezra, D. & Strobel. G. (2003) *Plant Sci.* **165:** 913-922.

Strobel, G. & Daisy, B. (2003) *Microbiol Mol Biol Rev.* **67:** 491-502.

Tan, R. X. & Zou. W. X. (2001) *Nat. Prod. Rep.* **18:** 448-459.

Tsai, H. F., Siegel, M. R. & Schardl, C. L. (1992) *Curr. Genetics* **22:** 399-406.

Van-Heeswijck, R. & McDonald, G. (1992) *Aus. J. Agr. Res.* **43:** 1683-1709.

Varma, A., Verma, S., Sudha, S., Sahay, N., Bütehorn, B. & Franken, P. (1999) *Appl. Environ. Microbiol.* **65:** 2741-2744.

Wakelin, S. A., Warren, R. A., Harvey, P. R. & Ryder, M. H. (2004) *Biol. Fertil. Soils* **40:** 36-43.

Waller, F., Achatz, B., Baltruschat, H., Fodor, J., Becker, K., Fischer, M., Heier, T., Hückelhoven, R., Neumann, C. & von Wettstein, D. (2005) *Proc Natl Acad Sci, USA* **102:** 13386-13391.

Wang, J., Machado, C., Panaccione, D. & Schardl, C. (1999) *Fungal Genetics Newslett.* **46:** 120.

Webber, J. (1981) *Nature* **292:** 449-451.

White, J. F., Sullivan, R. & Moy, M. (2001) In: *Proc. 4th Int. Neotyphodium/Grass Interactions Symp.* (Eds. Dapprich, P. D. & Paul, V. H.), Fachbereich Agrarwirtschaft, Soest, Germany, pp. 17-30.

Will, M. E. & Sylvia, D. M. (1990) *Appl. Environ. Microbiol.* **56:** 2073-2079.

Young, C., McMillan, L. & Scott, B. (1999) *Fungal Genetics Newslett.* **46:** 121.

13

# Microorganism-Mediated Management of Plant Pathogens

**Vishal Prasad, B. K. Singh, K. S. Kumar, M. Ahmed and R. S. Upadhyay**
*Laboratory of Plant Pathology and Microbial Technology, Centre of Advanced Study in Botany, Banaras Hindu University, Varanasi (U.P.)*

## ABSTRACT

Plants, like humans and other animals, also become sick, but they have evolved a sophisticated defence system against pathogens based on a combination of constitutive and inducible responses which can be localized or spread throughout plant organs and tissues. Apart from foe pathogens, there are several friendly microbes present in the soil which directly or indirectly protect plants from such deleterious pathogens. These microorganisms employ several mechanisms to kill or suppress the pathogens. These microorganisms also induce resistance in the plants against pathogens. Seeing the importance of such microorganisms in the protection of plant health, new biotechnological approaches are employed regularly to develop newer and much better microbial agents for management of the pathogen. This practice has led to the development of many such agents, several of which are also available commercially. This review describes general mechanisms of pathogen management and their commercial status.

## INTRODUCTION

Plants in their environment face potential deleterious organisms such as fungi, bacteria, viruses, nematodes, etc. Many of them are able to cause plant diseases, responsible for important losses in crop production worldwide. But, often the outcome of these interactions is not disease, since plants have developed multiple mechanisms to protect themselves against the pathogen attack. Moreover, beneficial microorganisms are common in the soil, improving plant growth and reducing the effects of deleterious organisms (Lucy *et al.*, 2004). While chemical control of plant diseases is usually expensive and may have a negative impact on the environment and on public health, the use of microorganisms to manage plant pathogens, known as biological control, is accepted as a durable and environment-friendly alternative in plant disease management. Microbial populations can be used directly for management of plant and animal pathogen populations based on negative population interactions. Identifying, understanding and utilizing microorganisms or microbial products to manage plant diseases and enhance crop production are integral parts of sustainable agriculture. Biological management has the potential to control crop diseases while causing no or minimal detrimental environmental impact. Some of the benefits of utilizing microorganisms include:

(i) Reduced dependence on chemical pesticides. This is important because of the increasing restrictions on chemical usage due to environmental and public concerns.

(ii) Lack of development of pathogen resistance to biocontrol organisms. This is important due to observed increase in resistance to many chemical controls.

(iii) Selective action against specific groups of pathogens and not against beneficial organisms.

(iv) Biodegradability of microbial pesticides and the by-products of their manufacture.

(v) Lack of danger to humans or animals.

(vi) Improvement of soil and enhancement of agricultural sustainability.

Biocontrol involves harnessing disease-suppressive microorganisms to improve plant health. Disease suppression by biocontrol agents is the sustained manifestation of interactions among the plant, the pathogen, the biocontrol agent, the microbial community on and around the plant, and the physical environment. Even in model laboratory systems, the study of biocontrol involves interactions among a minimum of three organisms. The complexity of these systems has influenced the acceptance of biocontrol as a means of controlling plant diseases in two ways. First, practical results with biocontrol have been variable. Thus, despite some stunning successes with biocontrol agents in agriculture, there remains a general skepticism born of past failures (Weller, 1988). Second, progress in understanding an entire system has been slow. Recently, however, substantial progress has been made in a number of biocontrol systems through the application of genetic and mathematical approaches that accommodate the complexity.

Several modes of action have been described in biological control. Direct effects of the biocontrol agent over the pathogen include inhibition by antimicrobial compounds (antibiosis), competition for colonization sites and nutrients, degradation of pathogenicity factors and parasitism. Indirect mechanisms include improvement of plant nutrition and damage compensation, changes in the root system anatomy, microbial changes in the rhizosphere and activation of plant defense mechanisms, leading to enhanced plant resistance. It is common that an effective biocontrol agent acts through the combination of different mechanisms (Whipps, 2001).

## INTERACTION WITH THE PATHOGEN

### Antibiosis

Biocontrol is often attributed to antibiosis. In many biocontrol systems that have been studied, one or more antibiotics have been shown to play a role in disease suppression (Raaijmakers *et al.*,

2002). The fact that antibiosis is a common mechanism of biocontrol may be due to a bias in choice of organisms for study. Alternatively, it may be due to the attractiveness of the antibiosis hypothesis, or antibiosis may be simply a highly effective mechanism for suppressing pathogens in the rhizosphere. Genetic analyses have been particularly informative in determining the role of antibiotics in biocontrol, in part because mutants can be screened easily *in vitro* for changes in antibiotic accumulation, providing the means to conduct thorough genetic analyses.

### *Role of antibiotics*

A number of highly effective disease-suppressive agents are found among the fluorescent pseudomonads, making this group of bacteria the most widely studied group of antibiotic producers in the rhizosphere (Ligon *et al.*, 2000). The first antibiotics clearly implicated in biocontrol by fluorescent pseudomonads were the phenazine derivatives that contribute to disease suppression by *Pseudomonas fluorescens* strain 2-79 and *P. aureofaciens* strain 30-84, which controlled take-all of wheat (Weller and Cook, 1983; Ownley *et al.*, 2003). Evidence for the role of phenazines includes an analysis of transposon insertion mutants that lack the ability to produce phenazine-1-carboxylate and are reduced in disease suppressive capabilities (Thomashow and Weller, 1988; Pierson and Thomashow, 1992). *P. fluorescens* strain CHAO produces hydrogen cyanide, 2,4-diacetylphloroglucinol, and pyoluteorin, which directly interfere with growth of various pathogens and contribute to disease suppression (Voisard *et al.*, 1989; Keel *et al.*, 1990,1992; Maurhofer *et al.*, 1994b; Blumer and Haas; 2000). Studies have established that mutants deficient in production of the antimicrobial substances show reduced ability to suppress certain diseases.

The genes for the biosynthesis of many of the metabolites involved in disease suppression by fluorescent pseudomonads have been isolated, and their regulation has been studied (Pierson and Thomashow, 1992; Pierson *et al.*, 1995; Bangera and Thomashow, 1996). An emerging theme in the fluorescent pseudomonads is that global regulatory elements coordinate production of secondary metabolites (Ovadis, 2004). For example, biosynthesis of phenazine derivatives in *P. aureofaciens* is regulated by a quorum sensor, PhzR, that perceives cell population density through the concentration of an autoinducer (Pierson *et al.*, 1994). Interestingly, mutants that lack the ability to produce the auto-inducer can use similar molecules produced by other rhizosphere inhabitants, suggesting that the presence of significant population of other bacteria could influence phenazine production by *P* aureofaciens in the rhizosphere (Pierson and Pierson, 1996; Wood and Pierson, 1996). Additionally, the environmental sensors ApdA and GacA influence the production of numerous secondary metabolites involved in biocontrol by pseudomonads (Laville *et al.*, 1992; Gaffney *et al.*, 1994; Corbell and Loper, 1995; Koch *et al.*, 2002; Chancey *et al.*, 1999).

Sigma factors also regulate antibiotic production in fluorescent pseudomonads. Sigma factors are subunits of RNA polymerase that direct transcription in bacteria. Each sigma factor has promoter specificity and regulates a distinct set of genes, thereby playing a key role in gene regulation in bacteria. In *Escherichia coli*, gene expression in stationary phase involves sigmaS (Kolter *et al.*, 1993). A homolog of sigmaS is significant in biocontrol by *P. fluorescens* through its role in expression of stationary phase genes required for the production of certain secondary metabolites and survival on plant material (Sarniguet *et al.*, 1995). The ratio of the housekeeping sigma factor sigma$^{70}$ and sigma$^{s}$ appears critical in the regulation of various metabolites involved in disease suppression (Schnider *et al.*, 1995).

Although bacilli have received less attention as potential biocontrol agents than the pseudomonads, evidence indicating that they may promote effective disease suppression is accumulating (Ryu, 2004). The bacilli are particularly attractive for practical use because they produce stable endospores, which can survive the heat and desiccation conditions that may be faced by biocontrol agents (Turner and Backman, 1991; Lumsden *et al.*, 1995; Osburn *et al.*, 1995; Schisler *et al.*, 2004). One well-studied example is *Bacillus cereus* strain UW85, which suppresses diseases caused by the Oomycetes, a group of fungi that cause severe plant diseases. Analysis of mutants of *B. cereus* shows a significant quantitative relationship between disease suppressiveness and the production of two antibiotics, zwittermicin A and kanosamine (Silo-Suh *et al.*, 1994; Milner *et al.*, 1996). Zwittermicin A is an aminopolyol representing a new class of antibiotic, and kanosamine is an aminoglycoside. The purified antibiotics suppress disease and inhibit development of Oomycetes by stunting and deforming germ tubes of germinating cysts.

*Trichoderma* (Harman *et al.*, 2004) and *Gliocladium* are closely related fungal biocontrol agents. Each produces antimicrobial compounds and suppresses disease by diverse mechanisms, including the production of structurally complex antibiotics, gliovirin and gliotoxin (Howell *et al.*, 1993). Mutants of *Gliocladium virens* that do not produce gliotoxin are reduced in their ability to control *Pythium* damping-off (Wilhite *et al.*, 1994). Mutants with increased or decreased antibiotic production show a corresponding effect on biocontrol (Howell and Stipanovic, 1983).

## *Antibiotic resistance*

One of the goals of using biocontrol in agriculture is to avoid the pitfalls associated with synthetic pesticides, including the development of resistance in pathogen populations. An attractive feature of biocontrol strategies is that population of pathogens resistant to antibiotics produced by biocontrol agents are likely to develop slowly. There are two reasons why this may be so. First, most biocontrol agents produce more than one antibiotic, and resistance to multiple antibiotics should occur only at a very low frequency. Second, total exposure of the pathogen population to the antibiotics is low because, in general, the populations of biocontrol agents are localized on the root, therefore, selection pressures are minimized. Nevertheless, if sufficient selection pressure is applied to pathogens, the appearance of strains that are not controlled by biocontrol agents is inevitable. The use of antimicrobial agents in human medicine and agriculture has shown that selection pressures drive the evolution of resistance faster than we expect or hope for. For example, the development of multiple-drug resistance derived from spontaneous mutations was not predicted to occur, but now it is recognized as a common mechanism of antibiotic resistance in bacteria (Hachler *et al.*, 1991; Cohen *et al.*, 1993).

Resistance to antibiotics usually arises in a sensitive population by spontaneous mutation or by horizontal gene transfer (Davies, 1994; Nikaido, 1994; Spratt, 1994; Duffy *et al.*, 2003). Mutations conferring resistance may affect antibiotic uptake or target sensitivity. Resistance genes that can be transferred encode antibiotic modifying enzymes, resistant target molecules, or efflux pumps. Such genes, when carried by antibiotic-producing bacteria to prevent them from committing suicide when they produce the antibiotic, are known as self-resistance genes (Cundliffe, 1989;

Davies, 1994). Self-resistance genes could be transferred to target pathogens in the soil, therefore, it is essential to understand mutations conferring resistance in target pathogens as well as self-resistance genes in organisms producing them. The importance of self-resistance is illustrated in the crown gall biocontrol system. *Agrobacterium radiobacter* effectively controls crown gall, which is caused by *A. tumefaciens*, largely through the action of the antibiotic agrocin 84. Efficacy was threatened when agrocin 84-resistant strains of the pathogen were isolated from galls on *A. radiobacter* treated plants (Panagopoulos *et al.*, 1979). These strains were likely the result of transfer of plasmid pAgK84, which carries both agrocin 84 production and resistance genes, from the biocontrol strain to the pathogen (Vicedo *et al.*, 1993; Stockwell *et al.*, 1996). A deletion in the plasmid encoded functions for conjugal transfer generated a non-mobilizable derivative of pAgK84, which provided more stable disease suppression (Jones *et al.*, 1988). This example demonstrates both the possibility of antibiotic resistance in target pathogens and the importance of understanding the genetic basis for antibiotic resistance in the design of more robust strategies for biocontrol.

## Iron Competition

Biocontrol can involve suppression of the pathogen by depriving it of nutrients. The best understood example of this mechanism is iron competition. Iron is abundant in earth crust, but most of it is found in the highly insoluble form of ferric hydroxide, thus, iron is only available to organisms at concentrations at or below $10^{-18}$ M in soil solutions at neutral pH. This presents a challenge for bacteria, which require iron at micromolar concentrations for growth. Bacteria have evolved high-affinity iron uptake systems to shuttle iron into the cell (Neilands, 1981; Neilands and Nakamura, 1991). The typical system involves a siderophore and an uptake protein, which transports the siderophore into the cell. Siderophores are low molecular weight, high affinity iron(III) chelators that transport iron into bacterial cells. When grown under low iron conditions, fluorescent pseudomonads produce yellow-green fluorescent siderophores (of the pyoverdin type) and membrane receptor proteins that specifically recognize and take up the siderophore iron complex. (Duijff *et al.*, 1994a; Buysens *et al.*, 1996). An interesting aspect of siderophore biology is that diverse organisms can use the same type of siderophore. Microorganisms may use each other's siderophores if they contain the appropriate uptake protein (Koster *et al.*, 1993; Raaijmakers*et al.*, 1995a), and plants can even acquire iron from certain pseudobactins (Duijff *et al.*, 1994b).

Siderophores have been shown to be involved in the suppression of formae speciales of *Fusarium oxysporum, G. graminis* var. *tritici*, *Pythium* spp. and deleterious rhizosphere microorganisms (DRMO). Since siderophores sequester the limited supply of iron (III) in the rhizosphere, they limit its availability to pathogens and ultimately suppress their growth. The availability of iron (III) in the soil declines logarithmically with increasing soil pH. Thus, siderophore-mediated suppression should be greater in neutral and alkaline soils than in acidic soils. Pathogens are thought to be sensitive to suppression by siderophores for several reasons:

(a) They produce no siderophores of their own.

(b) They are unable to use siderophores produced by the antagonist or by other microorganisms in their immediate environment.

(c) They produce too little siderophore or a siderophore with a lower affinity for iron than those of the antagonist.

(d) They produce a siderophore that can be used by the antagonist, but they are unable to use the antagonist's siderophore.

Several recent reviews have summarized the evidence supporting a role siderophore in biocontrol. The most convincing evidence has been the fact that siderophore minus mutants are less suppressive to pathogens in the rhizosphere than parental strains.

## Parasitism

In addition to antibiosis and iron deprivation, certain biocontrol agents also reduce plant diseases by parasitizing pathogens. For example, *Trichoderma* spp. parasitize fungal plant pathogens. The parasite extends hyphal branches towards the target host, coils around and attaches to it with appressorium-like bodies, and punctures its mycelium (Chet *et al.*, 1981; Goldman *et al.*, 1994). These events require specific interactions between the parasite and fungal host, including the detection of chemical gradients and mycelial surface features. The specificity is illustrated by the observation that *Trichoderma* coils around *Pythium ultimum* hyphae but not plastic threads of a similar diameter (Dennis and Webster, 1971). Digestion of host cell walls is accomplished by a battery of excreted enzymes, including proteases, chitinases, and glucanases. These enzymes often have antifungal activity individually and are synergistic in mixtures or with antibiotics (Di Pietro *et al.*, 1993; Lorito *et al.*, 1993, 1994). Mycoparasitism has been suggested as a mechanism of biocontrol by *Trichoderma* spp. and *G. virens*, but its contribution to disease suppression remains uncertain, and clarification will be achieved with mutants lacking the cell wall-degrading enzymes. These mutants are difficult to generate because *Trichoderma harzianum* produces at least three distinct chitinases (de la Cruz *et al.*, 1992) as well as proteolytic and glucanolytic enzymes.

## Substrate Competition and Niche Exclusion

Competition for nutrients supplied by root and seed exudates probably occurs in most interactions between bacteria and pathogens on the root and is responsible at least to some small degree for the observed biocontrol by the introduced bacteria (Elad and Baker, 1985; Elad and Chet, 1987). Large populations of bacteria established on planting material and roots become a particular sink for nutrients in the rhizosphere, thus reducing the amount of carbon and nitrogen available to stimulate spores of fungal pathogens or for subsequent colonization of the root (Elad and Baker, 1985; Elad and Chet, 1987). Fluorescent pseudomonads are especially suited to "mopping up" nutrients, since they are nutritionally versatile and grow rapidly in the rhizosphere (Weller, 1988).

As suggested by Suslow, niche exclusion is potentially an important mechanism of antagonism of deleterious rhizobacteria (DRB) by plant growth promoting rhizobacteria (PGPR) (Suslow, 1982; Suslow and Scroth, 1982). Certain areas on the root, such as cell junctions and points of emergence of lateral roots, appear to be favoured for colonization by many kinds of bacteria, including DRB because root exudates are abundant there. Inoculating planting material with PGPR presumably prevents or reduces the establishment by DRB at these sites (Suslow, 1982). Although *Agrobacterium* strain 84 suppresses *A. tumefaciens* mainly by production of agrocin 84 (Kerr 1980), physical blockage of infection sites may also contribute to biocontrol (Cooksey and Moore, 1982; Plessis *et al.*, 1985).

## Other Modes of Management

Arbuscular mycorrhizal fungi (AMF), which form symbiotic associations with root systems of almost all plants, also reduce root diseases caused by several soil-borne pathogens, mainly through indirect mechanisms. The AMF penetrates the root system, improving plant nutrition and growth, altering the anatomy and architecture of the root system. These changes, together with the activation of the plant defence mechanisms, seem to be responsible for the reduction of the disease (Azcón-Aguilar *et al.*, 2002; Pozo *et al.*, 2002a). For example, colonization of tomato roots by *Glomus mosseae* reduces disease development in plants infected with *Phytophthora parasitica* and the involvement of plant defence mechanisms has been pointed out to be responsible (Pozo *et al.*, 1996; Cordier *et al.*, 1998; Pozo *et al.*, 1998; Pozo *et al.*, 1999; Pozo *et al.*, 2002b).

# INTERACTIONS WITH THE PLANT

## Colonization

It seems logical that a biocontrol agent should grow and persist, or "colonize" the surface of the plant it protects, and colonization is widely believed to be essential for management (Weller, 1983; de Weger *et al.*,1987; Parke, 1991). However, colonization, or even the initial population size of the biocontrol agent, has been shown to be significantly correlated with disease suppression in only a few instances (Paulitz and Baker,1987; Parke, 1990; Bull *et al.*, 1991). For example, in the suppression of damping-off of peas by *P. cepacia* (renamed as *Burkholderia cepacia*), there is a significant relationship between population size of the biocontrol agent and the degree of disease suppression (Parke, 1990). Also, suppression of take-all of wheat is correlated with colonization of roots by *P. fluorescens* strain 2-79 (Bull *et al.*, 1991). However, even in interactions that require colonization for disease suppression, the biocontrol agent may not be required at high population density. It is intriguing that certain highly effective biocontrol agents, such as *P. cepacia* and *B. cereus*, achieve only modest populations on roots of field-grown plants and appear to replace, not augment, the population of indigenous members of their species (Halverson *et al.*, 1993; King and Parke, 1996). These results suggest that roots have a carrying capacity, or a limit on the size of a population they can support, for certain species of bacteria (Goddard *et al.*, 2001). The structure of rhizobacterial communities is determined by the plant species, and differences in the composition and the amounts of root exudates probably account for the differences in microbial populations.

Understanding the interplay between the sizes of the populations of the biocontrol agent and the pathogen has progressed significantly due largely to a theoretical modeling approach introduced by Johnson (1994), and augmented by empirical verification by others (Raaijmakers, 1995b; Montesinos and Bonaterra, 1996). The application of mathematical models provides a depth of analysis that was not achieved in past work on biocontrol. From these works, it is evident that careful attention must be paid to the dose-response relationship with both microbial partners, because biological outcomes are grossly different when observations are made at very high or very low disease pressure. Elaboration of these mathematical models will introduce a time variable, describing the relationship between disease severity or incidence and populations of the biocontrol agent and pathogen during plant development.

The bacterial characteristics that contribute to active or passive spread on roots are not well understood (Parke, 1991). However, under field conditions, percolating water probably plays an essential role in the passive distribution of bacteria on roots (Parke *et al.*, 1986; Liddell and Parke, 1989). Nevertheless, motility is important for biocontrol in some plant-bacterial pairs under some but not all conditions (de Weger *et al.*, 1987; Bowers and Parke, 1993), and osmotolerance is correlated with colonization ability (Loper *et al.*, 1985). Cell surface characteristics influence attachment to roots, which may be necessary for colonization (Vesper, 1987; Anderson *et al.*, 1988). Certain mutations that affect accumulation of secondary metabolites also influence colonization of plant material in field soil (Mazzola *et al.*, 1992; Pfender *et al.*, 1993; Natsch *et al.*, 1994; Carroll *et al.*, 1995). A promising approach that will likely broaden the array of traits considered to be important for colonization is to screen mutants directly for increased or decreased ability to colonize roots (Lam *et al.*, 1991). Mutants of *Pseudomonas* strains of both phenotypes have been identified, and analysis of these mutants indicates that prototrophy for amino acids and vitamin 61, rapid growth rate, utilization of organic acids, and lipopolysaccharide properties contribute to colonization ability.

The study of root colonization by fungal biocontrol agents is more complex. Defining a fungal unit to quantify and the difficulty of conducting genetic analyses are challenges in studying the ecology of fungi. One study focussed on mutants of *Trichoderma* that are resistant to the fungicide, benomyl. These mutants are dramatically increased in root colonization and biocontrol ability, even in the absence of benomyl (Ahmad and Baker, 1987, 1988a). Benomyl esistance correlates with several phenotypes, including increased cellulase production and altered morphology, making it difficult to determine the basis for increased colonization (Ahmad and Baker, 1988b; Peterbauer *et al.,* 1992). It is possible that increased cellulase production enhances root colonization by enabling *Trichoderma* to utilize plant cell debris and that increased colonization enhances biocontrol.

## Microorganism-Mediated Induced Resistance

Some biocontrol agents induce a sustained change in the plant, increasing its tolerance to infection by a pathogen, a phenomenon known as induced resistance (Bakker *et al.*, 2003). In some cases, it is clear that induced resistance by biocontrol agents involves the same suite of genes and gene products involved in the well documented plant response known as systemic acquired resistance (SAR), but this is not always the case. SAR is typically a response to a localized infection or an attenuated pathogen, which is manifested in subsequent resistance to a broad range of other pathogens (Ross, 1961; Uknes *et al.*, 1992). The best understood examples of induced resistance occur in the biocontrol of above-ground diseases. The idea that biocontrol agents might induce resistance in the host was first suggested on the basis of experiments showing that bacterial treatments protected potato tubers from subsequent infection by *P. solanacearum* (Kempe and Sequeira, 1983). Then, it was shown that the biocontrol agent *P. fluorescens* strain CHAO (Maurhofer *et al.*, 1994a) induces SAR-associated proteins, confers systemic resistance to a viral pathogen, and induces accumulation of salicylic acid, which plays a role in signal transduction in SAR (Gaffney *et al.*, 1993; Maurhofer *et al.*, 2002). Mutants of CHAO that do not produce the siderophore pyoverdin do not induce SAR, suggesting a novel role for bacterial metabolites in disease suppression (Maurhofer *et al.*, 1994a; Iavicoli *et al.*, 2003). Another fluorescent pseudomonad, *P. putida,* induces expression of the gene encoding PRla, which is associated with the classical SAR response (Zdor and Anderson, 1992). Other strains of *P. fluorescens* do not induce expression of the gene products associated with the classic SAR response but appear instead to induce a functionally analogous response (Hoffland *et al.*, 1995).

Another line of evidence for induced resistance, which may or may not involve SAR, is that some biocontrol agents suppress disease when they are applied far from the site of infection by the pathogen, and they cannot be found at the infection site (Wei *et al.*, 1991; Zhou and Paulitz, 1994; Liu *et al.*, 1995). Furthermore, in suppression of *Fusarium* wilt by *P. fluorescens*, preparations of lipopolysaccharides from the bacterial cell surface induce resistance as effectively as the living bacteria, demonstrating that biocontrol is not necessarily due to transport of the bacteria or an antibiotic through the plant (Leeman *et al.*, 1995a, 1995b). Similar to fluorescent pseudomonads, specific strains of *Bacillus* sp. can provide plant protection by way of induced systemic resistance (Kloepper *et al.*, 2004). Generalizations about the signal-transduction pathways that are involved in induced systemic resistance are a bit complicated by the fact that the induced systemic resistance response to a given biocontrol agent depends on the plant species and cultivar.

## GENETIC MANIPULATION OF MICROORGANISMS FOR BIOCONTROL

Before biocontrol can become an important component of plant disease management, it must be effective, reliable, consistent and economical to meet the criteria of superior strain, together with delivery system that enhances biocontrol activity, must be developed. Existing biological control attributes can be enhanced by improving existing, known biological agents, with genetic manipulation. Genetic manipulations of biocontrol agents not only can enhance their activity, but also expand their spectrum. There are number of different ways in which bicontrol agent might be improved by genetic engineering. Some of these strategies are relatively straightforward requiring the addition of only a single functional gene, while others are more complex and will require biocontrol agents to be transformed with several genes at one time.

Consistent with the involvement of siderophores in disease suppression it was shown that a mutant strain of *Pseudomonas aeruginosa* that was selected for its lack of siderophore production no longer had the ability, shown by the wild type, to protect tomato plants against *Pythium* damping-off (Buysens *et al.*, 1996). In one case, increasing the amount of iron present in the soil to 40 mmol $Fe^{3+}$/L caused a parallel decrease in both the amount of fluorescent siderophore produced and the inhibitory effect against *Gaeumannomyces graminis* var. *tritici,* a destructive pathogen of wheat (Elsherif and Grossmann, 1994). Similarly, a single Tn5 insertion into the genome of *Alcaligenes* sp. strain MFA1 resulted in simultaneous loss of the ability of the bacterium to grow in the absence of iron, and to inhibit microconidial germination and germination-tube elongation of the pathogen *Fusarium oxysporum* (Martinetti and Loper, 1992).

Since each siderophore is encoded by a number of different genes, genetically engineering bacteria to produce modified siderophores is not a simple matter. On the other hand, it is possible to improve biocontrol agent by extending the range of iron-siderophore complexes that a particular strain can utilize, so that a genetically altered biocontrol strain could take up and use siderophores synthesized by other soil microorganisms, thereby giving it a competitive advantage. This was done by cloning the genes for iron-siderophore receptors from one biocontrol agent and introducing them into other strains (Marugg *et al.*, 1989)

One of the most effective mechanisms which biocontrol agents employ to manage phtyopathogen proliferation is the synthesis of antibiotics. Hence, the utility of biocontrol agents may be improved by providing them with genes that encode the biosynthesis of antibiotics normally produced by other bacteria. In this way it should be possible to extend the range of phytopathogens that a biocontrol agent is able to manage. The insertion of phenazine biosynthetic genes in the non-producing or poor strains enhances their biocontrol ability (Woeng *et al.*, 2001). Another way in which the activity of a biocontrol agent can be enhanced is by genetic manipulation to increase the amount of antibiotic that the biocontrol agent synthesizes. While some increase in the amount of antibiotic produced by a particular biocontrol agent might be obtained by conventional mutagenesis and selection, more extensive manipulation of antibiotic production will, in all likelihood, only be obtained through the use of recombinant DNA technology. Using this approach, when an antibiotic-producing (wild type) strain of *Pseudomonas fluorescens* was genetically manipulated to overproduce the antibiotics pyoluteorin and 2,4-diacetylphloroglucinol, the resultant strain protected cucumber plants against fruit rot caused by *Pythium ultimum* to a greater extent than did the wild-type strain (Maurhofer *et al.*, 1992; Schnider *et al.*, 1995).

Since many of the enzymes from biocontrol agents that have been found to lyse fungal cells, including chitinases and β-glucanases, are encoded by a single gene, it should be relatively straightforward matter to isolate some of these genes and then transfer them to other biocontrol agents, thereby constructing biocontrol agents that, for example, can produce both antibiotics and fungus degrading enzymes. In one series of experiments, a chitinase gene was isolated from the bacterium *Serratia marcescens* and then transferred into *Trichoderma harzianum* and *Rhizobium meliloti* cells (Chet and Inbar,1994). In both cases, the transformed microorganisms expressed the chitinase and subsequently displayed increased antifungal activity. When the *Serratia marcescens* chitinase gene was into the strain of *Pseudomonas fluorescens* that acts as a biocontrol agent, the transformant stably expressed and secreted active chitinase and was more effective biocontrol agent to manage the pathogen *Rhizoctonia solani* (Koby *et al.*, 1994).

Sun *et al.*, (1995) reported that *Pseudomonas putida* GR12-12 secretes antifreeze protein(s) into the surrounding medium when the bacterium is grown at low temperatures. This protein(s) may regulate the formation of ice crystals outside of the bacterium, thereby protecting it from damage that might otherwise occur at freezing temperatures. By using genetic engineering, the addition of antifreeze proteins synthesizing capability of biocontrol agents that are otherwise unable to permit and proliferate at cold temperatures may make a bacterium more effective by permitting it to thrive under these seemingly adverse conditions.

Another approach that is developed for the management of a variety of phytopathogenic agents include the development of transgenic plants that are resistant to one or more pathogenic agents. Towards this goal, Lorito *et al.*, (1998) have transferred the *Trichoderma atroviride ech 42* gene into tobacco and patato. This resulted in nearly complete resistance in tobacco and potato against to *Rhizoctonia solani*, *Alternaria solani*, *Alternaria alternata* and *Botrytis cinerea*. This strategy, also confirmed by subsequent studies (Boler *et al.*, 2000), is now used in many laboratories to improve disease resistance of tobacco, potato, cabbage, broccoli, apple, tomato grapes, lemon, wheat, rice, forest trees and several other plants (Lorito and Scala, 1999).

## INTERACTION OF BIOCONTROL AGENTS WITHIN MICROBIAL COMMUNITY

The interaction of the biocontrol agent with the microbial community may provide clues to explain why many organisms suppress disease effectively in the laboratory but fail to do so in the field. Biocontrol organisms may be affected by microbial communities, and they may influence the communities they enter. In some cases, they may enhance components of the community that work in concert to suppress disease. Evidences suggests that deliberate manipulation of microbial communities may be a highly robust and effective form of biocontrol (English and Mitchell, 1988; Boehm and Hoitink, 1992). The ability of *P. fluorescens* to suppress *Fusarium* wilt in radish seems partly due to its effects on the fungal community, in particular on the non-pathogenic strains of *Fusarium oxysporum* (Schippers, 1993). Certain fluorescent pseudomonads have also been shown to displace resident fungi and bacteria, in some cases reducing populations of deleterious microorganisms (Kloepper and Schroth, 1981; Yuen and Schroth, 1986). Treatments that enhance plant health can also increase the frequency of manganese-reducing bacteria in the rhizosphere community, thereby increasing manganese availability to the plant, which may in turn enhance resistance to disease (Huber and Wilhelm, 1988; Elmer, 1995). Introduction of the biocontrol agent *B. cereus* strain UW85 can induce dramatic changes in the composition of culturable bacterial communities on soybean roots in the field (Gilbert *et al.*, 1993).

The change results in a community that more closely resembles a bacterial community from non-rhizosphere soil than does a non-treated root community. This finding, coupled with substantial support from previous work on the effects of host resistance and soil amendments on rhizosphere communities, suggested the "camouflage" hypothesis, which proposes that a mechanism for protecting plant roots from attack by pathogens is to make the roots "look" more like soil by enhancing populations of typical soil inhabitants and reducing populations of typical root-associated microorganisms (Gilbert *et al.*, 1994). The camouflage hypothesis has not been tested directly, but it may provide a useful basis for the study of community processes that affect biocontrol of root-associated pathogens. There are three challenges associated with studying microbial changes resulting from the introduction of biocontrol agents. First, it is difficult to determine whether a community change plays a role in disease suppression or whether it is simply an unrelated outcome of altering the rhizosphere microflora. Second, the data generated in community analysis require new mathematical tools to deal with the complex communities and their multiple levels of interaction. Caution must be used in interpreting results from purely descriptive studies. The power of such studies can be enhanced by application of multivariate statistical modeling (Pfender and Wootke, 1988; Gilbert *et al.*, 1996). Finally, all of the work addressing the effects of biocontrol organisms on microbial communities has relied on culturing to describe communities. Since <1% (and perhaps far less than that) of the bacteria in soil are culturable (Faegri *et al.*, 1977; Torsvik *et al.*, 1990a, 1990b), studies based on culturing bacteria undoubtedly ignore some key organisms and interactions. The application of molecular analyses to describe non-culturable communities in both extreme and familiar environments (Ward *et al.*, 1990; Weisburg *et al.*, 1991; Amann etal., 1994, 1995) provides a powerful set of tools for the study of microbial interactions that influence biocontrol (O'Connell *et al.*, 1996).

## COMMERCIAL STATUS

Over the past hundred years, research has repeatedly demonstrated that phylogenitically diverse microorganisms can act as natural antagonists of various plant pathogens. Interactions that lead to management of pathogens by microorganisms include antibiosis, competition, induction of host resistance and predation (discussed earlier). Intensive screening of microorganisms has yielded numerous candidate organisms for commercial development. Some of the microbial taxa that have been successfully commercialized in the United States include bacteria belonging to the genera *Agrobacterium, Bacillus, Pseudomonas* and *Streptomyces* and fungi belonging to the genera *Ampelomyces, Candida, Coniothyrium* and *Trichoderma.* Biocontrol products are either marketed as stand-alone products or formulated as mixtures with other microbials (e.g. High stick N/T) or synthetic fungicide (e.g., Kodiak). Some products are sold as plant strengtheners or growth promoters without any specific claims regarding disease control.

In India the popularity of managing plant pathogens by biological means has increased in recent years, as extensive and systematic research has greatly demonstrated their effectiveness. Also, techniques for the mass production, storage, transport and application of biocontrol agents have been improved in recent years. There has been tremendous breakthrough in this aspect, especially on standardization of production techniques of *Trichoderma, Gliocladium, Paecilomyces, Pseudomonas, Trichogramma* and *Bacillus* to use them against many insect pests and plant pathogens. In India the major biological control agents used are against pest and insects, although some biocontrol agents have also been used against plant pathogens.

There are a number of instances where biocontrol agents have been successfully employed in India (Prasad, 2001). Some examples are as follows:

1. Growth of Lantana weed was controlled by using the bug *Telonemia scrupulosa.*
2. Sugarcane pyrilla has been successfully controlled in a number of states by the introduction of its natural enemy *Epricania melanoleuca* and *Tetrastictus pyrillae.*
3. Trichogramma, which feeds on the eggs of sugarcane borers, has been used against the borers in the states of Tamil Nadu, Rajasthan, UP, Bihar and Haryana.
4. Similarly, *Trichogramma*, *Bracon*, *Chelonus* and *Chrysopa* spp. are being used for the control of cotton bollworms. *Trichogramma* has also been used against rice stem borer and leaf folder.
5. The sugarcane scale insect has been controlled with the help of predatory Coccinelid beetles in UP, West Bengal, Gujarat and Karnataka.

Apart from being used against pests, several fungi and bacteria have been employed as biocontrol agents for plant pathogens (Prasad, 2001).

Fungi—(a) *Trichoderma* spp. against damping-off of peas, tomato and tobacco, root rot of ground-nut.

(b) *Gliocladium virens* against *Pythium*, *Sclerotium* and *Rhizoctonia* diseases.

Bacteria—*Pseudomonas fluorescens* and *P. putida* control soft rot of potato.

In India the present commercial production of microbial products is able to meet the demand of only 1% cropped area. There exists a wide gap, which can only be bridged by setting up of more units for production of biocontrol agents. This requires large-scale investment and private participation. Some of the local small-scale industries have already started production and marketing of *Trichoderma viride* (against few fungal diseases) and *Trichogramma* (against sugarcane early shoot borer). There is a great scope to enhance production and use of microbial management products in the days to come as the demand is on the increase every year.

## FUTURE PERSPECTIVES

Successful biocontrol of plant disease requires studies involving an intricate array of interactions. Understanding these interactions at the molecular and ecological levels will make possible the rational development of biocontrol for agriculture. Application of genetic analysis to microorganisms involved in biocontrol has led to substantial progress in understanding the microbial metabolites and regulatory genes involved in biocontrol. Ecological analyses have also described the responses of microbial communities to introduction of biocontrol agents.

The complexity of the interactions involved in biocontrol and the variability in environmental conditions found globally in agriculture make it unlikely that any one strain will suppress even a single disease in all settings. The genetic diversity of microorganisms with disease-suppressive potential remains a powerful yet largely untapped resource for biocontrol of plant disease (Kerr and Htay, 1974). There is a need to seek new biocontrol strains, particularly strains adapted to the site where they will be used (Cook, 1993; Stabb *et al.*, 1994), but if random bacterial isolates from each site are screened for disease suppressiveness as has been done for years, the effort will continue to be labour intensive and will not make use of existing knowledge of biocontrol mechanisms.

The integrated use of genetic, molecular, and ecological approaches will form the basis for significant future advances in biocontrol research. In particular, additional effort in three areas will be essential for developing a more complete understanding of biocontrol and for making practical use of biocontrol strategies for agriculture.

First, understanding mechanisms of pathogen resistance to the action of biocontrol agents is critical to sustain disease suppression with long-term use. Strategies to minimize resistance and prevent its spread should be designed. The second area that is ripe for study is genetic diversity within species of both biocontrol agent and host plant. Exploitation of genetic variation among members of a microbial species that suppress disease may provide a solution to the variability across space and time that has been observed with many biocontrol agents. The genetics of the host should be exploited for supportiveness of biocontrol, and hospitality to biocontrol agents should be enhanced through directed breeding or genetic modification of the host plant. The third, and most challenging area of research needed to explain the biological context for biocontrol is microbial community ecology. A better understanding of the microbial interactions that enhance or detract from biocontrol will determine the long-term success of biocontrol. In particular, attention needs to be paid to non-culturable members of the root and soil associated communities, because these microorganisms may be numerically dominant and have not been studied. Molecular methods developed for the study of microorganisms in their environments are key tools for the study of the influences of the microbial community on biocontrol.

## References

Ahmad, J.S. & Baker, R. (1988a) *Can. J. Microbiol.* **34**: 229-234.

Ahmad, J.S. & Baker, R. (1987) *Phytopathology* **77**: 182-189.

Ahmad, J.S. & Baker, R. (1988b) *Can. J. Microbiol.* **34**:694-696.

Ahmed, A. S., Sanchez, C. P. & Candela, M. E. (2000) *Eur. J. Plant Pathol.* **106**: 817–824.

Amann, R.I., Ludwig, W. & Schlelfer, K. H. (1994) *ASM News* **60**: 360-365.

Amann, R.I., Ludwlg, W. & Schlelfer, K.H. (1995) *Microbiol. Rev.* **59**: 143-169.

Anderson, A.J., Hablbzadegah-Tarl, P. & Tepper, C.S. (1988) *Appl. Environ. Microbiol.* **54**: 375-380.

Azcón-Aguilar, C., Jaizme-Vega, M.C. & Calvet, C. (2002) In: *Mycorrhizal Technology in Agriculture: From Genes to Bioproducts* (Eds. S. Gianinazzi, H. Schuepp, K.Haselwandter & . Barea, J.M.) ALS Birkhauser Verlag, Basel pp. 187-197.

Bakker, P. A. H. M., Ran, L. X., Pieterse, C. M. J. & van Loon, L. C. (2003) *Can. J. Plant Pathol.* **25**: 5–9.

Baldani, V.L.D., Alvarez, M.A.B., Baldani, J.I. & Dobereiner, J. (1986) *Plant Soil* **90**: 35-46.

Baldani, V.L.D. & Doberelner, J. (1980) *Soil Biol. Biochem.* **12**: 433-439.

Bangera, M.G. & Thomashow, L.S. (1996) *Mol. Plant-Microbe Interact.* **9**: 83-90.

Bargabus, R. L., Zidack, N. K., Sherwood, J. W. & Jacobsen, B. J. (2002) *Physiol. Mol. Plant Pathol.* **61**: 289-298.

Bargabus, R. L., Zidack, N. K., Sherwood, J. W. & Jacobsen, B. J. (2004) *Biological Contr.* **30**: 342-350.

Birgit K., Tommy H. Nielsen, D. S., Jens, B. A., Carsten, C. Søren, M. Michael, G. Jan S. & Ole N. (2002) *Appl. Environ. Microbiol.* **68**: 4509–4516.

Blumer, C. & Haas, D. (2000) *Arch. Microbiol.* **173**: 170–177.

Boehm, M.J. & Hoitink, H.A.J. (1992) *Phytopathology* **82**: 259-264.

Bowers, J.H. & Parke, J.L. (1993) *Soil Biol. Biochem.* **25**: 1693-1701.

Bull, C.T., Weller, D.M. & Thomashow, L.S. (1991) *Phytopathology* **81**: 954-959.

Buysens, S., Heungens, K., Poppe, J. & Hofte, M. (1996) *Appl. Environ. Microbiol.* **62**: 65-871.

Carroll, H., Moenne-Loccoz, Y., Dawling, D.N. & O'Gara, F. (1995) *Appl. Environ. Microbiol.* **61**: 3002-3007.

Chancey, S.T., Wood, D. W. & Pierson, L. S. (1999) *Appl. Environ. Microbiol.* **65**: 2294-2299.

Chet, I. & Inbar, J. (1994) *Appl. Biochem. Biotechnol.* **48**: 37-43.

Chet, I., Harman, G.E. & Baker, R. (1981) *Microb. Ecol.* **7**: 29-38.

Chin-A-Woeng, T. F. C., Thomas-Oates, J. E., Lugtenberg, B. J. J. & Bloemberg, G. V. (2001) *Mol. Plant Microbe Interact.* **14**: 1006–1015.

Choong-Min R. Mohamed A. F. Chia-Hui H. Munagala S. R. Joseph W. K. & Paul W. P. (2004) *Plant Physiol.* **134**: 1017–1026.

Cohen, S.P., Hachler, H. & Levy, S.B. (1993) *J. Bacteriol.* **175**: 1484-1492.

Cook, R.J. (1993) *Phytopathology* **31**: 5340-5351.

Cook, R.J., Thomashow, L.S., Weller, D.M., Fuiimoto, D., Mazzola, M., Bangera, G. & Kim, D. (1995) *Proc. Natl. Acad. Sci.* **92**: 4197-4201.

Cooksey, D.A. & Moore, L.W. (1982) *Phytopathology* **72**: 919-21.

Corbell, N. & Loper, J.E. (1995) *J. Bacteriol.* **177**: 6230-6236.

Cordier, C., Pozo, M.J., Barea, J.M., Gianinazzi, S. & Gianinazzi-Pearson, V. (1998) *Molecular Plant Microbe-Interactions* **11**: 1017-1028.

Cundliffe, E. (1989) *Annu. Rev. Microbiol.* **43**: 207-233.

Davies, J. (1994) *Science* **264**: 375-382.

de la Cruz, J., Hidalgo-Gallego, A., Lora, J.M., Benitez, T., Pintor Toro, J.A. & Llobell, A. (1992) *Eur. J. Biochem.* **206**: 859-867.

de Weger, L.A., van der Vlugt, C.I.M., Wijfjes, A.H.M., Bakker, P.A.H.M., Schippers, B. & Lugtenberg, B. (1987) *J. Bacteriol.* **169**: 2769-2773.

Dennis, C & Webster, J. (1971) *Trans. Br. Mycol. Soc.* **57**: 363-369.

Di Pletro, A., Lorito, M., Hayes, C.K., Broadway, R.M. & Harman, G.E. (1993) *Phytopathology* **83**: 308-313.

Dowling, D.N. & O'Gara, F. (1994) *Trends Biotechnol.* **12**: 133-141.

Du Plessis, H.J., Hattingh, M.T. & Van Vuuren, H.J.J. (1985) *Plant Dis.* **69**: 302-305.

Duffy, B., Schouten, A. & Raaijmakers, J. M. (2003) *Annu. Rev. Phytopathol.* **41**: 501-538.

Duijff, B. J., Gianinazzi-Pearson, V. & Lemanceau, P. (1997) *New Phytol.* **135**: 325-334.

Duijff, B.J., Bakker, P.A.H.M. & Schippers, B. (1994a) *Biocontrol Sci. Technol.* **4**: 279-288.

Duijff, B.J., de Kogel, W.J., Bakker, P.A.H.M. & Schippers, B. (1994b) *Soil Biol. Biochem.* **26**: 1681-1688.

Elad, Y. & Baker, R. (1985) *Phytopathology* **75**: 1053-59.

Elad, Y. & Chet, I. (1987) *Phytopathology* **77**: 190-95

Elmer, W.H. (1995) *Phytopathology* **85**: 1461-1467.

Elsherif, M. & Grossmann, F. (1994) *Microbiol. Res.* **149**: 371-377.

English, J.T. & Mitchell, D.J. (1988) *Phytopathology* **78**: 1484-1490.

Faegri, A., Torsvik, V.L. & Goksoyr, J. (1977) *Soil Biol. Biochem.* **9**: 105-112.

Gaffney, T.D., Friedrich, L., Vernooij, B., Negrotto, D., Nye, G., Uknes, S., Ward, E., Kessmann, H. & Ryals, J. (1993) *Science* **261**: 754-756.

Gaffney, T.D., Lam, S.T., Ligon, J., Gates, K., Frazelle, A., Di Maio, J., Hill, S., Goodwin, S., Torkewitz, N., Ailshouse, A.M., Kempf, H.J. & Becker, O.J. (1994) *Mol. Plant-Microbe Interact.* **7**: 455-463.

Gilbert, G.S., Clayton, M.K., Handelsman, J. & Parke, J.L. (1996) *Microb. Ecol.* **32**: 123-147.

Gilbert, G.S., Handelsman, J. & Parke, J.L. (1994) *Phytopathology* **84**: 222-225.

Gilbert, G.S., Parke, J.L., Clayton, M.K. & Handelsman, J. (1993) *Ecology* **74**: 840-854.

Goddard, V. J., Bailey, M. J., Darrah, P., Lilley, A. K. & Thompson, I. P. (2001) *Plant Soil* **232**: 181-193.

Goldman, G.H., Hayes, C. & Harman, G.E. (1994) *Trends Biotechnol.* **12**: 478-482.

Hachler, H., Cohen, S.P. & Levy, S.B. (1991) *J. Bacteriol.* **173**: 5532-5538.

Halverson, L.J., Clayton, M.K. & Handelsman, J. (1993) *Soil Biol. Biochem.* **25**: 485-493.

Harman, G. E., Howell, C. R., Viterbo, A., Chet, I. & Lorito, M. (2004) *Nature Rev. Microbiol.* **2**: 43-56.

Hoffland, E., Pieterse, C.M.J., Bik, L. & van Pelt, J.A. (1995) *Physiol. Mol. Plant Pathol.* **46**: 309-320.

Homma, Y., Kato, Z., Hirayama, F., Konno, K., Shirahama, H. & Suzui, T. (1989) *Soil Biol. Biochem.* **21**: 723-728.

Howell, C. R. & Stipanovic, R. D. (1980) *Phytopathology* **70**: 712-715.

Howell, C. R., Hanson, L. E., Stipanovic, R. D. & Puckhaber, L. S. (2000) *Phytopathology* **90**: 248-252.

Howell, C.R. & Stipanovic, R.D. (1983) *Can. J. Microbiol.* **29**: 321-324.

Howell, C.R., Stipanovic, R.D. & Lumsden, R.D. (1993) *Biocontrol Sci. Technol.* **3**: 435-441.

Huber, D.M., & Wilhelm, N.S. (1988) In: *Manganese in Soils and Plants* (Eds. Graham, R.D., Hannam, R.J., & Uren, N.C.) Kluwer Academic Publishers, Boston, MA pp. 155-173.

Iavicoli, A., Boutet, E., Buchala, A. & Métraux, J.P. (2003) *Mol. Plant Microbe Interact.* **16**: 851-858.

Islam, T. M., Hashidoko, Y., Deora, A., Ito, T. & Tahara, S. (2005) *Appl. Environ. Microbiol.* **71**: 3786-3796.

James, M. L., Dwight, S. H., Philip, E. H., Nancy, R. T., Dirk. H., Hans-Joachim K. & Karl-Heinz van Pée (2000) *Pest Manag. Sci.* **56**: 688-695.

Johnson, K.B. (1994) *Phytopathology* **84**: 780-784.

Jones, D.A., Ryder, M.H., Clare, B.G., Farrand, S.K. & Kerr, A. (1988) *Mol. Gen. Genet.* **212**: 207-214.

Keel, C., Schnider, U., Maurhofer, M., Voisard, C, Laville, J., Burger, U., Wirthner, P., Haas, O. & Defago, G. (1992). *Mol.Plant-Microbe Interact.* **5**: 4-13.

Keel, C., Wirthner, P., Oberhansli, T., Voisard, C., Burger, U., Haas, D. & Defago, G. (1990) *Symbiosis* **9**: 327-341.

Kempe, J. & Sequeira, L. (1983) *Plant Dis.* **67**: 499-503.

Kerr, A. (1980) *Plant Dis.* **64**: 25-30.

Kerr, A. & Htay, K. (1974) *Physiol. Plant Pathol.* **4**: 37-44.

King, E.B. & Parke, J.L. (1996) *Soil Biol. Biochem.* **28**: 307-312.

Kloepper, J. W, Ryu, C. M. & Zhang, S. (2004) *Phytopathology* **94**: 1259-1266.

Kloepper, J.W. & Schroth, M.N. (1981) *Phytopathology* **71**: 1020-1024.

Koby, S., Schickler, H., Chet, I. & Oppenheim, A.B. (1994) *Gene* **147**: 81-83.

Koike, N. *et al.*( 2001) *Eur. J. Plant Pathol.* **107**: 523-533.

Kolter, R., Siegele, D.A. & Tormo, A. (1993) *Annu. Rev. Microbiol.* **47**: 855-874.

Koster, M., van de Vossenberg, J., Leong, J. & Weisbeek, P.J. (1993) *Mol. Microbiol.* **8**: 591-601.

Koumoutsi, A., Chen, X. H., Henne, A., Liesegang, H., Gabriele, H., Franke, P., Vater, J. & Borris, R. (2004) *J. Bact.* **186**: 1084-1096.

Lam, S., Ellis, DA. & Ligon, J.M. (1991) In: *The Rhizosphere and Plant Growth* (Eds. Keister, D.L. & Cregan, P.B.) Kluwer Academic Publishers, Boston, MA pp. 43-50.

Laville, J., Voisard, C., Keel, C., Maurhofer, M., Defago, G. & Haas, D. (1992) *Proc. Natl. Acad. Sci.* **89**: 1562-1566.

Leclere, V., Bechet, M., Adam, A., Guez, J. S., Wathelet, B., Ongena, M., Thonart, P., Gancel,F., Chollet-Imbert, M. & Jacques, P. (2005) *Appl. Environ. Microbiol.* **71**: 4577-4584.

Leeman, M., van Pelt, J.A., den Ouden, F.M., Heinsbroek, M., Bakker, P.A.H.M. & Schippers, B. (1995b) *Phytopathology* **85**: 1021-1027.

Leeman, M., van pelt, J.A., Hendreckx, M.J., Scheffer, R.J., Bakker, P.A.H.M. & Schippers, B. (1995a) *Phytopathology* **85**: 1301-1305.

Liddell, C.M. & Parke, J.L. (1989) *Phytopathology* **79**: 1327-1332.

Liu, L., Kloepper, J.W. & Tuzun, S. (1995) *Phytopathology* **85**: 1064-1068.

Loper, J.E., Haack, C. & Schroth, M.N. (1985) *Appl. Environ. Microbiol.* **49**: 416-422.

Lorito, M., Harman, G.E., Hayes, C.K., Broadway, R.M., Tronsmo, A., Woo, S.L. & Di Pietro, A. (1993) *Phytopathology* **83**: 302-307.

Lorito, M., Peterbauer, C., Hayes, C.K. & Harman, G.E. (1994) *Microbiology* **140**: 623-629.

Lucy, M., Reed, E. & Glick, B. R. (2004) *Antonie van Leeuwenhoek* **86**: 1-25.

Lumsden, R.D., Ridout, C.J., Vendemia, M.E. & Harrison, D.J. (2004) *J. Bacteriol.* **186**: 4986-4993.

Martinetti, G. & Loper, J.E. (1992) *Can. J. Microbiol.* **38**: 241-247.

Marugg, J.D., de Weger, L.A., Nielander, H.B., Oorthuizen, N., Recourt, K., Lugtenberg, B., van der Hofstad, G.A.J.M. & Weisbeek, P.J. (1989) *J. Bacteriol.* **170**: 1812-1819.

Maurhofer, M., Hase, C., Meuwly, P., Metraux, J.P. & Defago, G. (1994a). *Phytopathology* **84**: 139-146.

Maurhofer, M., Keel, C., Haas, D. & Defago, G. (1994b) *Eur. J. Plant Pathol.* **100**: 221-232.

Maurhofer, M., Keel, C., Schnider, U., Voisard, C., Haas, D. & Defago, G. (1992) *Phytopathology* **82**:190-195.

Maurhofer, M., Reimmann, C. Schmidli-Sacherer, P., Heeb, S., Haas, D. & Défago G. (1998) *Phytopathology* **88**: 678-684. .

Mazzola, M., Cook, R.J., Thomashow, L.S., Weller, D.M. & Pierson, L.S. (1992) *Appl. Environ. Microbiol.* **58**: 2616-2624.

Meziane, H., Van der Sluis, I., Van Loon, L.C., Hofte, M. & Bakker, P. A. H. M. (2005) *Mol. Plant Pathol.* **6**: 177-185.

Milner, J.L., Silo-Suh, L.A., Lee, J.C., He, H., Clardy, J. & Handelsman, J. (1996) *Appl. Environ. Microbiol.* **62**: 3061-3065.

Milner, J.L., Stohl, E.A. & Handelsman, J. (1996) *J. Bacteriol.* **178**: 4266-4272.

Montesinos, E. & Bonaterra, A. (1996) *Phytopathology* **86**: 464-472.

Moyne, A. L., Shelby, R., Cleveland, T. E. & Tuzun, S. (2001) *J. Appl. Microbiol.* **90**: 622-629.

Natsch, A., Keel, C., Pfirter, H.A., Haas, D. & Defago, G. (1994) *Appl. Environ. Microbiol.* **60**: 2553-2560.

Neilands, J.B. (1981) *Annu. Rev. Biochem.* **50**: 715-731.

Neilands, J.B. & Nakamura, K. (1991) In: *CRC Handbook of Microbial lron Chelates* (Ed. Winkelmann, G.) CRC Press, London pp. 1-14.

Nikaido, H. (1994) *Science* **264**: 382-388.

O'Connell, K.P., Goodman, R.M. & Handelsman, J. (1996) *Trends Biotechnol.* **14**: 83-88.

Ongena, M., Duby, F., Rossignol, F., Fouconnier, M. L., Dommes, J. & Thonart, P. (2004) *Mol. Plant-Microbe Interact.* **17**: 1009-1018.

Osburn, R.M., Milner, J.L., Oplinger, E.S., Smith, R.S. & Handelsman, J. (1995) *Plant Dis.* **79**: 551-556.

Ownley, B. H., Duffy, B. K. & Weller, D. M. (2003) *Appl. Environ. Microbiol.* **69**: 3333-3343.

Panagopoulos, C.G., Psallidas, P.G. & Alivizatos, A.S. (1979) In: *Soil-Borne Plant Pathogens* (Eds. Schippers, B. & Gams, W.) Academic Press New York pp. 569-578.

Parke, J.L. (1990) *Phytopathology* **80**: 1307-1311.

Parke, J.L. (1991) In: *The Rhizosphere and Plant Growth* (Eds. Keister, D.L. & Cregan, P.B.) Kluwer Academic Publishers Boston, MA pp. 33-42.

Parke, J.L., Moen, R., Rovira, A.D. & Bowen, G.D. (1986) *Soil Biol. Biochem.* **18**: 583-588.

Paulitz, T. & Belanger, R. (2001) *Annu. Rev. Phytopathol.* **39**: 103-133.

Paulltz, T.C. & Baker, R. (1987) *Phytopathology* **77**: 335-340.

Peterbauer, C.K., Heidenreich, E., Baker, R. & Kubicek, C. (1992) *Can. J. Microbiol.* **38**: 1292-1297.

Pfender, W.F. & Wootke, S.L. (1988) *Microb. Ecol.* **15**: 95-113.

Pfender, W.F., Kraus, J. & Loper, J.E. (1993) *Phytopathology* **83**: 1223-1228.

Pierson Ill, L.S. & Pierson, E.A. (1996) *FEMS Microbiol. Lett.* **136**: 101-108.

Pierson Ill, L.S. & Thomashow, L.S. (1992) *Mol. Plant-Microbe Interact.* **5**: 330-339.

Pierson Ill, L.S., Gaffney, T., Lam, S. & Gong, F. (1995) *FEMS Microbiol. Lett.* **134**: 299-307.

Pierson Ill, L.S., Keppenne, V.D. & Wood, D.W. (1994) *J. Bacteriol.* **176**: 3966-3974.

Pozo, M.J., Azcón-Aguilar, C., Dumas-Gaudot, E. & Barea, J.M. (1999) *Plant Science* **141**: 149-157.

Pozo, M.J., Cordier, C., Dumas-Gaudot, E., Gianinazzi, S., Barea, J.M. & Azcón-Aguilar, C. (2002b) *J. Exp. Bot.* **53**: 525-534.

Pozo, M.J., Dumas-Gaudot, E., Azcón-Aguilar, C. & Barea, J.M. (1998) *J. Exp. Bot.* **49**: 1729-1739.

Pozo, M.J., Dumas-Gaudot, E., Slezack, S., Cordier, C., Asselin, A., Gianinazzi, S., Gianinazzi-Pearson, V., Azcón-Aguilar, C. & Barea, J.M. (1996) *Agronomie* **16**: 689-697.

Pozo, M.J., Slezack-Deschaumes, S., Dumas-Gaudot, E., Gianinazzi, S. & Azcón-Aguilar, C. (2002a) In: *Mycorrhizal Technology in Agriculture: From Genes to Bioproducts* (Eds. Gianinazzi, S., Schuepp, H., Haselwandter, K. & Barea, J.M.) ALS Birkhauser Verlag, Basel pp. 103-111.

Press, C. M., Loper, J. E. & Kloepper, J. W. (2001) *Phytopathology* **91**: 593-598.

Prasad, S.S. (2001) Country Report – India. Annual meeting of Programme Advisory Committee, Ayutthaya, Thailand. November, 2001.

Raaijmakers J.M. & Walter, J.F. (1992) *Can. J. Microbiol.* **38**: 1274-1280.

Raaijmakers, J. M., Vlami, M. & de Souza, J. T. (2002) *Antonie van Leeuwenhoek* **81**: 537-547.

Raaijmakers, J.M., Leeman, M., van Oorschot, M.M.P., van der Sluils, I., Schippers, B. & Bakker, P.A.H.M. (1995b) *Phytopathology* **85**: 1075-1081.

Raaijmakers, J.M., van der Sluis, I., Koster, M., Bakker, P.A.H.M., Weisbeek, P.J. & Schippers, B. (1995a) *Can. J. Microbiol.* **41**: 126-135.

Reimmann, C., Valverde, C., Kay, E. & Haas, D. (2005) *J. Bacteriol.* **187**: 276-285.

Ross, A.F. (1961) *Virology* **14**: 340-358.

Sandra, A. I., Wright, C. H., Zumoff, L. S. & Steven, V. B. (2001) *Appl. Environ. Microbiol.* **67**: 282-292.

Sarniguet, A., Kraus, J., Henkels, M.D., Muehlchen, A.M. & Loper, J.E. (1995) *Proc. Natl. Acad. Sci.* **92**: 12255-12259.

Schippers, B. (1993) *Phytoparasitica* **21**: 275-279.

Schisler, D. A., Slininger, P. J., Behle, R. W. & Jackson, M. A. (2004) *Phytopathology* **94**: 1267-1271.

Schnider, U., Keel, C., Blumer, C., Troxier, J., Defago, G. & Haas, D. (1995) *J. Bacteriol.* **177**: 5387-5392.

Shanahan, P., O'Sullivan, D. J., Simpson, P., Glennon, J. D. & O'Gara, F. (1992) *Appl. Environ. Microbiol.* **58**: 353-358.

Silo-Suh, L.A., Lethbridge, B.J., Raffel, S.J., He, H., Clardy, J. & Handelsman, J. (1994) *Appl. Environ. Microbiol.* **60**: 2023-2030.

Smith, K. P., Havey, M. J. & Handelsman, J. (1993) *Plant Dis.* **77**: 139-142.

Spratt, B.G. (1994) *Science* **264**: 388-393.

Stockwell, V.O., Kawalek, M.D., Moore, L.W. & Loper, J.E. (1996) *Phytopathology* **86**: 31-37.

Sun, X., Griffith, M., Pasternak, J.J. & Glick, B.R. (1995) *Can. J. Microbiol.* **41**: 776-784.

Suslow, T.V. (1982). In: *Phytopathogenic Prokaryotes* (Eds. Mount, M.S. & Lacy, G.H.). Acdemic London pp. 187-223.

Suslow, T.V. & Schroth, M.N. (1982) *Phytopathology* **72**: 111-115.

Thomashow, L. S., Weller, D. M., Bonsall, R. F. & Pierson, L. S. III. (1990) *Appl. Environ. Microbiol.* **56**: 908-912.

Thomashow, L.S. & Weller, D.M. (1988) *J. Bacteriol.* **170**: 3499-3508.

Timms-Wilson, T. M. *et al.* (2000) *Mol. Plant Microbe Interact.* **13**: 1293-1300.

Torsvik, V., Goksoyr, J. & Daae, F.L. (199Oa) *Appl. Environ. Microbiol.* **56**: 782-787.

Torsvik, V., Kare, S., Sorheim, R. & Goksoyr, J. (1990b) *Appl. Environ. Microbiol.* **56**: 776-781.

Troxler, J., Berling, C.-H., Moënne-Loccoz, Y., Keel, C. & Défago, G. (1997) *Plant Pathol.* **46**: 62-71.

Turner, J.T. & Backman, P.A. (1991) *Plant Dis.* **75**: 347-353.

Uknes, S., Mauch-Manl, B., Moyer, M., Potter, S., Williams, S., Dincher, S., Chandler, D., Slusarenko, A., Ward, E. & Ryals, J. (1992) *Plant Cell* **4**: 645-656.

Van Peer, R. & Schippers, B. (1992) *Neth. J. Plant Pathol.* **98**: 129-139.

Van Wees, S. C. M., Pieterse, C. M. J., Trijssenaar, A., Van't Westende, Y. & Hartog, F. (1997) *Mol. Plant-Microbe Interact.* **10**: 716-724.

Vesper, S.J. (1987) *Appl. Environ. Microbiol.* **53**: 1397-1405.

Vicedo, B., Penalver, R., Asins, M.J. & Lopez, M.M. (1993) *Appl. Environ. Microbiol.* **59**: 309-315.

Voisard, C, Keel, C., Haas, D. & Defago, G. (1989) *EMBO J.* **8**: 351-358.

Ward, D.M., Weller, R. & Bareson, M.M. (1990) *Nature* **345**: 63-65.

Wei, G., Kloepper, J.W. & Tuzun, S. (1991) *Phytopathology* **81**: 1508-1512.

Weisburg, W.G., Barns, S.M., Pelletier, D.A. & Lane, D.J. (1991) *J. Bacteriol.* **173**: 697-703.

Weller, D.M. (1983) *Phytopathology* **73**: 1548-1553.

Weller, D.M. (1988) *Annu. Rev. Phytopathol.* **26**: 379-407.

Weller, D.M. & Cook, R.J. (1983) *Phytopathology* **73**: 463-469.

Whipps, J.M. (2001) *J. Exp. Bot.* **52**: 487-511.

Wilhite, S. E., Lunsden, R. D. & Strancy, D. C. (2001) *Appl. Environ. Microbiol.* **67**: 5055-5062.

Wilhite, S.E., Lumsden, R.D. & Straney, D.C. (1994) *Phytopathology* **84**: 816-821.

Wood, D.W. & Pierson Ill, L.S. (1996) *Gene* **168**: 49-53.

Yuen, G.Y. & Schroth, M.N. (1986) *Phytopathology* **76**: 176-180.

Zdor, R.E. & Anderson, A.J. (1992) *Plant Soil* **140**: 99-107.

Zhou, T. & Paulitz, T.C. (1994) *J. Phytopathol.* **142**: 51-63.

14

# Retrospects and Prospects of Tomato Leaf Curl Virus Research in India

**J.P. Tewari and N.A. Ansari**

*Plant Pathology Research Lab, Department of Botany, M.L.K. P.G. College, Balrampur- 271201 (U.P.) India*

**ABSTRACT**

Tomato *Lycopersicon esculentum* Mill family Solanaceae is one of the most popular and extensively grown vegetables in the country. The crop is subjected to attack by a number of biotic agents like fungi, viruses, bacteria, mycoplasma and nematodes. Among the different diseases attacking tomato, leaf curl diseases (TOLCD) is the most important, as sometimes it leads to cent percent crop losses. This chapter presents an overview of the tomato leaf curl virus as timely pathogens in the context of their distribution, symptoms, host range, incidence, losses, molecular approaches of characterization, virus vector interaction, various factors leading to epidemics, metabolic changes in infected plant, histopathology of infected plant and various approaches of disease control. Future prospect on different aspects of this virus in India has been also speculated.

## INTRODUCTION

Tomato, *Lycopersicon esculentum* Mill. is one of the most important vegetable crops cultivated in India. Tomatoes are a good source of vitamins in addition to their flavouring characters for other foods. Medicinally, fresh tomatoes are rich source of vitamin ‘c’, which is a safeguard against several diseases. Besides, tomato also contains ‘lycopene’ which is chiefly responsible for red colour and helpful in checking cancer and cardiovascular diseases. It is also effective in reducing cholesterol. Tomato contains also other nutrients essential for our body, hence, it is largely consumed as protective food. Tomatoes are used as salad, soup, sauce, ketchup, juice, paste, pickles and mixed vegetables (Sachan, 2004).

In India, tomato is grown nearly throughout the year because of favourable agro-climatic conditions and also being day neutral plants in one part or another. India is the second largest producer of tomato after China with an annual production of 7.42 million tonnes from 0.456 million ha area, productivity being 16.32 t/ha (Chhabra, 1992; Varghese and Chandramony, 2003). Tomato is cultivated especially in autumn in East and North India, whereas in the South India it remains available in fields throughout the year. Continuous growing of tomato has resulted in building up the population of insect pests and disease vectors and in the perpetuation of inoculum of a large number of diseases. Thirty-six virus diseases of tomato crops are reported in the world (Martyn, 1968), of which eight are found in India. Diseases like tomato leaf curl, tomato mosaic, tomato necrosis, tomato flavescence and tomato spotted wilt virus infect this crop in different parts of the country. Among them, tomato leaf curl virus (TLCV) is the most serious disease, taking a heavy toll of the crop. Tomato leaf curl disease, first reported in the Sudan (Cowland, 1932), differs from the other virus diseases of tomato, such as tomato yellow leaf curl (Cohen and Nitzany, 1966) and yellow mosaic (Verma *et al.,* 1975). The only reported difference between tomato yellow leaf curl virus (TYLCV) and TLCV is that TYLCV cannot infect *Nicotiana tabacum* or *Hibiscus esculentum* (Cohen and Nitzany, 1966; Nitzany, 1975), whereas the latter can infect these two plants (Yassin and Nour, 1965). Tomato yellow leaf curl disease producing almost similar symptoms is caused by another begomovirus, which has not been reported from the Indian subcontinent (Rishi, 2004).

Vasudeva and Somraj reported the occurrence of tomato leaf curl virus in India in 1948. In nature, the disease is characterized by bushy plant growth, reduction in leaf size accompanied by curling, rolling, twisting, puckering of the leaves and deformation of the flowers. The disease is transmitted by whitefly, *Bemisia tabaci* Genn. It affects all stages of the plant causing 40-90% loss, depending on the stage of the crop at the time of infection (Datar, 1984; Dasgupta *et al.,* 2003; Ansari *et al.,* 2007b).

This chapter presents an overview of the retrospects and prospects of tomato leaf curl virus research in India.

## SYMPTOMS AND HOST RANGE

Virus-infected plants show vein clearing, pronounced dwarfing, puckering and curling of leaves, dwarfing of flowers and poor fruit setting (Vasudeva and Somraj, 1948; Nair and Wilson, 1969). Another strain of TLCV causes enations on the lower side of the leaves (Nariani, 1968) and thickening of the veins (Ansari and Tewari, 2004). Tomato leaf curl disease is caused by three strains of tobacco leaf curl virus (Singh and Lal, 1964). The first is a typical strain known as enation leaf curl virus of tomato. The second strain causes slight leaf curl and twisting of the leaflets, but no reduction in leaf size. The third one is a golden yellow strain which is characterized by upward rolling of leaf margins and islands of golden colour scattered in green tissues. The viral nature of the disease has been established beyond doubt through grafting, dodder and transmission by the whitefly, *Bemisia tabaci* Genn. (Reddy and Yaraguntaiah, 1979).

Five isolates of TLCV have been reported on the basis of differences in symptoms (Reddy *et. al.,* 1981; Ansari and Tewari, 2004). Symptoms of TLCV on tomato plants differ, depending upon environmental condition. When average daily temperature was low, early symptoms on infected plants were leaf curling with reduction in leaf size (Fig. 1A). Under similar condition some plants had typical mosaic symptoms, which varied from mild to severe, with thickening of main veins and vein lets but no reduction in leaf size (Fig. 1A). With the rise in the temperature some infected

**Fig.** 1A. Healthy, 1B. Diseased

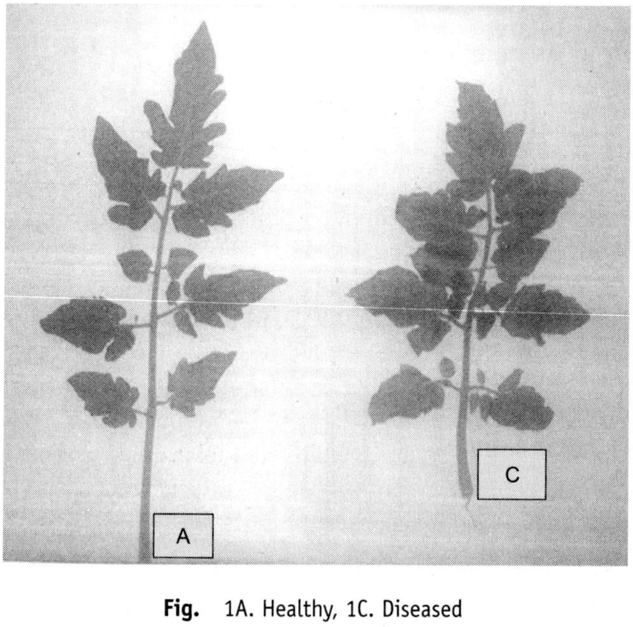


**Fig.** 1A. Healthy, 1C. Diseased

**Fig.** 1A. Healthy, 1B. Diseased

**Fig.** 1B. Healthy, 1E. Diseased

**Fig.** 1C. Healthy, 1F. Diseased

tomato plants, infected by golden yellow strain, were characterized by upward rolling and yellow golden coloration scattered in green tissue in the older leaves (Fig. 1D). The reduction in size of the leaves is more pronounced in successive leaves and is accompanied by shortening of the internodes with the result that the leaves are bushy, crowded, puckering, curling and stunting (Fig. 1E). Puckering of the leaflet is a common symptom of the disease and is severe in some plants. In the case of late infection, the plant bears less number of fruits but in earlier infections the symptoms are much more pronounced and the few flowers that are formed bear reduced floral parts and never attained normal size. They seldom open and soon drop off. The flowers that develop after infection are invariably sterile. The diseased plants usually develop purple patches, specially on the older leaves, and these vary in intensity in different plants (Fig. 1F). Fruit production is usually dependent upon disease severity and the growth stage of the tomato plants at the time of infection (Rataul and Brar, 1989).

The TLCV is readily transmitted to tomatoes, chillies, apaya and *Datura stramonium*. Besides infecting these plants, it is also known to infect other host plants such as *Abutilon indicum, Althea rosea, Ageratum conyzoides, Acanthospermum hispidum, Croton bonplandianum, Crotolaria juncea, Euphorbia hirta, Euphorbia geniculata, Jatropha curcas, Launia asplenifolia, Nicotiana glutinosa, Nicotiana tabacum, Galinosoga parviflora, Nicandra physaloides, Nicotiana plumbaginifolia, Rumex dentatus, Salvia plebia, Solanum nigrum, Sida moysoriansis, Sida rhomboidea, Sida rhobamdifolia, Sonchus arvensis, Sonchus asper, Scoparia dulcis, Shizanthus sp., Physalis peruviana, Petunia hybrida, Vernonia cineria Xanthium strumarium* and *Zinnia elegans* (Rataul and Butter, 1977; Reddy and Yaraguntaiah, 1981c, Saikia and Muniyappa, 1989; Ansari *et al.,* 2004a, 2006a; Ansari and Tewari, 2005b). However, symptoms of the disease in the following plant species could not be observed, although systemic infection was established by positive transmission tests, *Lens esculenta* Moench; *Lycopersicon peruvianum* (L) Mill., *Lycopersicon pimpinellifolium* Mill., *Malva nicaensis* All., *Nicotiana glutinosa* L., *Nicotiana tabacum* L and *Phaseolus vulgaris* L (Cohen and Nitzany, 1966).

The whitefly, *Bemisia tabaci* Genn. has a wide host range. Besides tomatoes and chillies, a large number of plants such as Bhindi, Mungbean, Urdbean, Pigeon pea, Soybean, Cassava and Cotton have been shown to be hosts in experimental studies. In nature, it has been observed on 142 plant species belonging to 23 different families (Saikia and Muniyappa, 1989). Previous researches indicate that weeds play an important role in the spread of viruses (Duffus, 1971; Sastry, 1984; Ansari *et al.*, 2004b). Thus, some of the weed plants recorded as host of both the whitefly or virus are *Acanthospermum hispidum, Amaranthus spinosus, Amaranthus viridis, Cleome viscosa, Blainvillea rhomboiden, Flavaria aurtralasica, Malvastrum coromandaliarum, Parthenium hysterophorus, physalis minima, Trianthema portulacastrum* (Table 1). Forty-five plants served either as hosts or as food plants (Mariyappan and Narayanswamy, 1986).

**Table 1** *Transmission test of tomato leaf curl virus from tomato to different plant species by whitefly (Bemisia tabaci)[#].*

| *Family/Host species* | *Symptoms** | *Incubation period (days)* |
|---|---|---|
| **Asteraceae** | | |
| *Ageratum conyzoides* | VT+LC+T+RLS | 25 |
| *Chrysanthemum* sp. | NI | - |
| *Dahlia* sp. | NI | - |
| *Helianthus annuus* | NI | - |
| *Parthenium hysterophorus* | SC+VT+B+S+RLS+D | 25 |
| *Sonchus arvensis* | VT+LC+UC | 20 |
| *Sonchus asper* | SC+B+S+E+D | 20 |
| *Tagetes erecta* | NI | - |
| *Vernonia cineria* | CP+LC | 21 |
| *Xanthium strumarium* | VT+LC+DC | 25 |
| *Zinnia elegans* | LVT+LC | 20 |
| **Acanthaceae** | | |
| *Acanthospermum hispidum* | LC+YG+CP | 25 |
| *Ruellia prostrata* | NI | - |
| **Amaranthaceae** | | |
| *Achyranthus aspera* | NI | - |
| *Gomphrena globosa* | NI | - |
| **Chenopodiaceae** | | |
| *Chenopodium album* | NI | - |
| *C. amaranticolor* | NI | - |
| *C. murale* | NI | - |
| *C. quinoa* | NI | - |
| **Cruciferae** | | |
| *Brassica oleraceae* var. *capitata* | NI | - |
| *Brassica rapa* | NI | - |
| **Capparidaceae** | | |
| *Cleome gynandra* | NI | - |
| *Cleome viscosa* | NI | - |

*Table 1 Contd.*

*Table 1 Contd.*

| | | |
|---|---|---|
| **Euphorbiaceae** | | |
| *Acalypha indica* | NI | - |
| *Croton bonplandianum* | LC+T+RLS | 28 |
| *Euphorbia geniculata* | VC+LC | 24 |
| *Euphorbia hirta* | VC+YG+LC | 24 |
| *Jatropha curcas* | VC+LC+YG | 35 |
| *Jatropha gossypifolia* | NI | - |
| *Ricinus communis* | NI | - |
| **Fabaceae** | | |
| *Clitoria ternatia* | NI | - |
| *Crotalaria juncea* | VT+E+LC | 18 |
| *Lens esculenta* | VT+LC | 18 |
| *Phaseolus vulgaris* | VT+CP | 21 |
| *Vigna sinensis* | NI | - |
| **Malvaceae** | | |
| *Abelmoschus esculentus* | VT+LC+YG | 15 |
| *Abutilon indicum* | VT+LC | 18 |
| *Althea rosea* | VT+LC | 16 |
| *Gossypium hirsutum* | VT+E+LC | 14 |
| *Hibiscus rosa sinensis* | LVT+LC | 18 |
| *Sida mysoriansis* | VT+E+LC | 18 |
| *Sida rhombadifolia* | VT+LC+E | 18 |
| **Oxiladaceae** | | |
| *Oxalis corniculata* | NI | - |
| **Solanaceae** | | |
| *Capsicum annuum* | VT+LC+B+D+BL | 20 |
| *Capsicum fruitiscence* | VT+LC | 20 |
| *Datura stramonium* | LC+E+VT | 22 |
| *Lycopersicon esculentum* | LC+VT+RLS+CP+ D+VC+P+SVT+MVT | 18 |
| *Nicandra physaloides* | VT+E+LC | 25 |
| *Nicotiana glutinosa* | VT+LC+E | 18 |
| *N. tobacum* | VT+MLC+E | 18 |
| *N. plumbaginifolia* | VT+DC | 24 |
| *Physalis minima* | VT+LC+E | 21 |
| *Solanum nigrum* | LC+E | 22 |
| *Solanum melongena* | NI | - |
| **Scrophulariaceae** | | |
| *Majus pumilus* | NI | - |
| **Lamiaceae** | | |
| *Ocimum sanctum* | NI | - |
| *Salvia plebia* | LC+E | 21 |
| *Salvia officinalis* | NI | - |
| **Polygonaceae** | | |
| *Polygonum sp.* | NI | - |
| *Rumex dendatus* | LC+E | 21 |

*VT-Vein thickening, LC-Leaf curling, T-Twisting, RLS-Reduction in leaf size, LVT-Light vein thickening, UC-Upward curling, SC-Severe curling, B-Bushy, S-Stunted, E-Enation, D-Dwarfing, CP-Cupping, DC-Downward curling, YG-Yellow golden, VC-Vein clearing, MVT-Thickening of main veins, BL-Blistering of leaves, SVT-Thickening of small veins, P-Puckering, MLC-Mild leaf curling, NI-No infection (only whitefly infested).

#Based on Ansari *et al.* (2006a)

## INCIDENCE AND LOSSES

Low incidence of leaf curl in late planted tomatoes compared with the 90% incidence in the early planted tomato crops was first reported in India regarding seasonal variations in the disease incidence (Vasudeva and Somraj, 1948). Besides several other factors the extent of loss due to leaf curl infection depends upon the stage of the plant at the time of infection. The plants infected within 20 days of transplanting remain stunted in growth and produce fewer leaves and fruits than those infected 35 to 50 days following transplanting (Sastry and Singh, 1973). Early infection resulted in 92.3% less fruits as compared with 74.1 to 28.9% loss in plants infected 35 to 50 days after transplanting. The lesser leaf curl disease incidence was reported in the crop planted in October (35%) as compared to the crop planted in February (69%). They, however, did not record the population of whitefly. It could be possible that the pattern of whitefly population build up may be different. The population of whitefly varies from year to year and place to place (Saklani and Mathahi, 1977; Varma, 1993).

In Punjab, the incidence of leaf curl is greater in the winter crop than in summer. The virus affects all stages of growth of tomato plants, causing 17.6 to 99.7% loss in yield depending upon the stage of the crop at the time of infection (Butter and Rataul, 1981). The disease incidence in Karnataka varied from 6.4 to 52.2% in summer and 52.5 - 100% in crops grown in the winter seasons (Saikia and Muniyappa, 1989). In Northern India, a higher degree of disease incidence (83%) was found in the winter crop as compared to the summer crop (Varma, 1993; Tripathi and Varma, 2002). In general, TLCV incidence and vector population were high in crop planted between late December and May, and low in crop planted from the end of June to early December. A high positive correlation was obtained between disease incidence and whitefly population (Hayati and Varma, 1984; Saikia and Muniyappa, 1989; Borah and Bordoloi, 1998; Ansari and Tewari, 2004; Ansari *et. al.*, 2006 b).

## MOLECULAR APPROACHES ON TOMATO LEAF CURL DISEASE

The molecular research of whitefly-transmitted geminiviruses belonging to genus begomovirus, which causes tomato leaf curl disease (ToLCD), has attracted much attention owing to the growing recognition of their importance. This virus has become an important deterrent to the tomato crop as well as productions of food and industrial crops during the past twenty years, as cropping practices have changed to meet the growing demand for agricultural commodities (Muniyappa, 1980; Brown, 1990, 1994, 2001; Brown and Bird 1992). Researches on molecular level attempt to explain the processes that lead to a successful state of disease in a host plant, primarily to enable identification of a weak link in the biotic and molecular interactions or processes that may be capitalized upon to achieve disease control. A successful viral pathogen must necessarily multiply, a stage of the disease process that requires a "replication-susceptible" host plant, while the second stage of the process requires a host permissive to invasion of cells adjacent to the primary infection court and ultimately to systemic spread.

With the recognition of diversity within begomoviruses has come the development of standard approaches to identify new and resurgent viruses. For most plant viruses, serology has traditionally been the method of choice for identification of new and resurgent viruses. However, virus-specific antiserum for begomoviruses (ToLCV) is not available because the capsid protein (CP) is extremely conserved across all species. DNA-based diagnostics, including polymerase chain reaction (PCR) and nucleotide sequencing of viral amplicons, have supplemented serology for detection, identification and classification of this group (Rybicki, 1994; Padidam *et al.,* 1995).

Begomoviruses (ToLCV) have circular, single-stranded DNA (ssDNA) genomes that replicate through double-stranded DNA (dsDNA) intermediates by a rolling-circle mechanism. The genomes of those are approximately 2.6 to 2.8 kb in size. Bipartite genomes share a common region (CR) of approximately 200 nucleotides that is highly conserved among cognate components of a viral species, while the analogous region in monopartite viruses is referred to as the large intergenic region (LIR). The CR and LIR contain modular *cis*-acting elements of the origin of replication and promoter elements (Lazarowitz, 1992; Bisaro, 1994; Hanley-Bowdoin *et al.,* 1999). Four to six open reading frames (ORFs) capable of encoding proteins >10 kDa in size are present on the A component or monopartite genome. The viral capsid protein is encoded by the ORF *AVI*, which is the most highly conserved begomovirus gene (Harrison, 1985; Padidam *et al.,* 1995). The capsid protein of this virus is required for encapsidation of ssDNA and for whitefly-mediated transmission. The capsid protein is also necessary for systemic spread of all monopartite and some bipartite viruses, depending upon the degree of virus-host adaptation (Pooma *et al.,* 1996; Hou *et al.,* 1998). The *AC1* ORF encodes Rep, a replication initiation protein and specificity is mediated through sequence-specific interactions with *cis*-acting elements of the *ori*. The *AC2* ORF encodes a transcription activator (TrAP) required for rightward gene expression and the *AC3* ORF encodes a replication enhancer (REn) (Lazarowitz, 1992; Bisaro, 1994; Hanley-Bowdoin *et al.,* 1999). Movement and systemic infection in bipartite viruses is accomplished by B component ORFs *BV1* and *BC1*, which are responsible for nuclear transport of ssDNA and cell-to-cell movement function (Sanderfoot and Lazarowitz, 1996). For monopartite viruses, several ORFs are involved in movement functions, including the *V1 (CP), V2* and translation products of the *C4* ORF, which exert host-specific effects on symptoms severity, virus accumulation and movement (Krake *et al.,* 1998; Brown, 2002).

The first conclusive aetiology of ToLCD in India as a geminivirus was reported by Muniyappa *et al.* (1991) and full length sequencing of ToLCV by Srivastava *et al*., (1994). Several species and strains of begomovirus causing ToLCD have been reported (Fauquet *et al.,* 2003). In Indian subcontinent, six species are known to be associated with the disease. These are *Tomato leaf curl Bangalore virus* (ToLCBV), *Tomato leaf curl Bangladesh virus* (ToLCBDV), *Tomato leaf curl Gujarat virus* (ToLCGV), *Tomato leaf curl Karnataka virus* (ToLCKV), *Tomato leaf curl New Delhi virus* (ToLCNDV), and *Tomato leaf curl Sri Lanka virus* (ToLCVSLV). ToLCNDV includes some isolates which cause only mild symptoms and isolates that cause leave distortion and mosaic in cucurbits. In India, ToLCNDV, ToLCGV reported from northern India are bipartite that requires both DNA-A and DNA-B for causing infection in the host plant, whereas ToLCBV, ToLCBV, ToLCKV from southern India are monopartite with only DNA-A (Muniyappa *et al.,* 2000).

Chakraborty *et al*., (2003) cloned and sequenced DNA-A and DNA-B of components of two isolates collected from Varanasi and Mirzapur and identified them as ToLCGV-(Var) and ToLCGV-(Mir), respectively. The two genomic components of ToLCGV-(Var) showed only 60% identity in the common region. In addition, another A or B components were found, which led to the conclusion that the two components of ToLCGV-(Var) are the cognate pairs that induced severe symptoms. DNA-B of ToLCGV-(Var) showed 80% sequence identity with that of ToLCNDV. Partial tandem dimeric constructs of DNA-A and DNA-B of ToLCGV-(Var) when coinoculated by bombardment produced symptoms in *N. benthamiana, N. tabacum,* tomato and pepper. DNA-A clone was also infectious but it produced milder symptoms and required longer incubation period. Thus, ToLCGV-(Var) DNA-A had properties of both monopartite and bipartite begomovirus. When coinoculated with ToLCGV-(Var) DNA-B, enhanced symptom severity and decreased incubation period were seen. ToLCGV-(Var) is sap transmissible and infected pepper (*Capsicum annuum* cv. California Wonder). No ToLCV isolate from India is known to infect pepper.

Sinha *et al.,* (2004) collected ToLCVD samples of tomato from four distinct zones in the country viz., New Delhi (ToLCNDV-IARI), Jabalpur (ToLCV-Jb), Raipur (ToLCV-Rp) and Dharwad (ToLCV-Dh). Replication initiator protein (Rep) gene of these isolates was amplified using gene specific primers. Sequence analysis of obtained amplicons revealed two distinct subgroups—ToLCV-Nde with bipartite genome having 94-95% homology and tomato leaf curl from Bangalore (ToLCBV) with monopartite genome that showed only 73-75% homology. Phylogenetic analysis showed that all the four isolates collected from distinct zones of the country belonged to subgroup I *i.e.*, ToLCV-Nde. The Rep gene nucleotide sequence of Dharwad isolate (ToLCV-Dh) had more similarity with ToLCV-Nde as compared to only 75% similarity with subgroup I i.e., ToLCBV from Bangalore (Sinha *et al.,* 2004). This supports independent origin of these virus isolates and that the geographical divergence in the country had no influence (Varma and Malathi, 2003). Alignment of ToLCNDV-IARI Rep protein governing Rep oligomerization, DNA binding and DNA cleavage with other viruses inducing ToLCVD showed that sequences were conserved in $\alpha$-helix involved in DNA cleavage, three motifs and P-loop for ATP binding inspite of high degree of variability at extreme N-and C-terminal (Dasgupta *et al.,* 2004).

In the quest to find sources of resistance to ToLCV, 122 varieties, lines and wild accessions of *Lycopersicon* were screened under heavy inoculum pressure for two years in all the three environmental seasons. High degree of resistance was found in *L. hirsutum* f. *glabratum* ('B6013') and *L. hirsutum* f. *typicum* ('A1904'). In addition, five accessions of *L. peruvianum* were also highly resistant. *L. pimpinellifolium* ('A1921') showed ToLCV symptoms after 90 days on terminal leaves. Inheritance of resistance in interspecific crosses between resistant line [*L. pimpinellifolium* ('A1921')] and five susceptible varieties of *L. esculentum* ('HS101', 'HA102', 'HS110', Pusa Ruby and 'Punjab Chhuhara') indicated to be monogenic in *L. pimpinellifolium* ('A1921') and incompletely dominant over susceptibility (Banerjee and Kallo, 1987a). In similar studies with resistant *L. hirsutum* f. *glabratum* ('B6013'), resistance was derived by two epistatic genes, one from the wild parent and one from the cultivated susceptible parent that resulted in a 13: 3 segregation in the $F_2$ generation (Banerjee and Kallo, 1987). Resistance in *L. hirsutum* f. *glabratum* ('B6013') was mapped at AVRDC, Taiwan to an introgression located on the lower end of chromosome 11 (Hanson *et al.,* 2000). The ToLCD resistant inbred line named H24 (Kallo and Banerjee, 1990) was used at AVRDC in the breeding program. In Southern India ToLCV resistant lines TLB 111, TLB 130 and TLB 182 developed at AVRDC using H24 were tried in field at five locations for two years. Yield of TLB111 and TLB130 was twice higher than those of susceptible Arka Vikas and Rashmi (Muniyappa *et al.,* 2002). For development of ToLCV resistant tomato transgenic, transgene (replicase gene of *Tomato leaf curl virus*) was characterized and its conserved functional motifs were identified (Dasgupta *et al.,* 2004). Gene silencing assays demonstrated the RNA mediated resistance through antisense suppression of the viral genome.

Gene silencing resulting into recovery of *Tomato leaf curl virus* infected tomato plants was achieved using homologous replicase gene constructs that produce RNAs duplex (sense (viral origin) and antisense (transgene)). This production of dsRNA template acts to produce sequence specific degradation of viral genome and recovery of infected plants (Praveen *et al.,* 2005).

## EPIDEMIOLOGY

Epidemiology is the study of environmental factors affecting the outbreak or increase and spread of infectious disease within plant population. The environmental factors that influence plant disease are: moisture, temperature, wind and the nature of the soil and so forth. The other factors include those, which affect host parasite relationship and determine the spread of disease through a host population. The important biological factors (Verma, 2003) are:

(i) **Physical ability of viruses and concentration reached:** The survival and spread of certain viruses appear to depend largely on a high degree of stability and the large amounts produced in infected tissue. This virus has a high degree of stability and is produced in large concentrations in the host.

(ii) **Rate of movement and distribution within host plant:** Viruses that move rapidly spread efficiently. Seed-born viruses have an advantage in spread and survival.

(iii) **Severity of the disease:** Virulent strains are self-eliminating as compared to moderate disease which allow the host plant to survive and reproduce effectively.

(iv) **Mutability and strain selection:** A virus that is able to mutate to produce strains that can cope effectively with changes in environment are able to survive and disperse in better way. Several new strains of this virus have been recorded and mutants are being reported from different places that indicate wide mutability.

(v) **Plant host range of virus:** A diversity of hosts gives a virus much greater opportunities to maintain itself and spread widely. Viruses that have perennials as hosts have become wide spread around the world.

From the epidemiological studies on TLCV carried out at Bangalore from 1969 to 1974 (Sastry *et al.,* 1978a) and Eastern Uttar Pradesh from 2002 to 2003 (Ansari and Tewari, 2004). It is seen that tomatoes planted during December to May are subjected to low rainfall, low humidity and high temperature. Such conditions help in the development of high population of whitefly and high TLCV incidence, thus resulting in low tomato yields. In contrast, tomato planted during July to November are subjected to high rainfall, high humidity and low temperature, resulting in low whitefly populations and a lower incidence of TLCV with better tomato yields (Saikia and Muniyappa, 1989; Roychoudhury and Jain, 1996 ).

## VIRUS-VECTOR RELATIONSHIP

Spread of tomato leaf curl virus between tomato plants by contact seems unlikely as the disease agent(s) have not been mechanically transmitted from tomato to tomato. In case of tomato yellow leaf curl virus, a few successful mechanical transmission to tomato from *D. stramonium, N. glutinosa* and *S. penelli* (Makkouk *et al.,* 1979). However, this is unlikely to be significant in the spread of tomato yellow leaf curl virus and tomato leaf curl virus in tomato fields. Transmission between tomato plants by natural root grafting has not been reported. TLCV or TYLCV was not transmitted through seed and tomato plants were not infected when grown in soil collected from around TLCV infected plants (Makkouk, 1978)

The virus is transmitted by whitefly, *Bemisia tabaci* Genn., from chilli to tomato and vice versa. The leaf curl symptoms on tomato develop in 9-12 days in August and 60-90 days in winter after inoculating the plant with whitefly. The virus acquisition and inoculation threshold in whitefly are 31 and 32 minutes, though such short time result in low transmission. If these periods are increased up to 24 hours each, virus transmission is increased to 30% and 24%, respectively. Pre and post – acquisition fasting of 1 and 2 hours enhance the virus transmission up to 36% and 40%. Further increase in these periods, however, causes a reduction in virus transmission efficiency (Butter, 1976; Reddy and Yaraguntaiah, 1981a; Ansari *et al.,* 2005). Young, true leaves of tomato are better sources of the virus than the cotyledons (Butter and Rataul, 1977a, b).

The inoculation period of the virus in the vector body is 6 hours (Reddy and Yaraguntaiah, 1981b). However, there was no latent period in the body of the whitefly vector. Whitefly females were efficient as males as vectors of tomato yellow leaf curl virus in Israel (Cohen and Nitzany, 1966) and tomato leaf curl virus in India (Pruthi and Samuel, 1939; Chakraborty, 2003; Ansari *et al.,* 2005). Minimum acquisition and inoculation feeding periods were 15-30 minutes. The latent period in the vector was at least 21 hours. The virus persisted in the vector for about 20 days, but not throughout the lifespan of the insect. Although the virus could be acquired by the larval stage of the insect it was not transmitted to the progeny. The virus could not be transmitted mechanically.

Once the whitefly has become viruliferous they remains infective for 14 days (Butter, 1973) and the virus is retained throughout the life span of the vector (Reddy and Yaraguntaiah, 1981a; Ansari *et al.,* 2005). A single whitefly is able to transmit the virus and the females are more efficient vectors of tomato leaf curl virus (TLCV) than the males. So for no whitefly-borne virus has been reported to be transovarially transmitted, and this is also the case with TLCV, where all results of transovarial transmission experiments were negative. The virus does not affect the life span of the vector.

## METABOLIC CHANGES IN VIRUS-INFECTED PLANTS

Leaf curl virus-infected tomato leaves showed a marked reduction in chlorophyll, xanthophyll and carotene content and in the rate of photosynthesis. Infected leaves showed an increase in respiratory rates in peroxidase and in catalase activity. Sugar and organic acid contents were not affected. Ascorbic acid oxidase and polyphenol oxidase activities increased, but ascorbic acid and phenolic contents decreased due to infection. Nitrite, protein, nitrogen contents increased, while there was a reduction in the nitrate, nitrogen contents in the infected leaves. Nitrate reductase increased in infected leaves. The amino acid content decrease while, RNA and DNA concentration increased due to infection. Phosphorus, magnesium and iron concentration decreased, while potassium, calcium, sodium, manganese and copper content increased in infected leaves plants. Fruits from the diseased plants have more acidity, protein, pectin, and ascorbic acid contents and less sugars and phenolics (Lodh *et al.,* 1971; Sasikumaran *et al.,* 1979).

## HISTOPATHOLOGICAL STUDIES

Histopathological studies of virus-infected plants play an important role in the development of relationship between the virus and tissues of their hosts. El-Shaieb *et al.,* (1981) observed phloem degeneration after 50 days of inoculation in tomato infected with tomato leaf curl virus. The infected leaves has less number of larger stomata, palisade tissues were compact in comparison to healthy. The epidermal cell and tissue of the midrib region especially on the dorsal surface of the infected leaves have been found degenerated due to infection. The walls of palisade cells were irregular in size and shape. Overall, the degeneration of cells starting from epidermal to spongy regions has been reported in infected leaves. The average stomatal frequency/mm$^2$ (719.73) was significantly lower in diseased plants than healthy. No significant differences were found in stomatal index, stomatal length, breadth in leaves of infected plants (Srivastava *et al.,* 2005)

## CONTROL

Tomato leaf curl virus cause serious loss of yield and quality of tomato. There is a range of control measures to reduce these losses. Direct method for the control of virus disease is not yet available. Most of the procedures that can be used effectively involve measures designed to reduce sources of infection inside and outside the crop to limit spread by vectors and to minimize the effect of infection on yield. The direct control measures for checking plant viruses are not practically convenient as viruses are obligate parasites and depend solely on host cells for nucleic acid and protein synthesis. The best way to control a virus disease is to keep it out of an area through quarantine inspection and certification systems. The effective methods of virus control, however, are to grow resistant varieties, cross protection and control of vectors. Various measures adopted so for to get rid of infection of this virus can be summarized as below:

1. **Use of insecticides:** Controlling the whitefly population can check the spread of TLCV. Use of correct insecticides at appropriate times not only reduces the whitefly population, but also checks the spread of the disease (Ansari *et al.,* 2007a). Chemical control of this disease has been attempted earlier with varying degrees of success (Sastry and Singh, 1971; Sastry *et al.,* 1974; Singh *et al.,* 1975, 1979; Naik and Muniyappa, 2004). Vector control during the early stages of the crop is very important. The tomatoes should be protected in the nursery as well as in the field with carbofuran, if higher yields are to be obtained. The next best procedures are nursery treatment and root dipping of seedling in carbofuran at the time of transplanting (Reddy and Yaraguntaiah, 1981b).

   Effect of different insecticides on the whitefly population in relation to TLCV incidence is also studied the (Sastry and Singh, 1971). Dimethoate (0.05%) proved most effective in controlling the vector and reduced the whitefly population from 65 to 13/75 leaves and disease incidence from 16.5 to 1.6%. Four sprays of dimethoate (0.05%) were recommended (Sastry and Singh, 1974). Rataul and Butter (1976) obtained similar results with dimethoate (0.025%), fenthion (0.05%), fentrothion (0.05%), dicrotophos (0.025%)

and monocrotophos (0.025%). Saklani and Mathai (1978) achieved good control of the disease through foliar sprays of tetrachlorvinphos (0.5,0.75 and 1.0 kg. a.i./ha) followed by endosulfan (0.35 and 0.525 kg. a.i/ha).

2. **Use of oils:** Mineral oils sprays are also effective in preventing the disease and increasing the yield of tomatoes. Three mineral oils, EWOS, BSOE, and E-9267, each at a concentration of 2% were tested and proved effective in preventing the disease (Butter and Rataul, 1973; Singh *et al.*, 1973, 1975). Krishi oil sprays are also most effective in control of the whitefly population and reducing the disease incidence. Soil systemic granular insecticides used in three split dosages at 20 days intervals beginning from the transplanting of tomato crop, were also effective in reducing whitefly population (Rataul and Butter, 1975; Sastry *et al.*, 1978b).
3. **Use of mulches:** The effect of straw mulching on the spread of TLCV and the influence of straw on the behaviour of the whiteflies was studied in Israel (Cohen and Madjar, 1974). The use of polythene mulching along with raising nursery under protective conditions was also recommended (Tripathi and Varma, 2002; Ansari *et al.*, 2007c).
4. **Use of growth regulators:** The use of gibberellic acid (GA) on tobacco was useful in reducing the severe curling and puckering of tobacco leaves, but spraying of GA (150 ppm) on tomatoes did not prove to be effective in the case of TLCV (Nariani and Maramorosch, 1957). However, Mukherjee and Raychaudhari (1966) recommended that root dipping of tomato seedlings in GA and 2 - thiouracil (50 ppm) was useful in reducing the severity of leaf curl.
5. **Planting density:** The relationship of insecticides to plant population density in relation to the incidence of TLCV is also studied (Sastry *et al.*, 1976). The results showed that increased plant density combined with insecticides might be effective as a control measures.
6. **Planting dates:** Studies have been carried out to find a suitable planting time to avoid excessive whitefly population and to minimize disease incidence during the cropping period. In this method, the plant may escape the infection or become infected at a later stage of growth, which does not adversely affect yield. October to mid December was found to be the most appropriate time for planting tomatoes followed by January to the first week of March. Planting of tomatoes during mid March to mid-September was badly affected with the disease, and the disease symptoms appeared very early (25-45 days after transplanting). There was, however, a delayed appearance of symptoms in October to mid December planting tomatoes (Saklani and Mathai, 1977; Borah and Bordoloi, 1998).
7. **Barrier cropping:** Studies on barrier cropping in relation to TLCV reveal that the disease incidence is reduced by planting six rows of either sun hemp or maize as barrier crops (Sastry *et al.*, 1977).
8. **Use of netting barrier:** The use of netting barrier on tomato was useful in reducing the disease incidence of TLCV (Govindappa *et al.*, 2003). Netting barrier (1.5 m height) with yellow cloth strip (40 cm width) 20 cm above the ground reduced the migration of *Bemisia tabaci* into tomato crop. Very low population of whitefly (1.1/plant) was observed in netted plot when compared to check (4.11/plant) after six week transplanting. Drastic

reduction in tomato leaf curl virus disease was observed with maximum of 44.6 after ten weeks of transplanting. However, in check pots the disease spread was very rapid during initial weeks and reached 100% by 6th week itself. The netting of tomato plots resulted significant increased in the crop yield.

9. **Use of plant extracts:** For the management of TLCV disease, *Clerodendron* leaf extract has been attempted (Baranwal and Ahmad, 1997). *Azardirachta indica*, *Clerodendron aculeatum*, and *Parthenium hysterophorus* leaf extracts also checked the vector *Bemisia tabaci* of the tomato leaf curl virus in UP and reduced the disease of incidence (Ansari and Tewari, 2005a). Spraying with neem seed kernel extracts and leaf extracts of *Pinus, Thuja, Araucaria, Cupressus* and *Cycas* proved effective in reducing the disease incidence, whitefly population and also in increasing the yield (unpublished data).
10. **Use of antibiotics:** Antibiotics viz. abomycin A and validomycin A showed marked anti-tomato leaf curl virus effect and plant growth stimulants and increased the yield per plant (Hayati and Varma, 1985). The plants sprayed with abomycin A (0.01%) produced maximum number (48.33%) of fruits per plant yet the maximum yield (302.40g) per plant was obtained by spraying 0.02% validomycin-A. Comparable results were reported using DPB antibiotic against tomato leaf curl virus (Thirumalachar *et al.,* 1973). Effect of 8-azaguanine on tomato leaf curl virus infection was also studied (Varma and Poonam, 1977).
11. **Use of resistance varieties:** Besides, integrated management practices like manipulation of date of sowing and application of insecticides for control of whitefly population, used of resistance cultivar are the most effective way to control of the disease (Varma *et al.,* 1980; Joshi and Chaudhury, 1981; Banerjee and Kallo, 1987, 1991). The pure lines showing resistance are recommended for varietal improvement to the breeders (Nariani and Vasudeva, 1963; Nainar and Pappiah, 2002 a, b).

## CONCLUSION

In the endeavour to feed the ever increasing population, intensive and extensive agricultural practices were adopted and high yielding varieties were introduced. Indiscriminate use of newer pesticides including synthetic pyrethroids resulted in enhancing the whitefly population, appearance of whitefly in newer areas and on additional hosts. As a result, large number of begomovirus diseases appeared and some of them became economically very important (Saikia and Muniyappa, 1989; Green and Kallo, 1994, Czosnek and Laterrot, 1997; Briddon and Markham, 2000; Varma and Malathi, 2003). Mapping of emerging and reemerging diseases on the regular basis should be faithfully followed so as to take timely measures to prevent the recurrence of these diseases. For the management of devastating this virus diseases, information on inheritance of resistance to most of the viruses is lacking. In addition epidemiological studies are required to address interrelationship of yellow mosaic viruses, golden mosaic viruses and leaf curl viruses infecting all the crop and weed plants. Weeds play very important role in survival of plant pathogens and vectors in the absence of their main hosts-plants during off season. Synthetic pesticides have largely failed to address the problem of whitefly population in field. Eco-friendly management alternatives are thus the answer as in the case of heavy population of biotype B of whitefly, which were successfully controlled using its natural enemies (Kirk *et al.,* 2000; Brown and Czosnek, 2002). There are several begomoviruses in India infecting economically important food, horticultural, ornamental crops and weeds, which need to be fully characterized. Understanding the phylogenetic relationship between them will help in developing management strategy.

Each of the measures investigated so for provides only partial control. In future, more work is needed on the epidemiology of the disease, their distribution, natural hosts, histopathological and cytological studies, transmissions by vectors and their relationship with weather in reference to epidemics, so that better control strategies can be developed. Much work is also required on the etiology of the disease.

The seriousness of TLCV in India requires collaboration and inter disciplinary research to exploit all possible control measures to produce acceptable tomato yields in areas where epidemics due TLCV are common.

## References

Ansari, N.A., Gautam, A., Tewari, H.D., Srivastava, G.P. & Tewari, J.P. (2007a) *J. Liv-World* **14**: accepted.

Ansari, N.A., Mishra, V.A. & Tewari, J.P. (2006a) *Ann. Pl. Protec. Sci.* **15**: 509-510.

Ansari, N.A., Pandey, A., Tewari, J.P. & Srivastava, G.P. (2004a) *J. Liv-World* **11**: 30-32.

Ansari, N.A., Pandey, A., Gupta, P.K., Srivastava, G.P. & Tewari, J.P. (2004b) *Geobios* **31**: 121-124.

Ansari, N.A., Pandey, A. & Tewari, J.P. (2005) *J. Liv-World* **12**: 1-8.

Ansari, N.A., Pathak, M., Shukla, A. & Tewari, J. P. (2007b) *Bioved* **17**: accepted.

Ansari, N.A., Srivastava, G., Srivastava, A.K., Pandey, A., Shukla, D., Chaturvedi, R.C., Shukla, S., Srivastava, G.P. & Tewari, J.P. (2006b) *J. Liv-World* **13**: 48-55.

Ansari, N.A. & Tewari, J.P.(2004) *Vegetos* **17:** 43-50.

Ansari, N.A. & Tewari, J.P. (2005a) *Geobios* **32**: 203-205.

Ansari, N.A. & Tewari, J.P. (2005b) *J. Liv-World* **12**: 22-26.

Ansari, N.A. Tewari, H.D., Chaturvedi, R.C., Srivastava G.P. & Tewari, J.P. (2007c) *Ind. J. Plant Pathol.* **25**: accepted (unpublished data).

Banerjee, M.K. & Kallo, G. (1987) *Euphytica* **36**: 581-588.

Banerjee, M.K. & Kallo, G. (1987a) *Theor. Appl. Genet.* **73**: 707-710.

Banerjee, M.K. & Kallo, G. (1991) *Vegetable Science* **18**: 102-112.

Baranwal, V.K. & Ahmad, N. (1997) *Ind. Phytopath.* **50 (2)**: 297-299.

Bisaro, D.M. (1994) In: *Homologous Recombination and Gene Silencing in Plants* (Eds. Paszkowski, J.), Kulwer, Dordrecht, The Netherlands, pp. 39-60.

Borah, R.K. & Bordoloi, D.K. (1998) *Ind. J. Virol.* **14**: 71-73.

Briddon, R.W. & Markham, P.G. (2000) *Virus Research* **71**: 151-159.

Brown, J.K. (1990) *FAO Pl. Prot. Bull.* **39**: 5-23.

Brown, J.K. (1994) *FAO Pl. Prot. Bull.* **42**: 3-23.

Brown, J.K. (2001) *Virus Research* **72**: 233-260.

Brown, J.K. (2002) In: *Plant Viruses as Molecular Pathogens* (Eds. Khan, J.A. & Dijkstra, J.), CBS Publishers & Distributors New Delhi, pp. 279-315.

Brown, J.K. & Bird, J. (1992) *Pl. Dis.* **76**: 220-225.

Brown, J.K. & Czosnek, H. (2002). *Ad. Bot. Res.* **36**: 65-100.

Butter, N.S. (1973) *M.Sc. Thesis*, Punjab Agric. Univ. Ludhiyana, pp. 72.

Butter, N.S. (1976) *Ph.D. Thesis*, Punjab Agric. Univ. Ludhiyana, pp. 87.

Butter, N.S. & Rataul, H.S. (1977a) *Phytoparasitica* **5**: 173-186.

Butter, N.S. & Rataul, H.S. (1977b) *Entomon* **2**: 163-164.

Butter, N.S. & Rataul, H.S. (1981) *Ind. J. Ecol.* **8**: 299-300.

Butter, N.S. & Rataul, H.S. (1973) *Curr. Sci.* **42**: 864-865.

Chakraborty, S., Pandey, P.K., Banerjee, M.K., Kallo, G. & Fauquet, C.M. (2003) *Phytopathology* **93**: 1485-1495.

Chhabra, N. (1992) *Ind. Hort.* **37**: 11-13.

Cohen, S. & Madjar, V.M. (1974) *Bull. Ent. Res.* **64**: 193-197.

Cohen, S. & Nitzany, F.E. (1966) *Phytopathology* **56**: 1127-1131.

Cowland, J.W. (1932) *Rep. Gezera Agric. Resi. Serg.* Sudan.

Czosnek, H. & Laterrot, H. (1997). *Archives of Virology* **142**: 1391-1406.

Dasgupta, A., Sinha, S.K. & Praveen, S. (2004) *Pl. Sci.* **166**: 1063-1067.

Dasgupta, I., Malathi, V.G. & Mukherjee, S.K. (2003) *Curr. Sci.* **84**: 341-354.

Datar, V.V. (1984) *Indian Phytopath.* **37**: 540-543.

Duffus, J.E. (1971) *Ann. Rev. Phytopath.* **9**: 319-349.

El-Shaieb, M.K.Z., Fahim, M. & Zaher, N.A. (1981) *Agri. Res Rev.* **59**: 1-19.

Fauquet, C.M., Bisaro, D.M., Briddon, R.W., Brown, J., Harrison, B.D., Rybicki, E.P., Stenger, D.C. & Stanley, J. (2003) *Archives of Virology* **148**: 405-421.

Govindappa, M.R., Muniyappa, V., Colvin, J., Hanson, P. & Kesav Murthy, K.V. (2003) *Ind. Phytopath.* **56**: 336 p. (Abstract).

Green, S.K. & Kallo, G. (1994) *AVRDC Tech. Bull.* **21**: 1-51.

Hanley-Bowdoin, L., Settlage, S.B., Orozco, B.M., Nagar, S. & Robertson, D. (1999). *Crit. Rev. Pl. Sci.* **18**: 71-106.

Hanson, P.M., Bernaeehi, S., Green, S., Tanksley, S.D., Muniyappa, V. & Padmaja, A.S. (2000) *J.Am. Soc. Hort. Sci.* **125**: 15-20.

Harrison, B.D. (1985) *Ann. Rev. Phytopath.* **23**: 55-82.

Hayati, J. & Varma, J.P. (1984) *Ind. Phytopath.* **37**: 223-227.

Hayati, J. & Varma, J.P. (1985) *Ind. J.Virol.* **1 (2)**: 152-156.

Hou, Y.M., Paplomatas, E.J. & Gilbertson, R.L. (1998). *Mol. Plant-Microbe Interact.* **11**: 208-217.

Joshi, G.C. & Chaudhury, D.(1981) *Vegetable Science* **8**: 15-50.

Kallo, G. & Banerjee, M.K. (1990) *Plant Breeding* **105**: 156-159.

Kirk, A.A., Lacey, L.A., Brown, J.K., Ciomperlik, M.A., Goolsby, J.A., Vacek, D.C., Wendel, L.E. & Napompeth, B. (2000) *Bull. Entomol. Res.* **90**: 317-327.

Krake, L.R., Rezaian, M.A. & Dry, I.B. (1998) *Mole. Plant-Microbe Interact.* **11**: 413-417.

Lazarowitz, S.G. (1992) *Crit. Rev. Pl. Sci.* **11**: 327-349.

Lodh, S.B., Ravel, P. & Singh, S.J. (1971) *Ind. J. Exp. Biol.* **9**: 281-283.

Makkouk, K.M. (1978) *Pl. Dis. Reporter* **62**: 259-262.

Makkouk, K.M., Shehab, S. & Majdalani, S.E. (1979) *Phytopathalogische Zeitschrift* **96**: 263-267.

Mariappan, V. & Narayanswamy, P. (1986) In: *National Seminar on whitefly transmitted plant virus disease*, IARI, New Delhi June 25-27, 1986, pp. 28.

Martyn, E.B. (1968) "*Plant Virus Names" Phytopath. Paper No.* 9 & *Supp. No.*1, C.M.I. kew Surrey, England.

Mukherjee, A.K. & Raychaudhari, S.P. (1966) *Ind. Phytopath.* **19**: 399-401.

Muniyappa, V. (1980) In: *Vectors of Plant Pathogens'* (Eds. Harris, K.F. & Maramorosch, K.) Academic press New York, pp. 39-85.

Muniyappa, V., Swanson, M.M., Duncan, G.H. & Harrison, B.D. (1991) *Ann. Appl. Biol.* **118**: 595-604.

Muniyappa, V., Venkatesh, H.M., Ramappa, H.K., Kulkarni, R.S., Zeidan, M., Tarba, C.Y., Ghanim, M. & Czosnek, H. (2000) *Archives of Virology* **145**: 1583-1589.

Muniyappa, V., Padmaja, A.S., Venkatesh, H.M., Sharma, A. & Chandrashekhar, S. (2002) *Hort. Science* **37**: 603-606.

Naik, R.G. & Muniyappa, V. (2004) *J.Mycol.Pl.Pathol.***34**: 216-219.

Nair, M.D. & Wilson, K.I. (1969) *Agric. Res. J. Kerala* **7**: 123-126.

Nainar, P. & Pappiah, C.M. (2002a) *South Indian Horticulture* **50**: 266-269.

Nainar, P. & Pappiah, C.M. (2002b) *South Indian Horticulture* **50**: 270-272.

Nariani, T.K. (1968) *Pl. Dis. Reptr.* **52**: 595-596.

Nariani, T.K. & Maramorosch, K. (1957) *Ind. Phytopath.* **10**: 101-102.

Nariani, T.K. & Vasudeva, R.S. (1963) *Ind. Phytopath.* **16**: 238-239.

Nitzany, F.E. (1975) *Phytopath. Medit.* **14**: 127-129.

Padidam, M., Beachy, R.N. & Fauquet, C.M. (1995) *J. Gen. Virol.* **76**: 249-263.

Pooma, W., Gillette, W.K., Jeffery, J.L. & Petty, I.T.D. (1996) *Virology* **21**: 264-268.

Praveen, S., Mishra, A.K. & Dasgupta, A. (2005) *Pl. Sci.* **168**: 1011-1014.

Pruthi, H.S. & Samuel, C.K. (1939) *Ind. J. Agric. Sci.* **9**: 223-275.

Rataul, H.S. & Brar, J.S. (1989) *Trop. Sci.* **29**: 111-118.

Rataul, H.S. & Butter, N.S. (1975) *J. Res. Punjab Agric. Univ.* **12:** 383-386.

Rataul, H.S. & Butter, N.S. (1976) *J. Res. Punjab Agric. Univ.* **13**: 303-307.

Rataul, H.S. & Butter, N.S. (1977) *Ind. Hort.* **21**: 15.

Reddy, K.S. & Yaraguntaiah, R.C. (1979) *Ind. Phytopath* **32**: 563.

Reddy, K.S. & Yaraguntaiah, R.C. (1981a) *Ind. Phytopath.* **34**: 310-313.

Reddy, K.S. & Yaraguntaiah, R.C. (1981b) *Ind. Phytopath.* **34**: 291-295.

Reddy, K.S. & Yaraguntaiah, R.C. (1981c) *South Indian Horticulture* **30**: 25-26.

Reddy, K.S., Yaraguntaiah, R.C. & Sastry, K.S. (1981) *Z.Pfi. krankh. Pfi. Path. Pfi. Schutz.* **88**: 400-404.

Rishi, N. (2004*) Ind. Phytopath.* **57**: 396-407.

Roychoudhury, R. & Jain, R.K. (1996) *Ind. J. Plant Pathol.* **14**: 20-25.

Rybicki, E.P. (1994) *Archives of Virology* **139**: 49-77.

Sachan, V.K. (2004) *Agrobios Newsletter* **2**: 35-36.

Saikia A.K. & Muniyappa, V. (1989) *Trop. Agri.* **66**: 350-354.

Saklani, U.D. & Mathai, P.J. (1977) *Ind. J. Hort.* **34**: 64-68.

Saklani, U.D. & Mathai, P.J. (1978) *Pesticides* **12**(8): 17-20.

Sanderfoot, A.A. & Lazarowitz, S.G. (1996) *Trends in Cell Biol.* **6**: 353-358.

Sasikumaran, S., Kandaswamy, T.K. & Vidyasekharan, P. (1979) *Indian Phytopath.* **32**: 352-359.

Sastry, K.S. (1984) In "*Virus Ecology*" (Eds. Mishra, A. & Polosa, H.), South Asian Publishers, New Delhi, pp. 59-90.

Sastry, K.S.M. & Singh, S.J. (1971) *Ind. J. Hort.* **28**: 304-309.

Sastry, K.S.M. & Singh, S.J. (1973) *Ind. J. Mycol. Pl. Path.* **3**: 50-54.

Sastry, K.S.M. & Singh, S.J. (1974) *Ind. J. Hort.* **31**: 178-181.

Sastry, K.S., Sastry, K.S.M. & Singh, S.J. (1974) *Pesticides* **8**: 41-42.

Sastry, K.S.M., Singh, S.J. & Sastry, K.S. (1976) *Indian J. Mycol. Pl. Path.* **6**: 8-13.

Sastry, K.S.M., Singh, S.J. & Sastry, K.S. (1977) *Ind. J. Hort.* **34**: 319-322.

Sastry, K.S.M., Singh, S.J. & Sastry, K.S. (1978a) *Ind. J. Hort.* **35**: 269-277.

Sastry, K.S.M., Singh, S.J. & Sastry, K.S. (1978b) *Pesticides* **12**: 28-29.

Singh D.V. & Lal S.B. (1964) *Proc. Natl. Acad. Sci. India* **34**: 179-187.

Singh, S.J., Sastry, K.S.M. & Sastry, K.S. (1973) *Ind. J. Agric. Sci.* **43**: 669-672.

Singh, S.J., Sastry, K.S.M. & Sastry, K.S. (1975) *Pesticides* **9**: 45-46.

Singh, S.J., Sastry, K.S. & Sastry, K.S.M. (1979) *Zeitsch. Pflanz. un. Pflanzen.* **86**: 253-256.

Sinha, S.K., Dasgupta, A., Baranwal, A. & Praveen, S. (2004) *J. Plant Biochem-Biotech* **13**: 43-66.

Srivastava, K.M., Hallan, V., Raizada, R.K., Chandra, G., Singh, B.P. & Sane V.P. (1994) *J. Virol. Meth.* **51**: 297-304.

Srivastava, S., Srivastava, G., Singh, S.V., Srivastava, G.P. & Tewari, J.P. (2005) *Flora and Fauna* **11**: 45-48.

Thirumalachar, M.J., Govindu, H.C. & Yaraguntaiah, R.C. (1973) *Ind. Phytopath.* **26**: 330-332.

Tripathi, S. & Varma, A. (2002) *Ind. Phytopath.* **55**: 473-478.

Vasudeva, R.S. & Somraj, J. (1948) *Phytopath.* **38**: 364-369.

Varma, A. (1993) In: *Crop Protection and Sustainable Agriculture Wiley Chichester, Ciba Foundation Symposium* **177**: 140-157.

Varma, A. & Malathi, V.G. (2003) *Ann. Appl. Biol.* **142**: 145-164.

Varma, J.P. & Poonam (1977) *Ind. J. Exp. Biol.* **15**: 458-459.

Varma, J.P., Hayati, J. & Poonam (1980) *Z. Pfl. Krankh. u. Pft. Sch.* **87**: 137-144.

Varghese, S. & Chandramony, D. (2003) *Geobios* **30**: 53-56.

Verma, H.N., Srivastava, H.N. & Mathur, A.K. (1975) *Pl. Dis. Reptr.* **59**: 494-498.

Verma, H.N. (2003) In: *Basics of Plant Virology,* Oxford & IBH Publishing Co. Pvt. Ltd. New Delhi.

Yassin, A.M. & Nour, M.A.(1965) *Annal Appl. Biol.* **56**: 207-217.

# 15

# Biotechnological Application of Microorganisms in the Management of Viral Diseases of Crops

**L.P. Awasthi, P. Kumar and S. Singh**

*Department of Plant Pathology, N.D. University of Agriculture & Technology, Kumarganj, Faizabad-224 229 (U.P.)*

## ABSTRACT

The viral diseases of plants are of immense importance as they cause extensive crop damage and severe losses in term of yield and quality of produce. Hence, their management has been of great interest to the researchers in the field of plant virology. Further, because of the peculiar nature and characteristics of plant viruses in the host system, any therapeutic method to control them has not been found very successful. Where resistant varieties are available, long-term and effective protection has been achieved but for a number of crops no resistant variety is available. Recently, a great success has been achieved by the application of microorganisms in various strategies to prevent and/or control the infection of plant viruses.

A few techniques have been devised that are collectively called 'genetic engineering'. These have allowed genetic material to be moved between organisms and expressed in them. These techniques were used initially to move genes from viruses, prokaryotes or eukaryotes into bacteria. However, foreign genes have also been transferred into other organisms and there is an increasing number of reports of success in moving genes into plants, mostly dicotyledons. Thus, the time is ripe to examine what use might be made of these techniques to make crop plants genetically resistant to virus infection.

## INTRODUCTION

Plant viruses are estimated to cause economic losses worldwide of about $15 billion per annum. Average yield losses, in many crops, have been estimated from 5 to 20%. However, in some crops losses have been reported to 80-90%. The development of effective control strategies is dependent on the availability of reliable methods for detection, identification, elimination of viruses from diseased propagules and development of virus resistant crops. The control of virus diseases of plants cannot be achieved along the lines followed for viral infections of animals and humans. Plants possess no immune system and usually they do not recover once they have become infected with a virus.

Several reports are available on the effectiveness of abiotic and biotic agents as elicitors of resistance in susceptible plants hosts. These agents appear to act by sensitizing the plant for activation of resistance mechanism. Natural plant substances have also been shown to protect plants against infections from viruses and also to prevent the establishment of viruses in the host (Verma *et al.*, 1979, 1980, 1982, 1985; Verma and Awasthi, 1978, 1979, 1980; Awasthi, 1981; Awasthi *et al.*, 1984; Kumar and Awasthi, 2004).

Naturally occurring genetic resistance to virus infection is known for many plant species including crop species (Russell, 1978). However, little progress has been made in identifying the mode of action of the genes responsible, even though many have been usefully incorporated into, or transferred between, important crop cultivars by conventional plant breeding techniques. Eventually, it may be possible to identify and transfer these plant resistance genes from one plant species to another by genetic engineering methods, but an alternative, that is already feasible, is to render plants resistant to viruses using those viral genes that are responsible for various types of viral interference, namely cross-protection, anti-sense (mic RNA) control and satellite RNA interference.

With the use of genetic engineering, insertion of a resistance gene will be the most fascinating approach. But to get a gene of interest is rather very difficult. In 1985, Beachy and his coworkers used a unique approach to get rid of virus infection and subsequent development of viral symptoms. They made transgenic plants expressing coat protein (CP) gene of TMV. After challenge inoculation symptoms were delayed by a few days. Besides CP, a number of other approaches like satellite mediated resistance, antisense RNA and ribozyme have been tried for conferring viral resistance in different host virus combinations.

## CROSS-PROTECTION

Cross-protection is a protective measure against severe viral strains. It requires infection of a plant with mild strain of virus which protects it further from infection from severe strains. It has been shown that tobacco plants infected with a green mosaic strain of TMV developed no further symptoms when inoculated with a yellow mosaic strain of TMV. Similarly, tobaccos inoculated with mild strain of PVX were immune from severe strains of PVX after only 5 days of inoculation (Matthews, 1970). There are many drawbacks in this method which made it unacceptable in most of the cases. For example, mild strains often reduce yield. The protected but infected plant may act as a reservoir, dominant strain may change to a more severe type in due course of time or in some plants, severe disease may result from mixed infection.

Cross-protection is not however confined to plant viruses, and examples of similar specific 'microbial territoriality' have been reported, for example, for viroids of plants (Fernow, 1967; Niblett, 1978); for viruses of insects especially alpha viruses (Davey *et al.*, 1979; Davey *et al.*, 1982; Eaton, 1979; Igearashi *et al.*, 1977; Peleg and Stollar, 1974) for bacteriophages (Hayes, 1968; Hutchisons and Sinsheimer, 1971; Sternberg and Hoess, 1983) and for bacterial plasmids (Duckworth *et al.*, 1981; Novick, 1969). Similar phenomena have been described for viruses of vertebrate animals, however, for them, cross-protection is probably often obscured by other forms of interference resulting from virus infection. For example, vertebrate hosts usually develop cellular or humoral immunity, produce and respond to interferon (Friedman, 1977), may lose cell receptors (Bratt and Rubin, 1968; Crowell 1966; Kimura *et al.*, 1976), and may produce defective interfering particles (Stollar, 1980). Nonetheless, there are some reports of intracellular interference between related viruses of vertebrates that seem to be analogous to cross-protection between plant viruses. These include reports for poliovirus (Cooper, 1977), foot-and-mouth disease virus (De La Torre *et al.*, 1985) and mucosal disease pestivirus (Gillespie *et al.*, 1962), as well as that for togaviruses, where there is evidence that cross-protection may be one of the factors influencing the geographical distribution of viruses (Davey *et al.*, 1982).

## Facts of Cross-protection Among Viruses of the GEA Super-group

**(A) Protection may be reciprocal or unilateral**—Viruses may cross-protect reciprocally (i.e., virus A protects against virus B and vice versa), but some, possibly the less closely related ones, may only protect 'unilaterally' (i.e., virus A protects against virus B, but virus B does not protect against virus A) (Fulton, 1978).

**(B) Protection is usually not complete**—Most tests for cross-protection merely assess whether the challenge virus induces additional symptoms. However, (Thung, 1931) also reported that he could not detect the challenge virus in the challenged plant.

**(C) Isolated infected cells show cross-protection**—Isolated cultured cells sequentially inoculated with related viruses exhibit cross-protection, though, like the intact host organism from which they were obtained, they may become infected with related viruses when inoculated simultaneously. This has been shown for TMV using tobacco protoplasts (Otsuki and Takebe, 1977), and for togaviruses in mosquitoes and their cultured cells (Davey *et al.*, 1979; Davey *et al.*, 1982; Pelej and Stollar, 1974; Stollar and Shenk, 1973).

**(D) Cross-protection develops rapidly after infection**—Experiments with isolated cells have shown that cross-protection develops rapidly after infection, probably before progeny virus particles are produced (Otsuki and Takebe, 1976).

**(E) Cross-protection involves only part of the viral genome**—Richardson, 1982 found that BHK cells became resistant to challenge inoculation after infection by a single SFV particle, however UV-irradiation studies indicated that translation of only one-third of its unipartite genome was required to establish cross-protection.

## ELIMINATION OF SOURCES OF INFECTION

As a layman thought, if a crop is free of virus and when there is no source of infection in the field or adjacent to the crop field, there will be no virus problem. In theory it seems very feasible to eliminate most of the sources (annual or perennial, weeds, ornamental plants, unrelated crops, volunteer plants) of infection from the field and crop is protected. In practice, it is rather difficult to eliminate, particularly where mixed cropping is being done.

Plant parts remain in the soil and can harbour mechanically transmitted stable viruses like TMV. In such cases it becomes very difficult to eliminate the source of inoculum from soil. Roguing of infected plants sometimes reduces the disease incidence, provided the virus in question is slow spreading. However, roguing may also increase disease incidence by disturbing the vectors and then it becomes more difficult as the disease spreads in non-persistent manner by vectors particularly in case of potyviruses.

## VIRUS-FREE PROPAGATING MATERIAL

Use of virus-free propagating material e.g., bulbs, corms, suckers or seeds, etc. and subsequent follow-up of hygienic practices minimize the chance of virus infection. In cases where mother stock of propagating material is infected, virus transmission occurs horizontally and are carried to their progenies. This situation is very common in vegetatively propagated crops. Such transmission may be an important factor of infection, since it introduces the virus into the crop at early stage which spreads to other plants while they are still young. Biotechnology offers a number of different strategies to obtain virus-free plant material. Considerable progress has been made in the application of plant tissue and cell culture techniques for the production of virus-free plants through meristem tip culture. Cell culture also affords an opportunity for study of aspects of the cellular and molecular basis of this meristem resistance which is difficult to investigate in the whole plant system. Meristem tip culture for production of virus-free plants is well established but it is worthwhile to mention that nucleus stocks should be checked with most sensitive serological methods e.g., ELISA, ISEM, DIBA (Singh and Raizada, 1988; Singh *et al.*, 1990) before distribution to nurserymen, extension agencies and other prospective growers.

## VECTOR CONTROL

It is an indirect approach of disease control. The viruses, which are transmitted by vectors viz., aphids, whitefly, mites and leafhoppers are the major insect vectors. Nematodes and fungi also disseminate some viruses. Aphids, whitefly and mites are called air-borne vectors while nematode and fungi are soil-borne. Use of insecticides, fungicides and nematicides is important for reducing the population of the vectors. The number of pesticide spray can be reduced to make it cost-effective when epidemiological and meteorological data are studied in relation to vector population dynamics, flight/movement direction, etc.

New control measures that obviate pesticidal treatment have been introduced. Aphids are attracted towards yellow and green colour. Yellow polyethylene sheets covered with glue for trapping in flying viruliferous aphids have been used in several cases. Similarly, some Petunia varieties have been identified for monitoring presence of mites in greenhouses. Now anti-virus nets/screens are available where a particular mesh size can be used for controlling the insect vector. Soil mulching is another way to reduce vector population. It is not effective with all viruses and varies with each host virus-vector combination. The use of oil sprays also prevents the transmission of viruses by aphids. Neem oil is highly effective for vector control (Srivastava *et al.*, 1986). Spraying of neem oil is a useful method for controlling vectors and it has been successfully used in controlling aphid-borne viruses in bulbous ornamentals (Zaidi and Mukherjee, 1989). The inhibitory effect on the spread of viruses in lilies viz., tulip breaking virus (non-persistently aphid-borne, potyvirus), lily symptomless virus (non-persistently aphid-borne, carlavirus), and lily virus X (a potexvirus on unknown etiology) was studied with a few brands of mineral oil, synthetic pyrethroid insecticides and a nematicide in crops in which virus-infected plants were present as virus sources. The spread of lily virus X was reduced by the insecticides and not effectively by the mineral-oil spray indicating the mode of transmission by vector as persistent or semi-persistent (Asjes, 1991). Mixture of mineral oil and pyrethroid was more effective in the reduction of spread of tulip breaking virus and lily symptomless virus than either components tested alone. Weekly sprays were effective between 90-95% in case of oil. However, pyrethroids alone gave 60-70% control (Asjes, 1991).

## MODIFICATION OF AGRICULTURAL PRACTICES

Crop rotation and delayed ploughing of fields may reduce the number of virus-infected reservoirs and thus diminish the disease incidence in the new crop. Isolation of infected crops or weeds from growing crops also gives an effective control measure for several virus diseases. It has been observed that occurrence of BYMV in gladiolus crop was reduced after elimination of Melilotus alba, a weed host of bean yellow mosaic virus (Srivastava *et al.*, 1990).

## SELECTION OF RESISTANT VARIETIES—THE VIRUS RESISTANCE GENES

Breeding programmes aim to render the plants more resistant to vectors of plant viruses. Hypersensitivity, which provides field resistance to the disease, is often controlled by a single dominant gene. Resistance to infection may be controlled by one or many genes. These genes are generally equally effective against strains of a virus.

Conventional plant breeding have been successfully used for developing resistant crop varieties against viral diseases. Search for resistant genotypes is still on with a worldwide approach. Towards that direction, protoplast culture may also help in improvement of viral resistance (Larkin and Scowcraff, 1981). Somaclonal variation may exhibit genetic changes and it has been used to select heritable virus resistance from tomato cultures (Barden *et al.*, 1986). Therefore, somaclonal variation could be an unexpected source of the resistant phenotypes by methods involving cell cultures. Breakdown in resistance or tolerance is always to be feared as the result of the appearance of new viral strains though its frequency is very low.

## DEVELOPMENT OF TRANSGENIC PLANTS

Progress in this exciting area of research for production of transgenic plants has been so spectacular that by the turn of the century, we hope to be growing crops which have been tailored to market specifications by the addition, subtraction or modification of genes. Transgenes will also be important in increasing the efficiency of crop production systems. For instance, transgenic plants resistant to herbicides, insects, viruses and a host of other stresses have already been produced.

### (a) Gene for Virus Coat or Capsid Protein (CP) from Positive Strand RNA Viruses

Coat protein gene from tobacco mosaic virus (TMV), classified as a positive strand RNA virus, has been transferred to tobacco and in the transgenic plant, expression of coat protein (CP) was observed. Further, when inoculated with TMV, the infection in the transgenic plants was very low and delayed relative to control plants that were not transformed. This provides a new approach for producing virus resistant plants complementing the efforts of classical plant breeding in producing disease resistant crop varieties. However, in such transgenic plants, coat protein gene should be constitutively expressed and may thus have effects on the nutritional value of plants. Subsequently, this approach has been applied to a range of crops (tomato, alfalfa, tobacco, potato, melons, rice) for developing resistance against a broad spectrum of positive RNA strand plant viruses (e.g. alfalfa mosaic virus, potato virus X = PVX, potato virus Y = PVY and potoato leaf roll virus). Potatoes have been produced which have coat protein genes and are tolerant in field tests to both PYX and PVY. DNA coding for a component of TMV replicase enzyme, was also transferred to tobacco plants conferring resistance to TMV.

### (b) Gene for Nucleocapsid (N) Protein from Tomato Spotted Wilt Virus (TSWV)

In earlier reports of resistance against viruses in transgenic plants, a number of positive (+) strand RNA viruses were involved. No report of resistance against any negative (–) strand virus was available. An important negative strand virus is tomato spotted wilt virus (TSWV), which causes considerable yield losses in crops like tomato, tobacco, lettuce, groundnut, pepper and in ornamentals like Impatiens, Ageratllill and Chrysanthemum. In this virus, genomic RNA is tightly associated with nucleocapsid (N) protein. This protein helps in wrapping of viral RNA and also in regulation of transcription-to-replication switch during infection cycle.

For producing transgenic tobacco plants, N gene was associated with CaMV 35S promoter and a leader sequence (to enhance expression). The transgenic plants showed expression of N-gene and also significant resistance to TSWV. A correlation was also observed between the amount of N protein and the level of resistance. The mechanism, involved in imparting resistance against negative strand RNA virus like TSWV, may differ from that involved in positive strand RNA virus, where coat protein (CP) gene has been used.

### (c) Satellite RNA and its use for Transformation

Satellite RNAs are species of RNA associated with specific strains of some plant RNA viruses, although it is not necessary for their replication. Replication of this satellite RNA depends on the virus, so that it gets packaged with it to cause infection elsewhere. Therefore, satellite RNA depends on virus for its replication and transmission, even though it is unrelated to viral genome. Presence of sat-RNA leads to reduction in severity of disease symptoms, and therefore has been used for developing resistance against specific viruses.

Using the above principle, transgenic tobacco plants have been produced which carried a DNA fragment, which when transcribed gives a species of satellite RNA which can reduce the severity of disease. The following attempts were made:

(i) In one case, transgenic tobacco plants using DNA for RNA satellite of cucumber mosaic virus (CMV) were produced and when satellite free aggressive CMV strains were used for infection, the presence of satellite RNA (transcribed from DNA) in transgenic plants led to reduction in disease symptoms. These transgenic plants are also protected against another related virus, i.e., tomato aspermy virus.

(ii) In another example involving tobacco ringspot virus (TobRV), transgenic tobacco plants could be produced, which carried DNA version of satellite RNA of tobacco ringspot virus (STobRV). These transgenic plants are protected against the deleterious effect of subsequent TobRV infection. The advantage of using satellite RNA strategy over coat protein is that satellite RNA is not constitutively expressed like coat protein, but is expressed only after virus infection.

## COAT PROTEIN GENE

The use of plant virus coat protein gene to produce virus resistant plants was first described by Powell-Abel *et al.*, (1986), which led to similar experiments for a number of plant viral coat protein genes. Golembeski *et al.*, (1990) used a non-structural coding region from tobacco mosaic virus to produce virus resistant plants. Plants with viral resistance have also been produced by expressing satellite RNAs or antisense RNAs complementary to viral positive-sense sequences. Ribozymes may be useful for controlling viral replication. Alternatively, plant nuclear genes encoding resistance or involved in defence mechanisms may be used for controlling viral diseases with the help of genetic engineering technology. These genes are, however, routinely being used by conventional plant breeders to develop genetically virus-resistant cultivars.

Genetically engineered coat protein mediated protection has been shown to be a promising approach for the control of a number of plant viral diseases. In some aspects, coat protein-mediated protection is similar to 'crops protection' to some extent (Sherwood, 1987). Various tools of plant genetic transformation results into the introduction of desired gene/genes into the plants to confer novel phenotypes. Such plants are referred as transgenic plants. They have a great potential for improvement of agricultural crops. By expression of the coat protein of a virus in transgenic plants, a delay or even complete inhibition of infection with the corresponding virus, has been observed. This has been reported for viruses belonging to different taxonomic groups by different workers viz., tobamovirus (Powell-Abel *et al.*, 1986, 1990; Nelson *et al.*, 1987); tobraviruses (Van Dun and Bol, 1988); potexviruses (Hemenway *et al.*, 1988; Hoekema *et al.*, 1989); potyvirus (Lawson *et al.*, 1990); Ilarvirus (Loesch-Fires *et al.*, 1987; Turner *et al.*, 1987); cucumoviruses (Cuozzo *et al.*, 1988); luteoviruses (Vander wilk *et al.*, 1991); carlaviruses (Mackenzie and Tremaine (1990); furoviruses (Kallerhof *et al.*, 1990).

However, the extent of the resistance achieved by CP gene expression varies and probably depends on a number of factors relating to the accumulation of the CP gene product, including its degree of expression, cellular localization, aggregation state and the tissue and/or developmental specificity of its expression. In addition, resistance may be affected by other unidentified factors inherent to the biology of the specific virus, vector, host plant and their interactions. Several studies have indicated that the coat protein of the protecting virus has an important role in systemic cross-protection. Sherwood and Fulton (1982) suggested that the cross-protection of TMV is the result from an inhibition of uncoating of the challenge virus thereby preventing the release of infectious RNA. Dodds *et al.* (1985) also reported similar results from studies of cross-protection of tomato plants infected with cucumber mosaic virus (CMV). They observed that CMVRNA could superinfect leaves infected with strains of CMV. Although the challenge virus RNA could infect healthy plants systemically. The infection resulting from the challenge RNA in previously infected leaves remained localized.

Recently developed techniques of gene manipulation and plant genetic engineering now provide a method for determining whether the expression of individual viral genes can mimic the cross-protection. The plants expressing coat protein of TMV were slower to develop systemic symptoms following inoculation with TMV than were sibling plants that did not contain TMV coat protein. However, 10-60% of the expressing plants never developed symptoms. It was suggested that the plants expressing coat protein have fewer sites susceptible to infection than do the non-expressing plants (Powell-Abel *et al.*, 1986).

In transgenic tobacco expressing coat protein of alfalfa mosaic virus (AMV), the coat protein made up nearly 0.1% of total protein in the highest expressers. However, infected plants contain >100 times this amount. Expression of the AMV coat protein gene appeared to provide virus resistance to the transgenic plants. Progeny of transforms that expressed the highest levels of coat protein were resistant to virion of two strains of AMV (Loesch-Fries *et al.*, 1987). The expressors, however, were much more susceptible to infection by AMV-RNA than they were to infection by AMV virion. Lastly, the expressers were highly susceptible to infection by the unrelated virus, i.e. TMV. This indicates that the transgenic plants behaved as if they had been protected from infection by AMV. Anderson *et al.* (1989) reported in case of transgenic plants that express the coat protein genes of TMV or AMV interfere with disease development of some unrelated viruses (PVX, PVY, CMV). The experiment indicates that transgenic tobacco plants that express different coat protein genes have a low but significant degree of protection against other viruses.

Since, aphid inoculation is the natural route to infection in the field, it is very important to test usefulness of the resistance phenotypes. Lawson *et al.* (1990) in their transgenic CP potato. Lines has used aphids for challenge inoculation. Only one of CP + lines (line 303) was resistant to inoculation by the insect vector tested. Similarly, mixed infection is common in several plant species which makes desirable to genetically engineer with multiple coat protein genes from different viruses. Lawson *et al.* (1990) produced transgenic Russett Burbank potato plants expressing both the PVX and PVY CP genes. The transgenic line 303 also showed complete resistance to PVX and PVY in mixed infection, demonstrating that multiple resistance can be achieved by genetically engineering with coat protein genes.

The mechanism by which coat-protein gene mediate resistance (Srivastava *et al.*, 1993) is not very clear. However, at least two distinct mechanisms seem to be involved.

(i) Experimental evidence indicates that possibly the coat-protein gene expression interferes with an early event in infection at least in some cases, i.e. uncoating. The resistance in some instances is largely overcome by inoculation with viral RNA rather than virions

(LoeschFries *et al.*, 1987; Nelson *et al.*, 1987; Van Dun *et al.*, 1988). In plants expressing PVX coat protein, however, the resistance was not broken by challenge inoculation with viral RNA (Hemenway et al., 1988).

(ii) Interference in movement of virus in coat protein expressing plant is another way by which symptoms could be delayed. It has also been suggested that TMV coat protein may play role in regulating the replication of TMV-RNA (Osbourn *et al.*, 1989). In case of bacteriophages, replication of several RNA containing phages and expression is affected by their coat protein.

## Satellite-mediated Resistance

Satellite RNA species have been found in association with a number of viruses. They are the RNA molecules like defective interfering (D) RNA molecule and multiply in association with some viruses, but are not strictly part of the virus genome. Satellite RNA molecules are usually small and occur naturally in some virus cultures. They depend on helper viruses for their multiplication but have no appreciable sequence homology with them (Fritsch and Mayo, 1989). However, satellite RNAs of CMV and tobacco ringspot virus (TRSV) have been used as sources of resistance. Transformation of tobacco with DNA complementary to satellite RNA variant that ameliorate the symptoms induced by CMV (Baulcombe et al., 1986; Harrison et al., 1987a) or TRSV (Gerlach *et al.*, 1987, 1988). These plants showed either milder symptoms or accumulated less virus than control plants after inoculation with the homologous virus. A secondary effect of infection of satellite-transformed plants was that virus transmitted from the transgenic plants containing satellite RNA and was less pathogenic for non-transgenic hosts (Harrison *et al.*, 1987b). Field trials performed satellite containing transgenic tomato, tobacco or pepper (Tien *et al.*, 1990 a, b), revealed that the disease symptoms become mild and less virus accumulated in transgenic plants than in control plants. After the introduction of such cDNA of satellite RNA yield was increased upto 14% in case of tomato. Satellite RNA may cause a decrease in helper virus, perhaps, by competing with viral RNA for the replicase activity (Burgyan *et al.*, 1989). However, there are a number of disadvantages or potential risks in using plants genetically engineered with satellite RNA copies. These relate to the fact that satellite RNAs are transmissible agents and are similar to the problem encountered with classical cross-protection. These include:

- Satellite RNAs that are ameliorative in one species may endanger other crops species. For example, CMV D-satRNA which is ameliorative in tobacco, yet lethal in tomato (Waterworth *et al.*, 1979).
- Satellite RNAs mutate rapidly and sequence variants have been observed within as few as two to four serial passages (Kurath and Paulkaitis, 1990). Furthermore, a single nucleotide change introduced by *in vitro* mutagenesis can alter an ameliorative satellite, so that it becomes necrogenic (Sleat and Paulkaitis, 1990).
- Recombination has been observed between satellite RNAs (Cascone *et al.*, 1990). After more understanding satellite RNAs molecular biology, there is possibility to manipulate the sequences of these agents, so that they lose their transmissibility from the transgenic plant and yet retain their efficacy in ameliorating plant viral diseases. Thus, it may lead to the use of satellite RNA sequences as virus-resistance genes in a more attractive way.

## Resistance Through Antisense RNA

Cuozzo *et al.* (1988) showed that plants expressing antisense CMV RNA were resistant to infection. In such plants movement of virus was slower and accumulation was less than in control plants. Effect was, however, only detected when the inoculum concentration was low. Similar observations have been made by Hemenway *et al.* (1988) and Powell *et al.* (1989) with antisense PVX RNA and TMV RNA, respectively. However, expression of TRSV RNA did not enhance the resistance of plants to TRSV infection (Baulcombe *et al.*, 1987; Angement *et al.*, 1990). Moreover, a sophisticated understanding of plant RNA metabolism as well as viral life cycles is needed to get success in designing antisense RNA molecules which can be effectively utilized in viral disease management.

## Ribozymes Activity for Target Viral Genome

Genome of most of the plant viruses are RNA type. Ribozymes (Haseloff and Gerlach, 1988) are also RNA molecules that have stretches of sequence complementary to the target RNA, and which therefore bind to the target RNA by hybridization. As the nucleotide sequence in the centre of the ribozyme between the regions that bind to target RNA has catalytic activity, the result of hybridization to a ribozyme is a cleavage of the target RNA. The cleavage is independent of protein and is the same mechanism as that whereby some satellite RNA molecules oligomers self cleave during replication. This ribozyme cleaves its target RNA (Potato leaf roll and virus RNA) *in vitro*, but there is no information about its activity *in vivo* (Lamb and Hay, 1990). However, transgenic expression of a ribozyme in mammalian cells has been shown to cause cleavage of human immunodeficiency virus RNA (Sarver *et al.*, 1990).

Ben-Tahar (1993) attempted to obtain melon resistant to CMV by transforming with 2 different gene, the coat protein and ribozyme. Three independent melon transform ants were found to be resistant at 80 to 100% in the T2 progeny. They concluded coat protein expressing plants simply are more tolerant. The most likely difficulty with this strategy for controlling plant virus replication is that ribozyme-containing control agent will be synthesized in the nucleus, while replication of plus strand RNA viruses occurs in the cytoplasm. Hence, the stability and transport of the ribozyme reagent may limit its potential for plant virus control.

Before the wealth of naturally occurring virus-resistance genes can be used to modify plants by genetic engineering, further advances in plant gene cloning technology and the understanding of plant virus resistance must be made. Advances are being made in genome mapping and in some cases molecular markers linked to major disease resistance genes have been found. Complementary developments are also being made in genome library construction, chromosome walking and automation of nucleic acid sequencing. Application of these technologies may permit direct isolation of plant genes, including those for virus resistance.

# CONCLUSION

Plant virus diseases can affect both yield and quality of crops, and even result in complete crop failure. Various eco-friendly means of controlling viral diseases are being practised, including virus free seeds, avoiding and controlling insect vectors, controlling weed species that act as virus hosts, using cultures bred for virus-resistance. These techniques and plant tissue culture along with immunodiagnostics are now playing an important role in the production of virus-tested plants and reducing the virus inoculum from the environment.

With the development in the novel approaches and modifications in the use of ELISA, new antibody labelling techniques, preparation and use of monoclonal antibodies, application of immunoblotting and immunoelectron microscopy and the use of synthetic peptides for producing antipeptide antibody reagents, rapid indexing with more accuracy for the presence of viruses has become easier compared to earlier days. Hence, the quality control system at national and international levels will be benefited to a great extent with these advancements in the indexing techniques.

Cell and tissue culture, however, could be used for studying virus host resistance mechanism. They may be useful for early screening of genes of interest to breeders. Attention should also be given to the possible direct utilization of somaclonal variation for selecting new morphological or physiological types (including resistant genotypes to the somatic hybridization with other cross incompatible species in the same genus or in different genera, and to the possibilities afforded by developing recombinant DNA techniques.

Biotechnology or more precisely molecular plant breeding offers a number of other different strategies to obtain virus resistant material. One possibility to introduce into a crop different types of genes, at will, offers new ways for the control of viral diseases. At present the most successful approach seems to be the introduction of genes leading to expression of viral coat proteins. Transgenic plants that accumulate virus coat proteins (CP) are protected from infection by the virus from which the CP gene was taken and closely related viruses. The potential of a number of other approaches such as gene encoding antisense RNAs, satellite or interfering RNAs, RNA-scrambling-RNAs, resistance factors, vector repellents and vector-toxic factors are needed to be exploited. Possible combination of genes leading to long lasting and preferably for a wide range of virus resistance is perhaps required for ultimate management of viral diseases in fields.

## References

Anderson, E. J., Stark, D. M., Nelson, R. S., Powell, P. A., Tomer, N. E. & Beachy, R. N. (1989) *Phytopathology* **79**: 1284-1290.

Angenent, G. C, Van Den Ouweland, J. M., Wand Bol, J. F., (1990) *Virology* **175**: 192-198.

Asjes, C. J. (1991) *J. Pl. Path.* **97**: 129-138.

Awasthi, L.P. (1981) *Archives of virology* **70**: 215-223.

Awasthi, L.P., Chowdhary, B. & Verma, H.N. (1984) *J. Trop. Plant. Dis.* **2**: 41-44.

Barden, K. A, Schillersmith, S.& Murakishi, H. H. (1986) *Plant Sci.* **45**: 209-213.

Baulcombe, D. C., Hamilton, W.D., Mayo, M. A. & Harrison, B. D. (1987) In: Plant Resistance to Viruses (Eds., Evered D & Harnett S), Chichester, UK, John Viley and Sons, pp 170-184.

Baulcombe, D. C., Saunders, G. R., Bevan, M. W., Mayo, M. A. & Harrison, B. D. (1986) *Nature* **321**: 446-449.

Ben-Tahar, S. (1993) In: Indo-French Symposium on Molecular Plant Pathology and Plant-Pathogen Interactions, held at IARI, New Delhi, pp. 39 (Abstr:).

Bratt, M. A. & Rubin, H. (1968) *Virology* **35**: 395-407.

Burgyan, J., Grieco, F. & Russo, M. (1989) *J. gen. Virol.* **70**: 235-239.

Cascone, P.J., Carpenter, C. D., Li, X. H. & Simon, A. E. (1990) *EMBOJ.* **9**: 1709-1715.

Cooper, P. D. (1977) In: Comprehensive Virology, Eds, H. Fraenkel-Conrat and R. R. Wagner, Vol. 9. Plenum Press, New York.

Crowell, R. L. (1966) *J. Bact.* **91**: 198-204.

Cuozzo, M., Connell, K. M., Kaniewski, W., Fang, R. X., Chua, N. H. & Tumer, N. E. (1988) *Biotechnology* **6**: 549-557.

Davey, M. W., Mahon, R. W. & Gibbs, A. J. (1979) *J. gen. Virol.* **42**: 641-642.

Davey, M. W., Mahon, R. W. & Gibbs, A. J. (1982) In: Viral diseases in South-East Asia and the Western Pacific, (Ed. Mackenzie, J. S.) 81-84. Academic Press, Sydney.

De La Torre, J. C., Davila, M., Sobrino, F., Ortin, J. & Domingo, E. 19B,. *Virology* **145**: 24-35.

Dodds, J. A., Lee, S. Q. & Tiffany, M. (1985) *Virology* **144**: 301-309.

Duckworth, D. H., Glenn, J. & McCorquodale, D. J. (1981) *Microbiol. Rev.* **45**: 52-71.

Eaton, B. T. (1979) *J. Virol.* **30**: 45-55.

Fernow, K. H. (1967) *Phytopathology* **57**: 1347-1352.

Fraser, R. S. A. (1990) *Annu. Rev. Phytopathol.* **28**: 79-200.

Friedman, R. (1977) *Bacl. Rev.* **41**: 543-567.

Fritsch, C. & Mayo, M. A. (1989) In: Plant Viruses, Vol. I, Structure and Replication (Ed., Mandahar C L, Boca Raton F L, USA: CRC Press, pp. 290-321.

Fulton, R.W. (1978) *Virology* **85**: 1-8.

Gerlach, W. L., Llewellyn D. & Haseloff, J. (1988) *J. Cell. Biochem. Suppl.* **12**: 239.

Gerlach, W. L, Llewellyn, D, & Haseloff, J. (1987) *Nature* **328**: 802-805.

Gillespie, J. H., Madin, S. H. & Darby, N. B. (1962) *Proc. Soc. Exp. Bioi. Med.* **110**: 248-250.

Golemboski, D. B., Lomonosolf, G. P. & Zaitlin, M. (1990) *Proc. Natl. Acad. Sci. USA* **87**: 6311-6315.

Harrison, B. D., Barker, H., Baulcombe, D. C., Bevan, M. Wand Mayo, M. A. (1987b) In: Positive Strand RNA Viruses (Eds. Brinton, M. A. & Rueckert, R. R.) New York, USA: Alan R, Liss, pp. 575-585.

Harrison, B. D., Mayo, M. A. & Baulcombe, D.C. (1987a) *Nature* **328**: 799-802.

Hasloff, J. & Gerlach, W. L. (1988) *Nature* **334**: 585-591.

Hayes, W. (1968) The genetics of bacteria and their viruses. Second Edition. Wiley, New York.

Hemenway, C., Fang, R.X., Kaniewski, J. J., Chua, N.H. & Tumer, N. E. (1988) *EMBO J.* **7**: 1273-1280.

Hoekema, A., Huisman, M. J., Molendijk, L., Van Dun, E. P. J. M. & Corelissen, B. J. C. (1989) *Biotechnology* **7**: 273-278.

Hutchisons, C. A. & Sinsheimer, R. L. (1971) *J. Virol.* **8**: 121.

Igarashi, A., Koo, R. & Stollar, V. (1977) *Virology* **82**: 69-83.

Kallerhof, J., Perez, P., Bouzoubae, S., Ben Tahar, S. & Perre, T. J. (1990) *Plant Cell Reports* **9**: 224-228.

Kimura, Y., Norrby, E., Nagata, I., Ito, Y., Shimokata, K. & Nishiyama, Y. (1976) *J. gen. Virol.* **33**: 333-343.

Kumar, P., Awasthi, L.P. & Khan, M.N. (2004) Antiviral efficiency of Boerhaaria diffusa against cucumber green mottle mosaic virus in muskmelon, (Cucumis melo). New Botanist. Published by Today & Tomorrow's Printers and Publishers New Delhi (Ed. Mishra, S.K.) pp. 39-43.

Kurath, G. & Palukaitis, P. (1990) *Virology* **176**: 8-15.

Lamb, J. W. & Hay, R. T. (1990) *J. gen Virol.* **71**: 2257-2264.

Larkin, P. J. & Scowcraft, W. R. (1981) *Theor. Appl. Genet.* **60**: 197-214.

Lawson, C., Kaniewski, W., Haley, L., Rozman, R. & Newell, C. et al. (1990) *Biotechnology* **8**: 127-134.

Loesch-Fries, L. S., Merlo, D., Zinnero, T., Burhop L, Hall K, Krau K, Jarvis N, Nelson, S. & Halk, E. (1987) *EMBOJ.* **6**: 1845-1851.

MacKenzie, D. J. & Tremaine, J. H. (1990) *J. gen. Virol.* **71**: 2167-2170.

Matthews REF (1970) Plant Virology. Academic Press, New York, 778 pp.

Nelson, R. S., Powell, A. P. & Beachy, R. N. (1987) *Virology* **158**: 126-132.

Niblett, C. L., Dickson, E., Fernow, K. H., Horst, R. K. & Zaitlin, M. (1978) *Virology* **91**: 198-203.

Novick, R.P. (1969) *Bact. Rev.* **33**: 210-235.

Osbourn, J. K., Watts, J. W., Beachy, R. N., & Wilson, M. A. (1989) *Virology* **172**: 370-373.

Otsuki, Y. & Takebe, I. (1976) In: Biochemistry and cytology of plant-parasite mteraction, (Eds. Tomiyama, K., Daly, J. M., Uritani, I., Oku, H. & Ouchi, S.) Elsevier, Amsterdam.

Peleg, J. & Stollar, V. (1974) *Arch. Ges. Virusforsch.* **37**: 54-61.

Powell-Abel, P., Nelson, R. S., De B, Hoffman, N., Rogers, S. G., Fraley, R. T. & Beachy, R.N. (1986) *Science* **232**: 738-743.

Powell-Abel, P., Stark, D. M., Sanders, P. R. & Beachy, R. N. (1989) *Proc. Natl. Acad. Sci. USA* **86**: 6949-6952.

Richardson, M. A. (1982) Ph.D. thesis.Australian National University, Canberra, Australia.

Russell, G. E. (1978) *Nature Lond.* **131**: 468.

Sarver, N., Cartin, E. M., Chang, P. S., Zaia, J. A., Landane, P. A., Stephens, D. A. & Rossi, J. J. (1990) *Science* **247**: 1222-1225.

Sastry K. S., Prabhakar, D. & Zaidi, A. A. (1992) In: Farming Systems and Integrated Pest Management (Eds. Verma, J. P. & Varma, A.), Malhotra Pub. House, New Delhi, pp 245-258.

Sherwood, J. L. (1987) In: Plant Resistance to Viruses (Eds. Evered, D. & Harnett, S.), Chichester UK, John Willey and Sons, pp. 136-150.

Sherwood, J. L. & Fulton, R. W. (1982) *Virology* **119**: 150-158.

Singh, B. P. & Raizada, R. K. (1988) In: Current Trends in Physiological Plant Pathology (Eds. Janardhanan, K. K. & Singh, B. P.), pp. 198-219.

Singh, B. P., Srivastava, K. M. & Raizada, R. K. (1990) *Vatika* **1**: 24-31.

Sleat, D. E. & Palukaitis, P. (1990) *Proc. Natl. Acad. Sci. USA* **87**: 2946-2950.

Srivastava, K. M., Raizada, R. K. & Singh, B. P. (1993) In: Genetic Engineering, Molecular Biology and Tissue Culture for Crop Pest and Disease Management (Ed. Vidhyasekaran, P.), Daya Publishing House, Delhi, pp. 31-41.

Srivastava, K. M., Rana, N. S., Dwadash Shreni, V. C. & Singh, B. P. (1986) *Indian Phytopath.* **39**: 20-25.

Sternberg, N. & Hoess, R. (1983) *Annu. Rev. Genet.* **17**: 123-154.

Stollar, V. (1980) In: The Togaviruses: biology, structure, replication, (Ed. Schlesinger, W.) 427-458. Academic Press, New York.

Stollar, V. & Shenk, T. (1973) *J. Virol.* **11**: 592-595.

Thung, T.H. (1931) Semstof en plantencel bij enkele virusziekten van de tabaksplant. Z. Ned. Indisch Natuurweyensch. Congr. Bandoeng, Java, pp. 450-463.

Tien, P., Li, C. L., Jiang, Z. R., Zhao, S. Z., & Yie Yand Liu, Y. Z. (1990a) Transgenic pepper plants resistant to cucumber mosaic virus expressing its satellite cDNA. Abstr. VIIIth Int. Congr. Virology, Berlin, Germany.

Tien, P., Zhao, S. Z., Wang, X., Wang, G. J., Zhang, C. X. & Wu, S. X. (1990b) Virus resistance in transgenic plants that express the monomer gene of cucumber mosaic virus satellite RNA in green house and field. Abstr., VIIIth Int. Congr. of Virology, Berlin, Germany.

Tumer, N. E., O'Connell, K. M., Nelson, R. S., Sanders, P. R., Beachy, R. N., Fraley, R. T. & Shah, D. M. (1987) *EMBO J.* **6**: 1181-1188.

Van Dun, C. M. P. & Bol, J. F. (1988) *Virology* **167**: 649-652.

Van Dun, C. M. P., Overduin, B., Vun Vloten-Dotting, L. & Bol, J. F. (1988) *Virology* **164**: 383-389.

Vander Wilk, F., Willink, D. P. S., Dursman, M. J., Duttinga, D, & Goldbach, R. (1991) *Plant. Mol. Biol.* **17**: 431-439.

Verma, H.N. & Awasthi, L.P. (1979) *Can J. Bot.* **57**: 926-932.

Verma, H.N. & Awasthi, L.P. (1978) *New Botanist* **5**: 49-59.

Verma, H.N. & Awasthi, L.P. (1980) *Can J. Bot.* **58**: 2141-2144.

Verma, H.N., Awasthi, L.P. & Mukerjee, K. (1982) *Indian J. Bot.* **5**: 161-165.

Verma, H.N., Awasthi, L.P. & Mukerjee, Ketaki (1979) *Z. Pflkrankh Pflschutz* **86**: 735-740.

Verma, H.N., Awasthi, L.P. & Mukherjee K. (1982) In: Advancing Frontier of Mycology and Plant Pathology. Published by Today & Tomorrow's Printers and Publishers, New Delhi (Eds. Bilgrani, K.S., Mishra, P.S. & Mishra, R.S.), pp. 255-264.

Verma, H.N., Awasthi, L.P. & Saxena, K.C. (1979) *Can. J. Bot.* **57**: 1214-1217.

Verma, H.N., Khan, M.M.A.A. & Dwivedi, S.D. (1985) *India J. Plant Pathol.* **3**: 13-20.

Verma, H.N., Mukerjee, K. & Awasthi, L.P. (1980) *Naturwissenschaften* **67**: 364.

Waterworth, D. E., Kaper, J. M. & Tousignant, M. E. (1979) *Science* **204**: 845-847.

Zaidi, A. A. & Mukherjee, D. (1989) Studies on the efficacy of neem oil for controlling virus disease of some ornamental bulbous plants. Abstr. Intn. Conf. on the Recent Advances in Medicinal Aromatics and Spice Crops. ISMAP, New Delhi, Jan. 28-31, 1989.

Zaidi, A. A., Ram, R., Zaidi, S. M. D. & Mukherjee, D. (1992) In: Plant Quarantine in the 90s and beyond. Proceedings of the International Plant Quarantine Congress, 1991, Malaysia, P L Manalo et al. editors, pp. 259-264.

# 16

# *Trichoderma* : Biology and Biotechnology

**V. K. Gupta[1], A. Gupta[2], N. Modi[3] and D.R. Modi[3]**
*[1]CISH, Lucknow, [2] CDRI, Lucknow and [3] B.B.A. University, Lucknow.*

## ABSTRACT

Chemicals are highly toxic to the environment and living organisms have also affected non-target species including some beneficial organisms, which balance nature and thereby have changed the natural ecological system. Most of the modern agrochemicals are selective/specific in their action and provide protection for short periods or till they are chemically dissociated. The success of *Trichoderma* strains as biocontrol agents is due to their high reproductively capacity, ability to survive under very unfavorable conditions, efficiency in utilization of nutritional capacity to modify the rhizosphere, strong aggressiveness against phytopathogenic fungi and efficiency in promoting plant growth and defense mechanism. *Trichoderma reesei* is a biotechnically important filamentous fungus used commercially in enzyme production and is also one of the best known cellulolytic organisms, producing readily and in large quantities a complete set of extracellular cellulases for the degradation of crystalline cellulose.

## INTRODUCTION

*Trichoderma* is a widely distributed ubiquitous hyphomycete fungi occurring in nearly all soil types and habitats, especially those rich in organic matter. Different species, in presence of conducive micro-climates, are capable of producing enzymes, which in turn degrade the target pathogenic fungi or in adverse conditions produce a number of toxins. In recent years, considerable success has been achieved through the use of fungal antagonists against a number of soil-borne diseases significantly affecting many important crops. Among several antagonists, *Trichoderma* sp. have been found of ubiquitous nature and promising in reducing soil-borne pathogens. The reliability of biological system can be improved by use of formulations that provide conducive environment for bioprotectants and fermentation system that economically produce propagules of high quality (Harman, 1991). The most common biotechnological agents of *Trichoderma* genus are strains of *Trichoderma viriens, Trichoderma viride, Trichoderma harzianum* and *Trichoderma reesei*. These properties made *Trichoderma* a ubiquitous genus present in any habitat and high population density. *Trichoderma* is more efficient in acidic soil than alkaline soils. Plant diseases, caused primarily by fungal and bacterial pathogens, cause loses of agricultural and horticultural crops. These result in reduced food supply and agricultural products of poor quality. Biological control involves the use of beneficial microorganisms, such as *Trichoderma* spp. to attack and control plant pathogens and diseases.

These fungi produce a number of enzymes such as cellulases, hemicellulases, chitinases and glucanases. The enzymes, chitinases and glucanases in their multiple forms are responsible for fungitoxic activity. It is reported that these fungi, when applied to the field, enhance plant growth as well. These days commercial preparations of these fungi are available in the market in different commercial forms, which are easy to use. In addition to *T. reesei*, a wide variety of other bacteria and fungi also produce cellulolytic enzymes. Cellulases originating from various organisms having different characteristics are used industrially in many applications, such as in the textile industry in finishing of denim fabric to impart a stonewashed appearance (biostoning) and in biofinishing of cotton. *T. reesei* strains producing significant amounts of homologous and heterologous cellulases and having defined cellulase profiles were constructed for specific industrial applications, *i.e.*, biostoning and biofinishing of cotton.

## BIOLOGY

### Occurrence and Distribution

The abundance of *Trichoderma* in various soils, coupled with their ability to degrade various organic substrates due to cellulolytic character, their metabolic versatility allow them to survive in many adverse ecological niches. It has been proved that some of the conidia of *Trichoderma harzianum* added to soil without nutrient-supplying amendments were able to survive between 110 and 130 days, but the length of survival varied with the isolates (Papavizas and Lewis, 1981). *Trichoderma* is widely distributed all over the world and occurs in nearly all soil types and habitats, especially in those containing or consisting of organic matter.

Although the ecological importance of chlamydospores has not been given due importance, there is an evidence which indicates their role in survival in the soil. Caldwell (1958) was amongst the first to report that chlamydospores survival is better in soil than conidia. Lewis and Papavizas (1984) demonstrated the potential of various *Trichoderma* species to form chlamydospores readily and in great numbers in natural soil or in fragments of organic matter after the introduction of the fungus to the soil as conidia. They also suggested that introduced isolates have the potential and aggressiveness to colonize and establish themselves in organic matter in natural environments. If chlamydospores can be helped to survive in soil for greater periods, the ecological importance of these structures would be significant in field implementation of biological control. The carbon and energy requirements can be met with monosaccharides and disaccharides, complex polysaccharides, purines, pyrimidines, amino acids, condensed tannins, catechins, aldehydes and organic acids particularly long-chain fatty acids (Danielson and Davey, 1973; Engelkes *et al.*, 1997).

The growth, however, has been variable and appears to depend upon $CO_2$ concentration and pH. Reports suggest that it is relatively unaffected up to 10 per cent (Danielson and Davey, 1973; Macauley and Griffin, 1969). The success of *Trichoderma* strains as biotechnological agents is due to their high reproductive capacity, ability to survive under very unfavorable conditions, efficiency in the utilization of nutrients, capacity to modify the rhizosphere, strong aggressiveness against phytopathogenic fungi, and efficiency in promoting plant growth and defence mechanisms. These properties have made *Trichoderma* a ubiquitous genus present in any habitat, and at high population densities *Trichoderma* is more efficient in acidic than alkaline soils. Excellent results of integrated control have been attained with strains of *Trichoderma virens* and metalaxyl against *phytium ultimum* infecting cotton (Chet *et al.*, 1997). It was classified for the first time in 1969 by Rifai. Later Bissett (1991a) further classified the genus into five sections based on branching of conidiophores. This system includes morphologically similar, morphologically variable anamorphs of Hypocrea and some forms which were previously included in *Gliocladium.* Due to its cosmopolitan nature (as abundant as 100 to 10,000 propagules per gm of soil) and easy cultivation, most of the species have been utilized as agents for biological control of a number of important soil-borne plant pathogenic fungi. So, among the known species, *Trichoderma harzianum*, *Trichoderma viride* and *Trichoderma reesei* are most commonly used as biotechnological agents.

## Morphology of *Trichoderma*

Colony characters, growth rate in culture, pigmentation, arrangement of conidiophores and shape of conidia are the important morphological characters used to distinguish different species of *Trichoderma.* Conidia shape may vary from globose to ellipsoidal, obovoidal or short-cylindrical with the basal end truncate and tapering type. Size of conidia may also differ. There is also variation in the culture colour. Some species show dull yellow pigmentation, other reddish colour and still others produce no pigment at all. Strains of *Trichoderma viride* produce aromatic odours. *T. aureoviride* produces a type of crystals in the culture media. Some strains of *Trichoderma* produce musty odours. Conidia are either roughened or verrucose in *T. viride* and wing-like projections of the outer wall in *T. saturnispoerum* and *T. ghanense.* Branching of conidiophore is either regular verticillate or irregular. Conidiophores are erect or straggling, branched at regular intervals, branches divergent, solitary or aggregated into floccose tufts. Each successive branches apically and distally progressively shorter and narrower giving pyramidal appearance. Conidiogenous cells are phialides, borne singly or in clusters terminally, lageniform, ampulliform, sub-globose or cylindrical, conidium bearing neck. Conidia single-celled, may be colourless, green, grey or brownish; smooth-walled to roughened, wing-like projections, obovoid, oblong or short cylindrical, sub globose bound in gloeoid heads or may be enclosed in a sac. Chlamydospores may be intercalary or terminal, globose to ellipsoid, greenish, yellowish or colourless.

Morphologically, *Trichoderma* somewhat resembles *Verticillium* and *Gliocladium.* In *Trichoderma*, the branching pattern is less regular as compared to *Verticillium* where branching is more regular, verticillate or bicilliate, whereas in *Gliocladium* branching arrangement is terminal. The colonies can easily be identified by their rapid growth, white, yellow green or bright green, dull reddish colour; matted wooly with strong or faint odour (Bisett, 1991a).The culture characters of some of the species, which are very commonly used for biological control, are described below.

## *Trichoderma viride* Pers

Colonies fast growing (5-9 cm). Conidiation forming compact tufts or more effuse, glaucous to dark bluish-green. Reverse typically uncoloured, less often pale yellowish. Odour usually distinctly aromatic, as of coconut. Conidiophores usually not extensively branched and having a relatively loose arrangement, branches most often paired, or single or 3-verticillate, often appearing flexuous. Phialides frequently paired or arising singly or 3-verticillate, narrowly lageniform, 8-14 × 2.4-3.0 μm. Conidia globose to ellipsoidal, usually conspicuously warted, bluish-green to dark green, 4.0-4.8 × 3.5-4.0 μm.

## *Trichoderma harzianum* Rifai

Colonies growing rapidly (most isolates 7-9 cm). Conidiation predominantly effuse, appearing granular or powdery due to dense conidiation; rapidly turning yellowish-green to dark green, or producing tufts or pustules fringed by sterile white mycelium. Reverse colourless to dull yellowish, buff or drab. Odour indistinct or faintly earthy. Conidiophores as in the section, tending to be regularly verticillate forming a pyramidal structure. Phialides ampulliform to lageniform, usually 3-4 verticillate, occassionally paired, mostly 3.5-7.5 × 2.5-3.8 μm, terminal phialides up to 10 μm long. Conidia subglobose to obovoid, mostly (2.5-) 2.7-3.5 × 2.1-2.6 (-3.0) μm, smooth-walled, subhyaline to pale green.

This species, even in the strict sense (Gams and Meyer, 1998), comprises the strains often used in biological control of plant-pathogenic fungi. Its optimum temperature for growth is 30°C with a daily increment of 3.3 cm on malt agar. Maximum growth temperature is 36°C.

## *Trichoderma virens*

Colonies fast growing (6-7 cm). Conidiation forming compact tufts or more effuse, dark-bluish green. Reverse colourless, slowly developing dull yellowish to amber shades. Odour usually indistinct. Conidiophores usually not extensively branched and having a relatively loose arrangement, branches most often paired, or single or 3-verticillate, often appearing flexuous. Phialides frequently paired or arising singly or 3-6 closely appressed phialides, narrowly lageniform, 4.5-10 × 2.8-5.5 μm. Conidia broadly ellipsoidal to obovoid, smooth walled (minutely warted at high magnifications), dark green, 3.5-6.0 × 2.8-4.1 μm.

## *Trichoderma reesei*

Colonies growing rapidly (5.5-5.7 cm). Conidiation trady and scattered in tufts, pale yellow-green. Conidiophores typical of the section, rarely rebranched. Phialides cylindrical, or slightly inflated, mostly 5.5-8 × 2.0-3.7 μm. Conidia pale green, ellipsoid, 3.5-4.5- × 2.3-3.0 μm.

## *Isolation of Trichoderma*

*Trichoderma species* are readily isolated from soil by all available conventional methods, largely because of their rapid growth and abundant conidiation. Due to the formation of chlamydospores and colonization of organic substrates, *Trichoderma sp. a*re also readily obtained by soil washing techniques. On wood, *Trichoderma* can frequently be observed as discrete colonies from which isolation into pure culture can be achieved in the field with moderate care. To increase recovery, selective methods have been devised by Askew and Laing (1993), Davet (1979), Elad and Chet (1983), Elad *et al.,* (1981), Johnson *et al.,* (1987), Papavizas and Lumsden (1982) and Smith *et al.,* (1990). Composition of selective medium devised by Papavizas and Lumsden (1982) is as follows:

Basal medium: V-8 juice - 200 ml, water - 800 ml; (g/l): agar-20; glucose-1; antimicrobial agents (μg/l): neomycin sulphate, bacitracin, penicillin G, and chloroneb - 100; chlortetracycline hydrochloride - 25, nystatin - 20, sodium propionate - 500; alkylaryl polyether alcohol is added to suppress Mucorales at 2.0 ml/l.

## *Cultivation of Trichoderma*

The methods described by Rifai (1969) are generally still followed. Good results can usually be obtained by point or streak inoculation on oat meal, corn meal or malt extract agars, and incubation under day light for 5 or a few more days at a temperature of 20-25°C.

## *Molecular taxonomy of Trichoderma*

Within the last 15 years, a new era of fungal systematics was initiated by the application of molecular techniques to answer questions of evolutionary patterns and processes. Despite the importance of *Trichoderma* in biocontrol and biotechnological applications, individual species of this genera are poorly defined and classification of strains at the species level is difficult. Although the genus *Trichoderma* is easily recognized and has been extensively studied (Bissett, 1984, 1991a-c; Rifai, 1969), it has been difficult to reliably define species because of the apparent paucity of morphological characters.

## *Characterization by isoenzyme analysis, RFLP, RAPD, PCR-based fingerprinting and DNA sequencing*

The first characterization of *Trichoderma* spp. by *isozyme* patterns was done by Zamir and Chet (1985). Twenty-three geographically diverse isolates of *T. harzianum* were grouped into five types according to the *isozyme* profiles. Results indicated that enzyme electrophoresis is useful for distinguishing *Trichoderma* at the intra-species level. Meyer (1991) used mitochondrial DNAs and plasmids as taxonomic characteristics for *T. viride*. He investigated 12 strains of the *T. viride* aggregate species and results of the molecular analyses were discussed in comparison to conidial ornamentation (Meyer and Plaskowitz, 1989). Morawetz *et al.,* (1992) and Kubicek *et al,.* (1996) produced restriction fragment length polymorphism (RFLP) patterns of a DNA fragment representing the cellobiohydrolase I-encoding gene cbh1 from different *Trichoderma* species that were species-specific. These studies illustrated the need to reclassify some species of *Trichoderma.* The polymerase chain reaction (PCR) technique has created new ways of revealing DNA polymorphisms among closely related genotypes with high sensitivity via a fast and easy-to-perform protocol. Schlick *et al*., (1994) and Kuhls *et al*., (1995) used PCR fingerprinting for

*Trichoderma* strain identification and detection of culture impurities. The characterization of potential biocontrol strains of *Trichoderma* by PCR fingerprinting was the aim of Arisan-Atac *et al.,* (1995), Fujimori and Okuda (1994), Turoczi *et al.*, (1996) and Zimand *et al.,* (1994). When analyzing soil isolates morphologically assigned to *T. harzianum, T. hamatum, T. viride* or *T. koningii,* all authors found two or more genetically distinct groups within each of these species. Turner *et al.,* (1996) used the *Random amplified polymorphic* DNA (RAPD) technique in an extensive study of molecular taxonomy of *Trichoderma* that included 145 strains of *Trichoderma* spp. Analysis of ITS-1 sequences of 128 strains of *Trichoderma* revealed 15 distinct ITS types with 34 unique sequences recognizing *T. harzianum* (3 taxa), *T. viride, T. virens, T. hamatum (2 taxa), T. aureoviride, T. citrinoviride, T. longibrachiatum, T. parceramosum, T. polysporum, T. piluliferum, T. saturnisporum and T. reesei* (Muthumeenakshi, 1996). Hence, molecular taxonomy has given new dimention for the study of identification of variation in the strains of *Trichoderma* spp.

## Biotechnology of *Trichoderma*

These versatile fungi are used commercially in a variety of ways i.e., plant growth promotion, protein sources, biocontrol agents, textiles industry and as gene source.

***Plant root colonization and biofertilization:*** *Trichoderma* strains must colonize plant roots prior to stimulation of plant growth and protection against infections. Colonization implies the ability to adhere and recognize plant roots, penetrate the plant, and withstand toxic metabolites produced by the plants in response to invasion by a foreign organism, whether pathogen or not. Mycorrhizal fungi interaction is modulated by plant flavonoids and fungal auxins, followed by morphogenetic events that include appressorium development. In addition, genes that encode hydrophobins and other cell wall structural proteins are specifically expressed, or their expression is up-regulated (Franlen *et al.,* 2002). Preliminary results have indicated the specific up-regulation of hydrophobins during colonization of *Trichoderma* strains and tomato plant roots. Some *Trichoderma* strains establish long-lasting colonization of plant roots and penetrate into the epidermis. There, they produce or release compounds that induce localized or systemic plant resistance responses. Plants react against fungal invasion by synthesizing and accumulating phytoalexins, flavonoids and terpenoids, phenolic derivatives, aglycones and other antimicrobial compounds. *Trichoderma* strains are generally more resistant to these compounds than most fungi; nonetheless, their ability to colonize plant roots strongly depends on the capacity of each strain to tolerate them. This resistance, considered an essential requirement for plant colonization, has been associated with the presence of ABC transport systems in *Trichoderma* strains (Harman *et al.*, 2004).

Root colonization by *Trichoderma* strains frequently enhances root growth and development, crop productivity, resistance to abiotic stress and the uptake and use of nutrients (Arora, 1992). Crop productivity in fields can increase up to 300% after the addition of *Trichoderma hamatum or Trichoderma koningii.* The same increase was observed when seeds were separated from *Trichoderma* strains by a cellophane membrane, which indicates that *Trichoderma* strains produce growth factors that increased the rate of seed germination. However, there are very few reports on strains that produce growth factors which have been detected and identified in the laboratory (auxins, citokinins and ethylene) (Arora, 1992). *Trichoderma* strains that produce cytokinin-like molecules, e.g., zeatyn and gibberelin GA3 or GA3-realted, have been recently detected. The controlled production of these compounds could improve biofertilization (Osiewecz, 2002). Root colonization by *Trichoderma* strains frequently enhances root growth and development, crop productivity, resistance to abiotic stress and the uptake and use of nutrients (Arora, 1992).

***Stimulation of Plant Resistance and Plant defence Mechanism:*** The ability of *Trichoderma* strains to protect plants against root pathogens has long been attributed to an antagonistic effect against the invasive pathogen (Chet *et al.*, 1997). However, these root-fungus associations also stimulate plant defensive mechanism. Strains of *Trichoderma* added to the rhizosphere protect plants against numerous classes of pathogens, e.g., those that produce aerial infections, including viral, bacterial and fungal pathogens, which points to the induction of resistance mechanisms similar to the hypersensitive response (HR), systemic acquired resistance (SAR), and induced systemic resistance (ISR) in plants (Harman *et al.*, 2004).

Plants genes respond to pathogens and elicitors for this reason. Plant defence mechanisms do not necessarily require stimulation by the living organism. The addition of *Trichoderma* metabolites that may act as elicitors of plant resistance, or the expression in transgenic plants of genes whose products act as elicitors, also results in the synthesis of phytoalexins, PR proteins and other compounds, and in an increase in resistance against several plant pathogens, including fungi and bacteria (Elad *et al.*, 2000, Dana *et al.*, 2001), as well as resistance to hostile abiotic conditions (Harman *et al.*, 2004). Activation of defense responses using elicitors could be a valuable strategy, as an alternative to the use of transgenic plants, to protect plants against pathogens.

***Trichoderma as protein source:*** *Trichoderma* strains have developed highly effective antagonistic mechanism to survive and colonize the competitive environment of the rhizosphere and spermosphere. One of its main mechanisms, mycoparasitism, relies on the recognition, binding and enzymatic disruption of the host-fungus cell wall. A major part of the *Trichoderma* antifungal system consists of a number of genes encoding for an astonishing variety of secreted lytic enzymes, including endochitinases, N-acetyl-beta-glucosaminidases, chitin 1, 4-beta-chitobiosidases, proteases, endo- and exoglucan beta-1, 3-glucosidases, endoglucan beta-1, 6-glucosidases, lipases, xylanases, mannanases, pectinases, pectin lyases, amylases, phospholipases, RNases, and DNases. Particularly useful for biocontrol application are chitinolytic and glucanolytic enzymes because of their ability to efficiently degrade the cell wall of plant pathogenic fungi by hydrolyzing biopolymers not present in plant tissues.

A substantial amount of work performed mainly during the past 7 years has indicated that cell-wall-degrading enzymes (CWDEs) from *Trichoderma* strains have great potential in agriculture as active components in new fungicidal formulations. This is because purified CWDEs from different strains of *Trichoderma harzianum* are highly effective in inhibiting spore germination and mycelial growth in a broad range of pathogens such as *Rhizoctonia, Fusarium, Alternaria, Ustilago, Venturia, Pythium, Phytophthora, Colletotrichum* and especially *Botrytis*. In contrast to plant enzymes, chitinases and glucanases from *Trichoderma* can degrade not only the immature wall at hyphal apices but also the strong chitin-glucan complexes of mature cell wall, as well as survival structures such as sclerotia and chlamydospores, which reduce not only disease symptoms but also pathogen spread. In particular, enzymes absent from plants such as beta-1, 6-glucanases can degrade important fungal cell wall structures such as beta-1, 6-glucans by linking chitin or beta-1,

3-glucans to cell wall proteins. *Trichoderma* enzymes have diverse structural and kinetic properties, which increase the probability of avoiding inhibitory mechanism. *Trichoderma harzianum* transformants over expressing Chit 33 chitinase constitutively inhibited the growth of *R. solani* under both repressing and depressing conditions; the antagonists' tests demonstrated that this chitinase also has an important role in mycoparasitism (Limon *et al.*, 1999). The strains producing the chimeric enzymes increased their specific chitinase activity. (Limon *et al.*, 2004). Biocontrol of *B. cinerea* and *Trichoderma harzianum* has been attributed in part to the action of proteases produced by the BCA that inactivate hydrolytic enzymes produced by this pathogen on bean leaves (Howell, 2003). Proteases involved in the degradation of heterologously produced proteins have been characterized (Delgado-Jarana *et al.,* 2003).The role of proteases in mycoparastism has been reinforced with the isolation of new protease-overproducing strains of *Trichoderma harzianum* (Szekeres *et al.*, 2004). Mutants with new profiles and higher quantities of secreted proteases were obtained by UV-irradiation. Some of these mutants have proved to be effective antagonists against plant pathogenic fungi such as *Fusarium culmurum* and *R. solani.* Synergism among lytic enzymes and between enzymes and antibiotics suggests formulations to test mixtures of *Trichoderma* transformants that produce different enzymes in order to improve the antagonistic effects of BCAs on phytopathogenic fungi, the effect of enzymes and secondary metabolites secreted by BCAs. Transformants were differently combined in order to test synergism among the enzymes secreted against several phytopathogens.

*Trichoderma* species have been investigated as biological control agents for over 70 years, but only relatively recently have strains become commercially available. This is largely a result of the change in public attitude towards the use of chemical pesticides, together with increased knowledge of their harmful side effects. Biocontrol agents are widely regarded by the general public as "natural" and therefore non-threatening products. *Trichoderma* species act against target organism in several ways. Strains are actively parasitic on host fungi through direct penetration of host hyphae and/or production of extracellular enzymes. In addition, species may produce antifungal antibiotics, and inhibition of pathogenicity-including hydrolytic enzymes has also been identified. The dual roles, antagonistic activity against plant pathogens and promotion of soil fertility, make *Trichoderma* strains appealing alternatives to hazardous fumigants and fungicides.

The antibiotic agent was exuded into the cultivation medium (Weindling, 1934). Pioneering efforts on interest in the secondary metabolites produced by *Trichoderma* species have been sustained and the isolation and characterization of a considerable number of metabolites has been achieved. Interestingly, gliotoxin and related compounds have recently attracted much interest in the medical field as potential immunomodulating agents that have the ability to suppress the immune system by triggering cell suicide (apoptosis) (Jiang *et al.*, 1993). An increasing number of metabolites from *Trichoderma* species are being discovered by the application of bioassay guided separation. With this technique, a biochemical assay is used to identify metabolites in an extract exhibiting a particular biochemical activity. In this way, biological activities other than antibiotic properties have been recognized for many compounds produced by *Trichoderma*. *Phytophthora parasitica* produces a cell wall glycoprotein (CBEL) that has a cellulose-binding domain (CBD) and elicits defence responses in the host plant. Infiltration of CBEL into non-host plant also induced defense reactions (McIntyre *et al.*, 2004). Expression in plants of fungal chitinases with CBDs, such as Chit 42CBD, has already increased antifungal activity (Limon *et al.*, 2004).

***Trichoderma as biocontrol agent:*** Chemical method of disease control has been in practise for long time, but after the disastrous effect of its over use was recognized, the importance of biological fungicides was emphasized. Cook and Baker (1983) defined 'Biological Control' as the reduction of the amount of inoculum or disease producing activity of a pathogen, accomplished by one or more organisms other than man. Though it is not possible to completely eradicate a pathogen or a disease by this method of control, but amount of inoculum build up can be maintained to a desired minimum level.

*Trichoderma* is one of the most commonly used biological control agents. Other important common biological control agents are *Aspergillus* and *Penicillium*. The most important quality of a bio-control agent is that it should have property of multiplying very fast in ordinary cheap medium, so that it can compete with disease causing agent. These can be easily grown in *in vitro* and *in vivo* conditions. In field conditions these are known to be quite aggressive in nature, growing on almost all types of substrates under a wide range of environmental conditions. In lab, they can be multiplied on cheap materials very easily.

These fungi either directly parasitize the pathogen or compete with other microorganisms and pathogens for food requirement. *Trichoderma* is known to have tremendous potential as bio-control agents. The selection of the species should be done on the basis of antagonistic traits as well as other behavioural traits.

Some strains are known to be plant growth stimulators; others induce resistant reaction in plants and few even cause diseases in plants. *T. viride* is known to attack 32 genera of plants (Far *et al.,* 1989). The high degree of ecological adaptability shown by strains within the genus *Trichoderma* is reflected in the fact that these fungi are common in soils all over the world, under differing environmental conditions, and live on various substrates. This considerable variation, coupled with their amenability to cultivation on inexpensive substrates, makes *Trichoderma* isolates attractive candidates for a variety of biological control applications. However, genetic characterization of this genus is as yet incomplete, and morphological classification is insufficient to predict behaviours of an unknown isolate. Therefore, although potential biocontrol agents with suitable antagonistic characteristics may readily be found, they must be screened carefully for other traits relevant to their use in a given application.

Even after careful screening, bio-protection candidates often show variable performance in field trials. The reliability of biological control systems can be improved by the use of formulations that provide conducive environments for the bio-protectant and fermentation systems that economically produce propagules of high quality (Harman, 1991). To become a successful bio-protectant, a *Trichoderma* strain must therefore not only show desirable antagonistic and ecological behaviour but must also be amenable to manipulation in relevant formulations and delivery systems. Nonetheless, the most appropriate criterion for selection of a biocontrol agent may ultimately be consideration of environmental requirements, both of the antagonist and of the pathogen which it is intended to control (Marois and Coleman, 1995).

It is observed that same pathogen can behave differently in field and in lab conditions. Though these are quite tolerant to the changes in the environmental conditions, the important abiotic factors affecting the behaviour, mode of action in the field are temperature, moisture and nutrition.

***Antagonistic mechanisms:*** The three most important antagonistic reaction processes of biocontrol agent are antibiosis, mycoparasitism and competition of nutrients. It is quite possible that a single antagonistic reaction may include all the three processes. There may be reactions other than these as suggested by Zimand *et al.,* (1996). According to this study, *T. harzianum* T39 checks/inhibits the pectin degrading enzymes produced by *B. cinerea*, there by reducing pathogenicity. Antagonistic mechanism is dependent upon type of strain, the organism intended to control and the environmental conditions.

***(a) Antibiosis:*** Chemical compounds (metabolites) of a microorganism that inhibits or kills other microorganisms are known as antibiotics. These chemical compounds are secondary metabolites that differ from primary metabolites, as these have no specific role to play in the organism that produces them. These are variously grouped-polypeptides, terpenes, phenols and alkaloids.

Luckner (1990) and Turner and Aldridge (1983) classified various secondary metabolites along with their structures, produced by *Trichoderma.* Secondary metabolites are produced by *Trichoderma* and other biocontrol agents in the form of volatile and non-volatile compounds that inhibit the growth processes of pathogen. Gliovirdin, virdin and gliotoxin are the most common and potent types as stated by Howell and Stipanovic (1983). 6-n-pentyl-2H-pyran-2-one (PPT) is characteristic of some species of *Trichoderma.* Antibiotic production varies with the strain type and environmental factors.

Polketides is the largest group of fungal secondary metabolites. These originate from the repetitive connections of short chain fatty acids e.g., acetate or propionate, by pathways very similar to those of fatty acid biosynthesis. Terpenes are the most abundant and widely distributed secondary metabolites in nature. According to Zeppa *et al.*, (1990) *Trichoderma* spp. form volatile terpenes. These are grouped as monoterpenes, sesquiterpene, diterpene, triterpene and tetraterpene.

***(b) Mycoparasitism:*** There are four stages in direct interaction between *Trichoderma* spp. and other fungi. These are:

(i) Chemotropic growth—A chemical stimulus produced by the target fungus attracts the antagonist.

(ii) Specific recognition—Lectins present on cell surfaces of both pathogen and antagonist are responsible for this process.

(iii) Attachment and coiling of *Trichoderma* around its host.

(iv) Secretion of lytic enzymes.

***Competition:*** When the antagonistic organism is applied to the field, it competes with the pathogen and other fungi for food. This competition leads to control of disease. *Trichoderma* spp. are aggressive in nature and can even survive as chlamydospores and conidia in the absence of sufficient nutrients. Ecological concepts of exploitation, competition and interference, competition to interactions between fungal populations was given by Lockwood (1981, 1992) and Wicklow (1992). According to Lockwood (1992), exploitation competition is depletion of a resource by any organism without directly limiting the resources for other organisms. In interference competition, either by chemicals or behavioural mechanisms, one organism directly limits the activity of another organism.

*Trichoderma* sp., when applied to the aerial parts of plants viz. leaves, are known to control pathogens that invade the dead tissues as base and from this base they enter into the healthy tissues.

*Botrytis* spp. and *Sclerotinia* spp. are best controlled by this method. *Trichoderma* is reported to control *Botrytis* and *Sclerotinia* on Strawberry and Apple (Tronsmo and Dennis, 1977; Tronsmo and Raa, 1977). Environmental factors play an important role in successful control of pathogen by this method. Dubos (1987) gave an account of some important factors that affect this very type of competition. According to him and his co-workers, selected *Trichoderma* spp., when sprayed on grape flowers at the time of blossoming are able to colonize tissues as it senesces, thereby delaying colonisation by *Botrytis*, and later disease level in the fruit. He laid emphasis on the importance of timing of applications and importance of repeated applications of the antagonist during flowering. The most important spray according to him is during late flowering (Dubos, 1987; Gullino, 1992 and Harman *et al.*, 1996).

***Trichoderma in textile industry:*** *Trichoderma reesei* is a biotechnically important filamentous fungus used commercially in enzyme production. *T. reesei* is also one of the best known cellulolytic organisms, producing readily and in large quantities a complete set of extracellular cellulases for the degradation of crystalline cellulose. In addition to *T. reesei*, a wide variety of other bacteria and fungi also produce cellulolytic enzymes. Cellulases originating from various organisms and having different characteristics are used industrially in many applications, such as in the textile industry in finishing of denim fabric to impart a stonewashed appearance (biostoning) and in biofinishing of cotton. The genus *Trichoderma* comprises a group of filamentous ascomycetes that are widely used in industrial applications because of their ability to produce extracellular lignocellulose-degrading hydrolases in large amounts. Enzymes secreted by *Trichoderma* have received widespread industrial interest, leading to commercial applications in the textile industry (Galante *et al.,* 1998a, Cavaco-Paulo and Gübitz, 2003, Nierstrasz and Warmoeskerken, 2003), the food and feed industries (Galante *et al.,* 1998b) and the pulp and paper industry (Buchert *et al.,* 1998). *Trichoderma reesei* has the capacity to secrete enzymes in high yields and this property can be exploited when using *T. reesei* as an industrial host for homologous and heterologous enzyme production (Mäntylä *et al.,* 1998, Penttilä, 1998, Penttilä *et al.,* 2004). However, there is a continuous need to improve enzyme production processes in order to enhance protein yields and improve the production economics. Cellulolytic enzymes hydrolyze cellulose and are produced by a wide variety of bacteria and fungi, *T. reesei* being one of the best-known cellulolytic organisms.

Cellulases are industrially important enzymes (reviewed by Schülein, 2000) with a current market value of about 190 million US$ (Nierstrasz and Warmoeskerken, 2003). In the textile industry, cellulase enzymes are tools for the fabric and garment finisher to produce higher value products, as cellulases clean fuzz and prevent formation of pills on the surface of cotton garments. Cellulases can also be used in denim finishing to create a fashionable stonewashed appearance in denim cloths in a process called biostoning. About 10% of textile finishing of cellulose materials is estimated to be performed by cellulases and approximately 80% of the 1.8 mrd pairs of denim jeans produced annually are finished with cellulases as an alternative to pumice stones (Buchert and Heikinheimo, 1998). Cellulases from the fungi *T. reesei* and *Humicola insolens* are widely used in the textile industry, and depending on the desired effect, different types of cellulases are utilized. In the detergent industry, cellulases are used to clean cotton garments or to brighten faded coloured garments by removing fuzz. In animal feed, cellulases are utilized together with other hydrolases in the degradation of non-starch polysaccharides to improve feed conversion rates (Galante *et al.,* 1998b). The food industry uses cellulases together with other plant cell wall-degrading enzymes in fruit and vegetable processing (Urlaub, 2002). Enzymatic hydrolysis of biomass to sugars for subsequent ethanol production has also been a major research area during recent years (Himmel *et al.,* 1999). Cellulases have potential in the pulp and paper industry, e.g., in deinking to release ink from fibre surfaces and in improving pulp drainage (Suurnäkki *et al.,* 2004). Within the forest industry cellulases have been shown to be effective in decreasing the energy consumption of mechanical pulping (Pere *et al.,* 2002). Cellulases have been found to increase the alkali solubility of treated pulp and directly alkali soluble cellulose has been obtained with specific cellulase compositions (Vehviläinen *et al.,* 1996, Rahkamo *et al.,* 1996). This property can be utilized in developing new environmentally beginning processes for manufacturing cellulosic articles such as films and fibres. The wide spectrum of industrial uses for cellulases establishes a need for commercial cellulase products containing different cellulose components and functioning optimally in different pH and temperature ranges.

***Trichoderma source of genes:*** Biocontrol microbes, almost by definition, must contain a large number of genes that encode products that permit biocontrol to occur. Several genes have been cloned from *Trichoderma* spp. that offer great promise as transgenes to produce crops that are resistant to plant diseases. These genes, which are contained in *Trichoderma* spp. and many other beneficial microbes, are the basis for much of "natural" organic crop protection and production. Several new general methods for biocontrol and for enhancement of plant growth have recently been demonstrated. In this processes of mycoparasitism, antibiosis, competition for the nutrients or space, tolerance to stress through enhanced root or plant development, solubilization and sequestration of inorganic nutrients, induced resistance and inactivation of the pathogen's enzymes. Transgenic expression of high levels of chitinolytic and glucanolytic *Trichoderma* enzymes does not affect plant morphology, development or yield, or infection by arbuscular mycorrhizal fungi.

Most of these genes have been patented and are commercially available, but a number are in development to be used in agricultural biotechnology. Cell wall-degrading enzymes, antibiotics, and other factors induced in the presence of the host fungus are thought to be directly involved (Lorito *et al*., 1996; Lorito 1998; Schirmböck *et al*., 1994). Chitinolytic enzymes secreted by *T. harzianum* have received much attention because of their lytic and inhibitory activity toward many phytopathogenic fungi ( Lorito *et al*., 1993, 1996). Although several exo- and endochitinases have been purified and their genes cloned from different *T. harzianum* isolates (Harman *et al*., 1993; Ulhoa *et al*., 1992), the role of these enzymes in mycoparasitism, antagonism, or biocontrol has not yet been established. The *T. harzianum* strain P1 is a well-studied isolate effective against foliar and post-harvest pathogens, such as *Botrytis cinerea* (Tronsmo, 1991). Strain P1 secretes several chitinolytic enzymes including at least one *N*-acetyl-b-glucosaminidase (CHIT72), at least one chitin 1,4-b-chitobiosidase (CHIT40), and a single 42-kDa endochitinase (CHIT42). The CHIT42 endochitinase exhibits a strong lytic and antifungal activity in vitro and is the only enzyme produced by P1 that is capable of rapidly degrading a suspension of colloidal chitin (Harman *et al*., 1993). Strain P1 contains a single copy of the gene encoding for CHIT42, *ech42* (previously named *ThEn42*; Hayes *et al*., 1994), which is expressed in the presence of chitin or fungal cell walls and in the early phases of the mycoparasitic interaction with *B. cinerea* and other fungi. Therefore, strain P1 appears suitable for gene disruption experiments to assess the role of CHIT42 in biocontrol (Lorito *et al*., 1996).

## CONCLUSION

*Trichoderma* and numerous other bioagents have been isolated from plant roots and there have been numerous attempts to use them for the biocontrol of certain root and foliar diseases resulting in failure as well as astounding success in specific pathogen and crops. Quantitative studies on the survival and establishment of a pathogen in plant rhizosphere and phylloplane need primary attention. The application of techniques for protoplast isolation and fusion and gene cloning will result in further development of new technologies for genetic manipulation of two fungi. With the help of latest tools of biotechnology, attempts should be made to develop Genetically Modified (GM) *Trichoderma* spp. Its mass multiplication methods must be simplified and have an acceptable shelf life and meet acceptable environmental and toxicological safety standards. It should be highly competitive with resident microbes and should be able to establish in soil for an acceptable period of time. It should be effective under broad range of environmental conditions (Cook 1993). Plant diseases, caused primarily by fungal and bacterial pathogens, causes forever losses of agricultural and horticultural crops every year.

For many diseases, traditional chemical control methods are not always economical nor are they effective, and fumigation as well as other chemical control methods may have unwanted health, safety, and environmental risks. Biological control involves the use of beneficial microorganism, such as specialized fungi and bacteria, to attack and control plant pathogens and the diseases they cause. Biological control offers an environmentally friendly approach to the management of plant disease and can be incorporated into cultural usage for an effective integrated pest management (IPM) system. Years after the discovery of gliotoxin and viridin, interest was renewed in toxic metabolites produced by other species. Improvement and new techniques in metabolite production have enabled its successful utilization in industrial microbiology. These improvements, aided with clarification of the taxonomy (Rifai, 1969), resulted in renewed interest in the ability of *Trichoderma* sp. as bioagent. Dennis and Webster (1971) demonstrated production of antibiotics different from gliotoxin and viridin, although they were unable to show production by other nine species aggregates. They demonstrated production of other chloroform-soluble antibiotics 'trichodermin' by *Trichoderma viride* and *Trichoderma polysporum* and other peptide antibiotics by *Trichoderma hamatum*.

Papavizas *et al.,* (1982) found that several UV-induced mutants of *Trichoderma harzianum* produced two unidentified metabolites one of which was heat-labile, and the other was heat-stable. 'Trichodermin', the least structurally complicated sesquterpenoid metabolite originally described by Godgredsen and Vangedal (1965) from *Trichoderma* should not be confused with trichodermin-4 as reported in literature (Fedorinchik *et al.*, 1975). This so-called antibiotic was crude preparation produced by growing *Trichoderma lignorum* on concentrated solid protein-carbohydrate media or liquid molasses media, drying the preparations, and grinding the biomass, including the solids from the media. Potential biocontrol agents were, at best, only marginally effective. Many strains were isolated because they were observed parasitizing plant pathogens or inhibiting their growth in culture. The enzymes and antibiotics produced by *Trichoderma* species that appear to be involved in biocontrol are strongly influenced by the substrate on which the fungus is grown, and conditions in the laboratory probably occur only rarely in nature or not at all. Temperature also has a profound effect on the production and activities of enzymes and antibiotics associated with biocontrol by

*Trichoderma* species. One mechanism of biocontrol is mycoparastism, and *Trichoderma virens* produces antifungal compounds to assist in killing its fungal targets. Peptides synthetases produce a wide variety of peptide secondary metabolites in bacteria and fungi. Many of these are known to possess antibiotic activities. Peptaibols form a class of antibiotics known for their high α-aminoisobutric acid content and their synthesis as a mixture of isoforms ranging from 7 to 20 amino acids in length. Industrially useful *T. reesei* strains producing significant amounts of homologous and heterologous cellulases and having defined cellulase profiles were constructed. Some recent results of the molecular mechanisms of cellulose gene regulation and protein secretion in *T. reesei* could possibly be utilized to increase the production levels of cellulases. The suitability of the cellulases in efficient conversion of biomass to sugars that can be fermented to ethanol should also be investigated. Cellobiohydrolases are the key components of the redesigned, highly synergistic cellulase mixtures required for such processes. *T. reesei* cellulases are useful in denim finishing to impart a certain type of stonewashed appearance, but they have disadvantages such as a tendency to promote backstaining and weakening of fabrics.

## References

Arisan-Atac, I., Heidenreich, E. & Kubicek, C.P. (1995). *FEMS Microbiol. Lett.* **126**: 249-256.

Arora, D.K., Elander, R.P. & Mukerji K.G. (1992). Handbook of applied mycology. *Fungal Biotechnol.* 1 4: Marcel Dekker, New York.

Askew, D.J. & Laing, M.D. (1993). *Plant Path.* **42**: 686-690.

Bissett, J. (1984) *Can J. Bot.* **62**: 924-931.

Bissett, J. (1991a) *Can. J. Bot.* **69**: 2357-2372

Bissett, J. (1991b) *Can.J. Bot.* **69**: 2373-2417.

Bissett, J. (1991c) *Can.J. Bot.* **69:** 2418-2420

Bliss DE (1951) *Phytopathol.* **41**: 665-683.

Buchert, J. & Heikinheimo, L. (1998) *Carbohydr. Eur.* **22**:32.34.

Buchert, J., Oksanen, T., Pere, J., Siika-aho, M., Suurnäkki, A. & Viikari, L. (1998) In: *Trichoderma* and *Gliocladium*, Enzymes, biological control and commercial applications (Eds. G.Harman & C.Kubicek), Vol. **2**: 343-364.

Cavaco-Paulo, A. & Gübitz, G. (2003). In: Textile processing with enzymes (Eds. A. Cavaco-Paulo & G. Gübitz). Woodhead Publishing Ltd., England. pp. 86.119.

Chet, I., Inbar J. & Hadar I. (1997) In: The Mycota IV: Enviornmental and microbial relationships (Eds. Dt. Wicklow & B. Soderstrom ). 165-184.

Cook, R.J. (1993) *Annu Rev Phytopathol.* **31**: 53-80.

Cook, R.J. & Baker, K.F. (1983). *American Phytopathological Society*, St. Paul, MN, 539 pp.

Dana., M.M., Limon, M.C., Mejias R., Mach, R.L., Benitez, T., Pintor-Toro, J.A. & Kubicek, C.P. (2001) *Curr Genet* **38**: 335-342.

Daneilson, R.M. & Davy, C.B. (1973) *Soil Biol Biochem.* **5**: 505-515.

Davet, P. (1979) *Ann. Phytopath.* **11**: 529-534.

Delgado-Jarana, J., Moreno-Mateos, M.A. & Benitez, T. (2003). *Euk Ccell* **2**: 708-717.

Dennis, C. & Webster J (1971) *Trans Br Mycol Soc.* **57**: 41-48.

Dubos, B. (1987) In: *Innovative approaches to plant disease control.* (Ed. I. Chet), John Wiley and Sons, New York, pp. 107-135.

Elad, Y., Freeman, S. & Monte, E. (2000) IOBC wprs Bulletin, vol 24. Sevilla, Espana.

Elad, Y. & Chet, I. (1983) *Phytoparasitica* **11**: 55-58.

Elad, Y., Chet, I. & Henis, Y. (1981) *Phytoparasitica* **9**: 59-68.

Engelkes, C.A., Nuclo, R.L. & Fravel, D.R. (1997) *Phytopathol* **87**: 500-505.

Farr, D.F., Bills, G.F., Chamuris, G.P. & Rossman, A.Y. (1989) APS Press, St Paul, MN.pp.999

Fedorinchik, N.S., Tarunina, T.A., Tyutyunnikov, M.G. & Kudryavtseva, K.I. (1975) *Plant Prot.* **3**: 67-72.

Franken, P., Khun, G. & Gianinazzi-Pearson V (2002) In: Molecular biology of fungal Development, (Ed. H.D. Osiewacz ) Marcel Dekker, New York, 325-348.

Fujimori, F. & Okuda, T. (1994) *J. Antibiotics* **47**: 173-182.

Galante, Y., De Conti, A. & Monteverdi, R. (1998a) In: *Trichoderma* and *Gliocladium*, Enzymes, biological control and commercial applications (Eds. G.Harman & C. Kubicek), Vol. **2**: pp. 311-326.

Galante, Y., De Conti, A. & Monteverdi, R. (1998b) In: *Trichoderma* and *Gliocladium*, Enzymes, biological control and commercial applications (Eds. G.Harman & C. Kubicek), Vol. **2**: pp. 327.342.

Godgredsen, W.O. & Vangedai, S. (1965) *Acta Chem Scand.* **19**: 1088-1102.

Gullino, M.L. (1992) In: *Biological control of plant diseases.* (Eds. E.C. Tjamos, G.C. Papavizas & R.J. Cook), Plenum Press, New York, pp. 125-132.

Harman, G.E., Howell, C.R., Viterbo, A., Chet, I. & Loritto, M. (2004) Nature Reviews **2**: 43-56.

Harman, G.E. (1991) *Crop Prot.* **10**: 166-171.

Harman, G.E., Latorre, B., Agosin, E., Martin, R.S., Riegel, D.G., Nielsen, P.A., Tronsmo, A., A. & Pearson, R.C. (1996) *Biol. Control* **7**: 259-266.

Harman, G.E., Hayes, C.K. & Lorito, M.(1993) In: Pest management biologically based technologies: proceeding of Beltsville symposium XVIII, (Eds. R.D. Lumsden & J.L.Vaughn), American Chemical Society, Washingatan, D.C., pp. 347-354.

Hayes, C.K., Klemsdal, S., Lorito, M., Di-pietro, A. Peterbauer, C. Nakas, J.P. Tronsmo, A. & Harman, G.E. (1994) *Gene.* **138**: 143-148.

Himmel, M. Ruth, M. & Wyman, C. (1999) *Curr. Opin. Biotechnol.* **10**: 358.364.

Howell, C.R. (2003) *Plant Dis* **87**(1): 4-10.

Howell, C.R. & Stipanovic, R.D. (1983) *Can. J. Microbiol.* **29**: 321-324.

Johnson, L. F., Bernard, E.C. & Peiyuan Quian, (1987) *Plant Dis.* **71**: 137-140.

Kubicek, C.P., Bolzlbauer, U.M., Kovacs, W., Kuhls, K., Lieckfeldt, E., Borner, T. & Samuels, G.J. (1996) *Fungal Gen. Biol.* **20**: 105-114.

Kuhls, K., Lieckfeldt, E. & Borner, T. (1995) *Microbiol. Res.* **150**: 363-371.

Limon, M.C., Chacon, M.R., Mejias, R., Delgado-Jarna, J., Rincon, A.M., Codon, A.C. & Benitez, T. (2004) *Appl Microbiol Biotechnol.* **64**: 675-685.

Lockwood, J.L. (1981) In: *The Fungal Community: its Organization and Role in the Ecosystem.* (Eds. D.T. Wicklow & G.C. Carroll) Marcel Dekker, New York, pp. 319-349.

Lockwood, J.L. (1992) In: *The Fungal Community: its Organization and Role in the Ecosystem*, 2nd edn (Eds. G.C. Carroll & D.T. Wicklow.), Marcel Dekker, New York, pp. 243-263.

Lorito, M., Hayes, C.K., Di-pietro, A. & Harman, G.E. (1993) *Curr. Genet.* **24**: 349-356.

Lorito, M., Woo, S.L., D'Ambrosio, M.D., Harman, G.E., Hayes, C.K., Kubicek, C.P. & Scala, F. (1996) *Molec. Plant-Microbe Interact.* **9**: 206-213.

Loritto, M. (1998) In: Trichoderma and Gliocladium, (Eds. Kubicek, C.P. & Harman, G.E.), vol 2. Taylor & Francis, London, pp. 73-99.

Loritto, M., Woo, S.L., Garcia Fernandez, I., Colucci, G., Harman, G.E., Pintor-Toro, J.A., Filippone, E., Mucciflora, S., Lawrence, C., Ziona, A., Tuzun, S. & Scala, F. (1998) *Proc Natl Acad Sci USA*. **95**: 7860-7865.

Luckner, M. (1990) Secondary Metabolism in Microganisms, Plants and Animals, 3rd ed. Springer-Verlag, Berlin.

Macauley, B.J. & Griffin, D.M.(1969) *Trans Br Mycol Soc.* **53**: 223-228.

Mäntylä, A., Paloheimo, M. & Suominen, P. (1998) In: Enzymes, biological control and commercial applications, (Eds. G. Harman & Kubicek) *Trichoderma* and *Gliocladium*, Vol. 2. pp. 291.310.

Marois, J.J. & Coleman, P.M. (1995) *Can. J. Bot.*, **73** (Suppl.1): S76-S82.

Mc Intyre, M., Nielsen. J., Arnau, J., Vander Brink, H., Hansen, K. & Madrid, S. (2004) Proceedings of the 7th European conference on fungal genetics. Copenhagen, Denmark.

Meyer, R. (1991) *Appl. Environ. Microbiol.* **57**: 2269-2276.

Meyer, R.J. & Plaskowitz, J.S. (1989) *Mycologia*, **81**: 312-317.

Morawetz, R., Gruber, F., Messner, R. & Kubicek, C.P. (1992) *Curr. Genet.* **21**: 31-36.

Muthumeenakshi, S. (1996) Ph.D. thesis, The Queen's University of Belfast.

Nierstrasz, V. & Warmoeskerken, M. (2003) In: Textile processing with enzymes (eds. Cavaco-Paulo, A. & Gübitz, G.), Woodhead Publishing Ltd., England. pp. 120.157.

Osiewecz, H.D. (2002) Molecular biology of fungal development. Marcel Dekker, New York.

Papavizas, G.C. & J.A.Lewis.( 1981) *Phytopathol* **71**: 247-248 (Abstr).

Papavizas, G.C. & Lumsden, R.D. (1982) *Plant Dis.* **66**: 1019-1020.

Papavizas, G.C., Lewis J.A. & Abd-El-Moity, T.H. (1982) *Phytopathol* **72**: 126-132.

Penttilä, M. (1998) In: *Trichoderma* and *Gliocladium*, Enzymes, biological control and commercial applications, (Eds. G.Harman & C.Kubicek.) Vol. 2. pp. 365.381.

Penttilä, M., Limon, C. & Nevalainen, H. (2004) In: Handbook of fungal biotechnology, (Ed. Arora, D). Marcel Dekker, Inc. pp. 413.427.

Pere, J., Ellmen, J., Honkasalo, J., Taipalus, P. & Tienvieri, T. (2002) In: Biotechnology in the Pulp and Paper Industry: 8th ICBPPI Meeting, Progress in Biotechnology, (Eds. Viikari, L. & Lantto, R.), Elsevier Science B.V. Vol. 21. pp. 281.290.

Rahkamo, L., Siika-Aho, M., Vehviläinen, M., Dolk, M., Viikari, L., Nousiainen, P. & Buchert, J. (1996) *Cellulose* **3**: 153.163.

Rifai, M.A. (1969) A revision of the genus *Trichoderma*. *Mycol. Pap.*, **116**: 1-56.

Rifai, M.A. (1969) *Commonw. Mycol. Inst. Mycol. Pap.*116: 56, 1969.

Schirmbock, M., Lorito, M., Wang, Y.L., Hayes, C.K., Arisan-Atac, I., Scala, F., Harman, G.E. & Kubicek, C.P. (1994) Appl. *Environ. Microbiol.* **60**: 4364-4370

Schlick, A., Kuhls, K., Meyer, W., Lieckfeldt, E., Borner, T. & Messner, K. (1994) *Curr. Genet.* **26**: 74-78.

Schülein, M. (2000) Protein engineering of cellulases. *Biochim. Biophys.* Acta. **1543**: 239.252.

Smith, V.L., Wilcox, W.F. & Harman, G.E. (1990) *Phytopathology* **80**: 880-885.

Suurnäkki, A., Niku-Paavola, M-L., Buchert, J. & Viikari, L. (2004) In: Enzymes in industry, (Ed. Aehle, W.) Weinheim, Wiley-VCH. pp. 232.244, 437.439.

Szekeres, A., Kredics. L., Antal, Z., Kevei, F. & Manczinger, L. (2004) *FEMS Microbiol. Lett.* **233**: 215-222

Tronsmo, A. (1991) *Biol. Control.* **1**: 59-62.

Tronsmo, A. & Dennis, C. (1977) *Neth.J. Plant Pathol.* **83**(Suppl.1): 449-455.

Tronsmo, A. & Raa, J. (1977) *Phytopathol. Z.* **89**: 216-220.

Turner, D., Kovacs, W., Kuhls, K., Lieckfeldt, E., Peter, B., Arisan-Atac, I., Strauss, J., Samuels, G.J., Borner, T. & Kubicek, C.P. (1996) *Mycol. Res.*, **101**: 449-459.

Turner, W.B. & Aldridge, D.C. (1983) Fungal metabolites II. Academic Press, London.

Turoczi, G., Fekete, C., Kerenyi, Z., Nagy, R., Pomazi, A. & Hornok, L. (1996) *J. Basic Mircorbiol.* **36**: 63-72.

Uloha, A.J., Vainsteun, M.H. & Peberdy, J.F. (1992) *Curr. Genet.* **21**: 23-26.

Urlaub, R. (2002) In: Enzymes in food technology, Whitehurst, (Eds. R. & Law, B.). Sheffield, Academic Press, CRC Press. pp. 145.183.

Vehviläinen, M., Nousiainen, P., Struszczyk, H., Ciechanska, D., Wawro, D. & East, G. (1996). In: The Chemistry and Processing of Wood and Plant Fibrous Materials, (Eds. Kennedy, J.F., Phillips, G.O. & Williams, P.A.), Woodhead Publishing Limited, Cambridge. pp. 197.204.

Weindling R. (1934) *Phytopathol.* **22**: 837-845.

Wicklow, D.T. (1992) In: *The Fungal Community: its Organisation and Role in the Ecosystem*, 2nd edn. (Eds. G.C. Carroll & D.T. Wicklow), Marcel Dekker, New York, pp. 265-274.

Zamir, D., and Chet, I. (1985) *Can. J. Bot.* **31**: 578-580.

Zeppa, G., Allegrone, G., Barbeni, M. & Guarda, P.A. (1990). *Ann. Microbiol. Enzim.* **40**: 171-176.

Zhang, J., Howell, C.R. & Starr JL. (1996) *Biocontrol Sci Technol.* **6**: 175-187.

Zimand, G., Elad, Y. & Chet, I. (1996) *Phytopathology* **86**: 1255-1260.

Zimand, G., Valinsky, L., Elad, Y., Chet, I. & Manulis, S. (1994) *Mycol. Res.* **98**: 531-534.

17

# *Trichoderma*: A Potential Biocontrol Agent for Biopesticide Industries

**H. B. Singh and Shishir Srivastava**

*Department of Plant Pathology, National Botanical Research Institute, Lucknow-226 001*

## ABSTRACT

Plant protection against pathogens, pests and weeds has been progressively reoriented from a therapeutic approach to a rational use of pesticide chemicals in which consumer health and environmental preservation prevail over any other productive or economic considerations. Microbial pesticides are being introduced in this new scenario of crop protection, and currently several beneficial microorganisms are the active ingredients of a new generation of microbial pesticides or the basis for many natural products of microbial origin. In such a scenario, the emergence of *Trichoderma* spp. as an effective biological control agent and growth enhancer has come as a boon to agriculture in India and other parts of the world. Since most of the crops are infected by the soil-borne plant pathogens that primarily attack the vulnerable seeds or seedlings, the antagonists can be applied directly to the targeted area i.e., to seeds or seedlings, and a single application using existing delivery systems (seed treatment, biopriming, furrow treatment) can significantly reduce crop losses. In addition, our studies have also indicated that the crop yield is enhanced both qualitatively and quantitatively. The development of a *Trichoderma* based biopesticide requires several steps addressed to its isolation in pure culture and screening by means of efficacy bioassays performed *in vitro*, *ex vivo*, *in vivo*, or in pilot trials under real conditions of application. For the commercial delivery of biopesticides, the biocontrol agent must be produced at an industrial scale, preserved for storage and formulated by means of biocompatible additives to increase survival and to improve the application and stability of the final product. A number of successful products based on different species of *Trichoderma* have been commercialized in India and elsewhere. Whatever limitations these biocontrol products may have, it can be addressed by enhancing biocontrol through manipulation of the environment using mixtures of beneficial organisms, physiological and genetic enhancement of the biocontrol mechanisms, manipulation of formulations, and integration of biocontrol with other alternative methods that alone do not provide adequate protection but in combination with biocontrol provide additive or synergistic effects. This review presents an overview of similar approaches adopted by Indian scientists and other workers all over the world.

## INTRODUCTION

A major problem in agriculture is the difficulty in managing different diseases caused by pathogens specially root pathogens, and even when chemical pesticides are available, questions arise concerning their efficiency when applied to soil in nurseries or in the field. It is likely that in the future most systems of agricultural crop production will be less intensive with significantly reduced chemical inputs, not only of fertilizers but also of pesticides. Pest resistance selection and biocontrol procedures, therefore, become the two essential approaches to guarantee an acceptable level of crop protection. Resistance can be selected through breeding or using tissue culture techniques to produce clones with desired traits. The management of plant diseases is an imperative need in the 21$^{st}$ century to meet the increasing demand for a continuous and healthy (devoid of residues of chemical pesticides) food supply for an ever-increasing human population. The practices in vogue for disease management nowadays are based largely on genetic resistance in the host plant (a long and time taking process), management of the plant and its environment (not always successful) and synthetic pesticides. The use of chemical pesticides in agriculture though have helped to maintain a low-cost, high quality food supply, their positive contribution has been challenged by individuals and groups suggesting that the negative environmental effects of these pesticides far outweigh their societal benefits. The US Environmental Protection Agency's (EPA) Science Advisory Board recommended, "The value of natural ecosystems is not limited to their immediate utility to humans. They have an intrinsic, moral value that must be measured in its own terms and protected for its own sake" (Anonymous, 1990). In this context the use of biological control is increasingly capturing the attention of imagination of plant pathologists, microbiologists and gaining the stature as a possible and practical agricultural method of controlling the soil-borne phytopathogens.

Though biological control of plant pathogens, as it stands now, is an established sub-discipline in plant pathology, its beginning can be traced back to 1920s when scientists demonstrated that control of potato scab, caused by *Streptomyces scabies* was due to the biocontrol activity of *Bacillus* sp. Weindling (1934, 1937) performed his pioneering work on *Trichoderma* and *Gliocladium*, the two antagonistic fungi which were going to hold the top position in the totem pole of biological control. These works were followed by blurred efforts in the area of biological control. The last few years, however, have witnessed a burst of activities in research efforts in this direction, and several review articles (Papavizas and Lumsden, 1980, Nautiyal, 2000 and Whipps, 2001) and books (Cook and Baker, 1983 Papavizas, 1981) discuss the use of specific microorganisms for the biocontrol of plant diseases. In the last decade, more than 60 biocontrol products have appeared in commercial market. Still they form a small fraction of total number and sales of chemical fungicides in the field. In a 1993 report, the sale of biofungicides represented less than \$ 2.5 million in sales whereas the sales of chemical fungicides were in excess of \$ 29 billion. Optimistic estimates projected that the sales of biocontrol products could reach \$ 15 million by the year 2000 (Powell and Justum, 1993). Therefore, it is preposterous to believe that the commercialized biocontrol agents, whether in the form of biopesticides, would be able to replace the chemical pesticides overnight as the foremost method for controlling plant pests viz. insects, fungi and bacteria. According to Schroth and Hancock (1982), the biocontrol of phytopathogens is a fascinating, challenging but elusive and frustrating area of study for plant scientists and microbiologists alike.

The emergence of biological control as a viable and eco-friendly alternative to chemical pesticides has much to do with the well-recognized observation that a substantial amount of biological control does operate in nature. In most of the cases, the biocontrol agents are microorganisms commonly found in nature and are well-known antagonists. It has also been observed that the induction of suppressiveness is related to factors that enhance their activity in a particular agricultural habitat. Most of the reports using biological control agents for disease management have been restricted to the study of different antagonists in controlling one or more diseases of a particular crop in micro-plot. However, very few reports are there for large-scale and successful application of biocontrol agents in disease management. There are still fewer examples of commercialization of biological control formulations despite the origin of concept of use of biocontrol agents for increasing plant yield over 100 years ago.

The major aim of biological control research is to provide additional/alternative tools for management of plant diseases, which are a major factor in reducing the quality and quantity of our agricultural produce. The enormity of the problem can be gauzed from the fact that more than 800 million people do not have adequate food; 1.3 billion live on less than $1 a day and at least 10% of global food production is lost to plant disease (James, 2003). To meet these challenges we have to place tools of biocontrol or better still the product of biocontrol research in the growers hands. This necessitates the commercialization of biocontrol products which would ensure wide reach of biocontrol strategies. This article discusses the process of commercialization and commercial products currently available in India. Also discussed is implementation of biocontrol, particularly use and testing of commercial products and testing new biocontrol agents under commercial or semi-commercial conditions.

Even though the long-term beneficial effects of many cultural practices, including crop rotation and green manures, as well as other practices such as solarization and biofumigation, are due to biological control, these are not discussed. In several important cropping systems, the replacement or reduction of chemical application has been achieved through use of biologically based pesticides, a concept included in the definition of biocontrol proposed by Cook and Baker (1983). They defined it as "Biological control is the reduction of the amount of inoculum or disease-producing activity of a pathogen accomplished by or through one or more organisms other than man". With such a limited definition of biocontrol, the overall contribution of biocontrol of plant diseases to plant health management is currently relatively small (Cook, 2000), representing about 1% of agricultural chemical sales (Lidert, 2001; Fravel, 2005). Nevertheless, the contributions of biopesticides are important because they offer different modes of action from chemical pesticides and can, therefore, be applied in rotation with pesticides or in isolation to reduce the possible development of pathogen resistance. It can also be used in situations where no control is available, where conventional pesticides cannot be used due to reentry or residue concerns, or where the product must be certified organic. In addition, biocontrol can also be used in combination with reduced rates of pesticides (Singh *et al.*, 2006).

Over the past century, the research has repeatedly demonstrated that phylogenetically diverse microorganisms can act as natural antagonists of various plant pathogens. The interactions between microorganisms and plant hosts can be complex. However, fewer isolates of microorganisms can suppress plant diseases under diverse conditions of cultivation and fewer still possess broad-spectrum activity against multiple pathogenic organisms causing immeasurable losses to various economically important crops. Nonetheless, intensive screening tests carried out all over the world have revealed a number of microorganisms with effective disease control and growth promotion properties and can be processed for commercialization. The screening for their efficacy is a critical step in the development of biocontrol agents. The success of all subsequent stages depends on the ability of a screening procedure to identify an appropriate candidate.

Many useful bacterial biocontrol agents have been found by observing inhibition zones in petri plates under *in vitro* conditions. However, this method does not identify biocontrol agents with other modes of action such as parasitism, induced plant resistance, or some forms of competition. The screening methods for parasitism should also include burying and retrieving propagules of the pathogen to isolate parasites. For competition, screening methods include looking for microbes that quickly colonize sterilized soil and have the ability to exclude other organisms that attempt to invade the space, and looking for microbes that colonize the infection court. Primary screening for new biocontrol microbes is still performed and it seems likely that continued prospecting will be required to diversify the potential applications of biocontrol as well as replace more widely used biocontrol products should resistance develop.

## DEVELOPMENT AND COMMERCIALIZATION OF BIOCONTROL AGENTS

The studies on biological control had started with the research on insects, whereas the biocontrol of plant diseases is a relatively new phenomenon. The first bacterium, *Agrobacterium radiobacter* strain K84, was registered with the United States Environmental Protection Agency (EPA) for control of crown gall in 1979. A decade after that, the first fungus, *Trichoderma harzianum* ATCC 20476, was registered with the EPA for control of a plant disease. A total of 14 bacteria and 12 fungi have been registered with EPA for control of plant diseases (Fravel, 2005). Sixty-five per cent of the EPA-registered organisms have been registered within the past 10 years, with 36% registered over the past 5 years. Several technological problems had to be conquered and shifts in thinking were required for these products to reach the shelves. Same has been the story in Indian context. The research on biocontrol using *Trichoderma* spp. started with minor studies being conducted under laboratory conditions for control of one of the widest and dreadful pathogen, *Sclerotium rolfsii*. Further, several workers studied the management of different soil-borne plant pathogens. This trend continued till the late eighties in nineteenth century, when several workers attempted to produce *Trichoderma* spp. on mass scale. It was only in the late nineties and early part of this century that efforts for commercialization of *Trichoderma* based biopesticides have been initiated and some success achieved.

The mass production of microorganism based pesticides was not regulated in India. Recently in 1999, the amendments were made in Insecticides Act 1968, making it mandatory that some bioagents viz. *Bacillus subtilis, Gliocladium* spp., *Pseudomonas fluorescens, Trichoderma* spp., *Beauvaria bassiana, Metarrhizium anisopliae, Verticillum lecanii, Grannulosis* viruses and

Nuclear Polyhedrosis Viruses (NPV) must be registered based on some specified guidelines before their commercialization. This paper presents an overview about the need and importance of registration and commercialization of biocontrol agents especially in Indian context.

**Table 1** List of fungal phytopathogens affecting economically important crops controlled by *Trichoderma* spp.

| *Crop* | *Disease* | *Pathogen* |
|---|---|---|
| **Chickpea (*Cicer arietinum* L.)** | Collar rot | *Sclerotium rolfsii* Sacc |
| | Damping-off | *Pythium debaryanum* Hesse |
| | | *Pythium ultimum* Trow |
| | Dry root rot | *Macrophomina phaseolina* |
| | Fusarium root rot | *Fusarium equiseti* (Corda) Sacc. |
| | | *Fusarium solani* (Mart.) Sacc. f. sp. *Eumartii* |
| | Fusarium wilt | *Fusarium oxysporum* Schlechtend: Fr. f. sp. *ciceris* |
| | Sclerotinia stem rot | *Sclerotinia sclerotiorum* (Lib.) de Bary |
| | Seedling or seed rot | *Aspergillus flavus Trichothecium roseum* |
| | Verticillium wilt | *Verticillium albo-atrum Verticillium dahliae* |
| | Wet root rot | *Rhizoctonia solani* Kühn |
| | Botrytis gray mold | *Botrytis cinerea* Pers.:Fr. |
| **Tomato (*Lycopersicon esculentum* Mill.)** | Charcoal rot | *Macrophomina phaseolina* |
| | White mold | *Sclerotinia sclerotiorum* (Lib.) de Bary |
| | Southern blight | *Sclerotium rolfsii* Sacc. |
| | Rhizoctonia damping-off and fruit rot | *Rhizoctonia solani* Kühn |
| | Pythium damping-off and fruit rot | *Pythium aphanidermatum* |
| | | *P. debaryanum* |
| | | *P. ultimum* |
| | Gray mold | *Botrytis cinerea* |
| | Fusarium wilt | *Fusarium oxysporum* Schlechtend.: |
| | | Fr. f. sp. *lycopersici* (Sacc.) |
| | | Fusarium crown and root rot |
| | | *Fusarium oxysporum* Schlechtend.: |
| | | Fr. f. sp. *radicis-lycopersici* |
| **Wheat (*Triticum* spp. L.)** | Foot rot, seedling blight, dryland | *F. graminearum* Schwabe |
| | root rot | *F. pseudograminearum* |
| | Karnal bunt | *Neovossia indica* |
| | Pythium root rot | *Pythium aphanidermatum* |
| | | *P. graminicola* |
| | Rhizoctonia root rot | *Rhizoctonia solani* Kühn |
| | Southern blight | *Sclerotium rolfsii* Sacc. |
| | Snow rot | *Pythium* spp |
| **Tobacco (*Nicotiana tabacum* L.)** | Charcoal rot | *Macrophomina phaseolina* (Tassi) |
| | Collar rot | *Sclerotinia sclerotiorum* (Lib.) de Bary |

*Table 1 Contd.*

*Table 1 Contd.*

| | | |
|---|---|---|
| | Damping-off and Stem rot of transplants | *P. aphanidermatum* |
| | Fusarium wilt | *Fusarium oxysporum* |
| | Gray mold | *Botrytis cinerea* |
| | Sore shin and damping-off | *Rhizoctonia solani* Kühn |
| | Southern stem rot | *Sclerotium rolfsii* Sacc. |
| **Sunflower *Helianthus annuus* L** | Botrytis head rot | *Botrytis cinerea* Pers.:Fr. |
| | Charcoal rot | *Macrophomina phaseolina* |
| | Fusarium stalk rot | *Fusarium equiseti* (Corda) Sacc. |
| | Fusarium wilt | *Fusarium moniliforme* J. Sheld. |
| | | *F. oxysporum* Schlechtend.:Fr. |
| | Phytophthora stem rot | *Phytophthora* spp. |
| | Pythium seedling blight and root rot | *P. aphanidermatum* |
| | | *P. debaryanum* |
| | Rhizoctonia seedling blight | *Rhizoctonia solani* Kühn |
| | Sclerotinia basal stalk rot | *Sclerotinia sclerotiorum* (Lib.) de Bary |
| | Sclerotinia basal stalk rot and wilt | *Sclerotinia minor* Jagger |
| | Sclerotium basal stalk and root rot (southern blight) | *Sclerotium rolfsii* Sacc. |
| **Sugar cane** | Fusarium sett and stem rot | *Fusarium moniliforme* |
| | Phytophthora rot of cuttings | *P. megasperma* |
| | Red rot | *Colletotrichum falcatum* |
| | Red rot of leaf sheath and sprout rot | *Sclerotium rolfsii* Sacc. |
| | Rhizoctonia sheath and shoot rot | *Rhizoctonia solani* Kühn |
| | Wilt | *Fusarium sacchari* |
| **Soybean** | Charcoal rot | *Macrophomina phaseolina* (Tassi) |
| | Damping-off | *Pythium aphanidermatum* |
| | Fusarium root rot | *Fusarium* spp. |
| | Phytophthora root and stem rot | *Phytophthora sojae* |
| | Rhizoctonia root and stem rot | *Rhizoctonia solani* Kühn |
| | Sclerotinia stem rot | *Sclerotinia sclerotiorum* (Lib.) deBary |
| | Southern blight | *Sclerotium rolfsii* |
| | Sudden death syndrome | *Fusarium solani* f. sp. glycines |
| **Pigeonpea (*Cajanus cajan* L.)** | Botrytis gray mold | *Botrytis cinerea* Pers.:Fr |
| | Collar rot | *Sclerotium rolfsii* Sacc |
| | Damping-off | *Pythium aphanidermatum* |
| | Dry root rot | *Rhizoctonia bataticola* |
| | Fusarium seedling rot | *Fusarium equiseti* |
| | Fusarium wilt | *Fusarium udum* |
| | Sclerotinia stem rot | *Sclerotinia sclerotiorum* (Lib.) de Bary |
| | Phytophthora root rot and stem blight | *Phytophthora drechsleri* |
| | Wet root rot | *Rhizoctonia solani* Kühn |
| **Peanut (*Arachis hypogaea* L.)** | Botrytis blight | *Botrytis cinerea* Pers.:Fr. |
| | Charcoal rot | *Macrophomina phaseolina* |

*Table 1 Contd.*

*Table 1 Contd.*

| | | |
|---|---|---|
| | Damping-off | *Fusarium* spp., *Pythium* spp., *Rhizoctonia* spp. |
| | Fusarium wilt | *Fusarium oxysporum* |
| | Pythium wilt | *Pythium myriotylum* Drechs |
| | Rhizoctonia foliar blight, peg and root rot | *Rhizoctonia solani* Kühn |
| | Sclerotinia blight | *S. sclerotiorum* (Lib.) de Bary |
| | Southern blight | *Sclerotium rolfsii* Sacc |
| **Mustard (*Brassica juncea* (L.)** | Damping-off | *Fusarium* spp. |
| | Head rot | *Rhizoctonia solani* Kühn |
| | Sclerotinia stem rot | *Sclerotinia sclerotiorum* (Lib.) deBary |
| | Southern blight | *Sclerotium rolfsii* Sacc. |
| | Yellows | *Fusarium oxysporum* |
| **Gladiolus** | Botrytis Blight | *Botrytis gladiolorum* |
| | Fusarium Yellows | *Fusarium oxysporum* f. sp. *gladioli* |
| | Stromatinia Dry Rot | *Stromatinia gladioli* |
| **Maize (*Zea mays* L.)** | Charcoal rot | *Macrophomina phaseolina* |
| | Fusarium ear and stalk rot | *Fusarium subglutinans* |
| | Pythium root rot | *Pythium* spp. |
| | Rhizoctonia root rot and stalk rot | *Rhizoctonia solani* Kühn |
| | southern blight | *Sclerotium rolfsii* Sacc |
| **Cucurbits** | Belly rot | *Rhizoctonia solani* Kühn |
| | Charcoal rot (vine decline and fruit rot) | *Macrophomina phaseolina* |
| | Crown and foot rot | *Fusarium solani* f. sp. *cucurbitae* |
| | Damping-off | |
| | Fusarium wilt | *Fusarium oxysporum* |
| | Gray mold | *Botrytis cinerea* Pers.:Fr. |
| | Phytophthora root rot | *Phytophthora* spp. |
| | Sclerotinia stem rot | *Sclerotinia sclerotiorum* (Lib.) deBary |
| | Southern blight | *Sclerotium rolfsii* Sacc |
| | Sudden wilt | *Pythium aphanidermatum* |
| | Web blight | *Rhizoctonia solani* Kühn |

## THE PROBLEM

Plant disease control as practised today is far cry for the judicious use in early part of pesticide use, which was used only when necessary. Scientifically, the chances that biocontrol will be successful are more in some pathosystems than others. Systems (or particular times in the cropping cycle) with low biological diversity and at least some degree of environmental control are logical targets. The pathogens with a limited window of opportunity for infection, as well as those with monocyclic disease cycles or slow rates of disease progress, have been targeted by different workers. The time and duration of the vulnerability of the pathogens should also be kept in mind while applying the biocontrol agents. Lidert (2001) has given some points which explain the reasons of failure of biocontrol agents to become products indicating that scientists outside industry do often:

(a) overestimate the power of environmental concerns as economic drivers;

(b) lack sufficient knowledge of grower needs, registration strategy and competitive forces;

(c) have naive ideas about positioning and market strategy; and

(d) underestimate registration costs and difficulties.

Lidert (2001) also reported that insufficient cost-performance and poor shelf life of the microbial biopesticides and biofertilizers are other reasons why promising microorganisms do not become successful products. The representatives from industry also make similar remarks, indicating that scientists choose problems for markets which are financially too small to justify development costs or problems which are "so expansive that it is impossible to clearly define what the product must do to be commercially acceptable" (Mintz and Walter, 1993). A "product concept" must be developed early in the overall research and development process (Mint and Walter, 1993).

## NEED FOR COMMERCIALIZATION OF BIOPESTICIDES IN INDIA

Our research in the field of agriculture immediately after independence has been mainly basic. It was only after the green revolution that the research in agricultural disciplines started to move towards applied areas and the scientists started to categorize the effective researches on the basis of their reach to the farmers, how easily it could be applied in the field and the cost-benefit ratio, etc. In our country and elsewhere N, P, K has been in use since ancient times through the use of cow dung, leftovers of the previous crops, etc. and the disease management practices were in vogue with the use of simple techniques of rouging and burning the diseased plant, use of $CuSO_4$, etc. But it was only after the chemical fungicides began to be produced on commercial scale that the production of crops burgeoned. Same is the case with biological control agents. The scientists have screened a large number of antagonists, selected the effective ones and tried them in field for controlling the diseases in micro-plots. In this process crude formulations of the effective antagonists were developed.

The next logical step of the above sequence should have been commercialization of the biocontrol products. The commercialization of biocontrol products is also necessary to overcome the negative and harmful effects of the chemicals, which are caused by their prevalence in our current schemes of disease management. This is more so in case of developing countries where the chemicals banned in most of the developed countries are still in use, e.g., pentachloronitrobenzene (PCNB), a highly toxic fungicide is still used for disease management in our country whereas it is banned in USA. It is anticipated that the successful development and commercialization of the biocontrol fungi will effectively reduce the use of hazardous chemicals and ultimately it may eliminate them. This will help us to provide our future generations with safer and cleaner soil, water and air than it is now. In India it is very difficult to convince largely illiterate farmers to use biocontrol products instead of hazardous chemical pesticides. Rough estimates by the New Delhi based Biotech Association of India indicate that the biocontrol products constitute less than 1 per cent of crop protection market which is in tune with the global scenario.

It has been said of biocontrol researchers that they are very good at moving the ball up and down the field but are not able at good finish, i.e., crossing the goal line. Worldwide, approximately 60 biocontrol products have made it over the goal line to become commercial products. The products are restricted in their use in the sense that they can control only one or two pathogens and have been tried on relatively few crops before being released into the market. The efforts made in the direction of commercialization of biocontrol agents may be categorized in three groups based on both the increasing level of difficulty and decreasing potential of repeat sales. The first group proposed to apply the antagonist directly and precisely to the infection court when and where needed, much like the use of soothing balm in medicine. This is done to preempt or swamp the pathogen by the application of high population of antagonist organism (Cook *et al.*, 1996) e.g. seed-applied antagonists for protection of germinating seeds against damping-off and antagonists applied to fruits for protection against fruit decay in storage and bacteria growing at below 0° C for protection of plants against frost injury (Lindow, 1983). Nearly all microbes commercialized to date fall into this group with the exception of *Gliocladium virens* GL-21 (Lumsden and Walter, 1995) which is applied in soil in close proximity to seed or plant or with sufficiently high inoculum density to assure instantly high population in the infection court. This group of bioagents forms the first logical step in their commercialization for use against plant diseases and is important for the group's significant contribution in increasing the experience and familiarity with microbes within the industry, growers and society.

Plant Growth Promoting Rhizobacteria (PGPR) form the backbone of second group which includes antagonists being applied at one place, e.g., on seeds with the intent that they will spread with the growth of the plants and protect their roots and shoots (Kloepper *et al.*, 1980 and Nautiyal, 2000). This group of antagonists may include cases exemplifying the strategy of inundative and augmentative application where the antagonist is expected to multiply, persist for part or all of the life of the plant and increase an inadequate population of a related or similar antagonist already present in the rhizosphere (Cook *et al.*, 1996 and Nautiyal, 2000). For example, atoxigenic *Aspergillus flavus* was used to spread from colonized wheat grains scattered on the soil surface on to the floral parts of cotton plants where they forestall the establishment of toxigenic strains of *A. flavus* (Cotty and Bayman, 1993). This category of examples presents a more challenging approach to biocontrol as the antagonists involved are subjected to greater vulnerability to the effects of competition and environmental conditions. No antagonist of this group has been commercialized to date.

The third category belongs to the antagonists, which are released into the infection court once or several times (Cook *et al.*, 1996). Most of the successful examples of biological control of perennial weeds and insects with microbes belong to this category. The reason for the success of this group of antagonists is that they are pathogens dependent on the target pest for their ability to multiply and persist on the host for a longer period of time. Despite all the success stories of effective biological control of plant diseases and pests, the rate or commercialization of the biocontrol agents has been slow and therefore their application in the field is restricted. Of the 33 commercial biocontrol products listed by Fravel *et al.*, (1999), over half of them have applications in nursery or greenhouses and many were developed specifically for soil-borne plant pathogens e.g., *Pythium* and *Rhizoctonia*. The world's total greenhouse area is 307,000 hectare including both plastic and glass (Gullino *et al.*, 1999), whereas the total land area under cultivation in 1998 was 1.51 billion hectare (FAOSTAT, 2000). The use of biocontrol is more prevalent in greenhouses and protected structures than in open field. This is another area of concern, which must be looked into before moving towards commercialization.

There are various problems associated with the slow rate of commercialization of the biocontrol products. Weller (1988) has reviewed the more important reasons for failure or inconsistent performance of the biological control in the rhizosphere with introduced microbes. They are:

(i) Low disease pressure for an effective test.

(ii) The treatment favours the increased damage from a non-target diseases.

(iii) Colonization of the roots affected by the introduced strain or the loss of ecological competence by the strain is variable.

(iv) The expression of production of antibiotics, wherever it is necessary for the effective action of antagonist, is either too late or too low in quantity to be effective for controlling the disease.

However, these limitations can be sorted out by selection of effective strains of antagonist and improvement in their activity through modern biotechnological methods. Some other challenges related to commercialization of biocontrol products are a mixture of institutional problems, technical limitations and unrealistic expectations from these products. However, we will have to overcome these obstacles if we are serious about the development of commercial products for biological control.

The rhizosphere competence of the formulation developed as a result of years of research varies with changing soil conditions (rhizosphere competent strain of an effective biocontrol agent/antagonist, all the research must be directed towards the development and production of the strain). The field trials are necessary supplements for the development of a product, which need to be carried out under different soil conditions in different field in different agro-climatic zones on different crops and various pathogenic fungi in order to develop a broad-spectrum product, which can effectively control a large number of pathogenic microbes under different conditions.

Another constraint is the selection of suitable substrate for the development of effective formulation. A number of substrates have been tried for the development of biocontrol formulations (Table 1). It would be economically viable to use the agricultural wastes viz., tea, coffee waste, rice hay, distilled wastes from oil yielding plants, wheat bran, etc. We have recently developed an economical method for mass multiplication of biocontrol fungi for field application (Singh *et al.*, 2007).

A number of commercialized products have been developed and are available in the world market for disease management but they are either location specific or disease specific or both. Not many products are available which are broad-spectrum one and used on large scale. As of today, a large number of diseases are still not covered by the commercialized biocontrol formulations. Thus, there is need for moving ahead with selecting the effective strains of biocontrol fungi and developing them as broad-spectrum formulations for use against a large number of soil-borne diseases in varying soil conditions. A large chunk of biocontrol products commercialized all over the world for managing soil-borne fungal diseases of plants are *Trichoderma*-based, e.g., in our country, about seventy manufacturers are producing *Trichoderma*-based formulations, most predominantly based on *Trichoderma viride*.

Economics is the key as well as an important barrier to commercialization of microbial agents for use in agriculture. This is manifested by the fact that the cost of development and registration (the policy varying in different countries) has to be paid by a very limited market e.g., the market represented by a single disease on single crop. In USA, providing the toxicological data to EPA greatly reduces the prospect for commercialization. Another major obstacle is an absence of infrastructure for scale up and commercialization of biocontrol products in most of the countries (Cook, 1993). A system similar to that developed for release of new cultivars by breeders must be developed for commercialization and dissemination of biocontrol agents to the farmers.

**Includes both provisional, u/s 9 (3B) and regular registration, u/s 9 (3)**

**Source http://cibrc.nic.in**

The ideal conditions required for development of a biocontrol agents and its subsequent commercialization are:

## SELECTION OF SUITABLE STRAIN

The relative ease of finding the right antagonist strain under laboratory and field conditions is an added advantage but it does ensure that the agent would work consistently and effectively in field. The search for an effective strain involves years of screening, testing and/or involvement of luck. For example Kim *et al.* (1997) screened 2000 strains of spore forming aerobic bacteria, from which strain L324-92 was selected. The discovery of phloroglucinol producing rhizosphere competent bacteria came after several years of research on phenazine-producing strains of fluorescent pseudomonads. It now forms the basis for the expectation that the right strain for biological control of take all disease of wheat is under development. The development of T-22 strain of *Trichoderma harzianum* by Harman and his associates took more than a decade and it was only in last decades of last century that the sale of its product, Root Shield, started to pick up from scratch (Harman, 2000).

Genetic engineering of effective strains or those strains, which do not have the desired traits, must also be tried. The demonstration by Huang *et al.*, (1997) of the activity against *Rhizoctonia* root rot in addition to the ability of phloroglucinol producing strain to produce phenazine-1-carboxylate shows that it may be possible to produce strains with ability to control more than one disease and possibly more than one crop. This may broaden the market potential needed to justify costs of development and registration of commercialized biological control products. There are some of the biological control agents like Plant Growth Promoting Rhizobacteria (PGPR) and some strains of *Trichoderma* and *Gliocladium*, which besides controlling various soil-borne diseases also enhance the yield of host crop (Windham *et al.*, 1986, MacKenzie and Starman, 1995 and Nautiyal, 2000). It would be a boon to the industries associated with development of biocontrol products, if a strain that possesses both these properties and can be produced successfully at large-scale.

Extensive performance trials should be conducted before accepting one strain for advancement towards commercialization. The release of new wheat cultivars is done after conducting experiments for 100 site years (number of sites X number of years) so as to produce enough data. The industries interested in developing a strain to a commercialized product conduct their own trials extensively but only after being sure about the potential of the strains.

## SHELF LIFE AND STORAGE

For the establishment of a product containing microbial antagonists in the market it is essential that it has long shelf life and can be stored at room temperature for at least one crop season and preferably well into the next season. Mathre *et al.*, (1999) suggests that a shelf life of 18 months is preferable for a commercial microbial product with a requirement of storage at room temperature. The last condition is essential, as the end users cannot afford to buy costly equipment to keep the products at any temperature other than room temperature. Besides, it will render the biocontrol product economically unviable due to high cost of maintenance.

## APPLICATION TECHNOLOGY

Generally, most biocontrol agents will be applied with the same equipment currently used to apply other seed treatment products, foliar sprays or granular formulations. Development of a new application technique will not be acceptable easily unless it is more economically viable than the existing techniques and/or fit into current production scheme. For example, the so-called 'hopper box' treatment of seeds whereby the seed is treated as a part of the operations of filling the drill or planting the seeds may well replace the current practice of treatment of the seed warehouse, by treating only the seeds that will be actually planted. Thus, no seeds will be treated which are not going to be used. We have developed a technique for mass production of biocontrol fungi by using simple and economically viable substrates.

## SCALE UP AND QUALITY CONTROL

The research on biological control all over the world is initially carried out on small-scale e.g., *in vitro* screening by dual culture, production on substrates in Erlenmeyer flasks, autoclavable bags etc. For example, the conditions required for growth of a bacterium in 250-ml Erlenmeyer flasks is much different from that which involves the production of 10,000 liters of cells depending mainly on the fermenter used. Likewise the growth of fungal biocontrol agents in flasks produces a small quantity of formulation, which can be used in microplots for disease control. When the production of biocontrol agents is scaled up for commercialization, (several tonnes of products instead of several grams/kilograms in laboratory) the basic problem is that of maintaining the quality of the material produced, which is outside the capability of most researchers. This can be sorted out by team work where public sector researchers collaborate with the scientist of industry. When growing in large quantities, the microbial agents may develop contamination or mutations at any of the various steps involved in the production.

There is another aspect regarding the use of fermentation technology for mass production of biocontrol agents. Technologically, liquid state fermentation of submerged cultivation of biocontrol agents is more superior in the sense that it is easier to control the process parameters and to guarantee homogenous conditions for development of the agent. However, solid state fermentation has its own advantages. The advantages associated with solid state fermentation are:

(i) simplicity of the technique;
(ii) smaller investment costs;
(iii) larger quantity of product per unit substrate;
(iv) reduction of energy consumption in comparison to liquid fermentation;
(v) the conditions are similar to those found in nature;

(vi) production of air conidia and spores that are more resistant and durable; and

(vii) better yields of pure and homogenous spores.

Diego *et al.*, (2001) compared static and dynamic solid state fermentation of *Trichoderma reesei* with the application of electromagnetic fields. They found that more spores per unit substrate were produced in dynamic state fermentation than those in static state fermentation. The minds from industrial and scientific fields will have to standardize and choose the better method for commercialization.

Cook *et al.*, (1996) have mentioned four adverse non-target effects of microbes used as introduced agents for biological control of pests and diseases. These are:

1. competitive displacement of a beneficial microorganism
2. allergenicity
3. toxigenicity
4. pathogenicity

Of the four non-target effects, all but allergenicity are also desired target effects depending on the mechanism involved in biological control. Since these non-target effects have been identified for all categories of biocontrol agents including insect and weed pathogens, there is a need to examine the potential of these effects on various aspects of biological control especially for antagonists of plant pathogens. Competitive displacement is one non-target effect of biological control which must be done away with. This kind of non-target effect would show up in the field as a negative rather that positive response of the treated crop. Allergenicity is another issue which was identified as worker safety issue but not as a public health issue (Cook *et al.*, 1996) and can be managed by exercising caution in workplace e.g., wearing of dust masks by the workers.

Toxigenicity becomes a major problem in biological control where the control mechanism involves the use of a secondary metabolite for inhibition of the target organism. However, this should not raise concern to workers unless the toxicant produced by the antagonist is a component of the product handled and applied to the soil or plant. Probably, the greatest hazard with secondary metabolites is their potential for toxigenicity to plants. This would be an agronomic problem but not an environmental or human health issue as is the case with the use of chemical pesticides.

Pathogenicity to non-target organisms is a major problem when the biocontrol agent used is an insect or weed pathogen. For antagonists, the concern would be for microbes with potential as opportunistic pathogens to humans, e.g., *Aspergillus ochraceus* and clinical strains of *Burkholderia cepacia*. Analysis of the elimination of these organisms should be done during the screening process. Every effort should be made to guarantee the safety of environment and human health while developing a microorganism on a commercial basis for controlling the plant diseases.

## REGISTRATION REQUIREMENTS IN INDIA

Several biocontrol agents have come under Insecticides Act, therefore it is imperative that these may be registered before they are produced on commercial scale. It is important that better quality biofungicide product be registered according to the norms set by Central Insecticide Board (The Gazette of India, 1999, Anonymous, 2003). To produce biofungicide in laboratory is one thing

(Singh *et al.*, 1999, Singh *et al.*, 2004a, b) and its large-scale production after registration from CIB is another thing. It requires certain specified parameters. Following data are needed before a biofungicide product based on *Trichoderma* be registered under 9(3B) and 9(3):

**A. Chemistry**

1. Analytical test report
2. Shelf life
3. Container content compatibility study

**B. Bioefficacy**

1. Laboratory Bioassay
2. Field trials (two location data)
3. Data on non-target organism

**C. Toxicological studies**

(a) (For formulated products to be directly manufactured)

1. Single dose oral- Rat (21 days) (Toxicity/Infectivity/Pathogenicity)
2. Single dose oral- Mouse (21 days) (Toxicity/Infectivity/Pathogenicity)
3. Single dose Pulmonary- Rat (14 days) (Toxicity/Infectivity/Pathogenicity)
4. Single dose Dermal- Rabbit (21 days) (Toxicity/Infectivity/Pathogenicity)
5. Single dose Dermal- Intraperitoneal (21 days) (Toxicity/Infectivity/Pathogenicity)
6. Primary skin irritation
7. Eye irritation

(b) Human Safety Records

**D. Environmental Toxicological Studies**

(For formulation only)

Non Target Vertebrates

1. Toxicity to chicken
2. Toxicity to pigeon
3. Toxicity to freshwater fish

A number of bioagents have been registered in India under Insecticides Act 1968. A list is given in Table 2.

## FUTURE PROSPECTS

There is a growing demand for sound, biologically-based pest management practices. Recent surveys of both conventional and organic growers indicate an interest in using biocontrol products, suggesting that the market potential of biocontrol products will increase in coming years. Applications of diverse biological control strategies have been successful in the greenhouse industry (where environmental design and control is greatest) and continue to increase. Funding for basic and applied research continues to be provided by different funding agencies in the country. Such investments will ensure that innovations in biological control research continue.

**Table 2** Different substrates used for mass production of biocontrol fungi in laboratory conditions

| *S.No.* | *Substrate* | *S.No.* | *Substrate* |
|---|---|---|---|
| 1. | Powdered rye grass seed | 17. | Soy flour and molasses |
| 2. | Diatomaceous earth granules + molasses | 18. | Sugar cane press mud |
| 3. | Wheat bran formulation | 19. | Used tea leaves |
| 4. | Wheat bran saw dust formulation | 20. | Pine needles |
| 5. | Molasses - yeast medium | 21. | Banana pseudo stem |
| 6. | Wheat bran peat formulation | 22. | Pine apple peelings |
| 7. | Sorghum seeds | 23. | Compost |
| 8. | Sugarcane and Maize straw | 24. | Fresh dried cow dung |
| 9. | Tapioca rind or thippi substrate | 25. | Rye grains |
| 10. | Talc formulation | 26. | Bajra grains |
| 11. | Coffee fruit skin and cherry husk | 27. | Jhigora grains |
| 12. | Alginate-Wheat bran | 28. | Maize grains |
| 13. | Coconut coir pith | 29. | Wheat grains |
| 14. | Alginate pellets | 30. | Peanut shells |
| 15. | Vermiculite wheat bran and cellulose granule formulation | 31. | Maize spent cob |
| 16. | Distilled waste of aromatic plants | 32. | Water hyacinth |

An upswing in commercial interests has also developed in the past few years, and prospects for increased growth are positive. Clearly, the future success of the biological control industry will depend on innovative business management, product marketing, extension education, and research. Increased demand for organic produce and participation in home gardening activities by pesticide-wary urban populations all over the world has enlarged the market for biocontrol products.

The field of plant pathology will contribute substantially to making the 21$^{st}$ century the age of biotechnology by the development of innovative biocontrol strategies. The development of such strategies will benefit a developing country like India which is rich in natural resources being made untenable by excessive and injudicious use of chemicals in agriculture.

## SUCCESS STORIES

The number of commercial biocontrol formulations based on *Trichoderma* spp. in India is given in Table 1. The preliminary trials for diseases management using biocontrol formulations involve the mass production of biocontrol fungi on different substrates. Our group has developed a process for successful mass production of biocontrol fungi on agri-waste and filed a US patent for the process (Singh *et al.*, 2003). Our group has also been awarded a US patent on a novel strain of *Trichoderma harzianum* having growth promontory effect, potential to control a large number of soil-borne fungal pathogens and also nematode inhibitory effect (Kalra *et al.*, 2002). We have also successfully controlled collar rot disease of mints using *Trichoderma and Gliocladium* sp. (Singh and Singh, 2004). Different scientists have tried a number of substrates (approximately 32) for producing bioagents *en* masse (Table 2).

**Table 3** List of companies registered with Central Insecticides Board & Registration Committee, INDIA for commercial production of *Trichoderma* based biopesticide.

| *S. No* | *Organism* | *Company* |
|---|---|---|
| 1. | *Trichoderma viride* 1% WP | *M/s* Agriland Biotech Limited |
| 2. | *Trichoderma viride* 1.5% WP | *M/s* Romvijay Bio Tech Pvt. Ltd. |
| 3. | *Trichoderma viride* 1% WP | *M/s* Multiplex Agricare Pvt. Ltd. |
| 4. | *Trichoderma viride* 1% WP | *M/s* Margo Biocontrol Pvt. Ltd. |
| 5. | *Trichoderma viride* 1% WP | *M/s* K.N. Bio Science (India) Ltd. |
| 6. | *Trichoderma viride* 1% WP | *M/s* Pragathi Bio-fertilizers |
| 7. | *Trichoderma viride* 1% WP | *M/s* Sri Biotech, Hyderabad (formerly M/s Vermigreen Bio-fertilizer) |
| 8. | *Trichoderma viride* 1% WP | *M/s* Vishva Mithra Bio Agro Pvt. Ltd., Guntur |
| 9. | *Trichoderma harzianum* 1% WP | *M/s* Vishwa Mithra Bio Agro Pvt. Ltd., Guntur |
| 10. | *Trichoderma harzianum* 1% WP | *M/s* Greentech Agro Services Pvt. Ltd., Coimbatore |
| 11. | *Trichoderma harzianum* 1% WP | *M/s* Sri Biotech, Hyderabad (formerly M/s Vermigreen Bio-fertilizer) |
| 12. | *Trichoderma harzianum* 1% WP | *M/s* Bio-pest Management Pvt. Ltd., Bangalore |
| 13. | *Trichoderma viride* 1% WP | *M/s Crop Health Products Ltd.* |
| 14. | *Trichoderma viride* 1% WP | *M/s International Panacea Ltd., New Delhi* |
| 15. | *Trichoderma viride* 1% WP | *M/s Pushpanjali Agri Input Technologies, Kurnool (AP)* |
| 16. | *Trichoderma viride* 1.15% WP | *M/s. Prathista Industries Ltd.* |
| 17. | *Trichoderma viride* 1% WP | *Maharashtra Manufacturers Association for Biocontrol Agents* |
| 18. | *Trichoderma viride* 1%WP | *Vidharbha Marathwada Agro Biotech Association* |
| 19. | *Trichoderma viride* 1%WP | *M/s Nafed Biofertilizer, Bharatpur (Rajasthan)* |
| 20. | *Trichoderma viride* 1% WP | *M/s Mitcon Consultancy Ltd., Pune* |
| 21. | *Trichoderma viride* 1% WP | *M/s Monarch Bio Fertilizers & Research Centre, Chennai* |
| 22. | *Trichoderma viride* 1% WP | *M/s Maa Bhagwati Biotech & Chemical* |
| 23. | *Trichoderma viride* 1% WP | *M/s K.N.S. Biotech, Degloor, Distt. Nanded (Maharashtra)* |
| 24. | *Trichoderma viride* 1% WP | *M/s Harit Bio-Control Lab., Yavatmal* |
| 25. | *Trichoderma viride* 1% WP | *M/s Sigma Bio Laboratory, Nagpur* |
| 26. | *Trichoderma viride* 1% WP | *M/s Sneha Biotech, Vijayawada* |
| 27. | *Trichoderma viride* 1% WP | *M/s Vaibavlaxmi Bio-control Lab., Wardha (MS)* |
| 28. | *Trichoderma viride* 1% WP | *M/s Advance Bio-tech Industries and Research Input, Indore* |
| 29. | *Trichoderma viride* 1% WP | *M/s Jain Biotech, Nagpur* |
| 30. | *Trichoderma viride* 1% WP | *M/s Ocean Agro (India) Ltd., Nandesari, Baroda* |
| 31. | *Trichoderma viride* 1% WP | *M/s Samridhi Bioculture Pvt. Ltd., Indore* |
| 32. | *Trichoderma viride* 1% WP | *M/s Vidarbha Biotech Lab., Yauatmal* |
| 33. | *Trichoderma viride* 1% WP | *M/s Ecophila Biotech, Ahmednagar* |
| 34. | *Trichoderma viride* 1% WP | *M/s Om Agro Organic, Yavatmal* |
| 35. | *Trichoderma viride* 1% WP | *M/s Universal Agro Bio-Tech., Nohar (Rajasthan)* |
| 36. | *Trichoderma viride* 1% WP | *M/s Durva Biotech, Nagpur* |
| 37. | *Trichoderma viride* 1% WP | *M/s Sheer Agro Bio-Fertilizer Creative Organization, Purba Medinipur (WB)* |
| 38. | *Trichoderma viride* 1% WP | *M/s Central Bio Tech, Nagpur* |

*Table 3 Contd.*

Table 3 Contd.

| | | |
|---|---|---|
| 39. | *Trichoderma viride* 1% WP | *Indore Biotech Inputs & Research (P) Ltd, Indore* |
| 40. | *Trichoderma viride* 1% WP | *M/s Yash Krishi Takniki Ewam Vigyan Kendra, Allahabad* |
| 41. | *Trichoderma viride* 1% WP | *M/s K.N. Bio Tech Pvt Ltd., Hyderabad* |
| 42. | *Trichoderma viride* 1% WP | *M/s Modern Biotech, Jalgaon (MS)* |
| 43. | *Trichoderma viride* 1% WP | *M/s Amit Bio Tech. Kolkata* |
| 44. | *Trichoderma viride* 1% | *M/s Jai Kissan Agro, Indore* |
| 45. | *Trichoderma viride* 1% WP | *M/s M.S. Industries, Amravati (MS)* |
| 46. | *Trichoderma viride* 1% WP | *M/s Ruchi Biochemicals, Gondia, Maharashtra* |
| 47. | *Trichoderma viride* 1% WP | *M/s International Panaacea Ltd., New Delhi* |
| 48. | *Trichoderma viride* 1% WP | *M/s Sudarshan Chemical Industries Ltd., Pune* |
| 49. | *Trichoderma viride* 1% WP | *M/s S&S Biotech, Nagpur* |
| 50. | *Trichoderma viride* 1% WP | *M/s J.R. Biocontrol Laboratories, Yavatmal* |
| 51. | *Trichoderma viride* 1% WP | *M/s Agri Life, Secunderabad* |
| 52. | *Trichoderma viride* 1% WP | *M/s Nomin Agri Bio Pvt.. Ltd., Pune* |
| 53. | *Trichoderma viride* 1% WP | *M/s R.B. Herbal Agro Satana, Distt. Nasik (Maharashtra)* |
| 54. | *Trichoderma viride* 1% WP | *M/s Parvara Agro Bio Tech, Sangamer (Maharashtra)* |
| 55. | *Trichoderma viride* 1% WP | *M/s Jai Bio Tech Ind.* |
| 56. | *Trichoderma viride* 1% WP | *M/s Esvin Advance Technologies Ltd., Chennai* |
| 57. | *Trichoderma harzianum* 2% WP | *M/s Gujarat Green Revolution Company Ltd.* |
| 58. | *Trichoderma viride* 1% WP | *M/s. Micoplex Biotech & Agrochem Pvt. Ltd., Wardha* |
| 59. | *Trichoderma viride* 1% WP | *M/s. Institute of Natural Organic Agriculture, Pune* |
| 60. | *Trichoderma viride* 1% WP | *M/s Sai Agrotech, Yavatmal* |
| 61. | *Trichoderma viride* 1% WP | *M/s Nirmal Organics Biotech P. Ltd., Mumbai* |
| 62. | *Trichoderma viride* 1% WP | *M/s Kalpavruksha Biosystems, Bangalore* |
| 63. | *Trichoderma viride* 1% WP | *M/s Govinda Agro Tech Ltd., Nagpur (Maharashtra)* |
| 64. | *Trichoderma viride* 1% WP | *M/s Maharashtra Research & Development Centre, Solapur (Maharashtra)* |
| 65. | *Trichoderma viride* 1% WP | *M/s Vasundhra Agrotech, Aurangabad (Maharashtra)* |
| 66. | *Trichoderma viride 1% WP* | *M/s Chhattisgarh Agro Biotech Lab., Raipur* |
| 67. | *Trichoderma viride* 1% WP | *M/s Sai National Rural Development & Research Institute, Allahabad* |
| 68. | *Trichoderma viride* 1% WP | *M/s Kan Biosys Pvt. Ltd., Pune* |
| 69. | *Trichoderma viride* 1% WP | *M/s Anjali Biotech, Amravati* |
| 70. | *Trichoderma viride* 1% WP | *M/s Deepa Farm Inputs (P) Ltd., Trivandrum* |

There has been a number of plant diseases controlled by using biocontrol formulations developed on these wastes e.g., wilt of chickpea, collar rot of *Mentha* spp., corm rot of gladiolus, etc. in addition to promotion of growth in some ornamental crops, legumes an cucurbitaceous crops (Singh *et al.*, 2004).

## FUTURE RESEARCH DIRECTIONS

A number of books have been written and innumerable research papers have been published on biocontrol of crops ranging from rice, wheat, cotton, tobacco, peanut, sugar cane, potato, soybean, tomato, apple, turfgrass, and postharvest citrus (Gnanamanickam, 2002). It is a fact that biological control still represents a very small portion of disease management, though the use of some microbes e.g., *Agrobacterium* for management of crown gall is comparably significant. The aim of the biocontrol research is to place biocontrol products in the growers' hands, therefore, there is a need for more communication between public funded researches and industry in the early stages of development. Over the past 30 years, the approach to biocontrol research has evolved towards being more ecologically holistic and more oriented toward both production systems and industry's concerns (Table 3). This evolution, most probably, will continue because the science is still young, and research is still needed in several areas. Particularly, the researches in production, formulation, and delivery can greatly assist in commercialization of biocontrol agents based on microorganisms. In addition, research is also needed in integrating the biocontrol agents into existing production systems e.g., in rotating or biocontrol with reduced quantity of chemical pesticides and in calculating these into forecast models to choose whether to apply a chemical pesticide or biocontrol. More research needs to be continued with concerted efforts in biocontrol for contribution to the movement towards sustainable agriculture and simply to ensure that alternatives are available if other management tools fail or are lost.

## References

Anonymous. (1990) In: *report of Science Advisory Board: Relative Risk Reduction Strategy Communicated to William K. Rilly,* Administrator, US EPA, 1990. pp. 25.

Cook, R.J. (2000) *Annu. Rev. Phytopathol.* **38:** 95-116.

Cook, R.J., Bruckart, W.L., Coulson, J.R., Goettel, M.S., Humber, R.A., Lumsden, R.D., Maddox, J.V., McManus, M.L., Moore, L., Meyer, S.F., Quimby, P.C., Stack, J.P. & Vaughn, J.L. (1996) *Biological Control* **7:** 333-351.

Cook, R. J. (1993) *Annu. Rev. Phytopathol.* **31:** 53-80.

Cook, R. J. (1996) *Can. J. Plant Pathol.* **18:** 439-445.

Cook, R.J. & Baker, K.F. (1983) In: *the Nature and Practice of Biological Control of Plant Pathogens. American J. Phytopathological Society*. St. Paul, M.N., 539 pp.

Cotty, P.J., & Bayman, P. (1993) *Phytopathology* **83:** 1283-1287.

FAOSTAT. (2000) Food and Agriculture Organization of United Nations. http: //apps.fao.org/

Diego, S.M.M., Morla, A.G., Sofia, M.C. & Pulles, D.C. (2001) In: *International Conference on New Horizons in Biotechnology*. Trivandrum, India. April 18-21, 2001. pp. 74.

Fravel, D.R., Rhodes, D.J. & Larkin, R.P. (1999) In: *Integrated Pest Management in Greenhouse Crops*. (Eds. Albajes, R., Lodovica Gullino, M., Van Lenteren, J. C., & Elad, Y.) Kluwer Academic Publishers, Dorcrecht, 545 pp. pp. 365-376.

Fravel, D.R. (2005) *Annu. Rev. Phytopathol.* **43:** 337-359.

Gnanamanickam, S.S.A. (2002) *Biological Control of Crop Diseases*. New York: Marcel Dekker. pp.468

Gullinos, M.L., Albajes, R. & van Lenteren, J.C. (1999) In: *Integrated Pest Management in Greenhouse Crops*. (Eds. Albajes, R., Lodovica Gullino, M., Van Lenteren, J. C. & Elad, Y.). Kluwer Academic Publishers, Dorcrecht, 545 pp. pp. 1-15.

Harman, G.E. (2000) *Plant Disease* **84(4):** 377-393.

Huang, Z., Thomashow, L.S., Mavrodi, D.V., Raijmakers, Weller, D.M. & Cook, R.J. (1997) In: transgenic strains of *Pseudomonas* for biocontrol of plant root diseases. (US Patent No. 6,277,625)

James, C. (2003) Global Status of Commercialized Transgenic Crops: Manila: Int. Serv. Acquis. Agri-Biotech Appl. (ISAAA).

Kalra, A., Singh, H. B., Pandey, R., Patra, N. K., Katiyar, N., Gupta, M. L., Dhawan, O. M. & Kumar, S. (2002) Strain of *Trichoderma harzianum* useful as nematode inhibitor, fungicide and plant growth promoter and a process for the isolation thereof. (US Patent No. 6,475,772).

Kim, D.S., Cook, R.J. &Weller, D.M. (1997) *Phytopathology* **87:** 551-558.

Kloepper, J.W., Schroth, M.N. & Miller, T.D. (1980) *Phytopathology* **70:** 1078-1082.

Lidert, Z. A. (2001) In: *Enhancing Biocontrol Agents and Handling Risks*, (Eds. Vurro, M., Gressel, J., Butt, T., Harman, G.E., Pilegram, A., *et al*.,) pp. 283. Amsterdam: IOS Press. 295 pp.

Lindow, S.E. (1983) *Annu. Rev. Phytopathol.* **21:** 363-384.

Lumsden, R.D. & Walter, J.F. (1995) In: *Benefits and Risks* (Eds. Hokkanen, M.T. & Lynch, J.M.) Cambridge University, Cambridge, UK.

MacKenzie, A.J., Starman, T.W. & Windham, M.T. (1995) *HortScience* **30(3):** 496-498.

Mathre, D.E., Cook, R.J. & Callan, N.W. (1999) *Plant Disease* **83(11):** 972-983.

Mintz, A.S. & Waltere, J.R. (1993) In: *Pest Management: Biologically Based Technologies* (Eds. Lumsden, R.D. & Vaughn, J.L.) American Chemical Society, Washington, DC, USA, pp. 398-404.

Nautiyal, C. S. (2000) In: Biocontrol Potential and its Exploration in Sustainable Agriculture (Eds. Upadhyay, R. K., Mukherji, K.G. & Chamola, B. P.) Kluwer Academic/Plenum Publishers, Dordrecht, New York, USA, pp. 9-23.

Papavizas, G.C. & Lumsden, R.D. (1980) *Annu. Rev. Phytopathol.* **18:** 389-413.

Papavizas, G.C. (1981) In: Biological control in crop production. Totowa, NJ: Allanhel, Osmun.pp. 461.

Powell K.A. & Justum A.R. (1993) *Pest. Sci.* **37:** 315-321.

Schroth, M.N. & Hancock, J.G. (1982) *Science* **216:** 1376-1381.

Singh. A. & Singh, H.B. (2004) *Current Science* **87(3):** 362-366

Singh, A., Srivastava, S. & Singh, H. B. (2007) *Bioresource Technology*, USA **98(2):** 470-473. (Available on line on www. sciencedirect.com).

Singh, H.B., Kalra, A., Patra, N.K., Kumar, S., Pandey, R., Khanuja, S.P.S. & Shasany, A. K. (2003) In: *Process for the preparation of novel growth media from distillation and other plant wastes for mass multiplication of bio-control fungi.* (US Patent No: 6,511,821).

Singh, H.B., Singh, A., Singh, S.P. & Nautiyal, C.S. (2004a) In: *Eco-agriculture with Bioaugmentation: An emerging concept* (Eds. Singh, S.P. & Singh, H.B.) Rohitashwa Printers, Lucknow, India.1-20.

Singh, H.B., Singh, Sarita, S., Singh, A. & Nautiyal, C.S. (2004b) In: *Eco-agriculture with Bioaugmentation: An emerging concept* (Eds. Singh, S.P. & Singh, H.B.) Rohitashwa Printers, Lucknow, India. 53-69.

Singh, H.B., Tripathi A., Srivastava, S., Vikas, Singh P.P. & A. Singh. (2006) *Journal of Eco-friendly Agriculture* **1(1):** 49-53.

The Gazette of India Part II-Section 3-Sub-section (I) published March 36, 1999.

Weinding, R. (1934) *Phytopathology* **24:** 1153-1179.

Weindling, R. (1937) *Phytopathology* **27:** 1175-1177.

Weller, D.M. (1988) *Ann. Rev. Phytopathol.* **26:** 379-407.

Whipps, J.M. (2001) *J. Exp. Bot.* **52:** 487-511.

Windham, M.T, Elad, Y. & Baker, R. (1986) *Phytopathology* **76:** 518-521.

18

# Potential of *Trichoderma* for Biocontrol and its Mass Production

**D. P. Singh**

*Crop Protection Programme, Directorate of Wheat Research, Karnal-132 001 (Haryana)*

**ABSTRACT**

*Trichoderma* spp. have been identified as potent biocontrol agents for a range of plant pathogens. More than 50 formulations of the fungus are available globally and increasing trend in its use has been witnessed during recent years. However, the great potential of *Trichoderma* shown against potent plant pathogens in *in vitro* has not been found true cent per cent under field conditions and affected by biotic and abiotic factors. The use of bioagent against foliar pathogens had been limited due to more adverse conditions for its establishment and multiplication than in case of rhizosphere. The most limiting factor in the uses of *Trichoderma* is its mass multiplication using cheaper substrates, proper formulation for enhancing its shelf life under tropical and subtropical conditions, and sensitiveness to higher pH and higher temperatures. The poor quality and substandard preparations of *Trichoderma* formulations marketed are also the cause of concern and adversely affecting its use by farmers.

## INTRODUCTION

During past two decades, research on *Trichoderma* spp. as potent biocontrol of plant diseases has intensified and about 50 industrial formulations are available in market with different trade names in different countries. The species of *Trichoderma* have emerged as most powerful bioprotectants for management of a wide variety of plant diseases. The fungus is being used as broad spectrum biopesticide against fungi, bacteria, viruses and nematodes and occupies the top position amongst biocontrol agents. The fungus when applied in the rhizosphere or as seed treatment induces hypersensitive response, systemic acquired resistance and induced systemic resistance. The low cost technology is likely to be the boon for seed industries also against a large number of seed and soil-borne diseases. However, major limiting factor in technology is the short shelf (4-6 months) life of formulations under tropical and subtropical conditions. The sensitiveness of *Trichoderma* to higher pH of >7 and higher temperatures are very important limiting factors restricting its application. Poor quality and substandard preparations of *Trichoderma* formulations marketed are also the cause of concern (Mukhopadhyay, 2005). The present review deals with potential of this fungus to manage plant diseases and prospects of mass production of inocula of its potent strains.

## BIOLOGY

The species of *Trichoderma* have been aggregated based on microscopic characters (Rifai, 1969). The studies on growth, morphogenesis and sporulation of fungus in *in vitro* have been done and information derived can be very useful in developing technologies for large-scale production of biomass for biological control. *Trichoderma* uses a wide range of compounds as source of carbon and nitrogen. Some of the carbon sources are mono, di and complex polysaccharides, purines, pyrimidines and amino acids, tannins and catechins, aldehydes, organic acids especially long chain fatty acids, methanol, methylamine and formate. Ammonia is the most readily used source of nitrogen by *Trichoderma* in buffered media. The vegetative growth is also supported by amino acids, nitrate and urea. The salts like magnesium increase the growth of *T. viride.* The effect of $CO_2$ has been found dependent with its concentration and medium pH and variations amongst species were observed. The rapid growth of fungus witnessed at high $CO_2$ concentrations in alkaline media explained slightly basic habitats of fungus (Papavizas, 1985).

Most of the *Trichoderma* spp. are photosensitive, sporulating readily on many natural and artificial substrates and concentric pattern of alternating rings corresponding to diurnal alternation of light and darkness, with conidia being produced during the light period can be seen. Exposure of culture for 20-30 seconds to light of 85-90 lux induces some sporulation. Best photo induction to sporulation was in daylight for three minutes or in NUV radiation for 10-30 seconds. It also induces the synthesis of DNA, RNA and protein. The sporulation is not hampered when cultures were exposed to light ranging from 254 to 1,100 nm. The compounds like actinomycin D, azaguanine, cyclohexamide, phenyl alcohol and ethidium bromide inhibit photo induced sporulation. The role of chlamydospores in biocontrol has not been fully investigated. The species namely, *T. hamatum*, *T. harzianum* and *T. viride* have been found producing chlamydospores in solid and liquid media, soil and its extracts as well as in natural plant debris. The washed conidia of this fungus do not germinate in distilled water and require an external source of nutrients. The acidic conditions favour more germination of conidia. The dried chlamydospores showed only 13-31% germination on nutrient agar (Papavizas, 1985). The information on possible role of such propagules in the survival of *Trichoderma* in soil under adverse conditions is not properly investigated (Mukhopadhyay, 1987). Kurian *et al.*, (2003) observed that under polythene cover, the temperature rises up to 51°C as compared to untreated soil having a temperature of 41°C. It reduced the population of pathogens and increased the population of *Trichoderma* due to less competition.

## PRODUCTION OF METABOLITES

Brian and McGrowan (1945) described the highly fungistatic antibiotic, viridin, produced by *T. viride*. Trichodermin, a sesquiterpenoid metabolite, described by Godtfredsen and Vangedal (1965) is also produced by *Trichoderma* spp. Besides, toxic metabolites, the fungus also produces various enzymes such as exo and endoglucanases, cellobiase and chitinase, which also play role in biocontrol. Thornton (2005) reported that *Trichoderma* spp. are efficient producers of extracellular hydrolytic and oxidative enzymes that enable them to colonize organic matter thereby preventing the saprotrophic spread of plant pathogens.

## ECOLOGY

Understanding the population dynamics of *Trichoderma* in soils and other habitats and their survival in soil and rhizosphere is useful for proper utilization of it as biocontrol agent. The fungus occurs in a variety of soils worldwide but prefers soils containing good amount of organic matter. *Trichoderma* is a secondary colonizer and frequent in well decomposed organic matter. It is also found on root surfaces of various plants, decaying bark, and on sclerotia and other propagules of other fungi. The population of fungus decreases under prolonged dry soil conditions. Certain strains of *T. hamatum* and *T. pseudokoningii* are adapted to excessive soil moisture. Likewise, *T. viride* and *T. polysoprum* are restricted to areas where low temperatures prevail, whereas *T. harzianum* is most commonly found in warm climatic regions. The iron content of the soil is an important determinant of microsite preference of *Trichoderma.* Isolation of *Trichoderma* using the selective medium helps to quantify the population in soil. Use of a new *Trichoderma* medium E (TME) with benomyl and alkylaryl polyether makes the selective isolation of mutant resistant to fungicide (Papavizas, 1985).

## POPULATION DYNAMICS

Survival in Soil: *Trichoderma* possesses the ability to survive in many ecological niches depending on prevailing conditions and the species or strain involved. The survival of conidia, when added to the soil without nutrient supplying amendments, was between 110-130 days and varies between strains. Most of the conidia probably lysed without geminating and at later stages in absence of adequate food bases. The original population added to soil could remain to its one-tenth level after a gap of two years. Hyphae also survive in soil. Caldwell (1958) observed that chlamydospores survive in soil better than conidia. Lewis and Papavizas (1984) demonstrated the potential of different species to form chlamydospores readily and in great numbers in natural soil. The sensitivity of conidia of *Trichoderma* to fungistatic is more in neutral or alkaline than acid soils. The organic matters added to soil partially or completely reverse fungistasis in case of acid soils. The conidia are more sensitive to fungistasis than chlamydospores and hyphae. The ammonia is fungistatic to conidia in alkaline soils. A possible approach of biocontrol will be to reduce fungistasis using energy sources and other organic matters in soil initially. *Trichoderma* populations do not change in soil in response to green compost but increase greatly in response to mature compost. The organic substrates may be colonized and exploited more rapidly by pathogenic or saprophytic fungi possessing high competitive saprophytic ability than by the antagonists. Kelly (1976) reported that excess nutrients added through molasses in granules containing *T. harzianum* can be used quite fast by *Phytophthora cinnamoni* and growth of antagonist is suppressed by lack of oxygen and resulted in net increase in the level of pathogen than of the antagonist. Lewis and Papavizas (1983) observed that *Trichoderma* and other biocontrol fungi proliferate abundantly in various neutral soils when added along with a food base like inoculated moist bran.

The proliferation and subsequent establishment in soil depend on inoculum age and how it is added in relation to the food base. The thorough colonization of the food base enables *Trichoderma* to grow relatively unimpeded through the soil. The alginate pallets with biomass of fungus thoroughly established in a food base also stimulate the population of *Trichoderma.* The pallets with conidia are less effective than those prepared with ground biomass containing mostly chlamydospores. The ground fermented biomass of antagonist added as a dry Pyrex (anhydrous aluminium silicate) formulation proliferated and increased more than conidia added to soil without nutrients. It may be due to less sensitivity of hyphae to fungistasis than spores. Lorito (2005) reported that *Trichoderma* can utilize a wide variety of nutrients sources, and effectively degrade some of them including hydrocarbon, chlorophenolic compounds, polysaccharides and xenobiotic pesticides. Many strains of *Trichoderma* are strong opportunistic invaders, fast growing, sporulation and powerful antibiotic producers. These properties make fungus ecologically quite successful and strains are found in agriculture, native prairie, forest, salt marsh and desert soil, in lake water, dead plant material, living roots of virtually any plant species, seeds and air.

The microbiological balance of the soil is subject to alterations mediated by various environmental factors including the application of pesticides or other chemical substances. The heat and chemical treatment of soil may cause complete or partial biological vacuum in the soil. It may favour certain indigenous or introduced strains of *Trichoderma.* It is known to possess the higher tolerance against broad spectrum biocides and ability to colonize the treated soil with carbon disulfide and formaldehyde more rapidly than other soil competitors. The application of sodium azide to soil resulted in the development of a large population of native *Trichoderma.* The rapid increase in treated soil is due to its ability to rapidly colonize the substrates in absence of competition from other microorganisms. The biocides also put stress, weaken and thus the pathogen propagules become more susceptible to attack by the antagonists. The sclerotia of *Sclerotium rolfsii* became more susceptible for invasion and degradation by *T. harzianum* due to sub-lethal dose of metham sodium as compared to fresh viable sclerotia (Papavizas, 1985).

Some fungicides like thiram and metalaxyl are not toxic to *T. harzianum* in *in vitro*, whereas benzimidazole fungicides (e.g., benomyl, thiobendazole, thophanate methyl) were toxic to fungus even at 0.5 mg/litre concentration. Indirect evidence of the non-target effect of benomyl in soil was provided by Backman and Rodriguez-Kabana (1975), and when it was applied on groundnut, it resulted in severe southern stem blight (*S. rolfsii*). Soil solarization and streaming may allow the proliferation of *Trichoderma.* Israel (2005) reported that the survival of *T. harzianum* was not affected when applied with Phorate up to 2000 ppm in dry soil and both can be combined. Shukla *et al.*, (2007) studied the tolerance levels of *Trichoderma* spp. to agrochemicals and reported that wettable sulphur, mancozeb and pendimethalin were the most compatible agrochemicals followed by zinc sulphate and carbofuran, whereas carbendazim even at 1 ppm was fungistatic to bioagent.

**Survival in the Rhizosphere:** The establishment of *Trichoderma* in the rhizophere is must for control of root diseases. However, it is affected due to lack of proper nutrients, presence of toxic substances in the root exudates and presence of antagonistic or competing microorganisms.

## MASS PRODUCTION AND PACKAGING

The most critical bottleneck in the use of *Trichoderma* in biocontrol has been the lack of knowledge of its mass culturing and packaging. Solid media for the experimental and greenhouse application have been used frequently. The other substrates include bark pellets, wheat bran plus peat, barley grain and pellet formulations containing *T. harzianum* and *T. polysporum*. There is a need to identify strains of *Trichoderma*, which can tolerate the effect of liquid fertilizers and pesticides and may be applied along with these in soil. Deep tank fermentation is most widely used in mass production of biocontrol microorganisms. Growth media may include readily available agricultural products such as glucose, starch, hydrolyzed corn and soya products, whey and molasses. The large amounts of biomass containing mycelia, chlamydospores and some conidia were produced in the liquid media using molasses and brewers yeast within 5-6 days. It can be air dried, grinded and diluted by using pyrax as carrier. The encapsulation of propagules using sodium alginate may be useful in future to improve delivery technology. Dutta *et al*., (2005) found paddy grain and husk as best locally available substrates for mass multiplication of *T. harzianum* in west Bengal. Khan *et al*., (2005) prepared mass culture of bioagent fungus *T. harzianum* using sawdust-soil-5% molasses moisture (15:1:1), whereas immobilization (carrier) was prepared in flyash-soil 5% molasses mixture (5:1:1). The composition of biopesticides was 1 part stock culture and 15 parts carrier. Using it, biopesticide 'Biowilt-X' was prepared which was found superior than other available formulations in market. The shelf life of bioagent was up to 32 weeks at ambient temperature. *Trichoderma* spp. are widely used in agriculture and industry and isolates of *T. reessei* are used as biopesticides and biofertilizers. It is due to the fact that propagules of fungus can be produced cheaply and in large quantity both in liquid and dry formulations and stored for months.

Methods have been developed to genetically modify this fungus in a very precise manner, which allows the improvement of their ability to secret desired enzymes, kill plant pathogens or stimulate plant growth and resistance to diseases (Lorito, 2005). Singh (2005) advocated that the research on *Trichoderma* can be fruitful only when the registration and commercialization of superior strains takes place in the country. The mass production of bioagent based formulation is important part in the development of technology. The substrates like wheat bran and bajra grains were found quite promising in terms of maximum colony forming units and shelf life. The cow dung inoculation technique has been found good at farmers' level. Two US patents, US PTO No. 6511821 and 6475772 have been obtained for this technique at NBRI, Lucknow, India.

## PRODUCTION OF POTENT STRAINS USING BIOTECHNOLOGICAL TOOLS

The purposeful induction of tolerance to pesticides and selection of stable biotypes for use in combination with pesticides may be suitable for better results in field. It may be done by selecting the tolerant strains after exposing to the pesticides. However, for better stability of tolerant biotypes, genome modification may be a morc promising approach to biocontrol. Mutagenesis is other way to develop new biotypes. The gene cloning will be of promise and is now getting lot of attention. Sible and Mukherjee (2005) reported that the stress resistant genes in *Trichoderma* spp. are Hsp70/DnaK, Hsp98, DnaJ, Mn-SOD, TrxB, Peroxiredoxin and Mannitol-1-phosphate dehydrogenase. The expression of the heat shock protein, Hsp70 was studied with heat stress (40°C for 1 h), cold shock (6°C for 1 h), $H_2O_2$ (5 mM) and copper sulphate (0.1 mm) and results revealed that the gene has a very high-level of constitutive expression.

Christian and Pandey (2005) estimated the protein contents of tolerant isolates of *T. harzianum* to pesticides (Metalaxyl MZ, Benomyl, carbendazim, thiram, Carbofuran, Metribuzin and propiconazole). A higher protein content of bioagent fungus was considered as an indicator of tolerance to pesticides of different isolates. A maximum of seven protein bands of 20-97 kDa obtained with isolates showed tolerance to 5000 ug/ml$^{-1}$. Mahato *et al.*, (2005) tested different species of *Trichoderma* for chitinase. Amongst ten isolates of *Trichoderma* (*T. harzianum* 1, 2, 3, *T. koningii*, *T. virens* 1, IABT-5, 6, 7, *T. viride* and *T. pseudokoningii*), *T. koningii* showed highest colloidal chitin hydrolysis. Endochitinase gene specific primers were designed for *T. koningii* and *T. virens* in BTI software GENETOOL with the sequence information available in database. A unique amplicon of 1.7 Kb obtained from *T. virens* 1 was cloned into pTZ57R/T vector and transformed into *Escherichia coli*. The transformation of clones was made through blue/white colony assay and PCR amplification and restriction analysis of plasmid. The gene was directionally cloned into plant transformation vector pHS 100 and transformed into *E. coli*. The microbes, including mycoparasitic fungi, are the rich sources of genes that can be utilized to genetically engineer crops for resistance against plant pathogens.

The gene *ech42* from *T. virens* encodes an endochitinase enzyme with high antifungal activity against a wide range of plant pathogenic fungi. Shah *et al.*, (2005) carried out such work and cDNA sequence of *ech42* gene cloned from an Indian isolate of *T. virens* was placed under the 35S promoter in the plant expression vector pCAMBIA 1301 having *hph* gene as a selectable marker and *uidA* as the reporter gene. Tobacco plants were transformed with this construct pCAMBIA-ech42 by using *Agrobacterium tumefaciens* LBA4404. Transformants were selected on the basis of resistance to hygromycin B and tested for GUS expression. Integration of the endochitinase in to tobacco genome was confirmed by PCR amplification using specific primers. Mukherjee (2005) reported that *T. virens* IMI 304061 produces a range of secondary metabolites including the antifungal compounds gliovirin, virdin and viridiol. The efforts to understand the molecular basis of its biocontrol properties has been initiated and a cDNA and cosmid library have been constructed and optimized the transformation protocols, thus developing this strain as a model organism for studying the genetics of biocontrol. The signal transduction gene, MAP kinase TmkA has been found to be involved in the mycoparasitism of the sclerotia of *S. rolfsii*.

## BIOCONTROL POTENTIAL OF *TRICHODERMA*

There are large reports of effectiveness of *Trichoderma* in controlling soil-borne diseases of wide range of crops. The indigenously present population of fungus may be affected with use of pesticides and it may increase in some cases, thus resulting in higher disease control. Purposeful adding of large amounts of *Trichoderma* with its food base to soil has been attempted by many workers using a range of media. The media used, however, should not help the pathogen to proliferate. The incorporation of conidia sometimes showed encouraging results to control soil-borne diseases. The soil pasteurization increases the colonization of *Trichoderma*. The plots treated with *T. viride* had low incidence of white rust and resulted in higher yield (Singh *et al.*, 2003b).

Seed treatment requires less amount of inocula and is now a exciting field to introduce the biocontrol agents in soil. The seed treatment has been tried for control of loose smut of wheat and was found effective, especially under low-level of embryo infection. The biocontrol agent must be able to multiply in the soil or rhizoshere to inhibit a given pathogen by competition, mycoparasitism or antibiosis. The correlations with *in vitro* antibiosis by *Trichoderma* against fungal pathogens, which actually happens in nature, have not been high. This is due to the fact that the level of production of metabolites by fungus reduced in case of soil. There is no evidence that *T. harzianum* produces any antibiotics. Chet and Baker (1981) showed that *Trichoderma* produces cellulose, Beta (1-3) glucanase, and chitinase and degrades the glucans in the walls of *Pythium* spp and the chitin and glucans in the walls of *Rhizoctonia solani*. Todkar *et al.*, (2005) studied the antagonistic effect of *T. viride* and *T. harzianum* on storage pathogens of grapes (*Aspergillus niger*, *A. fumigatus*, *Bipolaris taramira* and *Botrydiplodia theobromae*). Both were found quite effective and suppressed the growth of pathogens ranging from 52-75% under *in vitro* tests. Meena Devi and Paul (2005) reported that soil factors such as 35% moisture, 25 °C temperature and 6.6 pH were most suitable for the growth of JMA 4 isolate of *T. harzianum*. Amongst the fungicides tested, carbendazim was toxic to bioagent, whereas Captan @ 100 mg/litre could not affect bioagent much but suppressed the pathogen. Amongst different delivery systems tested, wheat bran base formulation and sodium alginate pallets were best against pathogens, *Rhizoctonia solani* and *Fusarium* spp. Sharma *et al.*, (2005) got control of Karnal bunt with propiconazole in case of two sprays of spore suspension of *T. viride* given at 40-47 growth stage (Zadoks' scale). Singh *et al.*, (2000) found good control of loose smut of wheat using *T. viride* seed treatment. Singh *et al.*, (2003a) also used *T. harzianum* and recorded effective against spot blotch caused by *Bipolaris sorokiniana* in wheat.

The registration, projected role and availability of *Trichoderma* inoculants is still a bottleneck and more research is needed to develop proper formulation with wide specificity and ability to control diseases of a wide range of crop plants. The genetic manipulation using biotechnological tools will definitely evolve better strains with wider adaptability and tolerance to different factors. There is a need to understand and predict the survival and proliferation of bioagent in soil and rhizosphere. Development of proper fermenting technology and delivery system will need long term commitment of researchers and industries. The information on the cost benefit ratio using bioagents and chemicals may be quite convincing to the farmers to integrate bioagents in crop protection technologies.

## References

Backman, P.A. & Rodriguez- Kabana, R. (1975) *Phytopathology,* **65**: 819-821.

Brian, P.W. & McGrowan, J.C.(1945) *Nature,* **156**: 144-145.

Caldwell, R. (1958) *Nature,* **181**: 1144-1145.

Chet, I. & Baker, R. (1981) *Phytopathology,* **71**: 286-290.

Christian, M.M. & Pandey, R.N. (2005) In: *Abstract of papers, Global conference II* (Eds. Khetrapal, R.K., Gupta, K., Chalam, C.V., Gour, H.N., Thakore, B.B.L. & Bhargava, S.), 25-29 Nov. 2005, Udaipur, India, pp. 93.

Dutta. S., Roy, M., Nath, R., Roy, A. & Roy, S. (2005) In: *Abstract of papers, Global conference II* (Eds. Khetrapal, R.K., Gupta, K., Chalam, C.V., Gour, H.N., Thakore, B.B.L. & Bhargava, S.), 25-29 Nov. 2005, Udaipur, India, pp. 199.

Godtfredsen, O.W. & Vangedal, S. (1965) *Acta Chem.Scand.* **19**: 1088-1102.

Israel, S., Bohra, A., Singh, M.P. & Lodha, S. (2005) In: *Abstract of papers, Global conference II* (Eds. Khetrapal, R.K., Gupta, K., Chalam, C.V., Gour, H.N., Thakore, B.B.L. & Bhargava, S.), 25-29 Nov. 2005, Udaipur, India, pp. 196.

Kelly, W.D. (1976) *Phytopathology,* **66**: 1023-1027.

Khan, M.R., Khan, S.M., Mohiddin, F.A., Khan, N. & Khan, M.M. (2005) In: *Abstracts of papers, National symposium and annual meet of IPS,* 16-18, Jan. 2007, RDV, Jabalpur, India, pp. 85.

Kurian, P.S., Backyiyarani, S. & Rajkumar, J. (2003) In: *Proceedings of 6th PGPR workshop,* (Eds. Reddy, M.N., Anandaraj, M., Eapen, S.J., Sarma, Y.R. & Kloepper, J.W.) 5-10 Oct. 2003, Calicut, India, pp. 47-52.

Lewis, J.A. & Papavizas, G.C. (1983) *Soil Biol.Biochem.* **15**: 351-357.

Lewis, J.A. & Papavizas, G.C. (1984) *Can. J. Microbiol.* **30**: 1-7.

Lorito, M. (2005) In: *Abstract of papers, Global conference II* (Eds. Khetrapal, R.K., Gupta, K., Chalam, C.V., Gour, H.N., Thakore, B.B.L. & Bhargava, S.), 25-29 Nov. 2005 Udaipur, India, pp. 162.

Mahato, U., Bhat, S., Krishnaraj, P.U. & Kuruvinashetti, M.S. (2005) In: *Abstract of papers, Global conference II* (Eds. Khetrapal, R.K., Gupta, K., Chalam, C.V., Gour, H.N., Thakore, B.B.L. & Bhargava, S.), 25-29 Nov. 2005 Udaipur, India, pp. 94.

Meena Devi & Paul, Y.S. (2005) In: *Abstract of papers, Global conference II* (Eds. Khetrapal, R.K., Gupta, K., Chalam, C.V., Gour, H.N., Thakore, B.B.L. & Bhargava, S.), 25-29 Nov. 2005 Udaipur, India, pp. 203-204.

Mukherjee, P.K. (2005) In: *Abstract of papers, Global conference II* (Eds. Khetrapal, R.K., Gupta, K., Chalam, C.V., Gour, H.N., Thakore, B.B.L. & Bhargava, S.), 25-29 Nov. 2005, Udaipur, India, pp. 163.

Mukhopadhyay, A.N. (1987). *Indian J. Mycol. Pl. Pathol.* **17**: 1-7.

Mukhopadhyay, A.N. (2005) In: *Abstract of papers, Global conference II* (Eds. Khetrapal, R.K., Gupta, K., Chalam, C.V., Gour, H.N., Thakore, B.B.L. & Bhargava, S.) 25-29 Nov. 2005, Udaipur, India, pp. 165.

Papavizas, G.C. (1985). *Ann. Rev. Phytopathol.* **23**: 23-54.

Rifai, M.A. (1969). *Commonw. Mycol.Inst. Mycol. Pap.* **116**, 56 pp.

Shah, M.R., Mukherjee, P.K. & Eapen, S. (2005) In: *Abstract of papers, Global conference II* (Eds. Khetrapal, R.K., Gupta, K., Chalam, C.V., Gour, H.N., Thakore, B.B.L. & Bhargava, S.), 25-29 Nov. 2005, Udaipur, India, pp. 96-97.

Sharma, A. K., Dhanbir Singh, Babu, K.S. & Kumar, K. (2005) In: *Abstract of papers, Global conference II* (Eds. Khetrapal, R.K., Gupta, K., Chalam, C.V., Gour, H.N., Thakore, B.B.L. & Bhargava, S.), 25-29 Nov. 2005, Udaipur, India, pp. 200-201.

Shukla, N., Chaudhary, R.G. & Singh, R.K. (2007) In: *Abstracts of papers, National symposium and annual meet of IPS,* 16-18 Jan. 2007, RDV, Jabalpur, India, pp. 37-38.

Sible, G.V. & Mukherjee, P.K. (2005) In: *Abstract of papers, Global conference II* (Eds. Khetrapal, R.K., Gupta, K., Chalam, C.V., Gour, H.N., Thakore, B.B.L. & Bhargava, S.), 25-29 Nov. 2005, Udaipur, India, pp. 6.

Singh, D.P., Nagarajan, S., Goel, L.B., Dharam Singh, Kumar, J., Singh, D.V., Amerika Singh, Srivastava, K.D., Rashmi Aggarwal, Beniwal, M.S., Tewari, A.N., Singh, K.P. & Grewal, A.S. (2000) *J. Biol. Control,* **14**: 35-38.

Singh, D.P., Sharma, A.K., Singh, K.P., Tewari, A.N., Satvinder Kaur, Singh, R.N., Singh, A.K., Singh, S.P., Kalappanawar, I.K., Hegde, Y., Chaudhury, A.K., Das, S.Y. & Pankaj Kumar. (2003a) In: *Abstracts of papers,* 25th annual conference and symposium of ISMPP, 8-10 Oct. 2003, Durgapura, Jaipur, India, pp. 40-41.

Singh, H.B. (2005) In: *Abstract of papers, Global conference II* (Eds. Khetrapal, R.K., Gupta, K., Chalam, C.V., Gour, H.N., Thakore, B.B.L. & Bhargava, S.), 25-29 Nov. 2005, Udaipur, India, pp. 162.

Singh, S., Kumar, S. & Singh, A. (2003b) In: *Proceedings of 6th PGPR workshop,* (Eds. Reddy, M.N., Anandaraj, M., Eapen, S.J., Sarma, Y.R. & Kloepper, J.W.), 5-10 Oct. 2003, Calicut, India, pp. 319-320.

Thorntron, C.R. (2005) In: *Abstract of papers, Global conference II* (Eds. Khetrapal, R.K., Gupta, K., Chalam, C.V., Gour, H.N., Thakore, B.B.L. & Bhargava, S.), 25-29 Nov. 2005, Udaipur, India, pp. 164.

Todkar, L., Sapal, R.T. & Sawant, D.M. (2005) In: *Abstract of papers, Global conference II* (Eds. Khetrapal, R.K., Gupta, K., Chalam, C.V., Gour, H.N., Thakore, B.B.L. & Bhargava, S.), 25-29 Nov. 2005, Udaipur, India, pp. 138.

19

# Rhizomediation in Soil for Sustainable Agriculture

**Meenu Saraf [1], Neelam Tank [1] and Rajendran Narayanan [2]**

[1] *Department of Microbiology, Gujarat University, Ahmedabad, Gujarat, India*

[2] *Molecular Biology, Department of Biology, Kentucky State University,* Frankfurt, Kentucky, USA.

## ABSTRACT

Rhizobial microbiology, plant nutrition, and soil science have diverged from their traditional science into new direction within the last decade. Likewise, the basic concepts on plant-microbe interaction, mineral-nutrient exchange, and soil acidity-alkalinity requirements have been customized to fit into the track of emerging "rhizomediation". The recent empirical observations on induced chemical signalling between microbes and plants as well as the acquisition of mineral nutrients in the rhizosphere have paved the way to understand the insight into this field. In addition, modern methods adapted from Molecular Biology and Biotechnology have already established the significance of rhizomediation as promising field of interest. In this review, accumulated knowledge on the management of rhizosphere, basic concepts and insights, application of modern techniques and major challenges in this field have been discussed to improve the crop yield for sustainable agriculture.

## INTRODUCTION

A decade ago, the success of the "green revolution" was totally dependent on the augmented contribution of fertilizer nitrogen. Approximately 90 million tonnes of fertilizers were applied annually to agriculture land (Sprent and Sprent, 1990). Since then the natural process of Biological Nitrogen Fixation (BNF) has played a significant role as an environmentally safe and sustainable farming practice. The increased use of BNF significantly reduces the need for fertilizer nitrogen and evades compounding recalcitrant environmental problems at large-scale proportions. Success of this innate process is dependent on the type of microorganisms used and in conjunction with the type of leguminous plants.

BNF is estimated to contribute $180 \times 10^6$ metric tonnes per year globally of which 80 per cent comes from free-living or associative biosystems (Tilak *et al.*, 2005). The expanded interest in microbial ecology has drawn attention to the fact that BNF is environmentally benign to reduce the use of fossil fuels and helps to restore distorted lands for high yield productivity. The axe has fallen simultaneously on the use of nitrogenous fertilizers due to the unacceptable levels of water pollution, toxic nitrates in drinking water, eutrophication of lakes and rivers they cause (Zahran, 1999). The ability of $N_2$ fixation by legumes is beyond doubt and represents a renewable source of $N_2$ for agriculture. However, this enormous quantity will have to be augmented as the human population increases from the current state of 6.5 billion and as the supply of nitrogenous fertilizer diminishes. This objective can be achieved through the development of superior crop varieties, improvement of agronomic practice and higher efficiency of nitrogen fixing processes through the better management of the symbiotic relationship between plants and bacteria.

In this review, we focus on the management of rhizosphere, role of rhizobia as a plant growth promoter and the mediatory qualities apart from nitrogen fixation. The plant-microbe interactions, major mediatory challenges involved at the volume of soil near the root and the application of modern techniques have also been discussed.

## MICROBIAL MANAGEMENT OF NITROGEN SOURCES

BNF is the reduction process of atmospheric N to ammonia ion ($N_2$ + 6e- + 8H+ gives rise to $2NH_4$+). The ammonium ion (the conjugate acid of ammonia, $NH_3$), the result of nitrogen-fixation process, is the form of nitrogen that is used by living systems. In another process, nitrates and nitrites released by decaying organic matter are converted to ammonium ions by nitrifying bacteria present in the soil. In a third process, denitrifying bacteria act on ammonia as well as nitrate compounds and convert to free nitrogen ($N_2$). This, eventually, is returned to the atmosphere by the denitrification process.

Agricultural soil management depends on the management of microbes involved and the level of nitrogen sources available. Nitrogen (N) is an essential soil nutrient needed in furthermost quantity by all living organisms. Most biological forms including plants and animals are unable to procure N gas, which comprises 80% of our atmosphere. Besides lightning, N can only be fixed by small and highly specialized groups of microorganisms that possess the enzyme, nitrogenase. The ability to reduce and siphon out such appreciable amounts of nitrogen from the reservoir of atmosphere and enrich the soil is confined to bacteria and archae. These include *(i)* symbiotic nitrogen-fixing forms such as *Rhizobium* (the obligate symbionts in leguminous plants) and *Frankia* (associate with the non-leguminous trees), and *(ii)* non-symbiotic nitrogen fixing forms like *Cyanobacteria, Azospirillum, Azotobacter, Acetobacter diazotrophicus, Azoracus*, etc. Good farming practice targets the free and biologically fixed nitrogen for many reasons. It is economically viable and environmentally prudent. This can be achieved by understanding the rhizosphere and by exploiting the BNF through genetically improved symbiotic ability of rhizobial population such as *Rhizobium*.

## INTERACTION OF RHIZOSPHERE AND LEGUME-RHIZOBIAL POPULATIONS

The rhizosphere is a unique term used to define the interaction between plant and its relative microbes. The term "rhizosphere" was first introduced in 1904 by Lorenz Hiltner and describes the relationship between bacteria and its legume, and now it can be redefined as the interactive ground between plant roots and microbes, at the degree of soil, around roots influenced by the soil nutrients and microbial activities. Rhizobacteria, which reside in the vicinity of root region, on the other hand find shelter at the root and sustenance within the host plant and compensate their host with a constant and renewable supply of nitrogen. Vessey *et al.*, (2005) have reviewed the root-based $N_2$ fixation symbiosis. They have compared and contrasted four symbiotic systems involved in $N_2$ fixation, namely legumes, actinorhizal plants, parasporia and cycads. Saraf (2001) has discussed nitrogen fixation in different ecosystems like agriculture fields, forests, aquatic ecosystems, etc. and the range of organisms involved. Regardless of genetic and functional diversity of the respective host as well as microsymbionts, the result of all the root symbiosis is well-designed.

In recent years, taxonomic status of such rhizobacteria has been considerably revised. Rhizobia are the members of the Rhizobiales, belonging to alpha-proteobacteria, which are capable of establishing symbiotic relationship with the members of the Leguminoceae family of flowering plants. Based on polyphasic taxonomy approach, Sahgal and Johri (2003) enlisted 36 species of bacteria distributed among seven genera (*Allorhizobium, Azorhizobium, Bradyrhizobium, Mesorhizobioum, Methylobacter, Rhizobium* and *Sinorhizobium*). Among them, *Rhizobium* nodulates more than one species of legumes and non-legumes plants such as *Parasporia* and *Tremma.* A few aquatic legumes such as *Aerchynmous, Neptunia* bear the stem nodules in addition to common root nodules. For example, *Neptunia natous* is nodulated by *Allorhizobium undicola* (Lajudie *et al.*, 1998), whereas *Azorhizobium caulinodons* is specific for stem nodulation in *Sesbania restrata* and *Phaseolus vulgaris* (Waelkans *et al.*, 1995). Recently, Kumar *et al.*, (2006) studied the genetic diversity of rhizobia isolated from the medicinal plant, *Mucuna pruriens*.

Biological $N_2$ fixation contributes directly through symbiotic relationships to productivity of forage and grain legumes. These symbiotic relationships involve bidirectional signal exchange between bacteria and plants culminating in the formation of root nodules. The amount of nitrogen fixed by two of the most important agricultural crops, cereals and legumes, is quite amazing. At least 139 million to 175 million tonnes of nitrogen fixed by BNF illustrates the magnitude (Sprent and Sprent, 1990). Legume inoculants, as commercial biofertilizers and a wealth of understanding of biological nitrogen fixation process, have been reviewed earlier (Saraf, 2001). Legumes are economically import crops and grown under different agroclimatic conditions. Out of 13,000 described species of legumes, approximately 3,000 species were examined for root nodulation. More than 90 percent of them form root nodules and about 20 per cent of food protein worldwide is derived from them (Sprent and Sprent, 1990). Significant legumes include dry bean (*Phaseolus vulgaris*), dry pea (*Pisum sativum*), chickpea (*Cicer arietinum*), broad bean (*Vicia faba*), pigeon pea (*Cajanus cajan*), cowpea (*Vigna unguiculata*), and lentil (*Lens culinaris*) (Agostini and Khan, 1986), peanut (*Arachis hypogaea*) and soybean (*Glycine max*). Most often legumes coexist with common nitrogen fixing *Rhizobium*. In an attempt to establish a database of such common *Rhizobium*, they were surveyed to determine the nativity and frequency of the effective types of rhizobial population as well as the fate of the introduced antigenic types (Dadarwal, 1978). The ecoserological assessment of chickpea in 13 major soils samples from India revealed three serological groups. Serogroup I was widely distributed. Serogroup II was limited to grey and brown soil types, and serogroup III which recognizes among strains of American origin did not occur in Indian soil (Tilak *et al.*, 2005). Many new species like *Sinorhizobium abri* from *Abrus precatorius* have been reported from time to time (Ogarawara *et al.,* 2003) and the genetic diversity of microsymbionts was documented. Currently, vast diversity of other rhizobacteria related to rest of the legumes that are wild or found in rare habitats, is under screening process.

## THE ROLE OF *RHIZOBIUM* IN RHIZOMEDIATION

The major metabolic functions of the root nodule are to mediate nitrogen from atmosphere with the help of an external biological entity, and convert it into ammonia to provide fixed nitrogen to plant. The role of *Rhizobium*, an efficient biological entity in the nodules, as nitrogen-fixer is well-established and thoroughly studied (Atkins, 1986). *Rhizobium*-legume symbiosis is important for global economy, since it is a major contributor to BNF. The global $N_2$ fixation through biological means is around $17.2 \times 10^7$ tonnes annually and 80% of it comes from legume-*Rhizobium* symbiosis alone (Dadarwal, 1997). Indeed, legume crops in US alone are responsible for the fixation of approximately 6 million tonnes of atmospheric nitrogen annually (Howarth *et al.*, 2002). The economic value of this process exceeds US$ 2.3 billion. *Rhizobium* based seed as soil inoculants are now sold worldwide reducing the need for expensive fertilizer by facilitating optimal nodulation of legumes and maximizing the levels of nitrogen fixation (Deaker *et al.*, 2004). However, several environmental conditions are limiting factors to the growth and activities of $N_2$ fixing plants. A principle of limiting factor states that "the level of crop production can be no higher than that allowed by the maximum factors" (Brockwell *et al.*, 1995). The positive correlation between nodulation and yield relationship is conditional and not very simple. If it was so simple, farmers would have readily accepted and adopted the use of bioinoculants leading to high-level of BNF (Saraf, 2001). Reasons contributing to this include inadequate demonstration of technology, presence of adequate native *rhizobia*, high soil mineral nitrogen levels, which suppress nitrogen fixation, and inadequate quality control of rhizobia inoculum (Bantilan and Johansen, 1995).

To fix the nitrogen, the rhizobial cells must undergo sequential and complex processes of metabolic and physiological differentiation into a mature state known as the bacteriod, which is enveloped by the plant cell. The metabolism of bacteriods, which persists in the nodule's low to microenvironment that is compatible with the key $O_2$-sensitive enzymes in the nitrogen-fixation process (Layzell *et al.* 1990), is overwhelmingly focused on the production of fixed nitrogen which is then transferred to the host plant. This process is fuelled by the host plant through the provision of large quantities of $C_4$-dicarboxylic acids such as malate (Salminen and Streeter, 1992). Most of the carbon from the plant is channelled into energy production to fuel nitrogen reduction. Similar to the rhizobial cells, plants also carry out a complex process (Ramalingam *et al.*, 2001). They symbiotically develop comprehensive iron uptake systems. In legume plants, it requires an integral strategy that takes into account the presence of the symbiotic system (Tang *et al.,* 1992). Legumes associated with *Rhizobium* form symbiotic tissues with the capacity to reduce $N_2$ to $NH_3$. $N_2$ + 6 e-

+ 8H+ → 2 $NH_4$+ (ammonium ion). For example, pea plant and its symbiotic *Rhizobium* amalgamate a number of different iron (FE) uptake systems. So-called "leghemoglobins" are found only in such nitrogen fixing legume root nodules. This hemoprotein regulates the oxygen tension in nodule tissues, which contains differentiated bacteria (Nadler *et al.,* 1990). A deficiency of Fe can limit the symbiotic fixation of $N_2$, modify the survival and growth of the rhizobia in the soil and restrict the initiation and development of the nodules. In the absence of Fe, growth restriction of nodule-meristems and the absence of nodule-function were found (Tang *et al.*, 1992). In *Rhizobium leguminosarum* a shortage of Fe causes reduced levels of cytochromes b and c (Nadler *et al.*, 1990).

## NITROGEN FIXATION IN RELATION TO RICE

The discovery of symbiotic nitrogen-fixation provided basic knowledge on plant-microbial interactions (Chung, King-Thom *et al.,* 1996), closely linked with the development of soil microbiology and phytopathology. This integrated the aspects of uptake and acquisition of nutrients by plant roots and nutrient-dynamics in soils. In *Azotobacter vinelandii*, the biological nitrogen fixation *depends on* three different pathways such as molybdenum (Mo)-dependent (*Nif involved),* vanadium (V)-dependent (Vnf involved) and neither Mo nor V dependent (Anf involved) (Shen, B. *et al.*, 1993). As reported earlier, electrons are required to reduce nitrogen. These are carried to nitrogenase by either flavodoxins or ferredoxins. Though in many cases the generation of low-redox potential electron carriers for nitrogenase reduction remains unclear, the immediate donation of electrons to nitrogenase reductase had been identified for many diazotrophs. For example, in Gram-positive bacteria and cyanobacteria [2Fe-2S], ferredoxins have been shown to supply electrons to nitrogenase. In Gram-negative bacteria 2[4Fe-4S], ferredoxins also do the job. However, some of the genes encoding these ferredoxins have been found to be localized in an operon with *nif* structural genes (other than nitrogenase genes) and cotranscribed, therefore, regulated in a *nif*-dependent manner. The structural genes for the nitrogenase enzyme complex are often co-transcribed in one operon in bacteria. Now we know what way the ferredoxins are involved in nitrogen-fixation in general.

*Azoarcus* sp. strain BH72 is an endophyte (like *Azospirillum*) capable of infecting roots (for example rice) and expressing the nitrogenase (*nif*) genes without causing symptoms of plant disease. Expression of structural genes (*nif HDK*) and nitrogenase activity requires low concentration of oxygen and ammonium (below 0.5 mM), but *Azoarcus* is a respiratory bacterium and the anaerobic condition does not permit nitrogen fixation. As an unusual case, *Azoarcus* has the following two advantages to overcome this problem: (i) it can shift into a state of "hyperinduction" under certain growth conditions that include extremely low oxygen concentrations (30nM) and (ii) *Azoarcus* has three copies of $P_{II}$-like proteins, the central signal transmitters of nitrogen metabolism (usually *Proteobacteria* has only two). So, nitrogenase (*nif*) genes of *Azoarcus* can be expressed endophytically in the aerenchyma of the roots seedlings suggesting that the interior of rice roots provides a microenvironment suitable for nitrogen fixation based on reporter gene studies. Hyperinduction of *Azoarcus* is characterized by *(i)* increased activity and efficiency of nitrogen fixation, *(ii)* appearance of intracellular membrane stacks (called diazosomes), and *(iii)* association of the iron-protein of nitrogenase with diazosome membranes. The association of nitrogenase with these membranes suggests that they are involved in efficient nitrogen-fixation, possibly by providing a more efficient electron flux to nitrogenase (Tanja Egner *et al.*, 2001).

## MANAGEMENT OF NUTRIENTS IN THE RHIZOSPHERE

Current knowledge on management of nutrients, rhizobia and rhizosphere is fragmentary and unbalanced. Though information on rhizobia as a nitrogen-fixer is available; its influences on abiotic stress conditions, parallel role of rhizobia in nitrogen-fixation and advancements in the molecular level are limited. The effect of symbiosis in different conditions admired by which the host plant and microsymbiont integrate to form the mutualistic relationship is important for the functioning of the rhizosphere. In the *Rhizobium*-legume symbiosis, the competitive and persistent rhizobial strain is not expected to express its full capacity. Many limiting factors impose limitations on the vigour of the host-legume relationship. They include salinity, unfavourable soil pH, nutrient deficiency, mineral toxicity, temperature extremes, soil moisture, disturbed physiological state of host, antagonism in soil, etc. In some cases single factor may also have more than one effect on host-microbe relationship. For example, salinity may also act as water stress, which affects both legume and host. Populations of *Rhizobium* species vary in their tolerance to major environmental factors; consequently screening of tolerant strains has been pursued (Saraf and Dhandhukia, 2005).

## PHOSPHATE SOLUBILIZATION

Phosphorus directly affects nitrogen-fixation, which is a principle yield-limiting nutrient in many types of soil in the world (Sindhu and Dadarwal, 2000). About 1% of phosphorus in the soil is immediately available to plants. Rest of it is in organic forms. Phytate, common organic soil phosphorus, is not readily utilized by plants. An enzyme phytase, available in plant, is needed to hydrolyze it. (Richardson *et al.*, 2005). Recent approach such as transgenic technique helps to understand the relationship between plant phytase secretion and soil phytate availability in the rhizosphere. This allows plants an improved access to soil phytate and to researchers a perceptive on the dynamics of organic P in the rhizosphere and unique insights into soil enzyme interactions. There are reports of *Rhizobium* being an efficient phosphate solubilizer. Halder and Chakraborty (1993) reported *Rhizobium meliloti* to be an efficient phosphate solubilizer. *Sinorhizobium* TR1 could also solubilize phosphate in both liquid and solid Pikovskaya's medium with a decline in pH (Tank and Saraf, 2003). Co-inoculations of *Rhizobium* with other phosphate solubilizing microorganisms in various legumes have shown synergistic effects resulting into better crop yields and phosphorus uptake. (Yadav and Dadarwal, 1997). Chabot *et al.*, (1996a) have constructed a strain of *R. leguminosarum* bv. *phaseoli* for phosphate solubilization for lettuce and maize crops. Root colonization of maize and lettuce using bioluminiscient phosphate solubilizing *R. leguminosarum* biovar *phaseoli* has been reported by Chabot *et al.*, (1996b). They have observed that maize growth promotion and colonization by PSM strain R1 lux $^{+}$of *R. leguminosarum* were not affected by soil P-fertilization and that P-solubilization is an effective mechanism of growth promotion of the strain. According to Klein *et al.*, (1990) inadequate root colonization is a major problem in potential biotechnological applications of beneficial strains. The use of highly efficient root colonizing bacteria like rhizobial strain R1 may, therefore, reduce the chances of failed nodulation. The observations made by Chabot *et al.*, (1998) that using strain *R. leguminosarum* R1Lux$^+$ colonized roots in both soils is better than the altered mutants suggests that P-solubilization has an important role to play in rhizosphere competitiveness.

## POLY β-HYDROXYBUTYRATE PRODUCTION

Rhizobia in soil are known to produce Poly β-hydroxybutyrate (PHB) when excess of carbon nutrient is available (Zevenhuizen, 1981). However, in non-carbon state, a limiting factor for growth, many bacteria accumulate the intracellular carbon storage compound PHB. This may serve as a buffer source of carbon and energy under carbon starvation conditions (Zevenhuizen, 1981). Pure culture of free-living *R. meliloti* may accumulate PHB up to 60% of its total cellular dry weight. Several species belonging to the genera *Rhizobium, Bradyrhizobium* and *Azorhizobium* accumulate PHB in free form (Tomboline and Nati, 1989) and in symbiosis (Karr *et al.*, 1984). Bergersen *et al.*, (1991) proved that stored PHB in bacteriods supports a limited nitrogen-fixation during darkness and a prolonged period of nitrogen-fixation. Cevallos *et al.*, (1996) have carried out the genetic and physiological characterization of *R. etli* mutants and proved that it is unable to synthesize Poly β-hydroxybutyrate. The capacity of *R. meliloti* to successfully compete with other soil microorganisms for the limiting nutrient resources in soil is an important determinant for the establishment of a successful symbiosis. The ability to synthesize and degrade PHB may influence the capability of bacterial cells to survive extended periods of starvation in the soil (Aneja and Trevor, 1999). These compounds have gained commercial significance since they can be developed as substitutes for fossil fuel derived polymers.

The cellular role of PHB, although not fully understood, is known to extend further by simply acting as an intracellular carbon store that can be mobilized to provide a bacterium with a competitive advantage over other soil microbes. PHB metabolism is also tightly linked to the redox state of the cell. Studies on *Rhizobium* ORS571 showed that accumulation of large amounts of PHB is induced under conditions of oxygen limitations (Stam *et al.,* 1986). Under low oxygen conditions such as those found in root nodule, the redox potential of the cell decreases as a result of aerobic metabolism leading to a concurrent rise in cellular NAD(P)H levels. This leads to inhibition of both NADP isocitrate dehydrogenase and citrate synthase, which in turn diverts Acetyl co-A and electrons away from the TCA cycle into PHB synthesis (Senior *et al.*, 1972). Trainer and Trevor (2006) have explored the progress made in determining biochemistry and genetics of PHB metabolism in their review and discussed various issues like elucidation of PHB cycle, variations in PHB metabolism among rhizobial species and the implications of these variations. They have proposed a model for the role of PHB metabolism and storage in symbiosis.

## INVOLVEMENT OF INDOLE ACETIC ACID

Nitrogen supply was initially thought to be the sole function of rhizobial root nodules, but later the hormone content of the root nodules drew attention due to involvement in the formation and development of the of the nodule (Kefford *et al.*, 1960). *Rhizobium* group has been showing production of IAA from tryptophan in culture (Dullart, 1970). The first report on nodular hormones of a tree legume was from *Pongamia pinnata* (Sinha and Basu, 1999). A preference to the use of L-tryptophan has been suggested by Bhattacharya and Pati (1999) for maximum growth and IAA production. However, bacteria could not sufficiently utilize D-tryptophan for their function because of its toxic effects (Vincent, 1974). Different carbon sources play an important role in the IAA production ability of rhizobia. A promotive effect of $Zn^{++}$ on IAA production has also been reported by Bhattacharya and Basu (1992). Tank and Saraf (2003) have also reported the IAA production related to different times of incubation in many different rhizobacteria including *Sinorhizobium meliloti*. By alteration in supply of metabolites within the nodules, the host plant might induce symbiosis to produce more IAA for its own use. Again there are also reports that hormone like auxin (Wheeler *et al.,* 1979), gibberllin (Evensen and Blevin, 1981) and cytokinin (Henson and Wheeler, 1976), which are produced within the root nodules, are transported to other plant parts as per requirements. The molecular cloning of gene for Indole-3-acetaminde hydrolase from *Bradyrhizobium japonicum*, enzyme which catalyses the biosynthesis of indole acetic acid (IAA), has been discussed by Sekine *et al.* (1989).

## INFLUENCE OF VITAMIN IN RHIZOSPHERE

Crop species capable of increasing the plant-available nutrient fraction exude organic (example carboxylates, phenolics, enzymes) and inorganic compounds (example protons, phosphate) to change chemistry and biology of rhizosphere. Amongst such compounds present in plant root exudates that can exert a positive chemostatic effect on microbes, vitamins influence the proliferation of beneficial bacteria in the vicinity of root systems considerably (Gaworizewska and Charlie, 1982). Some water-soluble vitamins like niacin, thiamine, riboflavin, pantothenic acid and biotin are liberated from legume roots (Rovira and Harris, 1961) but are also produced at biologically active levels by most bacteria and fungi isolated from the soil, rhizosphere or plant root surfaces (Mozafar, 1994). The higher incidence of vitamin producing bacteria and fungi in the root region suggests the assimilation of biologically synthesized vitamins by plants and a direct contribution of these compounds to plant development and rhizosphere interaction (Baya *et al.* 1985). *Rhizobium* strains are very demanding in their requirement for vitamins and growth factors and only a few of them are able to grow without any external additions (Elkan and Bunn, 1992). Sierra *et al.*, (1999) have reported the production of thiamine, niacin, riboflavin, pantothenic acid and biotin by two *Rhizobium* strains in chemically defined media on quantitative and qualitative *in vitro* productions of these vitamins. Steirt *et al.*, (1996) has demonstrated that wild type strain outcompetes a derived auxotrophic mutant, which was deficient in biotin production of *R. meliloti* nodulating alfa alfa roots.

## ROLE OF EXOPOLYSACCHARRIDES IN RHIZOSPHERE

In the rhizosphere, as plants produce organic and inorganic exudates, bacteria also produce exudates such as exopolysaccharrides (EPS). These exudates stimulate plant-bacterial activities amplifying the effect of bacterial EPS on soil properties. To be able to persist in a variety of soils and cause modulation, the rhizobia have been reported to produce variable exopolysaccharides. Besides, correlations between the quantity of EPS produced by a strain and its acid tolerance and metal resistance have been studied earlier (Cunninghan and Manns, 1984). Though the real mechanisms involved in the dynamic process were not well understood, the EPS-producer bacteria living in terrestrial ecosystems play an important role in soil aggregation. In a previous study, such a EPS producer *Rhizobium* sp., YAS34 was isolated from sunflower rhizosphere and inoculated to the plant to study the rhizosphere. It was observed that there is a change in the soil porosity and soil aggregate cohesion around root. Also it is noted that there is an increase in nutriment uptake and prevention from water stress (Alami *et al.*, 2000). In recent past, many reports revealed the significance of EPS in many bacteria and their role in rhizosphere. It plays a significant role in the legume infection process. Such biosynthesis process of EPS has been extensively studied in *S. meliloti* (Sobral *et al.,* 1991). There are two major types of EPS identified as EPS-I and EPS-II. The EPS-I consists of repeating octasaccharide containing seven glucose residues and one galactose with acetyl succinyl and pyruvyl residues, respectively. The EPS-II consists of alternating residues of glucose and galactose with 1-3-ß linkages with the sites of acetyl and pyruval substitutes. (Her *et al.,* 1990).

Regulation of EPS production is complex. In a wild type strain of *R. leguminoserum,* addition of ammonia results in a three-fold increase in the rate of EPS synthesis (Reeve *et al.*, 1997) but a 30-fold decrease in *R. meliloti* (Doherty *et al.,* 1988). Dilworth *et al*., (1999) have discussed the survival and exopolysaccharide production in *S. meliloti* in presence of calcium and acidic pH. On the other hand, Robertsen *et al*., (1981) have reported the structure of acidic EPS secreted by *R. leguminosarum* and *R. Trifolli* and determined the production of different types of acidic extracellular polysaccharides.

There are indications that the extracellular polysaccharides secreted by *Rhizobium* function in an early interaction with their host legumes (Bauer *et al.*, 1979), whereas Tank and Saraf (2003) have pointed out that exopolysaccharide production is an important trait of bacteria because it protects the cells against desiccation, phagocytosis, phage attack and helps in nitrogen fixation by preventing high oxygen tensions. Mutants that do not produce acidic exopolysaccharide have been described in *Rhizobium* and in most cases such mutants fail to nodulate or form empty non-nitrogen fixing nodules (Klein *et al.,* 1988). Long *et al.*, (1988) performed the genetic analysis of a cluster of genes required for synthesis of exopolysaccharide in *R. meliloti*, whereas Finan *et al.,* (1986) have demonstrated that *R. meliloti* SU47 has a symbiotic mega plasmid which included four loci involved in exopolysaccharide synthesis as well as two loci involved in thiamine biosynthesis.

## ANTAGONISTIC INFLUENCE IN RHIZOSPHERE

Antibiotics such as bacteriocins produced by rhizobia are one of the competitive factors responsible for dominance of particular strain in rhizosphere. The widespread occurrence of bacteriocins in bacteria especially cross inoculation groups of *Rhizobium* was first reported in 1967 by Roslycky. The antagonistic influence between rhizobia in variable cultures was reported by Schwinghamer (1971). Preliminary examinations of some of the bacteriocins by membrane filtration, dialysis and centrifugation indicated that they were all relatively high molecular weight substance but with different types of biological specificity, inducibility by UV light or adsorption to filter membrane (Bradley, 1967). Schwinghamer (1973) has reported a "phage-like" bacteriocin produced by *R. trifoli,* whereas Gross and Vidaver (1978) studied bacteriocins like substances produced by *Rhizobium japonicum* and other slow growing rhizobia. He concluded that these rhizobia representing genetic, serological, cultural and geographic diversity could be differentiated into three groups on the basis of production of bacteriocin. Schwinghamer and

Belkengren (1968) have suggested the use of bacteriocin producer strain to allow increase in nodulation by highly effective bacteriocin-resistant strain. Hodgson *et al.*, (1985) had employed these suggestions and were able to regulate nodulation of *Trifolium* inoculated by *Rhizobium trifolii*.

Rhizobitoxine (2-amino-4-2-(2-amino-3-hydropropoxy) trans-3-butenoic acid was found to be associated with rhizobia linked chlorosis reduction by Johnson *et al.,* (1959). The herbicidal potential of rhizobiotoxins was also studied by Omen (1973) and the potential as a pre-emergence weed control. In terms of biochemical and physiological functions, rhizobiotoxin is known to inhibit β-cysthathionase in the methionine biosynthetic pathway and ethylene formation (Omens *et al.*, 1972). Inhibition of hydrogenase synthesis in free-living *Bradyrhizobium japonicum* by rhizobiotoxine was reported by Minamisawa *et al.* (1990). One of the best characterized systems for genetic enhancement of nodulation and the competitiveness in root nodule bacteria is a cassette of genes that code for the production of a potent antirhizobacterial peptide, *Trifolium* (TFX). TFX is a ribosomally synthesized, post translationally modified peptide produced by *R. leguminosarum bv trifolli* (Briel *et al.,* 1993). Robelto *et al.* (1998) have shown that addition of the TFX production phenotype significantly increases nodule occupancy under field conditions without adverse effects on grain yield. Workers have isolated and sequence a set of genes that code for TFX production and resistance in *Rhizobium, Meso rhizobium, Sinorhizobium and Phyllobacteriamate* (Briel *et al.*, 1993, Triplett *et al.*, 1994).

## BIOLOGICAL CONTROL IN RHIZOSPHERE

Biocontrol mechanism in rhizosphere involves the synthesis of crystal proteins, antibiotics, HCN, siderophores and other biochemical agents by the producing bacteria (Narayanan Rajendran *et al.,* 1994). Though many soil bacteria such as *Pseudomonas*, *Rhizobium, Bacillus*, etc. have been extensively studied as plant growth promoting rhizobacteria (PGPR) (Narayanan Rajendran *et al.*, 1994), they also found to be the significant producer strains of imperative biocontrol substances. For instance, *Rhizobium* has been shown to reduce root rot of soybean caused by *Phytopthora megasperma* (Tu, 1978). Anton *et al.* (1978) studied the effect of 49 strains of *Sinorhizobium* towards *Fusarium oxysporum* and found them to be competent varying from 5-50% effectiveness. Malajcznk *et al.* (1984) found that rhizobia isolated from *Acacia* reduced the survival of zoospores of *Phytopthora cinnamoni*, whereas Chao (1990) tested six different *Rhizobium* strains for their bioactivity against 10 isolates of fungi and found that all the tested strains inhibited the growth of fungi. It was also reported that rhizobia and bradyrhizobia used as seed dressing or as soil drench were obtained to reduce the infection of *M. phaseolina, Rhizoctonia solani, Fusarium spp.* in both leguminous (Soyabean and mungbean) and non-leguminous (sunflower and Okra) plants in field conditions (Enteshamul *et al.,* 1993). Nautiyal (1997) observed that *Rhizobium* sp. NBRI19513 inhibited growth of *Fusarium oxysporum, R. bataticola and Phytium sp.* under *in vitro* conditions. Deshwal *et al.* (2003) have reported that *Bradyrhizobium* strains restrict growth of *Macrophomina phaseolina*. Synergistic action of plant defence chemicals also play a role in rhizosphere (Rajendran and Venkatesan, 1993).

Rhizobia are capable of colonizing the roots of non-legumes (Antoun *et al.*, 1998) such as radish and are found to increase its dry weight. Wiehe and Hoflich (1995) have demonstrated that *R leguminosarum bv trifolii* can survive and multiply under field conditions in the rhizosphere of *Brassica,* corn, rape and wheat. On the other hand, Skot *et al.* (1994) constructed transgenic *R. leguminosarum bv viciae* and *R. leguminosarum bv trifolii* in which the insecticidal crystal protein gene (*cryIIIA*) from *B. thuringiensis* subsp *tenebrionis*. The gene expression was found to be effective that these rhizobia protected their host plant from the damage of *Sitona flavescens* larvae.

The secondary metabolite HCN produced by several microorganisms has deleterious effect on the growth of some microbes (Knowler, 1996). Some rhizospheric microbes have been known to protect their host plants (Narayanan Rajendran *et al.,* 1992), through the inhibition of growth of the pathogen by HCN production. Rhizobia are relatively less efficient in HCN production. Among the rhizobia 12.5% and 3% strains were found to be HCN producers by Beaucheap *et al.* (1991) and Antoun *et al.* (1998), respectively. Rhizobia also appear to influence the plant defence mechanism by stimulating the production of phytoalexins by plants (Deshwal *et al.,* 2003). The involvement of 4-hydroxy-2,3,9-tromethoxy pterocarpan, a phytoalexin in resistance of pea plant against infection by *F. solani fsp pisi* has been recorded (Pueppkke and Etten, 1995) whereas Chakraborty and Chakrabory (1988) reported that *R. leguminosarum* protected pea root against infection by *F. solani* which could be due to increase production of phytoalexins by pea plants.

## IRON CHELATION AT RHIZOSPHERE

Soils that contain iron and other elements can be mediated by growing certain plants. These plants have the ability to augment the activities of bacteria and fungi in the rhizosphere. Many microbes have the ability to scavenge the iron and utilize them using high affinity ion transport system. This sophisticated system consists of two basic components. The synthesis and secretion of low molecular weight ferric (iron) specific chelating agent to solubilize iron termed as siderophores, is one component of the system. The second component consists of a membrane transport system that specifically recognizes the ferric complex of the particular siderophore produced by the microbe (Gill *et al.,* 1991). Siderophores are two types; 1) Catecholate and 2) hydroxamates (Neilands, 1981). Kloepper *et al.* (1990) demonstrated that siderophores also have antimicrobial activity against plant pathogens, whereas Modi *et al.* (1985) have isolated and characterized catechol like siderophore from cowpea *Rhizobium*.

Rhizobia have been reported to produce a number of siderophores. These includes rhizobactin from *S. meliloti* (Smith *et al.*, 1985), Citrate from *B. japonicum* (Guerinot *et al.,* 1990), anthranilate from *R. leguminosarum* (Patel *et al.,* 1988) and vicibactin from *R. leguminosarum bv viciae* (Dilworth *et al.,* 1998). The antagonistic action of *Rhizobium* and phytopathogens due to presence of siderophores has been studied by Sindhu *et al.*, (1997). Siderophore production by the rhizobia has been of particular interest in the view of the prominent role of iron enzymes in the nitrogen fixation and assimilation process. The iron enzymes and proteins involved include ferridoxin, hydrogenase, nitrogenase and leghemoglobin (Reigh and Connell, 1993). The siderophore iron complex is transported inside the cell by a specific energy dependent uptake system (Raymond *et al.*, 1984). In addition to the presence of siderophores, iron deficiency is also characterized by the appearance of one or more new major problems in the outer membranes formed iron regulated outer membrane proteins (IROMPs). The involvement of iron repressible outer membrane proteins in cowpea *Rhizobium* GN1 was studied by Jadhav and Desai (1994). The same researchers in 1997 reported the role of iron reductase in the release of iron from a ferri siderophore complex in GN1. To test the importance of a high affinity iron transport system the ability to *R. meliloti* 1021 to nodulate and efficiently fix nitrogen in plants, a number of Tn5 insertion mutants (Simon *et al,* 1983) where isolated which was defective in the production of transport and negative regulation of the production of siderophore. The identification of these mutants was accomplished using a moderate affinity iron binding dye that allows detection of siderophore production on agar medium *in situ* (Gill and Neilands, 1994) while siderophore producing ability of pesticide resistance mutants was studied by Saraf and Sood (2002). Rhizobia are well known for their ability to infect the root tissues and induce nodulation. However, presence of phytohormones ethylene has been known to inhibit nodulation (Lee and Larue, 1992). Decreased levels of nodulations have been observed after application of exogenous ethylene or 1-aminocyclopropane-1-Carboxilic acid (ACC) prior to or at the same time as the addition of rhizobia (Nukui *et al.,* 2000). Conversely nodulation can be promoted when plants are treated with ethylene inhibitors or antagonists (Yuhashi *et al.,* 2000). Low levels of ethylene is required for successful entry of the infection thread on the outermost layer of cortical cells, whereas higher levels of the hormone induce abortion of the infection thread by inducing cross linking of its matrix glycoproteins (Penmetsa and Cook, 1997) one rhizobial species, *Bradyrhizobium elkanii* appears to have done this by producing rhizobitoxine, and ethylene biosynthesis inhibition (Yuhashi *et al.,* 2000). The first report documenting the presence of ACC deaminase in *Rhizobium* spp, was provided by Wembo Ma *et al.* (2003a). They examined thirteen different strains of rhizobia out of which five tested positive.

## INFLUENCES OF ABIOTIC STRESS ON RHIZOMEDIATION

Rapid increase in human consumption of plant wealth, limited scope for increase in arable fertilizers, poor plant protection, continuous use of chemicals and high yielding fertilizers-responsive varieties resulted in an imbalance of nutrient status of soil, depletion of soil organic matter and deterioration of soil structure. It has also disturbed microbial equilibrium and a lot of non-target microbes like rhizobia. They face stress of every kind, which affects their physiological, morphological, biochemical and genetic activities. In this context the soil biologists have emphasized the role of the beneficial microbes in plant nutrition and pest management which are environmental-friendly and can be advantages only blended to replace or substitute a part of the energy intensive high cost agrochemicals. There are some reports on the use and manipulation of rhizosphere bacteria that keep cleaning up environmental pollution and help in growth of plants under such conditions. However, the activity of rhizobia or other PGPR may not be fully utilized to poor survival in stress conditions. These abiotic stresses are not only diverse but also very complex and rhizobia respond differently to stresses. These conditions range from levels of salinity, unfavorable soil pH, nutrient deficiency, mineral and heavy metal toxicity, temperature extremes, insufficient or excessive soil moisture, plant diseases, etc. which in turn imposes severe limitations on the vigor of host legumes (Brockwell *et al.,* 1995).

## INFLUENCE OF SALT AND OSMOTIC STRESS

Nearly 40% if the world's land surface can be categorized as having potential salinity problems. Increasing salt concentrations may have a detrimental effect on soil microbial populations as a result of direct toxicity as well as through osmotic stress (Tate, 1995). Variability in salt tolerance among crop legumes has been reported. For example, *Vicia faba, Phaseolus vulgaris* and *Glycine max* are more salt tolerant than others like *Pisum sativum*. The lipopolysaccharides profile using SDS-PAGE from various species of rhizobial legumes was found to be modified by salt (Tao *et al.*, 1992) with loss in nitrogenase activity at 200 mM salt concentration (Chien *et al.,* 1992). Salt stress inhibits the initial steps of *Rhizobium*-legume symbiosis. Soybean root hairs showed little curling or deformation when inoculated with *B. japonicum* in presence of 170 mM NaCl and complete suppression of nodulation at 210 mM NaCl (Tu, 1981). The reduction of nitrogen-fixing activity due to salt stress is usually attributed to a reduction in respiration of the nodules and a reduction in cytosolic protein production specifically leghemoglobin (Delgado *et al.*, 1994). Salt induced distortions in nodule structure, were observed by Sprent and Zahran (1988). Soybean and chickpea rhizobia were tolerant to 340 mM NaCl with fast growing strains being more tolerant than slow growing strains (Elskeikh *et al.*, 1990) rhizobia from woody legumes also showed substantial salt tolerance; strains from Acacia, Prosophis and Leucana are tolerant to 500 to 850 mM NaCl (Lal and Khanna, 1995). It has been reported (Jian *et al.*, 1993) that the growth of *R. meliloti* was severely inhibited by $Mg^{+2}$ ions where as $Na^{+2}$ and $K^{+}$ ions had little inhibitory effect. *Rhizobium* has been reported to accumulate osmolytes in saline conditions. In presence of high levels of salts, the level of intracellular free glutamate and/or K+ greatly increases in cells of *R. predu* (Fujihara and Yoheyama, 1993). While addition of glutamate was found to increase the growth of *S. meliloti* in 5% NaCl concentration (Saraf and Dhandhukia, 2005). Trehalose accumulates to higher levels in cells of *R. leguminosarum* under hypersaline conditions (Ghittoni and Bueno, 1996).

Soil osmotic potential is low at the rhizosphere due to the accumulation of minerals and other nutrients compared to other area of soil. This has profound effect on rhizobia and rhizomediation. As the result, microbial osmotic potential may be disturbed and the plant water uptake might be impeded. Busse and Bottomley (1989) reported irregular morphology of cells is potentials below –0.5MPa. This type of osmotic potentials analysis is very useful to improve the rhizomediation for the better yield. For example, the disaccharides sucrose and ectoine were used as osmoprotectants for *S. meliloti* (Gouffi *et al.*, 1999). The formation of osmotic-shock proteins was observed in 1999 using SDS-PAGE and was further studied using various biochemical techniques similar to that of heat-shock protein (Zahran (1999). On the other hand, Busse and Bottomley (1989) reported irregular morphology of cells is potentials below –0.5MPa. Similarly, the synthesis pattern in SDS-PAGE of lipopolysaccharides from various species of rhizobia from legumes was found to be modified by salt in the presence of which the length of side chains increased (Tao *et al.*, 1992) formation of ineffective nodules with loss in nitrogenase activity at 200 mM salt concentration was reported by Chien *et al.* (1992).

## EFFECT OF HIGH TEMPERATURE AND HEAT STRESS

Rhizosphere is a biologically very active area where microbes develop, interact with each other, and compete for space, nutrients and root exudates. Maintaining a constant temperature is critical for both microbes and plants. Temperature affects root hair infection, bacterial differentiation, nodule structure and functioning of root nodule of legumes (Roughley, 1970). The fluctuation in temperature also influences the nitrogen fixation and other biological activates. Critical temperature for $N_2$ fixation is 30°C for clover and pea, 35-40°C for soybean, guar and cowpea (Michiels *et al,* 1994). Nodulation of soybean was markedly inhibited between 42°C and 45°C at 12 hour and 9 hour, respectively (LaFavre *et al,* 1986). Piha and Munns (1987) reported that nodules formed at 35°C were small and had low nitrogenase activity. Hernandez-Armenta *et al.* (1989) found that transferring nodulated-bean plants from 26°C to 35°C inhibited nitrogen fixation. For most rhizobia the optimum temperature range for growth is 28-31°C, but strain adaptation at high temperature has been reported by Hartel and Alexander (1984). Screening of *R. leguminosarum* sp. *phaseoli* showed that some strains were able to nodulate *Phaseolus vulgaris* at high temperature (35-38°C), but nodules formed were very ineffective and plants did not accumulate nitrogen in shoots (Hungria and Franco 1993). Presence of some heat-shock proteins have been reported by Graham (1992). An increased synthesis of 14 heat-shock proteins in heat sensitive strains and 6 heat-shock proteins in heat tolerant strains were isolated between 40-45°C by Michelis (1994). However, *R meliloti* isolated from coastal arid regions of Gujarat in India showed optimum growth at 40°C (Saraf and Dhandhukia, 2005).

## INFLUENCE OF SOIL MOISTURE

Rhizobia are found to exist in soils where moisture level is low like desert soils (Jenkins *et al.,* 1987). Since rhizosphere is a complex resource centre, which sustains root exudates and harbours complex web of nutrients, rhizobia can uphold their life-activities even at the low moisture content. This extraordinary ability of rhizobia affects the plants directly through roots, and also indirectly by changing the physico-chemical environment of roots. It includes disease suppression, different chemical shift; adjust pH, change in redox potential, amendment of root exudation and enzymatic activities. The survival and phenomenal activity of rhizobia may depend on their distribution among microhabitats and influenced by the changes in soil moisture. Moderate water tension slowed the movement of *R. trifolii* (Hamdi, 1970) the migration of bacteria ceased when water filled pores in soil become discontinuous as a result of water stress. Mesquite *Rhizobium* and *R. meliloti* showed irregular morphology at low water potential (Busse and Bottomley, 1989), whereas Hunt *et al.* (1981) suggested low water content in soil leads to the lack of success of soybean inoculation in soils with a high indigenous population of *R. japonicum*. A favorable rhizosphere environment is vital to legumes-*Rhizobium* interaction. However, the magnitude of the stress plays an important role in the symptoms expressed. For example, mild water stress reduces only the number of nodules formed while moderate or severe water stress reduces both the number and size of nodules (Williams and De Mallerca, 1994). Reduction in nitrogen fixation is also associated with deficiency in soil moisture. Sellstedt *et al.* (1993) found that N derived from nitrogen-fixation was decreased by about 26% as a result of water deficiency when measured by acetylene reduction assay. *Rhizobium* can be optimized and selected with moisture stress tolerance. Optimization of soil moisture for growth of the host plant, which is generally more sensitive to moisture stress than bacteria, results in maximal development of fixed nitrogen inputs into the soil systems by *Rhizobium* legume symbiosis (Tate, 1995). Potassium is known to improve the resistance of plants to environmental stress. A report by Sangakkara *et al.* (1996) indicates that K can apparently alleviate the effects of water shortage in symbiotic nitrogen fixation of *V. faba* and *P. vulgaris*.

## INFLUENCE OF SOIL ACIDITY-ALKALINITY

The acidity and alkalinity activities of rhizobia in the rhizosphere are likely to be influenced by the plant genotypes and their symbiosis status. Soil acidity constrains symbiotic nitrogen-fixation in both tropical and temperate soils, limiting *Rhizobium* survival and persistence in soils and reducing nodulations (Ibekuwe *et al.,* 1997). Graham *et al.* (1982) has identified some *R. phaseoli* showing tolerance to acidity. It has been found that *R. loti* multiplied at pH 4.5 and can nodulate their legume host at pH 4.5 (Cooper *et al.*, 1995), *R. trifolii* and *R. loti* are moderately acid tolerant and while *R. meliloti* is very sensitive (Tiwari *et al.*, 1992) studies on the genetic basis of tolerance to low pH suggest that at least two loci of either megaplasmid or chromosomal location for pH genes are necessary for the growth of rhizobia at low pH (Chen *et al.*, 1993). On the other hand, availability of $Ca^{++}$ in acidic soils with a high Al content appears to be very important for nodulation. A low $Ca^{++}$ concentration at pH 4.5 greatly affects nodule number, nitrogenase activity and nodule ultra-structure of common bean (Vassileva *et al.*, 1997). Saraf and Dhandhukia (2005) have reported that *R. meliloti* shows growth optimum at pH 9.0 growth at pH 11.0 is 95% and it declines to 47% at pH 12. This tolerance to high pH may be due to saline alkaline soils from which the nodules have been isolated. Nour *et al.,* (1994) have reported pH tolerance of rhizobia isolated from chickpea to be 10.0.

## EFFECT OF SOIL AMENDMENTS ON RHIZOSPHERE

### Fertilizer Application

It is widely accepted that the capacity for nitrogen fixation by a nodulating or nodulated legume is influenced at least in two ways by mineral nitrogen in the soil in which it is grown. First, the process of nodulation may be promoted by relatively low levels of available $NO_3$ or $NH_4$ high concentrations of which always depress nodulation (Davidson and Robson, 1986). Second the rate of nitrogen-fixation by an actively growing and well nodulated legume is always suppressed by $NO_3$ ions (Sanginga *et al.,* 1996). Symbiotic nitrogen fixation in field legumes takes place against a changing background of mineral N availability as a consequence of mineralization, leaching and fertilizer applications (Davidson and Robson, 1986). It has been established that soil $NO_3$ inhibits root infection, nodule development and nitrogenase activity (Azcon and Elatrash, 1997). Furthermore, Danso *et al.* (1990) found that the inhibition of soybean nitrogen fixation at higher N levels (83 mg of N kg of soil$^{-1}$) was significantly reduced by a second inoculation. This treatment resulted in at least a doubling of the percentage and total amount of $N_2$ fixed by soybean plants after single slurry inoculation. However, rhizobia showed varied responses to combined N with regard to competitive abilities of 3 strains of *B. japonicum* nodulation of soybean at an increasing fertilizer N concentrations) upto 10 g of N m$^{-2}$) have been reported (Martensson *et al.*, 1989). While, Eaglesham (1984) reported that when $NO_3$ levels were sufficiently high to completely suppress nodulation on the original root, there was profuse nodulation and significant nitrogenase activity on the adventitious roots. The inhibitory effect of exogenous nitrate as nitrogen-fixation has variously hear attributed to a direct competition between nitrate reductase and nitrogenase for reducing power or to the fact that nitrite (a by product of nitrate reductase) inhibits the function of nitrogenase and leghaemoglobin (Becana and Sprent, 1987). To improve BNF by the legumes the soil N concentration must be managed through inclusion of appropriate nitrate tolerant high nitrogen-fixing legume crops.

## Pesticide Application

Due to vast applications and indiscriminate use of pesticides, the non-target organisms (rhizobia) present in the soil are destroyed. It has been reported that foliar applications of the herbicides bentazone and MCPA to red clover at the recommended rates altered the morphology of root hairs, reduced nodule number and nitrogenase activity. (Ljunggren and Martensson, 1980). On the other hand, some other herbicides like alachlor, fluazifop, sethoxydin and metalachor did not have detrimental effect on nitrogen-fixation or seed yields when added to recommended rates for weed control in soybean plantations (Kucey *et al.,* 1988). Fungicides like Thiram and Captan are harmful to nodulation and $N_2$ fixation of several grains and forage legumes (Heinonen *et al.,* 1982). Nematocides like carbaryl and Carbofuran at all the concentrations were inhibitory to the growth of *Rhizobium japonicum* (Maheshwari and Gupta, 1991). Certain strains of Rhizobia could resist high levels of pesticides by adaptations (Kanta *et al.*, 1986). Maheshwari *et al.* (1993) have observed that at extremely low doses of carbaryl and 2,4-D there is enhancement of growth, but it is inhibitory at higher levels while Maheshwari and Gupta (1991) reported stimulation in oxygen uptake with increase in concentration of organocarbamate pesticides. The dual behaviour of carbaryl and 2,4-D is attributed by them to a large amount of ATP hydrolyzing activity followed by uncoupling of reversible reaction for ATP generation. On the other hand, Saraf *et al.* (1994) have concluded the effects of carbaryl and 2,4-D nitrogenase and uptake hydrogenase activity in agar culture as well as root nodules formed by *B. Japonicum*. High concentrations of Gaucho and Uravax-300 clearly inhibited the growth of root nodule bacteria under laboratory conditions (Miettinan and Chegogen. 1996). Maheshwari and Saraf (1994) have concluded in their studies that changes ion the cell permeability of *Rhizobium* cell wall resulting from resistance to carbaryl and 2,4-d alter the ability of *Rhizobium* to develop nitrogenase and uptake hydrogenase activities. Meanwhile, Saraf and Sood (2002) have reported an increase in siderophore production of chickpea rhizobia treated with low doses of monocrotophos and substrate decline thereafter. The need of the hour is to tailor such rhizobia, which could tolerate some low doses of pesticides and exploit them as potential biofertilizers.

## INFLUENCE OF HEAVY METALS

Sewage sludge application to agricultural soils seemed to an efficient way of disposal. However, sewage sludge is contaminated with heavy metals and a variety of materials potentially toxic to rhizobacteria (Madariaga and Angle, 1992). Kinkel *et al.* (1987) examined two genera of soybean nodulating rhizobia to determine the level of resistance to eight different metals. Marked variations were found with several heavy metals, even for rhizobial strains belonging to the same species. Tang and Sadowsky (1994) reported that *Bradyrhizobium* strains were more resistant to heavy metals that were *Rhizobium* strain on the other hand *R. leguminosarum* was affected by copper treatment. The cells remain viable but non-cultivable Giller *et al.* (1989) suggested two possibilities to explain the mechanism by which the elevated metals concentration eliminated nitrogen-fixation. (i) one or more metals prevented formation of effective nodules or (ii) elimination of effective *Rhizobium* from metal contaminated soil. Obbard *et al.* (1994) suggested that important effects on the size of effective rhizobial populations were determined by soil pH, sludge type, addition rates and concentration of heavy metals present. Furthermore, the legume root nodules symbiosis can be used as an effective parameter for ecological evaluation of contamination soils (Wetzel and Werner, 1995). Meanwhile Farewell *et al.* (2006) have studied the use of transgenic canola and PGPR to enhance plant biomass at nickel contaminated site various bacterial strains containing the enzyme ACC deaminase have been found to increase plant growth following exposure to Cd (Belimov *et al.,* 2005), Cu (Reed and Glick, 2005) and Ni (Burd *et al.*, 1998). Rhizobia by virtue of it possessing the enzyme ACC deaminase could also be evaluated for its growth and potential at heavy metal contaminated sites.

## PLANT-RHIZOBIA INTERACTIONS AND MOLECULAR APPROACHES

Until recent past, improvement of analytical tools allowed us to characterize various rhizosphere-physico-chemical changes with some limitations. However, within the last decade the increase in technological innovations in tools and techniques has changed the entire field. Now the modern molecular approaches helped to understand the composition of root excretions, their functions in nutrient-acquisition, detoxification, plant microbial interactions, and plant-physiological approaches. The era of molecular microbial ecology has uncovered only a past of novel microbiota most of which is based on rRNA and rDNA analysis (Torsvik, 2002). Many molecular tools were employed to study regulatory processes involved in root-induced chemical changes and to characterize the diversity of rhizosphere-microbial communities. Since the rhizosphere and the host-rhizobial populations comprise a complex relationship, modern electron microscopy and other microbial ecological tools helped to localize the specific strain in their habitat. Direct visualization of the rhizosphere (*e.g.* resin-imbedding and thin sectioning, on-line observation chambers and three-dimensional imaging), fluorescence microscopy, fluorescent-antibody (FA) staining, single-cell microscopy, fluorescence *in situ* hybridization (FISH) technology, bioluminescence technology, green fluorescent protein tagging, microarray technique, cytometry, etc., are some of those techniques helps to understand rhizomediation process.

## MOLECULAR APPROACHES IN *RHIZOBIUM* STUDIES

As a system for the study of developmental biology *Rhizobium* root nodule formation provides unique experimental opportunities, each partner can be manipulated so as to permit genetic and molecular dissection of the developmental sequence. Recent results show the initiation nodulation steps can be characterized as a two-way molecular conversation. The host legume releases signal compounds that stimulate the coordinate expression of bacterial genes that are required for nodulation (*nod* genes). Those nod genes in turn encode enzymes involved in the synthesis of Nod factors substituted oligosaccharides that cause morphological changes in plant root. These nod factors compounds act as determinants for the host range exhibited by a specific bacterium. Echoing the structural similarity of nodules indifferent host bacterial systems the molecular signals in different host bacterial systems the molecular signals in the rhizobia-legume symbiosis are variations on a theme. Inducers from the plant are phenolics of the flavonoid group, with skeletal or side group modifications to distinguish the inducers from distinct plants; the morphogens synthesized by the bacteria are all oligomers of N-acetyl-glucosamine with different host specific modifications (Fisher and Long, 1992). The formation of effective nodules involves several stages of successive co-ordinations determined by different by different sets of genes both in the micro and macrosymbiont (Schulze *et al.,* 1994). The gene expression being temporarily and spatially coordinated the plant proteins that are specifically synthesized during the formation and functions of the nodule are called nodulins. Early nodulins are expressed at the onset of nitrogen-fixation while nodulins expressed at during $N_2$ fixation are termed as late nodulins and they participate in the function of the nodule by creating physiological conditions required for nitrogen-fixation, ammonium assimilation and transport (Munoz *et al.,* 1996).

The established legume-rhizobia symbiosis is initiated by the exudation of flavonoid and isoflavonoid compounds (eg genistein, naringenin, luteolin) from the host plant (Miklashevichs *et al.,* 2001). These substances act as both chemo-attractants and inducers of the nod genes such as regulatory Nod D genes and the structural Nod ABC genes, in the rhizobia. Other substances implicated in chemotaxis and proliferation of rhizobia at legume rhizosphere includes nutrients like amino acids, organic acids and sugars (Pandya *et al.,* 1999) and secondary metabolites example betaines (Phillips *et al.,* 1995) while some others like tree and shrub legumes produce a non-protein amino acid toxin, mimosine to enhance abundance of rhizobia in rhizosphere (Soedarjo and Borthakur, 1998). The interaction of the rhizobial regulatory Nod d protein with specific flavonoids is believed to be the first level of host specific recognition in the symbiosis. The consequence of nod gene expression is the bacterial synthesis of the nodule inducing "Nod factor". These factors may be lipo-chito-oligosachharides (LCOs) only certain LCO's can initiate nodulation in certain legumes (Miklashevichs *et al.* 2001). Recently, two putative Nod-factor receptor kinase genes (NFR1 and NFR5) have been identified in *Lotus japonicum* (Madsen *et al.,* 2003). Perception of Nod factor by the host legume results innumerous responses involved in infection and nodule formation including root hair deformation (Baker, 1981) partial dissolution of host cell wall (Turgeon and Bauer, 1985) formation of infection thread (Kijne, 1992) initiation of mitosis in root cortex (Dudley *et al.,* 1987) and induction of nodule specific genes expressed early in nodule development (Schultze and Kondorosi, 1998).

Aside from the Nod factor, there are plethora of other rhizobial cell surface ad excreted compounds implicated in symbiosis oriented signalling functions including polysaccharides (Price, 1999), phytohormones (Hirsch *et al.,* 1997) and low molecular weight proteins (nodulations outer proteins or Nops) (Marie *et al,* 2003). However, there is a considerable diversity of modes of infection and nodule development depending on the particular legume and the path of infection viz. intercellular or intracellular (Guinel and Geil, 2002). Intracellular infection occurs via infection threads, tubular cell wall-like structures constructed by the plant involving extensive cytoskeleton activity, the synthesis of which is initiated by bacterial Nod factor (Lhuissier *et al.,* 2001). The most extensively studied symbiosis involves crop legumes (*e.g.* commonbean, cowpea, fababeans, and soybeans) of *Papilionaceae* subfamily with intracellular infections leading to either indeterminate or determinate root nodules. Primordial of indeterminate nodules are initiated opposite to protoxylem poles while primordial of determinate nodules are initiated in outer root cortex. The bacteria move through the infection thread matrix and are released into the infected cells of developing nodule engulfed in a plant membrane (peribacteroid membrane) (Vessey *et al.,* 2005). The structure containing the bacteroids, the symbiosome is analogus to a plant organelle with a peribacteriod membrane (Day *et al.,* 2001). The symbiosome may contain one or several rhizobia that have undergone certain morphological changes associated with their new symbiotic lifestyle (Oke and Long, 1999). Depending on the legume, a single nodule may contain several strains of rhizobia, but in the majority of cases a nodule hosts a single strain of rhizobia (Martinez-Romero, 2003) and a single plant root can contain more than one strain of rhizobia among its nodules (Bromfield *et al.,* 2001).

A plethora of nodule specific plant genes are expressed during the development and functioning of legume nodules (Schultze and Kondorisoi, 1998). These genes have been characterized into two groups depending on whether the proteins they code for are synthesized early or late in nodule development late early and late nodulins respectively. Many of the early nodulins genes are associated with induction of root hair deformations, cortical cell divisions and cell wall modifications. The late nodulin genes are more closely associated with the mature nitrogen fixing nodule. Genes coding for leghaemoglobin are examples of late noduling (Heidstra *et al.*, 1997). Genes encoding proteins associated with carbon and nitrogen metabolism (Silvente *et al.*, 2003) and membrane transport (Kapranove *et al.,* 2001) in the nodules are also well-studied late nodulins. Pingret *et al.,* (1998) have investigated Nod factor signal transduction in the Medicago root epidermis by using pharmacological approach in conjunction with transgenic plants expressing the Nod factor, and found evidence for the participation of heterotrimeric G proteins in Nod factor signalling.

All diazotrophs utilize the same basic biochemical machinery for nitrogen fixation, which is carried out by nitrogenase enzyme. It is a promiscuous enzyme that catalyses the reduction of dinitrogen to ammonia, the reduction of protons to hydrogen, which appears to be an obligatory part of ammonia formation, and the reduction of small unsaturated molecules such as acetylene, azide and cyanide (Glick *et al.*,1999) and are designated as *nif* genes. They may be carried on plasmids in *Rhizobium*. Biochemistry and genetics of nitrogen fixation have been localized on plasmids in *R. leguminosarum* including symbiotic plasmid which gene essential for nodulation and nitrogen fixation. (Laguerre *et al.,* 2003) where as ABD and *Nif* genes are carried on megaplasmids (Trevor, 1990).

## OTHER MOLECULAR TECHNIQUES

### Stable Isotope Probing

Traditional techniques of rhizobial analysis including microsampling and microanalytical techniques have limitation to analyze the components of soil communities. However, modern molecular fingerprinting helps in rhizosphere community analysis in response to environmental change. Further, stable isotope probing helps to determine which members of a community are involved in the utilization of specific substrates. Here soil or plant communities are traced with 13C-labelled carbon substrates. Characterization of active populations is made following extraction of nucleic acids and separation of 13C-labelled (heavy) and 12C-labelled (light) nucleic acid pools by density gradient centrifugation. (Graeme *et al.*, 2004)

## Pyrolysis-Field Ionization Mass Spectrometry

Another technique in rhizosphere study for the quantification of rhizodeposition is pulse-labelling Py-FIMS. It is a rapid, versatile and highly sensitive method with broad application potential in rhizosphere research. Here leaching of rhizodeposits and their molecular-chemical characterization are carried out using solid-phase samples (Kuzyakov, 2003).

## Functional Genomics and *in situ* Studies

Microarray, cDNA array, PCR-based detection of genome polymorphisms, large-scale gene sequencing, genome analysis, analysis of gene expression, proteome analysis, metabolome are remarkably helping to understand the functional mechanisms of symbiosis and soil communities. The *in situ* techniques, especially molecular probes, are very useful for quantitative measurements at high spatial and temporal resolution. Molecular modelling and other techniques such as image analysis, sequence analysis, bioinformatics will be necessary to put together into action for the major challenges that result in the favourable management of the rhizosphere.

## Rhizomediation in Terms of Coinoculation

In general, beneficial rhizobacteria that stimulate plant growth are referred to as PGPR. A group that includes different bacterial species and strains belonging to genera such as *Gluconacetobacter, Azospirillum, Azotobacter, Bacillus, Burkholderia, Herbaspirillum,* and *Pseudomonas*. For the past 30 years, there has been an increasing effort in developing microbial inoculants (including rhizobia) for plant growth promotion in agriculture. The colonization of plant surfaces and rhizosphere ecology has also been investigated and significant advances in the elucidation of mechanisms such as role of biological nitrogen-fixation, production of phytohormones and biological control involved in plant growth promotion through rhizobial inoculation have been made specially using molecular biology approaches. Such interactions within microbial communities associated with plant roots are of considerable interest because such interactions might either enhance or inhibit the beneficial effect of the *Rhizobium* in the soils. It is therefore necessary to study the behaviour of *Rhizobium* as well as its competitiveness for colonization in presence of rhizosphere bacteria including other species of *Rhizobium* (Germida and Walley, 1996). Meyer and Lindermann (1986) showed that *P. putida* R-70 enhanced nodulation and nitrogen fixation by *Rhizobium* in clover. *Rhizobium* in presence of *P.fluorescens* also increased plant shoot weight, nodulation as well as its establishment in the rhizosphere. Nodulation was not only increased by the number, but also by the increased size and pink colour of nodules as compare to small and white. This could be due to the siderophore production by *Pseudomonas* and its effective iron scavenging and thus reducing deleterious microorganism populations, which competes with *Rhizobium* (Andrade *et al.*, 1998). The ability of rhizobacteria such as *Pseudomonas*, to complex Fe with siderophores is the key mechanism for suppressing pathogens. These bacteria intended to suppress diseases caused by soil pathogens enhance growth and yield ( Narayananan Rajendran and Jahn, 1998).

In combination with viable *Bacillus megatarium+Rhizobium*+legume symbiosis, interaction is enhanced but non-viable cell populations failed to enhance nodulation. Co-inoculation of *B. megatarium* and *R. etli* also enhanced root hair formation, which are infection sites for increasing nodulation. This can also be due to possible involvement of stimulatory substances produced by

*Bacillus,* which may be subject to regulation or vice versa (Srinivasan *et al.,* 1997). Another possibility is the overcoming of induction of 'nod-off' compounds implants which are produced as defence in presence of co-inoculation (Peter and Long, 1988; Mellor and Collinge, 1995). Combined inoculation of *Rhizobium, Pseudomonas and Bacillus* have shown increase dry matter, grain yield, $N_2$ and phosphorus uptake (Dhiman and Saraf, 2003) whereas Kundu and Gaur (1981) have observed that the use of composite cultures did not result in increase in PS activity. Total yield, nodule number and weight in chickpea decreased significantly due to co-inoculation of *Azotobacter* and *Rhizobium.* But seed bacterisation by mixed inoculums enhanced yield as compared to absence of *Rhizobium* or *Azotobacter* but was lesser than individual strains (Paul and Verma, 1999). Co-inoculation with *Azospirillum* also increases nodule number, yield, $N_2$ fixation and shoot dry matter (Rai, 1983; Iruthayana, 1983). But Plazinski and Rolff (1985) reported that such combination decreases acetylene reduction activity and thereby nitrogen fixation. Dhiman and Saraf (2003) also reported high total nodule number which was not higher than obtained by single *Rhizobium* inoculation. It was in the case of P solubilized and total weight of plants. On the contrary Rice *et al.,* (1995) reported increase P solubilization when *Rhizobium* was co-inoculated with *P. bilaii* in Cowpea and Pigeonpea. Seed bacterization of same pair of rhizobacteria enhanced crop production, shoot height, shoot length and dry weight but lower the germination ration (Dileepkumar *et al.*, 2001). Dashti *et al.* (1997) reported that the application of S*erratia liquifaciens* 2-68 or *Serratia proteamaclans* 1-102 along with *Bradyrhizobium japonicum* increased protein and organic matter as well as yield. Saraf *et al.* (2006) have studied combinations of 16 strains of rhizobacteria and concluded that a consortia of *Rhizobium, Pseudomonas* and *Bacillus* show the best results in terms of nitrogen and phosphorus solubilizations.

Dashti (1998) reported increased nodulation and $N_2$ fixation by nitrogen fixing bacteria and PGPR. Parmar and Dadarwal (1999) reported that a combination of *Pseudomonas* + *Bacillus* + *Rhizobium* consortium enhances nodule weight, root length, shoot length, biomass and promotes flavonoid like compound production and this increases plant productivity. Interaction between arbuscular mycorrhizae and *Rhizobium* in the rhizosphere of legumes often leads to an increase in nitrogen uptake by the plant (Smith *et al.*, 1979). This is thought because mycorrhizae corrects no only P deficiency but also any nutrient deficiency in plants that might be limiting to *Rhizobium* (O'Hara, 1988) thus making a greater proportion of photosynthates available to *Rhizobium* nodules (Harris *et al*, 1985) imparting larger and pinker nodules. Andrade *et al.,* (1998) showed that inoculation of DAPG positive mutants of *P. fluorescens* enhanced nodulation as well as mycorrhizal infaction of pea roots. Thus, although co-inoculation of *Rhizobium* with other rhizobacteria enhances over all vegetative parameters but it is still lesser than the increase obtained by individual inoculation. The only reason which comes out after all these reports is that co-inoculation must be increasing the competition for colonization between the soil microbial populations.

## Rhizomediation and Quorum Sensing

Rapid multiplication of rhizobia at rhizosphere indicates the availability of substrate. This helps a rich colonization process and enhances rhizosphere competence. The relationship between these two is a critical parameter. Based on this and other environmental parameters, many rhizobacteria respond through two regulatory systems, a secretion system and a density-dependant regulatory process known as quorum-sensing (QS) (Bassler, 2002). The latter system, that couples cell-density and gene expression, is presented in details because it exemplifies how plant and edaphic factors may interact with bacterial communication in the rhizosphere. The composition of rhizosphere differs greatly from that of the surrounding soils with the change in plant species and as a result of diverse plant-microbe interaction. Many bacteria control a variety of physiological functions in a cell density dependent manner via small diffusible signal molecules. Such mechanism is termed as 'Quorum Sensing' (Yijiang *et al.,* 2006; Fuqua *et al.,* 1994).

Quorum sensing is the regulation of gene expression in response to fluctuations in cell-population density (Melissa *et al.,* 2001). Here cell to cell communication is mediated through extracellular signalling molecules through which bacteria modulate their cellular functions like sporulation, biofilm formation, bacteriocin production, antibiotic production, enzyme secretion, plasmid transfer, symbiosis, etc. (Gonzalez *et al.*, 2003). *Rhizobium* modulates plant host interaction, nodulation and nitrogen fixation using such extracellular mechanism (Broughton *et al.,* 2000). Quorum sensing increases the number of rhizobia in and around the plant roots and nodules to infect the root hairs as well as forming plant host relationship. Quorum sensing relies heavily on production of low molecular weight signalling molecules called autoinducers *i.e. N*-acyl homoserine lactones which vary in their length and substituent at the third carbon of their N-acyl side chains (Whitehead *et al.,* 2001). Production of AHLs by rhizobial strains was identified by Cha *et al.,* (1998). Many of the gene product needed for this symbiosis are encoded by symbiosis (*sym*) plasmid. The QS in *R. leguminosarum* is regulated by chromosomally encoded *Cin*RI proteins (Sharma *et al.,* 2003,). The product of *cin* I gene is responsible for synthesis of N-(3-hydroxy-7-cis-tetrad-cenoyl)-L-homoserine lactone (htde-DHL) whose expression depends on Htde DHL expressed *Cin* R. *R. leguminosarum* possesses pRLIJI (syn plasmid) for producing AHL synthases whereas *R. meliloti* produces array of compounds with AHL like activity but *R. fredii* produced just one strongly non polar compound with such activity (Cha *et al.*, 1998). *S. meliloti* produces *Sin*R and *Sin*I, changes/mutations in which decreases number of pink nodules. *B. japonicum* has *NoI*A global regulatory component involved in repression of the nodulation genes. QS system also controls rhiABC genes important in rhizosphere establishment. Yijiang *et al.* (2006) have shown that autoinducers in *M. haukii* are not an acyl homoserine lactone but cyclic dipeptides readily detected by AHL bioassay and can use QS system. They key regulatory components of these signaling systems are LuxI-type proteins which as AHL synthases and *Lux*R-type proteins which serve as AHLreceptors and AHL dependent transcription factors (Waters and Bassler, 2005).

When QS defective mutants were compared for their growth pattern with wild type strains and were observed that the mutants showed decreased growth rate speculating that presence of autoinducers limits the growth of *M. haukii* thereby avoiding excessive bacterial growth that might induce plant defense response during plant root infection. QS deficiency also reduces symbiotic processes as well as nodulation by rhizobia (Yijiang *et al.,* 2006; Daniels *et al.,* 2002; Zheng *et al.,* 2006). QS varies from bacterium to bacterium due to the nature of the chemical signals, and the relay mechanisms. This allows bacteria to coordinate the gene expression (Melissa et *al.,* 2001). Most QS processes are idle when undertaken individually, but valuable when carried out simultaneously by a large number of cells (Christopher *et al.,* 2005).

## CONCLUSION

Rhizobia have to face various environmental conditions as free-living bacteria in soil that are capable of inducing the formation of nitrogen-fixing nodules on the roots of their leguminous host plants. Rhizobia are potential candidate as biofertilizer conferring diversity of plant growth promoting properties ranging from nitrogen fixation, phosphate solubilization, phytohormone production, siderophore, antimicrobials, ammonia and HCN production ability, etc. to benefit the plant. Future strategies are focused to clone such desired factors and incorporate most of them into those strains having the primary function of nitrogen fixation. The genes involved in plant growth regulation or suppression of pathogens are required to be identified and amplified and simultaneously field demonstrations are required to be carried out at larger scales with modifications in package practices if needed to introduce the microbial inoculants or their product in crop production.

The *Rhizobium*-legume symbiosis has tremendous potential to adapt to severe edaphic conditions and work as phytostabilizers in addition to biofertilizers. Several symbiotic systems of rhizobia and its host legumes, which are tolerant to different stresses, are prevalent in our environment today. This symbiosis represents the host source or the 'ideal' system to be applied in rehabilitation of degraded nutrient command great significance. Research in rhizobial genes, nodulins of plants, the biomolecules involved in nodule formation and the morphogenesis prompted to attempt nodule induction in non-legume crop plants like oilseeds and cereals. However, success has not been achieved in *nif* gene transfer to crop plants and, therefore, fertilizers nitrogen and phosphorus will continue to be vital for obtaining sustained crop yields. Rhizosphere management has been improved a lot in recent years. In the past 10 years, significant achievement has been made in nodulating roots of wheat and oilseed rape seedlings. Under defined laboratory conditions rhizobial infectivity was achieved, but effectivity in terms of nitrogenase expression in host is yet to be achieved. Efficient nitrogen fixation using biofertilizer is already a reality. Future rhizosphere research and biotechnological approaches are needed for sustainable development. Investment both financially at forming level and educationally at research level will further enhance the efficient use of nutrients at soil and improve crop yields.

## References

Agostini, B.B. & Khan, D. (1986) In: *World Crops: Cool Season Food Legumes*. (Ed. Summerfield, R.J.) Kluwer Academic, Boston. pp. 461-482.

Alami. Y., Achouak, W., Marol, C. & T., Heulin (2000). *Appl. Environ. Microbiol.* **66**: 3393-3398.

Andrade, G., Delelj, F. A. A. M. & Lynch, J. M. (1998) *Lett Appl Microbiol* **26**: 311-316.

Aneja, P. & Trevor, T. C. (1999) *J. Bacteriol.* **18 (3)**: 849-857.

Antoun, H., Bordeleau, L. M., Chabot, R., Groussand, N. & Lalande R. (1998*) Plant soil* **204**: 57-62.

Antoun, H., Bordeleau, L. M. & Gagnon, C. (1978) *J. Plant Sci.* **58**: 75-81.

Atkins, C. A., (1986) *Outlook on Agriculture* **15 (3)**: 128-134.

Azcon, R. & Elatrash, F. (1997) *Biol. Fertile. Soils.* **24**: 81-86.

Bantilan, M. C. S. & Johansen, C. (1995) *Plant soil.* **174 (1-2)**: 279.

Bassler, B. L. (2002) *Cell* **109**: 421-424.

Baker, W. D. (1981) *Ann. Revs. Plant. Physiol.* **32**: 407-409.

Bauer, W. D., Bhuvaneshwari, T. V., Mort, A. J. & Turgeon, G. (1979) *Plant Physiol.* **63**: 5-135.

Baya, A. M., Boethling, R. S. & Cormenzana, R. A. (1985) *Soil. Biol. Biochem.* **13**: 527-531.

Beauchamp, C. J., Dion, P., Kloepper, J. W. & Antoun, H. (1991) *Plant Soil* **132**: 273-277.

Becana, M. & Sprent, J. I. (1987) *Physiol. Planta.* **70**: 757-765.

Belimov, A. A., Hontzeas, N. Safronova, V. I., Demchinskaya, S. V. Piluzza, G., Bullitta, S. & Glick, B. R. (2005) *Soil. Biol. Biochem.* **37**: 241-250.

Bergersen, F. J., Peoples, M. B. & Turner, G. L. (1991) *Proc. Res. Soc. London* B **245**: 59-64.

Bhattacharya, R. N. & Basu, P. S. (1992) *Ind. J. Exptl. Biol.* **30**: 632-632.

Bhattacharya, R. N. & Pati, B. R (1999) *J. Microbiol. World* **1 (1)**: 25-31.

Briel, B. T., Ludden, P. W. & Triplett, E. W. (1993) *J. Bacteriol.* **175**: 3693-3702.

Bradley, D. E. (1967) *Bact. Rev.* **31**: 230-314.

Brockwell, J. P., Bottomley, P. J. & Thies J. E. (1995) *Plant soil* **174**: 143-180

Bromfield, E. S. P., Butler, G. & Barran, L. R. (2001) *Can. J. Microbiol.* **47**: 567-573.

Broughton, N. J., Jabbouri, S. & Perret, X. (2000) *J. Bacteriol.* **186**: 5641-5652.

Burd, G. I., Dixon D. G. & Glick B. R. (1998) *Appl. Environ. Microbiol.* **64**: 3663-3668.

Busse, M. D. & Bottomley, P. J. (1989) *Appl. Environ. Microbiol.* **55**: 2431-2436.

Cevallos, M. A., Sergio, E., Leija, A., Mora, Y. & Mora J (1996) *J. Bacteriol.* 1646-1654

Cha, C., Gao, P., Shaw, P. D. & Farrand, S. K. (1998) *Eur. J. Plant Pathol.* **104**: 569-582.

Chabot, R. Antain, H. & Cescas, M. P. (1996a) *Plant Soil* **184**: 311-321.

Chabot, R., Antoun, H., Kloepper, J. W. & Beauchaup, C. J. (1996b) *Appl. Environ. Microbiol.* **62**: 2767-2772.

Chabot, R., Beauchaup, C. J., Kloepper, J. W. & Antoun, H. (1998) *Soil. Biol. Biochem.* **30 (12)**: 1612-1618.

Chakraborty, V. & Chakraborty, B. N. (1988) *Can. J. Bot.* **67**: 1698-1701.

Chao, W. L., (1990) *Lett. Appl. Microbiol.* **10**: 213-219.

Chen, H. E., Gartner E. & Rolfe, B. G. (1993) *Appl. Environ. Microbiol.* **57**: 2005-2011.

Chien, C. T., Maundu, J., Covaness, L., Dandurand, L. M., Orser. & C. S. (1992) *FEMS Microbiol. Lett.* **90**: 135-140.

Christopher M. Waters & Bonnie L. Bassler (2005) *Ann. Rev. Cell Developmental Biol.* **21**: 319-346.

Chung, King-Thom & Ferris, Deam Hunter (1996) In: *Martinus Willem Beijerinck (1851-1931): Pioneer of general microbiology*. AMS News **62**: 539-543

Cooper, J. E., Wood, M. & Bjoursom, A. J. (1985) *Soil. Biol. Biochem.* **17**: 487-492.

Cunninghan, S. D. & Munns, D. N. (1984) *Soil. Sci. Soc. Am. J.* **48**: 1273-1276.

Dadarwal, K. R., (1997) In: *Biotechnological Approaches in soil Microorganisms for sustainable crop production*, Scientific Publishers Jodhpur, pp. 1-31.

Dadarwal, K. R., Prabha, S. & Tauro, P. (1978) *Proc. National Symp. Nitrogen Assimilation and Crop productivity*. IARI, New Delhi, India. pp. 235-239.

Daniels, R., Devos. D. F., Desair, J., Raedschelders, G., Luyten, E., Rosemeyer, V., Verreth, C., Schoeters, E., Vanderleyden, J. & Michiels, J. (2002) *J. Biol. Chem.* **277**: 462-468.

Danso, S. K. A., Zapata, F. & Awonaika K. O. (1990) *Appl. Environ. Microbiol.* **56**: 1793-1798.

Dashti, N., Zhang, F., Hynes, R. & Smith, D. L (1998) *Plant Soil* **200**: 205-213.

Dashti, N., Zhang, F., Hynes, R. & Smith, D. L. (1997) *Plant Soil* **188**: 33-41.

Davidson, I. A. & Robson, M. J. (1986) *Ann. Bot.* **57**: 331-338.

Day, D. A., Poole, P. S., Tyerman, S. D. & Rosendahl, L. (2001) *Cell. Mol. Life Sci.* **58**: 61-71.

Deaker, R, Roughley, R J. & Kennedy, I R., (2004) *Soil. Biol. Biochem.* **36 (8)**: 1275-1288.

Delgado, M. J., Ligero, F. & Lluch, C. (1994) *Soil. Biol. Biochem.* **26**: 371-376.

Deshwal, V. K., Dubey, R. C. & Mashewari, D. K. (2003) *Curr. Sci.* **84**: 443-448.

Deshwal, V. K., Pandey, P., Kang, S.C. & Maheswari, D. K. (2003) *Ind. J. Exptl. Biol.* **41**:1160-1164.

Dhiman, S. & Saraf, M. (2003) *Asian J. Microbiol. Biotech. Env. Sci.* **5(4)**: 495-498.

Dhiman, S. K., Dhandhukia, P. & Saraf M. S. (2006) *J. Phytopath. Res.* 63-69.

Dileepsumar, B. S., Breggren, I. & Martensson A. M. (2001) *Plant Soil* **229**: 25-34.

Dilworth, M. J., Carson, K. C., Giles, R. G. F., Byrne, L. T. & Glenn, A. R. (1998) *Microbiol* **144**:781-785.

Dilworth, M. J., Rynne, F. G., Cartekki, J. M. A., Marfisi, V. & Glenn, A. R. (1999) *Microbiol.* **145**: 1585-1593.

Doherty, D., Leigh, J. A., Glazebrook, J. & Walker, G. C. (1988) *J. Bacteriol.* **176**: 4249-4256.

Dudley, M. E. Jacobs, T. W. & Long, S. R. (1987) *Planta* **171**: 289-301.

Dullart, J, (1970) *Acta Bot. Neerl.* **19**: 573-618.

Eaglesham, A. R. J. & Ayanaba. (1989) In: *Current developments in Biological Nitrogen Fixation.* Edward Arnold Publishers, UK, 1-35.

El Sheikh, E. A. E. & Wood, M. (1990) *Soil. Biol. Biochem.* **22**: 343-347.

Elkan, G. H. & Bunn, C. R. (1992) In: *Prokaryotes 2nd edition A Handbook on the Biology of Bacteria Ecophysiology Isolation, Identification, Applications,* (Eds. Balows, A., Truper, H. G. Harder, W. & Schleifer, K. H.), Vol 3: pp. 2197-2213.

Enteshmul-Huque, S. & Gaffar, A. (1993) *J. phytopathol.* **138**: 157-163.

Evensen, K. B. & Blevins, P. G. (1981) *Plant Physiol.* **68**: 195-198.

Farewell, A. J., Versely, S., Nerov, Rodriguez, H., Shah, S., Dixon, G. D. & Glick, B. R. (2006) *Plant Soil* **288**: 309-318.

Finan, T. M., Kunkel, B., De vos, G. & Signer, E. R. (1986) *J. Bacteriol.* **167**: 66-72.

Fisher, R. F. & Long, S. R. (1992) *Nature* **357**: 655-659.

Fujihara, S. & Yokehama, T. (1993) *Appl. Environ. Microbiol.* **59**:1104-1109.

Fuqua, W. C., Winans, S. C.& Greenbery, E. P. (1994) *J. Bacteriol.* **176**: 269-275.

Gaworzewska, E. T. & Charlie, M. J. (1982) *J. Gen. Microbiol.* **128**: 1179-1188.

Germida, J. J. and walley, F. L. (1996) *Biol. Fertile. Soils* **23**: 113-120.

Ghittoni, N. E. & Bueno, M. A. (1996) *Symbiosis* **41**: 1021-1030.

Gill, P. R. & Neilands, J. B. (1989) *Mol. Microbiol.* **3**: 1183-1189.

Gill, P. R., Barton, L. L., Scoble, M. D. & Neilands, J. B. (1991) *Plant Soil* **130**: 211-217.

Giller, K. E., Mcgrath, S.P. & Hirsch, P. R. (1989) *Soil. Biol. Biochem.* **21**: 841-848.

Glick, B. R., Patten, C. L., Holguin, G. & Penrose, D. M. (1999) In: *Biochemical and genetic mechanisms used by plant growth promoting Bacteria*: Imperial College press. London.

Gonzalez, J. E. & Martenson, M. M. (2003) *Microbiol. Mol. Bio. Rev.* **67**: 574-592.

Gouffi, K., Pica, N., Pichereauv. & Blanco, C. (1999) *Appl. Environ. Microbiol.* **65**: 1491-1500

Graeme W. Nicol, Gordon Webster, L. Anne Glover & James I. Prosser (2004) *Environ. Microbiol.* **6 (8)**: 861

raham, P. H. (1992) *Can. J. Microbiol.* **38**: 475-484.

Graham, P. H., Viteri, S. E., Mackie, F., Vargas, A.T. & Palacios A (1982) *Crops Res.* **5**: 121-128.

Gross, D. C. & Vidaver, A. K. (1978) *Appl. Environ. Microbiol.* **36(6)**: 936-942.

Guerinot, M. J., Meidl, E. J. & Blessver, O. (1990) *J. Bacteriol.* **172**: 3598-3301.

Guinel, F. C. & Geil, R. D. (2002) *Can. J. Bot.* **80**: 695-720.

Halder, A. K. & Chakraborty, P. K. (1993) *Folia Microbiol.* **38**: 325-331.

Hamdi, Y. (1970) *Soil Biol. Biochem.* **3**: 121-126.

Harris, D., Pacovsky, R. S. and Paul, E. A. (1985) *New Phytologist* **101**: 427-440.

Hartel,P. & Alexander. (1984) *Can. J. Microbiol.* **30**:820-823.

Hartmann, A. (2005) In: *Rhizosphere 2004: perpectives and challenges—a tribute to Lorenz Hiltner* (Eds. Hartmann A., Schmid, M., Wenzel, W.W., Hinsinger, P.) Neuherberg, Germany, GSF-Forschungszentrum, Bericht 05/2005 (ISSN 0721-1694), pp. 1–4.

Heidstra, R., Nilson, G., martinez-Abarca, F., Kammer, V. A. & Bisseling, T. (1997) *Mol. Plant. Microbe. Interaction* **10**: 215-220.

Heinonen, T., Oros, G. & Keckas, M. (1982) *Acta Agric. Sham. Sci. Bull.* **28B**: 423-440.

Henson, J. E. & Wheeler, C. T. (1976) *New Phytol.* **76**: 433-439.

Her, G. R., Glazebrook, J., Walker, G. C. & Reinheld, V. N. (1990) *Carbo. res.* **198**: 305-312.

Hernandez, Armenta, R., Wein, H. C. & Eaglesham, A.R.J. (1989) *Crop Sci.* **29**: 1260-1265.

Hirsch, A. M. Fang, V. Asad, S. & Kapulnik Y. (1997) *Plant Soil* **194**: 171-184.

Hodgson, A. L. M., Roberts, W. P. & Waid, J. S. (1985) *Soil Biol. Biochem.* **4(17)**: 473-478.

Howarth, R. W., Boyer, E. W., Pabich, W. J. & Galloway N. (2002) *Ambio.* **31 (2)**: 88-96.

Hungria, H. & Franco, A. A. (1993) *Plant Soil* **149**:95-102.

Hunt, P. J., Nollum, A. G. & Matheny, T. A. (1981) *Agric. J.* **73**: 501-505.

Ibekuwe, A. M., Angle, J. S., Chang, R. L. & Vonberkum, P. (1997) *Agric. Ecosyst. Environ.* **61**: 103-111.

Iruthayana, E. E., Gunasekaran, S. & Vlassak, K. (1983) *Scientia. Hortie.* **20**: 231-240.

Jenkins, M. B., Virginia, R.A. & Jarrel, W. M. (1987) *Appl. Environ. Microbiol.* **33**: 36-40

Jhadhav, R. S. & Desai, A. (1994) *Fems. Microbiol. Lett.* **115**: 185-190.

Jhadhav, R. S. & Desai, A. (1997) *World J. Microbiol. Biotech.* **13**: 347-348.

Jian, W., Sushang, Y. & Jilun, L. (1993) *Acta Microbiol.* **33**: 260-267.

Johnson. H. W., Means, V. M., & Clark, F. E. (1959) *Nature* **182**: 308-309.

Kanta, K., Narula, N., Rabbi, S. & Gupta K. G. (1987) *Microbiology* **142**: 167-173.

Kapranov, P., Routt, S. M., Bankaitis, V. A., Bruijin, F. J. & Szczyglowski, K. (2001) *Plant Cell* **13**: 1360-1362.

Karr, D. B., Water, J. K., Suzaki, F. & Emerich, D. W. (1984) *Plant Physiol.* **75**: 1158-1162.

Kefford, N. P., Brockwell, J. & Zwer, J. A. (1960) *Aus. J. Biol. Sci.* **13**: 456-467.

Kijne, J. W. (1992) In: *Biological Nitrogen fixation* (Eds. Stacey G, Burris R H. & Evans H. J.) Chapman and Hall , NY, pp. 349-398.

Kinkel, B. K., Angle, J. S. & Keyser, H. H. (1987) *Appl. Environ. Microbiol.* **53**: 315-319.

Klein, D. A., Salwzmalel, J. L. & Dazzo, F. B. (1990) In: *Biotechnology of Plant-Microbe Interactions* (Eds Nakas, J. P. & Hagedorn, C.), McGraw Hill, New York, pp. 189-225.

Klein, S., Walker, G. C. & Signer, E. R. (1988) *J. Bacteriol.* **170**:1003-1006.

Kloepper, J. W., Leong, J., Teintze, M. & Schroth, M.N. (1990) *Nature* **286**: 885-886.

Knowler, C. J. (1996) *Bacteriol. Rev.* **40**: 652-657.

Kucey, R. M., Chaimonakupt, N. P., Arayang, Kool, P., Snitwonge, C., Siripaibool, P., Wadisirisuk, Bookerd, N. (1988) *Plant Soil* **108**: 87-92.

Kumar, B., Kumar, S. M., Annapurna, K. & Maheshwari, D. K. (2006) *Curr. Sci.* **91 (11)**: 1524-1529.

Kundu, B. S. & Gaur, A. C. (1981) *Haryana Agric. Univ. J. Res.* **11**: 559-566.

Kuzyakov, Y., P. Leinweber, D. Sapronov, K.U. Eckhardt. (2003) *J. Plant Nut. Soil. Sci.* **166:** 719-723.

La Favre. A. K. & Eaglesham, A.R.J. (1986) *Can. J. Microbiol.* **32**:22-27.

Laguerre, G., Louvirre, F., Allard, M. R., Amarger, N. C. (2003) *Appl. Environ. Microbiol.* **69 (4)**: 2276-2283.

Lajudie, de. P., Fubelle, L. E., Willams, A., Torch, U., Coopman, R., Collins, M. D., Kersters, K., Dreyfuss, B. & Gillis, M. (1998) *Int. J. Syst. Bact.* **48**: 1277-1290.

Lal, B. & Khanna, S. (1995) *World J. Microbiol. Biotech.* **10**: 637-639.

Layzell, D. B., Hurnst, S. & Palmer, G. R. (1990) *Plant Physiol.* **92**: 1101-1107.

Lee, K. H. & LaRue, T. A. (1992) *Plant Physiol.* **100**: 1759-1763.

Lhuissier, F. G. P., Ruijter, De N. C. A., Sieberer, B. J., Esseling, J. J. & Emons, A. M. C. (2001) *Ann. Bot.* **87**: 289-302.

Ljunggren, H. & Martenson, A. (1980) In: *Swedish weed conference*, pp. 99-108.

Long, S, Reed, J. W., Himavan, J. & Walker, G. C. (1988) *J. Bacteriol.* **170(9)**: 4239-4248.

Madariage, G. M. & Angle, J. S. (1992) *J. Environ. Qual.* **21**: 276-280.

Madsen, E. B., Madsen, L. H., Radutoiu, S., Olbryt, M., Rakwalska, M. Szczyglowski, K., Sato, S., Kareto, T., Tabato, S., Sandal, N., & Stongaard, J. (2003) *Mol. Plant Microbe Interact.* **16**: 743-751.

Maheshwari, D. K. & Gupta, M. (1991) *Biochem. Physiol. Pflanzen.* **187**: 316-322.

Maheshwari, D. K. & Gupta, M. (1991) *Zentrabl. Mikrobiol.* **146**:407-412.

Maheshwari, D. K. & Saraf, M. (1994) *J. Gen. Appl. Microbiol.* **40**: 569-574.

Maheshwari, D. K., Gupta, M., Sawhney, R. & Khandelwal, A. (1993) *Zentrabl. Mikrobiol.* **148**: 588-592.

Malaczak, N., Pearse, M. & Litchfield, R. T. (1984) *Trans. Br. Mycol. Soc.* **82**:491-495.

Marie, C., Deakin, W. J., Viprey, V., Kopcinska, J., Golinowski, W. Krishnan, H. B., Parret, X. & Broughton, W. J. (2003) *Mol. Plant Microbe Interact.* **16**: 743-751.

Martensson, A. M., Brutti, L. & Ljungeren, H. (1989) *Plant Soil* **117**: 219-225.

Martinez-Romero E. (2003) *Plant soil* **252**: 11-23.

*Melissa B. Miller & Bonnie L. Bassler* (2001) *Annu. Rev. Micro.* **55**.1.165

Mellor, R. B. & Collinge, D. B. ( 1995) *J. Exp. Bot.* **46**: 1-18.

Meyer, J. R. & Lindermann, R. G. (1986) *Soil Biol. Biochem.* **18**: 185-190.

Michiels, J., Verreth, C. & Vanderleydan, J. (1994) *Appl. Environ. Microbiol.* **60**: 1206-1212.

Miettinen. P. & Chegogen, P. E. (1996) *Food Sci. Finl.* **5**: 203-207.

Minamisawa, K., Eukai, K. & Sami, T. (1990) *J. Bacteriol.* **172(8)**: 4505-4509.

Miklashevicks E., Rohrig, H, Schell, J. & Schimdt, J. (2001) *Crit. Rev. Sci.* **20**: 373-394.

Modi, M., Shah, K. S. & Modi, V. V. (1985) *Arch. Microbiol.* **141**:156-158.

Mozafar, A. (1994) In: *Plant vitamins Agronomical, Physiological and Nutritional Aspects*, CRL Press, Boca Raton, Florida, USA.

Munoz, J. A., Palomares, A. J. & Ratet P. (1996) *J. Microbiol. Biotech.* **12**: 189-202.

Nadler, K.D., A.W. Johnston, J.W. Chen & John, T.R. (1990) *J. Bacteriol.* **172**: 670-677.

Nainawatee, H. S., Jain, V. & Singh, R. (1997) In: *Biotechnological Approaches in Soil Microorganisms for sustainable crop production,* Scientific Publisher, Jodhpur, India.

Narayanan, Rajendran & Sivamani, E. (1992) Protoplast fusion between BT and PS. Publisher: Oxford-IBH. Narayanan 147-164.

Narayanan Rajendran & Venkatesan, S.. (1993) In: *Chemical Ecology of Phytophagous Insects*. Pub: Oxford-IBH. N. 301-311.

Narayanan, Rajendran, & Dieter, Jahn (1998) *Lett. Applied. Microbiol.* **27**:1-6.

Narayanan Rajendran, *et al.* (1994) *FEMS Microbiol.Lett.***122**: 103-8.

Nautiyal, C. S. (1997) *FEMS Microbiol. Ecol.* **23**: 145-151.

Neilands, J. B. (1981) *Annual Rev. Biochem.* **5**: 715-731.

Nour, E. M., Marcel, J., Beck, D., Efforse, A. & Fernandes, M. P. (1994) *Can. J. Microbiol.* **40**: 345-353.

Nukui, N., Ezura, H., Yuhashi, K., Yasuta, T. & Minamisawa, (2000) *Plant Cell Physiol.* **41**: 893-897.

O'Hara, G. W., Boonkerd, N. & Dilworth, M. J. (1998) *Plant Soil* **108**: 93-110.

Obbard, J. P., Sauerbecl, D. R. & Jones, K. C. (1994) In: *Ecotoxicology of soil organisms*. Lewis Pub. UK. pp. 127-161.

Ogasawara, M, Suzuki, T., Muthch, I., Annapurna, K., Arora, N., Nishimura, Y. & Maheshwari, D. K. (2003) *Symbiosis* **34**: 53-68.

Oke, V. & Long, S.R. (1999) *Curr. Opin. Microbiol.* **2**: 641-646.

Omens, L. D. (1973) *Weed Sci.* **21 (1)**: 63-67.

Omens, L. D., Thompson, J. F., Pitcher, R. G. & Williams, T. (1972) *J. Chem. Soc. Chem. Commu.* 714.

Pandya, S., Iyer, P., Gaitonde, V., Parekh, T. & Desai, A. (1999) *Curr. Microbiol.* ***38***: 205-209.

Parmar, N. & Dadarwal, K. R. (1999) *J. Appl. Microbiol.* **86**: 36-44.

Patel, H. N., Chakraborty, R. N. & Desai, S. B. (1988) *FEMS Microbiol. Lett.* **56**: 131-136.

Paul, S. & Varma, O. P. (1999) *Ind. J. Microbiol.* **39**: 249-251.

Penmotsa, R. V. & Cook, D. R. (1997) *Science* **275**: 527-530.

Peters, N. K. & Long, S. R. (1988) *Plant Physiol.* **88**: 99-112.

Pueppkke, S. G. & Van, Elten. H. D. (1975) *J. Chem. Soc. Perkins. Trans.* **1**:946-948.

Phillips, D. A., Joseph, W. J., Jones, A. D. & Tueber, L. R. (1995) *Crop Sci* **35**: 805-808.

Piha, M. I. & Munns, D. N. (1987) *Plant Soil* **98**: 183-194.

Pingret, J. L., Journet, E. P. & Barker, D. G. (1998) *The Plant Cell* **s**: 659-671.

Plazinski, J. & Rolff, B. G. (1985) *Can. J. Microbiol.* **31**: 1026-1030.

Price, N.P.J. (1999) *Carbohy. Res.* **317**: 1-9.

Rai, R (1983) *J. agric. Soc. Comp.* **100**: 75-80.

Ramalingam, G. & Narayanan, Rajendran (2001) *J. Biosciences* **56**: 245-248.

Raymond, K. N., Muller, G. & Matzanke B. R. (1984) *Top. Curr. Chem.* **123**: 49-102.

Reed, M. L. E. & Glick, B. R. (2005) *Can. J. Microbiol.* **51**: 1061-1069.

Reeve, W. G., Dilworth, M. J., Tiwari, R. P. & Glenn, A. R. (1997) *Microbiol.* **143**: 1951-1958.

Reigh, G. & Connell, M. (1993) *J. Bacteriol.* **175**: 94-102.

Rice, W. A., Olsen, P. E. & Leggett, M. E. (1995) *Soil Biol. Biochem.* **27**: 703-705

Richardson, A.E., George, T.S., Hens, M., Simpson, R.J., (2005) In: *Organic Phosphorus in the Environment* (Eds. Turner, B.L., Frossard, E. & Baldwin, D.S.), CABI, Wallingford, UK. pp 165-184.

Robertsen, B. K., Aman, P., Darvill, A. G., McNeil, M. & Albersheim (1981) *Plant Physiol.* **67**: 389-400.

Robleto, E. A., Kmiecik, K., Optinger, E., Nienhuis, J. & Triplett, E. W. (1998) *Appl. Environ. Microbiol.* **64**: 2630-2633.

Roslycky, E. B. (1967) *Can. J. Microbiol.* **13**: 431-432.

Roughley, J. (1970) *Ann. Bot.* **34**: 631-646.

Rovira, A. D. & Harris, J. R. (1961) *Plant Soil* **14**: 199-213.

Sahgal, M. & Johri, B. N. (2003) *Curr. Sci.* **133**: 87-94

Salminen, S. & Streeter, J. G, (1992) *Plant Physiol.* **100**: 597-604.

Sangakarra, U. R., Hartwig, N. A. & Noesberg, J. (1996) *Plant Soil* **184**: 123-130.

Sanginga, N., Wirkom, L. E., Okogun, A. O., Akobunduo, Carsky, R. J. & Tian, G. (1996) *Biol. Fertil. Soils* **23**: 441-448.

Saraf, M. & Dhandhukia, P. (2005) *J. Microbiol. World* **7 (2)**: 250-257.

Saraf, M. & Sood, N. (2002) *Ind. J. Bot. Soc.* **81**: 313-316.

Saraf, M., (2001) In: *Innovative Approaches in Microbiology*. (Eds. Maheshwari D. K. & Dubey, R.C.), BSM Pal Publishers, Dehradun, India, pp. 53-81.

Saraf, M., Khandelwal, A., Sawhney, R. & Maheshwari, D. K. (1994) *Microbiol. Res.* **149**: 401-406.

Schultze, M & Kondorosoi A. (1998) *Ann. Rev. Genet.* **32**: 33-37.

Schulze, M., Kondorsoi, E, Patet, P., Buire, M. & Kondorosoi, A. (1994) *Intl. Review Cytol.* **156**: 1-75.

Schwinghamer, E. A. & Belkengren, R. P. (1968) *Arch. Mikrobiol.* **64**: 130-145.

Schwinghamer, E. A. (1971) *Soil Biol. Biochem.* **3**: 355-363.

Schwinghamer, E. A., Pankhurst, C. E. & Whitefield, P. R. (1973) *Can. J. Microbiol.* **19**: 359-368.

Sekine, M., Watanabe, K., & Syono, K. (1989) *J. Bacteriol.* **171**: 1718-1724.

Sellstedt, A., Stashl L., Mattsson, M., Johnsson, K. and Hoegberg, P. (1991) *J. Expt. Bot.* **44**: 1749-1755.

Senior, P. J., Beech, G. A., Ritchi, G. A. F. & Dames E. A. (1972) *Biochem. J.* **128**: 1193-1201.

Sharma, A., Manvika, S. & Johri, B. N. (2003*) Curr. Sci.* **85(8)**: 1164-1171.

Shen, B., Martin, A. E., Butt, J. N., Armstrong, F. A., Stout, C. D., Jensen, G. M., Stephens, P. J., LaMar, G. N., Gorst, C. M., & Burgess, B. K. (1993*) J. Biol. Chem.* **268**: 25928-25939

Sierra, S., Rodelas, B., Teledo, M. V., Pozo, C. & Lopez, J G (1999) *J. Appl. Microbiol.* **86**: 851-858.

Silvente S., Camas, A. & Lara, M. (2003) *J. Expt. Bot.* **54**: 749-755.

Simon, R., Priefer, V. & Puhler, A. (1983) *Biotechnol.* **1**: 784-791.

Sindhu, S. S., Suneja, S. & Dadarwal, K. R. (1997) *In: Biotechnological Approaches in Soil Microorganisms for Sustainable Crop Production* (Eds. Dadarwal, K R.), Scientific Publisher, pp. 149-191.

Sindhu, S. S. and Dadarwal, K. R. (2000) *Ind. J. Microbiol.* **40**: 211-246.

Sinha, B.K. & Basu, P. S. (1981) *Biochem. Physiol. Pflanzem.* **176**: 218-227.

Skot, L., Timms, E. & Mytton, L. R. (1994) *Plant Soil* **163**: 141-146.

Smith, M. J., Shoolery, J. N., Schwyn, B., Holden, I. & Neilands, J. B. (1985*) J. Am. Chem. Soc.* **107**: 1739-1745.

Sobral, B. W. S. Honeycutt, R. J., Atherly, A. G. & Noel, K.D. (1991) In: *The Biology and Biochemistry of Nitrogen Fixation* (Eds. Dilworth, M. J. & Glenn, A. R.), Elsevier, Amsterdam, pp. 229-258.

Soederjo, M. & Borthakur, D. (1998) *Soil Biol. Biochem.* **30**:1605-1613.

Sprent, J. I. & Sprent, P. (1990) In: *Pure and Applied Aspects*. Chapman and Hall, London, UK.

Sprent, J. I. & Zahran, H. H. (1988) In. *Nitrogen Fixation by legumes in Mediterrarean agriculture* (Eds. Beck, D. P. & Materan, L.), Martinus/ W Junk, Dordrecht, The Netherlands, pp. 145-151.

Srinivasan, M., Petersen, D. J. & Holl, F. B. (1997) *Can. J. Microbiol.* **43**: 1-8.

Stam, H., Verseneld, H. W., Vries, de W. & Stouhamer, A. H. (1986) *FEMS Microbiol. Lett.* **35**: 215-220.

Steirt, W. R., Joseph, C. M. & Phillips, D. (1996) *Mol. Plant-Microbe Interact.* **9**: 330-338.

Tang, C., Robson, A.D. & Dilworth , M..J. (1992) *J. Plant Nutr.* **15**: 2235-2252.

Egener, T., Martin, D. E., Sarkar, A. & Reinhold-Hurek, B. (2001) *J. Bacteriology* **183 (13)**: 3752-3760.

Tank, N. D. & Saraf, M. S. (2003) *Ind. J. Microbiol.* **43 (1)**: 37-40.

Tao, H., Brewin, N. J. & Noel, K. D. (1992) *J. Bacteriol.* **174**: 2222-2229.

Tate, R. L. (1995) *Soil Microbiology* (Symbiotic nitrogen fixation), John Wiley and Sons, NY, pp. 307-333.

Tilak, K.V.B.R., Ranganayaki N., Pal, K. K., De, R., Saxena, A. K., Nautiyal, S.C. Mittal, S., Tripathi, A. K., & Johri, B. N., (2005) *Curr. Sci.* **89 (1)**: 136-150.

Tiwari, R. P., Reeve, W. B. & Glean, A. R. (1992) *FEMS Microbiol. Lett.* **100**: 107-112

Tomboline, R. & Nuti, M. P. (1989) *FEMS Microbiol. Lett*: 299-304.

Tang, Z. & Sadowsky, M. J. (1994) *Appl. Environ. Microbiol.* **60**: 581-586.

Torsvik V. & Overseas, L. (2002) *Curr. Opin. Microbiol.* **5**:240-245

Trainer, M. A. & Trevor, C. T. (2006) *Appl. Microbiol. Biotech.* (in press).

Trevor, T. C., Singh, R. S. & Finan, T. M. (1990*) J. Gen. Microbiol.* **136**: 2497-2502.

Triplett, E. W., Briel, B. T. & Splitter, G. A. (1994) *Appl. Environ. Microbiol.* **60**: 4163-4166.

Tu, J. C. (1981*) Can. J. Plant Sciences* **61**: 231-239.

Tu, J. C. (1978) *Physiol. Plant Pathol.* **12**: 233-237.

Turgeon, B. G. & Bauer, W. D. (1985) *Planta* **163**: 328-349.

Vassileva, V. G., Milanov, G., Ignatov, G. & Nikolov, B. (1997) *J. Plant Nutrition* **20**: 279-294.

Vessey, J. K., Pawlowsky, K. & Berggman, B. (2005) *Plant Soil* **274**: 51-78

Vincent, J. M. (1974) In: *The Biology of Nitrogen Fixation,*(Eds. Quespel, A.) Amsterdam North Holland Publ. Co., pp. 265- 341.

Waelkens, F., Voets, T., Vlassak, K., Venderleyden, J., & Rhojn, V. P. (1995) *Mol. Plant Microbe Inter.* **8**: 147-154.

Waters, C. M. & Bassler, B. M. (2005) *Annu. Rev. Cell Dev. Biol.* **21**: 319-346.

Wenbo, Ma, Guinel, F. C. & Glick, B. R. (2003b) *Appl. Environ. Microbiol.* **69(8)**: 4369-4402.

Wenbo, Ma, Sebastianova, B. S., Sebestian, J., Burd, G. I., Guinel, F. C. & Glick, B. R. (2003a) *Antonie Van Leeuwenhoek* **83**:285-291.

Wetzel, A. & Werner, D. (1995) *Environ. Toxicol. Water Qual.* **10**: 127-134.

Wheeler, C. T., Henson, J. E. & Melanghlin, M. E. (1979) *Bot. Gaz.* **140**: 552-557.

Whitehead, N. A., Barnard, A. M., Slater, H., Simposn, N. J. & Salmond, G. P. (2001) *FEMS Microbiol. Rev.* **25**: 365-404.

Williams, P. M. & Demallorca, M. S. (1984) *Plant Soil* **90**: 267-283.

Yadav, K. S. & Dadarwal, K. R. (1997) In: *Biotechnological Approaches in Soil Microorganisms for Sustainable Crop Production.* Scientific Publishers, Jodhpur, pp. 279-293.

Yijiang, G., Zhong, Z., Sun, K.,Wang, H. & Zhu, J. (2006) *Plant Soil* **286**: 53-60.

Yuhashi, K I., Ichikama, N., Ezurah, Kao, S. A., & Minamisama (2000) *Appl. Environ. Microbiol.* **66**: 2658-2663.

Zahran, H. H. (1999) *Mol. Biol. Rev.* **63**: 968-989.

Zevenhuizen, L. P. T. M. (1981) *Antonie Van Leeuwenhoek* **47**: 481-497.

Zheng, H., Zhong, Z., Lai, X., Chen, W. X., Li, S. & Zhu, J. (2006) *J. Bacteriol.* **188**: 1943-1949.

20

# Strategy to Improve Ruminant Productivity Through Direct-Fed Microbials

**Ravinder Nagpal, A.K. Puniya[1], Monica Puniya[1], Tejpal Dhewa[2], J.P. Sehgal[1] and Kishan Singh**

*Dairy Microbiology Division, [1]Dairy Cattle Nutrition Division, National Dairy Research Institute, Karnal-132001, Haryana*

*[2]Dolphin (PG) Institute of Bio-Medical and Natural Sciences, Dehradun-248007, Uttarakhand.*

## ABSTRACT

Direct-fed microbials strategically increase the numbers of beneficial bacteria in the rumen, concomitantly stimulating the production of desired fermentation end products. Researchers have reported the positive effects of bacteria and fungi based direct-fed microbials on *in vitro* fibre digestion in cattle and buffaloes including interspecies transfer. Direct-fed microbials containing lactobacilli are quite common but claims for increased growth or decreased mortality are missing in absence of significant data back-up to support these effects. Of the various fungal preparations that have been tested so far, additives with yeasts have shown consistency in improving the milk production. Though, several bacteria other than lactobacilli have also been exploited to improve the performance of ruminants, yet none is commercially available due to different reasons. Consequently, direct-fed microbials may be a practically viable alternative of traditional antibiotic based growth promoters used in animal feeding practices. Hence, the aim of this chapter is to introduce the conception of direct-fed microbials to depict the potential mechanisms of their actions, and to confer the likely applications of living additives in ruminant nutrition and production.

## INTRODUCTION

Ruminant nutritionists and microbiologists have always been interested persistently in manipulating the microbial ecosystem of rumen to improve the production efficiency of domesticated animals. For this, a number of studies evaluated the effects of dietary modulations on feedstuff utilization and performance of ruminants (Burroughs *et al.,* 1960; Theurer *et al.,* 1963; Rust *et al.,* 1965). In recent past, the use of live microbial feed additives to improve the efficiency of ruminant's production has increased in light of demands for natural growth-promoting agents. On these lines, direct-fed microbials (DFM) have been used to improve animal performance either by manipulation of rumen environment or by directly altering the metabolic activities of microbes in the rumen microenvironment. Of late, the use of antibiotics and other growth stimulants in animal feeds has been of growing concern due to the potential risk of residues appearing in meat and milk. Hence, the need for a safe food supply to the consumers has instigated livestock producers to explore the safer alternatives to enhance the performance of livestock. DFM, the so-called animal probiotics, is one such natural preparation that has been incorporated into livestock diets in order to accomplish the objective of improved rumen efficiency. DFM, in particular filamentous fungi, have been found to stimulate the growth of rumen bacteria, in contrast to ionophores and antibiotics that are selectively toxic to bacterial community. Though, the total microflora is increased after DFM administration, yet the cellulolytic and lactic acid utilizing bacteria are selectively stimulated. For years, yeasts have been introduced in animal feeds as live supplements, mainly due to their high protein and vitamin content. However, when compared with the enzyme supplementations, very little research has been focused on evaluating the effects of DFM on rumen fermentation. In addition, scarce or no information exists that documents the effects of DFM on physiology of predominant rumen microflora.

## RUMEN MICROBIAL ECOLOGY

After birth, the digestive tract of newborn calves is naturally colonized by a variety of microbes from different environmental sources (Savage, 1987). Under normal conditions, microflora colonizes the rumen and lower gut, establishing a symbiotic relationship with host. Beneficial microorganisms supply nutrients to the host, aid in digestion of nutrients, and compete with potential pathogenic microbes in microenvironment. When young animals are raised under sterile conditions, their nutritional needs increase i.e., more vitamin K in diet, besides showing abnormal immune responses and ill-developed rumen. Such gnotobiotic animals are more susceptible to infections, probably due to establishment of pathogens in absence of competition from probiotics. This indicates that microbial colonization in the digestive tract is must for the normal development of rumen and hence, well being of livestock.

## LIVE MICROBIAL SUPPLEMENTS

The original concept of microbial additives in animal's diets involved the administration of beneficial microbes to livestock under stress, which was meant to prevent the establishment of pathogens or to reestablish normal gut microflora. To be an effective feed additive, a DFM should meet the following criteria:

- Easy to culture with no specific instrumental requirement.
- Economically feasible technology.
- Exert health benefits to the host.
- Genetically stable; non-pathogenic and non-toxic in nature.
- Viable at the time of administration to get a positive effect.
- Survive and remain active in the gut microenvironment.
- Maintain viability under storage and field conditions for longer durations.
- Should be categorized under "Generally recommended as safe".

The DFM and/ or probiotics have been defined as "a live microbial feed supplement that beneficially affects the host animal by improving its intestinal microbial balance" (Fuller, 1989). As per FDA, DFM is "a source of live (viable) naturally-occurring microorganisms", and this includes bacteria, fungi and yeast (Miles and Bootwalla, 1991). The claims that have been made in favour of DFM are detailed below:

- Better utilization of poorly digestible fibrous feeds.
- Enhanced milk production.
- Stimulation of immune system to improve animal's health.
- Improved resistance to infectious diseases.
- Increased growth rate of herd.

However, to support these claims, results are variable and inconsistent, and there is need to do more research in this direction before finally concluding the beneficial affects of DFM.

## MODES OF ACTION OF DFM

The most common explanation for improved animal health via addition of beneficial microflora is the prevention of colonization of pathogens in lower gut while competing for space and nutrients. The various approaches by which DFM benefits the animal are enlisted as follows (Fuller, 1989):

- By producing antibacterial compounds (lactic/acetic/formic acids, bacteriocins, antibiotics, hydrogen peroxide).
- Exclusion of undesirable microorganisms by competing for colonizing space and/or nutrients.
- Production of growth promoting substances (e.g., amino acids, vitamins).
- Production of fibre degrading enzymes (i.e., cellulases, hemicellulases, xylanases, proteases, etc.).
- Detoxification of undesirable compounds (i.e., amines, endotoxins, etc.)
- By stimulating host immune response.
- Production of nutrients (e.g., amino acids, vitamins) or other growth factors stimulatory to the host animal.

## DFM FOR RUMINANTS

During infancy of ruminants, when milk is the sole diet and the rumen is undeveloped, the food tends to bypass the rumen. The main purpose of microbial additives i.e., probiotics during neonatal life is the prevention of diarrhoea by positively modifying the flora of the small intestine in calves and lambs. Once the fermentation has fully established in the adult animal, the considerations for enhancing fibre breakdown becomes important. Research on the effects of DFM in young ruminants has mainly involved addition of *Lactobacillus* and *Streptococcus* as intestinal inoculants to suppress diarrhoea and to improve the growth rate. Efforts have identified specific microbial strains that stimulate ruminal bacteria more than others, and these have been selected for use to enhance production of meat or milk.

Most of the research work on DFM with lactating dairy cows has involved dietary supplementation with either *Aspergillus oryzae* or *Saccharomyces cerevisiae. A. oryzae* mainly increases cellulolytic bacteria, changes volatile fatty acids fermentation favorably and stabilizes rumen pH. *S. cerevisiae* reduces rumen ammonia and increases growth of rumen bacteria. Recently, experiments with feeding of anaerobic fungi have also been shown to improve the fibre degradation (Paul *et al.,* 2004; Dey *et al.,* 2004; Lee *et al.,* 2004; Tripathy, 2005; Thareja *et al.,* 2006). There were significant increases in dry matter intake, body weight, growth rate, milk production, and milk fat/ protein percentage in some trials. The cause of the variable response may be related to the type of DFM fed, and more research is needed to understand the exact mode of action of these microbials. There appears to be some relationship between the effectiveness of a DFM in eliciting a response and nutritional demands of an animal. The response to yeast culture is greater when supplemented to cows in early-lactation as opposed to late-lactation.

## BACTERIA AS DFM

Dairy producers have been inoculating indisposed ruminants with rumen fluid from a healthy animal to normalize rumen function and increase dry matter intakes. The efficacy of bacterial DFM has been extensively studied in the neonatal dairy calf. Bacteria-based DFM i.e., *Lactobacillus*, *Enterococcus*, *Streptococcus*, and *Bifidobacterium*, have been studied in young calves, (Newman and Jacques, 1995) and the importance of these when fed to young calves has been well established in maintaining normal rumen microflora. There is a number of bacteria-based DFM that are used in ruminant diets with targeted applications. Most DFM are probably beneficial because they have effects in the lower gut rather than in the rumen. For example, *L. acidophilus* produces lactic acid that lowers the pH in small intestine so as to inhibit the growth of pathogens. In contrast, a lack of beneficial effects of feeding DFM to calves has also been reported (Abu-Taroush *et al.,* 1996; Cruywagen *et al.,* 1996). High producing cows in early lactation would be the most appropriate host for products like DFM because these are in negative energy balance and maintained on diets that contain highly fermentable carbohydrates, sometimes leading to acidosis. Jaquette *et al.,* (1988) and Ware *et al.,* (1988) reported increased milk production from cows when fed *L. acidophilus* @ $1 \times 10^9$ cfu animal$^{-1}$ day$^{-1}$. Jeong *et al.,* (1998) fed *Lactobacillus* sp. and *Streptococcus* sp. to lactating cows and reported an improvement in milk production over control. Supplementation of *Lactobacilli* may be of special significance in dry periods of lactation when intake is depressed. Savoini *et al.,* (2000) reported that cows fed *Lactobacilli* during transition period produced more milk, and had lower non-esterified fatty acids and higher glucose in blood than controls.

Incidentally, where there has been no advantage of feeding bacterial DFM (Morrill *et al.,* 1977; Jenny *et al.,* 1991), calves generally experienced no health problems. In general, upon feeding DFM, increased milk yield has been a consistent response, whereas changes in milk composition were variable. Other studies claim that DFM might be efficacious for increasing milk production by lactating dairy cows (Gomez-Basauri *et al.,* 2001). Ghorbani *et al.* (2002) reported that steers receiving both lactate-utilizing *Propionibacterium* and *Enterococcus* had higher concentrations of ruminal acetate indicating a reduced risk of metabolic acidosis. The lactate-producers in rumen are thought to help the ruminal microflora adapting to the presence of lactic acid, whereas lactate-utilizers prevent accumulation of lactate (Nisbet and Martin, 1994; Kung and Hession, 1995). Nocek *et al.,* (2000) also reported decreased risk of acidosis for dairy cows receiving a combination of yeast and lactate-producers. There are potential DFM for ruminants but have not yet been commercialized for different reasons. For example, *Megasphaera elsdenii* is the major lactate-utilizer in rumen of cattle adapted to high grain diets. However, when cattle are abruptly shifted from a high-forage to high-concentrate diet, the numbers of these organisms are not sufficient to prevent lactic acidosis (Kung and Hession, 1995). Addition of a different strain of *Megasphaera elsdenii* has prevented lactic acidosis in steers (Robinson *et al.,* 1992). This organism could be used for high producing dairy cows in reducing sub-acute lactic acidosis but timing of administration and effective dose must be standardized first.

Propionibacteria, naturally found in the rumen of animals fed forage and concentrate diets, have the ability to convert lactic acid to acetic and propionic acid. Propionibacteria may be beneficial if inoculated into the rumen (Kung *et al.,* 1991) because higher ruminal propionate would improve the energy status of animal. Swinney-Flyod *et al.* (1999) and Huck *et al.* (1999) reported that feedlot cattle fed a diet containing Propionibacteria and *L. acidophilus* had better feed efficiencies. Kim *et al.* (2000) reported that *Propionibacterium* decreased propionic acid at the expense of acetic acid when fed to steers. *Propionibacterium* used in a commercial preparation also containing lactobacilli has improved weight gain in calves (Cerna *et al.,* 1991). Although propionibacteria can metabolize lactic acid, but they are slow growers and cannot prevent an acute lactic acidosis. As propionibacteria can also metabolize nitrates, it has been claimed that they reduce nitrate toxicity. DFM fed to ruminant livestock have been shown to decrease morbidity in newly weaned calves. It appears that adhesion, colonization, inhibitory action, and stimulation of immune function are all important for DFM to improve health status of animals. Depending on the species, DFM can increase ruminal propionate and decrease ruminal pH below sub-acute level, indicating the potential for efficient energy utilization and a reduction in acidosis.

## FUNGI AS DFM

The fungal feed additives and supplements, commonly categorized under animal probiotics, such as yeasts or fungi or their extracts, have been shown to affect the rumen fermentation pattern. These yeasts and fungi have been accepted as harmless commodity for human health (Wallace and

Newbold, 1993), as they are used since ages in food fermentations, and have also been shown to improve the health and productivity of animals. In comparison to antibiotics, fungal additives may not develop resistance in microbes or no accumulation in the ruminant's body will take place. *S. cerevisiae* and *A. oryzae* are the most prominent fungal probiotics. Aside from this, anaerobic fungi are also being tested as feed additives as they improve fibre degradation in rumen.

## Aerobic Fungi Based DFM

In contrast to most bacteria, fungal based-DFM appear to be more beneficial via favourable changes in ruminal fermentation, however, there is no definite evidence that fungal extracts affects digestion/metabolism in the lower gut. The total ruminal anaerobes (Dawson *et al.,* 1990; Newbold *et al.,* 1991) and cellulolytic bacteria (Harrison *et al.,* 1988) have been found to increase with fungal extracts. The possible mechanism for improvement in ruminal fermentation is that fungal DFM may prevent accumulation of excess lactic acid when ruminants fed high concentrate diets. Extracts of *A. oryzae* stimulated the uptake of lactic acid by rumen lactate-utilizer *Selenomonas ruminantium* (Nisbet and Martin, 1990) and *Megasphaera elsdenii* (Waldrip and Martin, 1993) by providing a source of malic acid. Similarly, Chaucheyras *et al.,* (1995a) reported that *S. cerevisiae* was able to prevent the accumulation of lactic acid by competing with *Streptococcus bovis* for glucose and by stimulating the uptake of lactic acid by *M. elsdenii* possibly by supplying growth factors i.e., amino acids and vitamins. However, the effects on pH are understated as yeast was unable to prevent acute lactic acidosis when fermentations were challenged with a diet rich in carbohydrates (Aslan *et al.,* 1995; Dawson and Hopkins, 1991). Higher ruminal pH may be the reason for increased numbers of rumen cellulolytic bacteria and hence, improvements in fiber digestion indirectly via fungal cultures addition (Arambel *et al.,* 1987).

Yeasts also stimulate acetogenic bacteria in presence of methanogens (Chaucheryas *et al.,* 1995b) that might result in a more efficient fermentation. *Aspergillus* extracts (Chang *et al.,* 1999) and yeast cultures (Chaucheryas *et al.,* 1995c) have also been shown to stimulate rumen fungi that may further improve fiber digestion. The other reason for improved ruminal fermentation may be due to the fact that yeasts scavenge excess oxygen from the rumen (Newbold *et al.,* 1996), thereby creating a more optimal environment for bacteria. Feeding of *S. cerevisiae* also increases rumen protozoa in steers fed straw-based diets (Plata *et al.,* 1994). Newbold *et al.* (1995) reported that the stimulation of rumen bacteria is different with different strains of *S. cerevisiae.*

Another hypothesis suggests that *Aspergillus* extracts may improve fibre digestion as these contain esterases (Varel *et al.,* 1993). Beharka *et al.* (1991) reported that young calves fed an *Aspergillus* extract weaned faster than controls and supplementation also increased numbers of rumen bacteria and volatile fatty acids. Williams *et al.* (1991), Wohlt *et al.* (1991), Kung *et al.* (1996) and Dann *et al.* (2000) reported that feeding of *S. cerevisiae* has increased dry matter intake and milk production but Arambel and Kent (1990), Schwartz *et al.* (1994) and Soder and Holden, (1999) reported differently. Similarly, feeding *A. oryzae* has improved milk production in some (Gomez-Alarcon *et al.,* 1991; Kellems *et al.,* 1990) but not in other studies (Bertrand and Grimes, 1997; Sievert and Shaver, 1993). No one reason can explain the inconsistent effects of fungal cultures on animal performance as they have multiple modes of action in the rumen ecology.

## Anaerobic Fungi Based DFM

Various researches of animal nutrition have suggested a positive relationship between ruminal anaerobic fungi and voluntary dry matter intake of low digestible herbage diets in ruminants (Akin *et al.*, 1983; Weston *et al.*, 1988; Gordon and Phillips, 1993). Few studies indicate that the intake of forage by early weaned calves was 35% higher in calves dosed with *Neocallimastix* sp. (Theodorou *et al.*, 1990) and the dosing of fungus-free sheep with *Neocallimastix* sp. resulted in a 40% increased intake of a low quality straw-based diet (Gordon and Phillips, 1993). These ruminal fungi also supply the proteins to the host through production of proteolytic enzymes and a proportion of the microbial protein synthesized in the rumen passes to abomasum and intestines for digestion and absorption (Gordon and Phillips, 1998).

The role of rumen fungi in the degradation of plant fibre has been examined extensively (Samanta *et al.*, 2001; Paul *et al.*, 2004; Dey *et al.*, 2004; Lee *et al.*, 2004; Thareja *et al.*, 2006). The rhizoids or bulbous holdfasts of vegetative thalli are better at penetrating plant tissue than are bacteria and protozoa, so they gain access to plant material that is not available to other rumen microorganisms (Orpin and Joblin, 1988). Bauchop and Mountfort (1981) suggested that such penetration leads to a faster and complete degradation of forage that enters the rumen. Degradation of lignified walls of plant-cells is an important characteristic of rumen fungi (Mountfort *et al.*, 1982; Akin and Benner, 1988). Zoospores of many species appear to colonize the lignified tissues preferentially and establish colonies localized on sclerenchyma and xylem cells. Species of *Piromyces, Neocallimastix, Orpinomyces* and *Anaeromyces* degrade fibre to a substantial degree. *Caecomyces* species degrade fibre less than the other genera (Gordon, 1990), perhaps because they lack an extensive rhizoidal system. In a study of Bermuda grass (*Cynodon dactylon*) stems and leaves, *Neocallimastix* species and *Orpinomyces* species were found to degrade plant material most effectively by weakening the textural strength of the residue (Akin *et al.*, 1990). The greater ability of rumen fungi, compared to rumen bacteria, to weaken forage fibre may be important in enhancing forage utilization by the host animal (Borneman and Akin, 1990).

Ito *et al.* (1994) studied sheep rumen fungi for digestibility of rice straw and found that there was a significant increase. Manikumar *et al.* (2002), Dey *et al.* (2004) and Thareja *et al.* (2006) indicated increased *in vitro* dry matter digestibility, and decreased cell wall contents of straws with the use of different anaerobic fungi viz, *Orpinomyces, Piromyces* and *Anaeromyces* in comparison to control. Gordon and Phillips (1998) reported a 7-12% increase in voluntary intake of straw based diet, when the sheep were dosed with monocentric fungi isolated from other herbivores. In contrast, no significant effect on feed intake was observed when growing crossbred calves were dosed with polycentric *Orpinomyces* sp. Yet, the growth rate and nutrient digestion improved in fungus administered group in growing cow calves (Dey *et al.*, 2004). Lee *et al.* (2000) studied the effect of administration of anaerobic fungi *Orpinomyces* from Korean native goat, or their enzymes on ruminal fermentation, microbial population, enzyme activities and nutrient digestion in sheep rumen.

## Genetically Modified DFM

The aim of genetic modification of rumen microorganisms is to increase the degradation of fibre components, and one of the possible strategies is the recombinant organisms expressing specific polysaccharidases at a high level. As cellulase and xylanase supplementation of diet increases the performance of ruminants (Yang *et al.,* 1999), it will generate interest in infusing genetically modified microorganisms in rumen so as to avoid the daily feeding with enzymes. The manipulation of predominant rumen bacteria such as *Prevotella ruminicola*, *Butyrivibrio fibrisolvens* or *Streptococcus bovis,* aims at the expression of heterologous polysaccharidase genes to increase fibrolytic activities in animals fed with concentrate-rich diets. Such diets decrease rumen pH, consequently inhibiting the native cellulolytic bacteria that are very sensitive to low pH (Russell and Wilson, 1988). Expression of gene from *Ruminococcus flavefaciens* in *S. bovis* (Whitehead and Flint, 1995), expression of an xylanase gene from *P. ruminicola* in another strain of the same species (Daniel *et al.,* 1995), and expression in *B. fibrisolvens* of a xylanase gene originating from the rumen fungus *Neocallimastix patriciarum* (Xue *et al.,* 1997) or from *Eubacterium ruminantium* (Kobayashi *et al.,* 1998) have been attempted in recent past.

Another area of research that has been identified is detoxification, as many plants produce toxic compounds that reduce productivity in ruminants. These include oxalic acid, fluoroacetate and a range of different tannin and related phenolic structures. The bacterial detoxification of these plant toxins may increase the range of vegetation used for grazing livestock. Bacteria that detoxify 3-hydroxy- 4(1H)-pyridone, the ruminal metabolite of mimosine, a toxic amino acid present in the leguminous shrub *Leucaena leucocephala*, were introduced and established in the rumen of Australian goats and steers which were then able to eat the plant (Jones and Megarrity, 1986). Similarly, inoculation in sheep of a novel bacterium isolated from the rumen of feral goats reduced the toxic effect of tannins of woody plants such as *Acacia aneura*. These successes have led to the studies of the possibility of potential detoxification using genetically modified bacteria (Brooker *et al.,* 1992; Gregg, 1995). The probiotics and their mode of actions are being determined. Since the genetics of *Saccharomyces cerevisiae* and *Aspergillus oryzae* is very well-known, it seems logical to modify them in order to increase their efficiency, or to use them as the vehicle of new activities by expressing added genes (Durand-Chaucheyras *et al.,* 1997).

An indirect approach for the enhancement of fibre digestion in ruminants is through the modification of silage inoculants. Ensiling involves the growth of lactobacilli on soluble sugars present in the plants. Inclusion of a number of hydrolytic enzymes i.e., amylase, cellulase or hemicellulase in silages with low carbohydrate contents has been shown to increase lactic acid by releasing easily accessible sugars for the growth of lactobacilli (Henk and Linden, 1992). Thus, inoculation of silage with the genetically modified bacteria to produce such enzymes has been proposed to pre-digest the plant material for better digestibility in the rumen subsequently. Recombinant *Lactobacillus plantarum*, a silage starter, was constructed to express $\alpha$-amylase, and cellulase or xylanase genes (Scheirlinck *et al.,* 1989; 1990). Considering thc genetic engineering tools already available, there is now a great potential to modify the lactic acid bacteria in order to increase their efficacy in silage making.

## Practical Applications of DFM

DFMs are available in different forms such as powder, paste, gel, and capsules. For usage, they may be mixed in feed, top-dressed, given as a paste, or mixed into the drinking water or milk replacer. However, use of DFM must be managed effectively as interactions with chlorine, water temperature, minerals, flow rate, and antibiotics can affect the viability of organisms. Non-hydroscopic whey is generally used as a carrier for bacteria based DFM. Bacterial DFM pastes are formulated with vegetable oil and inert gelling ingredients. Fungal DFM products are formulated with grain by-products as carriers. Some DFM are developed for one-time dosing while others are developed for feeding on a daily basis. Most DFMs contain live bacteria; however, some contain only bacterial or fungal extracts or fermentation by-products. The addition of DFMs to an animal's diet can assist in the replenishment of beneficial bacteria, resulting in a balanced intestinal microflora. The best response can be observed during the following situations: (a) when a newborn animal acquires beneficial bacteria from environment, (b) during weaning or dietary changes, (c) periods of stress i.e., shipping, vaccination, and other situations , and (d) antibiotic therapy.

The stability of DFMs is crucial because the microbes must be delivered live to the animal to be effective. For this, most DFMs require storage in a cool and dry area, away from heat, direct sunlight, and high levels of humidity. They must not only survive during processing and storage but also in the gut environment. The metabolites present in culture extracts have been suggested to be the "active" ingredients. Newbold *et al.* (1991) reported that autoclaving, but not irradiation, decreased the ability of an *A. oryzae* extract to stimulate rumen bacterial growth and activity. Dawson *et al.,* (1990) reported that the stimulatory effect of yeast on cellulolytic bacteria was negated when yeasts were autoclaved. Although yeasts were able to grow in continuous rumen cultures (Dawson *et al.,* 1990), *S. cerevisiae* did not multiply in sterile ruminal fluid, although they were metabolically active (Kung *et al.,* 1996). Durand-Chaucheyras *et al.* (1998) confirmed that added *S. cerevisiae* did not colonize the rumen of lambs and Kung *et al.* (1996) reported that yeasts were essentially washed out of rumen.

Lee and Botts (1988) reported that pulse dosing alone or pulse dosing with daily feeding of *Streptococcus faecium* resulted in improved performance of incoming feedlot cattle. Bacterial DFM should actually attach and colonize gut surfaces in order to have a beneficial effect; however, the argument is made that a DFM needs to produce its active component to be beneficial even without colonization. Hutchenson *et al.,* (1980) suggested that feeding more than $10^7$ cfu/ animal/ day may cause lower nutrient absorption due to overcrowded gut. Orr *et al.,* (1988) reported that feeding a continuous high dose of *L. acidophilus* to calves ($10^{10}$ cfu/ animal/ day) reduced feed efficiency when compared to feeding a lower dose ($10^6$). Tolerance of DFM to heat is important as many feeds are pelleted. In general, most yeast, *Lactobacillus*, *Bifidobacterium* and *Streptococcus* are destroyed by heat during pelleting. Genetic improvements may allow use of heat-sensitive organisms in pelleted feeds.

## FUTURE PROSPECTS AND CONCLUSION

DFM may be genetically modified to secrete essential amino acids or secrete high levels of growth factors. In near future, microbes capable of detoxification of compounds like alkaloids, oxalates, tannins, or mycotoxins may prove to be a better approach. However, commercial development of these organisms will not occur if they cannot be economically grown and stabilized, especially if daily feeding is not required. But, many questions with respect to type of enzymes, doses, interactions with maturity and moisture need to be answered. Herbal plant extracts may also have potentials as feed additives but information on their use for ruminants is lacking. Research is still needed to fully understand the mechanisms and specific requirements of predominant ruminal microorganisms under different feeding conditions before DFM could be developed to take advantage of improve feed digestion. The initial concept of feeding DFM to livestock was based on the potential for beneficial effects, including establishment of desirable gut microflora, prevention of pathogenic organisms and reducing ruminal acidosis. Although enough evidences are available with some uncertainty to warrant consideration for the use of a DFM in the feeding of various classes of livestock. As our understanding of this emerging technology is increasing, the use of DFMs in animal nutrition will most likely continue to increase.

## References


Abu-Taroush, H. .M., Al-Saiady, M.Y. & El-Din, A.H.K. (1996) *Anim. Feed Sci. Tech.* **57:** 39-49.

Akin, D.E. & Benner, R. (1988) *Appl. Environ. Microbiol.* **54:** 1117-1125.

Akin, D.E., Gordon, G.L.R. & Hogan, J.P. (1983) *Appl. Environ. Microbiol.* **46:** 738-748.

Akin, D.E., Rigsby, L.L., Lyon, C.E. & Windham, W.R. (1990) *Crop Sci.* **30:** 990-993.

Arambel, M.J. & Kent, B. (1990) *J. Dairy Sci.* **73:** 1560-1563.

Arambel, M.J., Weidmeier, R.D. & Walters, J.L. (1987) *Nutr. Repts. Intl.* **35:** 433-437.

Aslan, V.S., Thamsborg, M., Jorgensen, R.J. & Basse, A. (1995) *Acta. Vet. Scand.* **36:** 65-68.

Bauchop, T. & Mountfort, D.O. (1981) *Appl. Environ. Microbiol.* **42:** 1103-1110.

Beharka, A.A., Nagaraja, T.G. & Morrill, J.L. (1991) *J. Dairy Sci.* **74:** 4326-4336.

Bertrand, J.A. & Grimes, L.W. (1997) *J. Dairy Sci.* **80:** 1179-1184.

Borneman, W.S. & Akin, D.E. (1990) In: *Microbiol and Plant Opportunities to Improve Lignocellulose Utilization by Ruminants.* (Eds. Akin, D.E., Ljungdahl, L.G., Wilson, J.R. & Harris, P.J.) Elsevier, New York. pp. 325-339.

Brooker, J.D., Thomson, A.M. & Ward, H.W. (1992) *Australas. Biotechnol.* **2:** 288-291.

Burroughs, W., Woods, W., Ewing, S.A., Greig, J. & Theurer, B. (1960) *J. Anim. Sci.* **19:** 458.

Cerna, B., Cerny, M., Betkova, H., Patricny, P., Soch, M. & Opatrna, I. (1991) *Zivocisna Vyroba* **36:** 381-388.

Chang, J.S., Harper, E.M. & Calza, R.E. (1999) *J. Appl. Microbiol.* **86:** 389-398.

Chaucheryras, F., Fonty, G., Bertin, G. & Gouet, P. (1995b) *Appl. Environ. Microbiol.* **61:** 3466-3467.

Chaucheyras, F., Fonty, G., Bertin, G. & Gouet, P. (1995a) *Curr. Microbiol.* **31:** 201-205.

Chaucheyras, F., Fonty, G., Bertin, G., Salmon, J.M. & Gouet, P. (1995c) *Can. J. Microbiol.* **42:** 927-933.

Cruywagen, C.W., Jordaan, A. & Venter, L. (1996) *J. Dairy Sci.* **79:** 483-486.

Daniel, A.S., Martin, J., Vanat, I., Whitehead, T.R. & Flint, H.J. (1995) *J. Appl. Bacteriol.* **79:** 417-424.

Dann, H.M., Drackely, J.K., McCoy, G.C., Hutjens, M.F. & Garrett, J.E. (2000) *J. Dairy Sci.* **83:** 123-127.

Dawson, K.A. & Hopkins, D.M. (1991) *J. Anim. Sci.* **69:** 531.

Dawson, K.A., Neuman, K.E. & Boling, J.A. (1990) *J. Anim. Sci.* **68:** 3392-3398.

Dey, A., Sehgal, J.P., Puniya, A.K. & Singh, K. (2004) *Asian Aus. J. Anim. Sci.* **17:** 733-884.

Durand-Chaucheyras, F., Fonty, G. & Bertin, G.L. (1997) *Bull. Group Tech. Vét.* **5:** 35-52.

Durand-Chaucheyras, F., Fonty, G.; Bertin, G., Theveniot, M. & Gouet, P. (1998) *Reprod. Nutr. Dev.* **38:** 275-280.

Fuller, R. (1989) *J. Appl. Bact.* **66:** 365-378.

Ghorbani, G.R., Morgavi, D.P., Beauchemin, K.A. & Leedle, J.A.Z. (2002) *J. Anim. Sci.* **80:** 1977-1986.

Gomez-Alarcon, R.A., Huber, J.T., Higginbotham, G.E., Wiesma, F., Ammon, D. & Taylor, B. (1991) *J. Anim. Sci.* **69:** 1733-1740.

Gomez-Basauri, J., de Ondarza, M.B. & Siciliano-Jones, J. (2001) *J. Dairy Sci.* **84:** 283.

Gordon, G.L.R. (1990) In: *Microbial and plant opportunities to improve lignocellulose utilization by ruminants.* (Eds. Akin, D.E., Ljungdahl, L.G., Wilson, J.R. & Harris, P.J.) Elsevier, New York. pp. 301-309.

Gordon, G.L.R. & Phillips, M.W. (1993) *Lett. Appl. Microbiol.* **17:** 220-223.

Gordon, G.L.R. & Phillips, M.W. (1998) *Nutr. Res. Rev.* **11:** 1-36.

Gregg, K. (1995) *Trends Biotechnol.* **13:** 418-421.

Harrison, G.A., Hemken, R.W., Dawson, K.A., Harmon, R.J. & Barker, K.B. (1988) *J. Dairy Sci.* **71:** 2967-2975.

Henk, L.L. & Linden, J.C. (1992) *Enzyme Microbiol. Technol.* **14:** 923-930.

Huck, G.L., Kriekemeier, K.K. & Ducharme, G.A. (1999) *J. Anim. Sci.* **77:** 264.

Hutchenson, D.P., Cole, N.A., Keaton, W., Graham, G., Dunlap, R. & Pitman, K. (1980) *Proc. West. Sec. Amer. Soc. Anim. Sci.* **31:** 213.

Ito, K., Morita, Z., Kamel, H.E.M., Oura, R. & Sekine, J. (1994) *J. Faculty Agric.* **30:** 117-125.

Jaquette, R.D., Dennis, R.J., Coalson, J.A., Ware, D.R., Manfredi, E.T. & Read, P.L. (1988) *J. Dairy Sci.* **71:** 219.

Jenny, B.F., Vandijk, H.J. & Collins, J.A. (1991) *J. Dairy Sci.* **74:** 1968-1973.

Jeong, H.Y., Kim, J.S., Ahn, B.S., Cho, W.M.; Kweon, U.G., Ha, J.K. & Chee, S.H. (1998) *Korean J. Dairy Sci.* **20:** 247-252.

Jones, R.J., & Megarrity, R.G. (1986) *Aust. Vet. J.* **63:** 259-262.

Kellems, R.O., Lagerstedt, A. & Wallentine, M.V. (1990) *J. Dairy Sci.* **73:** 2922-2928.

Kim, S.W., Standorff, D.G., Roman-Roasario, H., Yokoyama, M.T. & Rust, S.R. (2000) *J. Anim. Sci.* **78:** 292.

Kobayashi, Y., Okuda, N., Matsumoto, M., Inoue, K., Wakita, M. & Hoshino, S. *FEMS Microbiol. Lett.* **163:** 11-17.

Kung, L. Jr., Hession, A.O., Tung, R.S. & Maciorowski, K. (1991) *Proc. 21st Biennial Conf. Rumen Function.* Chicago, IL, pp. 31.

Kung, L. Jr., & Hession, A.O. (1995) *J. Anim. Sci.* **73:** 250-256.

Kung, L. Jr., Kreck, E.M., Tung, R.S., Hession, A.O., Sheperd, A.C., Cohen, M.A., Swain, H.E. & Leedle, J.A.Z. (1996) *J. Dairy Sci.* **80:** 2045-2051.

Lee, S.S., Choi, C.K., Ahn, B.H., Moon, Y.H., Kim, C.H. & Ha, J.K. (2004) *Anim. Feed Sci. Tech.* **115:** 215-226

Lee, S.S., Ha, J.K. & Cheng, K.J. (2000) *Anim. Feed Sci. Tech.* **88:** 201-217.

Lee, R.W. & Botts, R.L. (1988) *J. Anim. Sci.* **66:** 460.

Manikumar, B., Puniya, A.K. & Singh, K. (2002) *Indian J. Microbiol.* **42:** 133-136.

Miles, R.D. & Bootwalla, S.M. (1991) In: *Direct-fed microbiols in animal production: A review of literature. Natl. Feed Ingred. Assoc.* West Des Moines, IA. pp. 117.

Morrill, J.L., Dayton, A.D., & Mickelsen, R. (1977) *J. Dairy Sci.* **60:** 1105-1109.

Mountfort, D.O., Asher, R.A. & Bauchop, T. (1982) *Appl. Environ. Microbiol.* **44:** 128-134.

Newbold, C.J., Brock, R. & Wallace, R.J. (1991) *J. Agric. Sci. Camb.* **116:** 159-162.

Newbold, C.J., Wallace, R.J. & McIntosh, F.M. (1996) *Brit. J. Nutr.* **76:** 249.

Newbold, C.J., Wallace, R.J., Chen, X.B. & McIntosh, F. (1995) *J. Anim. Sci.* **73:** 1811-1818.

Newman, K.E. & Jacques, K.A. (1995) In: *Biotechnology in Animal Feeds and Animal Feeding.* (Eds. Wallace, R.J. & Chesson, A.) VCH, Weinheim, Germany. pp. 247-258.

Nisbet, D.J. & Martin, S.A. (1990) *Appl. Environ. Microbiol.* **56:** 3515-3518.

Nisbet, D.J. & Martin, S.A. (1994) *J. Anim. Sci.* **72:** 1355-1361.

Nocek, J.E., Kautz, W.P., Leedle, J.A.Z. & Allman, J.G. (2000) *J. Dairy Sci.* **83:** 1242.

Orpin, C.G. & Joblin, K.N. (1988) In: *The rumen microbial ecosystem.* (Ed. Hobson, P.N.) Elsevier, London. pp. 129-150.

Orr, C.L., Ware, D.R., Manfredi, E.T. & Hutchenson, D.P. (1988) *J. Anim. Sci.* **66:** 460.

Paul, S.S., Kamra, D.N., Sastry, V.R.B. & Agarwal, N. (2004) *Anim. Feed Sci. Tech.* **115:** 143-157.

Plata, F.P., Mendoza, G.D., Barcena-Gama, J.R. & Gonzalez, S.M. (1994) *Anim. Feed Sci. Tech.* **49:** 203-210.

Robinson, J.A., Smolenski, W.J., Greening, R.C., Ogilvie, R.L., Bell, R.L., Barsuhn, K. & Peters, J.P. (1992) *J. Anim. Sci.* **70:** 310.

Russell, J.B. & Wilson, D.B. (1988) *J. Nutr.* **118:** 271-279.

Rust, J.W., Jacobson, N.L., McGilliard, A.D. & Hotchkiss, D.K. (1965) *J. Anim. Sci.* **24:** 156.

Samanta, A.K., Walli, T.K. & Singh, K.K. (2001) *Indian J. Anim. Sci.* **71:** 497-498.

Savage, D.C. (1987) *Die Nahrung.* **5:** 383-390.

Savoini, G., Mancin, G., Rossi, C.S., Grittini, A., Baldi, A. & Dell-Orto, V. (2000) *Obiettivi-e-Documenti-Veterinari.* **21:** 65-70.

Scheirlinck, T., DeMeutter, J., Arnaut, G., Joos, H., Claeyssens, M. & Michiels, F. (1990) *Appl. Microbiol. Biotechnol.* **33:** 534-541.

Scheirlinck, T., Mahillon, J., Joos, H., Dhaese, P. & Michiels, F. (1989) *Appl. Environ. Microbiol.* **55:** 2130-2137.

Schwartz, D.L., Muller, L.D., Rogers, G.W. & Varga, G.A. (1994) *J. Dairy Sci.* **77:** 3073-3080.

Sievert, S.J. & Shaver, R.D. (1993) *J. Anim. Sci.* **71:** 1032-1040.

Soder, K.J. & Holden, L.A. (1999) *J. Dairy Sci.* **82:** 605-610.

Swinney-Floyd, D., Gardner, B.A., Rehberger, T. & Parrot, T. (1999) *J. Anim. Sci.* **77:** 77.

Thareja, A., Puniya, A.K., Goel, G., Nagpal, R., Sehgal, J.P., Singh, P.K. & Singh, K. (2006) *Arch. Anim. Nutr.* **60:** 412-417.

Theodorou, M.K., Gill, M., King-Spooner, C. & Beever, D.E. (1990) *Appl. Environ. Microbiol.* **56:** 1073-1078.

Theurer, B., Woods, W. & Burroughs, W. (1963) *J. Anim. Sci.* **22:** 150.

Tripathi, V.K. (2005) Ph.D. dissertation. NDRI (Deemed University), Karnal.

Varel, V.H., Kreikemeier, K.K., Jung, H.J.G. & Hatfield, R.D. (1993) *Appl. Environ. Microbiol.* **59:** 3171-3176.

Waldrip, H.M., & Martin, S.A. (1993) *J. Anim. Sci.* **71:** 2770-2776.

Wallace, R.J. & Newbold, C.J. (1993) In: *Biotechnology in the Feed Industry.* (Ed. Lyons, T.P.). Alltech Technical Publications, Kentucky, pp. 173-192.

Ware, D.R., Read, P.L. & Manfredi, E.T. (1988) *J. Dairy Sci.* **71:** 219.

Weston, R.H., Lindsay, J.R., Purser, D.B., Gordon, G.L.R. & Davies, P. (1988) *Aust. J. Agric. Res.* **39:** 1107-1119.

Whitehead, T.R. & Flint, H.J. (1995) *FEMS Microbiol. Lett.* **126:** 165-169.

Williams, P.E.V., Tait, C.A.G., Innes, G.M. & Newbold, C.J. (1991) *J. Anim. Sci.* **69:** 3016-3026.

Wohlt, J.E., Finkelstein, A.D. & Chung, C.H. (1991) *J. Dairy Sci.* **74:** 1395-1400.

Xue, G.P., Johnson, J.S., Bransgrove, K.L., Gregg, K., Beard, C.E., Dalrymphe, B.P., Gobius, K.S. & Aylward, J.H. (1997) *J. Biotechnol.* **54:** 139-148.

Yang, W.Z., Beauchemin, K.A. & Rode, L.M. (1999) *J. Dairy Sci.* **82:** 391-403.

21

# Tripartite Symbiotic Association: Legume-Rhizobia-Mycorrhiza

**D. Harwani, Preeti Choudhary, Sangeeta Dhaker, Kamal Prasad and S. K. Mahna**

*INCO- DEV Research Laboratory, Department of Botany, Maharshi Dayanand Saraswati University, Ajmer, Rajasthan (India)-305009*

## ABSTRACT

Leguminous plants have the ability to fix atmospheric nitrogen through symbiosis with host specific Gram-negative soil bacteria, universally known as rhizobia. The vast majority of all land plants also form a symbiotic relationship with arbuscular mycorrhizal fungi (AMF), taxonomically grouped in the order Glomales. Recent researches have brought to light instances where such biological activities are markedly enhanced in two or three membered associations of organisms. Therefore, efforts are underway to investigate the influence of inoculation of more than one microbe for the improvement of plant health and overall crop production. In this context, rhizobia and mycorrhizal fungi are of paramount significance, which interact synergistically in improving the soil fertility level. In this review the interactions of rhizobia and mycorrhizal fungi with their host plant and environmental perspectives have been discussed. Subsequently, after describing the recent taxonomic status of rhizobia and AM fungi, the aspects of their exploitation for the sustainable improvement of agricultural systems and soil health have been reviewed.

## INTRODUCTION

Most of the legume plants possess two types of symbiotic microorganisms in their roots viz. atmospheric $N_2$-fixing bacteria and mycorrhizal fungi, thus establishing a tripartite association capable of supplying plants especially their nitrogen and phosphorus requirements (Silveira and Cardoso, 1987). The two symbioses typically act synergistically, resulting in greater nitrogen (N) fixation and phosphorus (P) acquisition. The interaction of legume plants with rhizobia and AM fungi has received considerable attention in recent years because of the relatively high phosphorus demand of nitrogen ($N_2$) fixation. Legumes are typically coarse-rooted and, therefore, inefficient in extracting phosphorus from the soil. In addition, a survey of Indian soils revealed that 98% of these need phosphorus fertilization either in the form of chemical or biological fertilizer (Tilak *et al.*, 2005). The AM fungi associated with legumes are an essential link for adequate phosphorus nutrition, leading to enhanced nitrogenase activity that in turn promotes root and mycorrhizal growth.

The nitrogen-fixation/carbon-fixation mutualism between leguminous plants and rhizobial bacteria is widespread, ancient and important for ecosystem function and human nutrition. Legumes and rhizobia can grow and reproduce independently, but neither usually fixes atmospheric $N_2$ alone (Sprent *et al.*, 1987). Enzymatic reduction of dinitrogen is energetically expensive and rhizobia rarely, if ever, fix nitrogen while free-living (Postgate, 1998). The successful plant symbiont must pay the energetic price of the reduction reaction, conduct a complicated signal exchange with rhizobia, produce leghaemoglobin, and form a novel organ-the nodule. In contrast to nodule formation higher plants have the ability to form arbuscular endomycorrhiza (AM), a symbiotic association of the plant root with fungi belonging to the Glomeromycota. Members of the Glomeromycota are the fungal partners of the AM symbiosis. Both fossil evidence and DNA sequence divergence within the Glomeromycota suggest that AM symbiosis is about 400 million years old, coinciding with the plant's colonization of land while symbioses between bacteria and plants apparently restricted to the angiosperms and the earliest bacterial-plant endosymbioses probably evolved several hundred million years later than the fungal ones (Parniske, 2000). Indeed, mycorrhizae may have facilitated the transition of plants to terrestrial habitats (Malloch *et al.*, 1980) and microbial symbioses remain critical determinants of global nutrient cycles, community structure and productivity (Vitousek and Walker, 1989; Rygiewicz and Andersen, 1994; Allen *et al.*, 1995; Bever *et al.*, 1997; van Der Heijden *et al.*, 1998; Klironomos *et al.*, 2000; Klironomos and Hart, 2001).

At first glance the interactions of plants with rhizobia and AM fungi seem to have little in common. The induced morphological responses of the host plants are different. Furthermore, both interactions are extremes in terms of host specificity. Whereas in AM formation there is a very little host specificity, the *Rhizobium*-legume interaction is highly specific. However, genetic studies have shown that several common steps are involved in establishing these symbioses (Bradbury *et al.*, 1991; Gianinazzi-Pearson, 1996).

In this review the interactions of rhizobia and mycorrhizal fungi with their host plant and their environmental perspectives have been discussed. Subsequently, after describing the recent taxonomic status of rhizobia and AM fungi, the aspects of their exploitation for the sustainable improvement of agricultural systems and soil health have been reviewed.

## SYMBIOSIS BETWEEN RHIZOBIA AND LEGUMINOUS PLANTS

The relevant aspects of the legume-rhizobia mutualism are described in detail by Denison (2000). Rhizobia-plant associations are referred as sympathogens (Spaink, 1995), prokaryotic plant parasites (Long and Staskawicz, 1993) and refined parasites of legumes (Djordjevic *et al.*, 1987). Their hypersensitive reactions or production of toxins renders them to be pathogens, while their dependency on plants for food and paying back plants as improvement in growth and development strongly suggest for their mutualistic nature. Hentschel *et al.,* (2000) reported that the bacterial interaction with its host is not necessarily different manifestation and the fundamental molecular mechanisms between symbiosis and pathogenesis are the same.

Only a few plant species have the ability to interact symbiotically with bacteria of the genera *Allorhizobium, Azorhizobium, Bradyrhizobium, Rhizobium, Mesorhizobium* and *Sinorhizobium*. Symbiosis between legumes and rhizobia is often mutually beneficial. If a plant is infected by a single strain of mutualistic rhizobia, it can obtain N needed for leaf growth; N that would otherwise require a larger and more expensive root system (Pate, 1986) by supplying the rhizobia with photosynthate. The symbiosis with rhizobia is apparently not an adaptation to a specialized ecological niche but rather depends upon some genetic peculiarity of legumes. Vincent's (1974) analysis showed that the rhizobia, which invade temperate zone legumes, tend to exhibit a narrow plant host range. In contrast, rhizobia that infect tropical legumes tend to be less restricted and can invade a broader range of tropical legumes. In 'effective' symbiosis, many of the rhizobia eventually differentiate into the bacteroid form. Nodules are the only plant organs, other than leaves, that have the ability to change their gas permeability rapidly and reversibly. A few aquatic legumes also bear stem nodules in addition to the normal root nodules. This peculiarity is restricted to fifteen species of *Aeschynomene*, one species of *Neptunia* and one species of *Sesbania* (Tilak *et al.,* 2005). Nodule can take several forms, although the best characterized are: (i) approximately spherical with determinate growth, with bacteroids, which throughout typically retain the ability to reproduce, and (ii) elongate with indeterminate growth, with undifferentiated rhizobia mainly in persistent infection threads near the tip. Bacteroids differentiate and fix nitrogen soon after nodule parenchyma has formed, when nodules are only a fraction of their potential size (whether determinate or indeterminate).

## SYMBIOTIC RECOGNITION AND EFFECTIVENESS

The development of the symbiotic relationship between rhizobia and legumes is a highly interactive process that includes (i) molecular communication between the free-living organisms, (ii) an infection phase when rhizobia enter root or stem nodules, and (iii) a final symbiotic phase when rhizobia occupying nodules and fix $N_2$. Bacterial recognition of the particular flavonoids produced by a host species provides the first opportunity for plant choice, as only the correct flavonoid/s induce symbiotic gene expression in a particular *Rhizobium* strain. This interaction starts in the soil with a specific plant-*Rhizobium* molecular signal exchange involving plant flavonoids released into the root exudates, which induce the expression of the *nod* operons in the rhizobia. These genes encode the enzymes for the biosynthesis and release of Nod factor (a lipochitooligosaccharide), which triggers nodule development in the inner root cortex for which the presence of active rhizobia is not required (Mylona *et al.,* 1995). Hence, the diversity of nod factors produced by rhizobia, and discrimination of these factors by plants, contribute the second level of specificity to the interaction and create an opportunity for partner choice by the plant (Perret *et al.*, 2000).

The nodule will provide the rhizobia with the nutrients and the micro aerobic environments required for $N_2$ fixation (Mylona *et al.,* 1995) and inside nodule rhizobia (differentiated cells-bacteroids) vary in number. For example, a single soybean nodule can contain $2.6 \times 10^5$ bacteroid-containing cells, each containing $3.8 \times 10^4$ bacteroids for a total of almost $10^{10}$ bacteroids per nodule. Moreover, one soybean nodule may contain as many rhizobia as 10-100 kg of soil (Bergersen, 1982). Though a plant gets nodulated with a number of rhizobial strains, effective nitrogen fixation takes place with relatively few strains (Turk and Keyser, 1992). In parallel with plant nodule organogenesis, rhizobia are chemoattracted to the root surface (Gulash *et al.,* 1984) than they attach to it in nonspecific (Smit *et al.,* 1992) as well as in bacterial lectin-mediated specific (Ho *et al.,* 1990) ways and penetrate the root hairs, forming characteristic structures known as infection threads. Thereafter, they finally invade the developing nodule (Stuurman *et al.,* 2000), where they subsequently differentiate into bacteroids, a distinct rhizobial form, which is capable of reducing atmospheric $N_2$ (Mylona *et al.,* 1995).

The process of infection and bacteroid differentiation is strongly dependent on the structure of the bacterial surface polysaccharide, which seems to play a role not only in recognition of rhizobia by the plant, but also as a signal to prevent the elicitation of plant defence activities arising against bacterial penetration (Becker *et al*., 1998). The sugars present on cell surfaces (i.e., exopolysaccharide, capsular polysaccharide, lypopolysaccharide) are essential for rhizobia to avoid host defense responses during the invasive phase (Brewin, 1998; Hadri and Bisseling, 1998; Albus *et al*., 2001). Albus *et al*., (2001) proposed that LPSs released from the bacterial surface function as specific signal to the plant, and promote symbiotic interaction and suppress defence responses. In addition to this, soybean lectin (SBL) are also released into the rhizosphere (van Rhijn *et al*.,1998) and specifically stimulate rhizobial adsorption and infection (Wall and Favelukes, 1991; Lodiero *et al.,* 2000). Compatibility at this stage depends, in part, on recognition by the plant of particular polysaccharides on the bacterial cell wall. Rhizobia vary in polysaccharide composition of their cell walls, which provides another opportunity for plant choice of partners using the "lock and key" cascade that determines the degree of plant and bacterial specificity (Perret *et al*., 2000).

Therefore, rhizobial adsorption, root hair infection, nodule formation and nitrogen fixation are the key steps of a complex process, each one contributing to a different level of symbiotic recognition and effectiveness.

## NITROGEN FIXATION PROCESS

However, a wide range of Bacteria and Archaea possesses nitrogenase and is capable of reducing dinitrogen to organic forms (Postgate, 1998). A substantial portion of the world's supply of organic nitrogen is fixed via the symbiosis between rhizobial bacteria and leguminous host plants (Postgate, 1998). Antonie Lavoisier named the element nitrogen, as "azote," meaning "without life," about 200 years ago, since it is a constituent of food, poisons, fertilizers and explosives (Schoot Uiterkamp, 1990). The atmosphere contains about $10^{15}$ tonnes of $N_2$ gas and the nitrogen cycle involves the transformation of some $3 \times 10^9$ tonnes of $N_2$ per year on a global basis (Postgate, 1982). However, transformation of nitrogen viz. $N_2$ fixation is not exclusively biological. Lightning probably accounts for about 10% of the world's supply of fixed nitrogen (Sprent and Sprent, 1990). The fertilizer industry also provides very important quantities of chemically fixed nitrogen.

For more than 100 years, BNF has been exploited extensively in agricultural practice (Dixon and Wheeler, 1986; Burris, 1994). The expanded interest in ecology has drawn attention to the fact that BNF is ecologically benign and that its greater exploitation can reduce the use of fossil fuels and can be helpful in reforestation and in restoration of misused lands to productivity (Sprent and Sprent, 1990; Burris, 1994). Nitrogen fixation (the reduction of atmospheric $N_2$ into $NH_3$) is catalyzed by the complex metalloenzyme nitrogenase, composed of two component proteins, which is dinitrogenase and dinitrogenase reductase. Until now three different forms of nitrogenase have been reported. All the three different nitrogenases share similar structural properties but differ in the heterometal present in the active site of the dinitrogenase unit. The most widespread nitrogenase contains iron-molybdenum cofactors at the active site and is encoded by the *nif* gene family. Second is vanadium containing and *vnf*-encoded (vanadium-dependent nitrogen fixation) nitrogenase that exhibits an iron-vanadium cofactor. A third so-called "alternative" nitrogenase contains an iron-only cofactor (FeFe-co) and is *anf*-encoded (alternative nitrogen fixation) (Rangaraj *et al.*, 2000). The rhizobial nitrogenase consists of two components, dinitrogenase (Fe-Mo protein) and dinitrogenase reductase (Fe-protein). The nitrogenase catalyzes the following reaction:

$$N_2 + 16\ MgATP + 8e^- + 8H^+ = 2NH_3 + 16\ MgADP + 16pi + H_2$$

Nitrogen fixation is an extremely energy-consuming reaction. Therefore, symbiotic $N_2$ fixers that benefit from plant photosynthesis mainly contribute biologically fixed $N_2$ on the earth (Kaminski *et al.*, 1998). In isolated bacteroids, 95% of the fixed $N_2$ was reported to be excreted into the medium as ammonia, the remainder being incorporated into bacteroids (Robertson and Farnden, 1980). Fixed $N_2$ is exported from bacteroid side across the symbiosome membrane to the plant cytoplasm as ammonium (Tyerman *et al.*, 1995) and as amino acids (Rosendahl *et al.*, 2001). From the nodules the fixed N is exported to the plant as amides (e.g., asparagine) or ureides (Boogerd and van Rossum, 1997).

Nitrogenase catalyses the reduction of $H_2$ which is an inhibitor molecule for $N_2$ fixation. With the help of uptake hydrogenase, the bacteroids of some rhizobial strains are capable of removing $H_2$ by oxidation, (Mellor and Werner, 1990; Bergersen, 1997). Nitrogenase is inactivated rapidly by excess $O_2$ (Bergersen, 1997). Therefore, several physiological and biochemical components regulate $O_2$ supply to nodules, such as a thin diffusion barrier in the inner nodule cortex, the presence of $O_2$-binding leghaemoglobins in the host cytosol, and respiration by large numbers of bacteroids and plant mitochondria (Bergersen, 1997; Kaminski *et al.*, 1998). Leghaemoglobin is the predominant plant protein of the nodule (up to 30%), giving usually a red colour characteristic to the effective nodule (Kaminski *et al.*, 1998).

## IMPORTANCE OF BIOLOGICAL $N_2$ FIXATION

BNF is an efficient source of nitrogen (Peoples *et al.*, 1995) and the total annual terrestrial contribution of N from BNF ranged 139 million—175 million tonnes as given by Burns and Hardy (1975) and Paul (1988). A wide range of organisms has the ability to fix nitrogen. However, only a very small proportion of species are able to do so, viz. about 87 species in 2 genera of archaea, 38 genera of bacteria, and 20 genera of cyanobacteria have been identified as diazotrophs or organisms that can fix nitrogen (Dixon and Wheeler, 1986; Sprent and Sprent, 1990; Zahran *et al.*,

1995). There are approximately 700 genera and about 13,000 species of legumes, only a portion of which (about 20% as reported by Sprent and Sprent, 1990) have been examined for nodulation and shown to have the ability to fix $N_2$.

The age of symbiotic nitrogen fixation is unknown, but may predate the origin of legumes in the Maastrichtian (late Cretaceous), some 65 million years before present (Herendeen *et al.*, 1992). Estimates are that the rhizobial symbiosis with nearly 100 agriculturally important legumes contribute nearly half of the annual quantity of BNF entering soil ecosystems (Tate, 1995). Legume symbioses contribute at least 70 million tonnes of N per year, approximately half deriving from the cool and warm temperature zones and the remainder deriving from the tropics (Brockwell *et al.*, 1995). Values estimated for various legume crops and pasture species are often impressive, commonly falling in the range of 200 to 300 kg of N $ha^{-1}$ $year^{-1}$ (Peoples *et al.*, 1995). Yield increase in crops planted after harvesting of legumes is often comparable to those expected from application of 30 to 80 kg of fertilizer N $ha^{-1}$. Inputs of fixed N for alfalfa, red clover, pea, soybean, cowpea, and vetch were estimated to be about 65 to 335 kg of N $ha^{-1}$ $year^{-1}$ (Tate, 1995) or 23 to 300 kg of N $ha^{-1}$ $year^{-1}$ (Wani *et al.*, 1995).

## Significance of Environmental Factors that Affect BNF

The process of $N_2$ fixation is strongly related to the physiological state of the host plant in $N_2$–fixing system (*Rhizobium*-legume symbiosis). Therefore, a competitive and persistent rhizobial strain is not expected to express its full capacity for nitrogen fixation if stress factors viz. salinity, unfavorable soil pH, nutrient deficiency, mineral toxicity, temperature and insufficient soil moisture impose limitations on the strength of the host legume (Brockwell *et al.*, 1995; Peoples *et al.*, 1995; Thies *et al.*, 1995). Typical environmental stresses faced by the legume nodules and their symbiotic partner (*Rhizobium*) may include photosynthetic deprivation, water stress, salinity, temperature and heavy metals (Walsh, 1995). A given stress may also have more than one effect e.g., salinity may act as a water stress, which affects the photosynthetic rate or may affect nodule metabolism directly. The most problematic environments for rhizobia are marginal lands with low rainfall, extremes of temperature, acidic soils of low nutrient status, and poor water-holding capacity (Bottomley, 1991). Populations of *Rhizobium* and *Bradyrhizobium* species vary in their tolerance to major environmental factors (Keyser, 1992; Slattery and Pearce, 2002; Mahobia, 2003; Harwani, 2006).

## BIODIVERSITY AND TAXONOMY OF RHIZOBIA

The name "rhizobia" describes root and stem-nodulating bacteria that live in symbiosis as $N_2$-fixing system mainly with leguminous plants; exceptions are species of non-legumes, *Parasponia* and *Tremma* (Akkermans and van Dijk, 1981). The atmospheric nitrogen fixing ability of legumes was discovered in 1838 by French agricultural chemist J.B. Boussingault and subsequent development in this field led to the discovery of symbiotic nitrogen fixation (SNF) in legumes by Hellriegel and Willfarth (1888). Beijerinck (1888) first isolated a bacterium from root nodules, which he named *Bacillus radicicola*. Although, rhizobia are soil-inhabitants, rhizobial species are not commonly isolated from sources other than nodules.

The taxonomy and nomenclature of the root nodule bacteria has been in constant review ever since. It was Frank (1889), who named the bacterium as *Rhizobium leguminosarum* and, thereafter, other species belonging to the same group were identified. Fred *et al.*, (1932) first recognized the taxonomic diversity of root nodule bacteria and recognized their relationship with agrobacteria. Further, they classified root nodule bacteria on the basis of growth rates viz. fast growers that acidify culture medium, and slow growers that do not acidify medium (Norris, 1965). Both groups were considered to be *Rhizobium* (rhizo = root, bios = living). Rhizobia were first classified on the basis of host specificity of each species of *Rhizobium* as a "cross-inoculation group", which nodulated plants of a particular cross-inoculation group (Fred *et al.,* 1932). Over the years, Allen and Allen (1939), Wilson (1944), Graham (1964), Norris (1965), Lange (1966), Trinick (1965, 1980) studied the host specificity of *Rhizobium* species and determined boundary jumping between the cross-inoculation groups and proved that cross-inoculation groups were not discrete and the host-based system of *Rhizobium* classification had several anomalies. Vincent (1974,1977) reported evidence of specific subgroups even within plant genera.

The $8^{th}$ edition of Bergey's Manual of Determinative Bacteriology (Jordan and Allen 1974) placed the legume nodulating group of bacteria, *Rhizobium,* in family Rhizobiaceae along with *Agrobacterium* because of their ability to cause cortical hypertrophies of different types on plants. Elkan (1984), in his review, quoted a new classification system for Rhizobiaceae proposed by the International Committee on Systematic Bacteriology of the International Union of Microbiological Societies, and divided *Rhizobium* into three recognized species: *R. leguminosarum* (biovars *trifoli*, *phaseoli* and *viceae*), *R. meliloti* and *R. loti.* Downie and Brewin (1992) signified the name *Azorhizobium caulinodans* (type species) that produces stem nodules on the stem of *Sesbania rostrata,* which reflected unusual characteristic. The so-called slow-growing and fast-growing groups of rhizobia were classified by Jordan (1984) as two different genera in the 1984 Bergey's Manual of Systematic Bacteriology. Thereafter, on the basis of different G+C DNA ratios, Werner (1992) recognized four genera of Rhizobiaceae as *Rhizobium*, *Bradyrhizobium*, *Agrobacterium* and *Phyllobacterium*. The members of the genera *Agrobacterium* and *Phyllobacterium* are crown, root, stem gall and leaf nodule-inducing bacteria in higher plants. The genera *Azorhizobium*, *Bradyrhizobium* and *Mesorhizobium* have been included in the Family Rhizobiaceae (van Berkum and Eardly, 1998), which are allocated in the second edition of the *Bergey's Manual of Systematic Bacteriology* (2001), in the new families Azorhizobiaceae, Bradyrhizobiaceae and Phyllobacteriaceae, respectively.

The current taxonomy of the Rhizobiaceae as a whole is dealt with comprehensively by Young and Haukka (1996), Sprent (2001), Young *et al.* (2001), Sahgal and Johri (2003) and Sawada *et al.* (2003). The system of classification as defined by these authors, plus more recent additions, is presented in Table 1. The rhizobia include the genera *Allorhizobium* (de Lajudie *et al.,* 1998a), *Azorhizobium* (Dreyfus *et al.,* 1988), *Bradyrhizobium* (Jordan, 1982), *Rhizobium* (Jordan, 1984), *Mesorhizobium* (Jarvis *et al.,* 1997) and *Sinorhizobium* (Chen *et al.,*1988; de Lajudie *et al.* 1994). These genera belong to three distinct phylogenetic branches within the α-2 subclass of Proteobacteria. Sy *et al.* (2001) discovered a strain of *Methylobacterium* sp. in nodules of *Crotalaria* spp. that constituted a fourth branch within the α-2 subclass. On the basis of recent observations of widespread occurrence of β -proteobacteria nodulating legume plants, rhizobia are now divided as α -rhizobia and β –rhizobia (Chen *et al.,* 2003; Verma *et al.,* 2004).

**Table 1** Nomenclature of Rhizobia

| *Genus* | *Specific name* | *Hosts (not necessarily exclusive)** | *Additional references* |
|---|---|---|---|
| *Rhizobium* | R. *leguminosarum* | | Frank (1889) |
| | bv. *trifolii* | Species of *Trifolium* | |
| | bv. *viciae* | Species of *Pisum, Lathyrus, Lens and Vicia* | |
| | bv. *phaseoli* | Species of *Phaseolus* | |
| | R. *etli* | | Segovia *et al.* (1993); Wang *et al.* (1999b) |
| | bv. *mimosae* | *Mimosa affinis* | |
| | bv. *phaseoli* | *Phaseolus vulgaris* | |
| | R. *gallicum* | Amarger *et al.* (1997) | |
| | bv. *gallicum* | P. *vulgaris* | |
| | bv. *phaseoli* | *P. vulgaris* | |
| | R. *giardinii* | | Amarger *et al.* (1997) |
| | bv. *giardinii* | P. *vulgaris* | |
| | bv. *phaseoli* | P. *vulgaris* | |
| | R. *galegae* | Lindstrom (1989) | |
| | bv. *giardinii* | *Galega officinalis* | |
| | bv. *phaseoli* | *G. orientalis* | |
| | *R. hiananense* | *Desmodium sinuatum* | Chen *et al.* (1997) |
| | *R. huautlense* | *Sesbania herbacea* | Wang *et al.* (1998) |
| | R. indigoferae | *Indigofera* | Wei *et al.* 2002 |
| | R. *mongolense* | *Medicago ruthenica* | van Berkum *et al.* (1998) |
| | *R. sullae* | *Hedysarum coronarium* | Squartini *et al.* (2002) |
| | R. *tropici* | *Phaseolus vulgaris, Leucaena esculenta, L. leucocephala* | Martinez-Romero *et al.* (1991) |
| | R. *yanglingense* | *Amphicarpaea trisperma, Coronilla varia, Gueldenstaedtia multiflora* | Tan *et al.* (2001) |
| *Bradyrhizobium* | B. *japonicum* | *Glycine max* | Jordan (1982) |
| | B. *elkanii* | *Glycine max* | Kuykendall *et al.* (1992) |
| | B. *liaoningense* | *Glycine max* | Xu *et al.* (1995) |
| | B. *yuanmingense* | Species of *Lespedeza* | Yao *et al.* (2002) |
| | B. *betae* | *Beta vulgaris* | Rivas *et al.* (2004) |
| | B. *canariense* | *Genistoid legumes* | Vinuesa *et al.* (2005) |
| | Several unnamed species of *Bradyrhizobium* | Genera of many species that are nodulated by slow-growing strains of rhizobia | Brockwell *et al.* (2005); Saxena *et al.* (2005) |
| *Sinorhizobium*[I] | S. *abri* | *Abrus precatorius* | Ogasawara *et al.* (2003) |
| | S. *indiaense* | *Sesbania rostrata* | Ogasawara *et al.* (2003) |
| | S. *meliloti* | Species of *Medicago, Melilotus* and *Trigonella* | Dangeard (1926) |
| | S. *medicae* | Species of Medicago | Rome *et al.* (1996a, 1996b) |
| | S. *adhaerens* | Not known | Willems *et al.* (2003); Young (2003) |

*Table 1 Cond.*

*Table 1 Cond.*

| | | | |
|---|---|---|---|
| | *S. arboris* | *Acacia senegal* | Nick *et al.* (1999) |
| | *S. fredii* | *Glycine max, Cajanus cajan* and *Vigna unguiculata* | Scholla and Elkan (1984) |
| | *S. kostiense* | Acacia Senegal | Nick *et al.* (1999) |
| | *S. kummerowiae* | Not known | cited from Young (2003) |
| | *S. morelense* | *Leucaena leucacephala*[III] | Wang *et al.* (2002) |
| | *S. saheli* | Species of *Acacia* and "a number of other tree genera" | de Lajudie *et al.* (1994) |
| | *S. terangae* | | de Lajudie *et al.* (1994) |
| | bv. *acaciae* | Species of *Acacia* | |
| | bv. *sesbaniae* | Species of *Sesbania* | |
| | *S. xinjiangense* | *Glycine max* | Peng *et al.* (2002) |
| *Mesorhizobium* | *M. loti* | Species of *Lotus, Anthyllis* and *Lupinus* | Jarvis *et al.* (1982) |
| | *M. amorphae* | *Amorpha fruticosa* | Wang *et al.* (1999a) |
| | *M. ciceri* | *Cicer arietinum* | Nour *et al.* (1994) |
| | *M. huakuii* | *Astragalus sinicus* | Chen *et al.* (1991) |
| | *M. mediterraneum* | *Cicer arietinum* | Nour *et al.* (1995) |
| | *M. plurifarium* | Species of *Acacia, Chamaecrista, Leucaena* and *Prosopis* | de Lajudie *et al.* (1998b) |
| | *M. septentrionale* | *Astragalus adsurgens* | Gao *et al.* (2004) |
| | *M. temperatum* | *A. adsurgens* | Gao *et al.* (2004) |
| | *M. tianshanense* | Various legumes including *Glycine max* | Chen *et al.* (1995) |
| *Azorhizobium* | *A. caulinodans* | Nodulates stems and roots of *Sesbania rostrata* | Dreyfus *et al.* (1988) |
| *Allorhizobium* | *A. undicola* | *Neptunia natans* | de Lajudie *et al.* (1998a) |
| *Blastobacter*[vi] | *B. denitrificans* | *Aeschynomene indica* | van Berkum and Eardly (2002) |
| *Burkholderia*[v] | *Burkholderia* sp | *Aspalathus carnosa* | Moulin *et al.* (2001) |
| | *B. caribensis* | Tropical legume(s) | Vandamme *et al.* (2002) |
| | *B. phymatum* | Tropical legume(s) | Vandamme *et al.* (2002 |
| | *B. tuberum* | Tropical legume(s) | Vandamme *et al.* (2002) |
| *Devosia*[vi] | *D. neptunii*[IV] | *Neptunia natans* | Rivas *et al.* (2002) |
| *Methylobacterium*[v] | *Methylobacterium* sp. | *Lotononis bainesii* | Jaftha *et al.* (2002) |
| | *M. nodulans* | *Species of Crotalaria* | Sy *et al.* (2001); Jourand *et al.* (2004) |
| *Ochrobactrum*[v] | *Ochrobactrum* sp | *Acacia mangium* | Ngom *et al.* (2004) |
| *Ralstonia*[II, vi] | *R. taiwanensis* | *Mimosa spp.* | Chen *et al.* (2001, 2003) |

*Genus names refer the better-known host legumes, nodulated by each species of rhizobia

[I] The generic name *Ensifer* (Young, 2003) may have nomenclatural priority over *Sinorhizobium*.

[II] The generic names *Wautersia* (Vaneechoutte *et al.*, 2004) and *Cupriavidus* (Vandamme and Coenye, 2004) have been proposed as alternative nomenclature for *Ralstonia*.

[III] Although the novel species *Sinorhizobium morelense* was isolated from nodules of *Leucaena leucocephala*, it did not form nodules when reinoculated on to the host plant (Wang *et al.*, 2002)

[IV] The specific name •e*neptunii*•f is tentative.

[v] Pseudo-*Bradyrhizobium* yet to be named and properly classified (Yonga, 1996).

[vi] These bacteria have been found to form effective root nodules on legumes. Analysis of their genomes indicated the presence of nodulation genes picked up from other rhizobia species.

On the basis of polyphasic taxonomy, there are four recognized and validated species belonging to the genus *Rhizobium*. These are *Rhizobium elti*, *Rhizobium galegae*, *Rhizobium leguminosarum* and *Rhizobium tropici*. Additional studies have revealed eight additional species, which include *Rhizobium hainanense* (Chen *et al.,* 1997), *Rhizobium huautlense* (Wang *et al.,* 1998), *Rhizobium giardinii* (*Phaseolus* genomic species 2 type H152), *Rhizobium gallicum* (*Phaseolus* genomic species 1 type R602) (Amarger *et al.,* 1997), *Rhizobium indigoferae* (Wei *et al.,* 2002), *Rhizobium mongolense* (van Berkum *et al.,*1998), *Rhizobium sullae* (Squartini *et al.,* 2002) and *R. yanglingense* (Tan *et al.,* 2001). Recent changes in the genus *Rhizobium* include the transfer of *Rhizobium loti* and *Rhizobium huakuii* to the new genus *Mesorhizobium*. The genus *Mesorhizobium* was proposed to house the meso-growing rhizobia related to *R. loti*. The present species belonging to this genus include *Mesorhizobium loti*, *Mesorhizobium huakuii*, *Mesorhizobium mediterraneum*, *Mesorhizobium ciceri*, *Mesorhizobium tianshanense* (Jarvis *et al.,* 1997), *Mesorhizobium plurifarium* (de Lajudie *et al.,* 1998b) and *Mesorhizobium amorphae* (Wang *et al.,* 1999a).

Chen *et al.*, (1988) proposed the genus *Sinorhizobium* to house fast-growing nodule-forming soybean rhizobia and contained *Sinorhizobium fredii*, *Sinorhizobium xinjiangense* (Chen *et al.,* 1988) and *Sinorhizobium meliloti*. However, this genus has expanded to include *Sinorhizobium terangae* and *Sinorhizobium saheli* (de Lajudie *et al.,* 1994), *Sinorhizobium kostiense* and *Sinorhizobium arboris* (Nick *et al.,* 1999) and *Sinorhizobium medicae* (Rome *et al.,* 1996b). Though *S. xinjiangense* was described some time ago, yet it is controversial due to the close similarity to *S. fredii* and lack of polyphasic data (Jarvis *et al.,* 1992). Jordan (1984) and Bradbury (1986) suggested that the members of the genus *Agrobacterium* are not capable of nodulating the roots of leguminous plants as caused by the six genera of rhizobia but instead are considered plant pathogens capable of causing gall disease and root hair disease.

*Bradyrhizobium japonicum*, which nodulates soybean, is the type species of *Bradyrhizobium*. *Bradyrhizobium japonicum* and *Bradyrhizobium elkanii* have been established species (Young and Haukka, 1996), but studies for extra slow-growing bradyrhizobia have proposed a new species, *Bradyrhizobium liaoningense* (Xu *et al.,* 1995), which exhibited unique phenotypic and genotypic characteristics. This genus currently includes another three (total six) species viz. *Bradyrhizobium yuanmingense* (Yao *et al.,* 2002) establish symbiosis with *Lespedeza*, *Bradyrhizobium betae* (Rivas *et al.,* 2004), produces nodules in *Beta vulgaris* and *Bradyrhizobium canariense* (Vinuesa *et al.,* 2005) that nodulates Genistoid legumes (Papilionoideae: Genisteae) from the Canary Islands. The *Bradyrhizobium* genus is phylogenetically closer to non-symbiotic bacteria like *Rhodopseudomonas palustris*, *Afipia*, *Nitrobacter* and *Blastobacter denitrificans*, than to *Rhizobium* (Willems and Collins, 1993; Molouba *et al.,* 1999; Yanagi and Yamasato, 1993).

Of special interest, photosynthetic *Bradyrhizobium* sp. strains were described, which were isolated from the stem nodules of *Aeschynomene indica*. It was shown to synthesise a pigment similar to the bacteriochlorophyll a (Eaglesham *et al.,* 1990; Evans *et al.,* 1990). Jaftha *et al.*, (2002) characterized the pink bacteria that nodulate *Lotononis bainesii* as *Methylobacterium*. At the same time there were reports of species of *Burkholderia*, belonging to the β-subclass of the Proteobacteria, isolated from root-nodules of legume. Moulin *et al.*, (2001) identified *Burkholderia* from nodules of *Aspalathus carnosa*, while Vandamme *et al.*, (2002) noted that *B. tuberum* and *B. phymatum* nodulated the roots of tropical legumes. More recently, Ngom *et al.*,

(2004) isolated bacteria from the root-nodules of *A. mangium* belonging to the *Ochrobactrum* clade. Still, it is an expanding field because more and more rhizobia are isolated and characterized everyday, especially from the tropical and the Mediterranean zones, where diversity is still poorly documented.

Nowadays the polyphasic approach (de Lajudie *et al.,* 1994; Jarvis *et al.,* 1997) seems to be best suited and has been adopted for rhizobial classification. At present the rhizobial taxonomy is undergoing frequent changes because strains with distinctive properties are being discovered as more and more sources of biodiversity are being studied. The diversity of rhizobia is assessed by an array of methods designed to gather data that are generated from phenotypic and genotypic characters. These include classical phenotypic analysis and numerical taxonomy, whole cell protein analysis, multilocus enzyme electrophoresis (MLEE), analysis of cellular fatty acids (FAME), DNA-base composition, Pulsed Field Gel Electrophoresis (PFGE), DNA-DNA reassociation, IS (Insertion sequence) typing, RFLP (Restriction Fragment Length Polymorphism) of 16S rDNA (Amplified Ribosomal DNA Restriction Analysis; ARDRA) conserved genes which includes rRNA genes and intergenic spacer regions profiling, whole genome fingerprinting using AFLP (Amplified Fragment Length Polymorphism), rep-PCR, RAPD (Randomly Amplified Polymorphic DNA), ap-PCR (arbitrarily primed PCR) and DAF (DNA Amplification Fingerprinting).

Methods that are intended to portray DNA fingerprints for genomic screening give a representative picture. Nevertheless, it is not usually possible to define which parts of the genome are varying or quantify the similarities or differences (Young, 1992). The DNA based techniques viz. random whole genome analysis, mobile genetic elements, specific gene variation have provided more information than the conventional phenotypic data. They also pointed out some questions that were previously not evident. Random whole genome analysis methods may apply for recent divergences, but where divergence may have occurred for sometime it is uncertain and whether these differences are linked to the stable part of the genome or the exchangeable genes or regions (Gurtler and Mayall, 2001). Unless they confirm using whole genome analysis, they have the same standing as phenotypic methods.

## MYCORRHIZA

### Fungus-Plant Root Symbiosis

Mycorrhizas are global associations including diverse morphological, functional and evolutionary categories (Smith and Read, 1997; Brundrett, 2002). Almost all the plant species of natural vegetation and the agricultural crop plants of the tropics live in mycorrhizal association with fungi (Gautam and Prasad, 2001). Vesicular arbuscular mycorrhizal associations (VAM), which are also called arbuscular mycorrhiza (AM) or glomeromycotan mycorrhiza (GM), are the most widespread and common root-fungus associations. It was Frank (1885), who gave the name "mycorhiza". Kelley (1931, 1950) incorporated the second "r" in the word "mycorrhiza".

### Vesicular Arbuscular Mycorrhiza (VAM)

According to Nehls *et al.*, (2001) and Pfeffer *et al.*, (2001), mycorrhiza differs primarily from other plant-fungus associations because it is close associations where exchange of materials takes place between living cells. Janse (1897) called the intramatrical spores "vesicules" and Gallaud (1905) called the other commonly observed intracellular structures "arbuscules". In general, arbuscules are used to define VAM associations. According to McGonigle *et al.*, (1990) and Toth *et al.*, (1990) their abundance is usually correlated with the degree of colonization of young roots by VAM fungi.

Arbuscules are short-lived structures that are often not in attendance or hard to see due to rootage and pigments in field-collected roots (Brundrett *et al.,* 1996). There is discrepancy in naming these mycorrhizas appropriately as arbuscular mycorrhizal association. The recognition that not all fungi formed vesicles led to the proposal that this symbiosis should be renamed arbuscular mycorrhiza. Nowadays, the term "arbuscular mycorrhiza" has become a trend, however, the term vesicular-arbuscular mycorrhiza (VAM) is as accurate as arbuscular mycorrhiza (AM).

## Discovery and Description

Earlier, mycorrhizal quantification was achieved by clearing the roots by heating in KOH and staining fungal cell walls with trypan blue in lactophenol (Phillips and Hayman, 1970). However, this stain does not differentiate between active and inactive or dead cells of the fungal hyphae (Singh, 2005). Probably, the most popular method for quantification of mycorrhization is based on the line intersect technique devised by Newman (1966). Giovannetti and Mosse (1980) compared various methods of mycorrhizal quantification, which led to greater acceptance of the line intersect method. Mycorrhizas are complex symbiosis and the fungi involved in it produce a variety of structures within the root. Quantification of these structures viz. hyphae, arbuscules, and vesicles was standardized by the method proposed by McGonigle *et al.*, (1990). The extraction of spores from soil is essential for their classification. It was made possible by wet sieving and decanting method, a method commonly used to extract nematodes from soil and adopted to arbuscular mycorrhizal fungi by Gerdemann (Gerdemann, 1955; Gerdemann and Nicolson, 1963).

## AMF Reproduction and its Morphological Traits

All AMF that have been described to date reproduce only by asexual means (Schubler *et al.,* 2001). There is no evidence that the Glomeromycota reproduces sexually. AMF reproduce clonally via spores, vesicles and hyphae. Furthermore, Kuhn *et al.* (2001), Pawlowska and Taylor (2004), Hijri and Sanders (2005) argued on the question, whether the nuclei in the mycelium and spores of one organism are genetically identical or not. Under favourable conditions glomeromycotan spores germinate, form appressoria on host roots and establish a new mycorrhizal symbiosis.

Mycorrhizal fungal hyphae are aseptate to sparsely septate viz., they are generally not divided by cell walls as described by Isaac (1992) and Schubler *et al.* (2001). This is unusual feature of the AMF, as most fungi, including members of the most phyla have septae (Isaac, 1992) and a lack of septae is a good first sign in identifying an AM fungus. To identify AM fungus at the lower taxonomic levels, certain hyphal traits have been used which include the number of hyphae per spore, and hyphal colour, shape and size, among others (Morton, 1988; Merryweather and Fitter, 1998). Different infection patterns viz., the arrangement of hyphae, vesicles and arbuscules are characteristic of different groups of species as described by Abbott (1982). Hence, the fine morphological details of AMF may help in species diagnosis. On the other hand, this can be difficult as vesicles, hyphae and particularly arbuscules often are visible as darker, blurry areas of stained root cortex cells. In addition, a single fungal species can have a different morphological appearance, which depend on the species of its host plant (Merryweather and Fitter, 1998).

## Species Concept in Relation to AMF

The biological species concept (Coyne, 1994) is irrelevant for asexual groups such as the AMF (Freeman and Herron, 1998). A typological species is usually delineated as a group of organisms that appear similar to one another and different to other groups of organisms. This species concept can be applied when the interbreeding criterion cannot be tested viz., in fossil organisms or in asexual lines such as the AMF (Futuyma, 1998) or when accurate phylogenies are not available. It is said that the typological species (morphospecies; Freeman and Herron, 1998) are convenient for describing particular syndromes of traits, but they do not inevitably bear a strong resemblance to consider being 'species'.

Furthermore, researchers have imposed a taxonomic system on AMF but it is not yet a 'natural systematics' (Schubler *et al.,* 2001). Since the AMF are asexual, we cannot describe AMF in terms of biological species. However, we could define them in terms of phylogenic species if accurate phylogenies were available at the fine scale, but until now this is not the case. We can define AMF in terms of typological species, but many biologists disagree with typological definitions because they do not generally reflect the evolutionary history of the organisms that are being classified. As per Wiley (1978) asexuality has been called "the bane of all proposed (species) definitions that are not obviously typological".

## Biological Network of Mycorrhiza

Beneficial microbial interactions involving arbuscular mycorrhiza is the omnipresent fungus–plant beneficial symbiosis (Barea *et al.,* 2005). The great majority of land plants are host for some type of mycorrhiza and members of most plant families form AM (Smith and Read, 1997). Other types of mycorrhiza are formed with fungi from the phyla Ascomycota or Basidiomycota viz., ectomycorrhiza of trees and shrubs, ericoid mycorrhiza of Ericales, orchid mycorrhiza and some others.

Only a few plant families are regarded as non-mycorrhizal, among them the Brassicaceae viz. cabbage, Caryophyllaceae (viz. carnation) and Chenopodiaceae (viz. spinach). AMF are best-known for their obligatory symbiotic relationship with plants. This exclusive association is found in virtually all terrestrial ecosystems and environments (Morton, 1988) and associated with the species in a majority of plant families (Rillig, 2004; Prasad, 2005).

Different AMF species are better/worse mutualists than others (van der Heijden *et al.,* 1998). Other fungi form symbiotic mutualisms viz., 'mycobionts' in lichens, including mutualisms with plants viz., endophytes, but the mycorrhizal mutualism formed by AMF and their host plants have a distinct biological network. The mycorrhizal mutualism is based primarily on the exchange of nutrients. Mycorrhizal fungi are better than roots at acquiring poorly mobile phosphate from the soil, which they exchange with their host plant for their sole source of carbon (Bolan, 1991; Chandrashekara *et al.,* 1995). In addition to their role in phosphate acquisition, AMF have other potential benefits to plants as described by Morton (1988), Prasad and Rajak (2002) and Rillig (2004), viz. the uptake of other important nutrients such nitrogen, protection against root pathogens, water acquisition and the mediation of pollution effects.

AMF also have some important ecological effects at the ecosystem scale viz., they have profound effects on plant species composition, diversity and productivity (van der Heijden *et al.*, 1998; Hart and Klironomos, 2002; Klironomos *et al.*, 2000; Manoharachary *et al.*, 2005). Rillig (2004) called AMF fungi as 'keystone mutualists' due to their vital importance for plant species composition, ecosystem properties, and all the indirect effects associated with it. According to Smith and Read (1997), host specificity of AM appears to be very low as many species were observed by them to colonize a wide range of host plants in the greenhouse study. Furthermore, plants are generally colonized by a mixture of AM fungal species, often within the same root (Helgason *et al.*, 1999).

Field studies using molecular identification methods have demonstrated that distinct fungal communities are associated with different hosts (Vandenkoornhuyse *et al.*, 2002). While according to Bidartondo *et al.* (2002) certain non-photosynthetic plants may cheat the fungus by obtaining all their nutrients from them, including carbohydrates.

## Effects of AMF on Host Plant

Generally, most researchers consider fungi, infecting plant tissues as pathogens or parasites and names as "Rhizophagus" (Dangeard 1900), which means "root eater". Rayner (1926–1927) accepted that ectomycorrhizal fungi might be beneficial to their hosts but she did not state the same for AM fungi. Using non-sterile soil as an inoculum, It was Asai (1943) who demonstrated that mycorrhizal plants grew faster than non-mycorrhizal plants. Nicolson (1967) and Gerdemann (1971) produced large volumes of AMF inoculum in pot cultures, which they produced from single isolates of fungal species. However, disagreement in the nature of mycorrhizal effects on the host has been also reported (Lohman, 1927; Janos, 1980; Johnson *et al.*, 1997).

Thus there are prominent cases of growth depression apparently caused by AMF in "non-host" species (Francis and Read, 1984) or in host species when phosphate availability in soil is high (Mosse, 1973; Peng *et al.*, 1993). Baylis (1959) reported first that the beneficial mycorrhizal effect was mediated by P uptake. Gerdemann (1964) also confirmed that non-mycorrhizal plants exhibited "severe phosphorus deficiency symptoms" and had significantly lower P concentrations and higher K and Mg concentrations than mycorrhizal plants.

Bowen and Rovira (1968) reported first that transfer of nutrients from fungus to host occurred across functional, intact arbuscules (Woolhouse, 1975). Marx *et al.* (1982) described the first biochemical evidence that made clear the role of the intact arbuscule in P transfer. They showed that the host plasma lemma, which invaginates around the arbuscular hyphae, had a very high ATP-ase activity, which later shown to be H+-ATPase (Gianinazzi-Pearson *et al.*, 1991, 2000), suggesting the presence of active transport mechanisms. Smith and Smith (1997) questioned whether arbuscules (alive or dying) are needed for P transfer, as intercellular hyphae may also be a site of P transfer.

AM fungi also absorb K, Ca, S, Fe, Mn, Cu, and Zn from the soil and translocate these nutrients to the host plants (Gerdemann, 1975; Hayman, 1982; Tinker and Gildon, 1983;

Newsham *et al.,* 1994). Moreover, they are also very useful to plant species that inherently lack morphological or physiological mechanisms for efficient P uptake (Manjunath and Habte, 1991; Koide and Schreiner, 1992). The early emphasis, placed on the role of mycorrhizal fungi in encouraging plant growth may have distracted us from an added, very significant role they play as stabilizers of soil structure (Clough and Sutton, 1976; Nicolson and Johnston, 1979; Tisdall and Oades, 1979; Miller and Jastrow, 2000) and as integral components of a very diverse soil biota (Bethlenfalvay and Schuepp, 1994; Franke-Snyder *et al.,* 2001).

Nevertheless, non-nutritional effects of AMF viz. root branching (Berta *et al.,* 1990), ethylene production (McArthur and Knowles, 1992; Besmer and Koide, 1999) or protection from pathogens (Prasad and Rajak, 2002) the suppression of plant diseases (Hooker *et al.,* 1994; Newsham *et al.,* 1994; Trotta *et al.,* 1996) including nematode infection (Cooper and Grandison, 1986; Habte, 1999), stimulate hormone production in plants which aid in improving soils structure (Wright and Upadhyaya, 1996; Bethlenfalvay, 1998; Wright and Upadhyaya, 1998), enhance leaf chlorophyll levels, (Tsang and Maun, 1999) and improve plant tolerance to water stress, salinity, soil acidity and heavy metal toxicity (Bethlenfalvay, 1992) are very important as well. Some of these functions may be the indirect effects of improved P nutrition (O'Keefe and Sylvia, 1991).

Several mechanisms have been proposed to explain the protection of AMF symbiosis such as changes in plant hormones (Allen *et al.,* 1982; Barea and Azcon-Aguilar, 1982; Danneberg *et al.,* 1992; Goicoechea *et al.,* 1995), increased leaf gas exchange and photosynthetic rate (Ruiz-Lozano *et al.,* 1996a), acidification of the rhizosphere (Bago and Azcon-Aguilar, 1997), increases in root phosphatase activity (Fries *et al.,* 1998), direct hyphal water uptake from the soil and transfer to the host plant (Hardie, 1985; Faber *et al.,*1991; Ruiz-Lozano and Azcon, 1995), enhanced activity of enzymes involved in antioxidant defense (Ruiz-Lozano *et al.,*1996b), nitrate assimilation (Ruiz-Lozano and Azcon, 1996), enhanced water uptake through improved hydraulic conductivity and increasing leaf conductance and photosynthetic activity (Koide, 1985; Dell-Amico *et al.,* 2002) and changes in cell-wall elasticity (Auge *et al.,*1987; Sanchez-Diaz and Honrubia, 1994).

## Biodiversity and Taxonomy of AMF

Gerdemann and Trappe (Gerdemann and Trappe, 1974) split the old *Endogone sensulato* into seven genera including *Endogone, Modicella,* and *Glaziella*, which were nonmycorrhizal genera, and four mycorrhizal genera including *Glomus, Sclerocystis* and two new genera *Gigaspora* and *Acaulospora* (Table 2). These were all grouped in the Endogonaceae, Endogonales and Zygomycetes. Another mycorrhizal genus Entrophospora was recognized by Trappe and Schenck (1982). Walker (1987) described, five arbuscular mycorrhizal fungal genera and dropped Sclerocystis and added Scutellospora in it. Morton and Benny (1990) placed the five genera of Walker (1987) into three families viz. Glomaceae, Acaulosporaceae, Gigasporaceae and two suborders viz. Glomineae and Gigasporineae, both of which were then placed in a new order, the Glomales. Later on, Morton and Benny (2001) identified two other families, the Archaeosporaceae and Paraglomaceae, with two new genera, *Archaeospora* and *Paraglomus* as illustrated below in Table 2.

The arbuscular mycorrhizal fungi composed an order, Glomales, within Zygomycota (Morton and Benny, 1990). These authors also corrected the incorrect ordinal name "Glomales" to "Glomerales". After isolating DNA from single spores of number of AMF taxa Schubler *et al.,*

**Table 2** Older Classifications of AM Fungi

| *Gerdemann and Trappe (1974)* | *Morton and Benny (1990, 2001)* |
|---|---|
| Zygomycetes | Zygomycetes |
| Endogonales | Glomales |
| Endogonaceae | Glomineae |
| *Endogone* | Glomaceae |
| *Sclerocystis* | *Glomus* |
| *Glomus* | *Sclerocystis* |
| *Gigaspora* | Acaulosporaceae |
| *Acaulospora* | *Acaulospora* |
| *Modicella* | *Entrophospora* |
| *Glaziella* | Gigasporineae |
| | Gigasporaceae |
| | *Gigaspora* |
| | *Scutellospora* |
| | Archaeosporaceae |
| | *Archaeospora* |
| | Paraglomaceae |
| | *Paraglomus* |

(2001) used primers specific to the small subunit ribosomal RNA (ssrRNA) (Schwarzott and Schubler, 2001). Their analysis suggested that AM fungi are clearly distinct from the other major groups of fungi and are more closely related to Ascomycota and Basidiomycota than the rest of Zygomycota with which they were grouped. Unlike the asexual members of the Glomeromycota, *Endogone* species reproduce sexually via zygospores, which indicate their phylogenetic link with the phylum Zygomycota. Phylogenetic analysis of the nuclear ssrRNA strongly suggests that *Endogone* (Endogonales) and the Glomeromycota do not form a clade (Gehrig *et al.,*1996). Further, Schubler *et al.,* 2001 reported that their current placement is within their own phylum, Glomeromycota, which is composed of a single order, Glomeromycetes. The Glomeromycota are a very old group with an estimated origin of at least 600 to 620 million years ago. Spores and hyphae of glomalean fungi were discovered in 460 million year old rocks from the Ordovician (Redecker *et al.,* 2000b). Moreover, Schubler (2000) suggested that all AMF are part of Glomeromycota, and all the known members of Glomeromycota form arbuscular mycorrhizas or their functional equivalent with non-higher plants.

The AMF are a very ancient lineage (Simon *et al.,* 1993) which might have diverged more than a billion years ago, before the evolution of the first land plants (Heckman *et al.*, 2001; Schubler, 2002). Schubler (2002) suggested that at the beginning, AMF might have been associated with cyanobacteria instead of plants and *Geosiphon* as is a single modern representative which produces bladders that harbour symbiotic cyanobacteria (Schubler *et al.,* 1994,1996). Nonetheless, molecular phylogenetic analysis revealed that *Geosiphon* is a member of the Glomeromycota (Schubler *et al.,* 2001). The mode of spore formation is not a useful diagnostic feature for some genera. Further, Redecker *et al.* (2000a) reported that the spores of the *Glomus* and *Acaulospora* types were produced by several distinct, deeply divergent lineages. They were described as two new genera *Archaeospora* and *Paraglomus* (Morton and Benny, 2001) and were placed in separate families.

Originally members of this genus *Archaeospora* were classified in separate families because some species were dimorphic (Morton *et al.,* 1997). *Glomus* groups A and B (Schwarzott *et al.,* 2001) are typified by the well-known species *Glomus mosseae* and *Glomus claroideum*, respectively. These two groups are genetically distant but form a monophyletic group in rDNA phylogenetic trees. Further, they suggested that the formation of a "sporiferous saccule" was characteristic for the Acaulosporaceae (*Acaulospora* and *Entrophosphora*), however, now it is known to occur in one additional lineage also viz., *Archaeospora*.

The members of Gigasporaceae viz., *Scutellospora* and *Gigaspora* are distinguished by the formation of their spores on a "bulbous suspensor" and which is well supported by molecular data also. Gigasporaceae and Acaulosporaceae form a clade in most rDNA phylogenies, which is in disagreement with previous morphology-based analyses that placed *Glomus* and Acaulosporaceae together (Morton and Benny, 1990). Redecker *et al.* (2000b) suggested that species which form complex sporocarps formerly placed in the genus *Sclerocystis* are in fact phylogenetically, nested within well-characterized *Glomus* species with simple spores.

**Table 3** Phylogenetic Relationship of AM Fungi (Schubler et al., 2001)

- Fungi
  - Glomeromycota
    - Glomeromycetes
      - Glomerales
        - Glomeraceae
          - *Glomus* gr. A & B (Schwarzott *et al.,* 2001)
      - Paraglomerales
        - Paraglomeraceae
          - *Paraglomus*
        - Pacisporaceae
          - *Pacispora* (Oehl and Sieverd, 2004)
        - Diversisporales
          - Diversisporaceae
            - Diversispora
          - Gigasporaceae
            - *Gigaspora*
            - *Scutellospora*
          - Acaulosporaceae
            - *Acaulospora*
            - *Entrophospora*
        - Archaeosporales
          - Archaeosporaceae
            - *Archaeospora*
          - Geosiphonaceae
            - *Geosiphon*

Oehl and Sieverding (2004) erected a new genus *Pacispora* that have characteristics intermediate between *Glomus* and the Gigasporaceae. The genus name *Gerdemannia* is a synonym of *Pacispora* which was published for the same fungal group after a few weeks (Walker *et al.*, 2004), and presented here as an illegitimate name based on the publication date. Another split off from *Glomus* (Morton and Benny, 1990) is a genus *Diversispora* (Table 3).

Walker *et al.* (2004) reported that the lower level taxonomy of the AMF is currently very plastic and major revisions continue to be made. Methods based on serology (Aldwell and Hall, 1987), isozyme variation as revealed by gel electrophoresis (Hepper, 1987) and fatty acid variation (Bentivenga and Morton, 1994) were introduced which will eventually make accurate species delineation more realistic. Furthermore, the ability to resolve species is being improved by DNA analyses as Redecker *et al.* (2003) claimed that "molecular identification techniques are about to revolutionize studies on ecology of arbuscular mycorrhiza" and systematists have come to depend increasingly on DNA-based methods (Cummings, 1990; Davidson and Geringer, 1990; Simon *et al.*, 1990, 1992; Redecker, 1997; Redecker, 2000). It will also make identification of AMF easier for those with access to specialized equipment viz., PCR machines, but more difficult for those without.

## AGRICULTURAL PERSPECTIVES AND SOIL FERTILITY

The ultimate goal of modern agricultural systems is high productivity within environmental and economic constraint. Today, the use of mineral fertilizers is considered the quickest and surest way of boosting crop production but their cost and other constraints deter farmers from using them in recommended quantities (Tilak *et al.*, 2005). Global utilization of N fertilizer has increased sevenfold since 1960 (Peoples and Ladha, 1995) and P fertilization use is forecasted to increase anywhere from 1·6-fold to 3·4-fold by 2050 (Tilman *et al.*, 2001). Nevertheless, in tripartite symbioses both the rhizobial and fungal microsymbionts improve the mineral nutrition (especially N/P requirements) of the host plant in exchange for assimilates provided by the later.

In most cases investigated, especially when both nitrogen and phosphate are limiting factors, VAM fungi and rhizobia appear to act synergistically since combined inoculation with mycorrhiza and rhizobia enhances plant growth and reproduction more than inoculation with either microsymbiont alone and also leads to a higher degree of host colonization by the two symbionts (Daft and El-Giahmi, 1974; Cluett and Boucher, 1983; Kawai and Yamamoto, 1986; Pacovsky *et al.*, 1986; Chaturvedi and Singh, 1989; Beukes, 2001).

Legume inoculation with rhizobia and bradyrhizobia is a long-established and successful practice, especially with particular crops. A few years after rhizobia were first described and cultured in pure medium, Nobbe and Hiltner received a patent on a process for inoculating legumes by coating the seed with a *Rhizobium* culture or, alternatively, mixing the culture with some soil, which was then mixed with the legume seed (United States Letter Patent 570,813, 1895). The inputs into terrestrial ecosystems of BNF from the symbiotic relationship between legumes and their rhizobia amount to at least 70 million tonns of N per year (Brockwell *et al.*, 1995). Further, this enormous quantity will have to be greater than before as the world's population increases and as the natural resources, which supply fertilizer-N, get smaller. The symbioses between *Rhizobium* or *Bradyrhizobium* and legumes are a cheaper and usually more effective agronomic practice for ensuring an adequate supply of N for legume-based crops than the application of fertilizer-N. Field trials conducted in India showed that nearly 50% of nitrogenous fertilizer can be saved through rhizobial inoculations with considerable increase in yield depending on the legume, soil and agroclimatic conditions (Rewari and Tilak, 1988; Tilak, 1993). Elimination or substantial reduction of N fertilizer is considered as a key factor in the development of eco-friendly agricultural system (Reis *et al.,* 2000).

As most crops are hosts for AM, this association is potentially an important resource for agriculture as described by Sieverding (1991). Asai (1944) was perhaps the first to describe that normal level of nodulation might depend on the presence of mycorrhizal fungi. The importance of AMF symbiosis has been attributed to high P requirements on the nodulation and $N_2$ fixation process in legume plants, which requires enhanced P uptake (Barea and Azcon-Aguilar, 1983). Improved P nutrition has been shown to increase in infertile and P fixing soils of the tropics (Dodd, 2000). Mycorrhizal fungi can also improve absorption of N fromss $NH_4^+$ -N mineral fertilizers, transporting it to the host plant (Ames *et al.,* 1983; Johansen *et al.,* 1993). Moreover, its transport and absorption can also increase biomass production in soils with low potassium, calcium and magnesium (Liu *et al.,* 2002).

Vesicular arbuscular mycorrhizal fungi (VAMF) can act as a potent biofertilizer but axenic cultures of VAMF are currently far from being realized. As the significant development of complete life cycle is only achieved in the presence of a host plant, the application of VAMF as a biofertilizer, from large-scale production, is also a far off vision, (Sahay *et al.,* 1998; Prasad and Kaushik, 2004). Most inocula have been produced in pot cultures using soil mixtures (Wood, 1985), but other technologies have been developed in the attempt to decrease costs and increase purity. These include nutrient film culture (Warner *et al.*, 1985), aeroponics (Hung and Sylvia, 1988) or expanded clay hydroponics (Dehne *et al.*,1987). There has been some notable research to develop practical on-farm inoculation production systems (Furlan, 1993), and recent large-scale production of *in vitro* mycorrhizas may also increase the practicality of using inoculum (Adholeya, 2003).

## CONCLUSION

We have come a long way in developing our understanding of rhizobial and mycorrhizal diversity and established new classification systems for them, but there are many milestones yet to meet. Progress in taxonomy is also due to an increasing number of available effective techniques in bacterial characterization, more and more accessible to an increasing number of laboratories. Molecular tools offer great opportunities to reveal the phylogeny among microorganisms (here bradyrhizobia and AMF) and on the basis of it, the recent taxonomic status of rhizobia and AMF have been provided. After considering the relative merits of rhizobia and vesicular arbuscular mycorrhizas, it is clear from the present review that legume symbiosis with rhizobia and AMF seems to be very promising for the production of eco-friendly, non-polluting, non-pathogenic and "ideal" bradyrhizobial and mycorrhizal biofertilizers for sustainable agriculture system and soil health. The three-dimensional symbiotic system of plant-rhizobia-AMF, when managed properly, in different agro-ecosystems is likely to enhance the production potential of nutrient deficient soils and minimize the use of deteriorating fertilizers. Moreover, elimination or substantial reduction of

N fertilizer and P fertilizer is a key factor in the development of environment-friendly agricultural systems. There are so many research and extension areas in which in-depth efforts are still needed, and numerous opportunities for improved $N_2$ fixation and nutrients acquisition (mainly P) likely to arise through advances in molecular biology.

## References

Abbott, L.K. (1982) *Aus. J. Bot.* **30**: 485-499.

Adholeya, A. (2003) In: *Proc. 4th International Conference on Mycorrhizae*, Montreal, Canada, pp. 240.

Akkermans, A.D.L. & Van Dijk, C. (1981) Non-leguminous root nodule symbioses with actinomycetes and rhizobia. In *Nitrogen Fixation* (Ed. Broughton, W.J.) Cambridge Univ. Press, Cambridge, pp. 57.

Albus, U., Baier, R., Holst, O., Puhler, A. & Niehaus, K. (2001)*New Phytol.* **151**: 597-606.

Aldwell, F.E.B. & Hall, I.R. (1987) In: *Mycorrhizae in the next decade: Practical applications and research priorities* (Eds. Sylvia, D.M., Hung, L.L. & Graham, J.H.) Institute of Food and Agricultural Sciences, University of Florida, Gainesville, Fla, pp. 305–307.

Allen, E.B., Allen, M.F., Helm, D.J., Trappe, J.M., Molina, R. & Rincon, E. (1995) *Plant Soil* **170**: 47–62.

Allen, M.F., Moore, T. S. Jr. & Christensen, M. (1982) *Can. J. Bot.* **60**: 468-471.

Allen, O.N. & Allen, E.K. (1939) *Soil Sci.* **47**: 63-76.

Amarger, N., Macheret, V. & Laguerre, G. (1997) *Int. J. Syst. Bacteriol.* **47**: 996-1006.

Ames, R.N., Reid, C.P.P., Porter, L. & Cambardella, C. (1983) *New Phytol.* **95**: 381-396.

Asai, T. (1943) *Jpn. J. Bot.* **12**: 359-436.

Asai, T. (1944) *Jpn. J. Bot.* **13**: 463.

Auge, R.M., Schekel, K.A. & Wample, R.L. (1987) *Plant Physiol.* **70**: 175-182.

Bago, B. & Azcon-Aguilar, C. (1997) *Zeitschrift Fur Pflanzen Und Bodenkunde* **160**: 333-339.

Barea, J.M. & Azcon-Aguilar, C. (1982) *Appl. Environ. Microbiol.* **43**: 810-813.

Barea, J.M. & Azcon-Aguilar, C. (1983) *Adv. Agron.* **36**: 1-54.

Barea, J.M., Pozo, M. J., Azcon, R. & Azcon-Aguilar, C. (2005) *J. Exp. Bot.* **56**: 1761-1778.

Baylis, G.T.S. (1959) *New Phytol.* **58**: 274-277.

Becker, A. & Puhler, A. (1998) In: *The Rhizobiaceae: Molecular biology of model plant associated bacteria* (Eds. Spaink, H.P., Kondorosi, A. & Hooykaas, P.J.J.) Kluwer Academic Publishers, Dordrecht, The Netherlands, pp. 97–118.

Beijernick, M.W. (1888) *Botanische Zeitung* **46**: 797-804.

Bentivenga, S.P. & Morton, J.B. (1994) *Mycol. Res.* **98**: 1419-1426.

Bergersen, F.J. (1982) Root nodules of legumes: structure and functions. Chichester: Research Studies Press.

Bergersen, F.J. (1997) *Plant Soil* **191**: 189-203.

Bergey's Manual of Systematic Bacteriology (2001) 2nd edn (Ed. George Garrity, M.) Springer-Verlag, New York.

Berta, G., Fusconi, A., Trotta, A. & Scannerini, S. (1990) *New Phytol.* **114**: 207–216.

Besmer, Y.L. & Koide, R.T. (1999) *Mycorrhiza* **9**: 161-166.

Bethlenfalvay, G.J. & Schuepp, H. (1994) In: *Impact of arbuscular mycorrhizas on sustainable agriculture and natural ecosystems* (Eds. Gianinazzi, S. & Schuepp, H.) Birkhauser, Basel, Switzerland.

Bethlenfalvay, G.J. (1992) In: *Mycorrhizae in Sustainable agriculture* (Eds. Bethlenfalvay, G.J. & Linderman, R.G.), ASA/CSSA/SSSA, Madison, Wisconsin, USA, pp. 1-27.

Bethlenfalvay, G.J., Cantrell, I.C., Mihara, K.L. & Schreiner, R.P. (1998) *Bio. Fert. Soil* **28**: 356-363.

Beukes, D.J. (2001) *Plant Prot. News* **59**: 16-17.

Bever, J. D., Westover, K. M. & Antonovics J. (1997) *J. Ecol.* **85**: 561-574.

Bidartondo, M.I., Redecker, D., Hijri, I., Wiemken, A., Bruns, T.D., Domínguez, L., Sersic, A., Leake, J.R. & Read D.J. (2002) *Nature* **419**: 389-393.

Bolan, N.S. (1991) *Plant and Soil* **134**: 189-207.

Boogerd, F.C. and van-Rossum, D. (1997) *FEMS Microbiol. Rev.* **5**: 21 -27.

Bottomley, P. (1991) In: *Biological Nitrogen Fixation* (Eds. Stacey, G., Burris, R.H. & Evans, H.J.), Chapman and Hall, New York, pp. 292-347.

Bowen, G.D. & Rovira, A.D. (1968) In: *Root growth* (Ed. Wittington, W.J.), Butterworth, London, pp. 170-199.

Bradbury, J. F. (1986) Agrobacterium- A Guide to Plant Pathogenic Bacteria, Farnham Royal, CAB International Mycological Institute, pp. 6-15.

Bradbury, S.M., Peterson, R.L. & Bowley, S.R. (1991) *New Phytol.* **119**: 115-120.

Brewin, N.J. (1998) In: *The Rhizobiaceae: Molecular biology of model plant associated bacteria* (Eds. Spaink, H.P., Kondorosi, A. & Hooykaas, P.J.J.), Kluwer Academic Publishers, Dordrecht, pp. 417-429.

Brockwell, J., Bottomley, P.J. & Thies, J.E. (1995) *Plant Soil* **174**:143-80.

Brockwell, J., Searle, S.D., Jeavons, A.C. & Waayers, M. (2005) Nitrogen fixation in acacias - An untapped resource for sustainable plantations, farm forestry and land reclamation, ACIAR Monograph No. 115, pp. 132.

Brundrett, M. & Abbott, L.K. (2002) In: *Microorganisms in plant conservation and biodiversity* (Eds. Sivasithamparam, K., Dixon, K.W. & Barrett, R.L.), Kluwer Academic Publishers, Dordrecht, pp. 151–193.

Brundrett, M., Bougher, N. Dell, B., Grove, T. & Malajczuk, N. (1996) Working with mycorrhizas in forestry and agriculture. Australian Centre for International Agricultural Research, Canberra.

Burns, R.C. & Hardy, R.W.F. (1975) Nitrogen fixation in bacteria and higher plants. Springer-Verlag, New York.

Burris, R.H. (1994) In: *Nitrogen fixation with non legumes* (Eds. Hegazi, N.A., Fayez, M. & Monib , M.), The American University in Cairo Press, Cairo, Egypt, pp. 1–11.

Chandrashekara, C.P. Patil, V.C. & Sreenivasa, M.N. (1995) *Plant Soil* **176**: 325-328.

Chaturvedi, C. & Singh, R. (1989) *Proc. Natl. Acad. Sci. India* **B59**: 443-446.

Chen, W.M., James, E.K., Prescott, A.R., Kierans, M. & Sprent, J.I. (2003) *Mol. Plant. Micro. Interac.* **16**: 1051–1061.

Chen, W.M. Laevens, S. Lee, T.M., Coenye, T., De Vos, P., Mergeay, M. & Vandamme, P. (2001) *Int. J. Sys. Evol. Microbiol.* **51**: 1729-1735.

Chen, W.X., Li, G.S., Qi, Y.L., Wang, E.T., Yuan, H.L. & Li, J. (1991) *Int. J. Syst. Bacteriol.* **41**: 275–280.

Chen, W.X.,Tan, Z.Y., Gao, J.L., Li, Y. & Wang, E.T. (1997) *Int. J. Syst. Bact.* **47**: 870-873.

Chen, W.X., Wang, E.T., Wang, S.Y., Li, Y.B., Chen, X.Q. & Li, J. (1995) *Int. J. Sys. Bact.* **45**: 153-159.

Chen, W.X., Yan, G.H. & Li, J.L. (1988) *Int. J. Syst. Bacteriol.* **38**: 392-397.

Chen, W.M., James, E. K., Prescott, A. R., Kierans, M. & Sprent, J.I. (2003) *Mol. Plant Micro. Interact.* **16**: 1051–1061.

Clough, K.S. & Sutton, J.C. (1976) *Can. J. Microbiol.* **24**: 326-333.

Cluett, H.C. & Boucher, D.H. (1983) *Oecologia* **59**: 405-408.

Cooper, K.M. & Grandison, G.S. (1986) *Ann. Appl. Biol.* **108**: 555-565.

Coyne, J. A. (1994) *Evolution* **48**: 19-30.

Cummings, B.A. (1990) In: *Abstracts, 8th North American Conference on Mycorrhizae* (Eds. Allen, M.F. & Williamsm, S.E.), University of Wyoming Agricultural Experiment Station, Laramie, Wyo, pp. 63.

Daft, M.J. & El-Giahmi, A.A. (1974) *New Phytol.* **73**: 1139-1147.

Dangeard, P.A. (1900) *Botaniste* **7**: 285-287.

Dangeard, P.A. (1926) In: *Recherches sur lestubercles radicauxdes legumineuses, Le Botaniste*, Series 16, Paris, pp. 270.

Danneberg, G. Latus, C. Zimmer, W. Hundeshagen, B. Schneider-Poetsch, H.J. & Bothe, H. (1992) *J. Plant Physiol.* **141**: 33-39.

Davidson, K. & Geringer, J.E. (1990) In: *Abstracts, 8th North American Conference on Mycorrhizae* (Eds. Allen, M.F. & Williams, S.E.), University of Wyoming Agricultural Experiment Station, Laramie, Wyo, pp. 70.

de Lajudie, P., Laurent-Fulele, E., Willems, A., Torck, U., Coopman, R. Collins, M.D., Kersters, K., Dreyfus, B. & Gillis, M. (1998a) *Int. J. Sys. Bacteriol.* **48**: 1277-1290.

de Lajudie, P., Willems, A., Nick, G., Moreira, F., Molouba, F., Hoste, B., Torck, U., Neyra, M., Collins, M.D., Lindstrom, K. Dreyfus, B. & Gillis, M. (1998b) *Int. J. Sys. Bacteriol.* **48**: 369-382.

de Lajudie, P., Willems, A., Pot, B., Dewettinck, D., Maestrojuan, G., Neyra, M. Collins, M.D. , Dreyfus, B., Kersters, K. & Gillis, M. (1994) *Int. J. Sys. Bacteriol.* **44**: 715–733.

Dehne, H.W., Backhaus, G.F. & Baltruschat, H. (1987) In: *Mycorrhizae in the next decade, practical applications and research priorities* (Eds. Sylvia, D.M., Hung, L.L. & Graham, J.H.), Institute of Food and Agricultural Sciences, University of Florida, Gainesville, Fla, pp. 272–273.

Dell-Amico, J., Torrecillas, A., Rodriguez, P., Morte, A. & Sanchez-Blanco, M.J. (2002) *J. Agric. Sci.* **138**: 387-393.

Denison, R.F. (2000) *Am. Nat.* **156**: 567-576.

Dixon, R.O.D. & Wheeler, C.T. (1986) In: *Nitrogen fixation in plants.* Blackie, Glasgow, United Kingdom.

Djordjevic, M. A., Gabriel, D.W. & Rolfe, B.G. (1987) *Annu. Rev. Phytopathol.* **25**: 145-168.

Dodd, J.C. (2000) *Outlook Agricul.* **29**: 55-62.

Downie, A. & Brewin, N. (1992) In: *The molecular genetic approach* (Eds. Russo,V.E.A., Brody, Cove, S. D. & Ottolenghi, S.), Springer Verlag, Berlin, pp. 257-270.

Dreyfus, B., Garcia, J.L. & Gillis, M. (1988) *Int. J. Syst. Bacteriol.* **38**: 89-98.

Eaglesham, A.R.J., Ellis, J.M., Evans, W.R., Fleischman, D.E., Hungria, M. & Hardy, R.W.F. (1990) In: *Nitrogen fixation : Achievements and Objectives* (Eds. Gresshoff, P.M., Roth, L.E., Stacey, G. & Newton, W.L.), Chapman and Hall, New York, 1990, pp. 805-811.

Elkan, G.H. (1984) In: *Biological nitrogen fixation ecology, technology and physiology* (Ed. Alexander, M.), Plenum Publishing Co. New York, pp. 1–38.

Evans, W.R., Fleischman, D.E., Calvert, H.E., Pyati, R.V., Alter, G.M. & Subba Rao, N.S. (1990) *Appl. Environ. Microbiol.* **56**: 3445–3449.

Faber, B.A., Zasoski, R.J., Munns, D.N. & Schakel, K. (1991) *Can. J. Bot.* **69**: 87-94.

Francis, R. & Read, D.J. (1984) In: *Management of mycorrhizas in agriculture, horticulture and forestry* (Eds. Robson, A.D., Abbott, L.K. & Malajczuk, N.), Kluwer, Dordrecht.

Frank, B. (1885) *Berichte der Deutschen Botanischen Gessellschaft* **3**: 128-145.

Frank, B. (1889) *Gesellschaft* **7**: 332-346.

Franke-Snyder, M., Douds, D.D. Jr., Galvez, L., Phillips, J.G., Wagoner, P., Drinkwater, L. & Morton, J.B. (2001) *Appl. Soil Ecol.* **16**: 35–48.

Fred, E.B., Baldwin, I.L. & McCoy, E. (1932) In: *Root nodule bacteria and leguminous plants.* University of Wisconsin Press, Madison, Wisconsin.

Freeman, S. & Herron J.C. (1998) In: *Evolutionary analysis*. Prentice-Hall, New Jersey.

Fries, L.L.M., Pacovsky, R.S., Safir, G.R. & Kaminsky, J. (1998) *Physiologia Plantarium* **103**: 152-71.

Furlan, V. (1993) In: *Abstracts, 9th NACOM* (Eds. Peterson, L. & Schelkle, M.), Guelph, Ontario, Canada, pp. 77.

Futuyma, D.J. (1998) In: *Evolutionary biology*. Sinauer Associates, Sunderland, Massachusetts.

Gallaud, I. (1905) In: E´tudes sur les mycorrhizes endophytes. *Revue General de Botanique* 17, pp. 5–48, 66–83, 123–136, 223–239,313–325, 425–433, 479–500.

Gao, J.L., Turner, S.L., Kan, F.L., Wang, E.T., Tan, Z.Y., Qiu, Y.H., Gu, J., Terefework, Z., Young, J.P.W., Lindstrom, K. & Chen, W.X. (2004) *Int. J. Sys. Evo. Microbiol.* **54**: 2003-2012.

Gautam, S.P. & Prasad, K. (2001) In: *Innovative approaches in microbiology* (Eds. Maheshwari , D.K. & Dubey, R.C.), Bishon Singh Mahendra Pal Singh, Dehradun, India, pp. 83-114.

Gehrig, H., Schubler, A. & Kluge, M. (1996) *J. Mol. Evol.* **43**: 71-81.

Gerdemann, J.W. & Nicolson, T.H. (1963) *Trans. Br. Mycol. Soc.* **46**: 235-244.

Gerdemann, J.W. & Trappe, J.M. (1974) *Mycol. Memoir.* **5**:1-76.

Gerdemann, J.W. (1955) *Mycologia* **47**: 619-632.

Gerdemann, J.W. (1964) *Mycologia* **56**: 342–349.

Gerdemann, J.W. (1971) In: *Proceedings of the first North American conference on mycorrhizae. Mycorrhizae* (Eds. Hacskaylo, E.), USDA Misc. Publ. 1189, pp. 9-18.

Gerdemann, J.W. (1975) In: *The development and functions of roots* (Eds. Torrey, J.G. & Clarkson, D.T.), Academic Press, London, UK, pp. 575-591.

Gianinazzi-Pearson, V. (1996) *Plant cell* **8**: 1871-1883.

Gianinazzi-Pearson, V., Arnould, C., Oufattole, M., Arango, M. & Gianinazzi, S. (2000) *Planta* **211**: 609-613.

Gianinazzi-Pearson, V., Smith, S.E., Gianinazzi, S. & Smith, F.A. (1991) *New Phytol.* **117**: 61-67.

Giovannetti, M. & Mosse, B. (1980) *New Phytol.* **84**: 489-500.

Goicoechea, N., Doleza, K., Antolin, M.C., Strand, M. & Sanchez-Diaz, M. (1995) *J. Exp. Bot.* **46**: 1543-1549.

Graham, P.H. & Parker, C.A. (1964) *Plant Soil* **20**: 383-396.

Gulash, M., Ames, P., Larosiliere, R.C. & Bergman, K. (1984) *Appl. Environ. Microbiol.* **48**: 149-152.

Gurtler, V. & Mayall, B.C. (2001) *Int. J. Syst. Evol. Microbiol.* **51**: 3-16.

Habte, M., Zhang, Y.C. & Schmidt, D.P. (1999) *Can. J. Bot.* **77**: 135-139.

Hadri, A.E. & Bisseling, T. (1998) In: *The Rhizobiaceae: Molecular biology of model plant associated bacteria* (Eds. Spaink, H.P., Kondorosi, A. & Hooykaas, P.J.J.), Kluwer Academic Publishers, Dordrecht, pp. 403-416.

Hardie, K. (1985) *New Phytol.* **101**: 677-684.

Hart, M.M. & Klironomos, J.N. (2002) In: *Mycorrhizal Ecology* (Eds. van der Heijden, M. G. A. & Sanders, I.), Springer-Verlag, Berlin, pp. 225-242.

Harwani, D. (2006) In: *Biodiversity and efficiency of bradyrhizobial strains and arbuscular mycorrhizal fungi of soybean cultivars grown in Haroti region of Rajasthan.* Ph.D. Thesis, Maharshi dayanand Saraswati University, Ajmer, India.

Hayman, D.S. (1982) *Phytopathology* **72**: 1119-1125.

Heckman, D. S., Geiser, D. M., Eidell, B. R., Stauffer, R.L., Kardos, N.L. & Hedges, S.B. (2001) *Science* **293**: 1129-1133.

Helgason, T., Fitter, A.H., Young, J.P.W. (1999) *Mole. Ecol.* **8**: 659-666.

Hellriegel, H. & Wilfarth, H. (1888) Untersuchungen uberdie steck stoffnahrungder Gramineen and Leguminoseen, Beilageheft Zuder Zeitschrift,Verein der Ruben zucker Industrie Deutschen Reichs 1888:234.

Hentschel, U., Steinert, M. & Hacker, J. (2000) *Trends Microbiol.* **8**: 226-231.

Hepper, C.M. (1987) In: *Mycorrhizae in the next decade: Practical applications and research priorities* (Eds. Sylvia, D.M., Hung, L.L. & Graham, J.H.), Institute of Food and Agricultural Sciences, University of Florida, Gainesville, Fla, pp. 308–310.

Herendeen, P.S., Crepet, W.L. & Dilcher, D.L. (1992) In: *Advances in legume systematics: The fossil record* (Eds. Herendeen, P.S. & Dilcher, D. L.), The Royal Botanic Gardens, Kew, pp. 303–316.

Hijri, M. & Sanders, I.R. (2005) *Nature* **433**: 160-163.

Ho, S.C., Wang, J.L. & Schindler, M. (1990) *J. Cell Biol.***111**: 1631-1638.

Hooker, J.E., Jaizme-Vegas, M. & Atkinson, D. (1994) In: *Impacts of arbuscular mycorrhizal on sustainable agriculture and natural ecosystems* (Eds. Gianinazzi, S. & Schuepp, H.), Birkhauser Verlag, Basel, Switzerland, pp. 191-199.

Hung, L.L. & Sylvia, D.M. (1988) *Appl. Environ. Microbiol.* **54**: 353-357.

Isaac, S. (1992) In: *Fungal-plant interactions*. Chapman and Hall, New York.

Jaftha, J.B., Strijdom, B.W. & Steyn, P.L. (2002) *Syst. Appl. Microbiol.* **25**: 440-449.

Janos, D.P. (1980) *Biotropica* 12 [Suppl]: 56-64.

Janse, J.M. (1897) *Ann. Jardin. Bot. Buitenzorg.* **14**: 53-201.

Jarvis, B.D.W., Downer, H.L. & Young J.P.W. (1992) *Int. J. Syst. Bacteriol.* **42**: 93-96.

Jarvis, B.D.W., van Berkum, P., Chen W.X., Nour, S.M., Fernandez, M.P., Cleyet-Jarvis, B.D.W., Pankhurst, C.E. & Patel, J.J. (1982) *Int. J. Syst. Bacteriol.* **32**: 378-380.

Jarvis, B.D.W., van Berkum, P., Chen, W.X., Nour, S.M., Fernandez, M.P., Cleyet-Marel, J.C. & Gillis, M. (1997) *Int. J. Syst. Bacteriol.* **47**: 895-898.

Johanssen, A., Jakobsen, I. & Jessen, E.S. (1993) *New Phytol.* **124**: 61-68.

Johnson, N.C., Graham, J.H. & Smith, F.A. (1997) *New Phytol.* **135**: 575–585.

Jordan, D.C. (1982) *Int. J. Syst. Bacteriol.* **32**: 136-141.

Jordan, D.C. (1984) In: *Bergey's manual of systematic bacteriology* (Eds. Krieg, N.R. & Holt , J.G.), Vol.1, Williams and Willkins, Baltimore, pp. 242-244.

Jordan, D.C. & Allen, O.N. (1974) In: *Bergey's manual of determinative bacteriology,* 8th edition, (Eds. Buchanan, R.E. & Gibbons, N.E.), Williams and Wilkins, Baltimore, Maryland, pp. 261-264.

Jourand, P., Giraud, E., Bena, G., Sy, A., Willems, A., Gillis, M., Dreyfus, B. & de-Lajudie, P. (2004) *Int. J. Syst. Evol. Microbiol.* **54**: 2269-2273.

Kaminski, P.A., Batut, J. & Boistard, P. (1998) In: *The Rhizobiaceae: Molecular biology of model plant associated bacteria* (Eds. Spaink, H.P., Kondorosi, A. & Hooykas, P.J.J.), Kluwer Academic Publishers, Dordrecht, pp. 431-460.

Kawai, Y. & Yamamoto, Y. (1986) *Plant Cell Physiol.***27**: 399-405.

Kelley, A.P. (1950) In: *Mycotrophy in plants*. Chronica Botanica, Waltham, Mass.

Kelley, A.P. (1931) In: *The concept of mycorrhiza. Mycologia* **23**: 147–151.

Keyser, H.H., Somasegaran, P. & Bohlool, B.B. (1992) In: *Soil Microbial Ecology: Applications in agricultural and environmental management* (Ed. Blaine-Metting, F.), Marcel Dekker, Inc., New York, pp. 205-226.

Klironomos, J.N. & Hart, M.M. (2001) *Nature* **410**: 651–652.

Klironomos, J.N., McCune, J., Hart, M. & Neville, J. (2000) *Ecol. Lett.* **3**: 137–141.

Koide, R.T. & Schreiner, R.P. (1992) *Annual. Rev. Plant Physiol.* **43**: 557-581.

Koide, R.T. (1985) *New Phytol.* **99**: 449-462.

Kuhn, G., Hijri, M. & Sanders, I.R. (2001) *Nature* **414**: 745-748.

Kuykendall, L.D., Saxena, B., Devine, T.E. & Udell, S.E. (1992) *Can. J. Microbiol.* **38**: 501–505.

Lange, R.T (1966) In: *Symbiosis* (Ed. Henry, S.M.), Academic press, London, pp. 99-170.

Lindstrom, K. (1989) *Int. J. Syst. Bacteriol.* **39**: 365–367.

Liu, A., Hamel, C., Elmi, A., Costa, C., Ma, B. & Smith, D.L. (2002) *Can. J. Soil Sci.* **82**: 271-278.

Lodeiro, A.R., Lopez-Garcýa, S.L., Vazquez, T.E. & Favelukes, G. (2000) *FEMS Microbiol. Letters* **188**: 177-184.

Lohman, M.L. (1927). *Studies Natural History* **11**: 33–58.

Long, S.R. & Staskawicz, B.J. (1993) *Cell* **73**: 921-935.

Mahobia, V. (2003) Ph.D. Thesis, Maharshi dayanand Saraswati University, Ajmer, India.

Malloch, D.W., Pirozynski, K.A. & Raven, P.H. (1980) *Proc. Natl. Acad. Sci. U.S.A.* **77**: 2113–2118.

Manjunath, A. and Habte, M. (1991) *Can. J. Bot.* **69**: 671-676.

Manoharachary, C., Sridhar, K., Singh, R., Adholeya, A., Suryanarayanan, T.S., Rawat, S.& Johri, B.N. (2005) *Curr. Sci.* **89**: 58-71.

Martinez-Romero, E., Segovia, L., Mercante, F.M., Franco, A.A., Graham, P. & Pardo, M.A. (1991) *Int. J. Syst. Bacteriol.* **41**: 417-426.

Marx, C., Dexheimer, J., Gianinazzi-Pearson, V. & Gianinazzi, S. (1982). *New Phytol.* **90**: 37–43.

McArthur, D.A.J. & Knowles, N.R. (1992) *Plant Physiol.* **100**: 341–351.

McGonigle, T.P., Miller, M.H., Evans, D.G., Fairchild, G.L. & Swan, J.A. (1990) *New Phytol.* **115**: 495–501.

Mellor, R.B. & Werner, D. (1990). In: *The molecular biology of symbiotic nitrogen fixation* (Ed. Gresshoff, M.), CRC Press, Boca Raton, Florida, pp. 111-129.

Merryweather, J. & Fitter, A. (1998) *New Phytologist* **138**: 117-129.

Miller, R.M. & Jastrow, J.D. (2000) In: *Arbuscular mycorrhizas: physiology and function* (Eds. Kapulnik, Y. & Douds, D.D.), Kluwer, Dordrecht, pp. 3–18.

Molouba, F., Lorquin, J., Willems, A., Hoste, B., Giraud, E., Dreyfus, B., Gillis, M., de Lajudie, P. & Masson-Boivin, C. (1999) *Appl. Environ. Microbiol.* **65**: 3084-3094.

Morton, J.B. & Benny, G.L. (1990) *Mycotaxon* **37**: 471-491.

Morton, J.B. & Benny, G.L. (2001) *Mycologia* **93**: 181-195.

Morton, J.B. (1988) *Mycotaxon* **32**: 267-324.

Morton, J.B., Bever, J.D. & Pfleger, F.L. (1997) *Myco. Research* **101**: 625-631.

Mosse, B. (1973) *New Phytol.* **72**: 127–136.

Moulin, L., Munive, A., Dreyfus, B. & Boivin-Masson, C. (2001) *Nature* **411**: 948–950.

Mylona, P., Pawlowski, K. & Bisseling, T. (1995) *Plant Cell* **7:** 869-885.

Nehls, U., Mikolajewski, S., Magel, E. & Hampp, R. (2001) *New Phytol.* **150**: 533-541.

Newman, E.I. (1966) *J. Appl. Ecol.* **3**: 139–145.

Newsham, K.K., Fitter, A.H. & Watkinson, A.R. (1994) *J. Ecology* **82**: 805-814.

Ngom, A., Nakagawa, Y., Sawada, H., Tsukahara, J., Wakabayashi, S., Uchiumi, T., Nuntagij, A., Kotepong, S., Suzuki, A., Higashi, S. & Abe, M. (2004) *J. Gen. Appl. Microbiol.* **50**: 17-27.

Nick, G., de Lajudie, P., Eardly, B., Soumalainen, S., Paulin, L., Zhang, X., Gillis M. & Lindstrom, K. (1999) *Int. J. Syst. Bacteriol.* **49**: 1359-1368.

Nicolson, T.H. & Johnston, C. (1979) *Trans. Br. Mycol. Soc.* **72**: 261-268.

Nicolson, T.H. (1967) Science Progress. *Oxford* **55**: 561–581.

Norris, D.O. (1965) *Plant Soil* **22**: 143-166.

Nour, S.M., Cleyet-Marel, J.C., Normand, P. & Fernandez, M.P. (1995) *Int. J. Syst. Bacteriol.* **45**: 640-648.

Nour, S.M., Fernandez, M., Normand, P. & Cleyet-Marel, J.C. (1994) *Int. J. Syst. Bacteriol.* **44**: 511-522.

Oehl, F. & Sieverding, E. (2004) *J. Appl. Bot. Angewandte. Bot.* **78**: 72-82.

Ogasawara, M., Suzuki, T., Mutoh, I., Annapurna, K., Arora, N.K., Nishimura, Y. & Maheshwari, D.K. (2003). *Symbiosis* **34**: 53-68.

O'Keefe, D.M. & Sylvia, D.M. (1991) In: *Handbook of applied mycology soil and plants* (Eds. Arora, D.K., Rai, B., Mukerji, K.G. & Knudseneds, G.R.), Vol. 1., Marcel Dekker, Inc., New York, pp. 35-53.

Pacovsky, R.S., Fuller, G., Stafford, A.E. & Paul, E.A. (1986) *Plant Soil* **92**: 37-45.

Parniske, M. (2000) *Curr. Plant. Biol.* **3**: 320-328.

Pate, J.S. (1986) In: *On the economy of plant form and function* (Ed. Givnish, T.J.), Cambridge University Press, Cambridge.

Paul, E. A. (1988) In: *Advances in nitrogen cycling in agricultural ecosystems* (Ed. J.R. Wilson), CAB International, Wallingford, United Kingdom, pp. 417–425.

Pawlowska, T. & Taylor, J.W. (2004) *Nature* **427**: 733-737.

Peng, G.X., Tan, Z.Y., Wang, E.T., Reinhold-Hurek, B., Chen, W.F. & Chen, W.X. (2002) *Int. J. Syst. Evol. Microbiol.* **52**: 457-462.

Peng, S., Eissenstat, D.M., Graham, J.H., Williams, K. & Hodge, N.C. (1993) *Plant Physiol.* **101**: 1063-1071.

Peoples, M.B., Herridge, D.F. & Ladha, J.K. (1995) *Plant Soil* **174**: 3-28.

Perret, X., Staehelin, C. & Broughton, W.J. (2000) *Microbiol. Mol. Biol. Rev.* **64**: 180–201.

Pfeffer, P.E., Bago, B.& Shachar-Hill, Y. (2001) *New Phytologist* **150**: 543-553.

Phillips, J.M. & Hayman, D.S. (1970) *Trans. Br. Myco. Soc.* **55**: 158-161.

Postgate, J. (1982) The fundamentals of nitrogen fixation. Cambridge University Press, Cambridge, United Kingdom.

Postgate, J. (1998) Nitrogen fixation. Cambridge University Press, Cambridge, UK.

Prasad, K. & Rajak, R.C. (2002) In: *Plant pest management* (Ed. Trivedi, P.C.), Aavishkar Publishers and Distributors, Jaipur, pp. 75-99.

Prasad, K. & Kaushik, S. (2004) In: *Biological diversity: Current trends* (Eds. Gautam, S.P., Bansal, Y.K. &. Pandey, A.K), Shree Publication and Distributors, New Delhi, pp.134-141.

Prasad, K., Kaushik, S. & Rajak, R.C. (2005) In: *Mycorrhiza* (Eds. Prakash , A. & Mehrotra, V.S.), Scientific Publication, India, Jodhpur, pp.281-289.

Rangaraj, P., Ruttimann, J.C., Shah, V.K. & Ludden, P.W. (2000) In: *Prokaryotic nitrogen fixation: A model system for analysis of a biochemical process* (Ed. Triplett, E. W.), Horizon Scientific Press, Wymondham, United Kingdom, pp. 55-79.

Rayner, M.C. (1926–1927) *New Phytol.* **25**: 1–50, 65–108, 171–190, 248–263, 338–372, **26**: 22–45, 85–114.

Redecker, D. (2000) *Mycorrhiza* **10**: 73-80.

Redecker, D., Morton, J.B. & Bruns, T.D. (2000b) *Mol. Phylo. Evol.* **14**: 276-284.

Redecker, D., Kodner, R. & Graham, L.E. (2000a) *Science* **289**: 1920-1921.

Redecker, D., Thierfelder, H., Walker, C. & Werner, D. (1997) *Appl. Environ. Microbiol.* **63**: 1756-1761.

Redecker. D., Hijri, I. & Wiemken, A. (2003) *Folia Geobotanica* **38:** 113-124.

Reis, V.M., Baldani, J.I., Baldani, V.L.D. & Dobereiner, J. (2000) *Critical Rev. Plant Sci.* **19**: 227-247.

Rewari, R.B. & Tilak, K.V.B.R. (1988) In: *Pulse crops* (Eds. Baldev, B. S. Ramanujam & H. K. Jain), Oxford and IBH, New Delhi, pp. 373–411.

Rillig, M.C. (2004) *Ecol. Letters* **7**: 740-754.

Rivas, R., Velazquez, E., Willems, A., Vizcaino, N., Subba-Rao, N.S., Mateos, P.F., Gillis, M., Dazzo, F.B. & Martinez-Molina, E. (2002) *Appl. Environ. Microbiol.* **68**: 5217-5222.

Rivas, R., Willems, A., Palomo, J.L., Garcia-Benavides, P., Mateos, P.F., Martinez Molina, E., Gillis, M. & Velazquez, E. (2004) *Int. J. Syst. Evol. Microbiol.* **54**: 1271-1275.

Robertson, J.G. & Farnden, K.J.F. (1980) In: *The biochemistry of plants: amino acids and derivates* (Ed. Miflin, B.J.), Academic Press, Inc., New York, pp. 65-112.

Rome, S., Brunel, B., Normand, P., Fernandez, M. & Cleyet-Marel, J.C. (1996a) *Arch. Microbiol.* **165**: 285–288.

Rome, S., Fernandez, M.P., Brunel, B., Normand, P. & Cleyet-Marel, J.C. (1996b) *Int. J. Syst. Bacteriol.* **46**: 972–980.

Rosendahl, L., Mouritzen, P. & Rudbeck, A. (2001) *Plant Soil* **230**: 31-37.

Ruiz-Lozano, J.M. & Azcon, R. (1996) *Agric. Ecosys. Environ.* **60**: 175-181.

Ruiz-Lozano, J.M. & Azcon, R. (1995) *Physiol. Plant* **95**: 472-478.

Ruiz-Lozano, J.M., Azcon, R. & Gomez, M. (1996a) *Physiol. Plant.* **98**: 767-772

Ruiz-Lozano, J.M., Azcón, R. & Plama, J.M. (1996b) *New Phytol.* **134**: 327-333.

Rygiewicz, P.T. & Andersen, C.P. (1994) *Nature* **369**: 58–60.

Sahay, N.S., Sudha, A.S. & Varma, A. (1998) *Ind. J. Exp. Biol.* **36**: 1069-1086.

Sahgal, M. & Johri, B.N. (2003) *Curr. Sci.* **84**: 44-48.

Sanchez-Diaz, M. & Honrubia, M. (1994) In: *Impact of arbuscular mycorrhizas on sustainable agriculture and natural ecosystems* (Eds. Gianninazi, S. & Schuepp, H.), Birkhauser Verlag, Basel, Switzerland. ISBN 3-7643-5000-8, pp. 167-178.

Sawada, H., Kuykendall, L.D. & Young, J.M. (2003) *J. Gen. Appl. Microbiol.* **49**: 155-179.

Saxena, A.K., Kingston, K.K., Mondal, S. & Shende, R. (2005) In: *Microbial diversity: Current perspectives and potential applications* (Eds. Satyanarayana, T. & Johri, B.N.), I.K. International Pvt. Ltd., New Delhi, pp. 427-456.

Scholla, M.H. & Elkan, G.H. (1984) *Int. J. Syst. Bacteriol.* **34**: 484-486.

Schoot Uiterkamp, A.J.M. (1990) In: *Nitrogen Fixation: Achievements and objectives*. (Eds. Gresshoff, P.M., Roth, L.E., Stacey, G. & Newton, W.E.), Chapman and Hall, NewYork, pp. 55–66.

Schubler, A. (2000) *Mycorrhiza* **10**: 15-21.

Schubler, A. (2002) *Plant Soil* **244**: 75-83.

Schubler, A., Bonfante, P., Schnepf, E., Mollenhauer, D. & Kluge, M. (1996) *Protoplasma* **190**: 53-67.

Schubler, A., Mollenhauer, D., Schnepf, E. & Kluge, M. (1994) *Botanica Acta* **107**: 36-45.

Schubler, A., Schwarzott, D. & Walker, C. (2001) *Mycol. Res.* **105**: 1413-1421.

Schwarzott. D., Walker. C. & Schubler, A. (2001) *Mol. Phylogenetics Evo.* **21**: 190-197.

Segovia, L., Young, J.P.W. & Martinez-Romero, E. (1993) *Int. J. Syst. Bacteriol.* **43**: 374-377.

Sieverding, E. (1991). Vesicular-arbuscular mycorrhiza management in tropical agrosystem, Deutsche Gesellschaft fur Technische Zusammenarbeit (GTZ) GmbH, Germany. ISBN 3-88085-462.

Silveira, A.P.D. & Cardoso, E.J.B.N. (1987) *Revista Brasileira de Ciência do Solo* **11**: 31-36.

Simon, L., Bousquet, J., Levesque, R.C. & Lalonde, M. (1993) *Nature* **363**: 67-69.

Simon, L., Lalonde, M. & Bruns, T. (1990) In: *Abstracts, 8th North American Conferenceon Mycorrhizae* (Eds. Allen, M.F. & Williams, S.E.), University of Wyoming Agricultural Experiment Station, Laramie, Wyo, p.265.

Simon, L., Lalonde, M. & Bruns, T. (1992) *Appl. Environ. Microbiol.* **58**: 291-295.

Singh, S. (2005) *Mycorrhiza News* **17**: 1-9.

Slattery, J. & Pearce, D. (2002) In: *Inoculants and nitrogen fixation of legumes in vietnam* (Eds. Herridge, D.), ACIAR Proceedings 109e. pp. 86-94.

Smit, G.S., Swart, B.J., Lugtenberg, J. & Kijne, J.W. (1992) *Mol. Microbiol.* **6**: 2897-2903.

Smith, F.A. & Smith, S.E. (1997) *New Phytol.* **137**: 373–388.

Smith, S.E. & Read, D.J. (1997) Mycorrhizal symbiosis, 2nd Edition. Academic Press, London.

Spaink, H. P. (1995) *Annu. Rev. Phytopathol.* **33**: 345-368.

Sprent, J.I. (2001) Nodulation in legumes. Kew, UK, Royal Botanic Gardens.

Sprent, J.I. and Sprent, P. (1990) Nitrogen-fixing organisms. Pure and applied aspects. Chapman and Hall, London, United Kingdom.

Sprent, J.I., Sutherland, J.M. & de Faria, S.M. (1987) *Biol. Sci.* **317**: 111-129.

Squartini, A., Struffi, P., Doring, H., Selenska-Pobell, S., Tola, E., Giacomini, A., Vendramin, E., Velazquez, E., Mateos, P.F., Martinez-Molina, E., Dazzo, F. B., Casella, S. & Nuti, M.P. (2002) *Int. J. Syst. Evol. Microbiol.* **52**: 1267-1276.

Stuurman, N., Bras, C.P., Schlaman, H.R.M., Wijfjes, A.H.M., Bloemberg, G. & Spaink, H.P. (2000) *Mol. Plant Microbe Interact.* **13**: 1163-1169.

Sy, A., Giraud, E., Jourand, P., Garcia, N., Willems, A.; de Lajudie, P., Prin, Y., Neyra, M., Gillis, M., Boivin-Masson, C. & Dreyfus B. (2001) *J. Bacteriol.* **183**: 214–220.

Tan, Z.Y., Kan, F.L., Pengm, G.X., Wang, E.T., Reinhold-Hurek, B. & Chen, W.X. (2001) *Int. J. Syst. Evo. Microbiol.* **51**: 909-914.

Tate, R. L. (1995) Soil Microbiology - Symbiotic nitrogen fixation, John Wiley and Sons, Inc., New York, pp. 307–333.

Thies, J.E., Woomer, P.L. & Singleton, P.W. (1995). *Soil Biol. Biochem.* **27**: 633–636.

Tilak, K.V.B.R. (1993) Bacterial Fertilizers, Indian Council of Agricultural Research, New Delhi, India, pp.4-33.

Tilak, K.V.B.R., Ranganayaki, N., Pal, K. K., De, R., Saxena, A.K., Nautiyal, C.S., Mittal, S., Tripathi, A.K. & Johri, B.N. (2005) *Curr. Sci.* **89**: 136-50.

Tilman, D., Fargione, J., Wolff, B., D'Antonio, C., Dobson, A., Howarth, R., Schindler, D., Schlesinger, W.H., Simberloff, D. & Swackhamer, D. (2001) *Science* **292**: 281–284.

Tinker, P.B. & Gildon, A. (1983) In: *Metals and micronutrients, uptake and utilization by plants* (Eds. Rob, D.A. & Pierpoint ,W.S.), Academic Press, London, UK, pp.21-32.

Tisdall, J.M. & Oades, J.M. (1979). *Aust. J. Soil. Res.* **17**: 429-441.

Toth, R., Doane, C., Bennett, E. & Alexander, T. (1990) *Mycologia* **82**: 519-522.

Trappe, J.M. & Schenck, N.C. (1982) In: *Methods and principles of mycorrhizal research* (Ed. Schenck, N.C.), American Phytopathological Society, St.Paul, pp. 1-10.

Trinick, M.J. (1965) *Aus. J. Sci.* **27**: 263-264.

Trinick, M.J. (1980) *J. Appl. Bacteriol.* **49**: 39-53.

Trotta, A., Varese, G.C., Gnavi, E., Fusconi, A., Sampo, S. & Berta, G. (1996) *Plant Soil* **185**: 199-209.

Tsang, A. & Maun, M.A. (1999) *Plant Ecol.* **144**: 159-166.

Turk, D. & Keyser, H.H. (1992) *Soil Biol. Biochem.* **25**: 69-74.

Tyerman, S.D., Whitehead, L.F. & Day, D.A. (1995) *Nature* **378**: 629-632.

van Berkum, P. & Eardly, B.D. (1998) In: *The Rhizobiaceae: Molecular biology of the modelplant-associated bacteria* (Eds. Spaink, H.P., Kondorosi, A. & Hooykaas, P.J.J.), Kluwer Academic Publishers, Dordrecht, The Netherlands, pp. 1-24.

van Berkum, P. & Eardly, B.D. (2002) *Appl. Environ. Microbiol.* **68**: 1132-1136.

van Der Heijden, M.G.A., Klironomos, J.N., Ursic, M., Moutoglis, P., Streitwolf-Engel, R., Boller, T., Wiemken, A. & Sanders, I.R. (1998) *Nature* **396**: 69–72.

van Rhijn, P., Goldberg, R.B., & Hirsch, A.M. (1998) *Plant Cell* **10**: 1233 1249.

Vandamme, P. & Coenye, T. (2004) *Int. J. Syst. Evol. Microbiol.* **54**: 2285-2289.

Vandamme, P., Goris, J., Chen, W.M., De Vos, P. & Willems, A. (2002) *Syst. Appl. Microbiol.* **25**: 507-512.

Vandenkoornhuyse, P., Husband, R., Daniell, T.J., Watson, I.J., Duck, J.M., Fitter A.H. & Young, J.P.W. (2002) *Mol. Ecol.* **11**: 1555-1564.

Vaneechoutte, M., Kaempfer, P., De Baere, T., Falsen, E. & Verschraegen, G. (2004) *Int. J. Syst. Evol. Microbiol.* **54**: 317-327.

Verma, S.C., Paulchowdhury, S. & Tripathi, A.K. (2004) *Can. J. Microbiol.* **50**: 312–322.

Vincent, J.M. (1974) In: *The biology of Nitrogen Fixation* (Ed. Quispel, A.). North Holland Publishing Amsterdam, pp. 265-341.

Vincent, J.M. (1977) In: *A treatise on dinitrogen fixation* (Eds. Hardy, R.H.F. & Silver, W.S.), section III, Biology, John Wiley, New York, pp.277-366.

Vinuesa, P., León-Barrios, M., Silva, C., Willems, A., Jarabo-Lorenzo, A., Pérez-Galdona, R., Werner, D. & Martínez-Romero, E. (2005) *Int. J. Syst. Evol. Microbiol.* **55**: 569-575.

Vitousek, P. M. & Walker, L. R. (1989) *Ecol. Monogr.* **59**: 247–265.

Walker, C. & Sanders, F.E. (1986) *Mycotaxon* **27:** 169-182.

Walker, C. (1985) In: *Proceedings of the 6th North American conference on mycorrhizae* (Ed. Molina, R.), Forestry Resarch Laboratory, pp. 193-198.

Walker, C. (1987) In: *Proceedings of the 7th NACOM. IFAS*, University of Florida, Gainesville, Fla.

Walker, C., Blaszkowski, J. & Schubler, A. (2004) *Myco. Res.* **108**: 707-718.

Wall, L.G. & Favelukes, G. (1991) *J. Bacteriol.* **173**: 3492–3499.

Walsh, K.B. (1995) *Soil Biol. Biochem.* **27**: 637–655.

Wang, E.T., Rogel, M.A., Santos, A.G., Martinez-Romero, J., Cevallos, M.A. & Martinez-Romero, E. (1999b) *Int. J. Syst. Bacteriol.* **49**: 1479-1491.

Wang, E.T., Tan, Z.Y., Willems, A.Y., Fernandez-Lopez, M., Reinhold-Hurek, B. & Martinez-Romero, E. (2002) *Int. J. Syst. Evol. Microbiol.* **52**: 1687-1693.

Wang, E.T., van Berkum, P., Beyene, D., Sui, X.H., Chen, W.X. & Martinez-Romero, E. (1998) *Int. J. Syst. Bacteriol.* **48**: 687-699.

Wang, E.T., van Berkum, P., Sui, H.X., Beyene, P., Chen, W.X. & Martinez-Romero, E. (1999a) *Int. J. Syst. Evo. Microbiol.* **49**: 51–65.

Wani, S.P., Rupela, O.P. & Lee, K.K. (1995) *Plant Soil* **174**: 29–49.

Warner, A., Mosse, B. & Dingemann, L. (1985) In: *Proceedings of the 6th North American conference on mycorrhizae* (Ed. Molina, R.), Forest Research Laboratory, Oregon State University, Corvallis, Ore, pp 85-86.

Wei, G.H., Wang, E.T., Tan, Z.Y., Zhu, M.E. & Chen, W.X. (2002) *Int. J. Syst. Evol. Microbiol.* **52**: 2231-2239.

Werner, D. (1992). The *Bradyrhizobium/Rhizobium* Fabales symbiosis. In: *Symbiosis of plants and microbes* (Ed. Werner, D.), Chapman and Hall, London, pp. 49-151.

Wiley, E.O. (1978) *Syst. Zool.* **27**: 17-26.

Willems, A. & Collins, M.D. (1993) *Int. J. Syst. Bacteriol.* **43**: 305-313.

Willems, A., Fernandez-Lopez, M., Munoz-Adelantado, E., Goris, J., De Vos, P., Martinez-Romero, E., Toro, N. & Gillis, M. (2003) *Int. J. Syst. Evol. Microbiol.* **53**: 1207-1217.

Wilson, J.K. (1944) *Soil Sci.* **58**: 61-69.

Wood, T. (1985) In: *Proceedings of the 6th North American conference on mycorrhizae* (Ed. Molina, R.), Forest Research Laboratory, Oregon State University, Corvallis, Ore, pp. 84.

Woolhouse, H. (1975) In: *Endomycorrhizas: Proceedings of a symposium held at the University of Leeds*, 22-25 July 1974. (Eds. Sanders, F.E., Mosse , B. & Tinker, P.B.), Academic Press, London, pp. 209-239.

Wright, S.F. & Upadhyaya, A. (1996) *Soil Sci.* **161**: 575-586.

Wright, S.F. & Upadhyaya, A. (1998) *Plant Soil* **198**: 97-107.

Xu, L.M., Ge, C., Cui, Z., Li, J. & Fan, H. (1995) *Int. J. Syst. Bacteriol.* **45**: 706-711.

Yanagi, M. & Yamasato, K. (1993) *FEMS Microbiol. Lett.* **107**: 115–120.

Yao, Z.Y., Kan, F.L., Wang, E.T., Wei, G.H. & Chen, W.X. (2002) *Int. J. Syst. Evo. Microbiol.* **52**: 2219-2230.

Yonga, M.M. (1996) PhD thesis, Australian National University, Canberra.

Yost, C.K. & Hynes, M.F. (2000) In: *Prokaryotic nitrogen fixation. A model system for the analysis of a biological process* (Ed. Triplett, E.W.), Horizon Scientific Press, Wymondham, UK, pp.237-250.

Young, J.M. (2003) *Int. J. Syst. Evol. Microbiol.* **53**: 2107-2110.

Young, J.M., Kuykendall, L.D., Martinez-Romero, E., Kerr, A. & Sawada, H. (2001) *Int. J. Syst. Evol. Microbiol.* **51**: 89–103.

Young, J.P.W. & Haukka, K.E. (1996) *New Phytol.* **133**: 87-94.

Young, J.P.W. (1992) In: *The nitrogen fixation and its research in China* (Ed. G. Hong), Springer-Verlag, Berlin, Heidelberg, pp.365-381.

Zahran, H.H., Ahmed, M.S. & Afkar, E.A. (1995) *Egypt J. Basic Microbiol.* **35**: 269-275.

# 22

# *Agrobacterium* as Natural Tool for Plant Genetic Engineering

R. Pati, [1] V.K. Gupta, [1] L. P. Yadava, [1] N. Srivastava, [1]A. Gupta[2], N. Modi[3] and D.R. Modi[3]
[1]Central Institute for Subtropical Horticulture, Lucknow-227107 (U.P.)
[2]Central Drug Research Institute, Lucknow
[3]Deptt. of Biotechnology, B.B.A. University, Lucknow

## ABSTRACT

Genetic engineering has got much attention for enhancement of plant quality, protection and production of new products with advances in agricultural biotechnology and plant molecular biology. During the last decade, using transgenic plants to express recombinant proteins for pharmaceutical use has become very attractive. Plants have the potential to provide a large amount of inexpensive and safe proteins for disease diagnosis and therapy. The aim of this chapter is to discuss recent advances and future prospects in the field of *A. tumefaciens*-mediated plant transformation. To achieve this end, an overview is provided of the molecular mechanisms involved in T-DNA generation, transfer, and integration in the host genome, together with an outline of the most recent developments in the identification of plant proteins interacting with T-DNA. These data provided insights into the value of *Agrobacterium* both as a model system and of its potential future importance in the area of transgenic crop production. This leads, in the second part of the review, to discussion of the controversy surrounding the release of GM crop plants into the environment.

## INTRODUCTION

Transgenic plants have been used to produce cytokines, hormones, monoclonal antibodies, industrial and therapeutic enzymes, and vaccines (Cramer *et al.*, 1999). The transgenic crop plants have been created through the use of molecular techniques that facilitate the introduction of new useful traits to target crop species. The recent techniques of plant genetic engineering have allowed scientists to introduce highly specific modifications in transgenic plants. Broadly speaking, three gene transfer techniques have been developed that can be recommended for developing transgenic crop plants. These are: (1) Agrobacterium-mediated gene transfer, (2) gene transfer by particle bombardment system, and (3) direct gene transfer to protoplasts through electroporation. Selectable marker genes or visible reporter genes need to be co-transformed with the gene of agronomic interest. The objective of expression of selectable marker gene(s) is essentially to identify the transgenic events among the large populations of untransformed cells. It may be mentioned here that gene transfer techniques, which yield thousands of transgenics with laboratory model plants species (tobacco or Arabidopsis), require an enormous investment in terms of human resource for the production of few transgenic plants of elitse cultivars of agriculturally important crops. Foreign genes are also being introduced with yet another method, that is, through chloroplast transformation, which prevents the flow of transgene to the neighbouring plant species. Several characteristic such as insect and disease resistance, herbicide tolerance, and improved nutritional quality are some of the changes brought about in important crops. Plants are likely to be used in future as bioreactors for production of pharmaceuticals, industrial oils, biodegradable plastics and other specialty chemicals through the use of transgenic technology. Twenty-five years ago, the concept of using *Agrobacterium tumefaciens* as a vector to create transgenic plants was viewed as a prospect and a "wish." Today, many agronomically and horticulturally important species are routinely transformed using this bacterium, and the list of species that is susceptible to *Agrobacterium*-mediated transformation seems to grow daily. In some developed countries, a high percentage of the acreage of such economically important crops as corn, soybean, cotton, canola, potato, tomato, rice (Cheng *et al.,* 1998), banana (May *et al.*, 1995), corn (Ishida *et al.*, 1996), wheat (Cheng *et al.*, 1997), sugar cane (Enríquez-Obregón 1998) and papaya (Fitch *et al.,* 1993) are transgenic; an increasing number of these transgenic varieties are/or will soon be generated by *Agrobacterium*-mediated, as opposed to direct transformation methods (polyethyleneglycol-mediated transfer, microinjection, protoplast and electroporation and gene gun technology). *Agrobacterium*-mediated transformation has remarkable advantages over direct transformation methods. It reduces the copy number of the transgene, potentially leading to fewer problems with transgene cosuppresion and instability. In addition, it is a single-cell transformation system and not forming mosaic plants, which are more frequent when direct transformation is used. In this chapter, we present update information about the mechanisms of gene transfer mediated by *A. tumefaciens* and the assessments for application of these methods in the transformation of dicotyledonous plants taking papaya as a model plant. There still remain, however, many challenges for genotype-independent transformation of many economically important crop species, as well as forest species used for lumber, paper, and pulp production. In addition, predictable and stable expression of transgenes remains problematic. In this chapter, we describe how scientists utilized knowledge of basic *Agrobacterium* biology to develop *Agrobacterium* as a "tool" for plant genetic engineering. We also explore how our increasing understanding of *Agrobacterium* biology may help extend the utility of *Agrobacterium*-mediated transformation. It is our belief that further improvements in transformation technology will necessarily involve the manipulation of these fundamental biological processes.

# AGROBACTERIUM MEDIATED TRANSFORMATION

## Biology of Agrobacterium

The genus *Agrobacterium* has been divided into a number of species. However, this division has reflected, for the most part, disease symptomology and host range. Thus, *A. radiobacter* is an "avirulent" species, *A. tumefaciens* causes crown gall disease, *A. rhizogenes* causes hairy root disease, and *A. rubi* causes cane gall disease. More recently, a new species has been proposed, *A. vitis*, which causes galls on grape and a few other plant species (Otten *et al.,* 1984). Plant transformation mediated by *Agrobacterium tumefaciens*, a soil plant pathogenic bacterium, has become the most used method for the introduction of foreign genes into plant cells and the subsequent regeneration of transgenic plants. *A. tumefaciens* naturally infects the wound sites in dicotyledonous plant causing the formation of the crown gall tumours. The first evidence indicating this bacterium as the causative agent of the crown gall goes back a century ago (Smith and Townsend, 1907). Since that moment, for different reasons, a large number of researches have focused on the study of this neoplastic disease and its causative pathogen. During the first and extensive period, scientific effort was devoted to disclose the mechanisms of crown gall tumour induction hoping to understand the mechanisms of oncogenesis in general, and to eventually apply this knowledge to develop drug treatments for cancer disease in animals and humans. When this hypothesis was discarded, the interest in crown gall disease largely decreased until it was evident that this tumour formation may be a result of gene transfer from *A. tumefaciens* to infected plant cells. *Agrobacterium tumefaciens* causes crown gall disease of a wide range of dicotyledonous (broad-leaved) plants, especially members of the rose family such as apple, pear, peach, cherry, almond, raspberry and roses. A separate strain, termed biovar 3, causes crown gall of grapevine.

## Crown Gall Disease

The disease gains its name from the large tumour-like swellings (galls) that typically occur at the crown of the plant, just above soil level. Although it reduces the marketability of nursery stock, it usually does not cause serious damage to older plants. Nevertheless, this disease is one of the most widely known because of its remarkable biology. Basically, the bacterium transfers part of its DNA to the plant, and this DNA integrates into the plant's genome, causing the production of tumours (Fig. 1) and associated changes in plant metabolism. *A. tumefaciens* is a Gram-negative, non-sporulating, motile, rod-shaped bacterium, closely related to *Rhizobium* which forms nitrogen-fixing nodules on clover and other leguminous plants. Strains of *Agrobacterium* are classified in three biovars based on their utilization of different carbohydrates and other biochemical tests. Most of the genes involved in crown gall disease are not borne on the chromosome of *A. tumefaciens* but on a large plasmid, termed the $T_i$ (tumour-inducing) plasmid. In the same way, most of the genes that enable *Rhizobium* strains to produce nitrogen-fixing nodules are contained on a large plasmid termed the *Sym* (symbiotic) plasmid (Table 1). Thus, the characteristic biology of these two bacteria is a function mainly of their plasmids and not of the bacterial chromosome. Plasmid is a circle of DNA separate from the chromosome, capable of replicating independently in the cell and of being transferred from one bacterial cell to another by conjugation. Plasmids encode non-essential functions. If the bacterium is grown near its maximum temperature (about 30°C),

**Fig. 1** Large crown galls are formed on the branches of *Madhuca india*.

then the plasmid is lost and pathogenicity (of *Agrobacterium*) is also lost. However, loss of the plasmid does not affect bacterial growth in culture—the plasmid-free strains are entirely functional bacteria.)

**Table 1** Chromosomal genes required for tumor formation

| *Chromosomal genes* | *Function in plants* |
|---|---|
| *pscA/exoC* | Biosynthesis |
| *chvB* | Modification of an extracellular polysaccharide required for attachment |
| *chvA* | Export |
| *att* | Bacterial attachment of plant cells |
| *chvD, miaA, ros* | Regulation of *vir* gene (in addition to *Vir* AG) |
| *chvG, chvI* | Two-component regulatory system: pH regulation of *vir* genes |
| *chvE* | Sugar transporter |
| *acvB* | Ti-DNA transport (unspecific DNA binding protein) |

In natural conditions, the motile cells of *A. tumefaciens* are attracted to wound sites by chemotaxis. This is partly a response to the release of sugars and other common root components, and it is found even in plasmid-cured strains. However, strains that contain the $T_i$ plasmid respond even more strongly, because they recognise wound phenolic compounds such as acetosyringone (Fig. 2), which are strongly attractive at even very low concentrations (100 nm). Thus, one of the functions of the $T_i$ plasmid is to code for additional, specific chemotactic receptors that are inserted in the bacterial membrane and enable the bacterium to recognise wound sites. Acetosyringone plays a further role in the infection process, because at higher concentrations (about 100 mM) than those that cause chemotaxis it activates the virulence genes (*Vir* genes) on the $T_i$ plasmid (Fig. 3). These genes coordinate the infection process and, in particular:

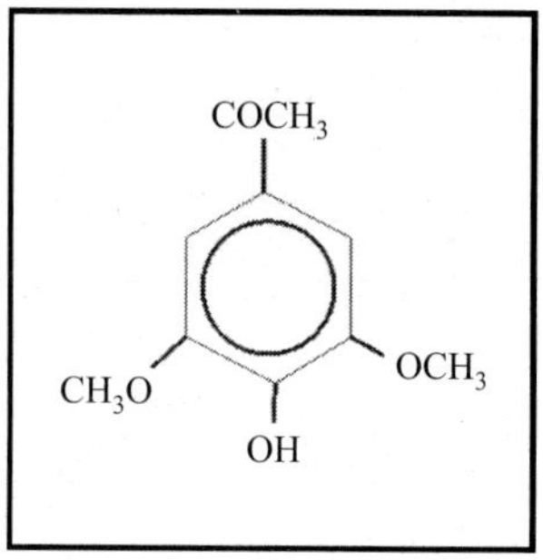


**Fig. 2** Acetosyringone.

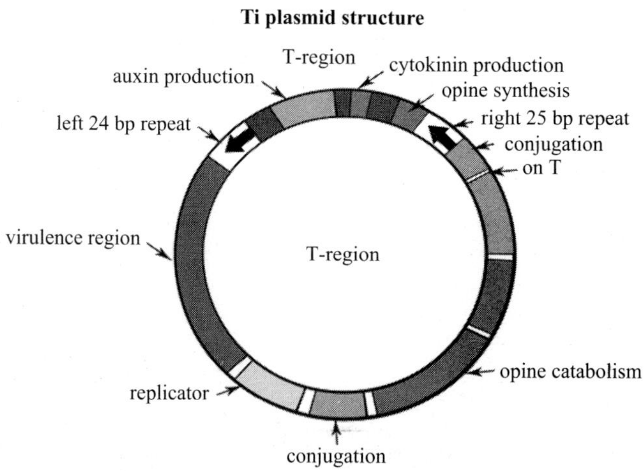


**Fig. 3** The octopine Ti plasmid. Regions involved in replication conjugation and opine catabolism are indicated as are the Vir region and the T region.

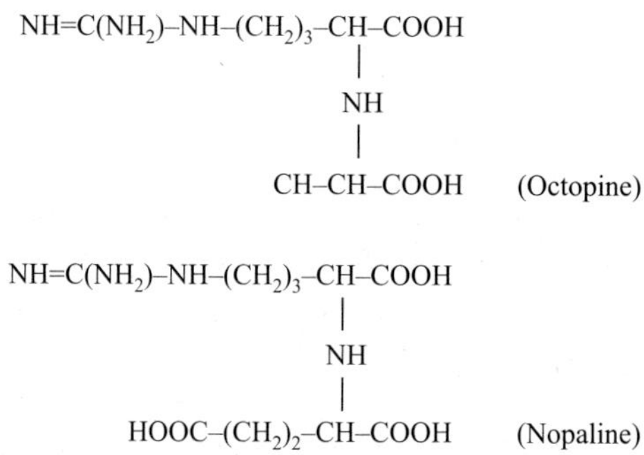


**Fig. 4** The chemical structure of the octopine (derived from arginine + pyruvate) and Nopaline (derived from arginine + a-ketoglutarate). Synthesis of octopine is directed by the Ti plasmids pTiA6 and synthesis of nopaline is directed by the Ti plasmids C58.

- lead to the production of proteins (permeases) that are inserted in the bacterial cell membrane for uptake of compounds (opines) that will be produced by the tumours (see later);
- cause the production of an endonuclease—a restriction enzyme—that excises part of the $T_i$ plasmid termed the T-DNA (transferred DNA).

## Ti Plasmid

Virulent strains of *A. tumefaciens* and *A. rhizogenes*, when interacting with susceptible dicotyledonous plant cells, induce diseases known as crown gall and hairy roots, respectively. These strains contain a large mega plasmid (more than 200 kb), which play a key role in tumour induction and for this reason it was named Ti plasmid, or Ri (root inducing) in the case of *A. rhizogenes*. Ti plasmids are classified according to the opines, which are produced and excreted by the tumours they induce. During infection, the T-DNA, mobile segment of Ti or Ri plasmid, is transferred to the plant cell nucleus and integrated into the plant chromosome. The T-DNA fragment is flanked by 25-bp direct repeats, which act as a *cis* element signal for the transfer apparatus.

The process of T-DNA transfer is mediated by the cooperative action of proteins encoded by genes determined in the Ti plasmid virulence region (*vir* genes) and in the bacterial chromosome. The Ti plasmid also contains the genes for opine catabolism produced by the crown gall cells, and regions for conjugative transfer and for its own integrity and stability. The 30 kb virulence (*vir*) region is a regulon organized in six operons that are essential for the T-DNA transfer (*virA, virB, virD,* and *virG*) or for the increase of transfer efficiency (*virC* and *virE*) (Hooykaas and Schilperoort, 1992; Jeon *et al.*, 1998). Different chromosomal-determined genetic elements have shown their functional role in the attachment of *A. tumefaciens* to the plant cell and bacterial colonization: the loci *chvA* and *chvB*, involved in the synthesis and excretion of the b -1, 2 glucan (Cangelosi *et al.*, 1989); the *chvE* required for the sugar enhancement of *vir* genes induction and bacterial chemotaxis; the *cel* locus, responsible for the synthesis of cellulose fibrils (Matthysse 1983); the *pscA* (*exoC*) locus, playing its role in the synthesis of both cyclic glucan and acid succinoglycan (Cangelosi *et al.,* 1991); and the *att* locus, which is involved in the cell surface proteins.

## Ti-plasmid Features

*Agrobacterium tumefaciens* carries a plasmid (Ti) with a number of important features.The *"vir"* or virulence functions are trans-acting elements that mobilize a plasmid containing the right border element (a cis-acting element). The transferred DNA is the region between the right and left border sequences, and includes genes that are tumorigenic (auxin and cytokinin production), and a gene that directs synthesis of specific opines (sugar derivatives that are not easily catabolized by other species). The genes for opine catabolism stay with the specific *A. tumefaciens* species, allowing that bacterium to benefit from the opine production of the plant. Engineered Ti vectors may simply delete the tumourigenic and opine-specific genes.

## T-DNA

The initial results of the studies on T-DNA transfer process to plant cells demonstrate three important facts for the practical use of this process in plants transformation. Firstly, the tumour formation is a transformation process of plant cells resulting from transfer and integration of T-DNA and the subsequent expression of T-DNA genes. Secondly, the T-DNA genes are transcribed only in plant cells and do not play any role during the transfer process. Thirdly, any foreign DNA placed between the T-DNA borders can be transferred to plant cells, no matter where it comes from. These well-established facts allowed the construction of the first vector and bacterial strain systems for plant transformation. *A. tumefaciens* has the exceptional ability to transfer a particular DNA segment (T-DNA) of the tumour-inducing (Ti) plasmid into the nucleus of infected cells where it is then stably integrated into the host genome and transcribed, causing the crown gall disease (Nester *et al.*, 1984). T-DNA contains two types of genes: the oncogenic genes, encoding for enzymes involved in the synthesis of auxins and cytokinins and responsible for tumour formation; and the genes encoding for the synthesis of opines. These compounds, produced by condensation between amino acids and sugars, are synthesized and excreted by the crown gall cells and consumed by *A. tumefaciens* as carbon and nitrogen sources. Outside the T-DNA are located the genes for the opine catabolism, the genes involved in the process of T-DNA transfer from the bacterium to the plant cell and the genes involved in bacterium-bacterium plasmid conjugative transfer.

**The process of gene transfer from *Agrobacterium tumefaciens* to plant cells implies several essential steps.**

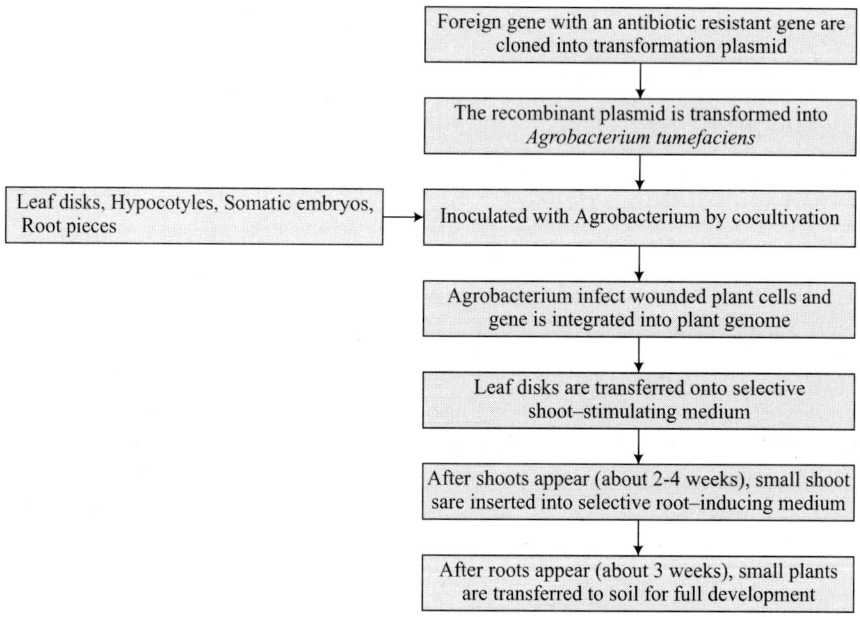


**Flow chart-1** Generation of transgenic plant using *Agrobacterium* – mediated transformation.

## BACTERIAL COLONIZATION

Bacterial colonization is an essential and the earliest step in tumour induction and it takes place when *A. tumefaciens* is attached to the plant cell surface. Mutagenesis studies show that non-attaching mutants loss the tumour-inducing capacity (Bradley *et al.,* 1997). The polysaccharides of the *A. tumefaciens* cell surface are proposed to play an important role in the colonizing process. The bacterial attachment could be prevented when lipopolysaccharides (LPS) solution from virulent strains is applied to the plant tissue before interaction with virulent bacteria. The LPSs are an integral part of the outer membrane and include the lipid A membrane anchor and the O-antigen polysaccharide in their composition. The chromosomal 20 kb *att* locus contains the genes required for successful bacterium attachment to the plant cell. This locus has been extensively studied using transposon insertion mutants. Insertions in the left 10 kb side of this region produced avirulent mutants that could restore its attachment capacity if the culture medium was previously conditioned by the incubation of wild-type virulent bacterium with plant cells.

This previous interaction resulted in the production and accumulation of a complement, absent in the mutant strain, conditioning the medium and allowing the attachment to the plant cell. In contrast, mutational insertion in the 10 kb right side of the *att* locus resulted in the irreversible loss of attachment capacity, which could not be restored by conditioned medium. These results suggest that genes, placed at *att* left side, are involved in molecular signalling events, while the right side genes are likely to be responsible for the synthesis of fundamental components. The *att* left side suggests an operon composed by nine open reading frames (ORF). Four of these ORF show homology to the genes involved in the so-called periplasmic binding protein dependent (or ABC) transport system (Ames *et al.,* 1990). The mutant analysis evidenced a failure in the production and accumulation of specific compounds essential for bacterial attachment. The ABC transporter encoding genes may be involved in the secretion of this substances or in the introduction into bacteria of some plant-originated activators of the synthesis of that compound specific for attachment.

## INDUCTION OF VIRULENCE SYSTEM

The T-DNA transfer is mediated by products encoded by the 30-40 kb *vir* region of the Ti plasmid. This region is composed by at least six essential operons (*vir A, vir B, vir C, vir D, vir E, virG* ) and two non-essential operons. (*virF, virH*). The number of genes per operon differs: *virA, virG* and *virF* have only one gene; *virE, virC, virH* have two genes while *virD* and *virB* have four and eleven genes respectively. The only constitutive expressed operons are *virA* and *virG*, coding for a two-component (*VirA-VirG*) system activating the transcription of the other *vir* genes. The *VirA-VirG* two component system has structural and functional similarities to other already described for other cellular mechanisms (Nixon, 1986).

*VirA* is a transmembrane dimeric sensor protein that detects signal molecules, mainly small phenolic compounds, released from wounded plants. The signals for *VirA* activation include acidic pH, phenolic compounds, such as acetosyringone (Winans, 1992), and certain class of monosaccharides which acts synergistically with phenolic compounds *VirA* protein can be structurally defined into three domains: the periplasmic or input domain and two transmembrane domains (TM1 and TM2). The TM1 and TM2 domains act as a transmitter (signalling) and receiver (sensor). The periplasmic domain is important for monosaccharide detection. Within the periplasmic domain, adjacent to the TM2 domain is an amphipatic helix, with strong hydrophilic and hydrophobic regions This structure is characteristic for other transmembrane sensor proteins and folds the protein to be simultaneously aligned with the inner membrane and anchored in the membrane. The TM2 is the kinase domain and plays a crucial role in the activation of *Vir*A, phosphorylating itself on a conserved His-474 residue (Huang *et al.*, 1990) in response to signaling molecules from wounded plant sites. Monosaccharide detection by *VirA* is an important amplification system and responds to low levels of phenolic compounds. The induction of this system is only possible through the periplasmic sugar (glucose/galactose) binding protein *ChvE*, which interacts with *VirA* ( Chang and Winans, 1992). Recent studies for determination of *VirA* regions, important for its sensing activity, suggested the position, which may be involved on TM1-TM2 interaction. This interaction causes the exposure of the amphipathic helix to small phenolic compounds and suggests a putative model for the *VirA-ChvE* interaction.

Activated *VirA* has the capacity to transfer its phosphate to a conserved aspartate residue of the cytoplasmic DNA binding protein *VirG*. *VirG* functions as a transcriptional factor regulating the expression of *vir* genes when it is phosphorylated by *VirA*. The C-terminal region is responsible for the DNA binding activity, while the N-terminal is the phosphorylation domain and shows homology with the *VirA* receiver (sensor) domain.

The activation of *vir* system also depends on external factors like temperature and pH. At temperatures greater than 32°C, the *vir* genes are not expressed because of a conformational change in the folding of *VirA* induce the inactivation of its properties. The effect of temperature on *VirA* is suppressed by a mutant form of *VirG*, which activates the constitutive expression of the *vir* genes. However, this mutant cannot confer the virulence capacity at that temperature to *Agrobacterium,* probably because the folding of other proteins that actively participate in the T-DNA transfer process is also affected at high temperature.

**Table 2** Virulence genes and their functions

| *Vir Gene* | *Function* |
|---|---|
| ***VirA*** | encodes a single protein which resembles a transmembrane chemoreceptor found in other bacteria; constitutively expressed; monocistronic |
| ***VirB*** | may play a role in directing T-DNA transfer events at the bacterial cell surface |
| ***VirC*** and ***VirD1& D2*** | site-specific endonuclease that cleaves at the 25 bp direct repeats borders of the T-DNA; produces a T-strand that is the intermediate molecule that is transported to the plant cell; polycistronic |
| ***VirE1& E2*** | a single-stranded DNA binding protein that appears to coat the T-strand during transfer to the plant cell; polycistronic |
| ***Vir F*** | cell-cycle regulation, elongation of S-phase |
| ***VirG*** | a positive regulatory protein which relays environmental information (when plant inducible factors are present) to other vir loci; requires *virA*; strongly affects *virB,C,D, E*: monocistronic |
| ***Vir J*** | T-DNA export |

## GENERATION OF T-DNA TRANSFER COMPLEX

The activation of *vir* genes produces the generation of single-stranded (ss) molecules representing the copy of the bottom T-DNA strand. Any DNA placed between T-DNA borders will be transferred to the plant cell as single strand DNA and integrated into the plant genome. These are the only cis acting elements of the T-DNA transfer system. The proteins *VirD*1 and *VirD*2 play a key role in this step, recognizing the T-DNA border sequences and nicking (endonuclease activity) the bottom strand at each border. The nick sites are assumed as the initiation and termination sites for T-strand recovery. After endonucleotidic cleavage, *VirD*2 remains covalently attached to the 5-end of the ss-T-strand. This association prevents the exonucleolytic attack to the 5-end of the ss-T-strand and distinguishes the 5-end as the leading end of the T-DNA transfer complex. *VirD*1 interacts with the region where the ss-T-strand is originated. Experiments *in vitro* evidenced that the presence of *VirD*1 is essential for the cleavage of supercoiled stranded substrate by *VirD*2 (Zupan and Zambryski, 1995). The simultaneous restoration of the excised ss-T-strand is evolutionarily related to other bacterial conjugative DNA transfer processes, which includes the generation of the single strand DNA.

Extensive mutation or deletion of the right T-DNA border is followed by almost completely loss of T-DNA transfer capacity, while at the left border results in lower transfer efficiency. This fact indicates that T-strand synthesis is initiated at the right border and it proceeds in the 5′ to 3′ direction. The termination process takes place even when the left border is mutated or completely absent, although with lower efficiency. Left border may act as a starting site for ssT-strand synthesis but the efficiency is much lower. The difference may be a consequence of the presence of an enhancer or "overdrive" sequence next to the right border. This enhancer has been found to be specifically recognized by Vir*C*1 protein. Deletion of *virC* operon is followed by attenuation of virulence of the *Agrobacterium* strains.

## T-DNA TRANSFER

The transferring vehicle to the plant nucleus is a ssT-DNA-protein complex. It must be translocated to the plant nucleus passing through the membranes—the plant cell wall and cellular spaces. According to the most accepted model, the ssT-DNA-*VirD*2 complex is coated by the 69-kDa *VirE*2 protein, a single strand DNA binding protein. This cooperative association prevents the attack of nucleases and, in addition, extends the ssT-DNA strand reducing the complex diameter to approximately 2 nm, making the translocation through membrane channels easier. However, that association does not stabilizes T-DNA complex inside *Agrobacterium* (Zupan *et al.*, 1996). *Vir*E2 contains two plant nuclear location signals (NLS) and Vir*D*2. This fact indicates that both proteins presumably play important roles once the complex is in the plant cell mediating the complex uptake to the nucleus. The deletion of NLS in one of these proteins reduces, but does not totally inhibit, the ssT-DNA transfer and its integration into plant genome, evidencing that the other partner can at least partially assume the function of the absent protein.

It is known that *VirE*1 is essential for the export of *VirE*2 to the plant cell, although other specific functions are still uncharacterized. Bacterial strains mutated in *virE1* cannot export *VirE*2, which is accumulated inside the bacterium. Such mutants can be complemented if coinfected with a strain that can export *VirE*2, indicating that this protein can be exported independently and that the transfer of *VirE*2 as part of the ss-T-DNA complex is not necessary for the transmission event being possible to transfer naked T-DNA to the plant cell (Sundberg *et al.*, 1996). Previous researches described the role of 9.5 kb *virB* operon in the generation of a suitable cell surface structure for the ssT-DNA complex transfer from bacterium to plant. The *VirD*4 protein is also required for the ss-T-DNA transport. The function of *VirD*4 is the ATP-dependent linkage of protein complex necessary for T-DNA translocation.

*VirB* are proteins that present hydropathy characteristics similar to other membrane-associated proteins. *VirD*4 is a transmembrane protein but predominantly located at the cytoplasmatic side of the cytoplasmic membrane. Comparative studies showed a high degree of homology between the V*irB* operon and transfer regions of broad host range (BHR) plasmids in genetic organization, nucleotide sequence and protein function. Both systems deliver non-self transmissible DNA-protein complex to recipient host cell. In addition, they have the capacity to DNA interkingdom delivery suggesting that the T-DNA transfer apparatus and conjugation systems are related and probably evolved from a common ancestral. The majority of *VirB* proteins are assembled as a membrane-spanning protein channel involving both membranes. Except for *VirB*11, they have multiple periplasmic domains. VirB1 is the only member of *VirB* proteins found in the extracellular milieu (Baron *et al.*, 1997), although it is possible that some of the other *VirB* proteins may be redistributed during the process of biogenesis and functioning of the transcellular conjugal channel, which could be the case of the *VirB*2, a protein with deduced extracellular functions. *VirB*2 is translated as a 12 kDa proprotein, which is later processed by proteolysis to its mature 7 kDa functional form.

*VirB*4 and *VirB*11 are hydrophilic ATPases necessary for active DNA transfer. *VirB*11 lacks continuous sequence of hydrophobic residues, periplasmic domains. Despite these structural characteristics, less than one-third of *VirB*11 constitutes its soluble fraction, while the rest of the protein remains associated with the cytoplasmic membrane. These characteristics are atypical for this type of protein and there are evidences for the possible dynamic coexistence of its different conformational forms *in vivo. VirB*4 tightly associates with the cytoplasmic membrane. It contains two putative extracellular domains conferring transmembrane topology to this protein, which presumably allows the ATP-dependent conformational change in the conjugation channel. Probably, the functional forms of *VirB*4 and *VirB*11 are homo- and heterodimers. The *VirB*4 synthesis is well correlated with the accumulation and distribution of *VirB*3. Other protein, *VirB*7, seems to be crucial for the conformation of the transfer apparatus. *VirB*7 interacts with Vir*B*9 forming heterodimers and probably higher-order multimeric complexes. The synthesis of *VirB*9 and its stable accumulation depends on heterodimer conformation, indicating that *VirB*9 alone may be unstable and requires the association with *VirB*7. In this intermolecular conformation, the monomeric subunits are joined by disulphide bridges. The *VirB*7-*VirB*9 heterodimer is assumed to stabilize other *Vir* proteins during assembly of functional transmembrane channels (Fernandez *et al.,* 1996).

Some of the initial steps of biogenesis of ssT-DNA complex apparatus have been recently identified. Firstly, *VirB*7 and *VirB*9 monomers are exported to the membrane and processed. They interact with each other to form covalently cross-linked homo- and heterodimers. Although the role of both types of dimers in the biogenesis of the transfer apparatus is widely accepted, it is likely that only heterodimers are essential. Subsequently, the *VirB*7-*VirB*9 heterodimer is sorted to the outer membrane. The sorting mechanism has not been elucidated. The next step implies the interaction with the other *Vir* proteins for assembling the transfer channel with the contribution of the transglycosidase *VirB*1. It is known that *Vir*B2 through *VirB*11 are essential for DNA transfer, suggesting that these proteins are fundamental component of the transfer apparatus, while *VirB*1 has a lesser contribution to this process. Two accessory *vir* operons, present in the octopine Ti-plasmid, are *VirF* and *VirH*. The *VirF* operon encodes for a 23 kDa protein that functions once the

T-DNA complex is inside the plant cells via the conjugal channel or independently, as it was assumed for *VirE*2 export. The role of *VirF* seems to be related with the nuclear targeting of the ssT-DNA complex but its contribution is less important than in the case of *VirF*. The *VirH* operon consists of two genes that code for *VirH*1 and *VirH*2 proteins. These *Vir* proteins are not essential but could enhance the transfer efficiency, detoxifying certain plant compounds that can affect the bacterial growth (Kanemoto *et al.,* 1989). If that is the function of *VirH* proteins, they play a role in the host range specificity of bacterial strain for different plant species.

## INTEGRATION OF T-DNA INTO PLANT GENOME

Inside the plant cell, the ssT-DNA complex is targeted to the nucleus crossing the nuclear membrane. Two *Vir* proteins have been found to be important in this step: *VirD*2 and *VirE*2, which are the most important, and probably *VirF*, which has a minor contribution to this process. The nuclear location signals (NLS) of *VirD*2 and *VirE*2 play an important role in nuclear targeting of the delivered ss-T-DNA complex, as described early. *VirD*2 has one functional NLS. The ssT-DNA complex is a large (up to 20 kb) nucleoprotein complex containing only one 5′-end covalently attached *Vir*D2 protein per complex. But the complex is coated by a large number of *VirE*2 molecules (approximately 600 per 20 kb T-DNA), and each of them has two NLS. The two NLS of *VirE*2 have been considered important for the continuous nuclear import of ss-T-DNA complex, probably by keeping both sides of nuclear pores simultaneously open. The nuclear import is probably mediated also by specific NLS-binding proteins, which are present in plant cytoplasm.

The final step of T-DNA transfer is its integration into the plant genome (Fig. 5). The mechanism involved in the T-DNA integration has not been characterized. It is considered that the integration occurs by illegitimate recombination. According to this model, paring of a few bases, known as micro-homologies, are required for a pre-annealing step between T-DNA strand coupled with *VirD*2 and plant DNA. These homologies are very low and provide just a minimum specificity for the recombination process by positioning *VirD*2 for the ligation. The 3'-end or adjacent sequences of T-DNA find some low homologies with plant DNA resulting in the first contact (synapses) between the T-strand and plant DNA and forming a gap in 3'-5' strand of plant DNA. Displaced plant DNA is subsequently cut at the 3'-end position of the gap by endonucleases, and the first nucleotide of the 5' attaches to *VirD*2 pairs with a nucleotide in the top (5'-3') plant DNA strand. The 3' overhanging part of T-DNA together with displaced plant DNA is digested away, either by endonucleases or by 3'-5' exonucleases. Then, the 5' attached to *VirD*2 end and other 3'-end of T-strand (paired with plant DNA during since the first step of integration process) joins the nicks in the bottom plant DNA strand. Once the introduction of T-strand in the 3'-5' strand of the plant DNA is completed, a torsion followed by a nick into the opposite plant DNA strand is produced. This situation activates the repair mechanism of the plant cell and the complementary strand is synthesized using the early inserted T-DNA strand as a template.

*VirD*2 has an active role in the precise integration on T-strand in the plant chromosome. The release of *VirD*2 protein may provide the energy containing in its phosphodiester bond, at the Tyr29 residue, with the first nucleotide of T-strand, providing the 5'-end of the T-strand for ligation to the plant DNA. This phosphodiester bond can serve as electrophilic substrate for nucleophilic 3'-OH from nicked plant DNA. When the mutant *VirD*2 protein is transferred

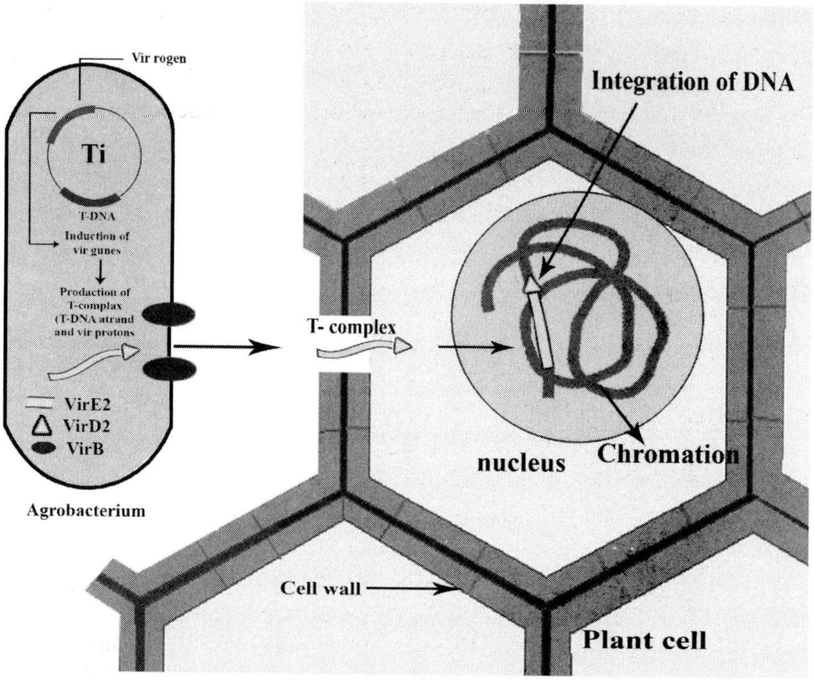


**Fig. 5** The path of T-DNA transfer and stable integration into the plant genome.

attached to the T-strand, the integration process take place with the loss of nucleotides at the 5'-end of the T-strand.

## *AGROBACTERIUM* MEDIATED TRANSFORMATION OF PAPAYA (*CARICA PAPAYA* L.)

Papaya (*Carica papaya* L.) belonging to family Caricaceae, has become an important fruit having great nutrition and commercial values. It is a polygamous fruit crop and both dioceous as well as gynodioceous varieties of papaya are being cultivated. Papaya Ring Spot and Papaya Leaf Curl are two major viral diseases that threaten papaya cultivation world over. An efficient *in vitro* regeneration protocol is therefore imperative for improvement of papaya through biotechnological approach. Somatic embryogenesis system using hypocotyls or immature zygotic embryos has been successfully utilized for *in vitro* regeneration and transformation in many countries. Micropropagation of papaya has been reviewed (Fitch *et al.,* 1993).

In this case study the procedure used is *Agrobacterium*-mediated plant transformation. Several factors have been considered in the design and implementation of any plant transformation study (Fig. 6).

**Explant to be Transformed:** The purpose of most plant transformation experiments in plant biotechnology is to produce whole transgenic plants. The explant used in transformation must be capable of producing whole plants by regeneration and should contain a high number of cells that are component of transformation.

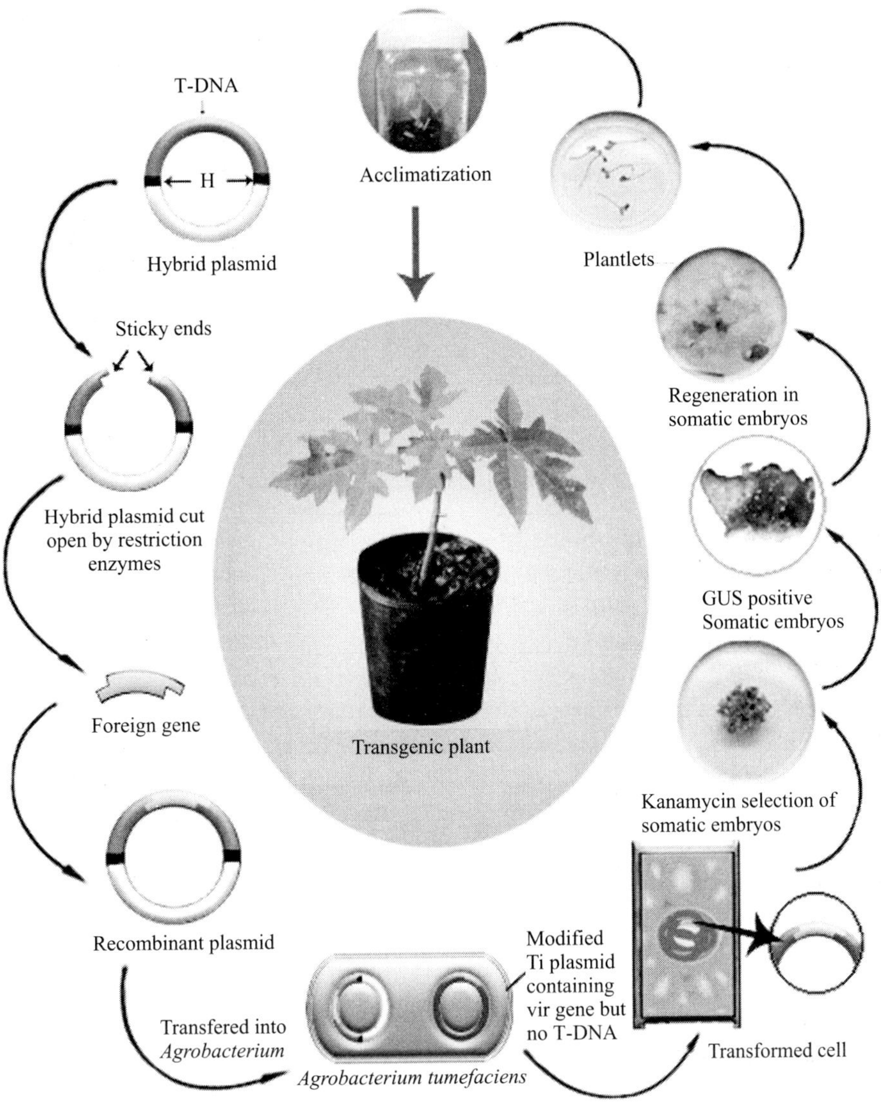


**Fig. 6** *Agrobacterium* mediated transformation in *Carica papaya* L.

**The Vector Used for Transgene Integration into Plant Genome:** Vectors used in *Agrobacterium*-mediated plant transformation are derivatives of the naturally occurring Ti plasmid. Generally, these are extensively modified so that the most features of a natural occurring Ti plasmid are removed, only the left and right border sequences being used to ensure transfer of the Ti-DNA region between them. The vector also contains a selectable marker (generally Kanamycin sulphate) on the T-DNA so that transformed plants can be separated from other non-transformed plants. Virulence genes required for transfer of the T-DNA are often located on a separate plasmid in the bacterium.

**Agrobacterium Strain to be used:** Several widely used *Agrobacterium* strains are available for plant transformation. For more recalcitrant plant species, the choice of strain is a major factor contributing to the success or failure of the experiment. A general protocol for *Agrobacterium*-mediated plant transformation in papaya is:

## STERILIZATION AND EXPLANT PREPARATION

Seeds from self-pollinated papaya are prepared for aseptic culture. The seeds are washed in running tap water and soaked in 0.1% Carbendazime (Bavestine) containing 1 drop of Tween 20 for surface sterilization for one hour, and then washed five times with sterile distilled water. Surface-sterilized seeds are shaken for 24 h at 27°C in 1 M $KNO_3$ and in sterile water at 32°C for 3 days or until the testate cracked. The seeds are germinated on 1% agar. After 10 days, seedlings with expanded cotyledons and 3-12 cm long hypocotyls are used as explant.

## CO-CULTIVATION

*A. tumefaciens* is grown in LB (Tryptone 10 gm/l, Yeast extract 5 gm/l and NaCl 10 gm/l) medium containing 50 mg/l kanamycin, in the dark at 28°C for 48 hours (0.8-1 OD at 600 nm). After this period, 10 ml of LB medium with antibiotics and 100 mM acetosyringone are added to each culture flask, and incubated for an extra period of 4-5 h. Bacterial cells are collected by centrifugation at 8000 rpm. The pellet is resuspended in 10 ml of MS medium, in which 100 mM acetosyringone was added prior to use. The hypocotyls are sectioned into 2-3 mm length and placed into a culture of *A. tumefaciens* (which contains the vector) for 30 minutes. To eliminate any bacterial excess, tissues are washed in MS liquid medium followed by blotting on sterile filter paper and placed on MS semisolid medium (no selective agent) for co-cultivation for two days to allow transfer of the T-DNA to the plant cells.

The explants are removed from the medium and washed in an antibiotic solution (such as cefotaxime 500 mg/l or carbenicillin 250 mg/l or rifimpicin 50 mg/l). Tissues contaminated with bacteria having apparent resistance to the antibiotics are discarded.

Histochemical GUS assay—Transient expression of GUS is histochemically assayed to screen for expression of B-glucuronidase (GUS) activity in genetically modified plants. The assay is carried out according to the method of Jefferson *et al.* (1987). Tissues are incubated in reaction buffer containing 50 mm Na $H_2PO_4$ (pH-7), 0.01% Tween-20, 10 mm $Na_2$ EDTA and 0.3% (w/v) 5- bromo 4- chloro 3-inolyl glucuronide (X-Glus) as a substrate for the enzyme. Tissue is incubated at 37°C overnight and screened for blue staining indicating GUS activity under a stereomicroscope on the third day after infection. Before assessing the number of blue spots in the explants, they are washed in methanol:acetone (3:1v/v). Each blue spot is considered as one transient GUS- expression focus.

## SELECTION AND REGENERATION

### Somatic embryo Induction and Multiplication

**Kanamycin Selection:** Seven days after the co-cultivation, the explants are transferred to fresh semi-solid medium supplemented with a selective agent (generally kanamycin, but this will depends on which selectable marker gene is present in the T-DNA of the vector) to prevent non-transformed plant cells. It also contains cefotaxime 500 mg/l or carbenicillin 250 mg/l (to prevent the growth of any *Agrobacterium* that are not killed by the initial treatment). Induction medium (½ strength MS medium supplemented with 400 mg/l glutamine, MS vitamins, 10 mg/l 2,4-d, 6% sucrose, and 0.8% agar of pH 5.8) contained 100 mg/l kanamycin. Four week later the kanamycin concentration is increased to 150 mg/l. The cultures are Incubated at 25°C ± 2°C in the dark for 4 weeks. Globular embryos become visible after 4 weeks, which form bipolar-shape. Subculture somatic embryos on this medium, which will result in large number of embryos.

The co-cultivation tissues are subcultured monthly. In contrast, putative transformed embryos emerge as pale yellow clusters that grow out among the mass of brownish embryogenic tissues.

### Somatic Embryo Maturation and Germination

After 4 weeks, several creamish white to light yellow thread like embryos appear. Transfer the clump of somatic embryos on half strength MS medium containing 400 mg/l glutamine, MS vitamins, 0.2 mg/l BAP and 0.1 mg/l NAA 3% sucrose and 0.8% agar (Regeneration medium). Cultures were incubated in growth room at 25 ± 2°C temperature for 16 h photoperiod with light at 40 μmolm$^2$ s$^{-1}$at 55% relative humidity. Allow the somatic embryos to germinate. If callusing starts at the base of the shoots, transfer the cultures on MS medium devoid of any hormone.

### Rooting and Acclimatization

Transfer regenerated plantlets in transparent culture bottles containing autoclaved vermiculite supplemented with ¼ MS liquid medium and cultures are incubated in growth room at 25 ± 2°C temperature for 16 h photoperiod with light at 40 μmolm$^2$ s$^{-1}$at 55% relative humidity for 2-3 weeks. Sometimes getting good root is a problem in such plants. In such plants add 1 mg/l IBA solution in vermiculite for better root growth. The caps of the bottles are loosened gradually and the plants are shifted to pots in poly house provided with 50% shade net and misting. Once plant develops 4-6 leaves (fully open), transfer it in polybags filled with 1:1:1 soil + sand + FYM.

## CONFIRMATION OF TRANSFORMATION

Transformation is confirmed by several means:

1. Transformed tissues are tested for their continued growth on the media with specific antibiotics (Kanamycin 150 mg/l).
2. Transformed calli, somatic embryos, leaf sections, roots and fruits sections from transformed plants are tested for GUS expression by a histochemical GUS assay (Jefferson *et al.,* 1987).
3. PCR analysis is performed (Ausubel *et al*., 1992) using DNA from transgenic and control plants prepared essentially as described (Dellaporta *et al.,* 1983).
4. For DNA hybridization, genomic DNA is isolated from the plants essentially as reported (Murray and Kennard, 1977). Fifty micrograms DNA is restricted with specific R.E. and separated on a 0.8% agarose gel. DNA is transferred onto a nylon membrane for hybridization with a cDNA fragment of specific promoter as a probe. The probe is designed to detect border sequences to give an estimation of T-DNA copies.

5. Double antibody sandwich enzyme-linked immunosorbent assay (DAS-ELISA) technique (Clark and Adam 1977), employing polyclonal and monoclonal antibodies is used to assay for presence of desired proteins (e.g. produced from PRV CP gene) and NPT II in putative transgenic leaves (Fitch et al., 1992).
6. ***Greenhouse test:*** Transgenic plants which contain specific gene sequences are screened for resistance. 3-4 young plant leaves are rubbed with carborandum and PRV virulent strain and/or the sap extract from the transgenic papaya leave of symptomless plants are applied to virus indicator plant ***Chenopodium quinoa*** for screening virus.
7. ***Field testing:*** Non-transformed and transformed plants are planted in the same field and the virulent virus strain of specific pathogen (e.g. PRV) is inoculated manually or naturally (aphids). Symptoms are monitored and PRV strains are confirmed by ELISA. Viral infections also test via cross-pollination between healthy, transgenic and infected plants.

**Table 3** Transgenic crop development and evaluation

| *Stage* | *Lab/Field* | *Analysis* | *Remark* |
|---|---|---|---|
| T0 | Lab testing or Green-house/Phytotron | • Molecular analysis<br>• Target trail data | Select a few promising plants |
| T1 | Greenhouse/Phytotron | • Molecular analysis of selected single plant progenies<br>• Trial segregation analysis | Identify individual positive plants |
| T2 | Greenhouse/Phytotron | • Molecular analysis of single plant progenies<br>• Target trial analysis | Identify non-segregating progenies |
| T 3 | Limited field trial | • Agronomic performance including DUS(distinctiveness, Uniformity and Stability) features<br>• Target trial expression<br>• Toxicity data<br>• Allergenicity data<br>• Environmental impact analysis<br>• Pollen flow study<br>• Soil data | • Tntegration of transgenic into IARI breading programme<br>• Identify promising progenies with desired level of target trait expression and advance to T4 generation |
| T4 and T5 | Multilocation field evaluation | • Agronomic data<br>• Target trait expression<br>• Pollen flow data | Dialogue with project Directors/<br>Coordinators of respective crops for VCU testing |
| T6 | VCU (Value for cultivation and use) testing in all India Coordinated Trials | • Seed multiplication | Variety identification and release |

[Source: A Scheme of Indian Agriculture Research Institute, New Delhi (India)]

## FUTURE OF GENETICALLY MODIFIED (GM) CROPS

Since the first reports on GM crops in the 1980s, the amount of debate surrounding the generation of transgenic plants and the political and economical ramifications of their development have seldom been far from the public eye. It could be said that attention on the use of trangenetically modified crop plants is substantially greater than any other area of plant science. To date, the majority of commercially produced GM crops have been created by ballistically delivering transgenes into embryonic tissue (Padgette *et al.,* 1995) However, as our understanding of the mechanisms of T-DNA integration grows and further information is reported on the proteins involved, the ballistic approach may well be superseded by a more sophisticated and "natural" solution to the problem.

*Agrobacterium tumefaciens* is a soil-borne bacterium that, in nature, is capable of inserting a defined fragment of its DNA into the genome of dicotyledonous plants (Tzfira and Citovsky, 2002; Gelvin, 2003). It has been the focus of research for a wide spectrum of biologists, from bacteriologists to molecular biologists to botanists, for a number of years. Moreover, the unique process by which it delivers trans-kingdom DNA into host plant cells has become a staple in almost all undergraduate plant science courses. However, it is important to remember that *A. tumefaciens* remains at the cutting edge of plant science. Our understanding of the molecular mechanisms of DNA integration has grown immeasurably in recent years, and in a climate of intense political debate surrounding the future of transgenic crops, the future role of *A. tumefaciens* in the controlled creation of transgenic crops or simply as the model type IV secretion system and as a research tool would appear to hang in the balance.

## BIOSAFETY REGULATIONS OF GMOS

The countries participating in the Earth Summit in 1992 have agreed upon the fact that biotechnology can offer indubitable benefits to sustainable development, world food supplies and economic prosperity. In order for this potential to be largely applied worldwide and with particular emphasis in developing countries and thus facilitating the alleviation of poverty, the countries joined their efforts in preparation of international rules, which would both ensure the further development of biotechnology for the benefit of the society at large and the conservation of genetic resources, especially in the centres of origin (mostly situated in the third world). The international rules, reflected both in the Cartagena protocol on biosafety and the WTO agreements are built on scientific basis and promote the case-by-case approach, i.e., every transgenic event should undergo separate risk assessment, and potential hazards specific to this event should be identified and specific risk management measures assigned. The specific international agreements that threat different aspects of the products of modern biotechnology (GMOs), however, as any other international instruments, are results of negotiations and compromises. In the last years, there are some tendencies in the negotiations of a few international instruments, e.g. the Cartagena protocol on biosafety and the Aarhus convention on public participation that diverge from the initial idea of the Earth Summit, by taking into consideration only the eventual negative effects that might be associated with the deliberate release into environment of the products of modern biotechnology. Some countries and nongovernmental organizations have expressed their willingness for stricter liability regimes that are in the position to hinder the development of public research in the countries, particularly developing countries and countries with economies in transition.

The policy makers in these countries should take into account the fact that public research is always oriented respond to a specific problem in the country' s agriculture or medicine and has a clear social benefit driven feature. Moreover, the scientific problem of the public research is not addressed by the multinational biotech companies, whose products are mostly in commodity crops that are able to bring fast and considerable profits. In the past years several international instruments that consider different aspects of the trade, transboundary movement and potential adverse effects for the environment of GMOs have been agreed. In most of the cases, closer interaction and cooperation, as well as further harmonization among these agreements would be recommendable. Unified stricter regimes may lose on flexibility and would not be able to satisfy the needs and interests of every country, particularly developing countries. Other than relying on international instruments e.g. the Cartagena Protocol on Biosafety, the countries all over the world are highly encouraged to develop their own national biosafety frameworks (GEF project) that would better reflect the countries' needs in terms of import-export of the products of modern biotechnology. There are several models of such national regulations, overviewed in this paper, that may be effective in building up a workable biosafety system. Which model to be chosen depends on the policy of the given country and should be in accordance with its international and regional obligations. It is commonly understood that international and regional harmonization, in addition to synchronizing the national regulatory frameworks, should focus on the issues of strengthening capacities and information sharing for biotechnological safety. Many countries, especially in the developing world, need to acquire the technology and the capacities necessary to sustainably handle the results of modern biotechnology. Therefore, public awareness, education and technology transfer play an important role. A number of international organizations such as FAO, WHO, UNEP, UNIDO, OECD, ICGEB and CGIAR, as well as the shown examples of regional cooperation are in the position to offer necessary assistance in capacity-building and dissemination of information on biosafety.

## TRANSGENIC PLANTS: ENVIRONMENTAL BENEFITS AND RISKS

Transgenic plants have been rapidly adopted globally in the past seven years. Planting transgenic plants bring huge benefits to the society and the environment by increasing yield and protecting environment, by reducing usage of toxic chemicals, efficient use of renewal resources, efficient use of arid land and improving environment, and monitoring, detecting detoxifying environmental pollution. However, like all other technologies, transgenic plants are not 100% safe. Potential environmental risks associated with gene flow, allergy or toxic to human being, beneficial insects or non-target organisms, directly switching transgenic plants to superweeds, increasing use of herbicides and the pest resistant to Bt plant are the major concerns of the day. Both the benefits and the risks of transgenic plants may vary spatially and temporarily on a case-by-case basis and to compare transgenic plants with traditional plants, and other agricultural practices for elucidating the relative benefits and risk of the transgenic plants.

## Box no. 1

### *Agrobacterium rhizogenes*

*Agrobacterium rhizogenes* is closely related to member of the biotype 2group of *A. tumefaciens* by physiological and DNA homology criteria. *A. rhizogenes* is distinguished from *A tumefaciens* because it induces hairy root disease, a disease characterized by extensive proliferation of roots from an infected plant wound. Plasmid in *A. rhizogenes* (Ri plasmids) strains have been characterized, and it has been shown that there are a number of different types, which can be classified based upon opine usage. Hormone biosynthetic genes are also present on some Ri plasmids. Homologues of the Ti-plasmid auxin-biosynthetic genes *tms1* and *tms2* have been identified on the agropine strain Ri plasmids. However, despite the well-established link between auxin and rooting, the genes are unnecessary for virulence. Rather, series of other open reading frames have been identified within the T-DNA of the plasmids, including the *rolB* and *rolC* genes that are involved in the metabolism of plant growth regulators and lead to the plant being sensitized to endogenous auxin. It is this increased sensitivity that leads to the root formation. As with *A. tumefactions*, vectors have been constructed that can be used in *A. rhizogenes* as binary systems.

Hairy roots are important in some area of plant biotechnology as they can be cultured *in vitro*. For many years they have been used as a system for the production of secondary metabolites and pharmaceutical proteins. *A. rhizogenes* transformation was, at one stage, consider an alternative strategy of *A. tumefaciens* for gene transfer as it lead to the production of the defined tissues (Hairy roots) that could be regenerated into whole plants. This strategy seems to have been discharged; however, as more efficient *A. tumefaciens* systems have been developed. *A. rhizogenes also* play an important role in production of roots in those plant species, which is difficult to root.

## Box no. 2

### *Gene silencing*

Gene silencing is a technique used to turn down or switch off the activity of genes. It is a powerful technology for gene discovery and determining gene function in humans, animals, plants and insects. It is also used to develop new crop varieties and holds tremendous promise as a therapeutic agent to control disease in humans and animals.

Two types of gene silencing are of interest to the plant biotechnologist: transcriptional gene silencing (TGS) and post-transcriptional gene silencing (PTGS), both of which are homology-dependent, gene silencing (HDGS) phenomena. The two types are distinguished according to the level of gene expression at which silencing occurs.

### Transcriptional Gene Silencing

TGS generally occurs when genes share homology in their promoter regions. It usually results in altered methylation patterns and altered chromatin confirmation, which results in gene silencing by repressing transcription.

## Post-transcriptional Gene Silencing

PTGS has been demonstrated that the expression of a homologous transgene can lead to the inhibition of expression of both the transgene and the endogenous gene, primarily by decreasing RNA stability. Thus instead of resulting in over expression of a particular gene, expression is actually reduced. The effect appears to increase with increases in the transcription of the transgene. Thus the use of very strong promoter to drive transgene expression may actually lead to more co suppression, and thus a lower level of expression, than the use of a weaker promoter. PTGS appears to result from the production, either in the nucleus or cytoplasm, ds RNA. This ds RNA can result from the presence of 'cryptic' transcription start sites producing antisense RNA, transcription through inverted repeats and through the activity of an RNA dependent RNA polymerase. This ds RNA is cut by a ds RNA endonuclease, resulting in the production of small (21-25 bases) stretches of sense and antisense RNA (termed 'siRNAs', i.e. small interfering RNAs) which can be transported throughout the plant. These siRNAs guide a different ribonuclease to the homologous ss mRNA, which is cut and then further degraded.

One useful 'spin-off' of gene silencing is the development of a technique known as RNAi (RNA interference) where ds RNA is deliberately made to silence gene expression in gene function studies.

## How Does Gene Silencing Work ?

Genes are found in living organisms and are passed on from one generation to the next. They are the coded instructions an organism uses to make proteins, and it is these proteins that make up the structures and perform the functions of living things. RNA is the courier that delivers the gene's instructions to make a protein. Gene silencing directs a natural mechanism to degrade the RNA instructions of a specified gene, preventing the gene from making its protein. Because of its effect on RNA, gene silencing is also referred to as RNA interference (RNAi). Gene silencing switches off the activity of only a targeted gene, so it is possible to determine the precise function of that gene. This specificity is important when using gene silencing to develop new traits where the activity of a targeted gene is switched off.

## Virus Resistant Plants

Gene silencing can be used to introduce virus resistance into a plant. Viruses are a significant problem in many of Australia's crop plants and developing virus resistant plants could provide many benefits to farmers and the environment. Introducing virus resistance to plants via gene silencing is done by inserting a piece of a DNA that has been modified to include one small fragment of a virus into a plant. This then effectively vaccinates the plant. When the plant is actually exposed to the virus it has a ready defense mechanism primed to destroy the now recognizable virus.

### Strategies to Eliminate Gene Silencing

Gene silencing can be effectively eliminated by a suitable plant breeding strategy. But several steps may be taken to minimize the likelihood of transgene silencing; some of these are discussed below.

(1) The transformation vector should not have duplicated sequences.

(2) Each gene construct in the vector should have a different promoter and polyadenylation signal.

(3) All the gene constructs in the vector should have the same orientation and should not be located adjacent to each other.

(4) The AT composition of the transgene should be similar to that of the host chromosome. In addition, the transgene sequence should be optimal for codon usage in plants and should not contain sequences that interfere with mRNA function/stability. The transgene construct must have appropriate 5' leader sequences and polyadenylation sites since they are essential for optimal gene expression.

(5) The transgene should be integrated in a single copy, and away from chromosomal locations having hypermethyation. Transgenes could be integrated into the large stretches 5' and 3' regions of endogenous genes to protect them from methylation due to the hypermethylation of flanking chromosomal sequences.

(6) Stability of transgene expression in homozygotes may be tested in tissue culture by exposing transgenic tissues to propionic or butyric acid. This test would eliminate plants in which silencing could occur due to environmental stress. But the plants showing stable transgene expression will have to the further evaluated under field conditions.

## SUMMARY

A variety of techniques for plant transformation are available to the plant biotechnologists. These techniques can be split into two groups: Agrobacterium mediated transformation; and direct gene transfer methods, of which the biolistics approach is probably the most widely used. These two groups of techniques are fundamentally different in mechanism, and are, in general, applied to different crops. Direct gene transfer methods and *Agrobacterium* mediated methods have there advantages and disadvantages. However, all the plant transformation methods can suffer from a problem known as gene silencing (Box no. 2). Gene silencing is impossible to predict precisely, although some precautions in vector design and transformation protocol can be taken to reduce its frequency. Gene silencing can prove a major hurdle to the commercialization of plant transformation products.

## ACKNOWLEDGEMENTS

Authors are grateful to Mr. Ramesh Pati for digital photography and redrawn of figures.

## References


Alexandrova, N., Georgieva, K. & Atanassov, A. (2005) R&D *Biotechnol. & Biotechnol.* Eq. **19**/Special Issue Page-153-172.

Ames, G.F.L., Mimura, C.S. & Shyamala, V. (1990) *FEMs Microbiological Review* **75**: 429-446.

Ausbel, F.m., Berent R., Kingston, R.E., Moore, D.D., Seidman, J.G., Smith, J.A. & Struhl, K. (1992) Current protocol in molecular biology, John Wiley, Media, PA.

Baron, C., Llosa, M., Zhou, S. and Zambryski, P.C. (1997) *Journal of Bacteriology* **179**: 1203-1210.

Bradley, L.R., Kim, J.S. & Matthysse, A.G. (1997) *Journal of Bacteriology* **179**: 5372-5379.

Cangelosi, G.A., Best, E.A., Martinetti, C. & Nester, E.W. (1991) Methods in Enzymology **145**: 177-181.

Cangelosi, G.A., Martinetti, G., Leigh J.A., Lee, C.C., Theines, C. & Nester, E.W. (1989) *Journal of Bacteriology* **171**: 1609-1615.

Chang, CH & Winans, S.C. (1992) *Journal of Bacteriology* **174**: 7033-7039.

Cheng, XY, Sardana, R., Kaplan, H. and Altosaar, I. (1998) *Proceedings of the National Academy of Sciences USA* **95**: 2767-2772.

Clark, M. & Adams, A. (1977) *J. Gen. Virol.* **34**:475-483.

Cramer, C. L., Boothe, J. G. & K. K. Oishi (1999) *Curr. Top. Microbiol. Immunol.* **240**: 95-118.

Dellaporta S.L., Wood J. & Hicks J.B. (1983) *Plant Mol. Biol. Reptr*. **1:** 19-21.

Enríquez-Obregón, G.A., Vázquez-padrón, R.I., Prieto-sansonov, D.L., de la Riva, G.A. & Selman-Housein,G.(1998) *Planta* **206**: 20-27.

Fernández, D., Spudich, G.M., Zhou, X.R., Berger, B.R. & Christie, P.J. (1996) *Journal of Bacteriology* **178**: 3168-3176.

Fitch, M. M., Manshardt, R. M., Gonsalves, D., Slinghtom, J. L. & Sanford, J. C. (1992) *Plant Cell Rep.* **12**:245-249.

Fitch, M.M., Manshardt, R. M., Gonsalves, D. & Slinghtom, J. L. (1993) *Plant Cell Rep.* **12**: 245-249.

Gelvin, S.B. (2003)*Trends Biotechnol* **21**: 95-98.

Gregorio G.H., Elide A. B., Lizbeth C. C. & Maria de L. M.H. (2006) *Plant Cell, Tissue and Organ Culture* **84**: 365-368.

Guideline for transgenic crop development and evaluation. Indian Agriculture Research Institute (ICAR), New Delhi.

Gustavo A. de la Riva., Joel González-Cabrera., Roberto Vázquez-Padrón & Camilo Ayra-Pardo. (1998) *Electronic Journal of Biotechnolog.,* Vol. **1**, No. 3, December,

Hooykaas, P.J.J. (2000) In: encyclopedia of Microbiology, vol, 1 2nd ed. (Ed. Lederberg, J.) academic Press, San Diago, USA, pp. 78-85.

Hooykaas, P.J.J. & Shilperoort, R.A. (1992) *Plant Molecular Biology* **19**: 15-38.

Huang, Y., Morel, P., Powell, B. & Kado, C.I. (1990) *Journal of Bacteriology* **172**: 1142-1144.

Ishida, Y., Saito, H., Ohta, S., Hiei, Y., Komari, T. & Kumashiro, T. (1996) *Nature Biotechnology* **4**: 745-750.

Jefferson R.A., Kavanagh T.A. & Bevan M. W. (1987) *EMBO J.* **6:** 3901-3907.

Jefferson, R.A. (1987) *Plant Mol Biol Rep* **5**:387-405.

Jeon, G.A., Eum, J.S. & Sim, W.S. (1998) *Molecules and Cells* **8**: 49-53.

Kanemoto, R.H., Powell, A.T., Akiyoshi, D.E., Regier, D.A., Kerstetter, R.A., Nester, E.W., Hawes, M.C. & Gordon, M.P. (1989) *Journal of Bacteriology* **171**: 2506-2512.

Lisa, V. (2003) *Plant Physiology* **133**:948-955.

Matthysse, A.G. (1983) *Journal of Bacteriology* **154**: 906-915.

May, G.D., Afza, R., Mason, H.S., Wiecko, A., Novak, F.J. & Arntzen, C.J. (1995). *Bio/Technology* **13**: 486-492.

Murray M.G. & Kennard W.C. (1977) *Biochemistry* **23**: 4225-4232.

Nester, E.W., Gordon, M.P., Amasino, R.M. & Yanofsky, M.F. (1984) *Annual Review of Plant Physiology* **35**: 387-413.

Nixon, B.T., Ronson, C.W. & Ausubel, F.M. (1986) *Proceedings of National Academy of Sciences USA* **83**: 7850-7854.

Otten, L., DeGreve, H., Leemans, J., Hain, R.; Hooykaas, P. & J. Schell. (1984) *Mol. Gen. Genet.* **195**:159-163

Padgette, S.R., Kolacz, K.H., Delannay, X., Re, D.B., LaVallee, B.J., Tinius, C.N., Rhodes, W.K., Otero, Y.I., Barry, G.F. & Eichholz DA. (1995) *Crop Sci* **35:** 1461-1467.

Sampson, J.A. (1986). Tropical Fruits. Second Ed. Longman Scientific and Technical, Longman Inc., New York, NY.

Smith, E.F. & Towsend, C.O. (1907) *Science* **25**: 671-673.

Stanton, B. (2003) *Microbiology and Molecular Biology Reviews* Vol.**67**, No. 1 16-37.

Sundberg, C., Meek, L., Carroll, K., Das, A. & Ream, W. (1996) *Journal of Bacteriology* **178**: 1207-1212.

Tzfira T. & Citovsky V. (2002) *Trends Cell Biol* **12:** 121-129.

Van Damme, P. (2003). Highland papayas. http://allserv.rug.ac.be/~tdirckx/caricaceae.htm. Accessed March, 2004.

Winans, S.C. (1992) *Microbiology Reviews* **56**: 12-31.

Xiaoping, Pan., Baohong, Zhang & George P. Cobb. (2005) *Physiol. Mol. Bio. Plants.* **11(1)**: 13-32

Zupan, J.R. & Zambryski, P.C. (1995) *Plant Physiology* **107:** 1041.1047.

Zupan, J.R., Citovsky, V. & Zambryski, P.C. (1996) *Proceedings of the National Academy of Sciences USA* **93**: 2392-2397.

23

# Coastal Sand Dune Vegetation and Microbial Resources: Benefits, Threats and Safeguards

**K.R. Sridhar[1] and A.B. Arun[2]**

[1]*Microbiology and Biotechnology, Department of Biosciences Mangalore University, Mangalagangotri-574 199 Mangalore, Karnataka, India*

[2]*College of Agriculture and Natural Resources, Department of Soil Environmental Science National Chung Hsing University, 250 Kuo-Khang Road Taichung, Taiwan 40227, Republic of China*

**ABSTRACT**

The coastal sand dunes (CSDs) undergo continuous alteration in topography as well as physico-chemical features mainly due to wind and wave action. In view of agriculturally and industrially important flora and microbes, the current overview provides a brief outline of ecological and economic value of dune vegetation followed by association and interaction of microflora (nitrogen-fixing bacteria, endophytic fungi and mycorrhizas) with dune vegetation and their importance in plant nutrition, defence and sand/soil aggregation. Importance of vegetation and microbes in dune stabilization, restoration and safeguards is also briefly discussed.

## INTRODUCTION

Although the coastal sand dunes (CSDs) are disturbance-prone, they support a variety of flora, fauna and microbes. Such disturbances result in spatial and temporal heterogeneity of community structure of vegetation (Watkinson and Davy, 1985). Common environmental fluctuations include temperature, desiccation, low moisture retention, sand erosion, sand accretion, salinity and salt spray (Maun, 1994). In distribution of vegetation on the dunes, Maun and Perumal (1999) indicated that the burial is a most important factor followed by salt spray. The nutrient input is usually intermittent through precipitation or salt spray (Kellman and Roulet, 1990). Organic matter serves as a major source of energy (John *et al.*, 1983) in CSDs in spite of deficiency of water, nitrogen, phosphorus, potassium and organic matter (Maun, 1994).

## VEGETATION

### Distribution

*Ammophila* is a major dune-building grass of temperate north-eastern US Atlantic coast, has the greatest influence on coastal geomorphology and development of vegetation (Maun and Baye, 1989). American beach grass (*Ammophila breviligulata*) is native to the East Coast and Grate Lakes of North America (Seabloom and Wiedemann, 1994). This pioneer plant species becomes senescent on stabilization of dunes and is often replaced by other plant species. American beach grass is declining due to root damaging nematodes (Selisker and Huettel, 1993) and nematode infection is substantially lowered in plants associated with arbuscular mycorrhizal (AM) fungi (Little and Maun, 1995). *Cakile edentula*, an annual and facultative perennial plant species grows along the shorelines of the coasts of Grate Lakes of Atlantic as well as Pacific of North America, Japan, New Zealand and Australia (Maun *et al.*, 1990). It is successful on the dunes due to resistance to salt spray, narrow life cycle, production of large number of fruits, dormant seeds and high seedling establishment. Similar to American beach grass, *Ammophila arenaria*, known as European beach grass (or marram grass), is adapted to CSD systems of European coasts (Seabloom and Wiedemann, 1994). *Desmoschoenus spiralis* and *Scirpoides nodosa* are native plants of CSDs of New Zealand and tolerant to high salinity (Sykes and Wilson, 1989). *Elymus farctus* is the most disturbance and salinity tolerant exotic grass on the New Zealand coasts. This species replaced the native sand dune vegetation along the coasts of South Island of New Zealand (Sykes and Wilson, 1991). The foredunes of New Zealand comprises cosmopolitan saline-tolerant prostrate plant species with long rhizomes of *Calystegia soldanella*, which grow in association with *E. farctus* and *Leymus arenarius* and establishes sand mounds.

Sea oats (*Uniola paniculata*), a semitropical C4 perennial grass dominant foredune sand binder of Southeastern Atlantic and US Gulf coasts (Hester and Mendelssohn, 1987). It is most abundant in Bahaman Islands and some parts of Cuba. Tropical shrubby legume, *Chamaecrista chamaecristoides* is endemic to the Mexican coastal dunes and partly to the Pacific coast (Martinez and Moreno-Casasola, 1998). These plants have adapted to mobile dunes, significantly increase the biomass on sand covering, fix nitrogen and support succession of associated flora. Greenhouse studies revealed that *C. chamaecristoides* provide sufficient nutrients to sustain the growth of associated fauna (Martinez and Rincon, 1993). Tropical coastal sand dunes are dominated by the members of Asteraceae, Convolvulaceae, Fabaceae and Poaceae (Arun *et al.*, 1999; Moreno-Casasola, 1988; Rao and Sherieff, 2002). *Ipomoea pes-caprae*, a stoloniferous perennial creeper, is most common strand vegetation of five continents including several tropical islands and warm temperate shores (St. John, 1970). It is restricted to the Indian Ocean, while *Ipomoea brasiliensis* is pantropical except for the Indian Ocean (Fosberg and Sachet, 1977). *Ipomoea imperati* is also pantorpical on the beaches of six continents and many islands of tropics and shores of subtropics (Leonard and Judd, 1999), adapted to low nutrients, withstands sand scouring, moderate burial and high temperature and promotes formation of embryonic dunes. Like *Ipomoea* spp., *Canavalia maritima* is also widely distributed throughout the tropical and subtropical coastal sand dunes and a common cover crop in arid lands of Australia as well as Africa. Biogeographic regions of coastal sand dunes of the Indian subcontinent have been divided into eight subdivisions (Pakistan, Kutch and north-west Kathiawar; southern Kathiawar-Gujarat; Konkan; Malabar; Coromandel-Circar; islands between India and Sri Lanka; Utkal and Bengal; Andaman and Nicobar Islands) (Rao and Meher-Homji, 1985; Rao and Sherieff, 2002). A variety of strand vegetation consisting of mat-forming creepers, prostrate/erect herbs/sedges, climbers, plants with perennating organs, scrubs and trees are common in the Indian subcontinent (Rao and Meher-Homji, 1985).

## Burial

Seeds, seedlings and established plants on the CSDs are adapted to chaotic conditions such as fluctuations in temperature and wind/wave action for seed germination, seedling emergence and survival (Maun, 1994). Burial of seeds at shallow depth is advantageous and increases soil contact and relative humidity, which enhances the chances of germination (Maun, 1998). Enforced dormancy increases on deep burial of seeds (Maun, 1998) and leads to establishment of seed banks. Large seeds emerge from deeper burial than small seeds (Maun and Lapierre, 1986) and the rate of germination is related to size of seeds (Jurado and Westoby, 1992). The CSDs of south-west coast of India receives considerable quantity of seeds from the mangroves (Arun *et al.*, 1999). For instance, seeds of *Canavalia cathartica*, *Derris triflorum, Pongamia pinata* and *Erythrina variegata* transferred from mangroves, grow successfully on the dunes. Seed germination, seedling emergence and recovery of five common CSD legumes (*C. cathartica, C. maritima, Crotalaria retusa, C. verrucosa* and *P. pinnata*) of the west coast of India under the influence of temperature, salinity and burial in greenhouse trials yielded interesting results (Arun *et al.*, 2001). Germination was highest at 28°C in 25% seawater at 2 cm burial, while the upper limit of germination was 42°C in 50% sea water at 5 cm burial. The upper limit of recovery of germination was 42°C in 100% sea water at 10 cm burial shows the adaptability of these plants to sand dunes. Dormancy of seeds at 42°C in 100% sea water at 10 cm burial reveals the formation of seed banks in CSDs. Seed decay was reduced in seawater in all plants except for *P. pinnata*. Moreno-Casasola *et al.* (1994) studied seed germination of nine tropical CSD legumes (*Acacia farnesiana, A. macracantha, Canavalia maritima, Chamaecrista chamaecristoides, Crotalaria incana, Indogofera suffruticosa, Macroptilium atropurpureum, Mimosa chaetocarpa* and *Schrankia quadrivalvis*) in the laboratory as well as field under fluctuating diurnal temperature gradient. Temperature fluctuations higher than 20°C increased the germination rate and diurnal wet-dry temperature fluctuations were responsible for seedling establishment. Burial and long exposure to wet-dry temperature treatment resulted in 50% seed permeability suggesting the forced seed dormancy.

## Economic Value

Besides stabilization (dune-building, sand-binding and erosion prevention), the CSD vegetation has several traditional, agricultural and pharmaceutical values (e.g., nutritional, medicinal, fibre/oil/

dye yielding, green manure, cover crop). Coastal dwellers are depending on CSD vegetation for a variety of traditional benefits for food, health, soil fertility and recreation. The CSD vegetation of south-west coast of India meets several traditional practices. The legumes are useful as green manure, mulch and cover crops (Arun *et al.*, 1999). The fibres of *Pandanus* leaves are useful in preparation of ropes, which also facilitate preservation of paddy in coastal Karnataka. *Canavalia maritima* serves as a common cover crop in Australia as well as Africa, as a potent cover crop to check soil erosion in dry and sandy areas of southwest coast of India. They are grown by farmers in agricultural fields as cover crops after harvest of main crops (e.g. paddy, sugar cane) to improve the soil nitrogen budget in coastal region of south-west coast of India.

*Cakile edentula* serves as forage, leaves are useful for salads and root powder blended with other flours facilitates preparation of bread (Maun *et al.*, 1990). Tender pods and seeds of *Canavalia cathartica* and *C. maritima*, seeds of *Alysicarpus rugosus, Mucuna pruriens* and *Sesbania bispinosa* are edible (Arun *et al.*, 1999). Raw seeds of *Canavalia* spp. possess 31.2-35.5% proteins, 50.5-61.4% carbohydrates, 1.3-1.86% fat, 1.7 and 10.2% fiber. The energy of seeds of *Canavalia* spp. (1520-1625 kJ/100 g) surpasses several commonly cultivated pulses (1358.3-1426.2 kJ/100 g). Potassium and manganese in seeds were higher than NRC/NAS (1989) recommended pattern for infants. All the amino acids in seeds meet over 50% of the whole egg protein amino acid profile. The essential amino acids of the raw seeds surpassed FAO/WHO (1991) recommended pattern for adults. The sum of polyunsaturated fatty acids of raw *C. cathartica* seeds exceeded wheat and soybean, so also *C. maritima* seeds against wheat. Thermal processing of the *C. cathartica* seeds showed a considerable increase in essential fatty acids.

The CSD plant species possess several traditional medicinal values. *Cakile edentula* is diuretic, antiscorbutic and purgative agent (Maun *et al.*, 1990). *Pandanus* fruits have insect repellent property. Extracts of *I. pes-caprae* constitute ingredients for medicines and hot water extract alleviate strain, fatigue and physical weakness on oral doses (Halberstein and Saunders, 1978). It is used as diuretic in French Guyana (Luu, 1975), to treat rheumatism in India (Christiansen and Reese, 1938) and dried leaves are used to treat arthritis in Nigeria (Iwu and Anyanwu, 1982). Australian aboriginals use *C. maritima* for medicinal purposes. In South America, Africa and Gulf of Mexico beans are ingested or smoked with dried leaves, which has active principle L-betonicine likely possesses hallucinogenic properties. Roots are useful in curing skin diseases in south-west coast of India. *Canavalia cathartica* is a wild ancestral form of *C. gladiata*, distributed throughout the tropical Asia and Africa and useful like *C. maritima* (Sridhar and Seena, 2006). Non-protein toxic amino acid, L-canavanine (L-amino-4-(guanidinooxy) butyric acid) (cav) is stored in many leguminous seeds (Rosenthal, 1991) serve as antiviral, antimicrobial and anti-insects (Rosenthal, 1992). It also possesses antitumour activity against Walker carcinoma cells, human melanoma or pancreatic cancer cells (Mattei *et al.*, 1992; Swaffar *et al.*, 1994), *in vivo* murine leukemia and rat colon tumor (Green *et al.*, 1980; Thomas *et al.*, 1986). The lectin in seeds, concanavalin A (con A) from *Canavalia* spp. has a wide range of medicinal value (antiviral, mitogenic, isolation of immunoglobulins, blood group substances and glycoprotein hormones) (Surolia *et al.*, 1973).

## MICROORGANISMS

Although CSD habitats are nutrient poor sandy soils with high salinity, they are favourable niches for several microorganisms (Sarig *et al.*, 1999). A variety of bacteria and fungi are common in the

CSDs and intertidal habitats. Their role in CSDs, particularly in nutrient cycling and symbiosis with dune vegetation needs special attention.

## Nitrogen Fixing Bacteria

Will and Sylvia (1990) inoculated the sea oats (*Uniola paniculata*) with CSD nitrogen-fixing bacteria isolated from the rhizosphere of Florida (*Alcaligenes denitrificans, Bacillus polymixa*) and other nitrogen fixing bacteria (*Azospirillum brasiliense, A. lipoferum,* and *Klebsiella pneumoniae*) alone or in combination with AM fungi. Root and shoot growth of sea oats increased by *K. pneumoniae* and it was further enhanced on addition of soluble P source ($CaHPO_4$). On inoculation of *K. pneumoniae* or *A. denitrificans* with *Glomus*, no synergistic effect on plant growth was seen. Bacterial inoculation increased root colonization of AM fungi and *K. pneumoniae* increased the spore germination and hyphal growth of *Glomus deserticola*. Will and Sylvia (1990) predicted that stimulation of hyphal extension by production of volatile substance by *K. pneumoniae* and non-volatile diffusible substance enhances the spore germination when *Glomus* spores and bacteria are contiguous. Recently, Chen *et al*., (2000) isolated 12 saline-tolerant rhizobia from the root nodules of *Canavalia maritima* grown on the CSDs of Southern Taiwan. These isolates belonged to the genus *Sinorhizobium* based on 16S rRNA and *nif* H gene sequencing. Several rhizobia isolated from the wild legumes of CSDs of south-west coast of India were fast growing, stress-tolerant (temperature, salinity and pH) and phosphate solubilizers (Arun and Sridhar, 2005). They tolerated or buildup high biomass at 30-40°C, 1-2.5% salinity and pH 5. Some strains showed peak of growth in acidic as well as alkaline pH range. Certain strains did not show phosphate solubilization, but in some strains, phosphate solubilizaton index ranged from 1.4-4.2. Interestingly, phosphate solubilization was associated with tolerance and elevated growth at alkaline pH (Arun and Sridhar, 2004). These stress-tolerant rhizobial strains were non-host specific and induced early flowering in edible legumes. The CSD rhizobia were also utilized a wide range of carbon sources as sole-carbon-sources and seem to be suitable as bioinoculants to agricultural soils (Sridhar *et al*., 2005).

Dune legumes profusely nodulate during post-monsoon season and fix atmospheric nitrogen (Arun and Sridhar, 2005). Physiological and biochemical adaptations of the dune legume along with their symbiotic associations provide additional strength to withstand the environmental perturbations of coastal dunes and in turn dune stabilization. On testing rhizobial strains of different geographical origin on *Phaseolus vulgaris*, native strains of rhizobia showed higher nitrogen-fixing efficiency than the strains of culture collections, indicating the importance of native rhizobia (Rodriguez-Navarro *et al*., 1999). Among the 10 CSD rhizobial isolates tested, five isolates performed better in saline dune sand (188 µg/g) and linked with there efficient phosphate solubilization (Arun and Sridhar, 2004, 2005). Environmental perturbations affect the colonization and nodulation of legumes by rhizobia (Alexander, 1984). Colonization of rhizobia with legumes on the CSDs reveals that suitable conditions prevail at least in specific seasons of the year. It is predicted that suitable conditions for rhizobial colonization will set up on the CSDs of south-west coast of India during the post-monsoon season, which allows rhizobial colonization. Cultivation of desired legume in unsterilized dune soil facilitates isolation of native rhizobia. It is possible to evolve a suitable technique to isolate efficient strains of native CSD rhizobia employing specific legumes as baits in unsterilized dune soils of different geographical origin (Arun and Sridhar,

2004). Interestingly, some edible legumes also establish on the CSDs of south-west coast of India (e.g. *Cajanus* sp., *Vigna* spp.) (Arun and Sridhar, 2005). There is a wide scope to understand the performance of these legumes to CSDs with rhizobia and AM fungi to develop elite varieties. Among the native legumes, cowpea (*V. unguiculata*) grown in large scale in the coastal belts of Karnataka and its improvement through suitable rhizobia and AM fungi will be of immense value. Although rhizobia of CSDs of southwest coast of India showed wide sole-carbon source utilization (Sridhar *et al.*, 2005), their activity depends mainly on the availability of organic matter on the dunes. Human interference particularly deforestation and stone fencing severely obstruct organic matter input to the dunes and in turn interferes the dune nitrogen and phosphorus cycles. It is important to screen native rhizobial isolates of CSDs, which are adapted to high temperature, salinity, pH with low moisture regimes as inoculants to the cultivated legumes.

Colonization of *Casuarina* by *Frankia* is common on the dunes of the west coast of India (Arun *et al.*, 1999). Batzli *et al.*, (2004) explained the abundance and distribution of three sympatric nitrogen-fixing shrubs (*Alnus*, *Myrica* and *Shepherdia*) and their symbiotic *Frankia* in CSD system along Lake Michigan. The results suggest that *Frankia* genotypes are not host specific, but are also associated with spatially and chronologically distinct sets of ecological conditions. Several *Nostoc* spp. associated with sand aggregates of CSDs reported by Forster and Nicolson (1981). Schiermonnikoog, island of the Netherlands show that mats of cyanobacteria consist of 15 species extended over several hectares of beach from foredune to high-tide mark (http://www.nioo.knaw.nl/science/progress/cemoMM.pdf). The sand becomes hard due to production of cyanobacterial extracellular polysaccharides.

## Endophytic Fungi

Mutualistic fungal inhabitants of live tissues of plants without pathological symptoms are called endophytic fungi. An assemblage of 31 endophytic filamentous fungi was reported from the roots of *I. pes-caprae* and *L. sarmentosa* and *Polycarpaea corymbosa* (Beena *et al.*, 2000a). The most common endophytic fungi belong to *Acremonium sp., Chaetomium globosum, Fusarium* sp. and *Torula caligans.* Although endophytic *Acremonium* is known to reduce the colonization and reproduction of AM fungi, no such inhibition was evident in *I. pes-caprae* and *L. sarmentosa*. Endophytic fungal assemblage consists of 46 taxa (33 mitosporic fungi, two zygomycetes and five sterile morphospecies in CSD *Canavalia* spp. have been reported by Seena and Sridhar (2004)). The highest endophytic fungi were recorded in the seedlings/mature plants than seeds. *Chaetomium globosum* was the most dominant species in segments of root, stem and leaf. Colonization of *C. globosum* increased from 5-12.5% in seeds up to 40-64.4% in seedlings or mature plants and may have protective role against herbivores and nematodes as it produces cytotoxic chaetoglobicin (Sekita *et al.*, 1976) and nematicidal flavipin (Chitwood, 2002).

## Endomycorrhizal Fungi

Arbuscular mycorrhizal (AM) fungi are common inhabitants of CSD vegetation (Sturmer and Bellei, 1994), and are involved in nutrient uptake, improvement of soil structure and facilitate formation of stable sand aggregates (Koske and Polson, 1984). Areas of temperate dunes with high vegetation cover by *Ammophila breviligulata* are associated with the greatest AM fungi as well as spore density (Koske and Halvorson, 1981). Species richness was non-linear in the rhizosphere of *Ammophila* and *Solidago* along the latitudinal temperature gradient, attributed to increased temperature from north to south and stimulatory effects of host plants in warm temperature towards south (Koske, 1987). The decline in species richness further south was predicted due to increased temperature stress on the host plants. The most conducive temperature for dune plant species and AM fungi was found in the belt of New Jersey to Virginia. *Gigaspora gigantea* was the most dominant AM fungus of US Atlantic Coast (Lee and Koske, 1994a). The AM fungal colonization was more extensive in *A. arenaria* of British sand dunes (Sutton and Sheppard, 1976). This grass grew on non-stabilized foreshores of Scotland on inoculation of *Glomus macrocarpum* and *G. fasciculatum* (Sylvia, 1986). In Italian CSDs, the members of Asteraceae, Fabaceae and Poaceae were heavily colonized by AM fungi (Giovannetti, 1985). The CSDs of Poland yielded several new species of AM fungi (e.g. *Glomus corymbiforme, G. gibbosum, G. minutum, A. gadanskensis, A. koskei, A. polonica* and *Entrophospora baltica*) (Blaszkowski, 1988, 1995a,b, 1997, Balszkowski *et al.*, 1998, 2000).

Association of the perennial grass, *Uniola paniculata* in Atlantic coast with AM fungi reduced the environmental stresses and resulted in rapid growth and dune stability (Sylvia and Will, 1988). Inoculation of two AM fungi (*Glomus deserticola* and *G. etunicatum*) of Florida dunes significantly increased the height, dry mass and plant cover of shoots of *U. paniculata* in greenhouse trials (Sylvia and Burks, 1988). Similarly, seedlings inoculated with *G. deserticola* and *G. macrocarpum* at Miami Beach showed elevated shoot dry mass, root length, plant height and number of tillers than control seedlings (Sylvia, 1989). Inoculation of nitrogen fixing bacteria, *Klebsiella pneumoniae* with *Glomus* spp. enhanced the AM fungal colonization (Will and Sylvia, 1990). The AM fungi colonized most plant species on the beach, embryo dunes, foredunes and stabilized the dunes of Gulf of Mexico (Corkidi and Rincon, 1997a). Colonization of the seedlings of *Chamaecrista chamaecristoides* by AM fungi in field bioassay showed elevated growth in the foredunes (Corkidi and Rincon, 1997a). Observations on seven early successional dune plant species suggested that the AM fungal colonization is crucial for the growth (Corkidi and Rincon, 1997b). The fixed dunes of Brazil showed the dominance of rhizomatous perennial grass, *Spartina ciliate*, which has been related to the high AM fungal diversity (Sturmer and Bellei, 1994). The AM fungal indices were significantly higher in endemic plant species than the alien plant species in Hawaiian Islands (Koske and Gemma, 1990). The high incidence of AM fungi was found in the strand habitats of Galapagos Islands (Schmidt and Scow, 1986), Heron Island and Hawaiian Islands (Peterson *et al.*, 1985). Up to 97% of plant species grown on the mobile to stable dunes of Gulf of Mexico were colonized by AM Fungi (Corkidi and Rincon, 1997a). *Scutellospora erythropa* in Bahamas, *Acaulospora scrobiculata* and *Gigaspora albida* in North America (Koske and Walker, 1984) and *Glomus* spp. in Japanese dunes (Abe *et al.*, 1994) are the most dominant species. In Hawaiian dunes, 23 vascular plant species colonized by AM fungi, with dominance of *Sclerocystis sinuosa* and *Glomus microaggregatum* (Koske and Gemma, 1990).

A few studies are available on the AM fungi of CSDs of Indian coasts. In south-east coast, 35 dune plant species were colonized by AM fungi (Mohankumar *et al.*, 1988). Rhizosphere of 12 plant species of the south-west coast yielded 16 AM fungi (Kulkarni *et al.*, 1997). Studies on AM fungi associated with CSD vegetation of Goa resulted in recovery of 17 species of AM fungi (Jaiswal and Rodrigues, 2001; Rodrigues and Jaiswal, 2001). Beena *et al.*, (2001) recovered 30

AM fungi from 28 dune plant species belonging to 14 families of the south-west coast. The dominant plant species, *Ipomoea pes-caprae* and *Launaea sarmentosa* consists of 41 and 28 AM fungi, respectively (Beena *et al*., 1997, 2000b). Colonization, richness and diversity of AM fungi were higher in moderately disturbed than severely disturbed dunes of south-west coast (Beena *et al*., 2000c). The species richness and spore density of AM fungi associated with *I. pes-caprae* strongly correlated with rhizosphere nitrogen. Three AM fungi (*Gigaspora gigantea, Glomus mosseae* and *G. dimorphicum*) were most common with *I. pes-caprae* in severely as well as moderately disturbed CSDs (Beena *et al*., 2000c). Similarly, *Gigaspora albida* and *Glomus microaggregatum* were common with *Canavalia cathartica* (Arun and Sridhar, 2006), *Gigaspora margarita* and *Glomus albidum* with *L. sarmentosa* (Beena *et al*., 1997). As these AM fungi are stress tolerant, they might be preferable candidates for future studies on dune stabilization and possibly serve as suitable biofertilizers.

In spite of a few surveys in tropical dunes, the species richness and diversity of AM fungi are higher than temperate dunes. The AM fungal colonization with dune vegetation influences plant distribution, sand aggregation and dune stabilization (Corkidi and Rincon, 1997b). Preferential association (John *et al*., 1983; Koske, 1975) and elevated growth (Joner and Jakobson, 1995) of AM fungi has been linked with high organic matter in soil. Organic matter supplies phosphorus for acid and alkaline phosphatases of AM fungi and the enzyme activity decreases in soil devoid of organic matter. The suborder Glomineae (*Glomus, Acaulospora* and *Entrophorpora*) is dependent on external mycelium to colonize roots, while Gigasporinae (*Gigaspora* and *Scutellospora*) is exclusively dependent on spores (Hart and Reader, 2004; Morton, 1993). Usually, in CSDs of south-west coast of India, the AM fungal colonization of roots attains a peak during post-monsoon, (October-January), while spore density and species richness in monsoon (June-September) (Beena *et al*., 1997). It is hypothesized that in CSDs of south-west coast of India, disturbance during monsoon and post-monsoon affect Gigasporinae and Glomineae, respectively. Detailed investigations may result in identification of disturbance-vulnerable or disturbance-resistant AM fungi as indicators of dune disturbance or stabilization.

The AM spores recovered in dune soil free from vegetation reveals dissemination of spores on the dunes to new sites by physical (e.g. wind blow and tidal activities) or biological (e.g. insects or small mammals) means. Sporulation of Gigasporaceae, the large-spored AM fungi increased significantly with increasing plant diversity (Burrows and Pfleger, 2002). However, large-spored (Gigasporaceae) as well as small-spored (Glomaceae) AM fungi were almost equal (4:3.5) on the CSDs of south-west coast (Bhagya *et al*., 2005). The AM fungi belongs to Glomineae primarily colonize by hyphal fragments, while Gigasporineae by spores (Hart and Reader, 2004). The soil disturbances less likely negatively affect Gigasporineae as they depend on spores for colonization than Glomineae, which are dependent on live mycelia for colonization (Hart and Reader, 2004). Interestingly, *Glomus mosseae* and *G. dimorphicum* were common in moderately as well as severely disturbed CSDs (Beena *et al*., 2000c). These AM fungi might have evolved on the CSDs as less vulnerable species of Glomineae and potential candidates adaptable to agricultural disturbances. Evaluation of high number of samples or plant species raises the number of AM spore recovery but not species (Kulkarni *et al*., 1997; Bhagya *et al*., 2005). Spore recovery elevated with increased number of samples, while species richness saturated at about 20 samples (Bhagya *et al*., 2005). Bhagya *et al*., (2005) hypothesized that AM spores might indicate the magnitude of disturbance and recovery of CSDs. The spore density was positively correlated with soil organic carbon, total nitrogen and available phosphorus on the CSDs of south-west coast of India. Thus, the AM spore load in CSDs might predict the magnitude of natural or anthropogenic disturbances, which influence input of organic matter or affect nitrogen and phosphorus cycles.

Gemma and Koske (1997) have discussed significance of AM fungi in planting stock in revegetation and restoration of CSDs of eastern US. Greipsson and El-Mayas (2000) Studied the AM fungi in natural and reclaimed areas of CSDs of Iceland and explained the importance of AM fungal inoculum in reclaimed sand dunes. For a successful restoration of CSD landscape as well as vegetation, attention should be focused on the below ground microbial communities (mycorrhizas, nitrogen fixers and phosphate solubilizers). Legumes are dependent on mycorrhizal fungi for efficient uptake of phosphorus. Alkaline phosphatase activity of mycorrhiza decreased in soil due to lack of organic matter (Joner and Jakboson, 1995). The phosphatase activity of AM fungi in acidic (pH, 5.2) as well as alkaline (pH, 8.5) soils was increased on amendment of organic matter (Joner and Jakobson, 1995) and amended organic carbon provides $P_0$ as substrate for the acid and alkaline phosphatases. Human disturbance, which prevents addition of organic matter into the CSDs (e.g. stone fencing) has detrimental impact on the functioning of AM fungi. Accumulation of organic debris (e.g. seaweeds, leaf litter, twigs, roots, seeds) on the dunes appears to be extremely important for nutrient enrichment and mineral cycling. Clearing such detritus for fuel, organic manure and recreation purpose severely affects dune vegetation as well as microbes.

## Ectomycorrhizal Fungi

Investigations on ectomycorrhizal (EM) fungi are scanty on the CSDs. Seventy-eight species belonging to 12 genera of EM fungi were reported from a dominant shrub of CSD (*Salix repens*) of Western Europe (van der Heijden *et al.*, 1999). On the Oregon CSDs, EM fungi are critical for establishment of pine under primary succession (Ashkannejhad and Horton, 2005). Due to introduction of European beach grass (*Ammophila arenaria*) at Oregon dunes about 40 years ago, shifting dunes have been stabilized (Ashkannejhad and Horton, 2005). Such dune stabilization by grass allowed the establishment of *Pinus contorta* var. *contorta* and provided new habitat for pines and their associated EM fungi. The EM fungal association is critical for pine establishment, particularly under primary succession. The primary succession of pinus is devoid of mycorrhizal inoculum and depends on the neighbouring forest for inoculum. The EM spore dispersal takes place by wind and animals. Laboratory raised seedlings in sterile soils were inoculated with fresh and one-year old dry black-tailed deer pellets (*Odocoileus hemionus*). The deer fecal pellet inoculant on an average yielded two EM fungi per seedling. *Suillus* and *Rhizopogon* species of EM fungi dominated in seedlings inoculated with fresh feces, while only latter in seedlings inoculated with one-year old dry feces. *Suillus* is known to produce resistant spore banks and acts as principle fungi supporting the seedlings on the sand dunes. Despite higher nitrogen concentrations, decomposition of EM fungal roots delayed in pine woodland (Langley *et al.*, 2006). Such delay in root decomposition promotes soil carbon accumulation and results in formation of stable soil organic matter. Further investigations on such impacts help in understanding of biogeochemical cycles of C and N in CSDs. Probably, the EM fungal inoculation to establish coastal silviculture might help to mitigate dune nourishment in view of depletion of organic matter accumulation by stone fencing.

## Lichens

Studies on lichens of CSDs are sparse. Forty-two macrolichens were reported from the CSDs of Lista (South-west Norway) (Tønsberg and Høiland, 1980). The dominant lichens preferred sand dune areas were *Cladonia conistea, C. furcata* sub sp. *subrangiformis, C. glauca* and *C. pityrea*.

# MICROBES IN SAND AND SOIL AGGREGATION

A wide range of microbes has been isolated from the maritime sand aggregates (bacteria, actinobacteria, cyanobacteria, filamentous fungi, mycorrhizal fungi and green algae) (Forster and Nicolson, 1981). *Bacillus* sp., *Pseudomonas* sp., *Nocardia* spp., *Streptomyces* spp. *Aspergillus fumigatus, Penicillium* sp. and *Glomus fasciculatus, Oedogonium* sp., *Ulothrix* sp. and *Nostoc* sp. are frequent in sand aggregates. Sand grains attach to the mucilage of *Nostoc* sp. and its nitrogen fixation increases microbial activity in sand dunes (Forster, 1980). It has been predicted that increase in number and diversity of microorganisms in unstable to stable dunes is responsible for the formation of stable aggregates. Mycorrhizal roots to stimulate specific groups of bacteria, which produce extracellular polysaccharides in soil (Bagyaraj and Menge, 1978; Meyer and Linderman, 1986).

A stable, well-aggregated soil structure maintains aeration and moisture essential for optimal growth of plants (Oades, 1984). The AM fungi are known to improve the soil structure as well as stability due to the formation of sand aggregates (Jakobson, 1994). They are responsible for three distinct phases of aggregates: a) hyphae entangle the soil particles; b) roots and hyphae create conditions congenial to form microaggregates in soil; c) roots and hyphae enmesh and bind the microaggregates to larger macroaggregates (Miller and Jastrow, 1990). Polysaccharide secretions of AM fungi firmly adheres the microaggregates (Tisdall, 1991). Similarly, AM fungi in sand dune stabilize the aggregates by the amorphous deposits of mycelia (Clough and Sutton, 1978). The members of Acaulosporaceae and Glomaceae contain β(1-3) glucan polymers (Lemoine *et al.*, 1995) and these polymers are present in the walls of external hyphae, hyphal coils and intercellular hyphae in the outer root tissues (Lemoine *et al.*, 1995), inner wall of external hyphae, germ tubes and spore wall (Gianinazzi-Pearson *et al.*, 1994). As an extracellular polysaccharide, β(1-3) glucan is involved in sand aggregation in CSDs. Glomalin, a stable glycoprotein produced by *Gigaspora gigantea* is known for soil aggregation (Wright and Upadhyaya, 1998). Insoluble sticky hydrophobic nature of glomalin coat protects aggregating materials such as soil minerals, organic matter along with microbes. The presence of glomalin is the best indication of soil aggregate stability, which has been linearly correlated with glomalin in soils. Iron deficiency in soils negatively affects the glomalin and in turn the aggregate formation (Kemper and Koch, 1966). The stability of soil aggregates is known as a function of organic matter, clay and free iron oxides for soils from the Western US and Canada (Kemper and Koch, 1966). The EM hyphae extend up to about 120 mm into the soil, covered by extracellular polysaccharides and hyphae often have particles of soil enmeshed between them (Foster, 1981). Microbial polysaccharides interact with cations or phenolic polymers in soil and protect from microbial degradation (Verma and Martin, 1976). The aggregate stability depends on the amount of organic matter or carbohydrates added to soil, the microbial biomass and length of fungal hyphae in soil (Tisdall,

1991). Experiments revealed that the stability of aggregates did not appear to persist once the organic matter disappeared (Tisdall, 1994).

Association of *Glomus* with *Phaseolus vulgaris* showed that the weight of sand aggregates in non-mycorrhizal plants was 10 g/kg, while it was 54 g/kg in mycorrhizal plants (Sutton and Sheppard 1976). The AM fungal association with mycorrhizal plants in CSDs stabilizes the dunes by binding the sand grains into wind-resistant aggregates. The amount of sand aggregation usually increases with distance from the sea due to decreasing salinity, elevated nutrients and establishment of higher plant species. The weight as well as size of the aggregates increases from mobile to fixed dunes. One gram of stable macroaggregate consists of about 50 m hyphal network (Tisdall, 1991). The aggregates of sand dunes remain intact even after the death of roots and hyphae (Koske and Polson, 1984). Although the extreme conditions result in the death of dune vegetation, possibly the microbes associated with sand aggregates favour the succession of vegetation.

The fungal hypahe are known to bring mineral particles and organic materials together to form stable microaggregates (<2 µm) and enmesh microaggregates into stable aggregates (>50 µm) (Tisdall *et al.*, 1997). A detailed study of sand aggregates of foredunes, mobile dune slopes, crest of dune ridges and early fixed dunes of Scotland was performed by Forster and Nicolson (1981). An increase in sand aggregate number (794 vs. 15,016/kg), size (>1-<1.4 vs. >2 mm) and weight (3.21 vs. 265.7 g/kg) was seen from the unstable foredunes to the early fixed dunes during summer. A significant difference was seen in foredune sites between the number of aggregates in three size ranges (>1-<1.4, >1.4-<2 and >2 mm). Seasonal study revealed increased average weight of aggregates in winter than summer in foredunes as well as mobile dune slopes (foredunes, 5 vs. 20 mg/kg; mobile dune slope, 40 vs. 70 mg/kg). But it was reverse for crest of dune ridges and early fixed dunes (crest of dune ridges, 300 vs. 290 mg/kg; early fixed dunes, 1260 vs. 950 mg/kg). Smaller sand aggregates (>1-<1.4 mm) possess fewer fungal species than the larger ones (>2 mm). Interestingly, foredune aggregates (>1.4-<2 mm) colonized by bacteria than fungi (Forster, 1979), probably the bacterial polysaccharides, might have important role in these habitats.

## THREATS AND SAFEGUARDS

The CSDs are vulnerable for a variety of natural disturbances like tidal action, wind blow, sand erosion and sand accretion. They are also under threat from human interference (e.g. encroachment, agriculture, waste disposal, road traffic). Increase in natural or human disturbance result in increased coastal dune vulnerability index and help as useful biological indicator of the degree of disturbance within the habitat (Garcia-Mora *et al.*, 2001). A case study of application of coastal dune vulnerability index in SW Iberian Peninsula resulted in classification of coastal dunes into suitable management units. Batanouny (1999) and Frihy (2001) documented the extent of human disturbances of Mediterranean CSDs in Egypt, considered as an endangered landscape.

Several projects were implemented to avoid beach erosion in Atlantic coast of US (Koske and Gemma, 1997; Sylvia and Will, 1988). Wilcock carter (1977) suggested mechanically moving sand and grading it into suitable slope and length to allow establishment of vegetation on the steep slopes. Such beach or sand dune nourishment protects from the sand deficit, erosion and enhances the activity of endangered species (e.g. sea turtles) (Jones and Mangun, 2001). Sand fences are effective means to trap windblown sand and reduce the velocity of wind in their immediate vicinity

(Woodhouse, 1978). Mats and netting practices are also effective methods to protect the sand dunes (Woodhouse, 1978). Coarse netting and mats are known to protect the transplanted dune grasses and new seedlings.

Garcia-Mora *et al.*, (2000) studied 30 dune systems of SE Portugal to SW Spain employing plant diversity measures for coastal dune monitoring. This study indicated that the native plant species have high potential to withstand environmental perturbations of the CSDs and useful in dune stabilization. Among the techniques of stabilizing CSDs, revegetation is the best means due to inexpensive, durable and self-repairing technique (Woodhouse, 1978). Planting beach grasses along the beaches initiated dune-building processes and reducing the erosion losses in Atlantic coast. Fertilizing the American beachgrass with nitrogen, phosphorus/potassium fertilizers and limestone showed success in Atlantic coast (Seliskar, 1995). Pioneer perennial dune grasses are employed to stabilize the coastal sand dunes in the United States: *Ammophila breviligulata* (North Pacific, North Atlantic and Great Lakes), *Ammophila arenaria* (North and South Pacific), *Panicum amarum* (South Atlantic and Gulf coasts), *Uniola peniculata* (South Atlantic and Gulf coasts) and *Apartina patens* (South Atlantic Gulf coasts) (Woodhouse, 1978). Several studies indicated that the success of dune grasses is dependent on AM fungal association. Mixed growth of grass stands with the perennial legume (*Lathyrus maitimus*) was effective in providing nutrients to the dune system in Pacific coast (Brown, 1948). In addition to grasses, tree species adapted to coastal habitats are immensely valuable in dune restoration. *Acacia sophorae* was more drought-tolerant and more plants established on the dunes of South Australia (Barron and Dalton, 1996). Establishment of the shrub, *Myrica cerifera* examined on a Virginia barrier island and found the success of tree seedlings on coastal soils, dependent on the presence of symbiont, *Frankia* (Wijnholds and Young, 2000).

## FUTURE OUTLOOK

The CSDs are of considerable ecological and economic importance throughout the world. They are critically important habitats for numerous terrestrial and aquatic organisms. Members of Poaceae dominate the temperate dunes, while tropical dunes are dominated by Asteraceae, Convolvulaceae, Fabaceae and Poaceae. Richness and diversity of vegetation constitute an authentic index of CSD vulnerability to natural and human disturbance. Dune plant species have high adaptive value and many of them are economically valuable (edible, green manure, medicinal). Dune stabilization and restoration is dependent on the development of vegetation in association with symbiotic microflora. Association of dune vegetation with mycorrhizas and nitrogen fixing bacteria determines the nutrient dynamics of dunes. Symbiotic association between legumes and rhizobia is one of the important biocatalytic links for the flow of nitrogen between the atmosphere and CSD habitats. Below ground diversity of AM fungi is one of the major factors governing the plant diversity and functioning of CSD ecosystem. Approximately one-third of the world's irrigated land is salt prone (400-950 $\times 10^6$ ha), about 40% of the world's land surface shows potential salinity problems and these areas are confined to tropical and Mediterranean regions. Native rhizobial and AM fungal strains are well adapted to stress conditions in association with legumes and help in reclamation of marginal lands as they enrich soil through symbiotic nitrogen fixation. The loss of AM fungal diversity decreases both plant equilibrium and ecosystem productivity, which in turn results in instability of dune ecosystems. Although microbial polysaccharides (bacteria, cyanobacteria and fungi) are known to play a major role in sand aggregation, knowledge on their importance in dune stabilization and application is inadequate. Deposition of organic matter on the dunes by waves and tides appears to be extremely important for the supply of phosphorus for vegetation and microbes. Phosphorus plays a major role in the legume growth and its deficiency prevents root nodulation of legumes. Most legumes are heavily dependent on mycorrhizas for active uptake of phosphorus. Among the dune plant-microbe interactions, association of plants with AM fungi gained considerable importance, while other symbiotic associations have been neglected. Human interference affects input of organic matter to the dunes, and in turn, the nitrogen and phosphorus cycles lead to destabilization of CSDs. Special attention has to be focused on the dynamics of nitrogen and phosphorus cycles of dunes. Coastal populations around the world are increasingly exerting pressure on coastal biodiversity. The CSDs are the black box of flora, fauna and microbes and sensitive species are always at risk due to human interference. Anthropogenic disturbance (e.g., agriculture, waste disposal, road traffic) accelerates the loss of CSD habitats. The loss of natural landscape means loss of natural resources (flora, fauna and microbes), cultural environments and traditional ecological knowledge. Stringent measures have to be enforced to protect and restore the coastal sand dune ecosystem. The CSDs being one of the extreme habitats provide excellent opportunities to understand the mutual interaction and adaptation of biota to the environmental stress. Stress-tolerant plant species and microbes of CSDs deserve special attention for their application in agriculture, forestry, bioremediation and environmental rehabilitation.

## References

Abe, J.P, Masuhara, G. & Katsuya, K. (1994) *Mycoscience* **35**: 233-238.

Alexander, M. (1984) In: *Biological Nitrogen Fixation: Ecology, Technology and Physiology*, (Ed. Alexander, M.), Plenum Press, New York, pp 39-50.

Arun, A.B. & Sridhar, K.R. (2004) *Biol. Fert. Soils* **40**: 435-439.

Arun, A.B. & Sridhar, K.R. (2005) *Int. J. Forest Usufructs Manage.* **6**: 60-68.

Arun, A.B. & Sridhar, K.R. (2006) *Mycorrhiza News* **18**: 13-17.

Arun, A.B., Beena, K.R., Raviraja, N.S. & Sridhar, K.R. (1999) *Current Science* **77**: 19-21.

Arun, A.B., Raviraja, N.S. & Sridhar, K.R. (2001) *Int. J. Ecol. Environ. Sci.* **27**: 23-29.

Ashkannejhad, S. & Horton, T.R. (2005) *New Phytol.* **169**: 345-354.

Bagyaraj, D.J. & Menge, J.A. (1978) *New Phytol.* **80**: 567-573.

Barron, P. & Dalton, G. (1996) *J. Coastal Res.* **12**: 1006-1008.

Batanouny, K.H. (1999) *Est. Coast. Shelf Sci.* **49**: 3-9.

Batzli, J.M., Zimpfer, J.F., Huguet, V., Smyth, C.A., Fernandez, M. & Dawson, J.O. (2004) *Can. J. Bot.* **82:** 700-709.

Beena, K.R., Raviraja, N.S. & Sridhar, K.R. (1997) *Kavaka* **25**: 53-60.

Beena, K.R., Ananda, K. & Sridhar, K.R. (2000a) *Sydowia* **52**: 1-9.

Beena, K.R., Raviraja, N.S. & Sridhar, K.R. (2000b) *J. Environ. Biol.* **21**: 341-347.

Beena, K.R., Raviraja, N.S., Arun, A.B. & Sridhar, K.R. (2000c) *Current Science* **79**: 1459-1466.

Beena, K.R., Arun, A.B., Raviraja, N.S. & Sridhar, K.R. (2001) *Tropical Ecol.* **42**: 213-222.

Bhagya, B., Sridhar, K.R. & Arun, A.B. (2005) *Int. J. Forest Usufructs Manage.* **6**: 1-18.

Blaszkowski, J. (1988) *Karstenia* **27**: 37-42.

Blaszkowski, J. (1995a) *Mycol. Res.* **99**: 237-240.

Blaszkowski, J. (1995b) *Mycologia* **87**: 732-737.

Blaszkowski, J. (1997) *Mycologia* **89**: 339-345.

Blaszkowski, J., Madej, T. & Tadych, M. (1998) *Mycotaxon* **78**: 165-184.

Blaszkowski, J., Tadych, M. & Madej, T. (2000) *Mycotaxon* **76**: 187-195.

Brown, R.L. (1948) *J. Soil Water Conser.* **3**: 69-74.

Burrows, R.L. & Pfleger, F.L. (2002) *Can. J. Bot.* **80**: 120-130.

Chen, W.M., Lee, T.M., Lam, C.C. & Cheng, C.P. (2000) *FEMS Microbiol. Ecol.* **34**: 9-16.

Chitwood, D.J., (2002) *Ann. Rev. Phytopath.* **40**: 221-249.

Christensen, B.V. & Reese, J.A. (1938) *J. Amer. Pharm. Assoc.* **27**: 195.

Clough, K.S. & Sutton, J.C. (1978) *Can. J. Microbiol.* **24**: 333-335.

Corkidi, L. & Rincon, E. (1997a) *Mycorrhiza* **7**: 9-15.

Corkidi, L. & Rincon, E. (1997b) *Mycorrhiza* **7**: 17-23.

FAO/WHO (1991) *Food and Agricultural Organization of the United Nations Food and Nutrition Paper # 51*, FAO, Rome, pp. 1-66.

Forster, S.M. (1979) *Soil Biol. Biochem.* **11**: 537-543.

Foster, R.C. (1981) *Science* **214**: 665-667.

Forster, S.M. & Nicolson, T.H. (1981) *Soil Biol. Biochem.* **13**: 205-208.

Fosberg, F.R. & Sachet, M.H. (1977) *Smithsonian Contributions to Botany* **36**: 1-34.

Frihy, O.E. (2001) *Ocean & Coastal Manage.* **44**: 489-516.

Garcia-Mora, M.R., Gallego-Fernandez, J.B. and Garcia-Novo, F. (2000) *J. Coastal Res.* **16**: 990-995.

Garcia-Mora, M.R., Gallego-Fernandez, J.B., Williams, A.T. and Garcia-Novo, F. (2001) *J. Coastal Res.* **17**: 802-811.

Gemma, J.N. & Koske, R.E. (1997) *J. Environ. Manage.* **50**: 251-264.

Gianinazzi-Pearson, V., Lemoine, M.C., Arnold, C., Gollotte, A. & Morton, J.B. (1994) *Mycologia* **86**: 478-485.

Giovannetti, M. (1985) *Trans. Br. Mycol. Soc.* **84**: 679-689.

Green, M.H., Brooks, T.L., Mendelsohn, L. & Howell, S.B. (1980) *Cancer Res.* **40**: 535-537.

Greipsson, S. & El-Mayas, H. (2000) *Restoration Ecol.* **8**: 144-150.

Halberstein, R.A. & Saunders, A.B. (1978) *Cul. Med. Psychiat.* **2**: 177.

Hart, M.M. & Reader, R.J. (2004) *Trop. Ecol.* **45**: 97-111.

Hester, M.W. & Mendelssohn, I.A. (1987) *Amer. J. Bot.* **74**: 1093-1101.

Iwu, M.M. & Anyanwu, B.N. (1982) *J. Phytother.* **9**: 125.

Jaiswal, V. & Rodrigues, B.F. (2001) *Current Science* **80**: 826-827.

Jakobson, I. (1994) *Plant and Soil* **159**: 141-147.

John, T.V.S., Coleman, D.C. & Reid, C.P.P. (1983) *Ecology* **64**: 957-959.

Joner, E.J. & Jakobson, I. (1995) *Plant and Soil* **172**: 221-227.

Jones, S.R. & Mangun, W.R. (2001) *Ocean & Coastal Manage.* **44**: 207-220.

Jurado, E. & Westoby, M. (1992) *J. Ecol.* **80**: 407-416.

Kellman, M. & Roulet, N. (1990) *J. Ecol.* **78**: 664-676.

Kemper, W.D. & Koch, E.J. (1966) Technical Bulletin # 1335, ARS, USDA, Washington DC.

Koske, R.E. (1975) *Can. J. Bot.* **53**: 668-672.

Koske, R.E. (1987) *Mycologia* **79**: 55-68.

Koske, R.E. & Gemma, J.N. (1990) *Amer. J. Bot.* **77**: 466-674.

Koske, R.E. & Gemma, J.N. (1997) *Amer. J. Bot.* **84**: 118-130.

Koske, R.E. & Halvorson, W.L. (1981) *Can J. Bot.* **59**: 1413-1422.

Koske, R.E. & Polson, W.R. (1984) *BioSci.* **34**: 420-424.

Koske, R.E. & Walker, C. (1984) *Mycologia* **76**: 250-255.

Kulkarni, S.S., Raviraja, N.S. & Sridhar, K.R. (1997) *J. Coastal Res.* **13**: 931-936.

Langley, J.A., Chapman, S.K. & Hungate, B.A. (2006) *Ecol. Lett.* **9**: 955-959.

Lee, P.J. & Koske, R.E. (1994a ) *Mycol. Res.* **98**: 453-457.

Lemoine, M.C., Gollotte, A. & Gianinazzi-Pearson, V. (1995) *New Phytol.* **129**: 97-105.

Leonard, R.I. & Judd, F.W. (1999) *J. Coastal Res.* **15**: 645-652.

Little, L.R. & Maun, M.A. (1995) *The Western J. Graduate Res.* **5**: 37-52.

Luu, C. (1975) *Pl. Med. Phytother.* **9**: 125.

Martinez, L.M. & Moreno-Casasola, P. (1998) *J. Coastal Res.* **14**: 162-174.

Martinez, L.M. & Rincon, E. (1993) *Acta Oecol.* **14**: 521-528.

Mattei, E., Damasi, D., Mileo, A.M., Delpino, A. & Ferrini, U. (1992) *Anticancer Res.* **12**: 757-762.

Maun, M.A. (1994) *Vegetatio* **111**: *59-70.*

Maun, M.A. (1998) *Can. J. Bot.* **76**: 713-738.

Maun, M.A.& Baye, P.R. (1989) *CRC Crit. Rev. Aquat. Sci.* **1**: 661-681.

Maun, M.A. & Lapierre, J. (1986) *Am. J. Bot.* **73**: 450-455.

Maun, M.A. & Perumal, J. (1999) *Ecology Lett.* **2**: 14-18.

Maun, M.A., Boyd, R.S. & Olson, L. (1990) *J. Coastal Res.* **6**: 137-156.

Meyer, J.R. & Linderman, R.G. (1986) *Soil Biol. Biochem.* **18**: 191-196.

Miller, R.M. & Jastrow, J.D. (1990) *Siol Biol. Biochem.* **22**: 579-584.

Mohankumar, V., Ragupathy, S., Nirmala, C.B. & Mahadevan, A. (1988) *Current Science* **57**: 367-368.

Moreno-Casasola, P. (1988) *J. Biogeogr.* **15**: 787-806.

Moreno-Casasola, P., Grime, J.P. & Martinez, M.L. (1994) *J. Tropical Ecol.* **10**: 67-86.

Morton, J.P. (1993) *INVAM Newslett. # 3.*

NRC/NAS (1989) In: *Recommended Dietary Allowances*. National Academic Press, Washington DC.

Oades, J.M. (1984) *Plant and Soil* **76**: 319-337.

Peterson, R.L., Ashford, A.E. & Allaway, W.G. (1985) *Aust. J. Bot.* **33**: 669-676.

Rao, T.A. & Meher-Homji, V.M. (1985) *Proc. Indian Acad. Sci. (Plant Sci.)* **94**: 505-523.

Rao, T.A. & Sherieff, A.N. (2002) In: *Coastal Ecosystem of the Karnataka State, India II. Beaches*, Karnataka Association for the Advancement of Science, Bangalore.

Rodrigues, B.F. & Jaiswal, V. (2001) *Indian J. Forestry* **24**: 18-20.

Rodriguez-Navarro, D.N., Santamaria, C., Temprano, F., & Leidi, E.O. (1999) *Eur. J Agron.* **11**: 131-143

Rosenthal, G.A. (1991) *Phytochem.* **30**: 1055-1058.

Rosenthal, G.A. (1992) In: *Insecticides: Mechanisms of Action and Resistance* (Eds. Otto, D. & Weber, B.), Intercept Ltd., Andover, England, pp. 35-46.

Sarig, S., Fliessbach, A. & Steinberger, Y. (1999) *Arid Soil Res. Rehabilitation* **13**: 75-80.

Schmidt, S.K. & Scow, K.N. (1986) *Biotropica* **18**: 236-240.

Seabloom, E.W. & Wiedermann, A.M. (1994) *J. Coastal Res.* **10**: 178-188.

Seena, S. & Sridhar, K.R. (2004) *Can. J. Microbiol.* **50**: 1015-1021.

Sekita, S., Yoshihira, K. & Natori, S. (1976) *Tetrahed. Lett.* **17**: 1351-1354.

Seliskar, D.M. (1995) *Restoration Ecol.* **3**: 54-60.

Sridhar, K.R. & Seena, S. (2006) *Food Chem.* **99**: 267-288.

Sridhar, K.R., Arun, A.B., Narula, N., Deubel, A. & Merbach, W. (2005) *Engineering in Life Sciences* **5**: 425-430.

St. John, H. (1970) *Botanische Jahrbucher Systematik* **89**: 563-583.

Sturmer, S.L. & Bellei, M.M. (1994) *Can. J. Bot.* **72**: 359-363.

Surolia, A., Prakash, N., Bishayee, S. & Bachhawat, B.K. (1973) *Indian J. Biochem. Biophys.* **10**: 145-148.

Sutton, J.C. & Sheppard, B.R. (1976) *Can. J. Bot.* **54**: 326-333.

Swaffar, D.S., Ang, C.Y., Desai, P.B. & Rosenthal, G.A. (1994) *Cancer Res.* **54**: 6045-6048.

Sykes, M.T. & Wilson, J.B. (1989) *Acta Bot. Neerl.* **38**: 173-182.

Sykes, M.T. & Wilson, J.B. (1991) *J. Veg. Sci.* **2**: 531-538.

Sylvia, D.M. (1986) *Mycologia* **78**: 728-734.

Sylvia, D.M. (1989) *J. Coastal Res.* **5**: 747-754.

Sylvia, D.M. & Burks, J.N. (1988) *Mycologia* **80**: 565-568.

Sylvia, D.M. & Will, M.E. (1988) *Appl. Environ. Microbiol.* **54**: 348-352.

Thomas, F.A., Rosenthal, G.A., Gold, D.V. & Dickey, K. (1986) *Cancer Res.* **46**: 2898-2903.

Tisdall, J.M. (1991) *Aust. J. Soil Res.* **29**: 729-743.

Tisdall, J.M. (1994) *Plant and Soil* **159**: 115-121.

Tisdall, J.M., Smith, S.E. & Rengasamy, P. (1997) *Aust. J. Soil Res.* **35**: 55-60.

Tønsberg, T. & Høiland, K. (1980) *Norw. J. Bot.* **27**: 131-134.

van der Heijden, E.W., de Vries, F.W. & Kuyper, Th. W. (1999) *Can. J. Bot.* **77**: 1821-1832.

Verma, L. & Martin, J.P. (1976) *Soil Biol. Biochem.* **8**: 85-90.

Watkinson, A.R. & Davy, A.J. (1985) *Vegetatio* **62**: 487-497.

Wijnholds, A.E. & Young, D.R. (2000) *J. Coastal Res.* **16**: 139-144.

Wilcock, F.A. & Carter, R.W.G. (1977) *Biol. Conser.* **77**: 279-291.

Will, M.E. & Sylvia, D.M. (1990) *Appl. Environ. Microbiol.* **56**: 2073-2079.

Woodhouse, W.W.Jr. (1978) *US Army Corp. of Engineers* **3**: 9-104.

Wright, S.F. & Upadhyaya, A. (1998) *Plant and Soil.* **198**: 97-107.

# Index


**A**

Abiotic stress, 375
  influences of, 375
  rhizomediation of, 375
ACC deaminases, 29
  biochemical characteristics of, 30
Acetosyringone, 440
Actinomycetes, 206
Agricultural practices,
  modification of, 309
Agricultural research and development,
  investment on, 135
Agrobacterium mediated transformation, 438
Agrobacterium rhizogenes, 455
Agrobacterium,
  biology of, 438
Algal biofertilizer,
  production of, 76
Alkaloids,
  roles of, 258
AM fungi,
  agricultural perspectives of, 423
  older classifications of, 421
  phylogentic relationship of, 422
  role in disease control of, 124
AMF,
  biodiversity of, 420
  effects of, 419
  reproduction of, 417
  species concept in relation to, 418
  taxonomy of, 420
Ammophila, 461
Amplified fragment length polymorphisms (AFLP), 93
Antagonist,
  formulation of delivery system for the, 160
  mechanisms of, 327
Antagonistic organisms,
  improvement of, 158
Antibiosis, 9, 105, 144, 181, 268, 328
Antibiotic interactions, 143
Antibiotic resistance, 270
Antibiotics,
  role of, 269
Antimicrobial metabolites, 209
Antisense RNA,
  resistance through, 314
Arbuscular mycorrhizae, 73
*Axadirachtin*, 127
*Azolla*, 72
  crop response to application of, 81
  production of, 78
*Azospirillum,* 71
  production of, 73
*Azotobacter,* 71
  production of, 73

**B**

Bacillus sphaericus, 118
  application of, 56
  benefit of, 57
  formulations of, 60
Bacillus thuringiensis (Bt), 117
  commercial formulations of, 118

Bacillus, 51
  classification of, 50
  colonization patterns of, 55
Bacteria, 201
  visualization of, 186
Bacterial colonization, 443
Bacterial determinants, 171
Bacterial factors, 173
Baculoviruses, 126
Baterial biocontrol agents, 117
Beauveria bassiana, 219
  as biocontrol agent, 222
  formulation of, 233
  production of, 228
  toxicity of, 224
BGA biofertilizers,
  application of straw based, 78
  crop response to application of, 81
  production technology of, 77
Bio-available iron, 180
Biocontrol agents, 133
  commercialization of, 133, 339
  development of, 339
  formulation of, 133
  inoculum preparation of, 133
  interaction of, 277
  mechanism of, 209
  plant derivatives used in, 127
Biocontrol fungi,
  substrates used for, 350
Biocontrol interactions via the plants, 107
Biocontrol mechanisms, 143
Biocontrol products, 14
Biocontrol,
  efficacy of, 131
Biodac formulation, 164
Biofertilization, 7, 172, 324
Biofertilizer production, 83
Biofertilizer technology, 68
Biofertilizers, 53, 68, 69
  crop response to, 80
  current status of, 81
  ISI standards for, 75
  types of, 70
Biological control agents, 197
  pseudomonas as, 109
  use of, 116
Biological control, 160, 249, 347
Biological $N_2$ fixation, 70
  importance of, 410
Biopesticides,
  commercialization of, 343
  performance of, 12
Bioprotectants, 52
Biostimulants, 54
Blue-green algae, 72
BNF,
  environmental factors and, 411

C

Capsid protein, 310
Carica papaya L., 448
  agrobacterium mediated transformation in, 449
Cell free culture filtrates,
  effect of, 150
Cell wall degrading enzymes, 11, 154
Cellophane agar diffusion technique, 150
Cellulolysis rate, 154
Chitinase assays, 159
Chitinases, 155
  production of, 156
Coastal sand dune, 460
Coat protein gene, 311
Coinoculation, 16
Colonization, 273
Competition, 328
  definition of, 145
Competitive colonization,
  bacterial traits involved in, 103
Competitive root colonization, 101
Conidial germination, 151
Conidial suspension,
  preparation of, 163
Crop management, 135

Cross-protection, 306, 307
Crown gall disease, 438
CSD vegetation, 462
 economic value of, 462
 microorganisms and, 463
 threats to, 470
CSDs,
 lichens of, 469
Culture filtrates,
 preparation of, 150
Cytochemical labeling, 157

**D**

Defensive mutalism, 257
DFM,
 aerobic fungi based, 399
 anaerobic fungi based, 400
 bacteria as, 397
 for ruminants, 397
 genetically modified, 400
 modes of action of, 396
 practical applications of, 401
DNA and RNA manipulation, 159
Drought and salt stress, 32
Dual culture experiment, 148

**E**

Endomycorrhizal fungi, 465, 468
Endophytes,
 distribution and diversity of, 246
 insecticidal activities of, 249, 251
 plant growth promotion and, 258
 products of, 251
Endophytic diazotrophs, 69
Endophytic fungi, 243, 465
 artificial inoculation of, 254
 classification of, 247
 dispersal of, 247
Epidemiology, 296
Exopolysaccharrides, 371

**F**

Fermenter, 74
Fertilizer application, 378
Flooding stress, 31
Fluorescent pseudomonas,
 biocontrol ability of, 101
 colonization ability of, 102
Functional genomics, 383
Fungal biocontrol agents, 119
Fungal cell wall lysing enzymes, 107
Fungal phytopathogens, 340
Fungi, 199
 as DFM, 398
Fungicides-bactericides, 120
Fungus-plant root symbiosis, 416
Fusarium, 88, 89
 molecular diversity, 90
 morphological characterization of, 89
 pathogenic characterization of, 90

**G**

Gene manipulation, 158
Gene silencing, 455, 456, 457
Gene transfer techniques, 437
Genetic engineering, 131
Genus bacillus, 50
GM crops,
 future of, 453
GMOS,
 biosafety regulations of, 453
Granuloviruses (GVs), 127
Grass endophytes, 248
Green manure, 208

**H**

HCN production, 10
Heavy metal stress, 32
Host plant, 419
Hydrolytic enzymes, 209

**I**

Indole acetic acid, 370
Induced systemic resistance (ISR), 53, 108, 209
Inoculants,
 monitoring of, 18

Insect control,
mechanisms of, 252
Insects,
pathogens of, 219
Inter simple sequence repeats (ISSR), 95
Interaction grades, 148
Interfungal interactions, 147
Iron competition, 271
ISR,
bacterial signals for, 108
mechanisms of, 108
rhizobacterial determinants of, 42

**L**

Live microbial supplements, 395
Lysis, 146
Lytic enzyme, 183

**M**

Marker technology,
applications of, 91
Market genes, 19
advantages of, 19
Microbes in sand and soil, 469
Microbial antagonists,
shelf life and storage of, 346
Microbial biocontrol agents, 117
Microbial communities,
effect of plant species on, 103
Microorganism-mediated induced resistance, 274
Microorganisms, 463
benefits of, 268
genetic manipulation of, 275
Mites, 208
Molecular markers, 91
Morphological traits, 417
Mycoinsecticide formulation, 236
Mycoparasitic interactions, 151, 152, 153
Mycoparasitism, 146, 328
Mycorrhiza, 416
biological network of, 418
Mycorrhizal fungi, 200

**N**

$N_2$ fixation, 173
Nematode control, 204
Nematodes,
as biological control agents, 125
Nematode-trapping fungi, 199
Nitrogen cycling, 259
Nitrogen fixation process, 173, 409
Nitrogen fixation,
in relation to rice, 368
Nitrogen fixing bacteria, 464
Nitrogen sources,
microbial management of, 365
Nitrogenous, 69
Nuclear polyhedron viruses (NPVs), 126
Nucleocapsid (N) protein, 310

**O**

Octopine ti plasmid, 440
Octopine, 440
Opportunistic fungi, 200

**P**

Parasitic bacteria, 202
Parasitism, 272
PCR amplification, 159
Pectinolytic activity, 155
Pesticide application, 379
PGPR, 2, 100
characterization of, 2, 187
classification of, 2
coinoculation of, 7, 8
commercialization of, 189
deaminase enzyme, 28
development of, 59
effect of, 11
field application of, 187
genetic diversity among, 184
genetic improvement of, 185
in abiotic stresses, 31
in management of biotic stresses, 33
in management of insects, 42

in management of nematodes, 43
in management of phytopathogens, 34
inducing plant systemic resistance (ISR), 40
influence on plant growth of, 4
producing antibiotics, 34
producing antifungal-metabolites, 36
producing defence enzymes, 37
producing siderophores, 38
Phosphate solubilization, 174, 369
Phosphatic biofertilizer, 69
Phytopathogens,
control of, 8
Phytostimulation, 8
Plant defence mechanism, 325
Plant derived biocontrol agents, 129, 131
Plant disease control, 342
Plant endophyte interaction, 244
Plant ethylene hormone and abiotic stresses, 28
Plant growth hormones, 176
Plant growth promotion activities, 7, 172, 259
Plant growth promotion,
mechanisms of, 52
Plant growth regulation, 173
Plant parasitic nematodes, 198
Plant resistance,
stimulation of, 325
Plant root colonization, 324
Plant-parasitic nematodes crop losses, 196
Plant-rhizobia interactions,
and molecular approaches, 380
Plants, 208
Plasmid construction, 159
Poly β-hydroxybutyrate production, 370
Post-transcriptional gene silencing, 456
Potent strains,
production of, 359
Primary screening, 147
Protozoans, 208
PSB,
production of, 73
Pseudomonads, 4, 119
P-solubilizing bacteria, 175
Pyrolysis-field lonization mass spectrometry, 383

**Q**
Quality control, 347
Quorum sensing, 384

**R**
Radial growth, 150
Random amplified polymorphic DNA (RAPD), 91
advantages of, 92
rDNA markers, 94
Recombinant DNA technology, 255
Restriction fragment length polymorphisms (RFLP), 92
Rhizobacteria,
biocontrol activity of, 179, 180
influence of heavy metals, 379
Rhizobia and leguminous plants,
symbiosis between, 407
Rhizobia,
biodiversity of, 411
nomenclature of, 413
taxonomy of, 411
Rhizobia, 3, 379
*Rhizobium* inoculants,
quality standards for, 76
Rhizobium studies,
molecular approaches in, 380
*Rhizobium,* 70
production of, 73
Rhizomediation, 384
and coinoculation, 383
the role of rhizobium in, 367
Rhizosphere and legume-rhizobial populations,
interaction of, 366
Rhizosphere colonization, 170
Rhizosphere competition, 36
Rhizosphere,
antagonistic influence in, 372
biological control in, 373
diversity and distribution in, 51
effect of high temperature and heat stress, 376
gene expression in the, 186
influence of vitamin in, 371
iron chelation at, 374

management of nutrients in, 369
microbial interactions in, 17
plant-microbe interactions in, 5
role of, 371
Rhizospheric bacteria, 202
Rice bran medium, 164
Rickettsias, 206
Root colonization, 17, 55, 171
Root exudates, 210
Rumen, 395

S

Salt and osmotic stress,
influence of, 375
Sand maize meal inoculum, 163
Satellite RNA, 310
Satellite-mediated resistance, 313
Seed germination, 11
Serial dilution technique, 148
Siderophore biosynthesis, 105
Siderophore, 9, 105, 178
Signal-transduction pathways, 41
Simple sequence repeats (SSR), 93
Site-specific recombinase, 171
Soil acidity-alkalinity,
influence of, 378
Soil amendments, 378
Soil fertility, 423
Soil fungi,
staling growth products of, 150
Soil microbial populations, 7
Soil moisture,
influence of, 377
Soil,
isolation of fungi from, 148
Soil-based BGA,
storage of, 77
Solid state fermentation,
advantages of, 347
Somatic embryo,
maturation and germination of, 451
Stable isotope probing, 382
Suitable strain,
selection of, 346

T

Target viral genome,
ribozymes activity for, 314
T-DNA integration of into plant genome, 447
T-DNA transfer complex, 444
T-DNA transfer, 445
T-DNA, 442
Ti plasmid, 441
Tomato leaf curl disease, 293
Tomato leaf curl virus, 298
Toxins, 210
Transcriptional gene silencing, 455
Transgenic crop development, 452
Transgenic plants, 454
benefits & risks of, 454
development of, 309
Tree endophytes, 247
Trichoderma harzianum rifai, 322
Trichoderma reesei, 322
Trichoderma spp., 119
Trichoderma virens, 322
Trichoderma viride pers, 322
Trichoderma, 3, 9, 320, 336, 355
as biocontrol agent, 327
as protein source, 325
biocontrol potential of, 360
biology of, 356
biotechnology of, 324
characterization of, 323
cultivation of, 323
ecology of, 357
in textile industry, 329
isolation of, 323
molecular taxonomy of, 323
morphology of, 321
occurrence and distribution of, 320
population dynamics of, 357
production and packaging of, 359
source of genes, 330

**V**

VAM,
  isolation of, 79
  production of, 79
Vector control, 308
Vesicular arbuscular mycorrhiza (VAM), 416
Vesicular mycorrhizae, 73
Viral biocontrol agents, 126, 128
Virulence factors, 107, 183
  degradation of, 107
  detoxification of, 107
Virulence genes and their functions, 444
Virulence system,
  induction of, 443
Virus resistance genes, 309
Virus resistant plants, 456
Viruses, 206
Virus-free propagating material, 308
Virus-infected plants,
  histopathological studies of, 298
  metabolic changes in, 297
Virus-vector relationship, 296
Viscometric method, 154

**W**

Western blot analysis, 160